T0215659

Lecture Notes in Computer Science   11304

Commenced Publication in 1973
Founding and Former Series Editors:
Gerhard Goos, Juris Hartmanis, and Jan van Leeuwen

More information about this series at http://www.springer.com/series/7407

Long Cheng · Andrew Chi Sing Leung
Seiichi Ozawa (Eds.)

Neural
Information Processing

25th International Conference, ICONIP 2018
Siem Reap, Cambodia, December 13–16, 2018
Proceedings, Part IV

 Springer

Editors
Long Cheng 🄳
The Chinese Academy of Sciences
Beijing, China

Seiichi Ozawa
Kobe University
Kobe, Japan

Andrew Chi Sing Leung
City University of Hong Kong
Kowloon, Hong Kong SAR, China

ISSN 0302-9743   ISSN 1611-3349   (electronic)
Lecture Notes in Computer Science
ISBN 978-3-030-04211-0   ISBN 978-3-030-04212-7   (eBook)
https://doi.org/10.1007/978-3-030-04212-7

Library of Congress Control Number: 2018960916

LNCS Sublibrary: SL1 – Theoretical Computer Science and General Issues

This Springer imprint is published by the registered company Springer Nature Switzerland AG
The registered company address is: Gewerbestrasse 11, 6330 Cham, Switzerland

Preface

The 25th International Conference on Neural Information Processing (ICONIP 2018), the annual conference of the Asia Pacific Neural Network Society (APNNS), was held in Siem Reap, Cambodia, during December 13–16, 2018. The ICONIP conference series started in 1994 in Seoul, which has now become a well-established and high-quality conference on neural networks around the world. Siem Reap is a gateway to Angkor Wat, which is one of the most important archaeological sites in Southeast Asia, the largest religious monument in the world. All participants of ICONIP 2018 had a technically rewarding experience as well as a memorable stay in this great city.

In recent years, the neural network has been significantly advanced with the great developments in neuroscience, computer science, cognitive science, and engineering. Many novel neural information processing techniques have been proposed as the solutions to complex, networked, and information-rich intelligent systems. To disseminate new findings, ICONIP 2018 provided a high-level international forum for scientists, engineers, and educators to present the state of the art of research and applications in all fields regarding neural networks.

With the growing popularity of neural networks in recent years, we have witnessed an increase in the number of submissions and in the quality of submissions. ICONIP 2018 received 575 submissions from 51 countries and regions across six continents. Based on a rigorous peer-review process, where each submission was reviewed by at least three experts, a total of 401 high-quality papers were selected for publication in the prestigious Springer series of *Lecture Notes in Computer Science*. The selected papers cover a wide range of subjects that address the emerging topics of theoretical research, empirical studies, and applications of neural information processing techniques across different domains.

In addition to the contributed papers, the ICONIP 2018 technical program also featured three plenary talks and two invited talks delivered by world-renowned scholars: Prof. Masashi Sugiyama (University of Tokyo and RIKEN Center for Advanced Intelligence Project), Prof. Marios M. Polycarpou (University of Cyprus), Prof. Qing-Long Han (Swinburne University of Technology), Prof. Cesare Alippi (Polytechnic of Milan), and Nikola K. Kasabov (Auckland University of Technology).

We would like to extend our sincere gratitude to all members of the ICONIP 2018 Advisory Committee for their support, the APNNS Governing Board for their guidance, the International Neural Network Society and Japanese Neural Network Society for their technical co-sponsorship, and all members of the Organizing Committee for all their great effort and time in organizing such an event. We would also like to take this opportunity to thank all the Technical Program Committee members and reviewers for their professional reviews that guaranteed the high quality of the conference proceedings. Furthermore, we would like to thank the publisher, Springer, for their sponsorship and cooperation in publishing the conference proceedings in seven volumes of *Lecture Notes in Computer Science*. Finally, we would like to thank all the speakers, authors, reviewers, volunteers, and participants for their contribution and support in making ICONIP 2018 a successful event.

October 2018

Jun Wang
Long Cheng
Andrew Chi Sing Leung
Seiichi Ozawa

ICONIP 2018 Organization

General Chair

Jun Wang   City University of Hong Kong,
            Hong Kong SAR, China

Advisory Chairs

Akira Hirose   University of Tokyo, Tokyo, Japan
Soo-Young Lee   Korea Advanced Institute of Science and Technology,
            South Korea
Derong Liu   Institute of Automation, Chinese Academy of Sciences,
            China
Nikhil R. Pal   Indian Statistics Institute, India

Program Chairs

Long Cheng   Institute of Automation, Chinese Academy of Sciences,
            China
Andrew C. S. Leung   City University of Hong Kong, Hong Kong SAR,
            China
Seiichi Ozawa   Kobe University, Japan

Special Sessions Chairs

Shukai Duan   Southwest University, China
Kazushi Ikeda   Nara Institute of Science and Technology, Japan
Qinglai Wei   Institute of Automation, Chinese Academy of Sciences,
            China
Hiroshi Yamakawa   Dwango Co. Ltd., Japan
Zhihui Zhan   South China University of Technology, China

Tutorial Chairs

Hiroaki Gomi   NTT Communication Science Laboratories, Japan
Takashi Morie   Kyushu Institute of Technology, Japan
Kay Chen Tan   City University of Hong Kong, Hong Kong SAR,
            China
Dongbin Zhao   Institute of Automation, Chinese Academy of Sciences,
            China

Publicity Chairs

Zeng-Guang Hou   Institute of Automation, Chinese Academy of Sciences,
                         China
Tingwen Huang   Texas A&M University at Qatar, Qatar
Chia-Feng Juang   National Chung-Hsing University, Taiwan
Tomohiro Shibata   Kyushu Institute of Technology, Japan

Publication Chairs

Xinyi Le   Shanghai Jiao Tong University, China
Sitian Qin   Harbin Institute of Technology Weihai, China
Zheng Yan   University Technology Sydney, Australia
Shaofu Yang   Southeast University, China

Registration Chairs

Shenshen Gu   Shanghai University, China
Qingshan Liu   Southeast University, China
Ka Chun Wong   City University of Hong Kong,
                         Hong Kong SAR, China

Conference Secretariat

Ying Qu   Dalian University of Technology, China

Program Committee

Hussein Abbass   University of New South Wales at Canberra, Australia
Choon Ki Ahn   Korea University, South Korea
Igor Aizenberg   Texas A&M University at Texarkana, USA
Shotaro Akaho   National Institute of Advanced Industrial Science
                         and Technology, Japan
Abdulrazak Alhababi   UNIMAS, Malaysia
Cecilio Angulo   Universitat Politècnica de Catalunya, Spain
Sabri Arik   Istanbul University, Turkey
Mubasher Baig   National University of Computer and Emerging
                         Sciences Lahore, India
Sang-Woo Ban   Dongguk University, South Korea
Tao Ban   National Institute of Information and Communications
                         Technology, Japan
Boris Bačić   Auckland University of Technology, New Zealand
Xu Bin   Northwestern Polytechnical University, China
David Bong   Universiti Malaysia Sarawak, Malaysia
Salim Bouzerdoum   University of Wollongong, Australia
Ivo Bukovsky   Czech Technical University, Czech Republic

Ke-Cai Cao	Nanjing University of Posts and Telecommunications, China
Elisa Capecci	Auckland University of Technology, New Zealand
Rapeeporn Chamchong	Mahasarakham University, Thailand
Jonathan Chan	King Mongkut's University of Technology Thonburi, Thailand
Rosa Chan	City University of Hong Kong, Hong Kong SAR, China
Guoqing Chao	East China Normal University, China
He Chen	Nankai University, China
Mou Chen	Nanjing University of Aeronautics and Astronautics, China
Qiong Chen	South China University of Technology, China
Wei-Neng Chen	Sun Yat-Sen University, China
Xiaofeng Chen	Chongqing Jiaotong University, China
Ziran Chen	Bohai University, China
Jian Cheng	Chinese Academy of Sciences, China
Long Cheng	Chinese Academy of Sciences, China
Wu Chengwei	Bohai University, China
Zheru Chi	The Hong Kong Polytechnic University, SAR China
Sung-Bae Cho	Yonsei University, South Korea
Heeyoul Choi	Handong Global University, South Korea
Hyunsoek Choi	Kyungpook National University, South Korea
Supannada Chotipant	King Mongkut's Institute of Technology Ladkrabang, Thailand
Fengyu Cong	Dalian University of Technology, China
Jose Alfredo Ferreira Costa	Federal University of Rio Grande do Norte, Brazil
Ruxandra Liana Costea	Polytechnic University of Bucharest, Romania
Jean-Francois Couchot	University of Franche-Comté, France
Raphaël Couturier	University of Bourgogne Franche-Comté, France
Jisheng Dai	Jiangsu University, China
Justin Dauwels	Massachusetts Institute of Technology, USA
Dehua Zhang	Chinese Academy of Sciences, China
Mingcong Deng	Tokyo University of Agriculture and Technology, Japan
Zhaohong Deng	Jiangnan University, China
Jing Dong	Chinese Academy of Sciences, China
Qiulei Dong	Chinese Academy of Sciences, China
Kenji Doya	Okinawa Institute of Science and Technology, Japan
El-Sayed El-Alfy	King Fahd University of Petroleum and Minerals, Saudi Arabia
Mark Elshaw	Nottingham Trent International College, UK
Peter Erdi	Kalamazoo College, USA
Josafath Israel Espinosa Ramos	Auckland University of Technology, New Zealand
Issam Falih	Paris 13 University, France

Bo Fan	Zhejiang University, China
Yunsheng Fan	Dalian Maritime University, China
Hao Fang	Beijing Institute of Technology, China
Jinchao Feng	Beijing University of Technology, China
Francesco Ferracuti	Università Politecnica delle Marche, Italy
Chun Che Fung	Murdoch University, Australia
Wai-Keung Fung	Robert Gordon University, UK
Tetsuo Furukawa	Kyushu Institute of Technology, Japan
Hao Gao	Nanjing University of Posts and Telecommunications, China
Yabin Gao	Harbin Institute of Technology, China
Yongsheng Gao	Griffith University, Australia
Tom Gedeon	Australian National University, Australia
Ong Sing Goh	Universiti Teknikal Malaysia Melaka, Malaysia
Iqbal Gondal	Federation University Australia, Australia
Yue-Jiao Gong	Sun Yat-sen University, China
Shenshen Gu	Shanghai University, China
Chengan Guo	Dalian University of Technology, China
Ping Guo	Beijing Normal University, China
Shanqing Guo	Shandong University, China
Xiang-Gui Guo	University of Science and Technology Beijing, China
Zhishan Guo	University of Central Florida, USA
Christophe Guyeux	University of Franche-Comte, France
Masafumi Hagiwara	Keio University, Japan
Saman Halgamuge	The University of Melbourne, Australia
Tomoki Hamagami	Yokohama National University, Japan
Cheol Han	Korea University at Sejong, South Korea
Min Han	Dalian University of Technology, China
Takako Hashimoto	Chiba University of Commerce, Japan
Toshiharu Hatanaka	Osaka University, Japan
Wei He	University of Science and Technology Beijing, China
Xing He	Southwest University, China
Xiuyu He	University of Science and Technology Beijing, China
Akira Hirose	The University of Tokyo, Japan
Daniel Ho	City University of Hong Kong, Hong Kong SAR, China
Katsuhiro Honda	Osaka Prefecture University, Japan
Hongyi Li	Bohai University, China
Kazuhiro Hotta	Meijo University, Japan
Jin Hu	Chongqing Jiaotong University, China
Jinglu Hu	Waseda University, Japan
Xiaofang Hu	Southwest University, China
Xiaolin Hu	Tsinghua University, China
He Huang	Soochow University, China
Kaizhu Huang	Xi'an Jiaotong-Liverpool University, China
Long-Ting Huang	Wuhan University of Technology, China

Panfeng Huang	Northwestern Polytechnical University, China
Tingwen Huang	Texas A&M University, USA
Hitoshi Iima	Kyoto Institute of Technology, Japan
Kazushi Ikeda	Nara Institute of Science and Technology, Japan
Hayashi Isao	Kansai University, Japan
Teijiro Isokawa	University of Hyogo, Japan
Piyasak Jeatrakul	Mae Fah Luang University, Thailand
Jin-Tsong Jeng	National Formosa University, Taiwan
Sungmoon Jeong	Kyungpook National University Hospital, South Korea
Danchi Jiang	University of Tasmania, Australia
Min Jiang	Xiamen University, China
Yizhang Jiang	Jiangnan University, China
Xuguo Jiao	Zhejiang University, China
Keisuke Kameyama	University of Tsukuba, Japan
Shunshoku Kanae	Junshin Gakuen University, Japan
Hamid Reza Karimi	Politecnico di Milano, Italy
Nikola Kasabov	Auckland University of Technology, New Zealand
Abbas Khosravi	Deakin University, Australia
Rhee Man Kil	Sungkyunkwan University, South Korea
Daeeun Kim	Yonsei University, South Korea
Sangwook Kim	Kobe University, Japan
Lai Kin	Tunku Abdul Rahman University, Malaysia
Irwin King	The Chinese University of Hong Kong, Hong Kong SAR, China
Yasuharu Koike	Tokyo Institute of Technology, Japan
Ven Jyn Kok	National University of Malaysia, Malaysia
Ghosh Kuntal	Indian Statistical Institute, India
Shuichi Kurogi	Kyushu Institute of Technology, Japan
Susumu Kuroyanagi	Nagoya Institute of Technology, Japan
James Kwok	The Hong Kong University of Science and Technology, SAR China
Edmund Lai	Auckland University of Technology, New Zealand
Kittichai Lavangnananda	King Mongkut's University of Technology Thonburi, Thailand
Xinyi Le	Shanghai Jiao Tong University, China
Minho Lee	Kyungpook National University, South Korea
Nung Kion Lee	University Malaysia Sarawak, Malaysia
Andrew C. S. Leung	City University of Hong Kong, Hong Kong SAR, China
Baoquan Li	Tianjin Polytechnic University, China
Chengdong Li	Shandong Jianzhu University, China
Chuandong Li	Southwest University, China
Dazi Li	Beijing University of Chemical Technology, China
Li Li	Tsinghua University, China
Shengquan Li	Yangzhou University, China

Ya Li	Institute of Automation, Chinese Academy of Sciences, China
Yanan Li	University of Sussex, UK
Yongming Li	Liaoning University of Technology, China
Yuankai Li	University of Science and Technology of China, China
Jie Lian	Dalian University of Technology, China
Hualou Liang	Drexel University, USA
Jinling Liang	Southeast University, China
Xiao Liang	Nankai University, China
Alan Wee-Chung Liew	Griffith University, Australia
Honghai Liu	University of Portsmouth, UK
Huaping Liu	Tsinghua University, China
Huawen Liu	University of Texas at San Antonio, USA
Jing Liu	Chinese Academy of Sciences, China
Ju Liu	Shandong University, China
Qingshan Liu	Huazhong University of Science and Technology, China
Weifeng Liu	China University of Petroleum, China
Weiqiang Liu	Nanjing University of Aeronautics and Astronautics, China
Dome Lohpetch	King Mongkut's University of Technology North Bangoko, Thailand
Hongtao Lu	Shanghai Jiao Tong University, China
Wenlian Lu	Fudan University, China
Yao Lu	Beijing Institute of Technology, China
Jinwen Ma	Peking University, China
Qianli Ma	South China University of Technology, China
Sanparith Marukatat	Thailand's National Electronics and Computer Technology Center, Thailand
Tomasz Maszczyk	Nanyang Technological University, Singapore
Basarab Matei	LIPN Paris Nord University, France
Takashi Matsubara	Kobe University, Japan
Nobuyuki Matsui	University of Hyogo, Japan
P. Meesad	King Mongkut's University of Technology North Bangkok, Thailand
Gaofeng Meng	Chinese Academy of Sciences, China
Daisuke Miyamoto	University of Tokyo, Japan
Kazuteru Miyazaki	National Institution for Academic Degrees and Quality Enhancement of Higher Education, Japan
Seiji Miyoshi	Kansai University, Japan
J. Manuel Moreno	Universitat Politècnica de Catalunya, Spain
Naoki Mori	Osaka Prefecture University, Japan
Yoshitaka Morimura	Kyoto University, Japan
Chaoxu Mu	Tianjin University, China
Kazuyuki Murase	University of Fukui, Japan
Jun Nishii	Yamaguchi University, Japan

Haruhiko Nishimura	University of Hyogo, Japan
Grozavu Nistor	Paris 13 University, France
Yamaguchi Nobuhiko	Saga University, Japan
Stavros Ntalampiras	University of Milan, Italy
Takashi Omori	Tamagawa University, Japan
Toshiaki Omori	Kobe University, Japan
Seiichi Ozawa	Kobe University, Japan
Yingnan Pan	Northeastern University, China
Yunpeng Pan	JD Research Labs, China
Lie Meng Pang	Universiti Malaysia Sarawak, Malaysia
Shaoning Pang	Unitec Institute of Technology, New Zealand
Hyeyoung Park	Kyungpook National University, South Korea
Hyung-Min Park	Sogang University, South Korea
Seong-Bae Park	Kyungpook National University, South Korea
Kitsuchart Pasupa	King Mongkut's Institute of Technology Ladkrabang, Thailand
Yong Peng	Hangzhou Dianzi University, China
Somnuk Phon-Amnuaisuk	Universiti Teknologi Brunei, Brunei
Lukas Pichl	International Christian University, Japan
Geong Sen Poh	National University of Singapore, Singapore
Mahardhika Pratama	Nanyang Technological University, Singapore
Emanuele Principi	Università Politecnica elle Marche, Italy
Dianwei Qian	North China Electric Power University, China
Jiahu Qin	University of Science and Technology of China, China
Sitian Qin	Harbin Institute of Technology at Weihai, China
Mallipeddi Rammohan	Nanyang Technological University, Singapore
Yazhou Ren	University of Science and Technology of China, China
Ko Sakai	University of Tsukuba, Japan
Shunji Satoh	The University of Electro-Communications, Japan
Gerald Schaefer	Loughborough University, UK
Sachin Sen	Unitec Institute of Technology, New Zealand
Hamid Sharifzadeh	Unitec Institute of Technology, New Zealand
Nabin Sharma	University of Technology Sydney, Australia
Yin Sheng	Huazhong University of Science and Technology, China
Jin Shi	Nanjing University, China
Yuhui Shi	Southern University of Science and Technology, China
Hayaru Shouno	The University of Electro-Communications, Japan
Ferdous Sohel	Murdoch University, Australia
Jungsuk Song	Korea Institute of Science and Technology Information, South Korea
Andreas Stafylopatis	National Technical University of Athens, Greece
Jérémie Sublime	ISEP, France
Ponnuthurai Suganthan	Nanyang Technological University, Singapore
Fuchun Sun	Tsinghua University, China
Ning Sun	Nankai University, China

Norikazu Takahashi	Okayama University, Japan
Ken Takiyama	Tokyo University of Agriculture and Technology, Japan
Tomoya Tamei	Kobe University, Japan
Hakaru Tamukoh	Kyushu Institute of Technology, Japan
Choo Jun Tan	Wawasan Open University, Malaysia
Shing Chiang Tan	Multimedia University, Malaysia
Ying Tan	Peking University, China
Gouhei Tanaka	The University of Tokyo, Japan
Ke Tang	Southern University of Science and Technology, China
Xiao-Yu Tang	Zhejiang University, China
Yang Tang	East China University of Science and Technology, China
Qing Tao	Chinese Academy of Sciences, China
Katsumi Tateno	Kyushu Institute of Technology, Japan
Keiji Tatsumi	Osaka University, Japan
Kai Meng Tay	Universiti Malaysia Sarawak, Malaysia
Chee Siong Teh	Universiti Malaysia Sarawak, Malaysia
Andrew Teoh	Yonsei University, South Korea
Arit Thammano	King Mongkut's Institute of Technology Ladkrabang, Thailand
Christos Tjortjis	International Hellenic University, Greece
Shibata Tomohiro	Kyushu Institute of Technology, Japan
Seiki Ubukata	Osaka Prefecture University, Japan
Eiji Uchino	Yamaguchi University, Japan
Wataru Uemura	Ryukoku University, Japan
Michel Verleysen	Universite catholique de Louvain, Belgium
Brijesh Verma	Central Queensland University, Australia
Hiroaki Wagatsuma	Kyushu Institute of Technology, Japan
Nobuhiko Wagatsuma	Tokyo Denki University, Japan
Feng Wan	University of Macau, SAR China
Bin Wang	University of Jinan, China
Dianhui Wang	La Trobe University, Australia
Jing Wang	Beijing University of Chemical Technology, China
Jun-Wei Wang	University of Science and Technology Beijing, China
Junmin Wang	Beijing Institute of Technology, China
Lei Wang	Beihang University, China
Lidan Wang	Southwest University, China
Lipo Wang	Nanyang Technological University, Singapore
Qiu-Feng Wang	Xi'an Jiaotong-Liverpool University, China
Sheng Wang	Henan University, China
Bunthit Watanapa	King Mongkut's University of Technology, Thailand
Saowaluk Watanapa	Thammasat University, Thailand
Qinglai Wei	Chinese Academy of Sciences, China
Wei Wei	Beijing Technology and Business University, China
Yantao Wei	Central China Normal University, China

Guanghui Wen	Southeast University, China
Zhengqi Wen	Chinese Academy of Sciences, China
Hau San Wong	City University of Hong Kong, Hong Kong SAR, China
Kevin Wong	Murdoch University, Australia
P. K. Wong	University of Macau, SAR China
Kuntpong Woraratpanya	King Mongkut's Institute of Technology Chaokuntaharn Ladkrabang, Thailand
Dongrui Wu	Huazhong University of Science and Technology, China
Si Wu	Beijing Normal University, China
Si Wu	South China University of Technology, China
Zhengguang Wu	Zhejiang University, China
Tao Xiang	Chongqing University, China
Chao Xu	Zhejiang University, China
Zenglin Xu	University of Science and Technology of China, China
Zhaowen Xu	Zhejiang University, China
Tetsuya Yagi	Osaka University, Japan
Toshiyuki Yamane	IBM, Japan
Koichiro Yamauchi	Chubu University, Japan
Xiaohui Yan	Nanjing University of Aeronautics and Astronautics, China
Zheng Yan	University of Technology Sydney, Australia
Jinfu Yang	Beijing University of Technology, China
Jun Yang	Southeast University, China
Minghao Yang	Chinese Academy of Sciences, China
Qinmin Yang	Zhejiang University, China
Shaofu Yang	Southeast University, China
Xiong Yang	Tianjin University, China
Yang Yang	Nanjing University of Posts and Telecommunications, China
Yin Yang	Hamad Bin Khalifa University, Qatar
Yiyu Yao	University of Regina, Canada
Jianqiang Yi	Chinese Academy of Sciences, China
Chengpu Yu	Beijing Institute of Technology, China
Wen Yu	CINVESTAV, Mexico
Wenwu Yu	Southeast University, China
Zhaoyuan Yu	Nanjing Normal University, China
Xiaodong Yue	Shanghai University, China
Dan Zhang	Zhejiang University, China
Jie Zhang	Newcastle University, UK
Liqing Zhang	Shanghai Jiao Tong University, China
Nian Zhang	University of the District of Columbia, USA
Tengfei Zhang	Nanjing University of Posts and Telecommunications, China
Tianzhu Zhang	Chinese Academy of Sciences, China

Contents – Part IV

Feature Selection

Multi-label Feature Selection Method Combining Unbiased
Hilbert-Schmidt Independence Criterion with Controlled
Genetic Algorithm.   3
    Chang Liu, Quan Ma, and Jianhua Xu

Anthropometric Features Based Gait Pattern Prediction Using Random
Forest for Patient-Specific Gait Training   15
    *Shixin Ren, Weiqun Wang, Zeng-Guang Hou, Xu Liang, Jiaxing Wang,
    and Liang Peng*

Robust Multi-view Features Fusion Method Based on CNMF.   27
    Bangjun Wang, Liu Yang, Li Zhang, and Fanzhang Li

Brain Functional Connectivity Analysis and Crucial Channel Selection
Using Channel-Wise CNN   40
    *Jiaxing Wang, Weiqun Wang, Zeng-Guang Hou, Xu Liang, Shixin Ren,
    and Liang Peng*

An Effective Discriminative Learning Approach for Emotion-Specific
Features Using Deep Neural Networks.   50
    Shuiyang Mao and Pak-Chung Ching

Convolutional Neural Network with Spectrogram and Perceptual Features
for Speech Emotion Recognition.   62
    *Linjuan Zhang, Longbiao Wang, Jianwu Dang, Lili Guo,
    and Haotian Guan*

Feature Selection Based on Fuzzy Conditional Distinction Degree.   72
    Qilai Zhang and Jianhua Dai

Multi-label Feature Selection Method Based on Multivariate Mutual
Information and Particle Swarm Optimization.   84
    Xidong Wang, Lei Zhao, and Jianhua Xu

Feature Selection Using Distance from Classification Boundary and Monte
Carlo Simulation.   96
    Yutaro Koyama, Kazushi Ikeda, and Yuichi Sakumura

Clustering

Approximate Spectral Clustering Using Topology Preserving Methods
and Local Scaling 109
  Mashaan Alshammari and Masahiro Takatsuka

Discovering Similarities in Malware Behaviors by Clustering
of API Call Sequences. 122
  Fatima Al Shamsi, Wei Lee Woon, and Zeyar Aung

A Storm-Based Parallel Clustering Algorithm of Streaming Data. 134
  Fang-Zhu Xu, Zhi-Ying Jiang, Yan-Lin He, Ya-Jie Wang,
  and Qun-Xiong Zhu

Iterative Maximum Clique Clustering Based Detection Filter 145
  Xinyu Zhang, Hao Sheng, Yang Zhang, Jiahui Chen, Yubin Wu,
  Guangtao Xue, and Quanrui Wei

Towards a Compact and Effective Representation for Datasets
with Inhomogeneous Clusters. 157
  Haimei Zhao, Zhuo Chen, Qiuhui Tong, and Yuan Bo

Adaptive Fuzzy Clustering Algorithm with Local Information
and Markov Random Field for Image Segmentation 170
  Jialiang Hu and Ying Wen

Efficient Direct Structured Subspace Clustering. 181
  Wen-ming Cao, Rui Li, Sheng Qian, Si Wu, and Hau-San Wong

Privacy-Preserving K-Means Clustering Upon Negative Databases 191
  Xiaoyi Hu, Liping Lu, Dongdong Zhao, Jianwen Xiang, Xing Liu,
  Haiying Zhou, Shengwu Xiong, and Jing Tian

Self-Paced Multi-Task Multi-View Capped-norm Clustering. 205
  Yazhou Ren, Xin Yan, Zechuan Hu, and Zenglin Xu

Shape Clustering as a Type of Procrustes Analysis 218
  Kazunori Iwata

Classification

Aspect-Level Sentiment Classification with Conv-Attention Mechanism. 231
  Qian Yi, Jie Liu, Guixuan Zhang, and Shuwu Zhang

Attention-Based Combination of CNN and RNN
for Relation Classification 244
  Xiaoyu Guo, Hui Zhang, Rui Liu, Xin Ding, Runqi Tian,
  and Bencheng Wang

Discrete Sparse Hashing for Cross-Modal Similarity Search 256
Lu Wang, Chao Ma, Enmei Tu, Jie Yang, and Nikola Kasabov

Solving the Double Dummy Bridge Problem with Shallow Autoencoders . . . 268
Jacek Mańdziuk and Jakub Suchan

Cross-Project Issue Classification Based on Ensemble Modeling in a Social
Coding World. 281
Yarong Zeng, Yue Yu, Qiang Fan, Xunhui Zhang, Tao Wang, Gang Yin,
and Huaimin Wang

Decision Tree Twin Support Vector Machine Based on Kernel Clustering
for Multi-class Classification 293
Qingyun Dou and Li Zhang

Machine Learning Techniques for Classification of Livestock Behavior 304
Natasa Kleanthous, Abir Hussain, Alex Mason, Jennifer Sneddon,
Andy Shaw, Paul Fergus, Carl Chalmers, and Dhiya Al-Jumeily

Two-Stage Attention Network for Aspect-Level Sentiment Classification 316
Kai Gao, Hua Xu, Chengliang Gao, Xiaomin Sun, Junhui Deng,
and Xiaoming Zhang

The Fuzzy Misclassification Analysis with Deep Neural Network
for Handling Class Noise Problem 326
Anupiya Nugaliyadde, Ratchakoon Pruengkarn, and Kok Wai Wong

A Neuronal Morphology Classification Approach Based on Deep
Residual Neural Networks 336
Xianghong Lin, Jianyang Zheng, Xiangwen Wang, and Huifang Ma

Privacy-Preserving Naive Bayes Classification Using Fully
Homomorphic Encryption 349
Sangwook Kim, Masahiro Omori, Takuya Hayashi, Toshiaki Omori,
Lihua Wang, and Seiichi Ozawa

Classification of Calligraphy Style Based on Convolutional
Neural Network 359
Fengrui Dai, Chenwei Tang, and Jiancheng Lv

Tropical Fruits Classification Using an AlexNet-Type Convolutional
Neural Network and Image Augmentation 371
Alberto Patino-Saucedo, Horacio Rostro-Gonzalez, and Jorg Conradt

Supervised and Semi-supervised Multi-task Binary Classification 380
Rakesh Kumar Sanodiya, Sriparna Saha, Jimson Mathew,
and Arpita Raj

Employ Decision Values for Soft-Classifier Evaluation
with Crispy References   392
   Lei Zhu, Tao Ban, Takeshi Takahashi, and Daisuke Inoue

Detection

Guide-Wire Detecting Based on Speeded up Robust Features
for Percutaneous Coronary Intervention   405
   Prasong Pusit, Xiaoliang Xie, and Zengguang Hou

New Default Box Strategy of SSD for Small Target Detection   416
   Yuyao He, Baoqi Li, and Yaohua Zhao

Weakly Supervised Temporal Action Detection with Shot-Based Temporal
Pooling Network.   426
   Haisheng Su, Xu Zhao, Tianwei Lin, and Haiping Fei

Drogue Detection for Autonomous Aerial Refueling Based on Adaboost
and Convolutional Neural Networks   437
   Yanjie Guo, Yimin Deng, and Haibin Duan

Deep Neural Network Based Salient Object Detection
with Image Enhancement.   444
   Lecheng Zhou and Xiaodong Gu

Brain Slices Microscopic Detection Using Simplified SSD
with Cycle-GAN Data Augmentation.   454
   Weizhou Liu, Long Cheng, and Deyuan Meng

Agglomeration Detection in Gas-Phase Ethylene Polymerization
Based on Multi-scale Convolutional Neural Network.   464
   Wenqian Zhang, Jing Wang, and Haiyan Wu

Intra-class Structure Aware Networks for Screen Defect Detection   476
   Chengchao Shen, Jie Song, Shuyi Song, Sihui Luo, Li Sun,
   and Mingli Song

Mobile Malware Detection - An Analysis of the Impact
of Feature Categories.   486
   Mahbub E. Khoda, Joarder Kamruzzaman, Iqbal Gondal,
   and Tasadduq Imam

Recurrent RetinaNet: A Video Object Detection Model
Based on Focal Loss   499
   Xiaobo Li, Haohua Zhao, and Liqing Zhang

Neural Causality Detection for Multi-dimensional Point Processes.   509
   Tianyu Wang, Christian Walder, and Tom Gedeon

Density-Induced Support Vector Data Description for Fault Detection
on Tennessee Eastman Process   522
   Yangtao Xue, Li Zhang, Bangjun Wang, and Baige Tang

ExtTra: Short-Term Traffic Flow Prediction Based on Extremely
Randomized Trees.   532
   Jiaxing Shang, Xiaofan Yan, Linhui Feng, Zheng Dong, Haojie Wang,
   and Shangbo Zhou

Actor Model Anomaly Detection Using Kernel Principal
Component Analysis   545
   Chunze Wang, Jing Wang, Chun Wang, and Qiwei Shen

Passive Detection of Splicing and Copy-Move Attacks in Image Forgery   555
   Mohammad Manzurul Islam, Joarder Kamruzzaman, Gour Karmakar,
   Manzur Murshed, and Gayan Kahandawa

Learning Latent Byte-Level Feature Representation for Malware Detection. . .   568
   Mahmood Yousefi-Azar, Len Hamey, Vijay Varadharajan,
   and Shiping Chen

Occlusion Detection in Visual Tracking: A New Framework
and A New Benchmark   579
   Xiaoguang Niu, Yueyang Gu, Zhifeng Lu, Zehua Hong, Yi Tian,
   Kuan Xu, Jie Yang, Xingqi Fang, and Yu Qiao

Attentional Payload Anomaly Detector for Web Applications   588
   Zhi-Quan Qin, Xing-Kong Ma, and Yong-Jun Wang

A Semantic Parsing Based LSTM Model for Intrusion Detection   600
   Zhipeng Li and Zheng Qin

Detecting the *Doubt Effect* and *Subjective Beliefs* Using Neural Networks
and Observers' Pupillary Responses   610
   Xuanying Zhu, Zhenyue Qin, Tom Gedeon, Richard Jones,
   Md Zakir Hossain, and Sabrina Caldwell

Driver Sleepiness Detection Using LSTM Neural Network.   622
   Yini Deng, Yingying Jiao, and Bao-Liang Lu

HTMTAD: A Model to Detect Anomalies of CDN Traffic
Based on Improved HTM Network   634
   Ning Zhao, Yongli Wang, Na Cao, and Xiaoze Gong

A Deep Learning Based Multi-task Ensemble Model for Intent Detection
and Slot Filling in Spoken Language Understanding   647
   Mauajama Firdaus, Shobhit Bhatnagar, Asif Ekbal,
   and Pushpak Bhattacharyya

An Image-Based Approach for Defect Detection on Decorative Sheets...... 659
  Boyu Zhou, Xin He, Zhongyi Zhou, and Xinyi Le

Complex Conditional Generative Adversarial Nets for Multiple Objectives
Detection in Aerial Images.................................... 671
  Dan Popescu, Loretta Ichim, and Andrei Docea

Facial Landmark Detection Under Large Pose...................... 684
  *Yangyang Hao, Hengliang Zhu, Zhiwen Shao, Xin Tan,
  and Lizhuang Ma*

Author Index 697

Feature Selection

Multi-label Feature Selection Method Combining Unbiased Hilbert-Schmidt Independence Criterion with Controlled Genetic Algorithm

Chang Liu, Quan Ma, and Jianhua Xu$^{(\boxtimes)}$

School of Computer Science and Technology, Nanjing Normal University,
Nanjing 210023, Jiangsu, China
liuchang_a@outlook.com, 356047012@qq.com, xujianhua@njnu.edu.cn

Abstract. In multi-label learning, some redundant and irrelevant features increase computational cost and even degrade classification performance, which are widely dealt with via feature selection procedure. Unbiased Hilbert-Schmidt independence criterion (HSIC) is a kernel-based dependence measure between feature and label data, which has been combined with greedy search techniques (e.g., sequential forward selection) to search for a locally optimal feature subset. Alternatively, it is possible to achieve a globally optimal solution using genetic algorithm (GA), but usually the final solution prefers to select about a half of original features. In this paper, we propose a new GA variant to control the number of selected features (simply CGA). Then CGA is integrated with HSIC to formulate a novel multi-label feature selection technique (CGAHSIC) for a given size of feature subset. The effectiveness of our proposed CGAHSIC is validated through comparing with four existing algorithms, on four benchmark data sets, according to four indicative multi-label classification evaluation metrics (Hamming loss, accuracy, F1 and subset accuracy).

Keywords: Multi-label learning · Feature selection
Hilbert-Schmidt independence criterion
Sequential forward selection · Genetic algorithm

1 Introduction

In traditional single label learning, a sample is only associated with one of predefined labels [1], whereas in multi-label learning (MLL), a sample may belong to multiple labels at the same time [2–4]. As lots of digital acquisition equipment and their post-processing software are widely available, it is unavoidable to produce many large-scale high-dimensional data in MLL applications. Usually,

This work was supported by Natural Science Foundation of China under grant No. 61273246.

L. Cheng et al. (Eds.): ICONIP 2018, LNCS 11304, pp. 3–14, 2018.
https://doi.org/10.1007/978-3-030-04212-7_1

high-dimensional sample vectors include some irrelevant and redundant features, which increase computational complexity and even degrade classification performance. Nowadays, this issue is dealt with via feature selection (FS) techniques to choose a subset of highly relevant and lowly redundant features from original ones [5,6].

Existing FS methods can be categorized into three sub-groups: filter, wrapper and embedded, according to alternative interaction with learning algorithm [5]. Filter methods evaluate the quality of features on the basis of their intrinsic characteristics and structures, without using any learning algorithm. Wrapper approaches select some discriminative features with high predicted classification performance from a pre-determined learning algorithm. Embedded methods consider FS as a part of training procedure of learning algorithm. The last two kinds of methods are time consuming due to their high computational costs, and furthermore their selected subsets are usually dependent on a specific learning algorithm. Due to effectiveness and efficiency, filter methods become the largest sub-group in multi-label FS field [5], which mainly consist of two ingredients: feature or subset evaluation criterion, and search strategy.

Feature or subset evaluation criterion is to assess the quality of individual feature or feature subset. To describe various characteristics of the data (e.g., distance, consistency, dependency and correlation), so far, many criteria have been applied, e.g., mutual information [7–10], information gain [11–13], Relief [14–16], Fisher statistics [17], and Chi-square statistics [7,18], biased Hilbert-Schmidt independence criterion [19], correlation-based index [20].

On the other hand, nowadays there are three widely-used search strategies to select a desired subset of features: (a) Simple ranking strategy, which evaluates and sorts the importance of individual feature in decreasing order, and then chooses a subset of top ranking features, as used in [8,14–18], without considering feature redundancy. (b) Greedy search strategy, which iteratively adds a feature to the selected feature subset, or removes a feature from the remained feature subset, and involves sequential forward selection (SFS) in [7,9,10,19] and hill-climbing method in [20]. (c) Stochastic search strategy, which aims at finding out a globally optimal subset of features according to various random operations, and covers genetic algorithm in [21] and memetic algorithm in [22]. It is still a challenging issue to find out a well-performed FS criterion and its corresponding search strategy in MLL FS field.

Hilbert-Schmidt independence criterion (HSIC) is a kernel-based dependency measure between two sets of random variables, which includes two versions: biased HSIC [23] and unbiased HSIC [24]. When HISC is applied to FS problem, it can describe the significant dependency between selected features and all labels. With linear kernels in both feature and label spaces, the biased HISC is combined with SFS to implement a multi-label FS method in [19]. In [24], the unbiased HSIC with non-linear kernels is integrated with SFS for single-label FS task. However, such two FS approaches only could find out a locally optimal subset of features due to their greedy search strategy.

In this paper, we apply unbiased HSIC to execute multi-label FS task, stemming from HSIC good performance in [24]. Furthermore, to find out a globally optimal FS subset, genetic algorithm is used as our search tool, rather than SFS in [19,24]. Additionally, we observed that GA-type methods prefer to choose a half of original features [25]. Therefore, we propose a GA variant to control the number of selected features, which is abbreviated as CGA. Finally, we combine CGA with HSIC to construct a new multi-label FS approach (simply CGAHSIC). The experimental results from four benchmark data sets illustrate our proposed CGAHSIC is superior to four existing FS methods (PMU [7], FIMF [8], FOHSIC [24] and FRHISC), according to four indicative multi-label classification evaluation metrics: Hamming loss, accuracy, F1 and subset accuracy.

2 Novel Multi-label Feature Selection Method

In this section, at first we review Hilbert-Schmidt independence criterion (HSIC) to evaluate the dependency between feature and label data. After summarizing traditional genetic algorithm, modified crossover and mutation operations are used to propose a controlled genetic algorithm (CGA). Finally, we integrate CGA with HSIC to constitute a new multi-label FS method: CGAHSIC.

2.1 Multi-label Classification and Feature Selection Settings

Assume that the feature and label data of a given training set consisting of N samples, D features and L labels are

$$
\begin{aligned}
\mathbf{X} &= [\mathbf{x}_1, ..., \mathbf{x}_i, ..., \mathbf{x}_N] = \left[\mathbf{x}^{(1)}, ..., \mathbf{x}^{(j)}, ..., \mathbf{x}^{(D)}\right]^T \in \Re^{D \times N}, \\
\mathbf{Y} &= [\mathbf{y}_1, ..., \mathbf{y}_i, ..., \mathbf{y}_N] = \left[\mathbf{y}^{(1)}, ..., \mathbf{y}^{(j)}, ..., \mathbf{y}^{(L)}\right]^T \in \{1, 0\}^{L \times N},
\end{aligned}
\tag{1}
$$

where the i-th sample is described using its column feature vector $\mathbf{x}_i = [x_{i1}, ..., x_{ij}, ..., x_{iD}]^T \in \Re^D$ and label vector $\mathbf{y}_i = [y_{i1}, ..., y_{ij}, ..., y_{iL}]^T \in \{0, 1\}^L$ ($y_{ij} = 1$ means that the j-th label is relevant), and $\mathbf{x}^{(j)} = [x_{1j}, ..., x_{ij}, ..., x_{Nj}]^T \in \Re^N$ and $\mathbf{y}^{(j)} = [y_{1j}, ..., y_{ij}, ..., y_{Nj}]^T \in \{1, 0\}^N$ represent the j-th feature vector and label vector, respectively.

The multi-label classification is to learn a classifier: $f(\mathbf{x}) : R^D \rightarrow \{0, 1\}^L$, which could predict the relevant labels for unseen samples. The multi-label feature selection (FS) is to choose d features from the original D ones ($d < D$) to remain those highly relevant and lowly redundant features. Formally, FS task is formulated as,

$$
\tilde{\mathbf{x}} = \mathbf{x} \circ \mathbf{p}, \tag{2}
$$

where $\circ$ represents Hadamard product of two vectors, and $\mathbf{p} = [p_1, ..., p_D]^T \in \{0, 1\}^D$ indicates a D-dimensional binary vector where '1' implies that the corresponding feature is chosen.

Algorithm 1. The pseudo-codes for CGAHSIC

Input: $\mathbf{X}$: feature data matrix; $\mathbf{Y}$: label data matrix; d: the number of selected features; G: the maximal number of generations; M: the number of chromosomes; Kernel function type and its parameters for label data.

Output: the optimal feature subset.

1: Initialize one population of size M, where each chromosome consists of d bits of 1's.
2: Calculate the kernel matrix for label data: $\mathbf{K}_y$.
3: Evaluate the HSIC value (HSIC$_j$) for each feature $\mathbf{x}^{(j)}$ ($j = 1, ..., D$).
4: Let the iteration index $g = 1$.
5: **while** $g \leq G$ **do**
6:     Calculate the HSIC value for each chromosome.
7:     Execute selection operation to produce $M/2$ offsprings.
8:     Run cross operation to create $M/2$ offsprings.
9:     Adjust the bits of 1's from all offsprings to be d.
10:    Apply mutation operation for two bits of 1 and 0.
11:    Obtain a new population.
12:    $g = g + 1$
13: **end while**
14: Determine one chromosomes with the highest HSIC value from the final population.
15: Constitute the best subset of features with the d bits of 1's.

2.2   Hilbert-Schmidt Independence Criterion

Hilbert-Schmidt independence criterion (HSIC) is a non-parametric dependence measure which considers all modes of dependencies between two sets of random variables. There are two versions: original form [23] and unbiased one [24].

Let $\mathbf{K}_x = [k_{ij}^x = k(\mathbf{x}_i, \mathbf{x}_j)|i, j = 1, ..., N]$ and $\mathbf{K}_y = [k_{ij}^y = k(\mathbf{y}_i, \mathbf{y}_j)|i, j = 1, ..., N]$ be two kernel matrices for feature and label data, respectively, in which $k(\cdot, \cdot)$ represents some kernel function [26]. The centering matrix is defined as $\mathbf{H} = \mathbf{I} - \mathbf{u}\mathbf{u}^T/l$, where $\mathbf{I}$ is the identity matrix and $\mathbf{u}$ is the column vector with all one elements.

The original empirical estimator for HSIC (simply HSIC^0) is defined as

$$\text{HSIC}^0(\mathbf{X}, \mathbf{Y}) = \frac{1}{(N-1)^2} tr(\mathbf{K}_x \mathbf{H} \mathbf{K}_y \mathbf{H}), \tag{3}$$

where tr is a trace operation of matrix. Due to the self-interaction terms, this estimator is biased, as proved in [24]. Therefore, an unbiased estimator is proposed [24], which removes those additional terms while ensuring proper normalization. Assume that $\tilde{\mathbf{K}}_x = [(1 - \delta_{ij})k_{ij}^x| i, j = 1, ..., N]$ and $\tilde{\mathbf{K}}_y = [(1 - \delta_{ij})k_{ij}^y|i, j = 1, ..., N]$, where $\delta_{ij} = 1$ if $i = j$, otherwise 0. This operation implies that the diagonal elements of two kernel matrices ($\mathbf{K}_x$ and $\mathbf{K}_y$) are set to zeros. In this case, the unbiased estimator of HSIC is formulated as

$$\text{HSIC}(\mathbf{X}, \mathbf{Y}) = \frac{1}{N(N-3)} \left(tr(\tilde{\mathbf{K}}_x \tilde{\mathbf{K}}_y) + \frac{\mathbf{u}^T \tilde{\mathbf{K}}_x \mathbf{u}\mathbf{u}^T \tilde{\mathbf{K}}_y \mathbf{u}}{(N-1)(N-2)} - \frac{2}{N-2} \mathbf{u}^T \tilde{\mathbf{K}}_x \tilde{\mathbf{K}}_y \mathbf{u} \right). \tag{4}$$

This HSIC has been combined with SFS to execute single-label feature selection task [24]. In this study, we consider the aforementioned HSIC as a feature selection criterion to depict the dependence between selected features and all labels.

Table 1. Statistics for four benchmark multi-label data sets.

Data set	#Domain	#Instances	#Features	#Classes	#Average labels
Emotions	Music	593	72	6	1.87
Image	Image	2000	294	5	1.24
Scene	Image	2407	294	6	1.07
Yeast	Biology	2417	103	14	4.24

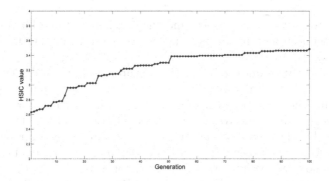

Fig. 1. The convergence analysis for CGAHSIC on Emotions.

2.3   Controlled Genetic Algorithm

In this subsection, we review basic genetic algorithm (GA) and then propose a new GA variant to control the number of selected features.

GA introduced originally by Holland [27] is a stochastic optimization algorithm which mimics natural evolution. Compared with traditional optimization techniques with a solution only during the entire procedure, GA creates a population of chromosomes or solutions during its evolution phase. To simulate the process of biological evolution, GA designs three genetic operations: selection, crossover and mutation [28]. In feature selection, each chromosome is coded as a string of D bits, i.e., 1 and 0, which imply selected and removed features, respectively. In this study, we apply GA to maximize the fitness function: unbiased HSIC (4).

Selection operation chooses some chromosomes for next generation according to fitness values. Generally speaking, those chromosomes with high fitness values are assigned a higher probability of survival. Crossover operation randomly select

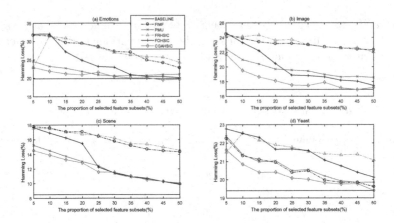

Fig. 2. The Hamming loss of five different algorithms on four data sets

several bits from two chromosomes, and then exchanges these bits each other. Mutation operation at random picks up one or several bits and then reverses their values (i.e., 1 to 0 or 0 to 1).

Basic GA firstly produces a population of chromosomes, evolutes these chromosomes gradually via three genetic operations, and stops evolution procedure after executing the preset number of generations. Finally the chromosome with the high fitness value is detected as our feature selection solution. The limitation of this GA is that the number of selected features could not be controlled strictly. Generally, GA-based FS methods prefer to choose a half of original features [25].

In this study, we design two control strategies to detect the pre-defined number of features strictly. We calculate HSIC values for all features alone ($\text{HSIC}_j = \text{HSIC}(\mathbf{x}^{(j)}, \mathbf{Y})|j = 1, ..., D$) and sort them in deceasing order. After crossover operation, when the number of selected features (h) is greater than d, we set $h - d$ bits of 1's to zeros according to the lowest ($\text{HSIC}(\mathbf{x}^{(j)}, \mathbf{Y})$). Reversely, if $h < d$, we add ($d - h$) bits of 1's using the highest ($\text{HSIC}(\mathbf{x}^{(j)}, \mathbf{Y})$) values. In mutation operation, we choose two bits with 1 and 0 and reverse them correspondingly. This GA variant is referred to as controlled GA or simply CGA, which could fix the number of selected features stringently during its evolution procedure.

2.4 Multi-label Feature Selection Method Based on HSIC and CGA

In this subsection, according to the previous work in Sects. 2.2 and 2.3, we summarize our new multi-label feature selection method (CGAHSIC) combining HSIC and CGA in Algorithm 1, whose key parameters include: (i) the number of generations (G); (ii) the number of chromosomes (M); (iii) the number of selected features (d); and (vi) the kernel type and coefficients for label data.

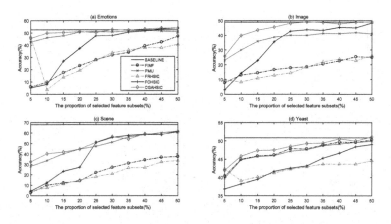

Fig. 3. The accuracy of five different algorithms on four data sets

3 Experiments

In this section, we introduce four multi-label data sets, four existing methods and their experimental settings, and finally provide detailed experimental results and analysis.

3.1 Four Benchmark Data Sets

In this paper, we downloaded four widely-used benchmark data sets: Emotions, Scene and Yeast from[1], and Image from[2], to evaluate and compare our algorithm and other existing feature selection methods, as shown in Table 1. This

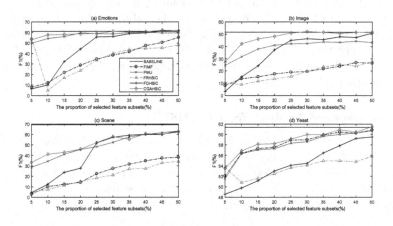

Fig. 4. The F1 of five different algorithms on four data sets

---

[1] http://mulan.sourceforge.net/datasets-mlc.html.
[2] http://cse.seu.edu.cn/PersonalPage/zhangml.

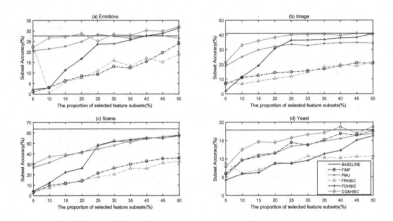

Fig. 5. The subset accuracy of five different algorithms on four data sets

table also shows some detailed descriptions including the numbers of samples, features, labels, and average labels. These data sets cover three different application domains: music, image and biology.

3.2   Compared Methods and Experimental Settings

In this study, we compare our CGAHSIC with FIMF [8], PMU [7], FOHSIC [24] and FRHSIC, where FRHSIC is to rank all features according to $HSIC_j$ $(j = 1, ..., D)$ values in decreasing order. Additionally, multi-label k nearest neighbour method (ML-kNN) with a recommended $k = 10$ [29] is chosen as a baseline classification technique to calculate four multi-label evaluation metrics: Hamming loss, accuracy, F1 and subset accuracy [2].

For FIMF and PMU, we accept their default settings in [7,8]. In FRHSIC, FOHSIC and our CGAHSIC, we apply linear kernel for feature data, and polynomial kernel of degree 4 (i.e., $k(\mathbf{y}_i, \mathbf{y}_j) = (\mathbf{y}_i^T \mathbf{y}_j + 1)^4$) for label data to describe the label interaction sufficiently. The number of generations is set to 100 and the size of population is 40. Finally, our experimental results come from ten-fold cross validation executing mode.

3.3   Convergence Analysis for CGAHSIC

Now we analyze the convergence for CGAHSIC experimentally. Here the HSIC value is defined as a function of the number of generations, as shown in Fig. 1 on Emotions data set. It is observed that as the number of generations increases, the HSIC value increases first and then tends to be stable. This shows that our CGAHSIC can find out an optimal solution for multi-label feature selection.

3.4   Experimental Results and Analysis

In this subsection, we compare our CGAHSIC with FIMF, PMU, FRHSIC and FOHSIC on four benchmark data sets. Due to the space limitation, we only provide the experimental comparison based on four indicative metrics (e.g., Hamming loss, accuracy, F1 and subset accuracy). These four metrics are regarded as functions of the proportion of selected features from 5% to 50% with an increment 5%.

Figure 2 shows the Hamming loss curves from five feature selection algorithms on four data sets. It is found out that, as the proportion of selected features increases, all curves decrease gradually. Attractively, our CGAHSIC perform the best at most of the proportional points among all compared methods.

Figures 3, 4 and 5 illustrate the accuracy, F1 and subset accuracy curves of five compared methods for four benchmark data sets. It is observed that, five curves ascend rapidly, when the more features are selected, for these three metrics. Again, our CGAHSIC is superior to the other four methods at most of the considered proportions.

To achieve a comprehensive comparison of five methods, we utilize win and rank indexes recommended in [30]. The win index counts how many times of each method to achieve the best results at different proportions on each data

Table 2. Win index of five methods from four data sets and four metrics.

Data set	CGAHSIC	FOHSIC	FRHSIC	PMU	FIMF
Hamming loss					
Emotions	**9**	0	0	1	0
Image	**10**	0	0	0	0
Scene	**7**	1	0	2	0
Yeast	**9**	0	1	0	0
Accuracy					
Emotions	**8**	0	1	1	0
Image	**10**	0	0	0	0
Scene	**7**	1	0	2	0
Yeast	**9**	0	1	0	0
F1					
Emotions	**7**	0	1	1	1
Image	**10**	0	0	0	0
Scene	**7**	1	0	2	0
Yeast	**10**	0	0	0	0
Subset accuracy					
Emotions	**9**	0	0	1	0
Image	**10**	0	0	0	0
Scene	**5**	2	0	3	0
Yeast	**9**	0	1	0	0
Overall	**136**	5	5	13	1

Table 3. Experimental results on four data sets from five methods with 30% features.

Data set	CGAHSIC	FOHSIC	FRHSIC	PMU	FIMF
Emotions					
Hamming loss	**20.43**(1)	22.95(3)	26.95(4)	20.60(2)	27.29(5)
Accuracy	**52.81**(1)	47.89(3)	33.81(4)	50.85(2)	31.81(5)
F1	**61.04**(1)	56.03(3)	40.21(4)	58.54(2)	38.41(5)
Subset accuracy	**28.34**(1)	23.79(3)	15.86(4)	27.47(2)	13.00(5)
Image					
Hamming loss	**17.46**(1)	18.80(2)	22.97(4)	19.43(3)	23.01(5)
Accuracy	**49.21**(1)	43.83(2)	18.38(5)	40.03(3)	18.56(4)
F1	**52.22**(1)	46.36(2)	19.62(4)	42.52(3)	19.56(5)
Subset accuracy	**40.35**(1)	36.40(2)	14.70(5)	32.70(3)	15.60(4)
Scene					
Hamming loss	11.55(3)	11.45(2)	16.22(5)	**11.38**(1)	15.77(4)
Accuracy	**56.47**(1)	55.81(2)	21.04(5)	51.25(3)	27.18(4)
F1	**57.84**(1)	57.21(2)	21.26(5)	52.35(3)	27.49(4)
Subset accuracy	**52.35**(1)	51.64(2)	20.40(5)	47.94(3)	26.26(4)
Yeast					
Hamming loss	**20.01**(1)	21.59(5)	21.53(4)	20.55(3)	20.49(2)
Accuracy	**49.30**(1)	43.20(4)	42.96(5)	47.62(3)	47.86(2)
F1	**59.94**(1)	54.49(4)	54.11(5)	58.61(3)	58.98(2)
Subset accuracy	**16.62**(1)	9.55(5)	10.75(4)	13.89(2)	13.73(3)
Average rank	**1.12**	2.88	4.50	2.56	3.94

set, as shown in Table 2. The last row lists the overall wins of each method, which demonstrate that our CGAHSIC significantly performs the best among all compared methods. With the fixed number of 30% features, we list four metrics for four data sets in Table 3, from which we estimate the rank index as shown in brackets. In the last row, we list the average rank of each method over four data sets and four metrics. It is illustrated that our proposed method is superior to four existing approaches.

4   Conclusions

For multi-label filter-type feature selection techniques, their key components are feature or subset evaluation criterion and corresponding search trick. In this paper, we proposed a new multi-label feature method based on unbiased Hilbert-Schmidt independence criterion and controlled genetic algorithm. The former is depict the significant dependency between the selected features and all labels, and the latter is to search for a globally optimal subset of fixed size. On four benchmark data sets, we evaluate and compare our proposed method and four existing FS methods, to demonstrate the effectiveness of our method. In future, we will validate our FS method via more benchmark data sets according to more evaluation measures.

References

1. Duda, R.O., Hart, P.E., Stork, D.G.: Pattern Classification, 2nd edn. Wiley, New York (2001)
2. Herrera, F., Charte, F., Rivera, A.J., del Jesus, M.J.: Multilabel Classification: Problem Analysis: Metrics and Techniques. Springer, Switzerland (2016). https://doi.org/10.1007/978-3-319-41111-8
3. Tsoumakas, G., Katakis, I.: Multi-label classification: an overview. Int. J. Data Warehouse Min. **3**(3), 1–13 (2007)
4. Zhang, M., Zhou, Z.: A review on multi-label learning algorithms. IEEE Trans. Knowl. Data Eng. **26**(8), 1338–1351 (2014)
5. Kashef, S., Nezamabadi-pour, H., Nipour, B.: Multilabel feature selection: a comprehensiove review and guide experiments. WIREs Data Min. Knowl. Discov. **8**(2), e1240 (2018)
6. Pereira, R., Plastino, A., Zadrozny, B., Merschmann, L.H.C.: Categorizing feature selection methods for multi-label classification. Artif. Intell. Rev. **49**(1), 57–78 (2018)
7. Lee, J., Kim, D.W.: Feature selection for multi-label classification using multivariate mutual information. Pattern Recogn. Lett. **34**(3), 349–357 (2013)
8. Lee, J., Kim, D.W.: Fast multi-label feature selection based on information-theoretic feature ranking. Pattern Recogn. **48**(9), 2761–2771 (2015)
9. Lee, J., Kim, D.W.: SCLS: multi-label feature selection based on scalable criterion for large label set. Pattern Recogn. **66**, 342–352 (2017)
10. Lin, Y., Hu, Q., Liu, J., Duan, J.: Multi-label feature selection based on max-dependency and min-redundancy. Neurocompting **168**, 92–103 (2015)
11. Spolaor, N., Chermana, E.A., Monarda, M.C., Lee, H.D.: A comparison of multi-label feature selection methods using the problem transformation approach. Eletronic Notes Theoret. Comput. Sci. **292**, 135–151 (2013)
12. Spolaor, N., Monard, M.C., Tsoumakas, G., Lee, H.D.: A systematic review of multi-label feature selection and a new method based on label construction. Neurocomputing **180**, 3–15 (2016)
13. Chen, W., Yan, J., Zhang, B., Chen, Z., Yang, Q.: Document transformation for multi-label feature selection text categorization. In: 7th IEEE International Conference on Data Mining (ICDM2007), pp. 451–456. IEEE Press, New York (2007)
14. Pupo, O.G.R., Morell, C., Soto, S.V.: ReliefF-ML: an extension of relieff algorithm to multi-label learning. In: Ruiz-Shulcloper, J., Sanniti di Baja, G. (eds.) CIARP 2013. LNCS, vol. 8259, pp. 528–535. Springer, Heidelberg (2013). https://doi.org/10.1007/978-3-642-41827-3_66
15. Reyes, O., Morell, C., Ventura, S.: Scalable extensions of the relieff algorithm for weighting and selecting features on the multi-label learning context. Neurocomputing **161**, 168–182 (2015)
16. Spolaor, N., Cherman, E., Monard, M., Lee, H.: Relief for multilabel feature selection. In: 2013 Brazlian Conference on Intelligent Systems (BRACIS2013), pp. 6–11. IEEE Press, New York (2013)

17. Kong, D., Ding, C., Huang, H., Zhao, H.: Multi-label relieff and f-statistics feature selection for image annotation. In: 2012 IEEE Conference on Computer Vision and Pattern Recognition (CVPR2012), pp. 2352–2359. IEEE Press, New York (2012)
18. Lewis, D.D., Yang, Y., Rose, T.G., Li, F.: RCV1: a new benchmark collection for text categorization research. J. Mach. Learn. Res. **5**, 361–397 (2004)
19. Xu, J.: Effective and efficient multi-label feature selection approaches via modifying Hilbert-Schmidt independence criterion. In: Hirose, A., Ozawa, S., Doya, K., Ikeda, K., Lee, M., Liu, D. (eds.) ICONIP 2016. LNCS, vol. 9949, pp. 385–395. Springer, Cham (2016). https://doi.org/10.1007/978-3-319-46675-0_42
20. Jungjit, S., Freitas, A.A., Michaelis, M., Cinatl, J.: A multi-label correlation based feature selection method for the classification of neuroblastoma microarray data. In: 12th Industrial Conference on Data Mining (ICDM2012): Workshop on Data Mining and Life Sciences (DMLS2012), pp. 149–157 (2012)
21. Jungjit, S., Freitas, A.A.: A new genetic algorithm for multi-label correlation-based feature selection. In: 23rd European Symposium on Artificial Neural Networks, Computational Intelligence and Machine Learning (ESANN2015), pp. 285–290 (2015)
22. Lee, J., Kim, D.W.: Memetic feature selection algorithm for multi-label classification. Inf. Sci. **293**, 80–95 (2015)
23. Gretton, A., Bousquet, O., Smola, A., Schölkopf, B.: Measuring statistical dependence with Hilbert-Schmidt norms. In: Jain, S., Simon, H.U., Tomita, E. (eds.) ALT 2005. LNCS (LNAI), vol. 3734, pp. 63–77. Springer, Heidelberg (2005). https://doi.org/10.1007/11564089_7
24. Song, L., Smola, A., Bedo, A.G.J., Borgwardt, K.: Feature selection via dependence maximization. J. Mach. Learn. Res. **13**, 1393–1434 (2012)
25. Yin, J., Tao, T., Xu, J.: A multi-label feature selection algorithm based on multi-objective optimization. In: 27th IEEE International Joint Conference on Neural Networks (IJCNN2015), pp. 1–7. IEEE Press, New York (2015)
26. Scholkopf, B., Smola, A.J.: Learning with Kernels: Support Vectors, Regulization, Optimization and Beyond. MIT Press, Cambridge (2001)
27. Holland, J.: Adaptation in Nature and Artificial Systems. MIT Press, Cambridge (1992)
28. Oh, I.S., Lee, J.S., Moon, B.R.: Hybrid genetic algorithms for feature selection. IEEE Trans. Pattern Anal. Mach. Intell. **26**(11), 1424–1437 (2004)
29. Zhang, M., Zhou, Z.: Ml-knn: a lazy learning approach to multi-label learning. Pattern Recognit. **40**(7), 2038–2048 (2007)
30. Demsar, J.: Statistical comparisons of classifiers over multiple data sets. J. Mach. Learn. Res. **7**, 1–30 (2006)

Anthropometric Features Based Gait Pattern Prediction Using Random Forest for Patient-Specific Gait Training

Shixin Ren[1,2], Weiqun Wang[1,2(✉)], Zeng-Guang Hou[1,3], Xu Liang[1,2], Jiaxing Wang[1,2], and Liang Peng[1,2]

[1] The State Key Laboratory of Management and Control for Complex Systems, Institute of Automation, Chinese Academy of Sciences, Beijing 100190, China
{renshixin2015,weiqun.wang,zengguang.hou,liangxu2013,wangjiaxing2016, liang.peng}@ia.ac.cn
[2] University of Chinese Academy of Sciences, Beijing 100049, People's Republic of China
[3] The CAS Center for Excellence in Brain Science and Intelligence Technology, Beijing 100190, China

Abstract. Using lower limb rehabilitation robots to help stroke patients recover their walking ability is becoming more and more popular presently. The natural and personalized gait trajectories designed for robot assisted gait training are very important for improving the therapeutic results. Meanwhile, it has been proved that human gaits are closely related to anthropometric features, which however has not been well researched. Therefore, a method based on anthropometric features for prediction of patient-specific gait trajectories is proposed in this paper. Firstly, Fourier series are used to fit gait trajectories, hence, gait patterns can be represented by the obtained Fourier coefficients. Then, human age, gender and 12 body parameters are used to design the gait prediction model. For the purpose of easy application on lower limb rehabilitation robots, the anthropometric features are simplified by an optimization method based on the minimal-redundancy-maximal-relevance criterion. Moreover, the relationship between the simplified features and human gaits is modeled by using a random forest algorithm, based on which the patient-specific gait trajectories can be predicted. Finally, the performance of the designed gait prediction method is validated on a dataset.

Keywords: Patient-specific gait · Anthropometric features
Random forest · Gait prediction

1 Introduction

Stroke is one of the common diseases that cause nervous system damage and even lead to death. Fortunately, due to timely treatment after stroke,

This research is supported by the National Natural Science Foundation of China (Grants 91648208) and Beijing Natural Science Foundation (Grant3171001).

L. Cheng et al. (Eds.): ICONIP 2018, LNCS 11304, pp. 15–26, 2018.
https://doi.org/10.1007/978-3-030-04212-7_2

the relative death rate has dropped rapidly [1]. However, the damages caused by stroke usually have long-term negative effects on patients' mobility, muscle control ability and gait patterns, and almost a half of stroke patients cannot walk without assistance [2].

Lower limb rehabilitation robots (LLRRs) have been developed to assist stroke patients to recover their walking ability in the last 20 years. Lokomat [3], ALEX [4] and Rewalk [5] are typical examples of the LLRR. Since gait training robots will be used by different patients, it is crucial to design personalized gait patterns for this kind of robot, which can be predicted by using the anthropometric parameters. However, the accurate relationship between anthropometric parameters and the gait pattern has not been well researched.

Luu et al. [8] proposed a gait trajectory generation method based on finite Fourier series (FFS) and modeled the relationship between the Fourier coefficients and gait feature, i.e., cadence and stride length. It can be seen from [8], human gait patterns can be represented by the associated Fourier coefficients. Koopman et al. [7] selected six key events to describe an individual's gait pattern in a gait cycle. Then a linear model was used to describe the relationship among the key events and human heights and walking speeds.

Luu et al. also adopted the multi-layer perceptron neural networks (MLPNN) model [6] and the general regression neural network (GRNN) model [9] for the gait pattern prediction, which are based on the gait parameters and four anthropometric features of the human legs. However, human gait trajectories are related to more factors, as is shown in [10]. Fourteen anthropometric features were used to estimate the hip, knee, and ankle joint angles, and the Gaussian process regression (GPR) method was used to design the estimation model [10]. The shortcoming of this method is that the estimation time is longer and the estimated gait patterns are inconvenient to be implemented on the platform of LLRRs.

This paper mainly focuses on developing a machine learning approach to predict personalized gait patterns. A Random Forest (RF) algorithm is designed to learn the relationship between the anthropometric features and the gait trajectories. To reduce calculation load, the gait trajectories are represented by Fourier coefficients which are used as the outputs of the RF algorithm. Fourteen anthropometric features which are same as [10] are used as the inputs of the RF algorithm. For the purpose of easy application on LLRRs, the minimal-redundancy-maximal-relevance (mRMR) criterion is adopted for the feature optimization, which is implemented based on the mutual information of inter-anthropometric features and the mutual information between the anthropometric features and Fourier coefficients. It can be found in the experiment that the modeling efficiency of the RF algorithm is higher than that of GPR. It can seen from the result of the anthropometric feature optimization of this paper and the comparison experiment that, the features used in [8,9] is insufficient for accurate prediction of personalized gait patterns, and the features used in [10] are too redundant for human gait prediction.

The remainder of this paper is organized as follows. Section 2 illustrates an LLRR developed at Institute of Automation, Chinese Academy of Sciences and the gait fitting method. The gait pattern prediction and the feature optimization are also given in Sect. 2. The results and discussion are presented in Sect. 3. This paper is summarized in Sect. 4.

2   Method

2.1   A Lower Limb Rehabilitation Robot

An LLRR has been developed at Institute of Automation, Chinese Academy of Sciences recently, based on which multiple training modes for patients at different rehabilitation phases can be provided.

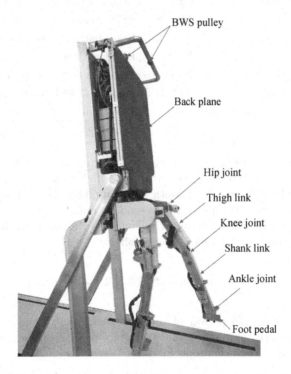

Fig. 1. Lower limb rehabilitation robot developed by Institute of Automation, Chinese Academy of Sciences.

As shown in Fig. 1, the LLRR has two leg exoskeletons, each of which has thigh and shank links and hip, knee and ankle joints. Each joint has one rotational degree of freedom in the sagittal plane, i.e., flexion/extension. The training effect can be improved by using a natural and personalized gait trajectory. On one hand, the lengths of thigh and shank links can be adjusted according to patients' shapes, which is one of the typical properties of the robot. On the other hand, how to design a personalized gait trajectory based on patients' features is one of the key points, which is investigated in the following text.

2.2 Anthropometric Features and Gait Trajectory Fitting

The human gait data which describes healthy subjects' gait kinematics and anthropometric features are obtained from [10]. Human age, gender and 12 body parameters are considered in this paper and given in Table 1.

Table 1. The associated human anthropometric features

Features	Ranges	Features	Ranges
Age (years old)	20–69	Bi-iliac width (cm)	26.1–35.8
Height (cm)	149–185	ASIS breath (cm)	20–30.6
Mass (kg)	43.3–99	Knee diameter (cm)	8.2–13
Gender	F/M	Foot length (cm)	20.5–28
Thigh length (cm)	27.5–41.6	Malleolus height (cm)	5.2–9
Calf length (cm)	30.5–46.3	Malleolus width (cm)	5.5–8
Bi-trochanteric width (cm)	28.8–38.6	Foot breath (cm)	6.4–11

The joint trajectories of human legs are continuous, smooth and periodic during walking. So, when patients participate in training on the LLRR, the output angle of LLRR must be smooth and compliant. Using Fourier series to fit the gait trajectories can generate the smooth and compliant output values. It has been proved that five terms of the Fourier series are enough for accurate fitting joint trajectories of human legs, as follows:

$$f(t_n) = a_0 + \sum_{n=1}^{5}(a_n \cos(n\omega t) + b_n \sin(n\omega t)), n = 1, \cdots, 5 \tag{1}$$

where $\omega = \frac{2\pi}{T}$ (T is the period of the gait pattern). a_n and b_n are Fourier coefficients. Hence, the human leg joint trajectory can be represented by a vector consisting of Fourier coefficients, as follows:

$$\mathbf{Y}_{(i,j)} = (a_{i,j}^0, a_{i,j}^1, b_{i,j}^1, a_{i,j}^2, b_{i,j}^2, a_{i,j}^3, b_{i,j}^3, a_{i,j}^4, b_{i,j}^4, a_{i,j}^5, b_{i,j}^5) \tag{2}$$

where $i = 1, 2, ..., P$; P is the number of subjects; $j = 1, 2, 3$, corresponding to the hip, knee and ankle joints, respectively.

2.3 Feature Selection Based on mRMR

Selecting the proper number of anthropometric features, which are strongly correlated with human gaits, is helpful for the performance improvement of the human gait prediction model. The mRMR method [11] is adopted to rank the importance of anthropometric features. It bases on the mutual information (MI) between the anthropometric features and the Fourier coefficients given in (3), which is a measure of the mutual dependence between the two variables.

By using the mRMR, an anthropometric feature subset can be obtained. The features in subset are not only highly related to Fourier coefficients, but also less redundancy among themselves. The mutual information, $I(X,Y)$, of two variables X and Y can be calculated in terms of their marginal probability functions $p(x)$, $p(y)$ and joint probability distribution function $p(x,y)$, as follows:

$$I(X,Y) = \sum_{y \in Y} \sum_{x \in X} p(x,y) \log \left(\frac{p(x,y)}{p(x)p(y)} \right). \tag{3}$$

A feature subset, where the redundancy of inter-features is minimum, can be obtained by:

$$\min W = \frac{1}{N^2} \sum_{i=1}^{N} \sum_{j=1}^{N} I(s_i, s_j), \tag{4}$$

where s_i and s_j are features in set S that contains N features.

Similarly, the features of subset having max-relevance with Fourier coefficients can be obtained by:

$$\max S = \frac{1}{N} \sum_{i=1}^{N} I(s_i, r_t), \tag{5}$$

where r_t is one of the Fourier coefficients.

The mRMR method combining the above two conditions can be described by:

$$\max_{s_i \in S - D_m} \left[I(s_i, r_t) - \left[\frac{1}{N-M} \sum_{j=1}^{N-M} I(s_i, s_j) \right] \right], \tag{6}$$

where D_m is an already selected feature subset with M features. Then one feature can be selected from the set $\{S - D_m\}$ into D_m by implement of (6) for one time. It can be seen that the features selected early into set D_m are more closely related to the Fourier coefficients.

2.4   The Random Forest Model

RF is a machine learning algorithm that combines the advantages of Bagging and Decision trees for classification or prediction [12]. The performance of RF has been verified in plenty of applications in the last ten years. Therefore, the RF algorithm is adopted to develop a model for describing the relationship between the anthropometric features and the Fourier coefficients. The RF algorithm designed in this paper is given by:

- Give a dataset containing X samples, where each sample contains L anthropometric features. X_{temp} $(X_{temp} < X)$ samples are randomly chosen from the original dataset to construct a subset X_{sub}^i. There will be T sample subsets, i.e. X_{sub}^i $(i = 1, ..., T)$, after T bootstrap iterations.
- Build T regression trees based on T sample subsets. A feature subset Θ_{sub} is formed by using K $(K < L)$ features which are randomly selected from the L anthropometric features. The best feature of the subset Θ_{sub} is used for division at each node of the tree. This is very effective to avoid the correlation of inter-trees. T trees can be built by this method to form a RF model.
- Let the response value of a tree to an input sample x is $f^t(x)$, the output value of the RF model can be given as follows:

$$Y(x) = \frac{1}{T} \sum_{t=1}^{T} f^t(x).$$ (7)

In this paper, two parameters of the RF model, which are highly related to the performance, are needed to be optimized: $ntree$, number of the trees, and $mfeature$, number of the features in the subset Θ_{sub}. The grid search method is used to find the optimal values of the two parameters. And the mean square error (MSE) is chosen to evaluate the predictive accuracy of the RF model.

3   The Results and Discussion

3.1   The Result of Fourier Series Fitting

The comparison between the reconstructed trajectories and actual trajectories of a sample are shown in Fig. 2. It can be seen that the actual trajectories are fitted very well by the Fourier coefficients.

3.2   Feature Selection and Optimization of the RF Model

The importance rankings of an anthropometric feature to each of the associated 11 Fourier coefficients can be obtained by the mRMR. Since a joint trajectory is represented by the 11 Fourier coefficients, the importance of each feature to a joint trajectory can be represented by the mean value of its 11 importance rankings corresponding to 11 Fourier coefficients. The importance rankings of 14 features are given in Table 2.

By adding one feature each time, 14 feature subsets can be obtained from the 14 features. Then the final optimal feature subset can be selected by using the optimal RF model. The specific steps are given as follows: Firstly, the first subset was obtained by using only the first feature in Table 2. Secondly, new subsets can be formed by adding one by one in the order of rankings until 14 subsets were obtained. Thirdly, the RF model was optimized for each feature subset. The grid search method was adopted to optimize the two parameters of each RF model: $ntree$ and $mfeature$ with ranges of 100–550 (in our experiments, the

MSE couldn't be decreased above 550 and below 100) and 1–12, respectively. The performance of each RF model was verified by using the 5-fold cross validation method [13]. For example, for the feature subset containing 12 features, the optimal $ntree(150)$ and $mfeature(3)$ can yield the lowest MSE for the ankle joint, as is shown in Fig. 3. Finally, the optimal feature subset can be obtained by selecting that with the lowest MSE of the RF model. It can be seen from Fig. 3 that, the best performance of the RF model for the ankle joint can be

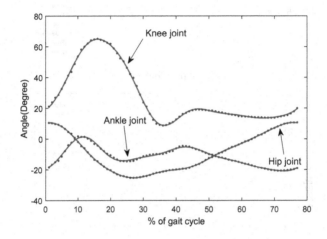

Fig. 2. Reconstructed joint trajectories by the Fourier coefficients and actual joint trajectories.

Table 2. The importance ranking of anthropometric features based on mRMR.

Features	Hip joint	Knee joint	Ankle joint	Mean
Mass	2.75	1.33	1.08	2.24
Age	4.17	2.58	2.67	3.73
Height	4.42	3.33	2.92	3.96
Calf length	6.33	5.42	4.08	5.56
Thigh length	6.25	5.83	6.33	6.28
ASIS breath	6.42	5.92	7.42	6.73
Bi-trochanteric width	7.58	8.17	7.5	7.63
Malleolus height	7.00	10.33	9.25	7.64
Bi-iliac width	8.17	7.25	8.00	8.21
Foot length	8.75	9.92	9.25	8.91
Malleolus width	9.25	12.25	12.08	10.09
Knee diameter	9.75	9.50	11.00	10.16
Foot breath	11.75	11	10.75	11.41
Gender	12.42	12.17	12.67	12.46

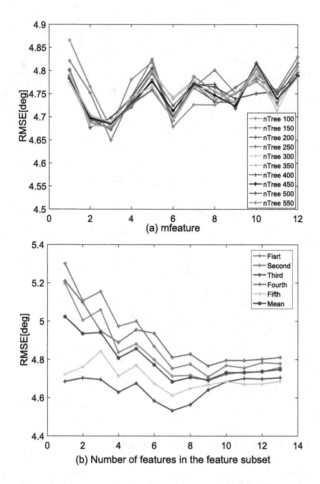

Fig. 3. (a) Optimaization of the RF models and (b) Feature selection for ankle joint.

obtained by using the feature subset, namely S_{OP}, containing the top seven features. Therefore, four anthropometric features are not enough for prediction, as is presented in the introduction section.

3.3 Predicted Performance of the RF Models

The predicted Fourier coefficients, obtained by using the optimized feature subset and the RF models, were used to reconstruct the joint trajectories. The reconstructed trajectories matched the actual trajectories well, which can be seen from Fig. 4. In order to validate the performance of the RF models, the 5-fold cross validation were used. The root mean squared error (RMSE) was adopted as the evaluation criterion. Meanwhile, the Pearson correlation coefficient was used to evaluate the similarity between the predicted and actual trajectories.

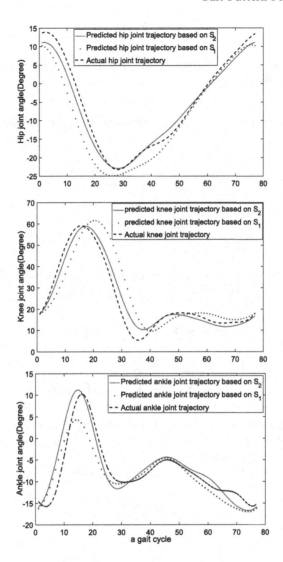

Fig. 4. Three joint trajectories comparison

Additionally, the feature dataset of [9], refered as S_2 in this paper, were also used to predict the gait trajectories, where the prediction model is the RF model as well. In addition to four anthropometric parameters, S_1 includes two gait parameters, namely stride length and cadence. They were not included in the gait dataset of this paper; however, they can be calculated by the gait period and walking speed as follows:

$$C = 2\frac{1}{T_{gait}},\tag{8}$$

$$L_{stride} = \frac{V_{walking}}{T_{gait}}, \tag{9}$$

where C is the cadence; T_{gait} is the period of a gait; L_{stride} is the stride length; $V_{walking}$ is the walking speed.

Table 3. The performance comparison by using the feature dataset S_1 and S_2

Joint	S_1		S_2	
	e	ρ	e	ρ
Hip	5.06	0.91	4.72	0.93
Knee	8.02	0.881	7.70	0.897
Ankle	5.15	0.74	4.92	0.75

The performance comparison by using the feature dataset S_1 and S_2 are given in Table 3, where the mean error, e, is defined by:

$$e = \frac{1}{5} \sum_{m=1}^{5} \left(\frac{1}{5} \sum_{n=1}^{5} e_{m,n} \right), \tag{10}$$

$$e_{m,n} = \frac{1}{P} \sum_{i=1}^{P} \sqrt{\frac{1}{Q} \sum_{j=1}^{Q} |v_i^*(j) - v_i(j)|}, \tag{11}$$

where i is the sample index; j is the time index; Q is the max time index; P is the total number of the test samples; v_i^* and v_i is the predicted and actual sample joint trajectory, respectively.

The mean value of Pearson correlation coefficients ρ is defined by:

$$\rho = \frac{1}{5} \sum_{m=1}^{5} \left(\frac{1}{5} \sum_{n=1}^{5} \rho_{m,n} \right), \tag{12}$$

$$\rho_{m,n} = \sum_{i=1}^{P} \frac{cov(v_i^*, v_i)}{\sigma_{v_i^*} \sigma_{v_i}}, \tag{13}$$

where $cov(v_i^*, v_i)$ is the covariance; $\sigma_{v_i^*}$ and σ_{v_i} are the standard deviations.

It can be found from Table 3 that the performance of the gait prediction model based on S_2 is better than that based on S_1 in the RMSE and correlation. Meanwhile, as shown in Fig. 5, the volatility of the RMSE and correlation of the gait prediction model based on S_2 is much smaller than that based on S_1. It indicates that the gait prediction model based on the optimized feature subset can generate relatively stable prediction results. Moreover, the performance comparison experiment between the RF and GRNN models was also carried out, from which it was found that the performance of these two models were similar to each other.

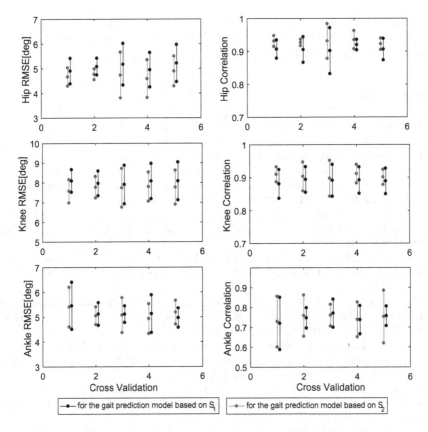

Fig. 5. RMSE and correlation coefficients of the predicted trajectories based on S_1 and S_2. Each bar is obtained from a 5-fold cross validation. The middle point in bar is the average, and two end points of the bar indicate the standard deviation.

Besides, the MLPNN model of [6] was also used to predict the gait trajectories of this paper. Lavenberg-Marquardt algorithm used in [6] was applied. It was found that, the volatility of RMSE for MLPNN was relatively large. It can be explained by that the performance of the MLPNN is unstable for gait dataset with small number of samples.

4   Conclusion

To generate the patient-specific gait trajectories for the LLRR, a RF algorithm is designed to learn the relationship between the anthropometric features and the Fourier coefficients, which are used to represent the gait trajectories. The anthropometric features are simplified by using an optimization method based on the criterion of mRMR. The tree and feature numbers of the RF model are optimized by the grid search method. The experiment results show that the performance of the proposed method based on the optimized feature subset is satisfactory. Patients will be included in experiments to further prove and improve the gait prediction method of this paper in the future.

References

1. Winstein, C.J., et al.: Guidelines for adult stroke rehabilitation and recovery: a guideline for healthcare professionals from the american heart association/american stroke association. Stroke **47**(6), e98 (2016)
2. Swinnen, E., Beckwe, D., Meeusen, R., Baeyens, J.P., Kerckhofs, E.: Does robot-assisted gait rehabilitation improve balance in stroke patients? a systematic review. Top. Stroke Rehabil. **21**(2), 87–100 (2014)
3. Jezernik, S., Colombo, G., Morari, M.: Automatic gait-pattern adaptation algorithms for rehabilitation with a 4-dof robotic orthosis. IEEE Trans. Robot. Autom. **20**(3), 574–582 (2004)
4. Banala, S.K., Kim, S.H., Agrawal, S.K., Scholz, J.P.: Robot assisted gait training with active leg exoskeleton (alex). IEEE Trans. Neural Syst. Rehabil. Eng. **17**(1), 2–8 (2009)
5. Esquenazi, A., Talaty, M., Packel, A., Saulino, M.: The rewalk powered exoskeleton to restore ambulatory function to individuals with thoracic-level motor-complete spinal cord injury. Am. J. Phys. Med. Rehabil. **91**(11), 911–921 (2012)
6. Luu, T.P., Lim, H., Qu, X., Hoon, K., Low, K.: Subject-specific lower limb waveforms planning via artificial neural network. In: 2011 IEEE International Conference on Rehabilitation Robotics, pp. 1–6. IEEE (2011)
7. Koopman, B., van Asseldonk, E.H., Van der Kooij, H.: Speed-dependent reference joint trajectory generation for robotic gait support. J. Biomech. **47**(6), 1447–1458 (2014)
8. Luu, T.P., Lim, H.B., Qu, X., Low, K.: Subject tailored gait pattern planning for robotic gait rehabilitation. In: 2010 IEEE International Conference on Robotics and Biomimetics, pp. 259–264, December 2010
9. Luu, T., Low, K.H., Qu, X., Lim, H.B., Hoon, K.H.: An individual-specific gait pattern prediction model based on generalized regression neural networks. Gait Posture **39**(1), 443–448 (2014)
10. Yun, Y., Kim, H.C., Shin, S.Y., Lee, J., Deshpande, A.D., Kim, C.: Statistical method for prediction of gait kinematics with gaussian process regression. J. Biomech. **47**(1), 186–192 (2014)
11. Peng, H., Long, F., Ding, C.: Feature selection based on mutual information: criteria of max-dependency, max-relevance, and min-redundancy. IEEE Trans. Pattern Anal. Mach. Intell. **27**, 1226–1238 (2005)
12. Breiman, L.: Random forests. Mach. Learn. **45**(1), 5–32 (2001)
13. Refaeilzadeh, P., Tang, L., Liu, H.: Cross-validation. In: Encyclopedia of Database Systems, pp. 532–538. Springer, New York (2009)

Robust Multi-view Features Fusion Method Based on CNMF

Bangjun Wang[1($\boxtimes$)], Liu Yang[2], Li Zhang[1], and Fanzhang Li[1]

[1] School of Computer Science and Technology & Joint International Research
Laboratory of Machine Learning and Neuromorphic Computing,
Soochow University, Suzhou, China
wangbangjun@suda.edu.cn

[2] School of Computer Science and Technology, Tianjin University, Tianjin, China

Abstract. Multi-view feature fusion should be expected to mine implicit nature relationships among multiple views and effectively combine the data presented by multiple views to obtain the new feature representation of the object using a right model. In practical applications, Collective Matrix Factorization (CMF) has good effects on the fusion of multi-view data, but for noise-containing situations, the generalization ability is poor. Based on this, the paper came up with a Robust Collective Non-negative Matrix Factorization (RCNMF) model which can learn the shared feature representation of multi-view data and denoise at the same time. Based on several public data sets, experimental results fully demonstrate the effectiveness of the proposed method.

Keywords: Multi-view · CMF · Future fusion

1 Introduction

Due to the diversity and convenience of data acquisition channels, a large amount of multi-view data has been accumulated, and its rational use has attracted more and more attention in machine learning and pattern recognition. Often, multi-view data is characterized by different features which are homogenous or heterogeneous. For example, multi-view features in human body object recognition such as face image, fingerprint information, sound information and signature information belong to heterogeneous data sets. Although multi-view data has different feature descriptions, the semantics represented under certain conditions are consistent. It can therefore be assumed that they share the same implicit high-level semantic space [4,8,9,15]. Based on which, there have been many researches, CCA (Canonical Correlation Analysis) [10] is a multivariate statistical analysis method studied the correlation between two groups of variables and extended by Chaudhuri et al. [4] to multiple views and obtained a multi-datasets canonical correlation analysis (Multiset CCA). Considering the nonlinear situation, Hardoon et al. [9] further expanded this method to KCCA

© Springer Nature Switzerland AG 2018
L. Cheng et al. (Eds.): ICONIP 2018, LNCS 11304, pp. 27–39, 2018.
https://doi.org/10.1007/978-3-030-04212-7_3

(Kernel CCA) by adding kernels. Guo et al. [8] proposed a convex subspace representation learning method for unsupervised multi-view clustering. Li et al. [14] proposed a discriminative multi-view interactive image rearrangement algorithm that integrated users' feedbacks and intents to fully describe multiple features of an image. Zhang et al. [19] proposed a multi-view dimension collaboration reduction approach considering the complementary of different views and similarities among the data points. The method enhances the correlation between different views and restrains the inconsistencies simultaneously with the kernel matching constraint based on the Hilbert-Schmidt independence criterion. Hou et al. [11] proposed a multi-view unsupervised feature selection algorithm using adaptive similarity and views weighting to overcome the problem of obtaining markup data in multi-view feature selection.

Non-negative matrix factorization (NMF) [13] shows good performance in single-view subspace clustering. Liu et al. [1,15] applied it to multi-view datasets. Zhao et al. [21] proposed a deep matrix factorization framework for multi-view clustering, using nonnegative matrix factorization to learn the hierarchical semantics of multi-view data in a hierarchical way. In recent years, the Collective Matrix Factorization (CMF) method [17] has become an important method for multi-view learning. This method can be used for finding the shared subspace of multiple views, so as to achieve the purpose of dimensionality reduction and feature fusion. CMF factorizes multiple matrices at the same time, and shares the subspace representation in the process of factorization, where $\mathbf{X}^1$ and $\mathbf{X}^2$ respectively denote the data matrix of the image and text, $\mathbf{W}^1$ and $\mathbf{W}^2$ are respectively the mapping matrix of the image and text, and $\mathbf{H}$ is the representation of the image and text data in subspaces. CMF has been successfully applied to many applications [16,18,22]. However, those CMF-based approaches did not consider scenes that the data contains noise in multi-view learning. In many practical applications, the data usually contain a lot of noise. In Fig. 1, for example, it is clear that the parts surrounded by the solid red boxes does not belong to the same class as the others, which are not helpful for the final clustering task. So the two parts can be regarded as noise.

For the NMF model in a single view, many researchers have considered the problem of how to remove the noise. These methods fall broadly into two categories. One kind of them draws on the idea of the robust PCA [3], introducing the error matrix, and adding sparse constraints to simulate the sparse noise in the data. Zhang et al. [20] applied this idea to the traditional NMF method. Such methods assume that noise is sparsely presented in the data so it's not suitable for processing data that contain dense noise points. The second category uses a different norm than Frobenius as the criterion for sample error in order to reduce the effect of noise on overall performance. The often criteria used are: ℓ_{21} distance [12], Manhattan distance [7], related entropy induction metric function [6], Alpha-Beta Divergence [5] and so on. Among them, the methods based on ℓ_{21} distance achieved good results. As shown in Fig. 2, the method like ℓ_{21}-NMF can weaken the effect of noise point on subspace learning well. Also, ℓ_{21} distances can easily be expanded from single view to multiple views. Yang et al. [18] used

ℓ_{21} distance to solve the problem of noise in migration learning, using two independent robust NMF models for the source and target respectively. While the limitation of this model is that it can only deal with isomorphic data, multi-view data is usually heterogeneous, so the model can not be applied to heterogeneous multi-view data.

To solve this noise issue, this paper proposes a Robust Collective Nonnegative Matrix Factorization (RCNMF) method based on nonnegative matrix factorization. RCNMF introduces the ℓ_{21} norm for the error of each view simultaneously and also for the new representation of multi-view data in the shared space which weakens the impact of noise on the overall performance. An iterative method is used to solve the objective function of RCNMF. Clustering tasks are performed to verify the performance of subspace fusion methods. By comparing with some existing methods on the real-world data-sets, the proposed method can effectively fuse the multi-view data and solve the noise problem.

Fig. 1. Display of noise in data

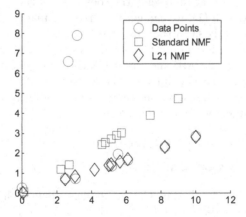

Fig. 2. The result of ℓ_{21}NMF model

The rest of this paper is organized as follows. Section 2 introduces the model of RCNFM proposed in this paper. Section 3 describes the related experiment and the result analysis. Section 4 makes a conclusion about this paper.

2   Robust Multi-view Subspace Fusion Method

This section introduces the traditional method of collective matrix factorization, then explains the robust collective non-negative matrix factorization model proposed in this paper, and finally shows the solution process.

Let $\{\{\mathbf{x}_i^j\}_{i=1}^n\}_{j=1}^m$ be the set of multi-view data, where $\mathbf{x}_i^j \in \mathbb{R}^{d_j}$ is a d_j dimension vector of the jth view, n is the number of samples, and m is the number of views. The task is to cluster the n samples into different groups. For simplicity, the data of each view is represented as a matrix $\mathbf{X}^j = [\mathbf{x}_1^j \cdots \mathbf{x}_n^j] \in \mathbb{R}^{d_j \times n}$, then the data of multi-view is $\{\mathbf{X}^j\}_{j=1}^m$. Since much of the data in multi-view is nonnegative, such as text and image data in the form of bag of words, so this paper assumes that each element is nonnegative.

2.1   CMF

CMF assumes that the representation of different views' data should be consistent in the new shared space, in which the same data matrix $\mathbf{H} \in \mathbb{R}^{r \times n}$ is shared where r is the feature dimension of the new space. Then we can learn the mapping matrix of each view to the shared space $\{\mathbf{W}^j\}_{j=1}^m \in \mathbb{R}^{d_j \times r}$. The related objective function is as follows:

$$\min_{\{\mathbf{W}^j\}_{j=1}^m, H} \sum_{j=1}^m \lambda_j \|\mathbf{X}^j - \mathbf{W}^j\mathbf{H}\|_F^2 \tag{1}$$

where λ_j is a parameter that represents the coefficient of each view. By optimizing the objective function (1), we can get the mapping matrix $\{\mathbf{W}^j\}_{j=1}^m$ and the new representation matrix $\mathbf{H}$ in the subspace.

Since this paper considers only nonnegative data, we can add non-negative constraints on the basis of CMF and call it CNMF. The collective non-negative matrix factorization model not only has the advantage of finding the essential components of the data, but also can reduce the solution space of the matrix factorization. The objective function becomes:

$$\min_{\{\mathbf{W}^j\}_{j=1}^m, H} \sum_{j=1}^m \lambda_j \|\mathbf{X}^j - \mathbf{W}^j\mathbf{H}\|_F^2$$
$$subject\ to\ \mathbf{W}^j, \mathbf{H} \geq 0, \quad j = 1, \cdots, m \tag{2}$$

Note that both the CMF and CNMF models assume that the high-level semantics of multi-view data are consistent, both of them map multi-view data to the same shared subspace, but do not consider the scene of noise.

2.2 RCNMF

In this section, we introduce a robust collective matrix factorization approach that is used primarily to reduce the effects of noise in data and to obtain more accurate subspace representation of features at the same time. The objective function is as follows:

$$\min_{\{\mathbf{W}^j\}_{j=1}^m, H} \sum_{j=1}^m \lambda_j ||\mathbf{X}^j - \mathbf{W}^j \mathbf{H}||_{21} + \alpha ||\mathbf{H}||_{21}$$
$$subject\ to\ \mathbf{W}^j, \mathbf{H} \geq 0, \quad j = 1, \cdots, m$$
(3)

where the parameter λ_j means the jth view's weight and α is the regularization coefficient. The norm ℓ_{21} of matrix $\mathbf{H}$ is defined as:

$$||\mathbf{H}||_{21} = \sum_{i=1}^n \sqrt{\sum_{k=1}^r \mathbf{H}_{ki}^2} = \sum_{i=1}^n ||\mathbf{h}_i||_2$$

where $\mathbf{h}_i$ is the ith column of $\mathbf{H}$. The norm ℓ_{21} of matrix $(\mathbf{X}^j - \mathbf{W}^j \mathbf{H})$ is defined as:

$$||\mathbf{X}^j - \mathbf{W}^j \mathbf{H}||_{21} = \sum_{i=1}^n \sqrt{\sum_{p=1}^{d_j} (\mathbf{X}^j - \mathbf{W}^j \mathbf{H})_{pi}^2} = \sum_{i=1}^n ||\mathbf{X}_i^j - \mathbf{W}^j \mathbf{H}_i||_2$$

and $(\mathbf{X}^j - \mathbf{W}^j \mathbf{H})$ takes the norm ℓ_{21} constraining errors.

The square term is no longer used to calculate each data point's error in this paper, which mainly hope to weak the impact of the larger error of the noise point on the entire data set. In extreme cases, if $||\mathbf{X}_i^j - \mathbf{W}^j \mathbf{H}_i||_2 = 0$, then the reconstructed ith sample and the original data are exactly the same. Thus the ith sample is less likely to be noise point. If the value of $||\mathbf{X}_i^j - \mathbf{W}^j \mathbf{H}_i||_2$ is large, then it means the reconstruction error is large and this sample is probably the noise point. Then we should weaken the effect of this sample while learning $\mathbf{W}$ and $\mathbf{H}$. Similarly, taking the ℓ_{21} norm for $\mathbf{H}$ is also expected weakening the effect of noise points. In addition, the norm ℓ_{21} regularization matrix performs the noise processing in a batch manner, so the mutual influence among the samples is considered in the denoising process at the same time.

2.3 Solution to RCNMF

Solving the RCNMF model (3) is a non-convex optimization problem, so we solved it iteratively. For each subproblems, we approximate it using an imprecise method. Existing methods for finding inaccurate solution include multiplicative update rule, projection ALS method and cyclic block coordinate gradient projection method. Among them, the multiplicative update rule method is relatively simple to calculate and widely used. Therefore, this paper uses the multiplicative update rule method to solve the objective function (3). Specific steps are as follows:

1. Update $\mathbf{W}^j$:

Fixing $\mathbf{H} = \mathbf{H}^\tau$ (where τ is the current iteration number), then $\mathbf{W}^j$ can be obtained by solving

$$\min_{\mathbf{W}^j} ||\mathbf{X}^j - \mathbf{W}^j\mathbf{H}||_{21}$$
$$subject\ to\ \mathbf{W}^j \geq 0, \quad j = 1, \cdots, m \tag{4}$$

The updating formula of $\mathbf{W}^j$ is:

$$(\mathbf{W}^j)^{\tau+1}_{pk} := (\mathbf{W}^j)^{\tau}_{pk} \frac{(\mathbf{X}^j\mathbf{D}^j\mathbf{H}^T)_{pk}}{((\mathbf{W}^j)^\tau\mathbf{H}\mathbf{D}^j\mathbf{H}^T)_{pk}} \tag{5}$$

where $\mathbf{D}^j$ is a diagonal matrix whose diagonal elements are $D^j_{ii} = 1/\sqrt{\sum_{k=1}^{d_j}(\mathbf{X}^j - \mathbf{W}^j\mathbf{H})^2_{ki}}$. Since $\mathbf{X}^j$ and $\mathbf{H}$ are known when solving $\mathbf{W}^j$ and the data of each views is independent, so all views' $\mathbf{W}^j$ can be calculated in parallel.

2. Update $\mathbf{H}$:

Fixing $(\mathbf{W}^j) = (\mathbf{W}^j)^\tau$, $\mathbf{H}$ can be solved by:

$$\min_{\mathbf{H}} \sum_{j=1}^m \lambda_j ||\mathbf{X}^j - \mathbf{W}^j\mathbf{H}||_{21} + \alpha||\mathbf{H}||_{21}$$
$$subject\ to\ \mathbf{H} \geq 0 \tag{6}$$

The updating formula of $\mathbf{H}$ is:

$$\mathbf{H}^{\tau+1}_{kn} := \mathbf{H}^\tau_{kn} \frac{\sum_{j=1}^m (\lambda_j(\mathbf{W}^j)^T\mathbf{X}^j\mathbf{D}^j)_{kn}}{(\sum_{j=1}^m (\lambda_j(\mathbf{W}^j)^T\mathbf{W}^j\mathbf{H}^\tau\mathbf{D}^j) + \alpha\mathbf{H}^\tau\mathbf{G})_{kn}} \tag{7}$$

where $\mathbf{G}$ is a diagonal matrix whose diagonal elements are $G_{ii} = 1/\sqrt{\sum_{k=1}^\tau H^2_{ki}}$. The objective function (3) can converge by iteratively applying formulas (5) and (7), similar to ℓ_{21}-NMF on the traditional single view whose convergence has been analyzed by Kong et al. [12].

2.4   Algorithm and Its Complexity Analysis

Algorithm 1 shows the algorithm of RCNMF. From the model (3), we can get the representation $\mathbf{H}$ of multi-view data in the subspace, which can be further used by all related tasks, such as clustering.

For each iteration, the computational complexity of updating $\mathbf{W}^j$ in (5) is $O(n^2d_j + n^2r)$. In general, $r \ll d_j$, so the complexity of updating $\mathbf{W}^j$ is about $O(n^2d_j)$. Similarly, the computational complexity of updating $\mathbf{H}$ in (7) is $O(mn^2d_j)$. Without loss of generality, let t be the number of iterations. Totally, the complexity of RCNMF is $O(tn^2m\sum_{j=1}^m d_j)$, where m is the number of views.

Algorithm 1. RCNMF

Inputs: A multi-view data $\{\mathbf{X}^j\}_{j=1}^m$, features dimension r of latent space, a threshold value σ, and parameters λ_j and α.

Outputs: H

1.while $\frac{\mathrm{Obj}^{\tau-1} - \mathrm{Obj}^\tau}{\mathrm{Obj}^{\tau-1}} >= \sigma$.

2. **for** $j = 1$ to m **do**

3.   using formula (5) updating $\mathbf{W}^j$

4. **end for**

5. using formula (7) updating $\mathbf{H}$

6.**end while**

---

3 Experimental Results and Analysis

This section introduces experiments on the real-word data-sets: Berkeley Drosophila Genome Project *BDGP* [2], *WebKB* and *Yale* and validate the performance of RCNMF on multi-view clustering. The data-sets used in experiments are described as:

1. *BDGP*: The Berkeley Drosophila Genome Project dataset consists of 2,500 embryo images of drosophila which are five categories, each corresponds to a stage of gene growth. Each image has two views: visual features (1750 dimensions) and text features (79 dimensions).

2. *WebKB:* This dataset includes 5 categories of documents: Course, Faculty, Student, Project and Staff. We select the website link collections of Cornell University, the sample number is 195. Each sample has two views. One is the property view, the feature number is 1703; the other is the relationship diagram between the samples, it is a matrix of 195×195.

3. *Yale:* The dataset is made up of 15 people's face images, each has 11 pictures, including different expressions or different perspectives. Totally, there are 165 images. Each picture has three views described by three types of features: intensity, LBP and Gabor, whose dimensions are 1024, 3304 and 6750 respectively.

3.1 Compared Methods

RCNMF proposed in this paper is to learn the fusion representation of multi-view data in latent space, based on which we can complete clustering tasks. The classic unsupervised learning method (K-means) is used as a benchmark. Moreover, we compare RCNMF with three re-representation methods, NMF [13], CCA and HTLIC [22], where NMF is tested on single view, and CCA maps two views' feature sets. The comparison algorithms are described as follows:

1. **K-means-best:** K-means-best means that we perform clustering directly on the data of each view and then pick up the best performance as its result.

2. **NMF-best:** NMF-best means that we reduce the dimension of each view with NMF separately, and then complete the clustering on the reduced data separately. The best performance is also taken as its final result.
3. **CCA:** CCA is to use canonical correlation analysis to learn shared subspaces, based on which we complete the clustering task. This method only applies to two views.
4. **HTLIC:** HTLIC is to use collective matrix factorization adding non-negative constraints and to learn the high-level feature subspace in which the clustering task is performed.

Among them, the K-means-best method works directly on the original data, NMF, CCA, HTLIC and RCNMF are all used to re-represent the data, and then K-means is used to cluster the newly represented data set. All the parameters in these methods are adjusted and we record the best result. The initial value of each variable are randomly selected because random initialization is relatively simple, and easy to calculate. In order to weaken the impact of random initialisation on the final clustering performance, each parameter in each method is randomly initialized 10 times and the average result is recorded. The termination criteria for all methods is:

$$\frac{\text{Obj}^{\tau-1} - \text{Obj}^{\tau}}{\text{Obj}^{\tau-1}} < \sigma$$

where Obj^{τ} is the objective function value in the τth loop, and σ is the threshold value. $\sigma = 10^{-4}$ in the following experiments. The performance criteria of clustering are ACC, NMI, AR, F-score, Precision and Recall. The larger the value, the better the clustering performance for all the criteria.

3.2   Effect of Parameters

This experiment takes the data set of *BDGP* to show the effect of parameters on the proposed model. *BDGP* has two views in total. View 1 is visual feature, view 2 is textual features.

1. The effect of subspace dimension r: We set the subspace dimension r to be a integer between 5 and 30 in steps of 5, Fig. 3 gives the corresponding clustering performance diagram when subspace dimension r changing in the set range. It can be seen that at $r = 20$, most of the indices of RCNMF (except for Recall) achieved the best performance in-scope. Which shows better clustering results can be obtained when the original visual features (500 dimensions) and the text features (1,000 dimensions) are fused and the dimensions are reduced to a lower level. In addition the complexity of the algorithm would also be reduced when the clustering is performed in the new space.

2. The effect of view factor λ: Fig. 4 shows the effect of the view factor λ on clustering performance. A small λ indicates that visual features are more important, a large λ indicates that text features are more important. It can be seen from Fig. 4 that the clustering performance is good when λ is close to the point of 0.8. This shows that the contribution of the text features is greater than the image.

3. The effect of regular parameter α: Fig. 5 shows the effect of the regular parameter α on the clustering result. The larger the value of α is, the greater the proportion of $\|\mathbf{H}\|_{21}$. It is observed that when α is 0.1, the clustering result is much better. It shows that it is necessary to add robust constraints on H.

To show the convergence performance of RCNMF, Fig. 6 gives the curve of the objective function (3) vs. iterations. Figure 6 demonstrates that the proposed model can converge to a local optimal value after several iterations on the *BDGP* dataset.

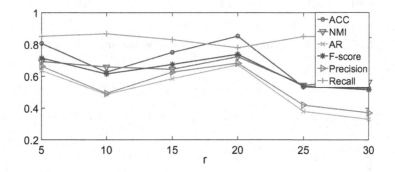

Fig. 3. Clustering performance vs. dimension of subspace r on *BDGP*

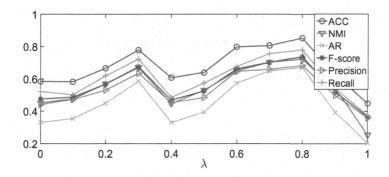

Fig. 4. Clustering performance vs. view coefficient λ on *BDGP*

3.3 Cluster Results Analysis

Tables 1, 2 and 3 show the clustering results on the selected three data-sets with the best results highlighted in black. Experimental results show that methods based on multi-view fusion are better than single-view clustering method (K-means-best and NMF-best). The main reason is that the fusion of multi-view data can obtain more information.

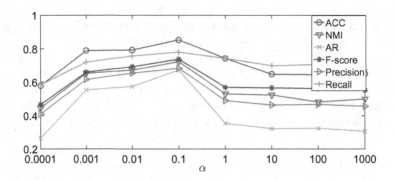

Fig. 5. Clustering performance vs. regularization parameter α on *BDGP*

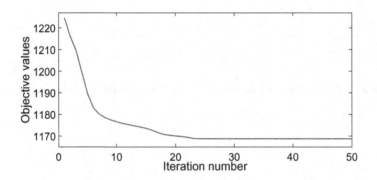

Fig. 6. Curve of convergence for (3) on *BDGP*

Table 1. Cluster results on the BDGP dataset

Method	K-means-best	NMF-best	CCA	HTLIC	RCNMF
ACC	0.526	0.624	0.657	0.772	**0.853**
NMI	0.480	0.511	0.719	0.705	**0.723**
AR	0.235	0.420	0.559	0.601	**0.672**
F-score	0.436	0.543	0.668	0.702	**0.738**
Precision	0.336	0.517	0.559	0.618	**0.683**
Recall	0.646	0.577	**0.852**	0.713	0.780

Table 2. Cluster results on the Cornell dataset

Method	K-means-best	NMF-best	CCA	HTLIC	RCNMF
ACC	0.564	0.564	0.567	0.583	**0.596**
NMI	0.405	0.354	0.366	0.385	**0.420**
AR	0.290	0.309	0.301	0.309	**0.334**
F-score	0.474	0.481	0.489	0.505	**0.510**
Precision	0.505	0.521	0.512	0.523	**0.531**
Recall	0.466	0.448	0.471	0.488	**0.504**

Table 3. Cluster results on the Yale dataset

Method	K-means-best	NMF-best	CCA	HTLIC	RCNMF
ACC	0.524	0.572	0.542	0.622	**0.642**
NMI	0.630	0.614	0.621	0.658	**0.669**
AR	0.371	0.362	0.402	0.420	**0.435**
F-score	0.415	0.405	0.415	0.460	**0.472**
Precision	0.364	0.366	0.380	0.414	**0.426**
Recall	0.484	0.454	0.490	0.520	**0.522**

In addition, the performances of RCNMF and HTLIC methods are better than that of CCA, this is because CCA treats all views equally and has orthogonal constraints. While RCNMF and HTLIC consider the different importance of different views, and relax the strong orthogonal constraint. Thus, the two methods can get a better performance. Since HTLIC does not consider the noise in multi-view data and RCNMF introduces the ℓ_{21} norm to effectively deal with the noise, RCNMF can weaken the influence of noise on the learning of subspace and makes the fused subspace more robust.

4   Conclusion

This paper presents a robust multi-view subspace fusion method RCNMF and applies it to clustering. In the proposed model, the multi-view features fusion is achieved by mapping multi-view features to a shared latent space. The proposed method adds the ℓ_{21} norm to the matrix factorization error and re-representation matrix based on CMF to eliminate the influence of noise. Clustering is completed on the re-representation of data in subspaces of all views. RCNMF is solved by an iteratively updated algorithm. After more than once iterations, RCNMF can converge. The performance of RCNMF is validated on three real data-sets. Through the clustering results, we can see that the proposed model can process the noise contained in the views while merging the features of multiple views into a shared subspace.

References

1. Akata, Z., Thurau, C., Bauckhage, C.: Non-negative matrix factorization in multimodality data for segmentation and label prediction. In: The 16th Computer Vision Winter Workshop (2011)
2. Cai, X., Wang, H., Huang, H., Ding, C.: Joint stage recognition and anatomical annotation of drosophila gene expression patterns. Bioinformatics **28**(12), 116–124 (2012)
3. Candès, E.J., Li, X., Ma, Y., Wright, J.: Robust principal component analysis. J. ACM (JACM) **58**(3), 11 (2011)
4. Chaudhuri, K., Kakade, S.M., Livescu, K., Sridharan, K.: Multi-view clustering via canonical correlation analysis. In: The 26th Annual International Conference on Machine Learning, pp. 129–136. ACM (2009)
5. Cichocki, A., Cruces, S., Amari, S.: Generalized alpha-beta divergence and their application to robust nonnegative matrix factorization. Entropy **13**, 134–170 (2011)
6. Du, L., Li, X., Shen, Y.: Robust nonnegative matrix factorization via half-quadratic minimization. In: The IEEE International Conference on Data Mining, pp. 201–210 (2012)
7. Guan, N., Tao, D., Luo, Z., Shawe-Taylor, J.: Mahnmf: manhattan non-negative matrix factorization. ArXiv Preprint ArXiv:1207.3438 (2012)
8. Guo, Y.: Convex subspace representation learning from multi-view data. In: The Twenty-Seventh AAAI Conference on Artificial Intelligence, pp. 387–393 (2013)
9. Hardoon, D., Shawe-taylor, J.: Convergence analysis of kernel canonical correlation analysis: theory and practice. Mach. Learn. **74**(22), 23–38 (2009)
10. Hotelling, H.: Relations between two sets of variates. Biometrika **28**, 321–377 (1936)
11. Hou, C., Nie, F., Tao, H., Yi, D.: Multi-view unsupervised feature selection with adaptive similarity and view weight. IEEE Trans. Knowl. Data Eng. **29**(9), 1998–2011 (2017)
12. Kong, D., Ding, C., Huang, H.: Robust nonnegative matrix factorization using l_{21}-norm. In: The International Conference on Information and Knowledge Management, pp. 673–682 (2011)
13. Lee, D., Seung, H.: Learning the parts of objects by non-negative matrix factorization. Nature **401**(6755), 788–791 (1999)
14. Li, J., Xu, C., Yang, W., Sun, C., Tao, D.: Discriminative multi-view interactive image re-ranking. IEEE Trans. Image Process. **26**(7), 3113–3127 (2017)
15. Liu, J., Wang, C., Gao, J., Han, J.: Multi-view clustering via joint nonnegative matrix factorization. In: The 13th SIAM International Conference on Data Mining, pp. 252–260 (2013)
16. Pan, W., Yang, Q.: Transfer learning in heterogeneous collaborative filtering domains. Artif. Intell. **197**, 39–55 (2013)
17. Singh, A., Gordon, G.: Relational learning via collective matrix factorization. In: The International Conference on Knowledge Discovery and Data Mining, pp. 650–658 (2008)
18. Yang, S., Hou, C., Zhang, C., Wu, Y.: Robust non-negative matrix factorization via joint sparse and graph regularization for transfer learning. Neural Comput. Appl. **23**(2), 541–559 (2013)
19. Zhang, C., Fu, H., Hu, Q., Zhu, P., Cao, X.: Flexible multi-view dimensionality co-reduction. IEEE Trans. Image Process. **26**(2), 648–659 (2016)

20. Zhang, L., Chen, Z., Zheng, M., He, X.: Robust non-negative matrix factorization. Front. Electr. Electron. Eng. **6**(2), 192–200 (2011)
21. Zhao, H., Ding, Z., Fu, Y.: Multi-view clustering via deep matrix factorization. In: The AAAI Conference on Artificial Intelligence, pp. 2921–2927 (2017)
22. Zhu, Y., et al.: Heterogeneous transfer learning for image classification. In: The AAAI Conference on Artificial Intelligence, pp. 1–6 (2011)

Brain Functional Connectivity Analysis and Crucial Channel Selection Using Channel-Wise CNN

Jiaxing Wang[1,2], Weiqun Wang[2(✉)], Zeng-Guang Hou[1,2,3], Xu Liang[1,2], Shixin Ren[1,2], and Liang Peng[2]

[1] University of Chinese Academy of Sciences, Beijing 100049, China
{wangjiaxing2016,zengguang.hou,liangxu2015,renshixin2015}@ia.ac.cn
[2] The State Key Laboratory of Management and Control for Complex Systems, Institute of Automation, Chinese Academy of Sciences, Beijing 100190, China
{weiqun.wang,liang.peng}@ia.ac.cn
[3] CAS Center for Excellence in Brain Science and Intelligence Technology, Beijing 100190, China

Abstract. Brain functional connectivity analysis and crucial channel selection, play an important role in brain working principle exploration and EEG-based emotion recognition. Towards this purpose, a novel channel-wise convolution neural network (CWCNN) is proposed, where every group convolution operator is imposed only on a separate channel. The inputs and weights of the full connection layer are visualized by using the brain topographic maps to analyze brain functional connectivity and select the crucial channels. Experiments are carried out on the SJTU emotion EEG database (SEED). The results demonstrate that positive and neutral emotions evoke greater brain activities than negative emotions in the left frontal region, which is consistent with the result from the power spectrum analysis in the literature. Meanwhile, 16 crucial channels, which are mainly distributed in the frontal and temporal regions, are selected based on the proposed method to improve emotion recognition performance. The classification accuracy by using the selected crucial channels is similar to that without channel selection. But the model with the 16 selected channels is more memory-efficient and the computation time can be reduced substantially.

Keywords: Channel-wise convolution neural network
Brain topographic maps · Full connection layer · Weights
Crucial channels

This work is supported in part by the National Natural Science Foundation of China (Grants 91648208 and 61720106012), Beijing Natural Science Foundation (Grants 3171001 and L172050).

L. Cheng et al. (Eds.): ICONIP 2018, LNCS 11304, pp. 40–49, 2018.
https://doi.org/10.1007/978-3-030-04212-7_4

1 Introduction

Correct recognition of emotions can make artificial intelligence better serve human beings in many aspects, such as human-machine interaction [1], emotional disorder diagnose and therapy [2,3], and lie detection [4]. In the earlier research on emotion recognition, physical signals were mainly used, which include facial expressions, gestures, voice information, etc. [5,6]. However, these external features are easy to be disguised and unstable, which will lead to inaccurate recognition results. On the contrary, physiological indicators [7], including galvanic skin response, electrocardiogram, and especially electroencephalogram (EEG), can directly reflect emotional changes with a high temporal resolution. Therefore, EEG-based emotion recognition has become an important research direction.

Many features (differential entropy, power spectral density, common spatial pattern, etc.) and classifiers (SVM, fuzzy logic, neural network, etc.) have been proposed for EEG-based emotion recognition [8–10]. However, basic research on brain functional connectivity under different emotions is still in the phase of infancy. Tandle et al. found that positive emotions could induce a higher theta power in the left hemisphere, while negative emotions could induce a higher theta power in the right [11]. Zheng et al. found that features from beta and gamma rhythms were more closely related to emotion recognition [12]. Zheng and Lu proposed a deep belief network based method to investigate crucial frequency bands and channels for emotion recognition, and the best classification accuracy with the selected channels was 86.65% on the SEED database [13]. However, there still exist several related problems, two of which concerned in this paper are given as follows:

- Firstly, the relationship between brain functional connectivity and emotions of different valence is still not well understood. Studying the influence of different emotions on the brain functional connectivity, and exploring topological properties of brain network, can provide a new evidence and perspective for brain functional network research.
- Secondly, in the process of multi-channel EEG data collection, signals from different channels are usually redundant, which has a negative effect on computation efficiency. Therefore, it is necessary to find crucial EEG channels for emotion recognition. Channel selection can make data processing more efficiently, while ensuring recognition rate with minimal loss.

To address the issues mentioned above, we propose a channel-wise convolution neural network (CWCNN), where every group convolution operator is only imposed on a separate input channel. Specifically, we divide the signals into 62 groups, of which the number is same to that of the original EEG channels. For each channel, different convolution kernels are allocated to extract characteristics of each channel, which ensures that the information from each channel is independent. The main contributions of this paper can be summarized as follows:

1. A novel channel-wise CNN is proposed, which can be used to analyze brain functional connectivity and select crucial channels.

2. The brain functional connectivity is analyzed based on the full connection layer's inputs, which are visualized by using the brain topographic maps.
3. The channel importance to emotion recognition is determined by analyzing inputs and weights of the full connection layer. 16 channels are selected as the crucial channels for emotion recognition. It is found that crucial channels are mainly concentrated in the frontal and both sides of temporal regions.
4. The emotion recognition performance before and after channel selection are compared. From the comparison, it can be seen that after channel selection the classification accuracy can be maintained while the computational efficiency can be improved substantially.

The remainder of this paper is organized as follows: Descriptions of short time Fourier transformation (STFT) and CWCNN model structure are presented in Sect. 2. Experiment setups and result analysis are given in Sect. 3. Finally, this paper is concluded in Sect. 4.

2   Methods

2.1   Short Time Fourier Transformation

Temporal-frequency spectra, which are obtained by STFT of the EEG signals, are used as the inputs of the emotion recognition model of this paper. The principle of the STFT is to extract the local signal by a sliding time window first, and then the Fourier transform is applied to the extracted signal to get the time-varying frequency spectrum. The sliding time window ensures that the Fourier transform only applies in a small range of the signal, which avoids the deficiency of the FFT's local analysis capability and makes Fourier transform have the ability to local orientation.

2.2   Design of the CWCNN Model

CNN has been developed in recent years and is widely applied in the field of pattern recognition. Compared with the fully connected network, CNN has two characteristics, namely weight sharing and local perception. Weight sharing refers to sharing of weights between certain neurons in the same layer. Local perception refers to the fact that the neurons are not fully connected, but local. Because of these two features, the number of parameters can be reduced greatly, thus reducing the complexity of the model.

The convolution operation is the most important operation in CNN, which is used to extract characteristics. In the convolution layer, different convolution kernels are set to operate on the input data to obtain different characteristics. Different characteristic expressions can be obtained by different convolution kernels, and the effective features will be strengthened in the continuous iterative training process. In this way, the feature extraction can be implemented.

In order to ensure information independence among channels, the collected data are grouped by different channels and convolution operation is carried out for each group in this paper, which means each kernel is imposed on one channel only.

Let the size of input data be N×M×T (channel numbers × frequency nodes × time nodes), and the class number of output data be 3, then the specific structure of the CWCNN model can be described by Fig. 1.

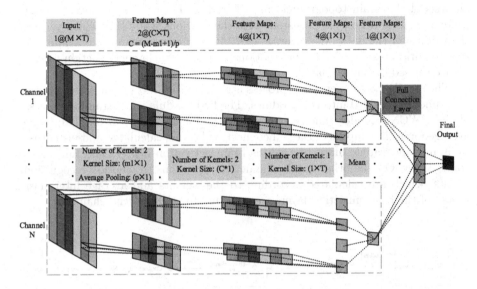

Fig. 1. The structure of the CWCNN model. The descriptions in yellow rectangles are parameter settings for each channel. The final mauve neurons, whose size are 1×1, are taken as the full connection layer's inputs.

From Fig. 1, it can be seen that all operations prior to the full connection layer are imposed on one channel. The average inter-class inputs of the full connection layer (**I**) and contributions of different channels to recognition results (**P**) can be calculated by:

$$i_c = \frac{1}{D_c} \sum_{d=1}^{D_c} (i(d)) \tag{1}$$

$$\mathbf{W_c} = diag(w_1, w_2, ..., w_N) \tag{2}$$

$$p_c = \mathbf{W_c} \cdot i_c \tag{3}$$

$$\mathbf{I} = [i_1, i_2, i_3] \tag{4}$$

$$\mathbf{P} = [p_1, \ p_2, \ p_3] \tag{5}$$

where D_c means the number of samples with the label of c, and $c \in \{1, 2, 3\}$, representing positive, neutral and negative emotions, respectively; $i \in R^{N \times 1}$ represents the inputs of the full connection layer; $i_c \in R^{N \times 1}$ means the average inter-class inputs of the full connection layer obtained by the samples whose original label is c. $\mathbf{W_c} \in R^{n \times n}$ is a diagonal matrix and represents the full connection layer's weights, which are related to the class c.

From the equations above we can see that, $\mathbf{I} \in R^{N \times 3}$ represents the average inter-class inputs of the full connection layer for positive, neutral and negative emotions. It can reflect the activities of different brain regions under different emotions, and therefore, the activation degrees of brain can be got roughly by analysis of the brain topographic maps based on $\mathbf{I}$. The analysis result about brain activation degrees is consistent with the previous study [11], which verifies the validity of the data. Moreover, $\mathbf{P} \in R^{N \times 3}$, represents channel contributions to recognition results of different emotions. The crucial channel selection can be realized based on the $\mathbf{P}$ matrix.

The detailed implementation process is given in Fig. 2. It can be seen that the experiment is divided into two modules. The first module is the training process of the CWCNN model. The second module is for analysis. On one hand, the average inter-class inputs of the full connection layer is calculated and visualized by brain topographic maps to analysis brain activation degrees and functional connectivity. On the other hand, channel contributions to emotion recognition results, which are calculated according to the weights and average inter-class inputs of the full connection layer, are analyzed to select crucial channels.

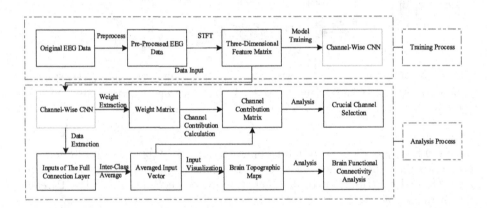

Fig. 2. The system block diagram for brain connectivity analysis and crucial channel selection.

3 Experiments and Results

3.1 Emotional Dataset

The dataset used in the experiments is the SJTU emotion database (SEED) [13], which contains 15 subjects' EEG data (7 males and 8 females). 15 Chinese film clips with three types of emotions, including positive, neutral and negative, were selected to arouse the subjects' inner feelings. All these films were simple, easy to understand, emotional, and the duration of each film was around 4 minutes.

The EEG data were collected by a 62-channel Neuroscan system at a sampling rate of 1000 Hz when they were watching each film clip. After each experiment, the subjects were asked to report their feelings by filling in a questionnaire to ensure that they had produced the same emotions as the film conveyed. To improve the reliability of the EEG signals, each subject was required to perform the experiment three sessions. The time interval between two sessions was at least one week. Therefore, there were 15 trials for each subject and each session, and 675 trials totally.

3.2 Data Processing and Model Training

The raw EEG signals were downsampled at 200 Hz, and band-pass filtered with a frequency band between 1 Hz and 100 Hz, which can cover all the important EEG rhythms. The preprocessed EEG signals were decomposed by the STFT with a 3-s sliding window (hamming window) and 1.5-s overlap. The resulting spectra were divided into 60 fragments without overlapping to increase training samples, and each sample inherited its parent's label with size of $62 \times 295 \times 2$ (channel numbers $\times$ frequency nodes $\times$ time nodes). Therefore, 40500 samples were obtained (675×60), and one fifth data chosen in random were used as the validation set, and others as the training set.

The CWCNN model was composed by three convolution layers and one full connection layer, which was same to that of Fig. 1. The first two convolution layers were used to extract the frequency domain information of each channel at each time node, and the first convolution layer was followed by an average pooling layer with size 2×1. The temporal domain characteristics were obtained by the last convolution layer. The group number was set equal to that of EEG channels. Kernel sizes for the convolution layers were 50×1, 123×1 and 1×2, respectively.

All experiments were established in Pytorch framework with the batch size of 32 [14]. The SGD method with 0.01 learning rate was used as the optimizer. Categorical cross entropy and Relu were used as the loss function and activation function, respectively.

3.3 Experiment Results

Once the final CWCNN model is determined, the weights of the full connection layer will be fixed, and the inputs of the full connection layer will be changed with the input data. The average inter-class inputs of the full connection layer can be calculated by (1) and (4). In order to analyze brain functional connectivity and select crucial channels, brain topographic maps associated with the average inter-class inputs and weights are shown in Figs. 3 and 4.

From Fig. 3, it can be seen that the overall connectivities of the brain are similar to each other for different emotional states. The areas with high activation are mainly distributed in the left side of prefrontal and bilateral frontal regions, while a small part of the parietal region is also highly activated. This means that the emotional processing unit of the brain is mainly distributed in the frontal

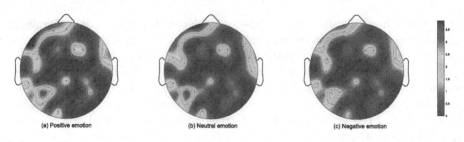

Fig. 3. Brain topographic maps of the inputs of the full connection layer under three emotions of different valence.

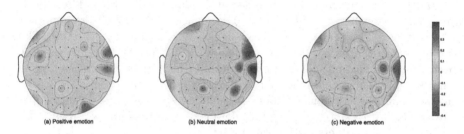

Fig. 4. Brain topographic maps of the weights of the full connection layer under three emotions of different valence.

region. In detail, compared with positive or neutral emotions, negative emotions evoke a lower activation in the left side of the brain, which is consistent with the results of previous study [11].

As shown in Fig. 4, for positive emotions, the channels with high weights are mainly distributed in both the left sides of the frontal and parietal regions. For neutral emotions, the channels with high weights are mainly distributed in right side of the temporal region. For negative emotions, channels with high weights are mainly distributed in the right frontal and temporal regions.

By comparing the high peak distributions in the above two figures, it can be seen that, the channels with high activation degrees are not necessarily of high importance to emotion recognition; on the contrary, the channels which are important to emotion recognition, are not necessarily high activated. The averaged channel contributions to emotion recognition results, which are calculated according to (2) and (3), are given in Table 1.

According to Table 1 and the channel distributions in brain (asymmetric properties in brain emotion processing), we selected 16 channels having relatively big contributions, as the crucial channels. The specific channel distributions in three different brain regions are given in Fig. 5.

In order to determine whether the selected channels have the ability to improve emotion recognition performance, an emotional classification neural network based on the general CNN, which consists of two convolution layers and two full connection layers, were designed. The first convolution layer aimed to

Table 1. Crucial channels and channel contributions to emotion recognition

Emotional valence	Channel ranking (Channel name/Contribution [%])						
Positive	FT8 / 5.0	F7 / 3.9	PO7 / 2.7	FT7 / 2.6	P7 / 2.4	FP1 / 2.2	TP7 / 2.0
Neutral	T8 / 3.2	FC5 / 2.5	P1 / 2.3	CP4 / 2.1	F2 / 2.1	P5 / 1.9	F4 / 1.8
Negative	P6 / 3.2	C6 / 3.0	FP1 / 2.6	F4 / 2.5	FT8 / 2.5	TP7 / 2.4	FPZ / 2.4

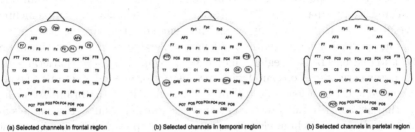

(a) Selected channels in frontal region   (b) Selected channels in temporal region   (b) Selected channels in parietal region

Fig. 5. The final selected channels which are distributed in frontal, temporal and parietal regions, respectively.

extract the characteristics in the frequency domain (kernel size: 50×1), and the second convolution layer was mainly used to extract the characteristics in the temporal and frequency domains (kernel size: 2×2). It should be noted that, owing that the coupling relationships among different channels are difficult to be reflected by the CWCNN, it is needed to design this emotion recognition model.

Performances of the classifiers including classification accuracy, model size, parameter number, and run-time performance are given in Table 2.

Table 2. Emotion recognition performance

Model name \ PARM	Accuracy [%]	Mode Size [M]	Parameter Number	Run-Time Performance [ms]
CNN-16	91.14±0.26	23.03	5,895,891	40.49
CNN-62	92.89±0.41	80.50	20,607,783	90.25

Note: "CNN-16" and "CNN-62" mean classifiers that input EEG data with the selected channels (16), and all the EEG channels (62), respectively. "Run-Time Performance" means that the time required for processing one sample.

From Table 2 we can see that, in terms of classification accuracy, there is basically no difference between the models of CNN-16 and CNN-62. But in terms of model size, number of parameters, and run-time performance, CNN-16 is much better than CNN-62, which verifies that the channel selection method of this paper can reduce the amount of memory and computational cost substantially.

4   Conclusion

In this paper, a channel-wise CNN is proposed, based on which the brain functional connectivity analysis and the crucial channel selection are implemented. The results show that connectivity for different emotions are similar to each other. The areas with high activation are mainly distributed in the left side of prefrontal and bilateral frontal regions, meanwhile, a small part of the parietal region is also highly activated. Compared with the positive or neutral emotions, the negative emotions evoke a lower activation in the left side of the brain. According to the channel contributions to the emotion recognition results, 16 channels are selected as the crucial channels. The comparison experiment between the two classifiers, the inputs of which were obtained respectively from the 16 selected channels and all of the channels, were carried out. The results show that similar classification accuracy can be obtained by the two classifiers. But in terms of model size and run-time performance, the classifier by using the selected channels is much better, which means the model with the selected channels are more memory-efficient and less time-consuming.

References

1. Guo, S., Zhao, X., Wei, W., Guo, J., Zhao, F., Hu, Y.: Feasibility study of a novel rehabilitation training system for upper limb based on emotional control. In: 2015 IEEE International Conference on Mechatronics and Automation (ICMA), pp. 1507–1512 (2015)
2. Sourina, O., Liu, Y., Nguyen, M.K.: Real-time EEG-based emotion recognition for music therapy. J. Multimodal User Interfaces 5(1–2), 27–35 (2012)
3. Datko, M., Pineda, J.A., Müller, R.A.: Positive effects of neurofeedback on autism symptoms correlate with brain activation during imitation and observation. Eur. J. Neurosci. 47(6), 579–591 (2017)
4. Hartwig, M., Bond, C.F.: Lie detection from multiple cues: a meta-analysis. Appl. Cogn. Psychol. 28(5), 661–676 (2014)
5. Jiang, M., Rahmani, A.M., Westerlund, T., Liljeberg, P., Tenhunen, H.: Facial expression recognition with sEMG method. In: 2015 IEEE International Conference on Computer and Information Technology; Ubiquitous Computing and Communications; Dependable, Autonomic and Secure Computing; Pervasive Intelligence and Computing, pp. 981–988 (2015)
6. Renjith, S., Manju, K.G.: Speech based emotion recognition in Tamil and Telugu using LPCC and hurst parameters #x2014; a comparitive study using KNN and ANN classifiers. In: 2017 International Conference on Circuit, Power and Computing Technologies (ICCPCT), pp. 1–6 (2017)

7. Wen, W., Liu, G., Cheng, N., Wei, J., Shangguan, P., Huang, W.: Emotion recognition based on multi-variant correlation of physiological signals. IEEE Trans. Affect. Comput. **5**(2), 126–140 (2014)

8. Li, H., Qing, C., Xu, X., Zhang, T.: A novel DE-PCCM feature for EEG-based emotion recognition. In: 2017 International Conference on Security, Pattern Analysis, and Cybernetics (SPAC), pp. 389–393 (2017)

9. Matiko, J.W., Beeby, S.P., Tudor, J.: Fuzzy logic based emotion classification. In: 2014 IEEE International Conference on Acoustics, Speech and Signal Processing (ICASSP), pp. 4389–4393 (2014)

10. Leslie, G., Ojeda, A., Makeig, S.: Towards an affective brain-computer interface monitoring musical engagement. In: 2013 Humaine Association Conference on Affective Computing and Intelligent Interaction, pp. 871–875 (2013)

11. Tandle, A., Jog, N., Dharmadhikari, A., Jaiswal, S.: Estimation of valence of emotion from musically stimulated EEG using frontal theta asymmetry. In: 2016 12th International Conference on Natural Computation, Fuzzy Systems and Knowledge Discovery (ICNC-FSKD), pp. 63–68 (2016)

12. Zheng, W.L., Zhu, J.Y., Peng, Y., Lu, B.L.: EEG-based emotion classification using deep belief networks. In: 2014 IEEE International Conference on Multimedia and Expo (ICME), pp. 1–6 (2014)

13. Zheng, W.L., Lu, B.L.: Investigating critical frequency bands and channels for EEG-based emotion recognition with deep neural networks. IEEE Trans. Auton. Ment. Dev. **7**(3), 162–175 (2015)

14. PyTorch deep learning framework. http://pytorch.org. Accessed 22 Aug 2018

An Effective Discriminative Learning Approach for Emotion-Specific Features Using Deep Neural Networks

Shuiyang Mao$^{(\boxtimes)}$ and Pak-Chung Ching

Department of Electronic Engineering,
The Chinese University of Hong Kong, Hong Kong SAR, China
maoshuiyang@hotmail.com

Abstract. Speech contains rich yet entangled information ranging from phonetic to emotional components. These different components are always mixed together hindering certain tasks from achieving better performance. Therefore, automatically learning a good representation that disentangles these components is non-trivial. In this paper, we propose a hierarchical method to extract utterance-level features from frame-level acoustic features using deep neural networks (DNNs). Moreover, inspired by recent progress in face recognition, we introduce centre loss as a complementary supervision signal to the traditional softmax loss to facilitate the intra-class compactness of the learned features. With the joint supervision of these two loss functions, we can train the DNNs to obtain separable and discriminative emotion-specific features. Experiments on CASIA corpus, Emo-DB corpus and SAVEE database show comparable results with that of state-of-the-art approaches.

Keywords: Speech emotion recognition · Deep neural networks
Hierarchical method · Centre loss

1 Introduction

Humans extensively use emotions to express their intentions through speech [1]. By incorporating appropriate emotions, the same phonetic information is encoded and conveyed with different semantics. Therefore, a system that can process emotions along with linguistic information must be developed. Over the past several years, speech emotion recognition has received increasing attention from researchers [2,3]. Despite the advances in this field, the systematic knowledge about the acoustic patterns that characterise different emotions remains lacking [4], whilst the feature selection for this task is far from being thoroughly investigated.

Deep neural networks (DNNs) have been proven as powerful sets of models for recognition tasks. Specifically, DNNs are well known for their capability to learn good representations from raw data with multiple levels of abstraction.

© Springer Nature Switzerland AG 2018
L. Cheng et al. (Eds.): ICONIP 2018, LNCS 11304, pp. 50–61, 2018.
https://doi.org/10.1007/978-3-030-04212-7_5

These methods have significantly improved the state-of-the-art in many tasks [5], including speech emotion recognition. For example, Han et al. [6] utilised a deep feedforward neural network to extract high-level features for each speech segment, and then used these high-level features to form utterance-level features for final classification. Recurrent neural networks (RNNs) and convolutional neural networks (CNNs) have also recently attracted increasing research interests in this area [7,8], for their ability to model the contextual dependencies that are embedded in speech data so as to improve the system performance.

Loss functions play a significant role in guiding neural networks towards capturing good data characteristics. However, in most of the available neural networks for speech emotion recognition, the importance of choosing the appropriate loss functions is rarely highlighted, and softmax loss is generally used by default. However, softmax loss only facilitates the inter-class separability of the learned features [9]. Intuitively, if we can further reduce intra-class variation, that is, if we make each feature cluster of the same class 'slim' in the feature space, then there will be less overlap amongst different feature clusters, hence further improving the recognition power. In order to enhance such discriminative power, contrastive loss [10] and triplet loss [11] have been recently and successfully applied to speech-related tasks [12–14]. They achieve their discriminative objective by mapping data pairs or triplets to a distance metric and by reducing or increasing the distance between similar or dissimilar pairs, respectively. However, both contrastive loss and triplet loss require the inconvenient construction of data pairs or triplets and may suffer from dramatic data expansion.

With the advances in face recognition, centre loss [9] has been recently proposed for face recognition tasks. It has been proven effective in enhancing the discriminative power of the deep features of input images [9,15]. Therefore, the same approach can be used to learn discriminative emotion salient features.

In this study we propose to use hierarchical classification approach to extract utterance-level features from frame-level acoustic features using DNNs. Meanwhile, we utilise softmax loss and centre loss to jointly supervise the learning process of DNNs instead of using softmax loss alone. Experiments on the CASIA Corpus, Emo-DB corpus [16] and SAVEE database [17] show comparable results with that of state-of-the-art approaches.

2   Overall Architecture

In this section, we describe the details of our classification architecture. Three stages are composed. In the first stage, we divide the signal into frames and extract the short-term frame-level features. Next, from these frame-level features, segment-level features are constructed to train a DNN for segment-level emotion classification. The trained DNN is used to make a decision about the emotional state distribution of every single segment. Lastly, these isolated decision values are combined to form the utterance-level features, which are then fed into a classifier to determine the emotional state of the whole utterance. Figure 1 presents an overview of the architecture, of which the details are outlined as follows.

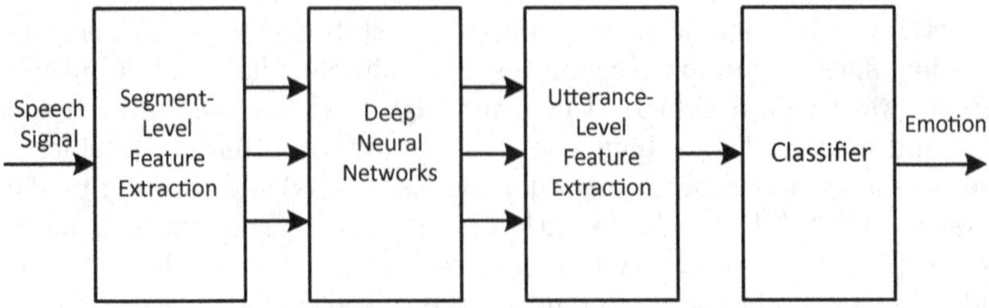

Fig. 1. Overview of our architecture for speech emotion recognition

2.1 Frame-Level Acoustic Feature Extraction

We firstly divide the input utterance into frames with overlapping windows and then compute 39 low-level descriptors (LLDs) for each frame, including zero-crossing rate (ZCR), energy, entropy of energy, 5 spectral-shape-related features (i.e., spectral centroid, spectral spread, spectral entropy, spectral flux and spectral rolloff), harmonic ratio, pitch, first 3 formant position, 13 log mel-filter banks, 12 chroma vectors and chroma deviation. We use the pyAudioAnalysis [18] library to extract these LLDs, of which the effectiveness in speech emotion recognition has been demonstrated in [19].

This process leads to a sequence of frame-level feature vectors and we denote the feature vector for each frame m as $\mathbf{F}(m) \in \mathbb{R}^{39 \times 1}$.

2.2 Segment-Level Emotion Classification

We firstly form the segment-level features $\mathbf{S}(m)$ by extracting relevant parametric statistics, namely mean, std, max and min over the aforementioned LLDs of neighbouring frames as follows:

$$\mathbf{S}(m) = \begin{bmatrix} \text{mean}\{\mathbf{X}(m)\} \\ \text{std}\{\mathbf{X}(m)\} \\ \text{max}\{\mathbf{X}(m)\} \\ \text{min}\{\mathbf{X}(m)\} \end{bmatrix} \in \mathbb{R}^{156 \times 1}, \tag{1}$$

where $\mathbf{X}(m)$ is formed by stacking the frame-level feature vectors of the neighbouring frames. It can be written as:

$$\mathbf{X}(m) = [\mathbf{F}(m-w), ..., \mathbf{F}(m), ..., \mathbf{F}(m+w)] \in \mathbb{R}^{39 \times (2w+1)}, \tag{2}$$

where w is the number of neighbouring frames in each side of frame m. In addition, since we use four kinds of parametric statistics, the dimension of $\mathbf{S}(m)$ is 156 (i.e., $39 \times 4 = 156$), as Eq. (1) shows.

We then train a DNN for segment-level emotion classification. The inputs to the DNN are these segment-level features and the training target is the label of the utterance. In other words, we assign the same label to all segments in one utterance. In addition, softmax loss and centre loss will be combined to jointly train the DNN. The DNN trained with these segments predicts a softmax distribution $\mathbf{U}$, which is regarded as a high-level feature vector for each single segment as follows:

$$\mathbf{U} = [p(E_1), p(E_2), ..., p(E_K)]^T \in \mathbb{R}^{K \times 1}, \tag{3}$$

where K denotes the number of emotion classes.

2.3 Utterance-Level Feature and Classification

From these segment-level emotional state distributions, the utterance-level features can be constructed by applying the same statistical functions as in Sect. 2.2, i.e., mean, std, max and min. Afterwards, these utterance-level features are fed into a relatively simple classifier (i.e., extreme learning machine (ELM) [6,20]) to determine the emotional state of the whole utterance.

3 Loss

In this section, we firstly review softmax loss and then present the formulation of the introduced centre loss. Afterwards, we show how to optimise total loss by using the gradient descent (GD) method, with which we can see that centre loss is trainable and easy to implement. We lastly summarise the algorithm employed in this work.

3.1 Softmax Loss

Softmax loss is formulated as follows:

$$\mathcal{L}_s = -log \frac{e^{\mathbf{W}_{y_i}^T \mathbf{x}_i + b_{y_i}}}{\sum_{j=1}^{K} e^{\mathbf{W}_j^T \mathbf{x}_i + b_j}}, \tag{4}$$

where $\mathbf{x}_i \in \mathbb{R}^d$ is the output of the last hidden layer of DNNs and denotes the deeply learned feature of the i-th sample, belonging to the y_i-th class. d is the deep feature dimension. $\mathbf{W}_j \in \mathbb{R}^d$ is the j-th column of the weights $\mathbf{W}_l \in \mathbb{R}^{d \times K}$ between the last hidden layer and the softmax loss layer. $\mathbf{b} \in \mathbb{R}^K$ is the bias term with K denoting the number of classes.

3.2 Combining with Centre Loss

Centre loss aims to minimise intra-class variations and can be formulated as follows:

$$\mathcal{L}_c = \frac{1}{2}||\mathbf{x}_i - \mathbf{c}_{y_i}||^2, \tag{5}$$

where $\mathbf{c}_{y_i}$ is the y_i-th class centre of deeply learned features and has the same dimension of the deep features. During the training process, $\mathbf{c}_{y_i}$ will be updated as the deep features change.

By combining Eqs. (4) and (5), we can obtain the total loss,

$$\mathcal{L} = \mathcal{L}_s + \lambda \mathcal{L}_c$$

$$= -log \frac{e^{\mathbf{W}_{y_i}^T \mathbf{x}_i + b_{y_i}}}{\sum_{j=1}^{K} e^{\mathbf{W}_j^T \mathbf{x}_i + b_j}} + \frac{\lambda}{2} ||\mathbf{x}_i - \mathbf{c}_{y_i}||^2, \qquad (6)$$

where λ is a scalar to balance the two loss functions. If λ is set to 0, then the joint supervision degrades to a conventional softmax loss.

Intuitively, softmax loss pushes apart the deeply learned features of different classes, while centre loss pulls the deeply learned features of the same class towards their corresponding centres. Therefore, both separability and discriminativity are obtained.

Figure 2 shows a DNN trained with the joint supervision of softmax loss and centre loss. In the figure, 'FC_ Layer' denotes the last hidden layer, of which the output is the deeply learned feature of the neural network that would be fed into the two loss functions.

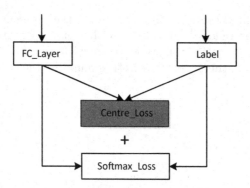

Fig. 2. DNNs with joint supervision of softmax loss and centre loss

3.3   Optimisation

In this part, we verify that the DNNs supervised by centre loss are trainable by calculating the gradients of the total loss function for backward operation. We omit the biases to simplify the analysis. Therefore, we have three variables that need to be updated during training, namely, the weight matrices $\mathbf{W}_l$ between the last hidden layer and the softmax loss layer, the weight matrices $\mathbf{W}_h$ in other hidden layers and centres $\{\mathbf{c}_j | j = 1, ..., K\}$.

The gradients of total loss with respect to $\mathbf{W}_l$ are the same as those of original softmax loss:

$$\frac{\partial \mathcal{L}}{\partial \mathbf{W}_l} = \frac{\partial \mathcal{L}_s}{\partial \mathbf{W}_l}, \qquad (7)$$

The gradients of total loss with respect to $\mathbf{W}_h$ can be obtained as follows by using the chain rule:

$$
\begin{aligned}
\frac{\partial \mathcal{L}}{\partial \mathbf{W}_h} &= \frac{\partial \mathcal{L}_s}{\partial \mathbf{W}_h} + \lambda \frac{\partial \mathcal{L}_c}{\partial \mathbf{W}_h} \\
&= \frac{\partial \mathcal{L}_s}{\partial \mathbf{x}_i} \frac{\partial \mathbf{x}_i}{\partial \mathbf{W}_h} + \lambda \frac{\partial \mathcal{L}_c}{\partial \mathbf{x}_i} \frac{\partial \mathbf{x}_i}{\partial \mathbf{W}_h} \\
&= (1 - g(\mathbf{x}_i)) \mathbf{W}_{y_i} \frac{\partial \mathbf{x}_i}{\partial \mathbf{W}_h} + \lambda (\mathbf{x}_i - \mathbf{c}_{y_i}) \frac{\partial \mathbf{x}_i}{\partial \mathbf{W}_h},
\end{aligned}
\tag{8}
$$

where

$$
g(\mathbf{x}_i) = \frac{e^{\mathbf{W}_{y_i}^T \mathbf{x}_i + b_{y_i}}}{\sum_{j=1}^{K} e^{\mathbf{W}_j^T \mathbf{x}_i + b_j}}
\tag{9}
$$

and calculating $\frac{\partial \mathbf{x}_i}{\partial \mathbf{W}_h}$ is same with the case that uses softmax loss alone.

The update rule for $\mathbf{c}_j$ is presented as follows:

$$
\Delta \mathbf{c}_j = \frac{\sum_{i=1}^{N} \delta(y_i = j)(\mathbf{c}_j - \mathbf{x}_i)}{1 + \sum_{i=1}^{N} \delta(y_i = j)},
\tag{10}
$$

where $\delta(y_i = j) = 1$ if the equation in the bracket holds, and $\delta(y_i = j) = 0$ otherwise, and the centres are updated once for every N input samples.

With these gradients of the weight matrices and the updating rule of the centres, we can optimise the DNNs by using the mini-batch gradient descent method. Algorithm 1 summarises our approach.

Algorithm 1. Optimising DNNs with joint supervision

Input: (a) Segment-level features and their corresponding labels as mentioned in Sect. 2. (b) Hyper parameters including learning rate μ^t, scalar α to control the updating rate of the centres and λ to balance the two loss functions. (c) The number of iterations $t \leftarrow 0$.

Initialise: Weight matrices $\mathbf{W}_l$ between the last hidden layer and the softmax loss layer. Weight matrices $\mathbf{W}_h$ in other hidden layers. Centres $\{\mathbf{c}_j | j = 1, ..., K\}$.

Repeat until convergence:

1:   Compute the deeply learned features, i. e., the output of the last hidden layer, $\{\mathbf{x}_i^t\}$

2:   Compute the total loss $\mathcal{L}^t = \mathcal{L}_s^t + \lambda \mathcal{L}_c^t$

3:   Compute the gradients of the total loss with respect to $\mathbf{x}_i^t$ for each i by $\frac{\partial \mathcal{L}^t}{\partial \mathbf{x}_i^t} = \frac{\partial \mathcal{L}_s^t}{\partial \mathbf{x}_i^t} + \lambda \frac{\partial \mathcal{L}_c^t}{\partial \mathbf{x}_i^t}$

4:   Update $\mathbf{W}_l$: $\mathbf{W}_l^{t+1} \leftarrow \mathbf{W}_l^t - \mu^t \frac{\partial \mathcal{L}^t}{\partial \mathbf{W}_l^t}$

5:   Update $\mathbf{c}_j$: $\mathbf{c}_j^{t+1} \leftarrow \mathbf{c}_j^t - \alpha \Delta \mathbf{c}_j^t$

6:   Update $\mathbf{W}_h$: $\mathbf{W}_h^{t+1} \leftarrow \mathbf{W}_h^t - \mu^t \sum_i \frac{\partial \mathcal{L}^t}{\partial \mathbf{x}_i^t} \frac{\partial \mathbf{x}_i^t}{\partial \mathbf{W}_h^t}$

7:   $t \leftarrow t + 1$

Output: $\mathbf{W}_l, \mathbf{W}_h$

---

4   Experiments and Results

We used three corpora of acted emotions to evaluate the validity and universality of our approach: a Chinese emotional database (CASIA), a German emotional corpus (Emo-DB) and an English emotional database (SAVEE), which are summarized as in Table 1.

Table 1. Overview of The Selected Emotion Corpora Used in The Experiments

	Language	Samples	Subjects	Quality	Sampling	Emotion
CASIA	Chinese	9,600	4 (2 female)	studio	16 kHz	anger, fear, happy, neutral, sad, surprise
Emo-DB	German	535	10 (5 female)	studio	16 kHz	anger, boredom, disgust, fear, joy, neutral, sad
SAVEE	English	480	4 (0 female)	studio	44.1 kHz	anger, disgust, fear, happy, neutral, sad, surprise

4.1   General Experimental Setting

In our experiments, each input signal is converted into frames by using a 25 ms window sliding at 10 ms each time. The size of each speech segment is set to 25 frames, including 12 frames in each side.

All utterances of each database are randomly divided into ten equal parts, among which eight parts are taken as the training data, one part is taken as the validation data, and the remained one is taken as the test data. This procedure is repeated by ten times, and the average classification results across all trials are computed.

The segment-level DNN consists of one input layer, 6 hidden layers, followed by one softmax loss layer. Network configuration (ordered from input to last hidden layer) is set to {156, 512, 512, 512, 512, 512, 256}, where '156' and '256' correspond to the dimension of the input segment-level features and the deeply learned features, respectively. A Sigmoid non-linearity is applied between two consecutive hidden layers. The mini-batch based gradient descent method is used to learn the weights in the DNN with batch size increasing during training (256→1024), which is equivalent to learning rate decay strategy [21]. We utilise the dropout regularisation technique [22] wherever possible with $p = 0.5$ to prevent neural network from overfitting. We also apply batch normalisation [23] between each pair of consecutive layers to accelerate the training process.

We use ELM to perform the utterance-level classification. For ELM training, the number of hidden units is set to 80 with no non-linearity being applied. The whole system is implemented using TensorFlow [24], an open source machine learning library developed by Google.

4.2 Experiment on CASIA Chinese Emotional Corpus

The Chinese emotional corpus was developed by Institute of Automation, Chinese Academy of Sciences (CASIA), which contains 9,600 utterances that are simulated by four subjects (two males and two females) in 6 different emotional states, namely, anger, fear, happiness, neutral, sadness and surprise. In our experiments, we only use 7,200 utterances that correspond to 300 linguistically neutral sentences with same statements. All the categories of emotions are selected. We compared our results with those of state-of-the-art approaches, as shown in Table 2.

Table 2. Speech Emotion Recognition Rate on CASIA Chinese Emotional Corpus (%)

	Anger	Fear	Happ.	Neutr.	Sadn.	Surpri.	Ave.
Wang et al. [3] (2015)	86.0	81.0	81.0	75.0	67.0	84.0	79.0
Sun et al. [25] (2015)	-	-	-	-	-	-	79.0
Yuan et al. [26] (2015)	100.0	-	89.5	81.8	83.8	-	88.8
Sun et al. [27] (2015)	-	-	-	-	-	-	85.1
Li et al. [28] (2017)	83.5	74.4	69.5	85.0	67.8	60.5	73.5
Liu et al. [29] (2018)	-	-	-	-	-	-	89.6
Liu et al. [30] (2018)	-	-	-	-	-	-	90.3
Prop.	89.2	77.5	86.7	98.3	85.0	93.3	88.3
Prop.(+ Centre Loss)	95.0	82.5	89.1	98.3	83.3	97.5	90.9

According to the results in Table 2, for the CASIA corpus, the average recognition rate (90.9%) obtained by the proposed hierarchical method using joint supervision of softmax loss and centre loss is higher than that (88.3%) obtained by using softmax loss alone under the same experimental conditions. This shows that the joint supervision can significantly enhance the discriminative power of deeply learned features, demonstrating the effectiveness of the centre loss. Meanwhile, it is observed that our approach also beats the other published methods by a significant margin.

4.3 Experiment on Berlin Emo-DB Corpus

Berlin Emo-DB German Corpus was collected by the Institute of Communication Science at the Technical University of Berlin. Ten professional actors (five males and five females) each produced ten utterances in German to simulate 7 different emotions. The number of spoken utterances for these 7 emotions in the Berlin Emo-DB is not equally distributed: 126 anger, 81 boredom, 46 disgust, 69 fear, 71 joy, 79 neutral, and 62 sadness. In our experiment, five categories of emotions are selected, i. e., anger, boredom, joy, sadness and neutral. Thus, a total number of 419 utterances are used in this experiment. The results are shown in Table 3.

Table 3. Speech Emotion Recognition Rate on Berlin Emo-DB German Corpus (%)

	Anger	Bored.	Joy	Neutr.	Sadn.	Ave.
Wang et al. [3] (2015)	98.3	71.5	92.9	87.6	91.2	88.3
Sun et al. [25] (2015)	-	-	-	-	-	88.9
Sun et al. [27] (2015)	-	-	-	-	-	89.3
Lim et al. [31] (2016)	86.9	92.0	81.2	93.8	90.7	88.9
Li et al. [28] (2017)	100.0	72.7	100.0	100.0	81.8	90.1
Sun et al. [34] (2017)	-	-	-	-	-	88.7
Human Perform.	-	-	-	-	-	84.3
Prop.	100.0	88.9	62.5	88.9	85.7	87.2
Prop.(+ Centre Loss)	100.0	83.3	68.8	94.4	92.9	89.4

For the Berlin Emo-DB corpus, we could see from Table 3 that our proposed hierarchical method using joint supervision of softmax loss and centre loss outperforms that using softmax loss alone, with average recognition rate 89.4% versus 87.2%. The improvement is also because of discriminative power of deeply learned features introduced by centre loss.

4.4   Experiment on SAVEE Database

The Surrey audio-visual expressed emotion database (SAVEE) consists of recordings from four male actors in 7 different emotions: anger, disgust, fear, happiness, sadness, surprise, and neutral. Each speaker produced 120 utterances. The sentences were chosen from the standard TIMIT corpus and phonetically-balanced for each emotion. In this experiment, all the categories of emotions are selected. In addition, for SAVEE database, 90% of the segments with the highest energy within each utterance are picked for training and test purposes, in order to remove the segments that are dominated by silence frames. The results are shown in Table 4. The pattern is similar to the results of experiments on CASIA and EMO-DB corpus.

4.5   Experiment on Hyper-parameters λ and α

The hyper-parameter λ balances the two loss functions and α controls the updating rate of the centres. Both of them are essential to our approach. So we conduct

Table 4. Speech Emotion Recognition Rate on SAVEE Database (%)

	Anger	Disgu.	Fear	Happ.	Neutr.	Sadn.	Surpri.	Ave.
Sidorov et al. [32] (2014)	-	-	-	-	-	-	-	48.4
Sun et al. [25] (2015)	-	-	-	-	-	-	-	77.4
Sun et al. [27] (2015)	-	-	-	-	-	-	-	75.6
Yogesh et al. [33] (2017)	-	-	-	-	-	-	-	76.2
Sun et al. [34] (2017)	-	-	-	-	-	-	-	76.3
Liu et al. [30] (2018)	-	-	-	-	-	-	-	76.4
Human Perform. [35]	-	-	-	-	-	-	-	66.5
Prop.	55.6	75.0	77.8	80.6	83.3	72.2	69.4	73.4
Prop.(+ Centre Loss)	50.0	86.1	75.0	80.6	83.3	80.6	77.8	76.2

several experiments to investigate the sensitivity of these two hyper-parameters. Due to the space limitation, only CASIA corpus is involved in this part.

We first fix $\alpha = 0.5$ and vary λ from 0 to 10^{-4}. The corresponding validation accuracies on CASIA corpus are shown in Fig. 3 (a). We can observe that only using the traditional softmax loss (i. e., λ is set to 0) is not a good choice and properly choosing the value of λ can boost the performance, which further demonstrate the effectiveness of centre loss. We then fix $\lambda = 10^{-5}$ and vary α from 0.001 to 1.0. The corresponding validation accuracies on the same dataset are illustrated in Fig. 3 (b). It can be observed that the validation performance remains stable across a wide range of α (from 0.2 to 0.6).

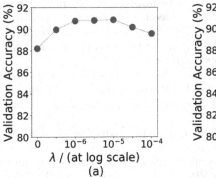
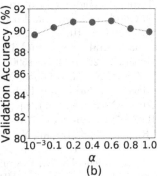

Fig. 3. Validation accuracies on CASIA Chinese Emotional Corpus. (a) Different λ was tested given fixed $\alpha = 0.5$. (b) Different α was tested given fixed $\lambda = 1 \times 10^{-5}$

5 Conclusions

In this paper, we proposed a hierarchical approach for speech emotion recognition. Moreover, in order to enhance the discriminative power of the deeply learned features, we introduced centre loss as a complementary supervision signal to softmax loss. We presented the definition and learning algorithm of centre loss and showed that such algorithm is trainable and easy to implement. Experiments on CASIA corpus (88.3%→90.9%), Emo-DB corpus (87.2%→89.4%) and SAVEE database (73.4%→76.2%) indicated that our approach using joint supervision of softmax loss and centre loss significantly outperforms systems that were trained with softmax loss alone.

References

1. Ververidis, D., Kotropoulos, C.: A state of the art review on emotional speech databases. In: 1st International Workshop on Interactive Rich Media Content Production (RichMedia 2003), Lausanne, Switzerland, pp. 109–119 (2003)
2. Rao, K.S., Koolagudi, S.G.: Emotion Recognition Using Speech Features. Springer, New York (2013). https://doi.org/10.1007/978-1-4614-5143-3
3. Wang, K., An, N., Li, B.N., Zhang, Y., Li, L.: Speech emotion recognition using fourier parameters. IEEE Trans. Affect. Comput. **6**(1), 69–75 (2015)
4. Banse, R., Scherer, K.R.: Acoustic profiles in vocal emotion expression. J. Pers. Soc. Psychol. **70**(3), 614–636 (1996)
5. Lecun, Y., Bengio, Y., Hinton, G.: Deep learning. Nature **521**(7553), 436–444 (2015)
6. Han, K., Yu, D., Tashev, I.: Speech emotion recognition using deep neural network and extreme learning machine. In: Interspeech 2014, Singapore (2014)
7. Lee, J., Tashev, I.: High-level feature representation using recurrent neural network for speech emotion recognition. In: Interspeech 2015, Dresden, Germany (2015)
8. Zhang, S., Zhang, S., Huang, T., Gao, W.: Speech emotion recognition using deep convolutional neural network and discriminant temporal pyramid matching. IEEE Trans. Multimed. **20**(6), 1576–1590 (2018)
9. Wen, Y., Zhang, K., Li, Z., Qiao, Y.: A discriminative feature learning approach for deep face recognition. In: Leibe, B., Matas, J., Sebe, N., Welling, M. (eds.) ECCV 2016. LNCS, vol. 9911, pp. 499–515. Springer, Cham (2016). https://doi.org/10.1007/978-3-319-46478-7_31
10. Hadsell, R., Chopra, S., LeCun, Y.: Dimensionality reduction by learning an invariant mapping. In: CVPR 2006, pp. 1735–1742. IEEE Press, New York (2006)
11. Schroff, F., Kalenichenko, D., Philbin, J.: Facenet: A unified embedding for face recognition and clustering. In: CVPR 2015, pp. 815–823. IEEE Press, Boston (2015)
12. Chen, K., Salman, A.: Extracting speaker-specific information with a regularized siamese deep network. In: NIPS 2011, pp. 298–306, Granada (2011)
13. Zheng, X., Wu, Z., Meng, H., Cai, L.: Contrastive autoencoder for phoneme recognition. In: ICASSP 2014, pp. 2529–2533. IEEE Press, Florence (2014)
14. Bredin, H.: Tristounet: triplet loss for speaker turn embedding. In: ICASSP 2017, pp. 5430–5434. IEEE Press, New Orleans (2017)
15. Wu, Y., Liu, H., Li, J., Fu, Y.: Deep face recognition with center invariant loss. In: Proceedings of the on Thematic Workshops of ACM Multimedia 2017, pp. 408–414. ACM, Mountain View (2017)

16. Burkhardt, F., Paeschke, A., Rolfes, M., Sendlmeier, W.F., Weiss, B.: A database of German emotional speech. In: Interspeech 2005, Lisbon (2005)
17. Haq, S., Jackson, P.J.B., Edge, J.: Speaker-dependent audio-visual emotion recognition. In: AVSP 2009, pp. 53–58. Norfolk (2009)
18. Giannakopoulos, T.: pyaudioanalysis: an open-source python library for audio signal analysis. PLoS ONE 10(12), 1–17 (2015)
19. Tsiakas, K., et al.: A multimodal adaptive dialogue manager for depressive and anxiety disorder screening: a wizard-of-oz experiment. In: Proceedings of the 8th ACM International Conference on PErvasive Technologies Related to Assistive Environments, p. 82. ACM, Corfu (2015)
20. Huang, G.B., Zhu, Q.Y., Siew, C.K.: Extreme learning machine: theory and applications. Neurocomputing 70(1–3), 489–501 (2006)
21. Smith, S.L., Kindermans, P.J., Le, Q.V.: Don't Decay the Learning Rate, Increase the Batch Size (2017). arXiv preprint arXiv:1711.00489
22. Srivastava, N., Hinton, G., Krizhevsky, A., Sutskever, I., Salakhutdinov, R.: Dropout: a simple way to prevent neural networks from overfitting. J. Mach. Learn. Res. 15(1), 1929–1958 (2014)
23. Ioffe, S., Szegedy, C.: Batch normalization: Accelerating deep network training by reducing internal covariate shift. In: ICML 2015, pp. 448–456. Lille (2015)
24. Abadi, M., et al.: Tensorflow: A system for large-scale machine learning. In: OSDI 2016, pp. 265–283. Savannah (2016)
25. Sun, Y., Wen, G.: Emotion recognition using semi-supervised feature selection with speaker normalization. Int. J. Speech Technol. 18(3), 317–331 (2015)
26. Yuan, J., Chen, L., Fan, T., Jia, J.: Dimension reduction of speech emotion feature based on weighted linear discriminate analysis. Image Process. Pattern Recognit. 8, 299–308 (2015)
27. Sun, Y., Wen, G., Wang, J.: Weighted spectral features based on local Hu moments for speech emotion recognition. Biomed. Signal Process. Control 18, 80–90 (2015)
28. Li, C.Z., Liu, F.K., Wang, Y.T., et al.: Speech Emotion Recognition Based on PSO-optimized SVM. In: 2nd International Conference on Software, Multimedia and Communication Engineering (SMCE). Shanghai (2017)
29. Liu, Z.T., Wu, M., Cao, W.H., et al.: Speech emotion recognition based on feature selection and extreme learning machine decision tree. Neurocomputing 273, 271–280 (2018)
30. Liu, Z.T., Xie, Q., Wu, M., Cao, W.H., Mei, Y., Mao, J.W.: Speech emotion recognition based on an improved brain emotion learning model. Neurocomputing 309, 145–156 (2018)
31. Lim, W., Jang, D., Lee, T.: Speech emotion recognition using convolutional and recurrent neural networks. In: APSIPA ASC 2016, pp. 1–4. IEEE Press, Jeju (2016)
32. Sidorov, M., Brester, C., Minker, W., Semenkin, E.: Speech-based emotion recognition: feature selection by self-adaptive multi-criteria genetic algorithm. In: LREC 2014, pp. 3481–3485. Reykjavik (2014)
33. Yogesh, C.K., Hariharan, M., Ngadiran, R., Adom, A.H., Yaacob, S., Polat, K.: Hybrid BBO_PSO and higher order spectral features for emotion and stress recognition from natural speech. Appl. Soft Comput. 56, 217–232 (2017)
34. Sun, Y., Wen, G.: Ensemble softmax regression model for speech emotion recognition. Multimed. Tools Appl. 76(6), 8305–8328 (2017)
35. Haq, S., Jackson, P.J.B.: Multimodal emotion recognition. In: Wang, W.W. (ed.) Machine Audition: Principles, Algorithms and Systems, pp. 398–423. IGI Global Press, Hershey (2010). Chapter 17

Convolutional Neural Network with Spectrogram and Perceptual Features for Speech Emotion Recognition

Linjuan Zhang[1], Longbiao Wang[1(✉)], Jianwu Dang[1,2(✉)], Lili Guo[1],
and Haotian Guan[3]

[1] Tianjin Key Laboratory of Cognitive Computing and Application,
College of Intelligence and Computing, Tianjin University, Tianjin, China
{linjuanzhang,longbiao_wang,liliguo}@tju.edu.cn
[2] Japan Advanced Institute of Science and Technology, Ishikawa, Japan
jdang@jaist.ac.jp
[3] Intelligent Spoken Language Technology (Tianjin) Co., Ltd., Tianjin, China
htguan@huiyan-tech.com

Abstract. Convolutional neural network (CNN) has demonstrated a great power at mining deep information from spectrogram for speech emotion recognition. However, perceptual features such as low-level descriptors (LLDs) and their statistical values were not utilized sufficiently in CNN-based emotion recognition. To solve this problem, we propose novel features to combine spectrogram and perceptual features in different levels. Firstly, frame-level LLDs are arranged as time-sequence LLDs. Then, spectrogram and time-sequence LLDs are fused as compositional spectrographic features (CSF). To fully utilize perceptual features and global information, statistical values of LLDs are added in CSF to generate rich-compositional spectrographic features (RSF). Finally, the proposed features are individually fed to CNN to extract deep features for emotion recognition. Bi-directional long short-term memory was employed to identify emotions and the experiments were conducted on EmoDB. Compared with spectrogram, CSF and RSF improve the unweighted accuracy by a relative error reduction of 32.04% and 36.91%, respectively.

Keywords: Speech emotion recognition · Spectrogram
Perceptual features · Convolutional neural network
Bi-directional long short-term memory

1 Introduction

The field of man-machine communication has witnessed a tremendous improvement in recent years. We still have difficulties in communicating with machines naturally. It is believed that speech emotion is particularly useful in human-computer interface, because the emotion carries the essential semantics and helps

© Springer Nature Switzerland AG 2018
L. Cheng et al. (Eds.): ICONIP 2018, LNCS 11304, pp. 62–71, 2018.
https://doi.org/10.1007/978-3-030-04212-7_6

machines better understand human speech [1]. However, speech emotion recognition is technically challenging because it is not clear what kinds of speech features are salient to efficiently characterize different emotions [2,3]. The aim of this study is to find new affect-salient features for speech emotion recognition.

Conventional speech emotion recognition approaches rely mostly on feature selection. Perceptual features have been intensively selected to estimate the emotion of speakers [2,4,5]. Perceptual features [6] consist of LLDs and statistical features, which are described in Table 1. LLDs are zero-crossing-rate (ZCR) from the time signal, root mean square (RMS) frame energy, fundamental frequency (F0), harmonics-to-noise ratio (HNR) by autocorrelation function, and mel-frequency cepstral coefficients (MFCC) 1–12. To each of these, the delta coefficients are additionally computed. Statistical features are statistical values of LLDs. Utilizing deep neural networks (DNN) to learn deep features from perceptual features is common in speech emotion recognition task. For example, DNN is utilized to obtain probability distribution of emotional state from perceptual features, and extreme learning machine (ELM) is used for classification [7]. Some studies proposed that the combination of bi-directional long short-term memory recurrent neural network (BLSTM-RNN) and full-connected neural network with a model of attention performs well when using perceptual features [8,9]. These studies highlight perceptual features based on deep networks for speech emotion recognition.

Table 1. The composition of perceptual features

LLDs (16*2)	Statistical Values
(Δ)ZCR (Δ)RMS energy	Mean, standard deviation, kurtosis, skewness,
(Δ)F0 (Δ)HNR	Extremes: value, rel. position, range
(Δ)MFCC(1-12)	Linear regression: offset, slope, MSE

Perceptual features are chosen by experience and not comprehensive. Thus it is uncertain whether the features selected from our prior knowledge are adequate for good performance in all the situations. Compare to perceptual features, using spectrogram for speech recognition proved to be successful [10–12]. It is recognized that emotional contents of utterances influence spectral energy in the frequency domain [13]. Deep networks with different structures based on raw spectrogram have shown significantly improvements of speech emotion recognition. In [14–16], they extracted deep features from raw spectrogram with CNN to find good results. These studies indicated that CNN can process spectrogram more effective and help identify emotions.

Comprehensive spectrogram can be obtained from the speech directly not by experience from the prior knowledge. If applying CNN on spectrogram alone, it is difficult to sufficiently learn the prior knowledge of perceptual features for automatic speech emotion recognition. In order to overcome this problem, we propose novel features to utilize the prior knowledge and comprehensive spectrographic information simultaneously. First, frame-level LLDs are arranged in timeline to make time-series LLDs. Then, segmental spectrogram and time-series LLDs are fused as CSF based on timeline. In [19], global features are thought to be important for speech emotion recognition. However, LLDs in CSF contain a wealth of local information. To solve this problem and fully utilize perceptual features, statistical features that are statistical values of LLDs are added in CSF manually to generate RSF. Finally, CNN is employed to extract deep features from our proposed features. It is the first work to combine perceptual features and spectrogram before feature extraction and treat them as 2-D images to be fed into the CNN model so as to extract deep features for emotional classification task.

The outline of this paper is as follows. The baseline system is described in Sect. 2. Section 3 introduces the proposed features and fusion methods. Sections 4 and 5 cover the experiments and conclusions.

2   Baseline System

In this section, our baseline system is described in Fig. 1 according to previous works [15–18]. First, speech signals are split into segments with a fixed length. Secondly, short-time Fourier transform (STFT) are used to transform segmental signals into amplitude spectrogram. When doing STFT, the FFT points are 256. Then, segmental spectrogram are fed to CNN to extract deep spectrogram. Finally, BLSTM is used to identify utterance-level emotions. The baseline system does not make use of the prior knowledge (e.g. F0).

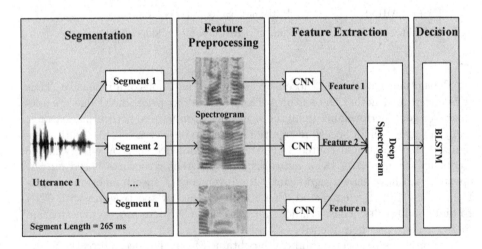

Fig. 1. Structure of the baseline system

The reasons why we primarily focus on CNN-BLSTM are:

(1) Since CNN models temporal and spectral local correlations [20], it is chosen first to extract deep features from the 2-D representations of our proposed features.

(2) Emotion is manifested in speech through a variable range of temporal dependencies. BLSTM is used to recognize the sequential dynamics in an utterance [21]. It is expected that BLSTM network captures long short-term dependent temporal details of the CNN-based features in a consecutive utterance.

3   CNN Based on Spectrogram and Perceptual Features

3.1   Motivation for Fusing Spectrogram and Perceptual Features

In this section, the motivation of the feature fusion is described from visual and theoretical perspective. Figures 2 and 3 show utterance-level spectrogram and time-sequence LLDs on different emotions with the same contents. Figure 2(a) describes the spectrogram of sadness emotion, and Fig. 2(b) depicts neutral emotion. The depth of reddish color implies the level of frequency energy. It is clear from Fig. 2 that the spectrogram of sadness and neutral standing for low-arousal emotions have similar patterns. In order to classify emotions with similar arousal, utilizing LLDs may be useful.

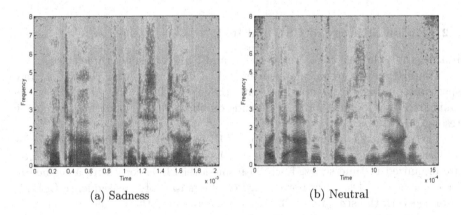

(a) Sadness   (b) Neutral

Fig. 2. Visualization of spectrogram

In Fig. 3, the horizontal axis represents time-domain of utterance. The vertical axis represents the 32-dimensional LLDs. Figure 3(a) and (b) are obviously different, which means the selected LLDs are easier to distinguish similar arousal emotions than spectrogram in this situation. However, it is unclear which kind of features are more effective to distinguish emotions in different cases. In order to adapt the features to various situations, our attempt is to fuse spectrogram and perceptual features for speech emotion recognition.

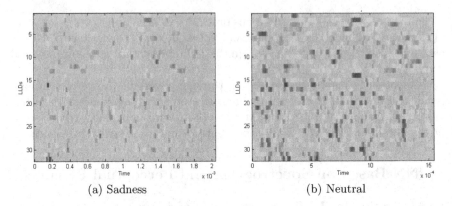

(a) Sadness   (b) Neutral

Fig. 3. Visualization of time-sequence LLDs

From a theoretical perspective, CNN is excellent at mining deep information from raw spectrogram. In the use of wide-band spectrogram, formants are emphasized but F0 is not, whereas F0 is known to compose the main vocal cue for emotion recognition [22]. Perceptual features can provide lots of prior knowledge (e.g. F0) that is useful for emotion recognition. In order to make use of spectrogram and perceptual features simultaneously, we propose novel features such as CSF and RSF for speech emotion recognition.

3.2   Fusion Strategy of CSF and RSF

In this study, the fusion Strategy of CSF and RSF consists of three steps.

The first step is to calculate segmental spectrogram. After the STFT, raw spectrographic matrix is obtained with the size of 25×129, where 25 is the number of time points, and 129 depends on the selected region and frequency resolution.

The next step is to utilize the openSMILE toolkit to get frame-level LLDs and segment-level statistical features that have been described in Table 1. LLDs are organized in time series. Each 25 frame-level LLDs constitute a segmental time-sequence LLDs. After normalization, the matrix of time-sequence LLDs is obtained with the size of 25×32, where 25 represents the number of frames in a segment, and 32 is the dimension of LLDs.

The third step is the feature fusion. Based on the timeline, segmental spectrogram and time-sequence LLDs are spliced together as CSF, where the size of CSF is 25×161. CSF vector of the j-th segment in the i-th utterance can be formulated as:

$$CSF_{ij} = [S_{ij}, L_{ij}], \tag{1}$$

where the S_{ij}, L_{ij} correspond to spectrogram vector and time-sequence LLDs vector of the j-th segment in the i-th utterance, respectively.

In order to splice spectrogram, time-sequence LLDs and statistical features, 384-dimensional statistical features are firstly reduced to 375 dimensions using PCA. Then, the resized statistical features are reshaped as 25 × 15. Finally, segmental spectrogram, time-sequence LLDs and statistical features are spliced together as RSF, where the size of RSF is 25 × 176. RSF vector of the j-th segment in the i-th utterance can be formulated as:

$$RSF_{ij} = [S_{ij}, L_{ij}, C_{ij}], \tag{2}$$

where the C_{ij} represents statistical features vector of the j-th segment in the i-th utterance. Figure 4 depicts the detailed feature extraction of RSF.

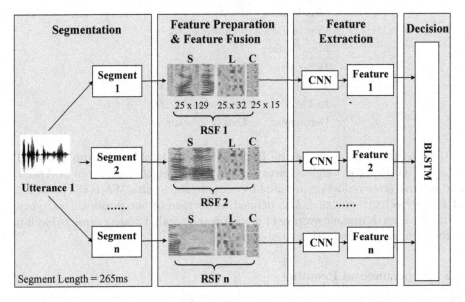

Fig. 4. Extraction of RSF. S represents spectrogram. L represents time-sequence LLDs. C represents statistical features.

4 Experiment

4.1 Experimental Setup

We choose speech materials from the EmoDB [23], which has seven categorical emotion types including disgust, sadness, fear, happiness, neutral, boredom and angry, where the number of utterances in each category are 46, 62, 69, 71, 79, 81 and 127, respectively. There are 535 simulated emotional utterances in German. All the utterances of approximately 2–3 s are sampled at 16000 Hz. The arousal is a descriptor of the intensity of the emotion. In terms of the arousal space [24], angry, fear, disgust and happiness belong to the high-arousal emotion, while, sadness, boredom and neutral belong to the low-arousal emotion.

According to [25], a speech segment contains sufficient emotional contents longer than 250 ms. In our experiment, the utterances are split into segments with a 265-ms window size. Each segment is divided into 25 frames using a 25-ms window, shifting 10 ms each time. About 50,000 segments are collected in this way. Table 2 depicts the detail of the network. Other parameters are also tested for experiments, and the configuration of Table 2 resulted the best performance.

Table 2. Parameters of the CNN-BLSTM network

Layers	Parameters
Convolution 1	32 filters of 5×5
Max-Pooling	2×2
Convolution 2	64 filters of 5×5
Max-Pooling	2×2
Dense layer	Length 1024
LSTM	Bi-directional, 200
LSTM	Bi-directional, 200
Dense layer	Length 7, softmax

Due to the limited size of the Berlin Emotion Database, we run a 10-fold cross validation. The weighted accuracy (WA), unweighted accuracy (UA), F1 and relative error reduction are used to evaluate the results. WA is the accuracy of all the test utterances. UA is defined as average of per emotional category recall. F1 is the harmonic average of precision and recall. Relative error reduction is the ratio of error reduction to original error.

4.2   Experimental Results

This section shows the classification results of our proposed features. From Table 3, we conclude: (1) Compared with spectrogram, the proposed time-sequence LLDs improve the WA and UA by a relative error reduction of 11.23% and 10.29%, respectively. One of the reasons is that time series information of LLDs is used more adequately by BLSTM. Another reason is that selected LLDs perform better than raw spectrogram on a small amount of training data. (2) CSF outperforms spectrogram with 33.76% and 32.04% relative error reduction in terms of WA and UA, respectively. RSF outperforms spectrogram with 38.06% and 36.91% relative error reduction in terms of WA and UA, respectively. The results reveal that spectrogram and perceptual features are complementary. Moreover, our proposed features are significantly effective. (3) RSF performs better than CSF. The results indicate that it is useful to add additional statistical features into CSF. And it is effective to reshape the statistical features and treat them as an additional graph for CNN to learn.

Table 3. WA and UA of different features with CNN-BLSTM

ID	Features	Components	Size	WA (%)	UA (%)
1	Spectrogram (baseline)	Spectrogram	25 × 129	86.73	86.40
2	Time-sequence LLDs (proposed)	LLDs	25 × 32	88.22	87.80
3	CSF (proposed)	Spectrogram + LLDs	25 × 161	91.21	90.76
4	RSF (proposed)	Spectrogram + LLDs + Statistical features	25 × 176	**91.78**	**91.42**

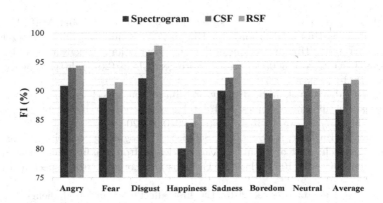

Fig. 5. The F1(%) of different features on different emotions

Figure 5 shows the contribution of proposed features on classifying different kinds of emotion in comparison to spectrogram (baseline features). (1) The results of CSF and RSF are both better than spectrogram on all kinds of emotions, especially on boredom emotion. (2) RSF performs better than CSF on most kinds of emotions. However, when classifying boredom and neutral, CSF performs better than RSF. We assume that there is no noticeable changes on LLDs in both boredom and neutral utterances. Therefore, it is unnecessary to add extra statistical features in this situation. (3) On average F1 of seven emotions, CSF and RSF significantly outperform spectrogram by relative error reduction of 33.68% and 38.80%, respectively. Overall, both CSF and RSF are effective for identifying different categories of emotions.

5   Conclusions and Future Works

In this paper, we first proposed time-sequence LLDs, CSF and RSF for speech emotion recognition. Then, the proposed features were individually fed into the CNN model to extract deep features. Finally, the BLSTM was employed to do final classification. It is the first work to combine spectrogram and perceptual features simultaneously for speech emotion recognition. Our results indicated that spectrogram and perceptual features were complementary and our proposed features were effective.

For future work, we will evaluate our proposed features on other large datasets and consider integrating speaker, gender and linguistic features in our experiment.

Acknowledgments. The research was supported by the National Natural Science Foundation of China (No. 61771333 and No. U1736219) and JSPS KAKENHI Grant (16K00297).

References

1. Kołakowska, A., Landowska, A., Szwoch, M., Szwoch, W., Wróbel, M.R.: Emotion recognition and its applications. In: Hippe, Z.S., Kulikowski, J.L., Mroczek, T., Wtorek, J. (eds.) Human-Computer Systems Interaction: Backgrounds and Applications 3. AISC, vol. 300, pp. 51–62. Springer, Cham (2014). https://doi.org/10.1007/978-3-319-08491-6_5
2. El Ayadi, M., Kamel, M.S., Karray, F.: Survey on speech emotion recognition: features, classification schemes, and databases. Pattern Recognit. **44**(3), 572–587 (2011). https://doi.org/10.1016/j.patcog.2010.09.020
3. Schuller, B., Batliner, A., Steidl, S., Seppi, D.: Recognising realistic emotions and affect in speech: state of the art and lessons learnt from the first challenge. Speech Commun. **53**(9–10), 1062–1087 (2011). https://doi.org/10.1016/j.specom.2011.01.011
4. Ringeval, F., et al.: Av+ ec 2015: the first affect recognition challenge bridging across audio, video, and physiological data. In: 5th International Workshop on Audio/Visual Emotion Challenge, pp. 3–8. ACM (2015). https://doi.org/10.1145/2808196.2811642
5. Valstar, M., et al.: Avec 2016: depression, mood, and emotion recognition workshop and challenge. In: 6th International Workshop on Audio/Visual Emotion Challenge, pp. 3–10. ACM (2016). https://doi.org/10.1145/2964284.2980532
6. Schuller, B., Steidl, S., Batliner, A.: The Interspeech 2009 emotion challenge. In: Tenth Annual Conference of the International Speech Communication Association (2009)
7. Han, K., Yu, D., Tashev, I.: Speech emotion recognition using deep neural network and extreme learning machine. In: INTERSPEECH, pp. 223–227 (2014). https://www.microsoft.com/en-us/research/publication/speech-emotion-recognition-using-deep-neural-network-and-extreme-learning-machine/
8. Huang, C. W., Narayanan, S. S.: Attention assisted discovery of sub-utterance structure in speech emotion recognition. In: INTERSPEECH, pp. 1387–1391 (2016). https://doi.org/10.21437/interspeech.2016-448
9. Mirsamadi, S., Barsoum, E., Zhang, C.: Automatic speech emotion recognition using recurrent neural networks with local attention. In: IEEE International Conference on Acoustics, Speech and Signal Processing (ICASSP), pp. 2227–2231. IEEE (2017). https://doi.org/10.1109/icassp.2017.7952552
10. Variani, E., Lei, X., McDermott, E., Moreno, I. L., Gonzalez-Dominguez, J.: Deep neural networks for small footprint text-dependent speaker verification. In: IEEE International Conference on Acoustics, Speech and Signal Processing (ICASSP), pp. 4052–4056. IEEE (2014). https://doi.org/10.1109/icassp.2014.6854363

11. Hannun, A., et al.: Deep Speech: Scaling up End-to-end Speech Recognition (2014). http://arxiv.org/abs/1412.5567
12. Amodei, D., et al.: Deep Speech 2: end-to-end speech recognition in English and Mandarin. In: International Conference on Machine Learning, pp. 173–182 (2016). http://dl.acm.org/citation.cfm?id=3045390.3045410
13. Nwe, T.L., Foo, S.W., De Silva, L.C.: Speech emotion recognition using hidden markov models. Speech Commun. **41**(4), 603–623 (2003). https://doi.org/10.1016/S0167-6393(03)00099-2
14. Huang, Z., Dong, M., Mao, Q., Zhan, Y.: Speech emotion recognition using CNN. In: 22nd ACM international conference on Multimedia, pp. 801–804. ACM (2014). http://doi.acm.org/10.1145/2647868.2654984
15. Lim, W., Jang, D., Lee, T.: Speech emotion recognition using convolutional and recurrent neural networks. In: Signal and Information Processing Association Annual Summit and Conference (APSIPA), pp. 1–4. IEEE, Asia-Pacific (2016). https://doi.org/10.1109/apsipa.2016.7820699
16. Satt, A., Rozenberg, S., Hoory, R.: Efficient emotion recognition from speech using deep learning on spectrograms. In: INTERSPEECH, pp. 1089–1093 (2017). https://doi.org/10.21437/interspeech.2017-200
17. Guo, L., Wang, L., Dang, J., Zhang, L., Guan, H.: A feature fusion method based on extreme learning machine for speech emotion recognition. In: IEEE International Conference on Acoustics, Speech and Signal Processing, pp. 2666–2670 (2018). https://doi.org/10.1109/icassp.2018.8462219
18. Guo, L., Wang, L., Dang, J., Zhang, L., Guan, H., Li, X.: Speech emotion recognition by combining amplitude and phase information using convolutional neural network. In: INTERSPEECH, pp. 1611–1615 (2018). https://doi.org/10.21437/interspeech.2018-2156
19. Hu, H., Xu, M.X., Wu, W.: Fusion of global statistical and segmental spectral features for speech emotion recognition. In: INTERSPEECH, pp. 2269–2272 (2007)
20. Yu, D., et al..: Deep convolutional neural networks with layer-wise context expansion and attention. In: INTERSPEECH, pp. 17–21 (2016). https://doi.org/10.21437/interspeech.2016-251
21. Lee, J., Tashev, I.: High-level feature representation using recurrent neural network for speech emotion recognition. In: Sixteenth Annual Conference of the International Speech Communication Association (2015). https://www.microsoft.com/en-us/research/publication/high-level-feature-representation-using-recurrent-neural-network-for-speech-emotion-recognition/
22. Petrushin, V. A.: Emotion recognition in speech signal: experimental study, development, and application. In: Sixth International Conference on Spoken Language Processing, pp. 222–225 (2000)
23. Burkhardt, F., Paeschke, A., Rolfes, M., Sendlmeier, W. F., Weiss, B.: A Database of German Emotional Speech. In: Ninth European Conference on Speech Communication and Technology, pp. 1517–1520 (2005)
24. Xie, B.: Research on Key Issues of Mandarin Speech Emotion Recognition [Ph.D. Thesis]. Hangzhou: Zhejiang University (2006)
25. Provost, E. M.: Identifying salient sub-utterance emotion dynamics using flexible units and estimates of affective flow. In: IEEE International Conference on Acoustics, Speech and Signal Processing (ICASSP), pp. 3682–3686. IEEE (2013). https://doi.org/10.1109/icassp.2013.6638345

Feature Selection Based on Fuzzy Conditional Distinction Degree

Qilai Zhang[1] and Jianhua Dai[2(✉)]

[1] School of Computer Science and Technology, Tianjin University,
Tianjin 300350, China
zql_wolf@tju.edu.cn
[2] Hunan Provincial Key Laboratory of Intelligent Computing and Language
Information Processing, College of Information Science and Engineering,
Hunan Normal University, Changsha 410081, Hunan, China
jhdai@hunnu.edu.cn

Abstract. Previous studies have shown that information entropy and its variants are useful at reducing data dimensionality. Yet, most existing approaches based on entropy exploit the correlations between features and labels, lacking of taking into account the relevance between features. In this paper, we propose a new index for feature selection, named fuzzy conditional distinction degree (FDD), based on fuzzy similarity relation by combining feature correlations with the relationship between features and labels. Different from existing approaches based on entropy, FDD considers the cardinality of the relation matrix instead of the similarity classes. Meanwhile, we encode the feature correlations into distance to measure the relevance of any two features. Some useful properties are discussed. Based on the FDD, a greedy forward algorithm for feature selection is presented. Experimental results on benchmark data sets denote the feasibility and effectiveness of the proposed approach.

Keywords: Feature selection · Fuzzy distinction degree
Dimension reduction

1 Introduction

In machine learning, data often contain redundant features, which can lead to heavy storage burden and high time-consuming [11]. The enormous pointless features may bring about performance degradation of the learning algorithms. To mitigate this problem, feature selection has become a necessity.

During the past few years, mutual information based approaches have been widely used in feature selection [1,10]. These methods employ a greedy scheme to select representative features one by one. It's noticeable that these approaches are prone to obtain suboptimal feature subset due to the computational complexity.

Generally, conditional entropy and its variants are extensively used for feature selection. Some kinds of conditional entropies do not have monotonicity.

© Springer Nature Switzerland AG 2018
L. Cheng et al. (Eds.): ICONIP 2018, LNCS 11304, pp. 72–83, 2018.
https://doi.org/10.1007/978-3-030-04212-7_7

Thus, they are not reasonable to be used as indices for evaluating the distinguishment ability of feature subset [2,4]. Compared with Yager's entropy [17] and its varieties [7,8], the computational complexity of using the cardinality of relations is smaller. Such as the neighborhood discrimination index proposed by Wang [13], which considers the cardinality of relations between the selected features and labels. However, it lacks of exploring the relevance between features.

Encoding the influence of feature correlations and guaranteeing the monotonic of the proposed index are the main motivations of this paper. Hence, we propose a fuzzy conditional distinction degree (FDD) based on fuzzy rough set [12,15] by computing the cardinality of relationship instead of the class, which measures the distinguishment ability of a feature subset. Based on FDD, a greedy scheme is employed to get the final feature subset. In particular, the advantages of FDD are summarized as follows:

- From the viewpoint of the cardinality of relation, we propose a new index for feature selection. The monotonicity has been proved theoretically and experimentally in this paper. Hence, the proposed FDD is reasonable and effective to be used as indexes for mutual information based approaches.
- We encode the influence of feature correlations into distance to quantify the relevance of any two features. This is a considerable difference from neighborhood discrimination index [13].

2   Preliminaries

In the following, a data set is called as a decision table, denoted as $T = \langle U, A, V, f \rangle$, where $U = \{u_1, u_2, \cdots u_n\}$ is a nonempty finite set of objects; $A = C \cup D$, where $C = \{a_1, a_2, \cdots a_m\}$ is feature set and $D = \{d\}$ is the set of labels. V is the union of feature domains, $V = \bigcup_{a \in A} V_a$, where V_a is the value set of feature a, called the domain of a; $f : U \times A \to V$ is an information function which assigns particular values from domains of feature to objects such as $\forall a \in A, x \in U, f(a, x) \in V_a$, where $f(a, x)$ denotes the values of feature a for object x. An equivalence relation (also called as indiscernibility relation) is defined as:

$$Ind(B) = \{(x, y) | f(a, x) = f(a, y), \forall a \in B\} \tag{1}$$

where $x, y \in U$ and $B \subseteq C$. According to the equivalence relation, the equivalence class of Ind(B) containing x can be denoted as $[x]_B$:

$$[x]_B = \{y | (x, y) \in Ind(B)\}$$

The family of all equivalence classes forms a partition, denoted as $U/D = \{P_1, P_2, \cdots P_r\}$. Meanwhile, $P = (p_{ij}) \in R^{n \times n}$ is an equivalence matrix induced by U/D, where $p_{ij} = 1$ if instances i and j satisfy $i, j \in P_k, \exists k \in \{1, 2, \cdots, r\}$, otherwise $p_{ij} = 0$. In what follows, $|\cdot|$ represents the cardinality of a set or matrix. In this paper, we use fuzzy similarity relation [9] to capture the correlations.

For a given decision table $T = \langle U, A, V, f \rangle$, where $U = \{u_1, u_2, \cdots u_n\}$, and $B \subseteq C$, $\tilde{R}_B$ is a fuzzy similarity relation induced by B defined on U if $\tilde{R}_B$ satisfies

- Reflectivity: $\tilde{R}_B (x, x) = 1$, for $\forall x \in U$.
- Symmetry: $\tilde{R}_B (x, y) = \tilde{R}_B (y, x)$, for $\forall x, y \in U$.
- T-transitivity: $\tilde{R}_B (x, x) \geq \left(\tilde{R}_B (x, y) \wedge \tilde{R}_B (y, x) \right)$.

Induced by $\tilde{R}_B$, an instance similarity matrix M_B is denoted as:

$$M_B = (m_{ij}) \in R^{n \times n} \tag{2}$$

where $m_{ij} = \underset{\forall a \in B}{\cap} \tilde{R}_a (i, j)$

$\tilde{R}_a (i, j)$ is the degree to which instances i and j are similar for feature a. Usually, there are many operators can be used to construct the similarity, for example

$$\tilde{R}_a (x, y) = 1 - \frac{|a(x) - a(y)|}{|a_{\max} - a_{\min}|} \tag{3}$$

$$\tilde{R}_a (x, y) = \exp \left(-\frac{(a(x) - a(y))^2}{2\sigma_a^2} \right) \tag{4}$$

where $a \in C$ and σ_a is standard deviation of feature a.

We expect to encode the feature correlations into distance to measure the relevance of any two features to improve the performance of learning algorithms. Based on this motivation, a new index, called fuzzy conditional distinction degree is introduced to compute the distinguishing ability of subset features.

3   Fuzzy Conditional Distinction Degree

In this section, fuzzy conditional distinction degree is proposed to identify salient features and some properties are discussed.

Given a decision table $T = \langle U, A, V, f \rangle$, and $B \subseteq C$, $\tilde{R}_B$ is a fuzzy similarity relation induced by B defined on U. Then, an instance similarity matrix on the universe can be denoted as M_B. M_B represents the minimum similarity between samples on the feature set B. In order to measure the relevance of any two features, we encode the feature correlations into distance by using a crisp similarity relation. A crisp similarity relation R_B on feature set B with respect to U can be represented as:

$$S_B = (s_{ij}) \in R^{m \times m}$$

where $|C| = m$. For arbitrary features i and j, the value of s_{ij} is calculated as follows:

$$s_{ij} = \begin{cases} 1, & i = j \\ \theta(i, j), & i, j \in B \wedge i \neq j \\ 0, & otherwise \end{cases}$$

and $\theta(i,j) = \sqrt{\sum_{k=1}^{n}(x_{ki} - x_{kj})^2}$ where x_{ki} represents the value of sample x_k at feature i and $|U| = n$. One thing to emphasize is that we define $s_{ii} = 1$ to avoid the situation that the cardinality of S_B is 0. That is to say, if $B = \emptyset$, the feature relevance matrix $S_B = I$ and I is the identity matrix. In addition, we only pay attention to the selected feature set B not all feature set C. It's obvious that S_B represents the strength of the relevance between features. S_B is a symmetric matrix with diagonal equal to 1. Especially, if only one feature a is selected, then the feature relevance matrix S_B is the identity matrix $I \in R^{m \times m}$, too.

According to the definitions of M_B and S_B, it's easy to get the following properties.

Theorem 1. *Let $T = \langle U, A, V, f \rangle$ be a decision table, $A = C \cup D$, B_1, B_2 are subsets of C, and $B_1 \subseteq B_2$, then*

$$|M_{B_2}| \leq |M_{B_1}| \quad and \quad |S_{B_1}| \leq |S_{B_2}|$$

This theorem shows that the instance similarity matrix M_B is monotonically decreasing with increasing the number of features. But in contrast with M_B, the feature relevance matrix S_B is monotonically increasing.

Theorem 2. *Let $T = \langle U, A, V, f \rangle$ be a decision table. B_1 and B_2 are two subset features. S_B is feature relevance matrix on feature set B with respect to U. Then*

$$S_{B_1 \cup B_2} = S_{B_1} \cup S_{B_2} \quad and \quad S_{B_1 \cap B_2} = S_{B_1} \cap S_{B_2} \qquad (5)$$

Obviously, the feature relevance matrix satisfies the law of combination. Especially, when $B_1 \subseteq B_2$, we have

$$S_{B_1 \cup B_2} = S_{B_2} \quad and \quad S_{B_1 \cap B_2} = S_{B_1}$$

It means that the discriminative power of core features is invariable as increasing feature sets. Meanwhile, expanding feature sets can increase the diversity of features.

Theorem 3. *Given a decision table T, M_{B_1} and M_{B_2} are fuzzy similarity matrices induced by B_1 and B_2, respectively. If $B_1 \subseteq B_2$, then*

$$M_{B_1} \cap M_{B_2} = M_{B_2} \quad and \quad M_{B_1} \cup M_{B_2} = M_{B_1}$$

From the definition of M_B and S_B, it's obvious that they are based on pairwise relations between features, not addressing the joint contribution of three or more features. It's the prior assumption of independent between features and instances in the proposed method.

Now, we introduce a new index, named fuzzy conditional distinction degree, to capture the intrinsic structure of feature subset by considering feature relevance and sample similarity.

Definition 1. *Let $T = \langle U, A, V, f \rangle$ be a decision table, $A = C \cup D$. For any subset $B \subseteq C$, M_B and S_B are instance similarity matrix and feature relevance matrix induced by B, respectively. P is an equivalence matrix induced by U/D. The fuzzy conditional distinction degree (FDD) is defined as:*

$$FDD(D|B) = \frac{|S_C|}{|S_B|} \log \frac{|P \cap M_B|}{|P \cap M_C|} \tag{6}$$

The fuzzy conditional distinction degree measures the discriminative power of feature subsets relative to all features. It indicates that $FDD(D|B)$ reveals the discriminative capability of B. In addition, let

$$F(B) = \frac{|S_C|}{|S_B|} \quad \text{and} \quad I(D|B) = \log \frac{|P \cap M_B|}{|P \cap M_C|}$$

where $F(B)$ represents the rate of feature relevance degree of B on C and $I(D|B)$ is the difference degree of B on C. We have $F(B) \geq 1$ and $I(D|B) \geq 0$. According to the Theorem 1, $F(B)$ and $I(D|B)$ are monotonically decreasing. In particular, $F(B) = 1$, $I(D|B) = 0$, and $FDD(D|B) = 0$ if $B = C$.

Theorem 4. *If $B_1 \subseteq B_2$, then*

$$FDD(D|B_2) \leq FDD(D|B_1) \tag{7}$$

Proof. If $B_1 \subseteq B_2$, due to Theorem 1 we have

$$F(B_2) \leq F(B_1) \quad \text{and} \quad I(D|B_2) \leq I(D|B_1)$$

According to the definitions of $I(D|B)$ and $F(B)$, the fuzzy conditional distinction degree can be rewritten as: $FDD(D|B) = F(B) I(D|B)$. From the perspective of function, by deriving the above formula, obviously

$$(FDD(D|B))' = (F(B) I(D|B))' \tag{8}$$

$$= F(B)' I(D|B) + F(B) I(D|B)' \tag{9}$$

What's more, $F(B)' \leq 0$ and $I(D|B)' \leq 0$ are apparent due to the monotonicity of $F(B)$ and $I(D|B)$. Meanwhile, $I(D|B) \geq 0$ and $F(B) \geq 1$ are established. Hence, $(FDD(D|B))' \leq 0$ is satisfied. The fuzzy conditional distinction degree $FDD(D|B)$ is decreasing with increasing the number of features. Therefore, we can get $FDD(D|B_2) \leq FDD(D|B_1)$. □

This theorem reveals the difference of data distribution under feature sets B and C. As increasing the number of features, the difference becomes smaller. In machine learning, finding a feature subset, which is relevant to the learning task, is a commonly used method. Supposed that $FDD(D|C)$ represents the structure information of the original data and $FDD(D|R)$ represents the structure information obtained by using the feature subset R, the goal of objective function is to find a subset feature R:

$$\min_{R} \| FDD(D|R) - FDD(D|C) \|$$

Because of the monotonicity of fuzzy conditional distinction degree, $DD(D|C) = 0$, it's easy to find that R^* is called a reduct of C relative to decision D:

$$R^* = \arg \min_{R \subseteq C} FDD(D|R)$$
$$= \arg \min_{R \subseteq C} F(R)I(D|R) \tag{10}$$

That is to say, if a subset of features denoted as R^* is a reduct of C relative to decision D, then it satisfies such properties in theory:

- $FDD(D|R^*) = FDD(D|C)$
- $\forall a \in R^* \Rightarrow FDD(D|R^* - \{a\}) > FDD(D|R^*)$

Hence, the less a decision attribute D has fuzzy conditional distinction degree with respect to a feature subset, the more important the feature subset is. Adding a new feature into the selected feature set, the fuzzy conditional distinction degree will decrease. The decrement of fuzzy conditional distinction degree reveals the increment of distinction ability produced by a new feature subset. In this case, the significance of a feature can be defined as follows.

Definition 2. *Supposed a decision table* $T = \langle U, A, V, f \rangle$, $A = C \cup D$, $B \subseteq C$ *and* $a \in C - B$, *the significance degree of feature* a *with respect to* B *and* D *is defined as:*

$$SIG(a, B, D) = FDD(D|B \cup \{a\}) - FDD(D|B) \tag{11}$$

Especially, we define $SIG(a, B, D) = DD(D|a)$ if $B = \emptyset$. Before we describe the algorithm, it's worth to well-considered the details of SIG, which represents the maximum step length of descent. Considering the convergence speed, we use the gradient ∇ as the termination condition instead of SIG.

$$\nabla(a, B) = \frac{SIG(a, B, D)}{FDD(D|B)}$$

According to the above definition, the algorithm is described as:

Algorithm 1. Feature Selection based on fuzzy conditional distinction degree with heuristic search strategy (DDFS)

Input: A decision table $T = \langle U, A, V, f \rangle$,
Output: One reduct R;
1: $R \leftarrow \emptyset$
2: $R \leftarrow a*$, where $a*$ maximises $FDD(D|a)$ in $C - R$
3: calculate the value of $\nabla(a, R)$ for all $a \in C - R$
4: $R \leftarrow a_k$, where a_k maximises $\nabla(a, R)$ for all $a \in C - R$
5: if $\nabla(a, B) > \delta$, then goto the Step 3. Otherwise, return the reduct R.

The parameter δ is a threshold to stop the loop when the change is small in each iteration. As described above, the DDFS algorithm terminates when the gradient $\nabla(a, B)$ of $FDD(D|B)$ is less than δ, which means that the addition of any remaining features does not obviously decrease the $FDD(D|B)$. If there are n samples and m features, the time complexity for computing the fuzzy similarity relations is $O\left(mn^2\right)$. By greedy scheme, the worst search time for a reduct is m^2. Hence, the overall time complexity of DDFS is $O\left(mn^2 + m^2\right)$.

4   Experiments

In this section, we conduct experiments to verify the effectiveness of the proposed DDFS method on several data sets. Meanwhile, we compare DDFS with other representative feature selection methods.

4.1   Datasets and Experimental Setup

Datasets: All datasets are available at the UCI[1] repository and Keukemiaa[2]. The number of features in all datasets varies from 15 to 11225. The characteristics of datasets are described in Table 1. The meaning of abbreviations is as follows.

NI: the number of instance; NF: the number of features; NC: the number of labels.

Table 1. The main details of used data sets in the experiments

Dataset	Abbre	NI	NF	NC
credit	credit	635	15	2
congressEW	congress	435	16	2
parkinsonsEW	park	195	22	2
wdbc	wdbc	569	30	2
splice	splice	3175	60	3
musk1	musk	476	166	2
pengcolonEW	colon	62	2000	2
Hepatocellular	hepat	33	7130	2
Leukemia	leu	72	11225	3

Comparison: We compare DDFS with five representative feature selection methods: CFS [6], FCBF [18], FRS [3], HANDI [13] and NFRS [14]. In addition to feature selection algorithms, we also select two different learning algorithms: KNN and C4.5 [16], to evaluate the accuracy of selected features. Each data

---

[1] https://archive.ics.uci.edu/ml/datasets.html.
[2] http://www.ntu.edu.sg/home/elhchen/data.htm.

set are divided into train and test data. After selecting the most representative features and training the learning algorithms on train data, we further use KNN and C4.5 to evaluate the performance on test data. Since the KNN algorithm is susceptible to initialization results, we performed a random initialization and repeat 20 times. The average results are recorded. In addition, we also use 10-fold cross-validation to ensure the accuracy of experimental results. During the result recording, we record the mean of all the algorithms under the evaluation criteria. To verify the effectiveness, a widely used evaluation metrics, classification accuracy, is employed to assess the performance. The larger metrics is, the better performance is.

Parameter setting: There are two parameters in HANDI [13] algorithm, ε and δ. The parameter δ is set as 0.001 for low dimensional data and 0.01 for high dimensional data. The parameter ε varies from 0 to 1 with a step of 0.05 to select an optimal feature subset for each data set. For the algorithm NFRS [14], we set ε to a value between 0.1 and 0.5 in steps of 0.05 and λ to a value between 0.1 and 0.6 in steps of 0.1. As for δ in FCBF [18], the value of δ is same as that in HANDI.

4.2 Monotonicity Experiment

Figure 1 shows the results of the monotonicity, where x axis represents the number of features and y axis denotes the corresponding values of $D(D|B)$, $I(D|B)$ and $F(B)$, respectively. From Fig. 1, it's noticeable that the values of $FDD(D|B)$, $I(D|B)$ and $F(B)$ are decreasing gradually with increasing the number of features. Meanwhile, it's obvious that $FDD(D|B)$ is equal to zero when all attributes are selected. In other words, all samples are distinguishable in this case. With only partial features selected, the value of $FDD(D|B)$ is almost equal to zero on some data sets. It is noticeable that with increasing the number of features, the uncertainty is reduced leading to the decrementing of

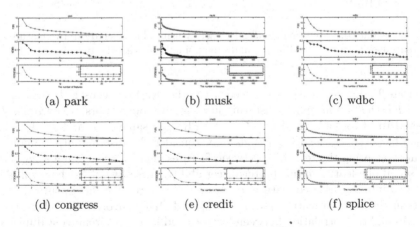

(a) park   (b) musk   (c) wdbc

(d) congress   (e) credit   (f) splice

Fig. 1. The monotonicity of $F(B)$, $I(B)$ and $FDD(D|B)$ on different data sets.

fuzzy conditional distinction degree. This property guarantees the validity and feasibility of the $FDD(D|B)$ in feature selection.

4.3   Redundancy Experiment

As mentioned above, the $FDD(D|B)$ takes into account the correlations between features and the similarity between instances and labels. Taking *wdbc* data set as an example. Let $T = \langle U, A, V, f \rangle$ represent *wdbc* and $A = C \cup D$, where U represents the samples and C is the feature set of *wdbc*. It has 30 features and two classes. Then, a copy of C is generated, named C'. Finally, we get a new data set $T' = \langle U, A', V, f \rangle$ and $A' = (C \cup C') \cup D$, where it satisfies $a \in C$, $a' \in C'$. In order to show the ability of $FDD(D|B)$ to distinguish redundant features, we present the distributions of the data in two-dimension feature space with the top two selected features by different algorithms in Fig. 2.

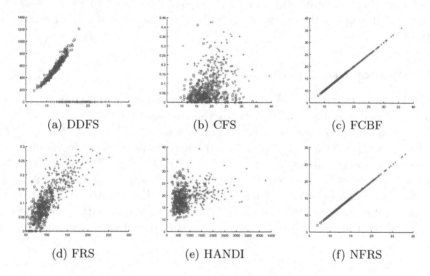

Fig. 2. The data distribution on *wdbc* data set by using the selected top two features of different algorithms. The red points and blue points represent the two classes in *wdbc*.

From Fig. 2, it's easily observed that the features chosen by the proposed $FDD(D|B)$ have best discernibility than its competitors. Meanwhile, the selected top two features of FCBF and NFRS are same in T and T'. It's due to the fact that the two algorithms only consider the discriminative power of features, ignoring the relevance between features. That is to say, in a redundant dataset, if one feature a has better discernibility, these two algorithms will add feature a into the selected feature set, even though a copy of this feature already exists in the selected feature set. The proposed DDFS involves the relevance of features and the correlation between features and labels to circumvent duplicate attributes.

4.4   Classification Experiment

Average accuracy(acc) and standard deviation(std) are calculated to represent the performance of classification by KNN and C45. The results are shown in Tables 2 and 3 in the form of acc±std(rank value), where the bold symbols highlight the highest classification accuracies among the selected features and the rank value represents the merits of learning performance. The smaller rank value is, the better performance is.

Table 2. The performance of KNN classifier on different algorithms

DataSet	Raw Data	HANDI	FCBF	FRS	NFRS	CFS	DDFS
credit	62.630.2273(7)	86.06±0.0382(3)	86.02±0.0239(4)	85.78±0.0402(5)	86.22±0.0361(2)	70.25±0.0224(6)	**86.71±0.0443(1)**
congress	87.36±0.2036(7)	95.49±0.0228(2)	94.92±0.0282(4)	94.87±0.0226(5)	95.3±0.0644(3)	88.56±0.0885(6)	**95.63±0.0125(1)**
park	84.14±0.0988(6)	86.84±0.055(2)	84.18±0.025(5)	85.56±0.1219(4)	86.6±0.0551(3)	79.98±0.1753(7)	**86.67±0.0217(1)**
wdbc	92.18±0.0161(6)	95.67±0.0181(4)	88.05±0.0559(7)	95.72±0.0549(3)	96.17±0.0294(2)	95.25±0.0266(5)	**96.18±0.0835(1)**
splice	79.99±0.0114(6)	89.64±0.0301(5)	94.62±0.0189(4)	94.82±0.0232(3)	94.83±0.02(2)	51.91±0.0514(7)	**95.19±0.0257(1)**
musk	57.8±0.195(7)	83.62±0.1353(4)	83.7±0.1037(3)	86.97±0.1197(2)	82.75±0.2515(5)	73.63±0.0706(6)	**87.27±0.097(1)**
colon	74.6±0.585(5)	77.6±0.0769(3)	**83.87±0.2229(1)**	64.52±0.1966(7)	77.4±0.7329(4)	64.52±0.1177(6)	83.54±0.1265(2)
hepat	63.64±0.2895(7)	**90.91±0.2873(1)**	89.39±0.0835(2)	72.43±0.6537(4)	69.81±0.1444(5)	63.64±0.1641(6)	81.17±0.2808(3)
leu	63.64±0.2895(7)	89.39±0.0835(2)	81.17±0.2808(3)	72.43±0.6537(4)	69.81±0.1444(5)	63.64±0.1641(6)	**90.91±0.2873(1)**

Table 3. The performance of C4.5 classifier on different algorithms

DataSet	Raw Data	HANDI	FCBF	FRS	NFRS	CFS	DDFS
credit	62.63±0.2912(7)	86.76±0.0628(3)	85.88±0.0098(4)	85.01±0.0421(5)	86.92±0.0535(2)	70.34±0.015(6)	**87.00±0.0702(1)**
congress	87.36±0.0473(7)	95.63±0.0493(2)	95.18±0.0047(3)	94.43±0.0219(5)	94.82±0.0211(4)	88.95±0.0302(6)	**95.83±0.0524(1)**
park	85.13±0.1265(6)	88.21±0.2747(2)	86.15±0.2401(5)	87.44±0.0339(3)	87.05±0.069(4)	80.51±0.0426(7)	**88.72±0.1456(1)**
wdbc	94.85±0.0354(5)	95.79±0.0323(3)	92.27±0.2802(6)	95.61±0.0524(4)	96.00±0.0356(2)	87.97±0.1072(7)	**96.29±0.0457(1)**
splice	51.91±0.2547(7)	89.65±0.0347(5)	94.03±0.0137(4)	94.54±0.0252(2)	94.52±0.0345(3)	79.94±0.0079(6)	**94.55±0.0202(1)**
musk	74.58±0.0715(6)	88.09±0.0723(3)	88.69±0.041(2)	84.9±0.1035(5)	87.43±0.0425(4)	60.50±0.2872(7)	**89.37±0.1023(1)**
colon	80.62±0.2356(5)	82.55±0.5331(3)	82.26±0.2377(4)	74.52±0.1967(6)	**84.02±0.0999(1)**	64.52±0.0107(7)	83.87±0.2878(2)
hepat	63.64±0.0293(7)	89.39±0.2444(2)	75.12±0.1897(3)	69.35±0.6595(5)	69.70±0.2740(4)	63.64±0.2717(6)	**90.91±0.0381(1)**
leu	63.64±0.0293(7)	89.39±0.2444(2)	75.12±0.1897(3)	69.35±0.6595(5)	69.70±0.2740(4)	63.64±0.2717(6)	**90.91±0.0381(1)**

Tables 2 and 3 give the classification performance of HANDI, FCBF, FRS, NFRS, CFS, and DDFS. Considering classification accuracy, we can conclude that DDFS is better than other algorithms with respect to the KNN classifier and C45 classifier except for colon and hepat. The slight differences between these two data sets may be due to the unbalanced of features and samples. Compared with HANDI algorithm, the classification performance is similar may due to the same motivation by considering the cardinality of a relation rather than similarity classes.

In addition, we perform Friedman test and the Bonferroni-Dunn test to show the statistical significance of the result [5]. The Friedman test is defined as:

$$\tau_F = \frac{12N}{k\left(k+1\right)}\left(\sum_{i=1}^{k} r_i^2 - \frac{k(k+1)^2}{4}\right)$$

where k is the number of algorithms, N is the number of data sets, and r_i is the mean value of algorithms i among all data sets. Then, by using the value of $F_F = \frac{(N-1)\tau_F}{N(k+1)-\tau_F^2}$, we can determine whether the performance of these algorithms are same. If the assumption that "all algorithms have the same performance" is rejected, the performance of the algorithms are significantly different, and post-hoc test is needed to further differentiate the algorithms. If the distance of the average ranks exceeds the critical distance: $CD_\alpha = q_\alpha \sqrt{\frac{k(k+1)}{6N}}$, where q_α is the critical tabulated value for post-hoc test [5]. Tables 2 and 3 demonstrate the ranking values of different algorithms under different classifiers. It's obvious that the average ranking value is different. That is to say, these algorithms are different. According to the [5], the critical value of F for $\alpha = 0.05$ is 2.4495. and $q_{0.05} = 2.850$. Friedman test shows that at the 0.05 significance level, the average accuracies of DDFS is best among all algorithms. For both of KNN and C4.5, the Bonferroni-Dunn tests reveal that DDFS is statistically better than its competitors.

5   Conclusion

In this work, we introduce a new index to measure the discriminative power of feature subsets. Under the proposed index, a greedy scheme algorithm is proposed for feature selection. Compared with the classic entropy approaches, the proposed fuzzy conditional distinction degree is defined on the cardinality of the relation matrix. In addition, it associates the relevance between feature space and the similarity of samples. Extensive experiments demonstrate that DDFS is more efficient than some popular existing algorithms in classification.

Future work should include the study of reducing the computational complexity of greedy scheme due to the high time-consuming.

Acknowledgments. This work was partially supported by the National Natural Science Foundation of China (Nos. 61473259, 61502335, 61070074, 60703038) and the Hunan Provincial Science and Technology Project Foundation (2018TP1018, 2018RS3065).

References

1. Battiti, R.: Using mutual information for selecting features in supervised neural net learning. IEEE Trans. Neural Netw. **5**(4), 537–550 (1994)
2. Dai, J., Wang, W., Xu, Q.: An uncertainty measure for incomplete decision tables and its applications. IEEE Trans. Cybern. **43**(4), 1277–1289 (2013)
3. Dai, J., Xu, Q.: Attribute selection based on information gain ratio in fuzzy rough set theory with application to tumor classification. Appl. Soft Comput. J. **13**(1), 211–221 (2013)
4. Dai, J., Xu, Q., Wang, W., Tian, H.: Conditional entropy for incomplete decision systems and its application in data mining. Int. J. Gen. Syst. **41**(7), 713–728 (2012)

5. Demsar, J.: Statistical comparisons of classifiers over multiple data sets. J. Mach. Learn. Res. **7**, 1–30 (2006)
6. Hall, M.A.: Correlation-based feature selection for discrete and numeric class machine learning. In: Proceedings of the Seventeenth International Conference on Machine Learning (ICML 2000), pp. 359–366 (2000)
7. Hu, Q., Yu, D., Xie, Z., Liu, J.: Fuzzy probabilistic approximation spaces and their information measures. IEEE Trans. Fuzzy Syst. **14**(2), 191–201 (2006)
8. Hu, Q., Zhang, L., Zhang, D., Pan, W., An, S., Pedrycz, W.: Measuring relevance between discrete and continuous features based on neighborhood mutual information. Expert Syst. Appl. **38**(9), 10737–10750 (2011)
9. Jensen, R., Shen, Q.: New approaches to fuzzy-rough feature selection. IEEE Trans. Fuzzy Syst. **17**(4), 824–838 (2009)
10. Tallón-Ballesteros, A.J., Riquelme, J.C.: Tackling ant colony optimization meta-heuristic as search method in feature subset selection based on correlation or consistency measures. In: Corchado, E., Lozano, J.A., Quintián, H., Yin, H. (eds.) IDEAL 2014. LNCS, vol. 8669, pp. 386–393. Springer, Cham (2014). https://doi.org/10.1007/978-3-319-10840-7_47
11. Tang, J., Liu, H.: Unsupervised feature selection for linked social media data. In: ACM SIGKDD International Conference on Knowledge Discovery and Data Mining, pp. 904–912 (2012)
12. Tiwari, A.K., Shreevastava, S., Som, T., Shukla, K.K.: Tolerance-based intuitionistic fuzzy-rough set approach for attribute reduction. Expert Syst. Appl. **101**, 205–212 (2018)
13. Wang, C., Hu, Q., Wang, X., Chen, D., Qian, Y., Dong, Z.: Feature selection based on neighborhood discrimination index. IEEE Trans. Neural Netw. Learn. Syst. **29**(7), 2986–2999 (2017)
14. Wang, C., Qi, Y., Shao, M., Hu, Q., Chen, D., Qian, Y., Lin, Y.: A fitting model for feature selection with fuzzy rough sets. IEEE Trans. Fuzzy Syst. **25**(4), 741–753 (2017)
15. Wang, C., Shao, M., He, Q., Qian, Y., Qi, Y.: Feature subset selection based on fuzzy neighborhood rough sets. Knowl.-Based Syst. **111**, 173–179 (2016)
16. Witten, I.H., Eibe, F., Hall, M.A.: Data Mining: Practical Machine Learning Tools and Techniques, 3rd edn. Morgan Kaufmann, Elsevier, Burlington (2011)
17. Yager, R.R.: Entropy measures under similarity relations. Int. J. Gen. Syst. **20**(4), 341–358 (1992)
18. Yu, L., Liu, H.: Feature selection for high-dimensional data: A fast correlation-based filter solution. In: Machine Learning, Proceedings of the Twentieth International Conference (ICML 2003), August 21–24, 2003, Washington, DC, pp. 856–863 (2003)

Multi-label Feature Selection Method Based on Multivariate Mutual Information and Particle Swarm Optimization

Xidong Wang, Lei Zhao, and Jianhua Xu[✉]

School of Computer Science and Technology, Nanjing Normal University,
Nanjing 210023, Jiangsu, China
1529980236@qq.com, xujianhua@njnu.edu.cn

Abstract. Multi-label feature selection has become an indispensable pre-processing step to deal with possible irrelevant and redundant features, to decrease computational burdens, improve classification performance and enhance model interpretability, in multi-label learning. Mutual information (MI) between two random variables is widely used to describe feature-label relevance and feature-feature redundancy. Furthermore, multivariate mutual information (MMI) is approximated via limiting three-degree interactions to speed up its computation, and then is used to characterize relevance between selected feature subset and label subset. In this paper, we combine MMI-based relevance with MI-based redundancy to define a new max-relevance and min-redundancy feature selection criterion (simply MMI). To search for a globally optimal solution, we add an auxiliary mutation operation to existing binary particle swarm optimization with mutation to control the number of selected features strictly to form a new PSO variant: M2BPSO. Integrating MMI with M2BPSO builds a novel multi-label feature selection method: MMI-PSO. The experiments on four benchmark data sets demonstrate the effectiveness of our proposed algorithm, according to four instance-based classification evaluation metrics, compared with three state-of-the-art feature selection approaches.

Keywords: Multi-label classification · Feature selection
Multivariate mutual information · Particle swarm optimization
Mutation operation

1 Introduction

Multi-label learning is a special supervised classification task, in which any instance is possibly associated with multiple labels simultaneously [1–4]. As lots of observed and estimated features are easily available, many real world application data sets inevitably involve some irrelevant and redundant features, which increases computational burdens and even deteriorates classification performance.

© Springer Nature Switzerland AG 2018
L. Cheng et al. (Eds.): ICONIP 2018, LNCS 11304, pp. 84–95, 2018.
https://doi.org/10.1007/978-3-030-04212-7_8

Therefore, feature selection (FS) has become a necessary pre-processing step to cope with this issue via selecting a subset of more discriminative features [5,6].

In traditional binary and multi-class learning, mutual information (MI) between two random variables is one of successful measures to develop various FS techniques [7,8]. In multi-label learning, MI and its multivariate form (MMI) [9] also have become popular FS criteria in the past several years. Their resultant FS techniques can be divided into four sub-groups: ranking, greedy, random and optimization, according to detailed search strategies.

After evaluating the relevance between single feature and all labels based on MI or MMI, the ranking-type FS methods sort significance of all features in deceasing order, and then choose a subset of top ranking features. In [10], integrating pruning problem transformation (PPT) with MI is to construct PPT-MI as a compared method, in which PPT converts the multi-label data sets into multi-class ones considering all possible label combinations and deletes some classes with a few instances, and then the significance for a given feature is averaged over all MI-based importance values between this feature and each label. After approximating MMI evaluation between a single feature and all labels via limiting two-label combinations, a fast MMI-based FS method (FIMF) is proposed in [11].

Greedy-based methods utilize the sequential forward selection strategy to find out a sub-optimal subset of discriminative features. Based on MI, max-dependency and min-redundancy FS criterion is defined in MDMR [12], which is similar to mRMR [13]. In MFNMI [14], MI is evaluated using k-nearest neighbour instances. To accelerate MMI computational procedure, the original MMI is approximated via considering three-degree interactions only in PMU [10]. Furthermore, MAMFS could be regarded as an extension of PMU, which takes some higher-degree interactions into account [15].

Random-type methods apply some stochastic optimization techniques (e.g., genetic algorithm, particle swarm optimization and so on) to search for a globally optimal subset of features. To the best of our knowledge, only memetic algorithm is directly used to optimize MMI in [16].

Optimization FS methods are to construct constrained optimization problems to maximize MI-based feature-label relevance (linear term) and to minimize MI-based feature-feature redundancy (quadratic term) at the same time. In [17] and [18], a quadratic programming (QP) problem with non-negative constraints is designed, which is solved via QP solver in [17] and by Nystrom low-rank approximation in [18]. In [19], a QP problem with a unit simplex constraint is considered and then is solved using Frank-Wolfe optimization technique.

In the above FS methods, according to MI, MDMR [12] maximizes feature-label relevance and minimizes feature-feature redundancy at the same time, but it neglects label-label correlations. In PMU [10], the relevance between the selected feature subset and the entire label set is described using approximated MMI, which essentially considers the correlations between one feature and two labels, and between two features and one label. In this paper, we define a new max-relevance and min-redundancy multi-label FS criterion via combining

MMI-based relevance in PMU and MI-based redundancy in MDMR, which is simply referred to as MMI. In order to find out a globally optimal subset of selected features, after applying binary particle swarm optimization with muta- tion operator (MBPSO), we add an additional mutation operation to control the number of selected features strictly to construct a new PSO variant (sim- ply M2BPSO). Finally, MMI is integrated with M2BPSO to propose a novel multi-label FS method: MMI-PSO. On four benchmark data sets, we validate the classification performance of MMI-PSO, according to four instance-based evaluation metrics [4], via comparing with three state-of-the-art FS methods (PPT-CH [10], PMU [10] and FIMF [11]).

2 Multi-label Feature Selection Algorithm Based on Multivariate Mutual Information and Particle Swarm Optimization

Let a training set of size N be

$$\{(\mathbf{x}_1, \mathbf{y}_1), ..., (\mathbf{x}_i, \mathbf{y}_i), ..., (\mathbf{x}_N, \mathbf{y}_N)\}, \tag{1}$$

where $\mathbf{x}_i = [x_{i1}, ..., x_{ij}, ..., x_{iD}]^T \in \Re^D$ and $\mathbf{y}_i = [y_{i1}, ..., y_{ij}, ..., y_{iC}]^T \in \{1, 0\}^C$ indicate the D-dimensional real feature vector and C-dimensional binary label vector, respectively, for the i-th instance. Additionally, we assume that $F = \{f_1, ..., f_D\}$ and $L = \{l_1, ..., l_C\}$ represent feature and label sets, respec- tively.

The feature selection is to choose a subset S of size d from the original set of features F ($d < D$). To achieve this task, we introduce a new max-relevance and min-redundancy criterion based on two-variate mutual information and multi- variate mutual information, and binary particle swarm optimization with two mutation operators, at first. Then a novel multi-label feature selection method is proposed in this section.

2.1 Max-relevance and Min-redundancy Criterion Based on Multivariate Mutual Information

Mutual information has been widely-used in single-label feature selection [7,8]. Given three random variables (x, y and z), assume that two-variate mutual information (MI) is denoted by $I(x, y)$, 3-variate MI by $I(x, y, z)$ and conditional MI by $I(x, y|z)$, respectively.

In binary feature selection (here $L = \{l_1\}$), min-redundancy and max- relevance method (mRMR) [13] is one of the most successful techniques based on MI, which originally is formulated as a two-objective optimization problem to maximize feature-label relevance and minimize feature-feature redundancy at the same time,

$$\max \frac{1}{|S|} \sum_{f_i \in S} I(f_i, L),$$
$$\min \frac{1}{|S|^2} \sum_{f_i \in S} \sum_{f_j \in S} I(f_i, f_j), \tag{2}$$

and then is converted into a single-objective task,

$$\max \frac{1}{|S|} \sum_{f_i \in S} I(f_i, L) - \frac{1}{|S|^2} \sum_{f_i \in S} \sum_{f_j \in S} I(f_i, f_j). \tag{3}$$

In multi-label feature selection, max-dependency and min-redundancy app-roach (MDMR) [12] extends the above mRMR to deal with multi-label case, which essentially optimizes

$$\max \frac{1}{|S|} \sum_{f_i \in S} \sum_{l_j \in L} I(f_i, l_j)$$
$$- \frac{1}{|S|^2} \sum_{f_i \in S} \left(\sum_{f_j \in S} I(f_i, f_j) - \sum_{f_j \in S} \sum_{l_k \in L} I(f_i, l_k | f_j) \right). \tag{4}$$

On the other hand, after a multivariate mutual information $I(S, L)$ [9] between the selected feature subset S and the entire label set L is approxi-mated via limiting 3-degree interactions to speed up its evaluating procedure, multivariate MI based method PMU [10] is to maximize the following criterion,

$$\max \sum_{f_i \in S} \sum_{l_j \in L} I(f_i, l_j) - \sum_{f_i \in S} \sum_{f_j \in S} \sum_{l_k \in L} I(f_i, f_j, l_k) - \sum_{f_i \in S} \sum_{l_j \in L} \sum_{l_k \in L} I(f_i, l_j, l_k). \tag{5}$$

Table 1. Four different MI-based terms involved in feature selection methods

Method	$I(f_i, l_j)$	$I(f_i, l_j, l_k)$	$I(f_i, f_j)$	$I(f_i, f_j, l_k)$
mRMR	O		O	
MDMR	O		O	O
PMU	O	O		O
MMI (ours)	O	O	O	O

From the above feature selection criteria, as shown in Table 1, there are four MI-based terms to be investigated: (a) the feature-label relevance: $I(f_i, l_j)$; (b) the relevance between one feature and two labels: $I(f_i, l_j, l_k)$, which is to depict the possible two label correlations; (c) the feature-feature redundancy: $I(f_i, f_j)$; (d) the redundancy between two features and one label: $I(f_i, f_j, l_k)$, which aims at characterizing the conditional feature-feature redundancy given a fixed label. In Table 1, we also point out which term is involved in mRMR, MDMR and PMU: (a) $I(f_i, l_j)$ is maximized in all three methods; (b) $I(f_i, f_j)$ is minimized in mRMR and MDMR; (c) $I(f_i, f_j, l_k)$ is investigated in MDMR and PMU, since $I(f_i, l_k | f_j) = I(f_i, f_j, l_k) - I(f_i, f_j)$; (d) $I(f_i, l_j, l_k)$ is considered in PMU.

To consider all four terms comprehensively, via combining the criterion (5) in PMU and feature-feature redundancy $I(f_i, f_j)$, we define a new FS criterion as follows,

$$\max f(S) = \sum_{f_i \in S} \sum_{l_j \in L} I(f_i, l_j) - \sum_{f_i \in S} \sum_{f_j \in S} \sum_{l_k \in L} I(f_i, f_j, l_k)$$
$$- \sum_{f_i \in S} \sum_{l_j \in L} \sum_{l_k \in L} I(f_i, l_j, l_k) - \frac{1}{|S|(|S|-1)} \sum_{f_i \in S} \sum_{f_j \in S, i \neq j} I(f_i, f_j). \tag{6}$$

It is worth noting that the feature self-redundancy $I(f_i, f_i)$ is removed in our definition. This criterion is to maximize MMI-based relevance and minimize MI-based redundancy at the same time, and is simply referred to as MMI in this paper.

2.2 Binary Particle Swarm Optimization with Two Mutation Operations

In mRMR, MDMR and PMU, the sequential forward selection (SFS) strategy is used to search a sub-optimal subset of features sequentially. In this paper, we apply and modify particle swarm optimization (PSO) to optimize our feature selection criterion (6) to search for a globally optimal feature subset.

PSO is a population-based stochastic approach for solving continuous or discrete optimization problems. In PSO, its individual is called as a particle, which moves in the search space of an optimization problem. The position of a particle denotes a possible solution to the optimization problem to be solved. Each particle searches for its better positions in the search space by changing its velocity according to some rules originally inspired by behavioral models of bird flocking [20].

Assume that: (a) the size of population is set to M; (b) the dimensions of solution is D; (c) the position of the i-th particle is denoted by $\mathbf{z}_i = [z_{i1}, z_{i2}, ..., z_{iD}]^T$ $(i = 1, ..., M)$; (d) the moving velocity for the i-th particle is depicted as $\mathbf{v}_i = [v_{i1}, v_{i2}, ..., v_{iD}]^T$; and (e) the optimal position of the i-th particle found is $\mathbf{p}_i = [p_{i1}, p_{i2}, ..., p_{iD}]^T$; and (f) the globally optimal position is $\mathbf{g} = [g_1, g_2, ..., g_D]^T$. After randomly creating the initial position and velocity for the i-th particle (i.e., $\mathbf{z}_i^{(1)}$ and $\mathbf{v}_i^{(1)}$), the original PSO updates its velocity and position vectors using the following rules at its $(t + 1)$-th iteration:

$$\mathbf{v}_i^{(t+1)} = w\mathbf{v}_i^{(t)} + c_1 r^{(t)}(\mathbf{p}_i - \mathbf{z}_i^{(t)}) + c_2 r^{(t)}(\mathbf{g} - \mathbf{z}_i^{(t)}), \tag{7}$$

$$\mathbf{z}_i^{(t+1)} = \mathbf{z}_i^{(t)} + \mathbf{v}_i^{(t+1)}, \tag{8}$$

where w is the inertia weight, which determines the particle's inheritance of the velocity from the previous iteration; c_1 and c_2 are positive learning factors, which reflect the self-confidence of the particle and its ability to learn from the excellent individuals in the population; $r^{(t)}$ is a uniformly distributed random number whose value is between 0 and 1, and t implies the iterative index.

For some applications (e.g., feature selection), the previous continuous solution $\mathbf{z}_i$ can be reduced into a binary one, where 1 or 0 represents that a feature is selected or not. Therefore binary particle swarm method (simply BPSO) was also proposed in [20]. BPSO randomly initializes the position and velocity of the i-th particle as follows,

$$z_{ij}^{(1)} = \begin{cases} 1, \text{ if } r^{(1)} > 0.5, \\ 0, \text{ otherwise,} \end{cases} \tag{9}$$

$$v_{ij}^{(1)} = -v_{\max} + 2r^{(1)}v_{\max}, \tag{10}$$

where $j = 1, ..., D$ and $v_{\max}$ indicates the maximal velocity value. With the same velocity formula (7), at the $(t+1)$-th iteration, the position update rule becomes:

$$z_{ij}^{(t+1)} = \begin{cases} 1, \text{ if } S(v_{ij}^{(t+1)}) > 0.5, \\ 0, \text{ otherwise,} \end{cases} \tag{11}$$

where $S(v_{i,j}^t)$ is the logistic function:

$$S(v_{ij}^{(t+1)}) = \frac{1}{1 + \exp(-v_{ij}^{(t+1)})}, \tag{12}$$

which determines the pseudo-probability to choose the j-th component to be set to 1. When the velocity $v_{ij}^{(t+1)}$ is large enough, $S(v_{i,j}^{(t+1)})$ tends to 1, which means that all components of $\mathbf{z}_i$ are possibly set to 1. Therefore, it is necessary to limit the range of maximal velocity v_{max}.

In [21], to maintain the diversity of particles and to discover the globally optimal solution with a high possibility, mutation operation originally in genetic algorithm (GA) is added to BPSO, to construct MBPSO. The mutation rule is designed as,

$$z_{ij}^{(t+1)} = \begin{cases} 1 - z_{ij}^{(t+1)}, \text{ if } r_m^{(t+1)} < p_{mut} \\ z_{ij}^{(t+1)}, \quad \text{ otherwise} \end{cases} \tag{13}$$

where $r_m^{(t+1)} \in [0,1]$ is a random number satisfying uniform distribution law, and p_{mut} is a probabilistic threshold to control the number of mutation operations.

Generally speaking, the aforementioned BPSO and MBPSO finally find out a solution with an optimal fitness function value. Due to random optimization strategy, the number of 1 components or selected features is unknown before running. In this paper, to choose a fixed size for selected feature subset, we apply the above mutation operation to adjust the number of 1 components to be a fixed value d. Suppose that the number of 1's from the i-th position is d_0. When $d_0 > d$, we randomly select $d_0 - d$ 1's to be converted into 0's. Correspondingly, if $d_0 < d$, $d - d_0$ randomly selected 1 components are reversed into 0's. We simply referred to our PSO version as M2BPSO.

2.3 Multi-label Feature Selection Method Based on MMI and M2BPSO

In this subsection, we apply our M2BPSO to optimize our MMI to build a new multi-label feature selection method (simply MMI-PSO), as shown in Algorithm 1. We execute Algorithm 1 to select an optimal subset of size d.

Algorithm 1. Multi-label feature selection method based on MMI and M2BPSO

Input

    $\mathbf{X}$ and $\mathbf{Y}$: feature data and label data.

    M: the size of population.

    I_{max}: the maximal number of iterations.

    d: the number of selected features.

    S_{rnd}: the random seed.

    p_{mut}: the mutation probability threshold.

    w: the inertia weight.

    c_1 and c_2: the learning factors.

Procedure

    To set the random seed as S_{rnd}.

    To initialize the binary position vectors and real velocity vectors:

        $\mathbf{z}_i$ and $\mathbf{v}_i((i = 1, ..., M))$.

    To evaluate the fitness values $f(\mathbf{z}_i)$.

    To set the optimal positions $\mathbf{p}_i = \mathbf{z}_i$.

    To find out the global optimal position: $\mathbf{g} = \mathrm{argmax}_{i=1,...,M} f(\mathbf{p}_i)$.

    to set $t = 1$.

    Repeat

        To create the random number $r^{(t+1)}$.

        To update the velocity and position vectors $\mathbf{v}_i^{(t+1)}$ and $\mathbf{z}_i^{(t+1)}$.

        To execute the mutation operation with p_{mut}.

        To adjust the number of selected features to be d with mutation operator.

        To evaluate the fitness values $f(\mathbf{z}_i^{(t+1)})$.

        To update the optimal positions $\mathbf{p}_i$.

        To update the globally optimal positions $\mathbf{g}$.

        t=t+1.

    Until $(t > I_{max})$

To detect an optimal subset S of selected features according to the highest MMI value.

Output:

    S: the subset of selected features.

---

3 Experiments

In this section, we compare and evaluate our HMI-PSO and three existing FS methods (PPT-CHI, PMU [10] and FIMF [11]). Before presenting the detailed experimental results, we introduce four benchmark data sets and experimental settings.

3.1 Four Benchmark Data Sets

In this study, we downloaded four widely-validated benchmark data sets: Image, Plant, Scene and Yeast form[1], as shown in Table 2. These data sets have been divided into two parts: training and testing subsets at our homepage randomly.

---

[1] http://computer.njnu.edu.cn/Lab/LABIC/LABIC_Software.html.

Table 2. Statistics for four benchmark multi-label data sets.

Data set	Domain	Training instances	Testing instances	Features (D)	Labels (C)	Average labeles
Image	Scene	1200	800	294	5	1.24
Plant	Biology	558	390	440	12	1.08
Scene	Scene	1211	1196	294	6	1.07
Yeast	Biology	1500	917	103	14	4.24

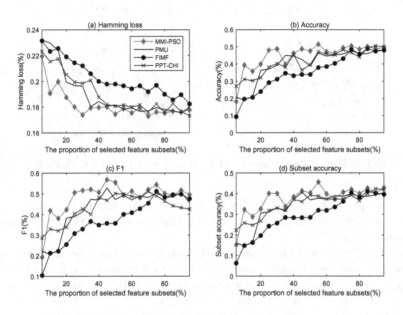

Fig. 1. Four instance-based metrics from four FS methods on Image

Table 2 also shows some important statistics for these sets, including the number of training and testing instances, dimension of features, the size of labels, average labels, and application fields.

3.2 Compared Methods and Experimental Settings

In this paper, we evaluate and compare our MMI-PSO and three existing FS methods (PPT-CHI [10], PMU [10], and FIMF [11]) experimentally. For three compared approaches, we accept their default settings. On our MMI-PSO, its key parameters are assigned as follows: $c_1 = 2$, $c_2 = 2$, $w = 0.5$, $v_{max} = 1$, $M = 100$, $I_{max} = 100$ and $r_{mut} = 1/D$.

In this study, four instance-based metrics are evaluated, including Hamming loss, accuracy, F1 and subset accuracy. Except for Hamming loss, the higher the other metric values are, the better the FS methods work. On their detailed

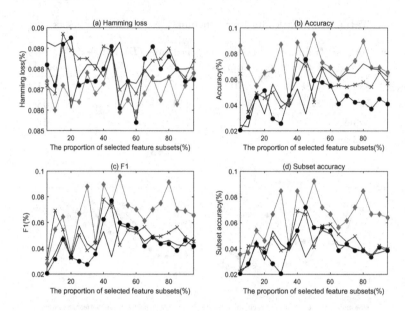

Fig. 2. Four instance-based metrics from four FS methods on Plant

definitions, please refer to [2,4]. For each data set and each method, these metrics are evaluated via multi-label k nearest neighbor method (ML-kNN) with a recommended $k = 10$ [22].

Additionally, we accept a training-testing mode to verify classification performance since each benchmark data set has been divided into training and testing subsets. Each FS method is firstly executed on training instances to choose a subset of features, then correspondingly original training and testing feature data are filtered to construct low-dimensional data, and finally four metrics are evaluated on testing instances using ML-kNN.

3.3 Performance Evaluation and Analysis

In this paper, to evaluate the classification performance of each FS method comprehensively, we regard each metric as a function of the proposition of selected features from 5% to 95 % with a step 5%, as shown in Figs. 1, 2, 3 and 4.

From Figs. 1, 2, 3 and 4, it is observed that at most of propositions of selected features, our MMI-PSO performs the best. To compare these four FS methods comprehensively, we use "Win" index [23] to count the number of the best results for each method and each metric across four data sets and 19 propositions (76 combinations), as shown in Table 3. It is obvious that our MMI-PSO achieves the maximal times on all four metrics. In the last row of Table 3, we sum up the number of wins for each method over all four metrics, and then sort four methods as MMI-PSO (247), PPT-CHI (21), FIMF (19), and PMU (17) in deceasing order. It is illustrated that our proposed method MMI-PSO is superior to the other three methods (PPT-CH, PMU and FIFM).

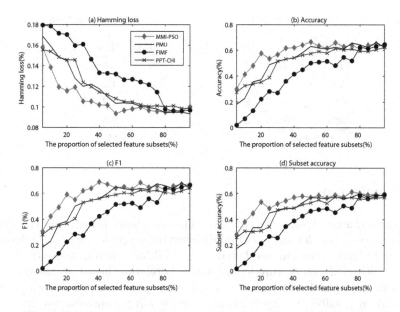

Fig. 3. Four instance-based metrics from four FS methods on Scene

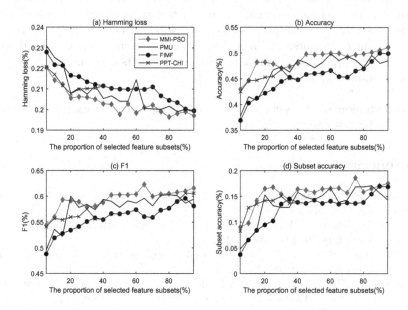

Fig. 4. Four instance-based metrics from four FS methods on Yeast

Table 3. The number of wins for each method and metric across four data sets

Metric	PPT-CHI	PMU	FIMF	MMI-PSO
Hamming loss	7	5	6	**58**
Accuracy	7	4	3	**62**
F1	5	2	7	**62**
Subset accuracy	2	6	3	**65**
Total wins	21	17	19	**247**

4   Conclusions

In this paper, a novel max-relevance and min-redundance feature selection criterion is proposed, which consists of four different terms: (a) feature-label relevance and feature-feature redundance based on mutual information, respectively; (b) feature-label-label and feature-feature-label correlation based on three-variate mutual information. Furthermore to search for a globally optimal feature selection solution possibly, we apply binary particle swarm optimization with two mutation operators to maximize our criterion. Through adding two mutation operations we both strengthen diversity of particles and control the number of selected features. The experiments from four widely-validated benchmark data sets demonstrate that our proposed feature selection algorithm is superior to three existing techniques. In future work, we will compare our method with more existing approaches on more benchmark data sets according to more performance evaluation metrics.

Acknowledgments. This work was supported by the Natural Science Foundation of China (NSFC) under Grant 61273246.

References

1. Tsoumakas, G., Katakis, I.: Multi-label classification: an overview. Int. J. Data Warehouse Min. **3**(3), 1–13 (2007)
2. Zhang, M., Zhou, Z.: A review on multi-label learning algorithms. IEEE Trans. Knowl. Data Eng. **26**(8), 1338–1351 (2014)
3. Gibaja, E., Ventura, S.: A tutorial on multilabel learning. ACM Comput. Surv. **47**(3), 1–38 (2015). Article 52
4. Herrera, F., Charte, F., Rivera, A.J., del Jesus, M.J.: Multilabel Classification: Problem Analysis, Metrics and Techniques. Springer, Switzerland (2016). https://doi.org/10.1007/978-3-319-41111-8
5. Kashef, S., Nezamabadi-pour, H., Nipour, B.: Multilabel feature selection: a comprehensiove review and guide experiments. WIREs Data Min. Knowl. Discov. **8**(2), e1240 (2018)
6. Pereira, R., Plastino, A., Zadrozny, B., Merschmann, L.H.C.: Categorizing feature selection methods for multi-label classification. Artif. Intell. Rev. **49**(1), 57–78 (2018)

7. Vergara, J.R., Estevez, P.A.: A review of feature selection methods based on mutual information. Neural Comput. Appl. **24**(1), 175–186 (2014)
8. Cai, J., Luo, J., Wang, S., Yang, S.: Feature selection in machine learning: a new perspective. Neurocomputing **300**, 70–79 (2018)
9. McGill, W.J.: Multivariate information transmission. Trans. IRE Prof. Group Inf. Theor. **4**(4), 93–111 (1954)
10. Lee, J., Kim, D.W.: Feature selection for multi-label classification using multivariate mutual information. Pattern Recognit. Lett. **34**(3), 349–357 (2013)
11. Lee, J., Kim, D.W.: Fast multi-label feature selection based on information-theoretic feature ranking. Pattern Recognit. **48**(9), 2761–2771 (2015)
12. Lin, Y., Hu, Q., Liu, J., Duan, J.: Multi-label feature selection based on max-dependency and min-redundancy. Neurocomputing **168**, 92–103 (2015)
13. Peng, H., Long, F., Ding, C.: Feature selection based on mutual information: criterion of max-dependency, max-relevance and min-redundancy. IEEE Trans. Pattern Anal. Mach. Intell. **27**(8), 1226–1238 (2005)
14. Lin, Y., Hu, Q., Liu, J., Chen, J., Duan, J.: Multi-label feature selection based on neighborhood mutual information. Appl. Soft Comput. **38**, 244–256 (2016)
15. Lee, J., Kim, D.W.: Mutual information-based multi-label feature selection using interaction information. Expert Syst. Appl. **42**(4), 2013–2025 (2015)
16. Lee, J., Kim, D.: Memetic feature selection algorithm for multi-label classification. Inf. Sci. **293**(293), 80–96 (2015)
17. Lim, H., Lee, J., Kim, D.W.: Multi-label learning using mathematical programming. IEICE Trans. Inform. Syst. **98**(1), 197–200 (2015)
18. Lim, H., Lee, J., Kim, D.W.: Low-rank approximation for multi-label feature selection. Int. J. Mach. Learn. Comput. **6**(1), 42–46 (2016)
19. Xu, J., Ma, Q.: Multi-label regularized quadratic programming feature selection algorithm with frank-wolfe method. Expert Syst. Appl. **95**, 14–31 (2018)
20. Poli, R., Kennedy, J., Blackwell, T.: Particle swarm optimization: an overview. Swarm Intell. **1**(1), 33–57 (2007)
21. Zhang, Y., Wang, S., Phillips, P., Ji, G.: Binary PSO with mutation operator for feature selection using decision tree applied to spam detection. Knowl.-Based Syst. **64**, 22–31 (2014)
22. Zhang, M., Zhou, Z.: ML-kNN: A lazy approach to multi-label learning. Pattern Recognit. **40**(7), 2038–2048 (2007)
23. Demsar, J.: Statistical comparisons of classifiers over multiple data sets. J. Mach. Learn. Res. **7**, 1–30 (2006)

Feature Selection Using Distance from Classification Boundary and Monte Carlo Simulation

Yutaro Koyama[1], Kazushi Ikeda[2], and Yuichi Sakumura[1,2(✉)]

[1] Aichi Prefectural University, Nagakute, Aichi 480-1198, Japan
saku@bs.naist.jp
[2] Nara Institute of Science and Technology, Ikoma, Nara 630-0192, Japan

Abstract. In binary classification, to improve the performance for unknown samples, excluding as many unnecessary features representing samples as possible is necessary. Of various methods of feature selection, the filter method calculates indices beforehand for each feature, and the wrapper method finds combinations of features having the maximum performance from all combinations of features. In this paper, we propose a novel feature selection method using distance from the classification boundary and a Monte Carlo simulation. Synthetic sample sets for binary classification were provided, and features determined by random numbers were added to each sample. For these sample sets, the conventional methods and the proposed method were applied, and it was examined whether the feature forming the boundary was selected. Our results demonstrate that feature selection was difficult with the conventional methods but possible with our proposed method.

Keywords: Feature selection · Support vector machine
Margin-based exploration · Monte Carlo method

1 Introduction

Preprocessing of data is an important step in data sciences as the performance of any algorithm highly depend on the quality of data. Most data preprocessing algorithms are aimed at reducing noise that is mixed with the signal. In binary classification problem, each class sample is represented as a vector wherein elements, which are the so-called features, are observed regardless of whether they have information regarding classification. Consequently, feature selection prior to binary classification is important to improve classification performance [1–6].

Feature selection algorithms used for data preprocessing are roughly divided into two types: filter and wrapper methods [7, 8]. In the filter methods, before applying binary classification, each feature is evaluated by the quantity representing the relationship with the class such as information entropy [9–12] and is removed from the feature set based on the quantity. The filter method uses several evaluation criteria [13]. Information gain (IG) [14], which is often used in the decision tree algorithm [15], is one criterion that represents feature importance for the sample. The relief algorithm

© Springer Nature Switzerland AG 2018
L. Cheng et al. (Eds.): ICONIP 2018, LNCS 11304, pp. 96–105, 2018.
https://doi.org/10.1007/978-3-030-04212-7_9

uses ranking based on the weight calculated from the distance between the samples [16–19]. The filter method incurs a low computational cost because it computes only feature importance from feature characteristics. However, as the filter method is independent of the classification algorithm, classification performance cannot be guaranteed.

In the wrapper methods [20–22], samples labeled by binary class are actually trained by a classifier to select features that contribute to classification accuracy. The accuracy is evaluated using generalization performance computed from cross-validation among samples. Hence, this method is equivalent to extracting combinations of features that maximize generalization performance. However, there are often combinations that have only a small difference in accuracy when compared with the optimal one. In this case, it is uncertain whether the optimal combination is truly optimal. It is impossible to exclude the possibility that a feature unrelated to classification happens to contribute to accuracy. Moreover, performing cross-validation on all combinations of subsets of features requires enormous computation cost; if we suppose n features, we have $\sum_{i=1}^{n} {}_nc_i$ combinations.

We propose a novel method to select features contributing to the determination of classification boundary by sequentially exploring pseudosamples using random perturbation so that the distance between the classification boundary and the sample increases. The problem of the wrapper method is that it detects unnecessary features by overlearning on given samples. The problem with the filter method is that classification performance is not guaranteed because classification boundaries are not used. In our proposed method, classification boundaries are constructed for various subsets of samples so as to avoid overlearning of all samples. Moreover, by randomly generating pseudosamples, statistical feature selection like that undertaken by the filter method is performed. Repeating Monte Carlo's sample generation in various subsets, informative features are likely to converge relatively. Comparing the performance of the feature selection methods using two synthetic sample sets, we demonstrate that the proposed method has the highest accuracy in selecting features.

2   Method

We use support vector machine (SVM) [23, 24] for binary classification. SVM defines a classification boundary so that the distance from the boundary to the closest sample of each class is maximized. If the boundary has high nonlinearity, the feature values are projected to a higher-dimensional space (kernel space [25, 26]) where the samples are classified by a flat boundary. In addition to the given sample, we introduce a generated pseudosample so that the distance from the boundary in the kernel space increases.

2.1   Pseudosample Exploration in Kernel Space

When samples undergo binary classification via SVM, samples as support vectors are distributed near both sides of the boundary. The classification boundary is calculated using all features of these samples, and the features that contribute less to boundary formation are squeezed while being projected to kernel space. Then, after computing a classification boundary, consider adding an independent perturbation to each feature value of the support vector. We call this a pseudosample. The pseudosample is accepted only when the distance between the pseudosample and the boundary in the kernel space increases compared to the original distance, and an independent perturbation is given to the feature values of the pseudosample again. On repeating this, the sample moves away from the classification boundary in the kernel space (Fig. 1). As a result, it is expected that the features contributing to boundary formation (necessary features) converge with a directional change, whereas the features that do not contribute to boundary formation (unnecessary features) continue with an omnidirectional change. A necessary or unnecessary feature can be determined by repeatedly exploring such pseudosamples through a Monte Carlo simulation and evaluating the variance of the feature values when at least one of the features converges.

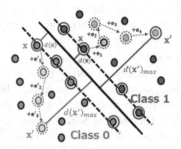

Fig. 1. Schematic of pseudosample exploration algorithm. The black line indicates the classification boundary that classifies classes 0 and 1. A pseudosample is generated by adding an independent perturbation (e_i) to the existing sample $(\mathbf{x})$ and held if the distance between the pseudosample and the boundary $(d(\mathbf{x}'))$ increases.

2.2 Preliminary Experiment for Pseudosample Exploration

In this section, we will confirm that there is a significant difference in variance of necessary or unnecessary feature at convergence via a Monte Carlo simulation of pseudosamples. For this purpose, we created simple synthetic data (Fig. 2a). In this data, a sample is represented by two-dimensional features (x, y) and the label is determined only by x; when $x < 0$ it is class 0 (blue), and when $x > 0$ it is class 1 (red).

First, the synthetic samples were learned via SVM with a Gaussian kernel and a classification boundary was provided. A perturbation was added to the sample of the support vector to generate a pseudosample that has a longer distance between the sample and the boundary. Then, the average and variance of the features at the time of convergence were calculated (Fig. 2b). The resultant variance of x is much smaller than that of y as expected, suggesting that x contributes to the construction of the boundary more than y.

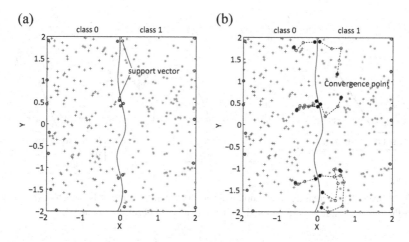

Fig. 2. (a) Test samples expressed by two features (*x*- and *y*- axes). They are classified by the boundary $x = 0$. (b) Sample trajectories of pseudosample explorations. The black and blue circles indicate start points (support vectors) and convergence points, respectively. The variances of coordinates x and y are 0.07 and 1.29, respectively, for class 0 and 0.04 and 1.17, respectively, for class 1. (Color figure online)

2.3 Core Procedure for Feature Selection

Subset of Samples

Nonlinear classification boundaries locally have normal vectors in various directions. The direction of the normal contains information of necessary features that form the boundary. Therefore, if you use a subset of samples that straddle the boundary, you will focus on the necessary feature that forms the boundary in the local region. If many such subsets are generated and pseudosample exploration is performed for each subset, there is a possibility that necessary features in each local region can be extracted. A subset of samples is created as follows:

1. Select coordinates randomly in sample space (Fig. 3a).
2. Select samples existing within the radius r from the selected coordinates (Fig. 3b).
3. Increase r until both of the following conditions are satisfied (Fig. 3c):
   a. The number of samples with in the circle is 10%–15% of all samples.
   b. The number of samples in each class is approximately equal.
4. Return to 1 if condition 3 is not satisfied.

Using the subset created by the above procedure, it is possible to narrow down to samples near the boundary and let the SVM learn them.

Discovering Unnecessary Features with Large Variance

In the subset near the boundary, if the pseudosample exploration is performed so as to be away from the boundary, the variance is likely estimated to be small for the necessary feature. Therefore, the necessary feature can be extracted by the following procedure:

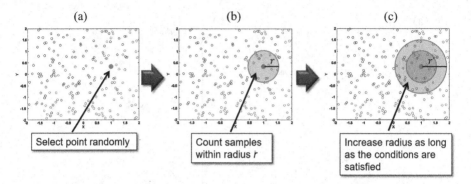

Fig. 3. Procedure for creating the sample subset. (a) a point is randomly selected, (b) samples within the circle of radius r are counted, and (c) a sample subset is completed.

1. Make sample subsets and select one.
2. Construct the classification boundary via SVM with the selected sample subset.
3. Repeat pseudosample exploration from support vectors until convergence.
4. Calculate the variance of converged pseudosample features for each class.
5. Repeat Steps 2–5 with another subset.
6. Extract features of the three lowest variance for each subset.
7. Exclude features that were extracted with the least count in all subsets.
8. Return to Step 1 with the samples that have no features excluded in Step 7.

Through this procedure, unnecessary features are deleted one by one, and it can be expected that necessary features will remain until the end.

3   Results

3.1   Dataset

To evaluate the performance of the feature selection algorithms, two sample sets comprising 17 features were used. Of these, 15 are unnecessary features and 2 are necessary features. The unnecessary features were determined by a uniform random number in $[-2: 2]$. The necessary features were made to have a clear boundary, and the samples divided by the boundary were labeled as 0 and 1, respectively. Two types of boundaries for necessary features were prepared (Fig. 4). One is a cubic function boundary (Fig. 4a), and the other is an XOR boundary (Fig. 4b). At each boundary, labels close to the boundary were randomly exchanged to other levels; label 0 to 1 and label 1 to 0. The total number of samples is 200, with 100 samples for each class. The feature selection algorithms were applied to these sample sets and we examined whether the two necessary features can be extracted in each set.

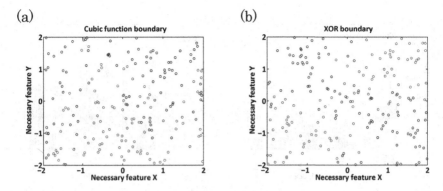

Fig. 4. Two sample sets applied for feature selection algorithms. (a) Cubic function boundary and (b) XOR boundary. Each sample sets have 17 features, two of which are shown as necessary features. The remaining 15 features are unnecessary and are determined by uniform random numbers in $[-2:2]$.

3.2   Performance Comparison of Algorithms

Using the sample sets in Fig. 4, the performance of the proposed feature selection algorithm was evaluated. We compared the performances of the filter method using information gain and relief weights and that of the wrapper method by all combinations of features and the Monte Carlo method proposed herein.

Filter Method
A feature with a large information gain or relief weight means that it likely has information on classification. When we computed the information gain and relief weights for the sample set in Fig. 4, it was observed that feature selection depends on the classification boundary (Fig. 5).

Using a sample set with a cubic function boundary (Fig. 4a), the two necessary features (NFX and NFY) have a large information gain and relief weight, and the other 15 unnecessary features become smaller (Fig. 5a). In this case, it is possible to select features for this sample set because there is a clear difference between the necessary and unnecessary features. However, if relief method is employed, the weight of NFX takes extremely small values to distinguish those of other unnecessary features (Fig. 5c). In the case of the sample set with the XOR boundary (Fig. 4b), the distribution of information gain changed dramatically while the relief method could extract the necessary features (Fig. 5d). The clear difference of the information gain disappeared as seen at the cubic function boundary, and the information gains of unnecessary features were estimated to be larger than those of necessary features (Fig. 5b). These suggest that stable feature selection is impossible for simple data sets in Fig. 4 by using information gain or relief weight.

Wrapper Method
We made all combinations of the 15 unnecessary features (32,767 combinations) with two necessary ones and calculated the accuracy by cross-validation for each combination. The histograms of all accuracy values for two boundaries (Fig. 4) are shown in Fig. 6.

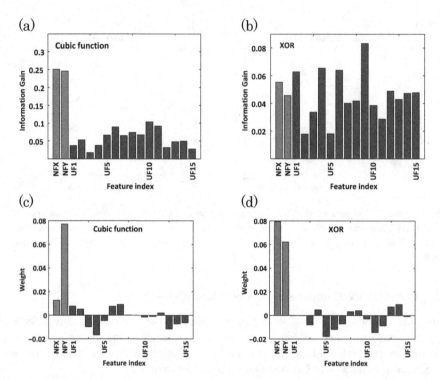

Fig. 5. (a, b) Information gains and (c, d) relief weights for features in sample sets in Fig. 4. (a) and (c) computed from the datasets in Fig. 4a, and (c) and (d) from that in Fig. 4b. Labels in axis indicate feature index (15 unnecessary features, UF*, and two necessary features, NFX and NFY).

The combination with optimal accuracy contains the unnecessary feature in the both boundaries (Fig. 6a and b). The accuracies of the combination comprising only necessary features was ranked 39 and 530 from the tops, respectively. The unnecessary features obey uniform random numbers and those in high rank happen to form some structure, which is considered to be the cause of high accuracy. However, it is due to the high classification ability of nonlinear SVM; hence, high accuracy cannot be demonstrated when an unknown sample is added. These facts suggest that the wrapper method is highly likely to pick up unnecessary features.

Proposed Method

In the proposed method, features are deleted one by one from all features to make it the last feature. Hence, if the necessary features (NFX and NFY) remain till the end, it is suggested that this method is effective for feature selection.

The order of deleted features is shown in Fig. 7, for the cubic function boundary (Fig. 4a) and the XOR boundary (Fig. 4b). In both sample groups, the necessary features remain till the end (red bars). These results demonstrate that feature selection, which is difficult using conventional methods, is possible with our proposed method.

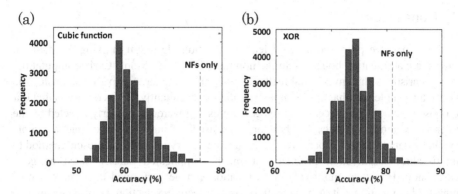

Fig. 6. Histograms of accuracy for all feature combinations of sample sets in Fig. 4. (a) and (b) correspond to Fig. 4a and b, respectively. The highest accuracies are (a) 79.0% and (b) 87.0%, which are realized by the combination containing unnecessary features. The accuracies only by the necessary features (NFs, red dashed lines) are 75.5% and 81.0%, which are ranked 39 and 530 from the tops, respectively. (Color figure online)

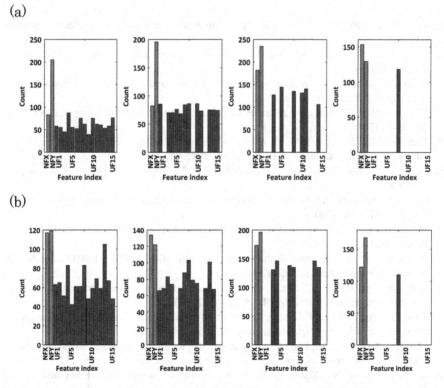

Fig. 7. Feature selection process by proposed method for (a) cubic and (b) XOR boundaries in Fig. 4. Red bars represent necessary features. The first, 4th, 10th, and 15th steps are shown from the left. (Color figure online)

4 Conclusion

In this paper, we proposed a novel feature selection algorithm utilizing the distance from the classification boundary in the nonlinear SVM and Monte Carlo computation. A comparison with conventional methods, filter and wrapper, was performed using the synthetic sample sets that have simple boundaries. Resultantly, it was observed that our proposed method can successfully select features that were not properly selected by the filter or wrapper methods. Furthermore, we used a binary classifier and a simple synthetic dataset in this paper. Practically, it is necessary to apply the present method to multi-value classification problems or much complicated datasets such as gene expression. However, multi-value classification can be performed by a repetition of binary classifications; it is meaningless unless feature selection is possible with a simple dataset. It is necessary to compare the filter method using other indices and to examine whether we need to select features in other sample sets that have complex boundaries.

References

1. John, G.H., Kohavi, R., Pfleger, K.: Irrelevant features and the subset selection problem (1994)
2. Almuallim, H., Dietterich, T.G.: Learning with many irrelevant features. In: AAAI, vol. 91 (1991)
3. Blum, A.L., Langley, P.: Selection of relevant features and examples in machine learning. Artif. Intell. **97**, 245–271 (1998)
4. Chandrashekar, G., Sahin, F.: A survey on feature selection methods. Comput. Electr. Eng. **40**, 16–28 (2014)
5. Vergara, J.R., Estévez, P.A.: A review of feature selection methods based on mutual information. Neural Comput. Appl. **24**, 175–186 (2015)
6. Li, Y., Li, T., Liu, H.: Recent advances in feature selection and its applications. Knowl. Inf. Syst. **53**, 551–577 (2017)
7. Guyon, I., Elisseeff, A.: An introduction to variable and feature selection. J. Mach. Learn. Res. **3**, 1157–1182 (2003)
8. Jain, A.K., Duin, R.P.W., Mao, J.: Statistical pattern recognition: a review. IEEE Trans. Pattern Anal. **22**, 4–37 (2000)
9. Shannon, C.: A mathematical theory of communication. ACM SIGMOBILE Mobile Comput. Commun. Rev. **5**, 3–55 (2001)
10. Bennasar, M., Hicks, Y., Setchi, R.: Feature selection using joint mutual information maximisation. Expert Syst. Appl. **42**, 8520–8532 (2015)
11. Zhao, G., Wu, Y., Chen, F., Zhang, J., Bai, J.: Effective feature selection using feature vector graph for classification. Neurocomputing **151**, 376–389 (2015)
12. Pes, B., Dessì, N., Angioni, M.: Exploiting the ensemble paradigm for stable feature selection: a case study on high-dimensional genomic data. Inf. Fusion **35**, 132–147 (2017)
13. Sánchez-Maroño, N., Alonso-Betanzos, A., Tombilla-Sanromán, M.: Filter methods for feature selection – a comparative study. In: Yin, H., Tino, P., Corchado, E., Byrne, W., Yao, X. (eds.) IDEAL 2007. LNCS, vol. 4881, pp. 178–187. Springer, Heidelberg (2007). https://doi.org/10.1007/978-3-540-77226-2_19
14. Mitchell, T.M.: Machine Learning, vol. 45. McGraw Hill, Burr Ridge (1997)

15. Quinlan, J.: Induction of decision trees. Mach. Learn. **1**, 81–106 (1986)
16. Kira, K., Rendell, L.A.: The feature selection problem: traditional methods and a new algorithm. In: AAAI, vol. 2 (1992)
17. Kononenko, I.: Estimating attributes: analysis and extensions of RELIEF. In: Bergadano, F., De Raedt, L. (eds.) ECML 1994. LNCS, vol. 784, pp. 171–182. Springer, Heidelberg (1994). https://doi.org/10.1007/3-540-57868-4_57
18. Liu, H., Motoda, H., Yu, L.: Feature selection with selective sampling. In: ICML (2002)
19. Kira, K., Rendell, L.A.: A practical approach to feature selection (1992)
20. Kohavi, R., John, G.H.: Wrappers for feature subset selection. Artif. Intell. **97**, 273–324 (1997)
21. Panthong, R., Srivihok, A.: Wrapper feature subset selection for dimension reduction based on ensemble learning algorithm. Procedia Comput. Sci. **72**, 162–169 (2015)
22. Mi, H., Petitjean, C., Dubray, B., Vera, P., Ruan, S.: Robust feature selection to predict tumor treatment outcome. Artif. Intell. Med. **64**, 195–204 (2015)
23. Vapnik, V.: Pattern recognition using generalized portrait method. Autom. Remote Control. **24**, 774–780 (1963)
24. Boser, B.E., Guyon, I.M., Vapnik, V.N.: A training algorithm for optimal margin classifiers, pp. 144–152 (1992)
25. Aizerman, M.A.: Theoretical foundations of the potential function method in pattern recognition learning. Autom. Remote Control **25**, 821–837 (1964)
26. Buhmann, M.D.: Radial Basis Functions: Theory and Implementations, vol. 12. Cambridge University Press, Cambridge (2003)

Clustering

Approximate Spectral Clustering
Using Topology Preserving Methods
and Local Scaling

Mashaan Alshammari$^{(\boxtimes)}$ and Masahiro Takatsuka

School of Information Technologies, The University of Sydney,
Sydney, NSW 2006, Australia
mals6571@uni.sydney.edu.au,
masa.takatsuka@sydney.edu.au

Abstract. Spectral clustering is the type of unsupervised learning that separates data based on their connectivity instead of convexity. However, its computational demands increase cubically with the number of points n. This triggered a stream of studies to ease these demands. An effective solution is to provide an approximated graph $G^* = (V^*, E^*)$ for the input data with a reduced set of vertices and edges. Recent similarity measures used to construct the approximated graph $G^* = (V^*, E^*)$ have some deficiencies such as: (1) weights on edges highly depend on the cluster density, and (2) larger memory footprint compared to conventional similarity measures. In this work, we employed topology preserving methods (e.g., neural gas) to obtain $G^* = (V^*, E^*)$ due to their ability to preserve input data topology. Then we used a conventional similarity measure to assign weights on the graph. The experiments reveal that graphs obtained through topology preserving methods and passed to a locally scaled similarity measure, produce performances comparable to the recent measures with a significantly smaller memory footprint.

Keywords: Spectral clustering · Topology preserving methods
Neural gas

1 Introduction

Spectral clustering uses the spectrum of the affinity matrix A to partition the connected components of the graph G that connects data points $\{x_1, x_2, \ldots, x_n\}$. Unfortunately, storing A requires $O(n^2)$ and performing eigendecomposition needs computations of $O(n^3)$ [1]. Due to the effectiveness of spectral clustering, many studies investigated minimizing its computational demands. Reducing the graph vertices V shrinks the size of A from $n \times n$ to $m \times m$, where $m \ll n$. On the other hand, reducing the number of edges E results in a sparse matrix [2] which reduces the overall computations. Nevertheless, achieving both reductions is not trivial and has been the subject for most of spectral clustering researchers [2].

The idea behind using m representatives out of n points is to perform spectral clustering on a reduced set of points then generalize the outcome. The selection of representatives could be obtained through sampling [2]. The more sophisticated the

© Springer Nature Switzerland AG 2018
L. Cheng et al. (Eds.): ICONIP 2018, LNCS 11304, pp. 109–121, 2018.
https://doi.org/10.1007/978-3-030-04212-7_10

sampling scheme, the more we are confident that small clusters are not left out. A more effective approach would be using a learning algorithm to place the representatives [1]. However, both approaches are incapable of constructing the graph $G(V, E)$ and we have to rely on the similarity measure to produce a sparse graph.

To construct the graph $G(V, E)$ we could use a similarity measure penalized by a global scale (σ) which controls the decay of the affinity as proposed in [3]. Nevertheless, σ is insensitive to local statistics [4] and difficult to tune. An alternative solution proposed in [4] is to use a local scale σ_i ($i \in \{1, 2, \ldots, m\}$) set as the distance to K^{th} neighbor. However, K is another parameter that needs tuning. Moreover, local scaling measure is unable to produce a sparse graph (i.e., eliminating some edges) and it is highly depending on the graph selection like *knn* or ε graphs. Recently, novel similarity measures were introduced [2]. On one hand, these measures can produce a sparse graph by eliminating weak edges. On the other hand, their computational demands are much higher than their conventional counterparts.

In this work, we propose the use of topology preserving methods and demonstrate that they are more efficient in placing representatives to approximate input data. Unlike other approximation methods such as KASP [1], topology preserving methods can produce a sparse graph, since edge drawing is part of their training. A topology preserving method has two components: vector quantization and establishing lateral connections [5]. We tested self-organizing maps (SOM) [6], neural gas (NG) [7], and growing neural gas (GNG) [8]. We found out that graphs obtained through topology preserving methods improve the accuracy of the local scaling measure with markedly smaller memory footprint than other measures.

2   Related Work

Two topics were the subject for most research on spectral clustering. The former relates to the question: how to draw edges between data points by quantifying their similarity. The second topic is approximate spectral clustering (ASC) which concerns of placing m representatives to cluster n points.

2.1   Similarity Measures for Spectral Clustering

Defining similarities between data points is crucial for spectral clustering accuracy. Those similarities are represented as weights on edges of the graph G which should be informative to highlight the optimal cut for spectral clustering. Ideally, large weights are assigned between the points in same cluster, and relatively small weights (or no edges at all) between points in different clusters. There are number of options for the similarity measures (see Appendix D in [9]). Globally scaled similarity which is the Euclidean distance between the compared points penalized by a Gaussian scale:

$$A_{ij} = \exp\left(\frac{-d^2(i,j)}{\sigma^2}\right) \tag{1}$$

However, this similarity measure ignores the local statistics around the compared points [4]. An advanced option introduced in [4], that penalizes the Euclidean distance by the product of local scales σ_i. and σ_j. The local scale of a particular point is set as the distance to its Kth neighbor:

$$A_{ij} = \exp\left(\frac{-d^2(i,j)}{\sigma_i\sigma_j}\right) \tag{2}$$

The local scale set as $\sigma_i = d(i, i_K)$. This measure is sensitive to local statistics and adjusts itself accordingly. However, setting the Kth neighbor is an issue for this measure. All aforementioned measures are unable to eliminate edges to produce a sparse affinity matrix. To avoid this, one could restrict the Euclidian distance to a certain threshold [9]. Unfortunately, this means one more parameter that needs tuning.

There are number of similarity measures that can produce sparse versions of the graph G. A measure called common-near-neighbor (CNN) is defined as the number of points in the intersection of two spheres with radius ε centered at i and j [10]. The weight of the edge in CNN graph is determined by the number of shared points. Intuitively, if there are no shared points between i and j, there will be no edge connecting them, hence it results in a sparse graph. Nevertheless, in large datasets computing CNN requires heavy computations [2]. Connectivity matrix (CONN) is suitable for vector quantization methods [2]. It utilizes the concept of induced Delaunay triangulation presented in [5]. $CONN(i, j)$ is defined as the number of points which i and j are their best-matching-unit and second-best-matching-unit. Like CNN, CONN can eliminate edges in case of no common neighbors. However, CONN is much faster than CNN since its computations are embedded in vector quantization training.

$$CONN(i,j) = \left|\left\{v \in V_{ij} \cup V_{ji}\right\}\right| \tag{3}$$

2.2   Approximate Spectral Clustering

High computational demands of spectral clustering stimulate a stream of studies addressing what is known as approximate spectral clustering (ASC) [1, 2]. The basic idea of ASC is to carry out the spectral clustering with m representatives, where $m \ll n$, then generalize the results to the entire dataset. However, the fundamental question is how to choose m representatives. Choosing m via random sampling, introduces the risk of missing small clusters. A different approach is to set a low rank approximation of the affinity matrix A (e.g., Nystrom method). However, some studies reported high memory consumption of such methods [1].

Vector quantization methods (e.g., $k.$ -means and SOM) proved to be an efficient way to approximate spectral clustering [1, 2]. Even though they require a training time, the m representatives inherent density information found in n points. Thus, the training step mitigates the risk of missing small clusters. The initial effort in this direction was by Yan et al. [1], where they used k-means to approximate n points. However, k-means is unable to preserve the topology of the input data. Topology preserving methods, namely self-organizing maps (SOM) and neural gas (NG), were used to approximate input data [2]. Nevertheless, the edges drawn by SOM and NG were replaced by the edges drawn via CONN [2]. In other words, the vector quantization component of SOM and NG was used, whereas the lateral connection component was not utilized. This means the training time spent on lateral connections was wasted.

The approximation process should place representatives and draw edges between them to justify its training time. If we could get edges through the approximation step, the similarity measure will weigh those edges and skip non-existing ones. This will result in a sparse affinity matrix A. Topology preserving methods are perfectly suited to this objective, since they produce a new graph with a reduced set of representatives.

3   Proposed Approach

The algorithm passes through four main steps to perform approximate spectral clustering with smaller memory footprint. (1) It approximates the input data using one of four methods: k-means, SOM, NG, or GNG, each of which is coupled with an edge drawing scheme. (2) The weights on edges were set using local σ [4] or *CONN* [2] followed by an eigendecomposition for the graph Laplacian L. (3) It passes through a cost function that automatically selects the dimensions needed to construct an embedding space where the clusters could be detected. (4) The number of clusters in the embedding space was estimated by another cost function.

3.1   Approximating Input Data

The computational bottleneck in the spectral clustering could be avoided by performing vector quantization and carry on with a reduced number of representatives. For this purpose, four vector quantization methods were used to produce an approximated graph, three of which contain a topology preserving component. A map M satisfies the topology preserving condition if points i and j which are adjacent in $\mathbb{R}^d$ are mapped to neurons w_i and w_j which are adjacent in M, and, vice versa [11].

k-means. k-means is a well-known method for vector quantization. Given data points $\{x_1, x_2, \ldots, x_n\}$ distributed around m centroids $\{\mu_1, \mu_2, \ldots, \mu_m\}$. It attempts to minimize the squared distances between data points and their closest centroids:

$$\min_{\{\mu_i\},\{c_{ij}\}} \sum_{j=1}^{m} \sum_{i=1}^{n} c_{ij} ||x_i - \mu_i||^2 \qquad (4)$$

where $c_{ij} \in \{0, 1\}$ such that $c_{ij} = 1$ if x_i was assigned to the cluster μ_j and $c_{ij} = 0$ otherwise. k-means was used to approximate input data. Unlike SOM, k-means is not equipped with a topology preserving component. Therefore, its centroids were connected using nearest neighbor graphs (NN).

Self-Organizing Map. SOM consists of set of neurons attached to lateral connections. During SOM training, the map attempts to capture input data topology. Its training starts by randomly selecting a data sample x_i. Its closest neuron is called the best matching unit w_b:

$$||x_i - w_b|| = \min_j \{||x_i - w_j||\} \tag{5}$$

This process is the competitive stage in which neurons compete to win x_i. Then, w_b pulls its topological neighbors to be closer in a process known as the cooperative stage:

$$w_j(t+1) = w_j(t) + \alpha(t)\eta(t)(x_i - w_j(t)) \tag{6}$$

where $\alpha(t)$ is the learning rate monotonically decreasing with time t and $\eta(t)$ is the neighborhood kernel. In our case, SOM was set to 2D hexagonal grid.

Neural Gas. One of SOM deficiencies is that the network adaption is performed based on grid connections regardless of neurons positions in the input space. Another problem is the fixed connections which may restrict its ability to capture data topology. These two problems were addressed by the neural gas (NG) [11]. NG initiates neurons in the input space without lateral connections. Initially, a random point x_i is introduced to neurons and its best matching unit w_b is identified. Other neurons are ordered ascendingly based on their distance from w_b and updated as per the following adaption rule:

$$w_j(t+1) = w_j(t) + \varepsilon \cdot e^{-k_j/\lambda}(x_i - w_j(t)) \tag{7}$$

where $j \in \{1, 2, \ldots, m\}$, k is the rank associated with w_j based on its distance from w_b. $\varepsilon \in \{0, 1\}$ controls the extent of the adaption and λ is a decaying constant. Next, NG applies the competitive Hebb learning (CHL) to connect best matching unit w_b^0 and the second best matching unit w_b^1, to maintain a perfectly topology preserving map (see Theorem 2 in [11] and the discussion therein). Due to neurons movement, these edges are allowed to age and perhaps removed if they are not refreshed. In our experiments, we set the stopping criteria to the stability in quantization error.

Growing Neural Gas. Although neural gas was an improvement over SOM, it kept some properties of original SOM. It uses a fixed number of neurons and relies on decaying parameters for adaption. These two deficiencies were overcome by growing neural gas (GNG). It starts by introducing a random x_i to the competing neurons and selects the best matching unit and computes its error using:

$$error_{t+1} = error_t + ||w_b - x_i||^2 \tag{8}$$

w_b and its topological neighbors are adapted using the following adaption rules:

$$w_b(t+1) = \varepsilon_b(x_i - w_b) \tag{9}$$

$$w_k(t+1) = \varepsilon_k(x_i - w_k) \tag{10}$$

where k indicates all direct topological neighbors of w_b. ε_b and ε_k are the fractions of change. Then, GNG applies competitive Hebb learning (CHL) to connect BMU and the second BMU and set the weight on the edge to zero. If the current iteration is an integer multiple of the parameter l, a new neuron is inserted in halfway between the neuron with the maximum accumulated error w_q and its topological neighbor with the largest error w_f. The newly inserted neuron is connected to both w_q and w_f, and the old edge connecting them is removed. The training continues until the stopping criteria is met, which was set in our experiments to the stability in quantization error.

3.2   Constructing the Affinity Matrix A

To construct the affinity matrix A, a similarity score should be set for each pair of neurons. We used two measures to set the similarity between neurons. The first is local σ [4] defined as:

$$A_{w_i w_j} = \exp\left(\frac{-d^2(w_i, w_j)}{\sigma_{w_i} \sigma_{w_j}}\right) \tag{11}$$

It cannot draw a sparse graph by itself, because its formula does not produce zero values. Therefore, we have to rely on edges in the graph obtained through approximation methods. This similarity measure can preserve the local statistics between the neuron w_i and w_j regardless of the density around them as illustrated in Fig. 1. It is obvious that the density of the points on the outer ring is much less than the middle ring. However, this similarity measure set similar weights on the outer ring edges as the middle ring.

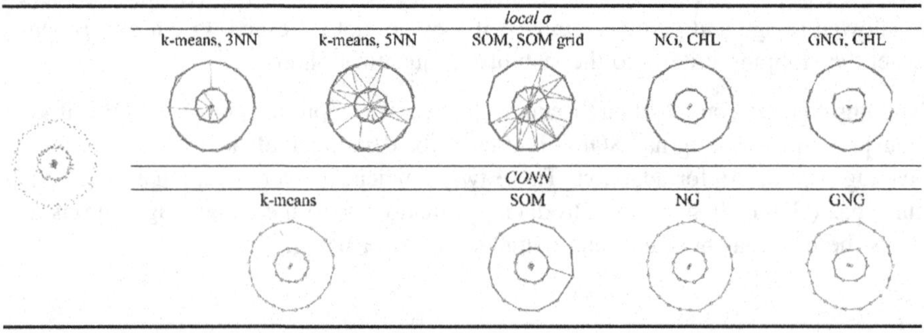

Fig. 1. Constructing an approximated graph $G^* = (V^*, E^*)$ using local σ and *CONN*

The second measure is *CONN* [2]. It is suitable for vector quantization since it measures the density between neurons. Unlike local σ, the weight of the edge is entirely

determined by the number of points in the Voronoi region $V_{w_i w_j}$ shared by w_i and w_j.
Examples of *CONN* graphs shown in Fig. 1

$$A_{w_i w_j} = \left| \left\{ v \in V_{w_i w_j} \cup V_{w_j w_i} \right\} \right| \tag{12}$$

Setting K for Computing Local σ. Selection of the Kth neighbor is a problem for the
use of local σ. Although some studies have used $K = 7$ [4, 9], it remains as an
empirical selection and have not been used in approximate spectral clustering. Performing vector quantization affects our decision to set K. It is probable that the selected
value has no actual edge in the approximated graph $G^* = (V^*, E^*)$.

Let K be the direct neighbor of the current neuron w_i (i.e., $K = 1$). For nearest
neighbor graphs, this edge is guaranteed to exist since each neuron has k edges (i.e., 3
or 5 edges). For SOM with a hexagonal grid the existence of this edge is guaranteed
since each neuron on that grid has at least 2 edges. For NG and GNG, the existence of
this edge is guaranteed by the definition of competitive Hebbian rule that connects best
matching neuron with the second best matching neuron. Therefore in all experiments
we set $K = 1$ that is the direct neighboring neuron.

Memory Footprint for Local σ and *CONN*. Computing each of the similarity
measures, requires a certain array needs to be present in the memory. That array
contains values needed to compute the similarities between neurons. Starting with local
σ, it requires a value attached to each neuron. That value represents the local scale σ_i.
Therefore, a one-dimensional array of size $1 \times m$ is needed to be present in the memory
to construct the affinity matrix A using the local scaling measure. For *CONN* similarity
measure, the edge between w_i and w_j is determined by the number of points where w_i is
the BMU and w_j is the second BMU, in addition to the points where w_j is the BMU and
w_i is the second BMU. To achieve that, a two dimensional array is needed.

3.3   Embedding Space Dimensions $\mathbb{R}^{m \times k}$

The graph Laplacian is computed as $L = I - D^{-\frac{1}{2}} A D^{-\frac{1}{2}}$. Then, eigendecompsition is
performed on L to obtain the largest eigenvectors that partition the graph. In the
original spectral clustering algorithm in [3], it is recommended to build an embedding
space using k eigenvectors then run k-means on that space to find clusters. In that work,
k was set manually, however, automating the selection of k would make the algorithm
more practical. The automation could be achieved by counting the number of eigenvalues of multiplicity zero. However, eigenvalues could deviate from zero due to noise
leaving this technique unreliable [4].

For an eigenvector v_i to be included in the embedding space $\mathbb{R}^{m \times k}$, it must be able
to sperate the neurons. An eigenvector that cannot separate the neurons could confuse
the clustering in the embedding space. v_i discrimination power could be measured by:
(1) a clustering index that gives a score on how well data is separated, and (2) its
corresponding eigenvalue λ_i. The closer λ_i to zero the more discriminative v_i becomes.
Therefore, we used the following formula to evaluate all eigenvectors $\{v_1, v_2, \ldots, v_m\}$:

$$\frac{\sum_{c=2}^{4} DBI_c(v_i)}{\lambda_i}, \qquad 2 \leq i \leq m \tag{13}$$

For every eigenvector v_i containing one dimensional data $\{w_1, w_2, \ldots, w_m\}$, we compute how well it can separate them into 2, 3, and 4 clusters ($c \in \{2,3,4\}$) using the Davies-Bouldin index (DBI). This quantity was penalized by the eigenvalue λ_i corresponding to the eigenvector v_i The Davies-Bouldin index is defined as:

$$\frac{1}{c} \sum_{i=1}^{c} \max_{i \neq j} \left\{ \frac{S_c(Q_i) + S_c(Q_j)}{d_{ce}(Q_i, Q_j)} \right\} \tag{14}$$

Given neurons $\{w_1, w_2, \ldots, w_m\}$ clustered into $\{Q_1, Q_2, \ldots, Q_c\}$ clusters. $S_c(Q_i)$ is within-cluster distances in cluster i, and $d_{ce}(Q_i, Q_j)$ is the distance between clusters i and j. After scoring all eigenvectors $\{v_1, v_2, \ldots, v_m\}$, the scores were fit into a histogram. The bin size was determined via Freedman-Diaconis rule, defined as $2Rm^{-1/3}$, where R is the inter-quartile range. The desired eigenvectors fall outside the interval $[\mu \pm \sigma]$, where μ is the mean and σ is the standard deviation.

3.4   Number of Clusters in the Embedding Space

The embedding space $\mathbb{R}^{m \times k}$ is spanned by eigenvectors qualified through the method explained in the previous subsection. In this space the connected components of the graph $G^* = (V^*, E^*)$ form convex clusters could be detected by k-means. However, the parameter k must be set prior to performing k-means.

One way to select k would be through a clustering index (like DBI in Eq. 14) then use k that yields the lowest score. However, this approach tends to favor large values of k for better separation. This could be avoided by examining the eigenvalues of the graph Laplacian L. As stated in [12] the number of connected components of the graph equals the number of eigenvalues with multiplicity zero. Therefore, we should penalize each value of k by the sum of eigenvalues it accumulates (Fig. 2).

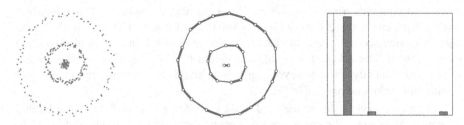

Fig. 2. (left) input data, (middle) $G^* = (V^*, E^*)$ via GNG, (right) histogram for eigenvectors scores, suggesting the top two eigenvectors are sufficient for clustering (best viewed in color).

$$DBI_k(X) + \sum_{i=1}^{k} \mu_i, \qquad 2 \le k \le m \qquad (15)$$

4 Experiments

The experimental design contains three experiments to test the competing methods. The experiments vary in the input type. The edges weights were assigned using local σ [4] and *CONN* [2]. The local scaling used all the 5 approximation graphs, whereas, *CONN* used only four. *CONN* only needs vector quantization without the edge drawing component. Apart from the second experiment, all methods used an automated selection of k discussed in Sects. 3.3 and 3.4. The experiments were run on a windows 10 machine (3.40 GHz CPU and 8 GB of memory) where methods were coded in MATLAB 2017b.

4.1 Synthetic Data

The competing methods were tested using synthetic data (shown in Fig. 3). The number of representatives m was set by running multiple values using k-means++ algorithm then select the one that represents an elbow point in quantization error curve.

Fig. 3. A collection of synthetic data.

Local σ outcomes in Table 1, suggest that 3NN, 5NN, and SOM are not very useful. However, it could be considerably improved by using edges obtained through NG and GNG. Since these two methods used the competitive Hebbian learning (CHL) [11], they can eliminate edges where the probability of x_i is discontinued. This was extremely helpful for local σ to assign weights on these edges and discard the ones that do not exist, resulting in a better clustering outcome. This demonstrates that the local σ outcome is highly influenced by the quality of the graph. The results were not surprising, it was anticipated by looking at Fig. 1, where 3NN, 5NN, and SOM added unnecessary edges between clusters. On the other hand, NG and GNG provided better graphs that clearly show the separation between clusters.

Moving to the other side of the table where graphs were weighed by *CONN*. It is observable that k-means struggled compared to its peers (i.e. SOM, NG, and GNG). This could be explained that in k-means centroids are moving independently from each other. Therefore, the concept of BMU and second BMU which is required by *CONN*

Table 1. Clustering accuracies for data in Fig. 3. All values are averages of 100.

	n	m	local σ					CONN			
			k-means, 3NN	k-means, 5NN	SOM, SOM grid	NG, CHL	GNG, CHL	k-means	SOM	NG	GNG
Spirals	1000	64	0.59±0.1	0.46±0.1	0.37±0.0	0.97±0.1	0.97±0.1	0.85±0.2	0.88±0.1	0.96±0.1	**0.97±0.1**
Rings	299	32	0.89±0.1	0.94±0.1	0.67±0.0	0.92±0.1	**0.97±0.1**	0.88±0.2	0.97±0.1	0.82±0.2	0.95±0.1
Lines	512	64	**0.99±0.0**	0.99±0.1	0.76±0.2	0.98±0.1	**0.99±0.0**	0.86±0.2	0.93±0.2	0.86±0.2	0.87±0.2
Smile	266	32	0.96±0.1	0.99±0.1	0.91±0.1	**0.99±0.0**	0.97±0.1	0.84±0.2	0.95±0.1	0.90±0.1	0.88±0.1
Moons	1000	64	0.89±0.2	0.89±0.2	0.67±0.2	0.95±0.1	0.76±0.2	0.81±0.2	**0.96±0.1**	0.91±0.1	0.89±0.2

was not fully implemented. For other methods, graphs obtained through SOM yields the best performance. While, NG and GNG performed in a similar manner with an advantage for GNG. This could be explained that SOM provided better neurons density than NG and GNG. Neurons density is determined by their adaptation to the input data using Eqs. (6), (7), and (9) for SOM, NG, and GNG respectively.

The outcome of this experiment emphasizes on the efficiency of the local scaling similarity measure if it was passed the appropriate graph. In addition to its small memory footprint, local σ produced the best accuracy in 3 datasets out of 5. Another advantage for local σ over CONN is that local σ weighs edges regardless of the density around neurons. This is clearly demonstrated by the performance drop in NG and GNG when they are weighted by CONN. On the other hand, local σ could perform poorly when the graph contains unnecessary edges as we saw in 3NN, 5NN, and SOM graphs.

4.2 UCI Datasets

In this experiment, we used datasets retrieved from UCI repository. The automatic estimation of σ discussed in Sects. 3.3 and 3.4 was disabled to reduce the volatility and provide better comparison. For graphs weighted by local σ in Table 2, SOM was the best performer, NG and GNG being closely comparable except for BC-Wisconsin dataset where SOM was considerably higher. On the other hand, graphs weighted by CONN did not produce exceptional performance especially in Wine dataset.

Table 2. Clustering accuracies on UCI datasets for 100 runs.

	n	m	local σ					CONN			
			k-means, 3NN	k-means, 5NN	SOM, SOM grid	NG, CHL	GNG, CHL	k-means	SOM	NG	GNG
Iris	150	16	0.83±0.1	0.85±0.1	0.78±0.1	0.86±0.1	0.85±0.1	0.86±0.1	**0.88±0.1**	**0.88±0.1**	0.87±0.1
Wine	178	16	0.82±0.1	0.82±0.1	**0.83±0.1**	0.81±0.1	**0.83±0.1**	0.75±0.1	0.72±0.1	0.76±0.1	0.75±0.1
BC-Wisconsin	699	16	0.77±0.1	0.79±0.1	**0.97±0.0**	0.84±0.1	0.83±0.1	0.96±0.0	0.96±0.0	0.95±0.0	0.95±0.0
Segmentation	2100	32	0.48±0.1	0.49±0.1	0.57±0.1	0.55±0.0	0.56±0.0	**0.58±0.1**	0.54±0.1	0.56±0.1	0.56±0.1
Pen Digits	10992	64	0.70±0.1	0.66±0.1	0.73±0.1	0.66±0.0	0.65±0.0	0.73±0.1	**0.75±0.1**	0.68±0.1	0.67±0.1

The memory footprint is where local σ and *CONN* are split apart. In Fig. 4, it is clear that the memory footprint of *CONN* is exponentially increasing with m. This was not the case with local σ where it maintained a linear increase with m. With local σ performing similar to *CONN* in terms of clustering accuracy, it represents a memory efficient option if it was coupled with a topology preserved graph (Fig. 5).

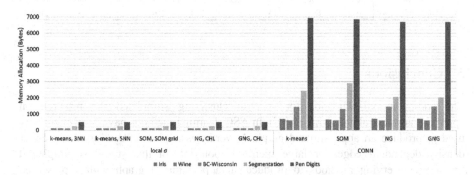

Fig. 4. Similarity measures average memory consumption (best viewed in color).

4.3 Berkeley Segmentation Dataset (BSDS500)

Berkeley Segmentation Dataset (BSDS500) contains 500 images. It uses 3 clustering quality metrics: segmentation covering (covering), rand index (RI), and variation of information (VI). It important to mention that the achieved segmentation results are lower than the numbers reported in the original study [13]. This mainly due to the absence of the edge detection component, which is beyond the scope of this work.

From local σ part in Table 3, one could observe that 3NN, 5NN, and SOM graphs, fluctuated over the clustering metrics. NG and GNG maintained a consistent performance over all measures compared to *CONN* graphs. In terms of covering, they deviated by 0.03 and 0.04 respectively from the best performer. For RI, they were off by 0.06 and 0.08 from the best performer. Finally, for VI, their performance was in line with *CONN* graphs. For *CONN* graphs (on the right side in Table 3), the best performer in terms of covering was NG and the others were not far away from its score. In terms of RI, all methods achieved an equal score. This emphasizes that *CONN* is highly influenced by the edges it draws rather than the vector quantization method used.

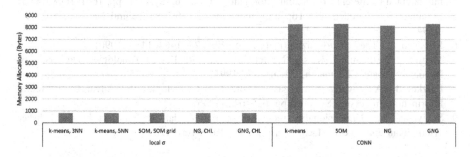

Fig. 5. Average memory needed to compute the similarity measure for methods in Table 3.

Table 3. Image segmentation evaluation using BSDS500

Evaluation metric	local σ					CONN			
	k-means, 3NN	k-means, 5NN	SOM, SOM grid	NG, CHL	GNG, CHL	k-means	SOM	NG	GNG
Segmentation covering	0.33	0.36	0.34	0.36	0.35	0.37	0.36	**0.39**	0.38
Rand index	0.59	0.58	0.64	0.62	0.6	**0.68**	**0.68**	**0.68**	**0.68**
Variation of information	3.06	**2.75**	3.03	2.91	2.94	2.91	2.97	2.81	2.83

5 Conclusions

The computational demands for spectral clustering stimulated the research on approximate spectral clustering (ASC). In ASC, learning methods have been effectively used to produce an approximated graph. Nevertheless, prior ASC efforts have either density dependent edges or large memory footprint. In this study we employed topology preserving methods to produce an approximated graph with a preserved topology. The edges were weighed via a locally scaled similarity measure (local σ) that is independent from the density around them. The experiments reveal that local σ coupled with topology preserving graphs could match the performance of recent ASC methods with less memory footprint.

Approximation methods for spectral clustering are known for long preprocessing time to obtain the approximated graph. For future work, we attempt to minimize the preprocessing computations required to produce the approximated graph.

References

1. Yan, D., Huang, L., Jordan, M.I.: Fast approximate spectral clustering. In: Proceedings of the 15th ACM SIGKDD International Conference on Knowledge Discovery and Data Mining, pp. 907–916 (2009)
2. Tasdemir, K.: Vector quantization based approximate spectral clustering of large datasets. Pattern Recogn. **45**, 3034–3044 (2012)
3. Ng, A.Y., Jordan, M.I., Weiss, Y.: On spectral clustering: Analysis and an algorithm. In: Advances in Neural Information Processing Systems (2002)
4. Zelnik-Manor, L., Perona, P.: Self-tuning spectral clustering. In: Proceedings of the 17th International Conference on Neural Information Processing Systems, pp. 1601–1608 (2004)
5. Martinetz, T., Schulten, K.: Topology representing networks. Neural Netw. **7**, 507–522 (1994)
6. Kohonen, T.: The self-organizing map. Proc. IEEE **78**, 1464–1480 (1990)
7. Martinetz, T., Schulten, K.: A "Neural-Gas" Network Learns Topologies. University of Illinois at Urbana-Champaign, Champaign (1991)
8. Fritzke, B.: A growing neural gas network learns topologies. In: Advances in Neural Information Processing Systems, pp. 625–632 (1995)
9. Sugiyama, M.: Dimensionality reduction of multimodal labeled data by local fisher discriminant analysis. J. Mach. Learn. Res. **8**, 1027–1061 (2007)

10. Zhang, X., Li, J., Yu, H.: Local density adaptive similarity measurement for spectral clustering. Pattern Recogn. Lett. **32**, 352–358 (2011)
11. Martinetz, T.: Competitive hebbian learning rule forms perfectly topology preserving maps. In: Gielen, S., Kappen, B. (eds.) ICANN 1993, pp. 427–434. Springer, London (1993). https://doi.org/10.1007/978-1-4471-2063-6_104
12. Von Luxburg, U.: A tutorial on spectral clustering. Stat. Comput. **17**, 395–416 (2007)
13. Arbelaez, P., Maire, M., Fowlkes, C., Malik, J.: Contour detection and hierarchical image segmentation. IEEE Trans. Pattern Anal. Mach. Intell. **33**, 898–916 (2011)

Discovering Similarities in Malware Behaviors by Clustering of API Call Sequences

Fatima Al Shamsi[1], Wei Lee Woon[2], and Zeyar Aung[2(✉)]

[1] Abu Dhabi Systems and Information Centre, Abu Dhabi, UAE
Fatima.AlShamsi@adsic.abudhabi.ae
[2] Khalifa University of Science and Technology, Masdar Institute, Abu Dhabi, UAE
{wei.woon,zeyar.aung}@ku.ac.ae

Abstract. New genres of malware are evading detection by using polymorphism, obfuscation and encryption techniques. Hence, new strategies are needed to overcome the limitations of current malware analysis practices. In this paper, we propose an unsupervised learning (clustering) framework to complement the supervised learning (i.e., classifier-based malware detection) approach. We cluster malware instances to discover similarities in their dynamic behaviors and to detect new malware families. For that, we utilize Application Programming Interface (API) call sequences to represent the behaviors of malware in dynamic runtime environment. We investigate three sequence comparison algorithms, namely, Optimal Matching (OM), Longest Common Subsequence (LCS), and Longest Common Prefix (LCP) for calculating sequence–sequence distances to be used for hierarchical clustering. Among the three algorithms, LCP is found to be both the most effective in terms of clustering quality and the most efficient in terms of time complexity (linear-time).

Keywords: Malware · API calls · Clustering · Malware patterns

1 Introduction

Nowadays malware becomes prevalent everywhere: from conventional servers and PCs to mobile and IoT devices [1]. Among an average of 1 million malware instances appear every day, most of them are variants of the existing ones [2]. Polymorphism, obfuscation and encryption techniques are applied by malware developers to evade detection by anti-malware programs [3,4]. They generate different malware instances from the same malware family to avoid traditional detection techniques. As a result, members of the same malware family are functionally similar to one another [5] although their binaries can be quite different. A recent study revealed that there was a decrease in the number of new malware families detected between the year of 2014 and 2015. However, there was noticeable increase with a percentage of 36% in the number of malware variants detected between 2014 and 2015 [6]. With the increased number of malware

© Springer Nature Switzerland AG 2018
L. Cheng et al. (Eds.): ICONIP 2018, LNCS 11304, pp. 122–133, 2018.
https://doi.org/10.1007/978-3-030-04212-7_11

instances, effective and computationally efficient methods are required to detect similar malware behaviors in order to accelerate the malware analysis process. Leveraging data mining techniques like clustering can help us fulfill this requirement.

In this paper, we conduct a clustering analysis of malware based on their dynamic behavioral patterns represented by Application Programming Interface (API) call sequences. Although supervised learning (classification) approach, such as [7–10], is the most common data mining method for malware analysis (detection), unsupervised learning (clustering), such as [11–14], is also a useful malware analysis tool that can complement supervised learning. The benefits of a clustering analysis are as follows.

1. Organizing malware into homogeneous clusters can help us better understand the malware activities [15]. Visualization of clustering results can provide us with the "bird-eye view" of the similarities/differences of malware in their behavioral patterns such as host infection procedures, attack routines, and dissemination mechanisms.
2. Clustering analysis enables "variant detection". It allows us to uncover behaviorally similar malware that should belong to the same family even though identified as distinct and sometimes even assigned to different malware categories by commercial anti-malware programs. This facilitates malware analysis experts to better sample malware variants of the same/different malware families when carrying an in-depth manual analysis process [16].
3. Clustering allow us to identify the emergence of new malware families by examining the formation of new clusters and/or sub-clusters. In fact, clustering is recommended as an adaptive technique to look for unknown patterns potentially corresponding to new types of attacks [15,17].

In our study, we first transform API call sequences into symbolic strings to allow us to detect similar patterns. Then, we calculate all-against-all pairwise similarities (and hence distances) of malware samples in our dataset in order to construct the distance matrix and subsequently perform a hierarchical clustering using the Weighted Pair Group Method with Arithmetic Mean (WPGMA) algorithm.

The new contributions of this paper are as follows.

1. We have investigated three different similarity/distance functions, namely Optimal Matching (OM), Longest Common Subsequence (LCS) (as in [14]), and Longest Common Prefix (LCP) and compare the results in terms of clustering quality (silhouette coefficient) and time complexity.
2. It turns out that LCP offers the best results with the best clustering quality and the lowest time complexity (linear-time). To our best knowledge, we are the first to apply the LCP function in API sequence-based malware analysis and demonstrate its usefulness. The results suggest that prefix (i.e., initial API subsequence) could be used as a quick but reliable indicator for overall similar behaviors in malware.

This paper is a summary version of the master's thesis [18] by the first author.

2   Related Work

Supervised learning (classification) and unsupervised learning (clustering) are the two main approaches used in data mining-based malware analysis. The works [7–10] are some examples of malware classification. Some examples of malware clustering are described below.

Bayer et al. [11] provided a scalable approach for malware clustering. They investigated malware applications that share similar behavioral traits through three phase model, namely, dynamic analysis phase, identification of behavioral signatures based on system calls phase, and the clustering phase. To suggest a feasible solution to the proximity search of nearest neighbor, the clustering algorithm they used in their study is based on locality sensitive hashing. Their experiment effectively and efficiently clustered samples sharing similar behavioral patterns.

Kim et al. [12] calculated the similarity between android malware applications by utilizing Androguard to construct control flow graphs. Similar behaviors were compared by reviewing the structural information collected from the flow graphs to match target applications. Moreover, clustering sub-families of each family and identifying representative malware samples help reduce the number of similarity calculation. The clustering algorithm used was DBSCAN.

Qiao et al. [13] clustered API sequences based on frequent pattern mining algorithms. The analysis framework consists of three steps, API call abstraction, frequent itemset mining, and analysis. For API call abstraction, dynamic analysis tools, namely, CWSandbox and Cuckoo Sandbox were used to extract behavioral reports of malware binaries. API pattern mining for frequent itemsets was identified using Apriori Program. Nevertheless, frequent API calls are used as an input to calculate the similarity between different malware binaries. The output served as an input for computing signature sequences. Instead of API calls, strings of opcodes representing malware behaviors are clustered using a suffix tree in the recent work by Oprişa et al. [19].

Zhong et al. [14] proposed analysis system, namely ARIGUMA, which is composed of four components, Sandbox, code analyzer, visualizer, and web User Interface (UI). The system use code analyzer to classify malware applications based on their behavior. Comparison between malware instructions was completed using LCS. Given that LCS algorithm requires polynomial time by dynamic programming, the computational time was reduced by filtering out the numeric features of codes.

3   Proposed Malware Clustering Framework

3.1   Generating Encoded API Call Sequences

Instead of looking directly into the characteristics of malware through its binaries (executable codes), we examine its Application Programming Interface (API) call information in order to better understand its behavioral pattern. This approach gives us three advantages. (1) It enables a dynamic analysis of malware though run-time API calls instead of a static analysis. (2) It is much more efficient to deal with API call sequences (whose lengths are typically less than 1000) than malware binaries (whose lengths are typically a few megabytes). (3) Two instances of malware from the same family may not have a detectable common signature in their binaries because of polymorphism and/or obfuscation; however, their behaviors in terms of API call sequences are expected to be quite similar.

The API-based malware detection system (APIMDS) dataset used in this study is obtained from Ki et al. [20]. It contains the API system call sequences for 6 nominal malware categories, namely, backdoor, packed, PUP (Potentially Unwanted Programs), Trojan, virus and worm. Those sequences were obtained by executing the malware samples using virtual machine environment such as VirtualBox. (The detailed process can be referred to in the paper [20].) From the whole dataset, we randomly select 1,797 of them for our clustering studies.

After obtaining the API call sequences of those selected malware, we preform "sequence encoding". Its purpose is to reduce the dimensionality of sequence unique states as well as to reduce the time complexity for calculating distances between the different sequences. API code sequence generation step is carried out by referring to MSDN library [21] for categorizing API system calls into 26 categories as given in Table 1.

Table 1. Codes for API call categories.

Category	Code	Category	Code
I/O Create	A	Window class functions	N
I/O Open	B	File I/O	O
I/O Write	C	Data exchange	P
I/O Find	D	Keyboard	Q
I/O Read	E	Volume management	R
I/O Access	F	Windows GDI	S
Load	G	Menus and Other resources	T
Debugging & Error handling	H	Interprocess communication	U
Memory management	I	Process and thread	V
Delete/Destroy	J	Networking and internet	W
Get info	K	System information/services	X
Windows	L	Device management	Y
COM	M	Other	Z

3.2 Generating Distance Matrix

The concept of "similarity" can be applied to a pair of encoded API call sequences representing dynamic behaviors of malware. Similar malware instances share a certain amount of behaviors in common. The most common algorithms that used to measure the similarity between the two given sequences are: (i) Optimal Matching (OM), (ii) Longest Common Subsequence (LCS), and (iii) Longest Common Prefix (LCP) [22,23].

The purpose of these algorithms are to extract the lengths of the "common patterns" of malware behaviors in their own unique ways. The longer the two sequences that follow a common pattern, the more similar they are and the lesser the difference (a.k.a. "distance") between them. Distance can be regarded as the inverse of similarity and can be easily derived from it, and vice versa.

Given a dataset of n malware samples, for each distance function, we need to construct an all-against-all (i.e., $n \times n$) distance matrix, where each matrix entry d_{ij} is the pairwise distance between two malware sequences indexed i and j. In fact, we only need the compute the distance values in the upper triangle of the matrix (whose size is $n(n-1)/2$), because the lower triangle is the mirror image of the upper one. This distance matrix is to be used as an input in the subsequent clustering procedure.

In our research, we use R's TraMineR package [23] to compute the OM, the LCS, and the LCP distances. Their brief descriptions are given below.

Optimal Matching (OM) Distance: OM produces edit distance by calculating the minimal amount of edits required to change a pair of malware behaviors X and Y into a different pair such that they are similar to some extent [14]. It is based on Needleman-Wunsch sequence alignment algorithm [24], which is commonly used in Bioinformatics. The insertion/deletion cost is single value to be specified by the user, while the substitution cost matrix (of replacing one API call category for another) can be produced by assigning constant values, or using the transition rate between sequences observed in the dataset. All the insertion/deletion/substitution costs are set to 1 in our experiment. Equation 1 defines the OM distance recursively [22].

$$d_{OM}(X_i, Y_j) = \min \begin{cases} d_{OM}(X_{i-1}, Y_j) + \text{deletion cost} \\ d_{OM}(X_i, Y_{j-1}) + \text{insertion cost} \\ d_{OM}(X_i, Y_{j-1}) + \text{substitution cost if } X_i \neq Y_j \end{cases} \quad (1)$$

The OM distance of two sequences of lengths $|X|$ and $|Y|$ can be computed in a quadratic $O(|X| \times |Y|)$ time [22,24]. It should be noted that OM directly computes the distance without needing to compute the similarity first.

Longest Common Subsequence (LCS) Distance: Given two sequences X, and Y, LCS calculates the largest number of common sequential elements between them. For example, let the two sequences be $X = $ **A-B-A-C-A-D** and $Y = $ **A-B-C-A-P-Q-R**. Then, their LCS is **A-B-C-A**. LCS measures the similarity and difference of instructions between sequences that have same relative order, but not necessarily adjacent. For the two sequences X_i, and Y_j, where i and j represent the length of the longest common subsequence present in both sequences, Eq. 2 defines the length of LCS recursively [22].

$$LCS(X_i, Y_j) = \begin{cases} 0 & \text{if } i=0 \text{ or } j=0 \\ LCS(X_{i-1}, Y_{j-1}) + 1 & \text{if } X_i = Y_j \\ \max(LCS(X_i, Y_{j-1}), LCS(X_{i-1}, Y_j)) & \text{if } X_i \neq Y_j \end{cases} \tag{2}$$

The LCS distance d_{LCS} is derived from the LCS length using Eq. 3.

$$d_{LCS}(X, Y) = |X| + |Y| - 2 \cdot LCS(X, Y) \tag{3}$$

LCS was also used in the previous malware clustering study by Zhong et al. [14]. Computing LCS of two sequences with $|X|$ and $|Y|$ elements using dynamic programming is of a quadratic $O(|X| \times |Y|)$ time complexity [14].

Longest Common Prefix (LCP) Distance: Given two sequences X, and Y, LCP calculates the length of the longest common prefix shared by them. For example, let the two sequences be $X = $ **A-B-A-C-A-D** and $Y = $ **A-B-C-A-P-Q-R**. Then, their LCP is **A-B**. LCP offers a quick first-order approximation of the similarity of two given sequences by looking at their prefixes only. For the two sequences X, and Y, Eq. 4 defines the length of LCP [22].

$$LCP(X, Y) = |\{u \neq \lambda : X^{|u|} = u = Y^{|u|}\}| \tag{4}$$

where λ is an empty string, and $X^{|u|}$ and $Y^{|u|}$ ($|u| \leq \min(|X|, |Y|)$) are the longest prefix sequences of length $|u|$ of X and Y respectively. Just like the LCS distance, the LCP distance d_{LCP} can be derived from the LCP length in the same manner as in Eq. 3.

LCP of two sequences with $|X|$ and $|Y|$ elements can be calculated in linear time, which is $O(\min(|X|, |Y|))$.

3.3   Clustering

Clustering enables grouping of similar encoded API sequences representing malware behaviors to detect repeating patterns. Moreover, sub-patterns in the same cluster can also be recognized and further analyzed. To evade binary code signature-based detection by anti-malware tools, malware developers make minor changes (polymorphism) or apply obfuscation to the source code of existing malware to produce new variants. This greatly increases the number of variants for the same malware with different signatures. API sequence-based clustering can help reveal those kinds of variants easily.

Clustering uses the distance matrix generated from the above step. In our case, we conduct 3 distinct clustering exercises using 3 distance matrices with OM, LCS, and LCP distance functions respectively. We employ a hierarchical clustering algorithm called Weighted Pair Group Method with Arithmetic Mean (WPGMA) [25] to generate clusters. (Note: we also tried k-means clustering, but the results are not as good as those by WPGMA. We do not show the k-means results in this paper because of the page limit.)

The WPGMA algorithm constructs a rooted tree (dendrogram) that reflects the structure present in a pairwise distance matrix. At each step, the nearest two clusters, say i and j, are combined into a higher-level cluster $i \cup j$. Then, its distance to another cluster k is simply the arithmetic mean of the distances between k and members of $i \cup j$ [25].

$$d_{(i \cup j),k} = (d_{i,k} + d_{j,k})/2 \tag{5}$$

Clustering n elements using WPGMA takes $O(n^2)$ space and $O(n^2)$ time. There are $n-1$ iterations, with $O(n)$ work in each one [26]. We use R's `Cluster` package [27] to perform WPGMA clustering.

4   Results and Discussions

In this section, we will discuss the results of our proposed framework on the 1,797 malware samples in our experimental dataset. The will identify the suspicious API calls and examine the suspicious call patterns based on the clustering results. Visualization and statistical validation of the clustering results are also performed along with a running time analysis.

4.1   Clusters of Similar Malware Behavioral Patterns

Clustering helps us to find similar groups among malware API sequences. Hierarchical clustering (WPGMA) algorithm is used to cluster sequences and identify similar patterns of malicious behavior. Hierarchical clustering divides data into clusters and produce a tree structure to display the results. As mentioned above, three distinct clustering tasks using OM, LCS, and LCP distance matrices are performed. However, only the results of LCP will be shown here as an example.

Figure 1 represents clustering results for LCP. By referring to the figure, we can notice that there is a clear distinction between the clusters, except for Cluster #1 which is not very homogeneous. Each cluster exhibits a different set of behaviors. For example, in Cluster #4, we can notice that there are 6 sub-clusters that are very similar in initial subsequences yet differ in overall sequence length. These sub-clusters follow the same pattern, namely, `Networking and Internet > Get Info > Debugging and Error Handling > I/O Create > File I/O > Get Info > I/O Open > Window Class`. This pattern is an example of the initial subsequence taken by all sub clusters within the same cluster.

Likewise, for a large and heterogeneous cluster like Cluster #1 (with 1278 members), we can also apply sub-clustering to generate more homogeneous sub-clusters. Figure 2 shows the sub-clustering result for Cluster #1. It should be noted that for the sub-clusters that are still heterogeneous and sufficiently large, such as Sub-Cluster #1 (with 984 members), it is still possible to further sub-divide them. (The sub-division results are not shown in this paper.)

One notable observation is that malware variants which belongs to different nominal categories (like virus, backdoor, worm, and Trojan) can exhibit similar behaviour and thus belong to the same cluster. An example can be seen in Fig. 3, which shows malware variants with different nominal categories having similar API sequence patterns—indicating a strong possibility that they are variants of each other, albeit differently categorized by the anti-malware companies.

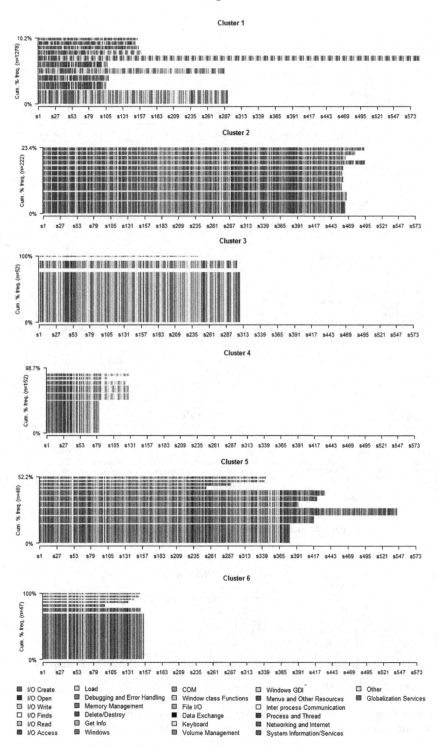

Fig. 1. Clustering results with LCP distance function (number of clusters = 6).

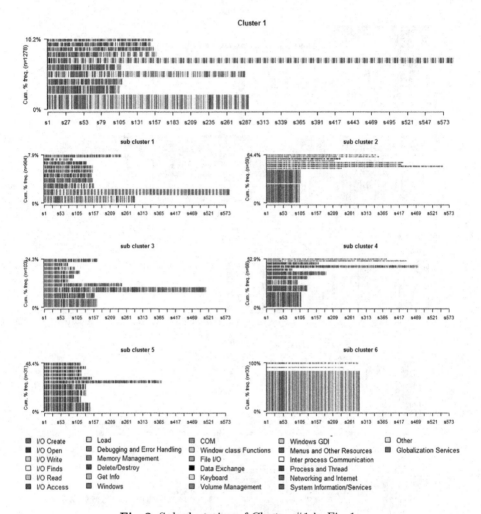

Fig. 2. Sub-clustering of Cluster #1 in Fig. 1.

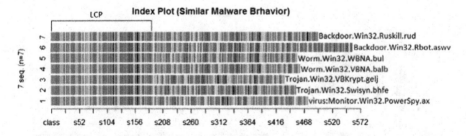

Fig. 3. Zooming into a subset of Cluster #2 of LCP clustering result, exhibiting different nominal categories of malware in the same cluster.

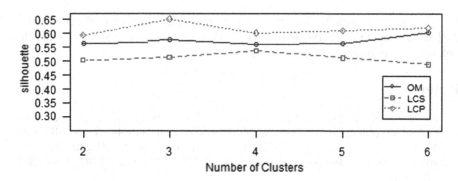

Fig. 4. Silhouette coefficients by 3 distance functions (OM, LCS, and LCP).

4.2 Clustering Quality

To statistically validate clustering results, silhouette coefficient ($\bar{s}$) is used to determine the optimal number of clusters which produces best clustering results. Given the distance matrix, the normalized distance m_{ij} between two sequences indexed i and j is calculated from the matrix element d_{ij} as: $m_{ij} = d_{ij}/d_{\max}$, where $d_{\max}$ is the maximum distance value in the matrix.

The average intra-cluster distance of the sequence i to other members of its own cluster C (donated as a), the average inter-cluster distance of the sequence i to the members in the nearest cluster C' (denoted as b), and finally the silhouette coefficient ($\bar{s}$) for the sequence i are defined in Eq. 6. The silhouette coefficient value always lies between -1 and 1. The general guideline is that a coefficient greater than 0.5 represents a meaningful clustering result [28]. The higher the coefficient, the better the clustering quality.

$$ a = \sum_{j \in C \backslash \{i\}} \frac{m_{ij}}{|C|-1}; \quad b = \min \left(\sum_{j \in C'} \frac{m_{ij}}{|C'|} \right); \quad \bar{s} = \frac{b-a}{\max(a,b)} \tag{6} $$

Figure 4 shows the average silhouette coefficients for 3 WPGMA clusterings results of the 1,797 malware samples based on 3 different distance functions of OM, LCS, and LCP. Here, LCP is found to offer the best result with the highest average silhouette coefficient of **0.65** when the number of clusters is 3.

5   Conclusion

In this paper, we have proposed a clustering framework to detect similar malware behaviors based on the comparison of API call sequences. We have investigated the performances of three sequence–sequence comparison functions, namely, OM, LCS, and LCP, as a basis for clustering. To the best of our knowledge, we are the first to apply the LCP function in API sequence-based malware analysis, which offers the most promising results in terms of clustering quality and time complexity (linear-time). This gives us a strong indication that prefix (i.e., initial API subsequence) can serve as a quick and reliable descriptor for overall similar behaviors in malware.

References

1. Alwahedi, S., Al Ali, M., Ishowo-Oloko, F., Woon, W.L., Aung, Z.: Security in mobile computing: attack vectors, solutions, and challenges. In: Agüero, R., Zaki, Y., Wenning, B.-L., Förster, A., Timm-Giel, A. (eds.) MONAMI 2016. LNICST, vol. 191, pp. 177–191. Springer, Cham (2017). https://doi.org/10.1007/978-3-319-52712-3_13
2. Lee, T., Kwak, J.: Effective and reliable malware group classification for a massive malware environment. Int. J. Distrib. Sens. Netw. (2016). Article ID 4601847
3. Cho, I.K., Im, E.G.: Extracting representative API patterns of malware families using multiple sequence alignments. In: Proceedings of the 2015 ACM Conference on Research in Adaptive and Convergent Systems (RACS), pp. 308–313 (2015)
4. Stamp, M.: Information Security: Principles and Practice, 2nd edn. Wiley, New York (2011)
5. Dinh, A., Brill, D., Li, Y., He, W.: Malware sequence alignment. In: Proceedings of the 2016 IEEE International Conferences on Big Data and Cloud Computing (BDCloud), Social Computing and Networking (SocialCom), Sustainable Computing and Communications (SustainCom), pp. 613–617 (2016)
6. Symantec Enterprise Security: 2016 symantec internet security threat report. Technical report (2016). http://www.symantec.com/security-center/threat-report
7. Narayanan, A., Chen, Y., Pang, S., Tao, B.: The effects of different representations on static structure analysis of computer malware signatures. Sci. World J. **2013**, 8 (2013). Article ID 671096
8. Kate, P.M., Dhavale, S.V.: Two phase static analysis technique for Android malware detection. In: Proceedings of the 3rd International Symposium on Women in Computing and Informatics (WCI), pp. 650–655 (2015)
9. Milosevic, N., Dehghantanha, A., Choo, K.K.R.: Machine learning aided Android malware classification. Comput. Electr. Eng. **61**, 266–274 (2017)
10. Al Ali, M., Svetinovic, D., Aung, Z., Lukman, S.: Malware detection in Android mobile platform using machine learning algorithms. In: Proceedings of the 2017 IEEE International Conference on Infocom Technologies and Unmanned Systems (Trends and Future Directions) (ICTUS), pp. 763–768 (2017)
11. Bayer, U., Comparetti, P.M., Hlauschek, C., Krügel, C., Kirda, E.: Scalable, behavior-based malware clustering. In: Proceedings of the 2009 Network and Distributed System Security Symposium (NDSS), pp. 1–18 (2009)
12. Kim, J., Kim, T.G., Im, E.G.: Structural information based malicious app similarity calculation and clustering. In: Proceedings of the 2015 ACM Conference on Research in Adaptive and Convergent Systems (RACS), pp. 314–318 (2015)
13. Qiao, Y., He, J., Yang, Y., Ji, L.: Analyzing malware by abstracting the frequent itemsets in API call sequences. In: Proceedings of the 2013 IEEE 12th International Conference on Trust, Security and Privacy in Computing and Communications (TrustCom), pp. 265–270 (2013)
14. Zhong, Y., Yamaki, H., Yamaguchi, Y., Takakura, H.: ARIGUMA code analyzer: Efficient variant detection by identifying common instruction sequences in malware families. In: Proceedings of the 2013 IEEE 37th Annual Computer Software and Applications Conference (COMPSAC), pp. 11–20 (2013)

15. Cordeiro De Amorim, R., Komisarczuk, P.: On partitional clustering of malware. In: Proceedings of the 1st International Workshop on Cyberpatterns: Unifying Design Patterns with Security, Attack and Forensic Patterns (CyberPatterns), pp. 47–51 (2012)
16. Perdisci, R., U, M.: VAMO: Towards a fully automated malware clustering validity analysis. In: Proceedings of the 28th ACM Annual Computer Security Applications Conference (ACSAC), pp. 329–338 (2012)
17. Monshizadeh, M., Yan, Z.: Security related data mining. In: Proceedings of the 2014 IEEE International Conference on Computer and Information Technology (CIT), pp. 775–782 (2014)
18. Al Shamsi, F.: Mapping, Exploration, and Detection Strategies for Malware Universe. Master's thesis, Masdar Institute of Science and Technology, Abu Dhabi, UAE (2017)
19. Oprişa, C., Cabău, G., Pal, G.S.: Malware clustering using suffix trees. J. Comput. Virol. Hacking Tech. **12**, 1–10 (2016)
20. Ki, Y., Kim, E., Kim, H.K.: A novel approach to detect malware based on API call sequence analysis. Int. J. Distrib. Sens. Netw. (2015). Article ID 659101
21. Microsoft: MSDN: Learn to develop with Microsoft developer network (2018). https://msdn.microsoft.com/
22. Elzinga, C.H.: Sequence analysis: Metric representations of categorical time series. Department of Social Science Research Methods. Technical report, Vrije Universiteit Amsterdam, The Netherlands (2006)
23. Gabadinho, A., Ritschard, G., Müller, N.S., Studer, M.: Analyzing and visualizing state sequences in R with TraMineR. J. Stat. Softw. **40**, 1–37 (2011)
24. Needleman, S.B., Wunsch, C.D.: A general method applicable to the search for similarities in the amino acid sequence of two proteins. J. Mol. Biol. **48**, 443–453 (1970)
25. Gonnet, G.H., Scholl, R.: Scientific Computation, 1st edn. Cambridge University Press, New York (2009)
26. Itsik, P.: WPGMA (2001). http://www.cs.tau.ac.il/~rshamir/algmb/00/scribe00/html/lec08/node21.html
27. Maechler, M., Rousseeuw, P., Struyf, A., et al.: Methods for cluster analysis (2018). https://cran.r-project.org/web/packages/cluster/index.html
28. Kaufman, L., Rousseeuw, P.J.: Finding Groups in Data: An Introduction to Cluster Analysis. Wiley, New York (2005)

A Storm-Based Parallel Clustering Algorithm of Streaming Data

Fang-Zhu Xu[1,2], Zhi-Ying Jiang[1,2], Yan-Lin He[1,2], Ya-Jie Wang[3], and Qun-Xiong Zhu[1,2(✉)]

[1] College of Information Science and Technology,
Beijing University of Chemical Technology, Beijing 100029, China
zhuqx@mail.buct.edu.cn
[2] Engineering Research Center of Intelligent PSE,
Ministry of Education of China, Beijing 100029, China
[3] Guizhou Food Safety Testing Engineering Technology Research Center Co.,
Ltd., Guizhou, China

Abstract. Aiming at solving the shortcomings of traditional Single-Pass clustering algorithms, such as low accuracy and large amount of computation, a novel Storm-based parallel Single-Pass clustering algorithm is proposed to discovery of hot events in the food field. In order to solve the problem of data inconsistency in parallel computing, a method of dynamically acquiring cluster increments and random delays is adopted to improve the Single-Pass algorithm. In order to validate the performance of the proposed method, a case study of news events classification is carried out. Simulation results show that the proposed algorithm can effectively improve the cluster repetition in clustering results and greatly improve the accuracy and efficiency of clustering compared with the traditional Single-Pass algorithm.

Keywords: Parallel clustering · Streaming data · Single-Pass algorithm

1 Introduction

With the rapid development of the Internet and the massive data brought about by information explosion, the era of big data is coming. However, in recent years, the traditional off-line data processing methods are far from satisfying the growth of today's data. Streaming data have gradually become the mainstream data form. Compared with traditional data, streaming data has four characteristics: (1) data arrives in real time; (2) data arrival order is independent and is not controlled by the application system; (3) the data scale is large and cannot be predicted; (4) once the data is processed, unless the data is intentionally saved, or the data cannot be taken out again [1]. Based on the above situations, new effective data processing methods for streaming data are particularly important.

Clustering is one of the most basic and important techniques in data mining. It is to divide data with similarities into one set from complex data. With the development and researches of streaming data, the cluster analysis of stream-oriented data has been widely developed [2]. Streaming clustering algorithms refer to the real-time processing

© Springer Nature Switzerland AG 2018
L. Cheng et al. (Eds.): ICONIP 2018, LNCS 11304, pp. 134–144, 2018.
https://doi.org/10.1007/978-3-030-04212-7_12

of data, which requires the continuous arrival of data sources. The timeliness requirements are extremely high. Classical streaming clustering algorithms include the Clu-Stream algorithm and the Single-Pass algorithm. The Clu-Stream algorithm implements a two-tier clustering framework, which divides the flow clustering process into online and offline parts [3]. Gu et al. [4] applied the Clu-Stream algorithm to the detection of turbine performance and effectively implemented real-time data processing. The Single-Pass clustering algorithm is to process sequential data sequentially, and add the most similar class or self-contained class according to the similarity of current data with other categories, and finally form a clustering result. Based on the idea of multi-Agent, Forestierod and others [5] used self-organizing strategies to improve the Single-Pass algorithm and cluster similar data points. However, in practical applications, when dealing with large-scale, high-dynamics, and high-dimensional data streams, the traditional single-machine serial processing method is often incapable of completing the task timely, efficiently, and accurately. Therefore, finding the way to improving the timeliness and scalability of mass flow data has become a hot spot and a difficult topic for scholars.

In recent years, with the rapid rise of clustering technology, parallel computing and distributed computing, the widespread application of the Hadoop distributed computing framework and the Storm computing framework have brought about the dawn of solving the real-time mining of large-scale streaming data. Among them, Hadoop focuses more on offline computing and is not very suitable for streaming data mining [6]. While Storm has advantages for streaming data clustering, both in terms of timeliness and efficiency [7]. Therefore, this article relies on Storm to parallelize the single-pass algorithm for the clustering of stream data. At the same time, an improved method is proposed to solve the problem of data inconsistency when the clustering algorithm is executed in parallel.

2 Clustering Algorithm

2.1 Data Preprocessing

The Single-Pass clustering algorithm actually operates on the word vector. Therefore, before clustering, the arrived text data should be pre-processed and converted into a directly-calculated vector. An essential step before the text vectorization is to process the text segmentation using the Word open source word segmentation tool which is based on JAVA and provides basic word segmentation functions.

The text vectorization, as the name implies, is to transform the existing text into a multidimensional vector, namely the following form:

$$D = \{(w_1, v_1), (w_2, v_2), (w_3, v_3) \ldots \ldots (w_n, v_n)\} \tag{1}$$

Among them, represents a feature item in the text content. The value is obtained by the word segmentation device. indicates the corresponding weight value of the feature item, which is obtained by a certain vectorization rule, and the text is vectorized to facilitate the numerical calculation [8].

At present, the most common text vectorization method is the TF-IDF algorithm. The main idea of TF-IDF is that if a word or phrase appears frequently in an article and rarely appears in other articles, the word or phrase will be considered to have a good class distinguishing ability and be suitable for classification. The weight of this feature is, where TF means the term frequency which is calculated by Eq. 2, and IDF means inverse document frequency which is calculated by Eq. 3.

$$f_T = m/M \tag{2}$$

$$f_{ID} = lg(N/n + 0.01) \tag{3}$$

where m represents the number of occurrences of a feature word in the text, M represents the total number of words in the text, N represents the total number of files, n represents the number of files containing the feature word, and 0.01 is used to prevent equals to 1. This algorithm only performs the TF calculation, and uses stop word table to filter the high-frequency stop words in the segmentation process.

The cosine similarity is selected for calculation. The cosine between two vectors can be calculated by Euclidean dot product (Eq. 4).

$$a \cdot b = \| a \| \| b \| \cos\theta \tag{4}$$

For two specified vector A and B, their cosine similarity can be calculated by dot product and the length of two vectors, as shown in Eq. 5:

$$\cos \theta = \frac{A \cdot B}{\| A \| \| B \|} = \frac{\sum_{i=1}^{n} A_i \times B_i}{\sqrt{\sum_{i=1}^{n}(A_i)^2} \times \sqrt{\sum_{i=1}^{n}(B_i)^2}} \tag{5}$$

where and represent the components of vector A and B respectively. The closer the cosine similarity to 1, the higher the similarity of two texts [9].

2.2   Single-Pass Clustering Algorithm Principle

The Single-Pass algorithm is a classical method of streaming data clustering. As shown in Fig. 1, the main idea is to read data from the data stream at a time, compare this data with the existing clusters, and calculate the degree of similarity. If the similarity value exceeds the preset threshold, the data falls into the cluster. If there is no similar cluster, then the new data will be treated as a new cluster. This process is repeated until no new data is read [10].

Single-Pass is a classical algorithm for streaming clustering. The advantages and disadvantages of this algorithm are also obvious. The advantage is that the algorithm is simple. The algorithm computes only once for the data, and does not require iterative calculations. But the disadvantage is also very prominent. Since the Single-Pass algorithm performs only one calculation and the cluster set is only the current result, the accuracy of the calculation is inferior to that of multiple iterations. Moreover, unlike K-Means and other algorithms, Single-Pass does not predefine the number of clusters. So, the calculation time increases as the clusters increase. If the text correlation is not

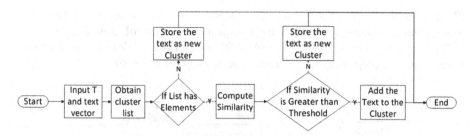

Fig. 1. Single-Pass clustering algorithm

large, the calculated time will increase sharply as the number of clusters increases. In order to solve the problem, a method of dynamically acquiring cluster increments and random delays is adopted to improve the Single-Pass algorithm. The following section will introduce the improved algorithm of this paper in detail and demonstrate the feasibility of the improved algorithm through experiments.

3   Storm-Based Parallel Clustering Algorithm

3.1   Storm Framework

The system uses Storm as a computing framework. Storm is an open source distributed streaming computing framework developed by Twitter in 2010. Data is calculated in the topology. Each topology is a directed acyclic graph consisting of a series of Spouts and Bolts. As shown in Fig. 2, Spout is the producer of Streams. Usually spout reads the data tuple from an external data source (Kafka, etc.) and sends it to the topology. Bolt handles all the operations and business logic in topology. If the logic is complex, it is usually solved by using multiple bolts. Bolt will subscribe to the spout or other tuple sent by the bolt, and the entire application may have multiple spouts and bolts. Together a map structure of topology can be formed [11]. Storm's streaming data clustering has the advantages in terms of timeliness and efficiency.

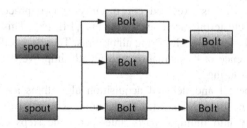

Fig. 2. Storm framework flow chart

Figure 3 shows the current system's calculation flow and concurrency settings. SqlSpout is the MySQL data read layer, SplitBolt is the word segmentation layer, and tfBolt is the text vectorization layer. The ClassifytBolt layer is the cluster layer, the addItem layer is responsible for the subsequent processing of clustering, and the SqlSave layer is responsible for writing MySQL data. In general, it can be divided into three parts, data preprocessing, clustering and subsequent processing.

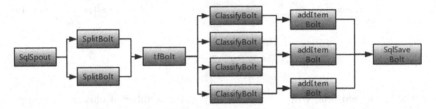

Fig. 3. System flow chart

3.2 Parallel Clustering Algorithm

When the entire clustering process is calculated by Storm, the parallel computing does not cause problems because the preprocessing and subsequent processing sections are independent processes that are not related to each other. However, in the clustering operation, the Single-Pass algorithm parallel computing will produce consistency problems [12]. Because clustering is based on the fact that each piece of text is strongly correlated, when comparing one piece of text, you need to get a collection of all the pieces; but each piece of text is processed independently in parallel computing, which creates data inconsistencies [13].

Specifically, concurrent execution is likely to have changed the set of clusters at the end of the execution; but the text does not detect changes to the cluster. For example, assume that event A and event B have a very high degree of similarity and have no correlation with other clusters, that is, normal calculation A and B will be divided into the same cluster, and the cluster contains only these two elements. When A and B perform the clustering operation at the same time, there is a possibility that none of the other parties acquire the cluster information, and thus two clusters are established. For a large number of data flow situations, similar information has a high probability of being close to each other. If it is executed concurrently in two processes, it will cause particularly serious clustering accuracy problems [14, 15]. This problem is often neglected in the field of parallel clustering algorithms. Therefore, effectively solving or improving the occurrence of this situation will greatly improve the accuracy and efficiency of parallel clustering.

Dynamic incremental and delayed acquisition algorithms are used to minimize concurrent consistency issues. The idea of dynamic acquisition is to divide the calculation of similarity into multiple parts, intersperse the steps of obtaining cluster increments, so as to reduce the time of occurrence of the consistency problem. Although the use of dynamic acquisition methods can reduce the problem of concurrency to a certain extent, it does not completely eliminate the consistency problem. The delay calculation is used to solve the remaining consistency problems. When obtaining the cluster increment, if no new cluster is available for calculation, then randomly delay a short period of time. Therefore, the random delay time is set to 30 ms. In this way, the consistency problem when creating a new cluster can be greatly improved, and the time efficiency of the clustering operation will not have an excessive impact.

The improved parallel computing process of Single-Pass clustering algorithm is shown in Fig. 4.

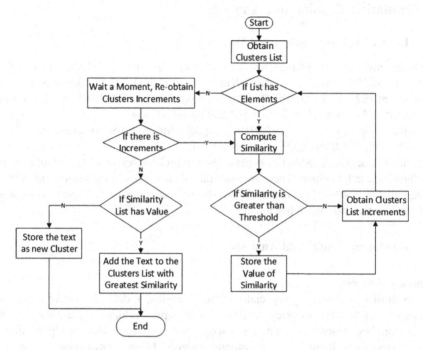

Fig. 4. The improved calculation process of single-pass

The first step is to obtain the current information in the cluster list; the second step is to judge whether there are elements in the list, and then divide it into two cases. If there are elements, then calculate the value of similarity, and then determine whether it is greater than the threshold, if yes, store the value of the similarity and obtain the cluster list until no increments. If not, directly obtain the cluster list. The next step can be performed when there is no increment in the list; The third step is to acquire cluster increments again after a period of random time. The fourth step is to judge whether there is increment in the list. At this time, it is divided into two cases. If there is, return to the calculated cosine similarity in the second step. If not, continue to the next step; The fifth step is to determine if there is a value in the similarity list; The sixth step, if it is not, then the text is stored as a new cluster; If it is, the text is stored in the cluster with the highest degree of similarity.

In this algorithm, when comparing the cosine similarity, the specific approach is to find the cluster that is larger than and closest to the threshold, add the text to the cluster and recalculate the center point of the cluster. Through the current element we can calculate a virtual center point and use it as the center point of the cluster. The calculation speed is faster because the only thing need to know is the current center point of the cluster and the number of elements in the cluster, therefore, the center point position can be obtained quickly.

4 Emulation Results and Analysis

4.1 Emulation Environment and Data Set

Hardware environment: Storm Cluster consists of five servers, including four Supervisors and one Nimbus. Among them, Nimbus is located at 192.168.0.101, Supervisor is located at 192.168.0.102-105, Redis is configured at 192.168.0.101, and MySQL is configured at 192.168.0.103. Zookeeper unified scheduling.

Software environment: jdk1.7.0_67, apache-Storm-0.9.5, apache-maven-3.3.9, IDEA, Hadoop 2.5.0-cdh5.3.9, zookeeper-3.4.6.

Data: The data set used in this paper is the food-related textual data provided by the GuiZhou Research Institute. There are more than 40 million articles and 40,000 of them are tested. These 40,000 data form a "strawberry" carcinogenic and Qingdao prawns, Harbin sky price fish and other topics.

4.2 Emulation Results and Analysis

Accuracy Analysis

The result of text clustering, especially for news events, is difficult to have a uniform evaluation standard for accuracy. Therefore, this paper evaluates the accuracy of the results from both macroscopic and microscopic perspectives so that the algorithm can be displayed more intuitively and comprehensively. In the macroscopic aspect, the overall situation of the 40,000 data is compared. Because of the large amount of data, a more general method of discriminating the accuracy of the clustering results is used, that is, the average distance between the elements in each cluster and its center point. It should be noted that the distance here is the cosine distance, which means that the closer the value to 1, the smaller the distance within the cluster, and the higher the clustering accuracy. Finally, we operate the accuracy analysis on the total average value we obtained. Figure 5 shows the comparison of the results of the stand-alone Single-Pass algorithm and the improved parallel Single-Pass algorithm in this paper. Both of them have a clustering calculation threshold of 0.3. It can be found that the clustering result of the proposed algorithm is much higher than 0.3, while the one of stand-alone Single-Pass algorithm only exceeds the threshold. The clustering achieved by the improved parallel Single-Pass algorithm is proved to be more compact.

To the micro perspective, 10 clusters of topics were selected from the clustering results and each of the texts was observed. Due to the small amount of data, the evaluation criteria of traditional text classification techniques can be used are: precision, recall, and a combination of both F-Measure and False.

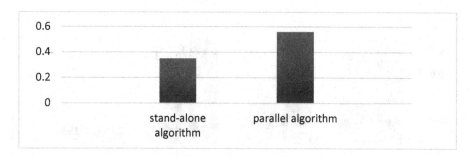

Fig. 5. Cluster distance comparison graph

(1) The precision rate P is the ratio that the number of correctly cluster texts occupies the number of total cluster texts from the clustering results:

$$P = \frac{d_{right}}{d_{all}} \tag{6}$$

(2) The recall rate R is the ratio of the number of related documents correctly clustered occupies the number of all related documents in the clustering result:

$$R = \frac{d_{right}}{d_{allRight}} \tag{7}$$

(3) The F value is the harmonic average of the precision and recall:

$$F = \frac{2PR}{P+R} \tag{8}$$

(4) The error rate False is the ratio of the number of incorrectly clustered texts in each cluster to the total number of all texts in the clustering result:

$$False = \frac{d_{wrong}}{d_{all}} \tag{9}$$

$$d_{right} + d_{wrong} = d_{all} \tag{10}$$

As shown in Fig. 6, based on the number of correct clustering texts and the number of missing text of each cluster in the stand-alone Single-Pass algorithm and the improved parallel Single-Pass algorithm respectively, we calculate and compare the precision rate, recall rate, F value and error rate. From the figure, it can be seen that all indicators of the improved algorithm increase by about 5%.

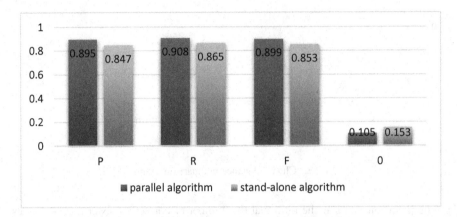

Fig. 6. Clustering results accuracy index comparison chart

Efficiency Analysis

The comparison of the algorithm efficiency is to see the increase in processing time by the increase in the amount of data. Figure 9 shows the comparison of the time of the stand-alone Single-Pass algorithm and the improved parallel Single-Pass algorithm running on Storm when processing the same data. From Fig. 7, it can be seen that as the amount of data processed increases, the running time of the stand-alone Single-Pass algorithm increases greatly, and the gap between the stand-alone Single-Pass algorithm and the Storm-based algorithm is increasing. When the amount of data reaches 10,000, the processing time of the stand-alone Single-Pass algorithm is almost 2 times the processing time of the improved parallel Single-Pass algorithm. This also shows that Storm-based parallel algorithms can significantly improve the computation efficiency and the larger the amount of data, the more obvious of the improvement. The reason lies in that although the improved method seems to increase the processing time, it actually greatly avoids the occurrence of duplicate clusters and reduces the number of cluster, leading to reduction of the time for calculating the similarity of each text. Moreover, the time of dynamically requiring cluster increments and random delays is

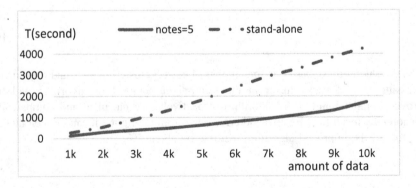

Fig. 7. Comparison of clustering efficiency

too small to affect the overall efficiency, so the efficiency will be significantly improved.

5 Conclusions

This article is based on the Storm platform, using the Single-Pass algorithm to achieve parallel clustering of streaming data. At the same time, for the problem of inefficient running of clustering and inaccurate clustering results caused by the inconsistency of parallel computing data, a method of dynamic incremental and random delay is proposed to improve the algorithm.

We cluster the news of 40,000 food fields, observe the clustering results, and compare it with the stand-alone Single-Pass algorithm. It can be found that the improved parallel Single-Pass algorithm can effectively improve the clustering accuracy. From the overall macroscopic aspect, by comparing the closeness of the clusters, the proposed method can increase the average distance of each text to the center point by about 20%. From the microscopic aspect, by comparing results for each cluster, the proposed method can improve 4 accuracy indexes by about 5%. After passing through a large amount of the same data in the system, observing and comparing the processing time, the method of dynamically acquiring incremental and random delay will not only affect the processing efficiency, which makes the algorithm more efficient than the stand-alone algorithm.

References

1. Cheng, X.Q., Jin, X.L., Wang, Y.Z., Guo, J., Zhang, T., Li, G.: Survey on big data system and analytic technology. J. Softw. **25**(9), 1889–1908 (2014)
2. Hengmin, Z., Weiwei, Z.: Study on web topic online clustering approach based on single-pass algorithm. Data Anal. Knowl. Discov. **27**(12), 52–57 (2011)
3. Wu, Y.: Network big data: a literature survey on stream data mining. JSW **9**(9), 2427–2434 (2014)
4. Gu, H., Si, F., Xu, Z.G.: Turbine performance monitoring method based on clustream data steam. China Acad. J. Electron. Publishing House **5**, 5180–5184 (2013)
5. Forestiero, A., Pizzuti, C., Spezzano, G.: A single pass algorithm for clustering evolving data streams based on swarm intelligence. Data Min. Knowl. Discov. **26**(1), 1–26 (2013)
6. Dong, X., Li, R., Zhou, W., Wang, C., Xue, Z., Liao, D.: Performance optimization and feature enhancements of hadoop system. J. Comput. Res. Develop. **50**(S2), 1–15 (2013)
7. Zhao, F., Lin, S., Gao, X., Computers, S.O.: The research and application of storm framework for large data. Microcomput. Appl. (2016)
8. Tu, S., Huang, M.: Mining microblog user interests based on textrank with tf-idf factor. J. China Univ. Posts Telecommun. **23**(5), 40–46 (2016)
9. Guo, Q.: The similarity computing of documents based on VSM. Comput. Softw. Appl. **5186**, 585–586 (2008)
10. Yan, D., Hua, E., Hu, B.: An improved single-pass algorithm for Chinese microblog topic detection and tracking. In: IEEE International Congress on Big Data, pp. 251–258 (2016)
11. Karunaratne, P., Karunasekera, S., Harwood, A.: Distributed stream clustering using micro-clusters on apache storm. J. Parallel Distrib. Comput. **108**, 74–84 (2017)

12. Yi, W., Teng, F., Xu, J.: Noval stream data mining framework under the background of big data. Cybern. Inf. Technol. **16**(5), 69–77 (2016)
13. Hassani, M., Seidl, T.: Clustering big data streams: recent challenges and contributions. IT Inf. Technol. **58**(4), 206–213 (2016)
14. Hyde, R., Angelov, P., Mackenzie, A.R.: Fully online clustering of evolving data streams into arbitrarily shaped clusters. Inf. Sci. **382**, 96–114 (2016)
15. Zheng, L., Huo, H., Guo, Y., Fang, T.: Supervised adaptive incremental clustering for data stream of chunks. Neurocomputing **219**, 502–517 (2017)

Iterative Maximum Clique Clustering Based Detection Filter

Xinyu Zhang[1], Hao Sheng[1(✉)], Yang Zhang[1], Jiahui Chen[1], Yubin Wu[1], Guangtao Xue[2], and Quanrui Wei[3]

[1] State Key Laboratory of Software Development Environment, School of Computer Science and Engineering, Beihang University, Beijing, People's Republic of China
{xinyu.zhang,shenghao,yang.zhang,chenjh,yubin.wu}@buaa.edu.cn
[2] Department of Computer Science and Engineering, Shanghai Jiao Tong University, Shanghai, People's Republic of China
gt_xue@sjtu.edu.cn
[3] The 15th Research Institute of China Electronics Technology group Corporation, Beijing, People's Republic of China
wquanrui@163.com

Abstract. Object detection is an important research field of computer vision, but getting accurate object detection from a large number of detection candidates has always been a challenge. The most current algorithms use an insufficient Greedy Non-Maximum Suppression (NMS) strategy which heavily relies on the confidence of the detection candidates. This paper proposes the Iterative Detection Filter (IDF) approach, which considers more information of the detection candidates, including overlapping, the confidence generated by the detector, and the ground position perception information of the scene. Through this approach, the detection candidates are mapped to more accurate detections. Our method achieves a significant improvement on the MOT16 and MOT17 datasets, which are widely used in video tracking and detection.

Keywords: Detection filter · Maximum clique · Iterative clustering
Detection candidate · Non-Maximum Suppression

1 Introduction

Object detection is an important research field in computer vision and has made remarkable progress in recent years. When a detector searches an object in a video or image, it simply judges whether or not to include the object in the given box (either via a sliding window or through object proposals) and provide it with confidence. In this case, the detector might find more potential objects, *i.e.* an object has multiple responses. It is a challenge that how to find accurate object detection from a large number of detection candidates with low reliability confidence description.

Non-Maximum Suppression (NMS) is a popular method that is used to remove redundant detections (*e.g.* [6,8,9,16]). It greedily deletes lower scoring detections and keeps higher scoring ones, if they overlap enough

© Springer Nature Switzerland AG 2018
L. Cheng et al. (Eds.): ICONIP 2018, LNCS 11304, pp. 145–156, 2018.
https://doi.org/10.1007/978-3-030-04212-7_13

(*e.g.* intersection-over-union $IoU > 0.5$); this is what we refer to as Greedy NMS in the following. Greedy NMS relies too much on the confidence of detection candidates; in fact, the greedy removal of any conflict detection is insensible, especially when considering its statistical significance. Although the effect of the detector is very good, it produces some inaccurate detection candidates without considering the scene information.

This paper focuses more on the detection candidates information, including overlapping, the confidence generated by the detector, and the ground position perception information of the scene. First, we cluster these detection candidates using the iterative maximum clique clustering approach based on overlapping information. Then, we introduce the filtering method of the conflict detections to the detection that is based on confidence. Finally, we use the ground-sensing filtering method to redefine the confidence of detections and remove the unreliable detections.

We used the DPM [6] algorithm, which is widely used in detecting pedestrians in videos [15], in order to get the original detection candidates. Our method was a significant improvement in the MOT16 and MOT17 datasets.

2   Related Work

Object detection is vital in the field of computer vision and machine learning. Early object detection algorithms used sliding window detection. Viola *et al.* [19] proposed 'Integral Image' feature and used AdaBoost learning algorithm which sped up calculations. Dalal *et al.* [3] used Histograms of Oriented Gradient (HOG) feature and adopted linear SVM approach to detect. Felzenszwalb *et al.* [7] proposed a discriminatively trained, multi-scale, deformable part model (DPM) for object detection. Later on, Felzenszwalb *et al.* [5] built cascade classifiers from DPM models, such as pictorial structures.

With the ongoing depth of the convolution neural network (CNN) layer, the network's abstraction ability – which is the ability to resist translation and the scale change – became stronger, and the sliding window detection approach was discarded. Some outstanding detection algorithms were proposed based on object proposal approach [2,17,21]. They used a lower computation to achieve an accurate positioning of the bounding box of the general object that was contained in the image. Girshick *et al.* [9] applied high-capacity convolutional neural networks (CNNs) to bottom-up region proposals and suggested the RCNN (Regions with CNN features) algorithm. With the development of the algorithm and the improvement of computing speed, the SPPNet [10], Fast-RCNN [8], Faster-RCNN [16], FPN [14] *et al.* algorithms that were based on CNN, were quickly developed and achieved good results.

In regards to the sliding window detection and object proposal methods, the mapping of the detection candidates to the detection outputs was an important step. Greedy NMS is widely used in these algorithms. Even in the case of end-to-end detection algorithms [11,18,20] used in recent years, they cannot completely remove the Greedy NMS strategy [12] from the final output of detections. Greedy

NMS relies heavily on the confidence of the detection candidates and does not make good use of redundant detections information. This resulted in a variety of wrong deletions (false negative, FN) and inaccurate positioning (false positives, FP).

In light of the above challenge, we propose the Iterative Detection Filter (IDF) algorithm, which focused more on the detection candidates information, and weighed detection candidates to form the final output using the iterative maximum clique clustering.

3 Iterative Clustering Based Detection Filter

Greedy NMS only uses the extent of overlapping among detections and confidence information to roughly delete redundant detections. The deleted detections do not have a role to play. Our results made better use of both of the aforementioned features. First, we cluster these detection candidates using the iterative maximum clique clustering. Then, we introduce the filter of the conflicting detection candidates to the final detection. Finally, we introduce the ground-sensing filtering method to refine confidence and remove any unreliable detections.

3.1 Iterative Maximum Clique Clustering

Traditionally, the Greedy NMS method selected a detection from the redundant candidate set to represent pedestrian, which did not make full use of the detector information. We discovered that it was more effective when multiple detection candidates were used to indicate an objects detection.

Our first priority was to classify the detections in the candidate set in order to represent an object with multiple detections. We found that there was a large number of detection candidates around a significant object, but fewer around insignificant features such as occluded pedestrians. We usually picked out the significant object detection first, so that the detection had a higher quality. A clique is a complete sub-graph in a graph. A maximal clique is a clique to which no more vertices can be added, *i.e.* maximal complete sub-graph. We noted a similarity in the two problems, so we proposed an iterative clustering algorithm based on the maximal clique model.

Our iterative maximum clique clustering has two main steps. In the first step, we built a graph with detection candidate as node. In the second step, the nodes are iteratively classified based on solving maximum cliques.

We built a graph $\mathcal{G} = \{\mathcal{V}, \mathcal{E}\}$ for all detection candidates in image, where the position information and confidence $v = \{x, y, w, h, c\} \in \mathcal{V}$ as nodes, the overlap between detections as the weight of the edge $e(v_i, v_j) = IOU(v_i, v_j) \in \mathcal{E}$. $\{x, y, w, h, c\}$ represents the upper left corner coordinates, width, height and confidence of a pedestrian detection. The overlap function is defined as follows:

$$IOU(v_i, v_j) = \frac{v_i \cap v_j}{v_i \cup v_j}. \tag{1}$$

Fig. 1. The process of iterative maximum clique cluster in MOT16-02, where **Iter-0** is the original detection candidates, and the **Iter-1**, **Iter-2**, **Iter-n** is the first, second, n-th iteration. **Iter-n** is the final output of the detection.

In order to transform the classification problem into solving the maximal clique problem, we map the weight of the edge to $\{0, 1\}$ by constructing the indication function:

$$e(v_i, v_j) = \begin{cases} 1 & IOU(v_i, v_j) \geq threshold \\ 0 & IOU(v_i, v_j) < threshold \end{cases} \tag{2}$$

Detection candidates are conflicting when the overlap is bigger than the threshold.

In the iterative classification, we found the maximum cliques in graph $\mathcal{G} = \{\mathcal{V}, \mathcal{E}\}$ obtained above, *i.e.*, finding first maximal complete sub-graph $\mathcal{G}_1$. We assigned these nodes the same label and then removed them from the graph $\mathcal{G} = \mathcal{G} \backslash \mathcal{G}_1$. We iterate over this method until there are no nodes in the graph, ending the step. We got a set $C = \{\mathcal{G}_1, \mathcal{G}_2, \ldots, \mathcal{G}_n\}$ composed of complete sub-graphs.

After clustering, we used the detection filter to get the new detection set with sub-graphs set (Sect. 3.2). We built a new graph with a new detection set using the first step. If there were any conflicting nodes in the graph, we continued to perform the second iteration clustering step. If there was no edge in the graph, the node that was obtained by the detection filter became the final detection result. As the Figs. 1 and 2 shows, the detection candidates converge to pedestrians over multiple iterations. Since imprecise detection candidates usually

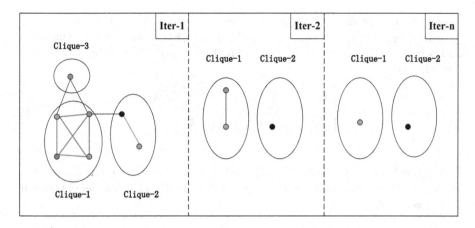

Fig. 2. The iterative maximum clique clustering process corresponding to two pedestrian detection candidates of Fig. 1 in the lower right corner. The two pedestrians are surrounded by multiple detection candidates in **Iter-0**. After several iterations, the detection gradually converges to the pedestrian. For the each iteration step, the color circle nodes indicate the detection candidate at present, the large ellipse is the maximum cliques obtained by iterative clustering. And the edges represents that the two nodes are conflicting, *i.e.* overlap is greater than the threshold.

have smaller confidence and contain little object information, they play a role without reducing accuracy when detection is generated.

3.2   Detection Filter

We found that directly removing conflict detections does not take full advantage of the detected information. In Sect. 3.1, we classify all the detections, *i.e.* the same label detections is conflicting and at most one detection is left. Therefore, we proposed a detection filter based on the confidence of the original detection candidates. Detections that belonged to the same class were mapped to a new detection.

The detections that were obtained through the detector contain a confidence which reflects the possibility of belonging to a object. We used the DPM detector in the experiment with a range of confidence of $[-\infty, \infty]$. We intercepted the detection of confidence greater than -0.5, as the original detection set.

In order to effectively use the detector information, we generated the detection by combining detection candidates which had the same label with certain weight. We found that confidence is a good metric that partially reflected the quality of the detection. However, the range of confidence of the original detection is $[-0.5, \infty]$, a range that is too large to balanced distinguish detection. Therefore, we proposed a weight mapping function which mapped the confidence $c \in [-0.5, \infty]$ to the weight $\lambda \in (0, 1)$. The definition of the weight mapping function is as follows:

$$\lambda = \frac{1}{1 + e^{-c}}. \tag{3}$$

For any sub-graph $\mathcal{G}_i$, it contains a series of detections $\{(x_1, y_1, w_1, h_1, c_1, \lambda_1),$ $(x_2, y_2, w_2, h_2, c_2, \lambda_2), \cdots , (x_m, y_m, w_m, h_m, c_m, \lambda_m\}$ that are assigned the same label, where confidence uses weight mapping function above. Our detection filter is as follows:

$$
\begin{cases}
x_{new} = \sum_{i=1}^{m} (x_i * \lambda_i)/\sum_{i=1}^{m} \lambda_i \\
y_{new} = \sum_{i=1}^{m} (y_i * \lambda_i)/\sum_{i=1}^{m} \lambda_i \\
w_{new} = \sum_{i=1}^{m} (w_i * \lambda_i)/\sum_{i=1}^{m} \lambda_i \\
h_{new} = \sum_{i=1}^{m} (h_i * \lambda_i)/\sum_{i=1}^{m} \lambda_i \\
c_{new} = max\{c_1, c_2, \ldots, c_m\}
\end{cases}
\tag{4}
$$

After such a transformation, each new detection fuses more information from the original detection candidates, which makes the detection of the extraction more reasonable.

3.3   Ground-Perception Filtering Method

When referring to pedestrian tracking, accurate access to the ground position provides a great advantage for the processing of abnormal pedestrians. In a video, it is very difficult to manually mark the ground position, especially in the case of a moving camera.

We found that it was more effective to use a series of continuous detection data in videos to perceive the ground position. The detections with the highest confidence usually had the best quality, so it was more feasible to use the points of their feet to represent the position of the ground. Although a few high confidence detections were inferior, the majority of the detections we chose were of higher quality, which made the ground position we obtained understandable. In the case of a motion camera, the ground had a slight shaking in the video, but we could still get a balanced ground position using our method.

In order to demarcate the range of pedestrian heights, we assumed that the height of a human obeyed normal distribution $\mathcal{N}(H, \sigma^2)$, where $H = 1.7\,\mathrm{m}$ and $\sigma = 0.3$. Hence for a pedestrian in video, the perspective height $\tilde{H}$, obeyed normal distribution $\mathcal{N}(\tilde{H}, \tilde{\sigma}^2)$, where $\tilde{\sigma} = \sigma * \tilde{H}/H$. The perspective height was used to refine the confidence of detection. When the height of detection is anomalous, it is given a low score. The refined function is defined as follows:

$$
c_i^{new} = 2\pi\tilde{\sigma} * p(\tilde{H}_i) * c_i, \tag{5}
$$

where $p(\tilde{H}_i)$ is the probability density function of normal distribution $\mathcal{N}(\tilde{H}, \tilde{\sigma}^2)$, c_i is the confidence of detection $\{x_i, y_i, w_i, h_i, c_i\}$, $h_i(i.e.\ \tilde{H}_i$ in Eq. 5) is perspective height of the detection and c_i^{new} is the refined confidence. Base on height that obeyed normal distribution, the of range new confidence is $[0, 1]$.

However, the current confidence optimization strategy has a disadvantage. Short pedestrians (especially children) are easily reset to lower confidence. We found that there were a lot of original detection candidates around pedestrians with significant features. After resetting the detection with low confidence and it corresponded to the original sub-graph with a large number of nodes, we handled it as a special detection. For example, when deleting low confidence detections, they were not actually deleted.

4   Experiments

We experimented on the MOT16, MOT17 [15] datasets to prove that our approach is an improvement for both precision and recall.

Datasets: In our experiments, we used the MOT16 and MOT17 [15] datasets, which covered several different types of scenes, including variable directions, different density of pedestrians and partial occlusions. The detection candidates came from publicly available datasets in the MOT website[1] (*i.e.* raw (no NMS) detections).

Metrics: Artificial label were used as the ground truth. We used classic comprehensive evaluation metrics recall (Rcll↑), precision (Prcn↑) and F-Score (F-Sc↑, the harmonic mean of precision and recall), where

$$F - Sc = 2 * Rcll * Prcn/(Rcll + Prcn). \tag{6}$$

We also introduced Multiple Object Detection Accuracy (MODA↑) and Multiple Object Detection Precision (MODP) [4] to evaluate detection.

4.1   Analysis of Iterative Clustering

In our clustering process, we defined detection conflict that the overlap was greater than threshold. Choosing an appropriate threshold to determine the conflict was an important issue. We chose the two kinds of ground truth (GT) standard; one was the minimum visible region greater than 0.5 as a pedestrian, which was generally used to evaluate the effectiveness of detection, and the other was the visible area greater than 0.05, which was generally used in the tracking because a bit of evidence should also be seen as pedestrians appearances. We selected different thresholds for analysis (results are shown in Tables 1 and 2).

Table 1. The quantitative detection results of our approach under different threshold with 0.05 as minimum visible region.

Threshold	Rcll	Prcn	FAR	TP	FP	FN	MODA	MODP	AP	F-sc
0.20	46.83	**87.50**	**1.23**	45609	**6518**	51777	40.14	76.99	0.45	61.01
0.30	49.49	86.17	1.46	48197	7735	49189	**41.55**	77.50	0.45	62.87
0.40	50.99	82.87	1.93	49653	10266	47733	40.44	77.78	**0.53**	**63.13**
0.50	52.36	71.43	3.84	50989	20391	46397	31.42	**77.92**	0.52	60.43
0.60	**54.52**	44.60	12.41	**53096**	65961	**44290**	−13.21	77.55	0.40	49.06

We selected multiple evaluation indicators to evaluate the detection. Among them, MODA, AP and F-sc are three important evaluation indicators, which

---

[1] https://motchallenge.net/data/MOT16/.

Table 2. The quantitative detection results of our approach under different threshold with 0.5 as minimum visible region.

Threshold	Rcll	Prcn	FAR	TP	FP	FN	MODA	MODP	AP	F-sc
0.20	63.37	**86.47**	**1.23**	41639	**6516**	24066	53.46	77.42	0.62	73.14
0.30	66.15	84.90	1.45	43466	7731	22239	**54.39**	77.99	**0.63**	**74.36**
0.40	67.37	81.19	1.93	44266	10255	21439	51.76	78.35	**0.63**	73.64
0.50	68.05	68.71	3.83	44714	20366	20991	37.06	**78.58**	0.61	68.38
0.60	**68.77**	40.67	12.40	**45183**	65907	**20522**	−31.54	78.52	0.46	51.11

comprehensively reflect the quality of detection. When combining these two tables, we saw that the weakening of conflict (*i.e.* increase of the threshold) caused a rise of the truth positive and brought on a rapid increase in the false positive. We found that when the threshold is 0.4, both in F-sc and MODA got an excellent value while Rcll, Prcn and MODP maintained a higher position.

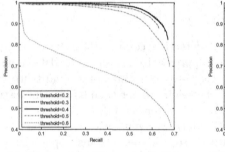
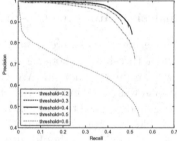

(a) Minimum visible region is 0.5   (b) Minimum visible region is 0.05

Fig. 3. The Rcll-Prcn curves results of our methods under different visual region and conflicting thresholds.

At the same time, we analyzed the Rcll-Prcn curves corresponding to different thresholds. As we all know, when a curve is completely surrounded by another, the approach is superior to the another. Figure 3 shows that we got the best results when the threshold was 0.4.

Under the threshold is 0.4, we analyzed the convergence (Fig. 4) of the detection candidate. As shown in Fig. 4(a), the number of detection candidates had dropped significantly after the first iteration, and all the frames converged within seven iterations. This proved that the algorithm has a very good convergence performance. As shown in Fig. 4(b), most frames would converge over 4 or 5 iterations.

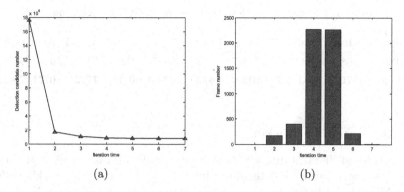

(a)   (b)

Fig. 4. Convergence analysis in MOT16 train_dataset. (a) The change in the number of detection candidates as the iteration progresses. After the first iteration, the number of detection candidates has dropped significantly, and all the frames converge within seven iterations. (b) The distribution of the number of iterations in order to reach convergence. For example, there are approximately 2300 frames that converge after four iterations.

4.2   Comparison with Baseline

We chose the publicly available detection candidate on the MOT website as the original dataset, which was obtain by DPM-v5 [6] algorithm. We analyzed the different NMS thresholds in the MOT16 train_dataset, where the NMS approach came from the DPM-v5 (Fig. 5). Our approach greatly outperformed the Greedy NMS method at all NMS values.

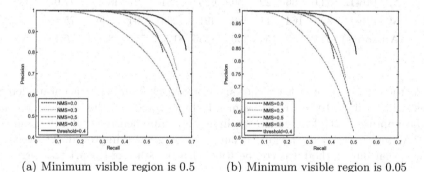

(a) Minimum visible region is 0.5   (b) Minimum visible region is 0.05

Fig. 5. The Rcll-Prcn curves results of DPM approach under different visual region and NMS values, where the black solid line is our method at the threshold of 0.4.

We submitted the results of the MOT17 test set on the MOT website. Table 3 shows that we exceeded the DPM algorithm based on the Greedy NMS in a lot of metrics. Our method reduced a large number of redundant detections (FP) while ensuring the increase of TP; in regards to integrated indicator MODA, AP, it increased by 25.9% and 10.0% respectively.

Table 3. Comparison of detection results of MOT17 Benchmark

Method	Rcll	Prcn	FAR	TP	FP	FN	MODA	MODP	AP	F-sc
DPM [6]	68.1	64.8	7.1	78007	42308	36557	31.2	75.8	0.61	66.41
IDF (ours)	**70.2**	**83.3**	**2.7**	**80395**	**16153**	**34169**	**56.1**	**77.2**	**0.71**	**76.19**

4.3   Application in Tracking

In regards to off-line tracking, the tracking-by-detection framework is the most popular, so the accuracy of detection is important. We tested the tracking effect on the MOT16 dataset with the detection obtained by our method. We experimented with a popular multiple hypothesis tracking (MHT) [13] algorithm and followed the current popular CLEAR MOT [1] metrics to evaluate the tracking performance (Table 4).

Table 4. Tracking results of MOT16 benchmark

Sequence	Rcll	GT	MT	ML	FP	FN	IDs	MOTA	MOTP
MOT16-01	42.5	23	6	9	176	3679	15	39.5	72.2
MOT16-03	62.1	148	41	26	6893	39677	241	55.2	76.3
MOT16-06	55.8	221	56	111	549	5102	42	50.7	75.2
MOT16-07	52.8	54	8	5	985	7704	66	46.4	75.1
MOT16-08	43.9	63	11	20	1616	9389	103	33.6	78.6
MOT16-12	54.0	86	22	30	625	3814	21	46.2	77.0
MOT16-14	32.9	164	11	89	491	12398	64	29.9	75.4
All (ours)	**55.2**	**759**	**155**	**300**	**11335**	**81763**	**552**	**48.6**	**76.1**

As the aforementioned table shows, our approach found more accurate detections, which allowed the FP to be massively elevated. At the same time, due to the improvement of detection quality, the comprehensive evaluation metric MOTA (multiple object tracking accuracy) was greatly improved. The results showed that our method was competitive with the state-of-the-art tracker.

5   Conclusion

This paper proposed an iterative detection filter approach based on maximum clique clustering and used it to complete the mapping from detection candidates to detections. Our approach considered more of the detection candidates information, including overlapping, the confidence generated by the detector, and the ground position perception information of the scene. We discovered more accurate detection, while severely reducing false positives. Our method had a significant improvement in the MOT16 and MOT17 dataset, which is widely used in video tracking and detection. In addition, our methods also achieved good results in tracking application.

Acknowledgement. This study is partially supported by the National Key R & D Program of China (No. 2016QY01W0200), the National Natural Science Foundation of China (No. 61472019), the Macao Science and Technology Development Fund (No. 138/2 016/A3), the Open Fund of the State Key Laboratory of Software Development Environment under grant SKLSDE-2017ZX-09, the Project of Experimental Verification of the Basic Commonness and Key Technical Standards of the Industrial Internet network architecture, and the Technology Innovation Fund of China Electronic Technology Group Corporation. Thank you for the support from HAWKEYE Group.

References

1. Bernardin, K., Stiefelhagen, R.: Evaluating multiple object tracking performance: the clear mot metrics. EURASIP J. Image Video Process. **2008**(1), 1–10 (2008)
2. Cheng, M., Zhang, Z., Lin, W., Torr, P.H.S.: BING: binarized normed gradients for objectness estimation at 300 fps. In: IEEE Conference on Computer Vision and Pattern Recognition, pp. 3286–3293 (2014)
3. Dalal, N., Triggs, B.: Histograms of oriented gradients for human detection. In: IEEE Conference on Computer Vision and Pattern Recognition, pp. 886–893 (2005)
4. Ellis, A., Ferryman, J.: Pets2010 and pets2009 evaluation of results using individual ground truthed single views. In: IEEE International Conference on Advanced Video and Signal Based Surveillance, pp. 135–142 (2010)
5. Felzenszwalb, P.F., Girshick, R.B., McAllester, D.A.: Cascade object detection with deformable part models. In: IEEE Conference on Computer Vision and Pattern Recognition, pp. 2241–2248 (2010)
6. Felzenszwalb, P.F., Girshick, R.B., McAllester, D.A., Ramanan, D.: Object detection with discriminatively trained part-based models. IEEE Trans. Pattern Anal. Mach. Intell. **32**(9), 1627–1645 (2010)
7. Felzenszwalb, P.F., McAllester, D.A., Ramanan, D.: A discriminatively trained, multiscale, deformable part model. In: IEEE Conference on Computer Vision and Pattern Recognition, pp. 1–8 (2008)
8. Girshick, R.: Fast R-CNN. In: IEEE International Conference on Computer Vision, pp. 1440–1448 (2015)
9. Girshick, R.B., Donahue, J., Darrell, T., Malik, J.: Rich feature hierarchies for accurate object detection and semantic segmentation. In: IEEE Conference on Computer Vision and Pattern Recognition, pp. 580–587 (2014)
10. He, K., Zhang, X., Ren, S., Sun, J.: Spatial pyramid pooling in deep convolutional networks for visual recognition. IEEE Trans. Pattern Anal. Mach. Intell. **37**(9), 1904–1916 (2015)
11. Henderson, P., Ferrari, V.: End-to-end training of object class detectors for mean average precision. In: Lai, S.-H., Lepetit, V., Nishino, K., Sato, Y. (eds.) ACCV 2016, Part V. LNCS, vol. 10115, pp. 198–213. Springer, Cham (2017). https://doi.org/10.1007/978-3-319-54193-8_13
12. Hosang, J.H., Benenson, R., Schiele, B.: Learning non-maximum suppression. In: IEEE Conference on Computer Vision and Pattern Recognition, pp. 6469–6477 (2017)

13. Kim, C., Li, F., Ciptadi, A., Rehg, J.M.: Multiple hypothesis tracking revisited. In: IEEE International Conference on Computer Vision, pp. 4696–4704 (2015)
14. Lin, T., Dollár, P., Girshick, R.B., He, K., Hariharan, B., Belongie, S.J.: Feature pyramid networks for object detection. In: IEEE Conference on Computer Vision and Pattern Recognition, pp. 936–944 (2017)
15. Milan, A., Leal-Taixé, L., Reid, I.D., Roth, S., Schindler, K.: MOT16: A benchmark for multi-object tracking. CoRR abs/1603.00831 (2016)
16. Ren, S., He, K., Girshick, R.B., Sun, J.: Faster R-CNN: towards real-time object detection with region proposal networks. In: Annual Conference on Neural Information Processing Systems, pp. 91–99 (2015)
17. van de Sande, K.E.A., Uijlings, J.R.R., Gevers, T., Smeulders, A.W.M.: Segmentation as selective search for object recognition. In: IEEE International Conference on Computer Vision, pp. 1879–1886 (2011)
18. Stewart, R., Andriluka, M., Ng, A.Y.: End-to-end people detection in crowded scenes. In: IEEE Conference on Computer Vision and Pattern Recognition, pp. 2325–2333 (2016)
19. Viola, P.A., Jones, M.J.: Rapid object detection using a boosted cascade of simple features. In: IEEE Conference on Computer Vision and Pattern Recognition, pp. 511–518 (2001)
20. Wan, L., Eigen, D., Fergus, R.: End-to-end integration of a convolutional network, deformable parts model and non-maximum suppression. In: IEEE Conference on Computer Vision and Pattern Recognition, pp. 851–859 (2015)
21. Zitnick, C.L., Dollár, P.: Edge boxes: locating object proposals from edges. In: Fleet, D., Pajdla, T., Schiele, B., Tuytelaars, T. (eds.) ECCV 2014, Part V. LNCS, vol. 8693, pp. 391–405. Springer, Cham (2014). https://doi.org/10.1007/978-3-319-10602-1_26

Towards a Compact and Effective Representation for Datasets with Inhomogeneous Clusters

Haimei Zhao, Zhuo Chen, Qiuhui Tong, and Yuan Bo[(⊠)]

Intelligent Computing Lab, Division of Informatics, Graduate School at
Shenzhen, Tsinghua University, Shenzhen 518055, People's Republic of China
{z-chen17, zhaohml7}@mails.tsinghua.edu.cn,
tongqh@126.com, yuanb@sz.tsinghua.edu.cn

Abstract. Due to the restriction of computing resources, it is often inconvenient to directly conduct analysis on massive datasets. Instead, a set of representatives can be extracted to approximate the spatial distribution of data objects. Standard data mining algorithms are then performed on these selected points only, which typically account for a small fraction of the original data, reducing the computational time significantly. In practice, the boundary points of data clusters can be regarded as a compact and effective representation of the original data, with great potential in clustering, outlier or anomaly detection and classification. As a result, given a complex dataset, how to reliably identify a set of effective boundary points creates a new challenge in data mining. In this paper, we present a boundary extraction technique similar to the method in SCUBI (Scalable Clustering Using Boundary Information). The key difference is that our technique exploits the clustering information in a feedback loop to further refine the boundary. Experimental results show that our technique is more robust and can produce more representative boundary points than SCUBI, especially on complex datasets with large inhomogeneity in terms of cluster density.

Keywords: Boundary · Extraction · Clustering · SCUBI

1 Introduction

In face of the ever increasing volume of datasets, efficiency is becoming a critical concern in data analytics. For example, the computational cost of many clustering algorithms [1, 2] such as K-means, DBSCAN, Affinity Propagation and Hierarchical Clustering is greatly affected by the number of data points (n) in the dataset. Among them, the time complexity of K-means is $O(tkn)$ where t is the iteration number [4]; the time complexity of DBSCAN is $O(n^{4/3})$ in the best situation and $O(n^2)$ on some datasets [5]; the time complexity of Affinity Propagation is $O(n^2 T)$ [6] and Hierarchical Clustering features $O(n^2 \log n)$ time complexity [7].

Consequently, it may be impractical to directly apply existing clustering algorithms on large datasets, due to the potentially intolerable computing time. Although it is possible to accelerate these algorithms using parallel or distributed platforms such as

© Springer Nature Switzerland AG 2018
L. Cheng et al. (Eds.): ICONIP 2018, LNCS 11304, pp. 157–169, 2018.
https://doi.org/10.1007/978-3-030-04212-7_14

Hadoop, an interesting idea is to tackle this challenge from a different perspective: whether it is necessary to involve all available data into the clustering process? Under some general assumptions (e.g., each cluster is a solid object without holes), the shape of each cluster can be determined solely by the data points on its surface while all interior points impose no influence on the clustering results. This observation can be significant as it implies that only data points on the surface of each cluster need to be clustered, which only account for a small fraction of the entire dataset.

SCUBI (Scalable Clustering Using Boundary Information) is a latest clustering scheme that exploits the idea of boundary information to achieve good scalability [8]. Firstly, it extracts boundary points from the original dataset. Next, boundary points are clustered using existing clustering algorithms. Finally, the boundary information and the clustering results are used to assign interior points to appropriate clusters. SCUBI is a general scheme and a variety of clustering methods such as DBSCAN, Affinity Propagation and Spectral Clustering can be plugged into the framework. Typically, less than 5% data points are extracted as boundary points and experimental results show that SCUBI can significantly improve the efficiency of existing clustering algorithms on massive datasets with little impact on the quality of clusters.

In this paper, we investigate the boundary information from a more general perspective: it can be regarded as a compact and generic representation of the dataset, which potentially plays a key role in many data mining tasks such as stream clustering, outlier/anomaly detection and classification. The major contribution of our work is an effective approach to the extraction of boundary points, which follows the general principle of boundary extraction in SCUBI but alleviates the issues of SCUBI on complex datasets with large inhomogeneity in terms of cluster density. The core idea is a strategy with an extra feedback loop: (i) select a relatively large set of boundary points initially to ensure robustness; (ii) conduct clustering on boundary points; (iii) use the cluster information as feedback to further refine the boundary set.

In Sect. 2, we give a brief review of existing boundary extraction techniques based on different principles. The details of our improved boundary extraction method are presented in Sect. 3. Comprehensive experimental results on 2D and higher dimensional datasets are shown in Sect. 4. This paper is concluded in Sect. 5 with some directions for future work.

2 Review of Boundary Extraction

The importance of boundary points comes from the fact that they inexplicitly specify the distribution of data points. For example, once the boundary of each cluster is known, all data points can be easily assigned to a certain cluster based on their relative locations. Similarly, in classification tasks, if two classes can be perfectly classified, it is likely that the decision boundary is only determined by those boundary points, in analogy to support vectors in SVM, instead of any other interior points.

Existing approaches to boundary extraction can be divided into four categories: concave theory based, diagram based, information entropy based and density based. The methods based on concave theory sequentially build the boundary from one point to another. Some typical methods include alpha shape [9], conformal alpha shape [10] and concave hull [11]. They tend to perform well on 2D datasets with evenly distributed data points and are widely used in image processing and machine vision [12]. However, they cannot be directly extended to high-dimensional datasets.

The methods based on diagram establish a Delaunay diagram of the dataset. Since the Delaunay diagram is unique for a fixed dataset, the boundary points that are found are also fixed. Furthermore, these methods do not require user defined parameters and can therefore be used when little *priori* knowledge of the original dataset is available. However, the complexity of building a Delaunay diagram is $O(n^2)$ and the computational cost is strongly influenced by the dimension of datasets. As a result, they are only used to process 2-D or 3-D data [13–15].

The methods based on information entropy are mainly used in the domain of point clouds [16]. They are useful for simplifying 3D models but are insufficient for processing high-dimensional datasets. Also, they suffer from high time complexity on datasets with multiple clusters, as the internal model needs to converge several times.

The core idea of density-based approaches is that a boundary point should be the point where the density difference on both sides of its tangent plane is large [17, 18] as shown in Fig. 1. However, in practice, it is not easy to calculate such a tangent plane directly. Instead, the normal vector, which is directly related to the tangent plane, is much easier to calculate [19–21]. The normal vector (direction of density gradient) of a point is conceptually defined as the average of the vectors pointing to its k nearest neighbors [22], as shown in Fig. 2. Note that as the direction instead of the length of the normal vector is important, in Eq. 1, we do not need the value of ρ.

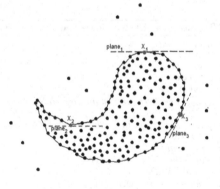

Fig. 1. Boundary point and tangent planes

Next, according to Eq. 2 with normalized *NV*, a score is calculated for each data point based on the cosine of the angle between each vector to its nearest neighbors and *NV*. The higher the score of a point, the greater the density gradient at that point, which means that it is more likely to be a boundary point (Algorithm 1).

$$NV(x) = \rho \cdot \frac{1}{k} \sum_{i=1}^{k} u_i \tag{1}$$

$$u_i(x) = x_i - x, \text{ where } x_i \in kNN(x)$$

$$\text{scores}(x) = \sum_{i=1}^{n} cos(u_i, NV(x)) \tag{2}$$

$$\text{where } NV(x) = \sum_{i=1}^{k} \frac{(x_i - x)}{|x_i - x|}$$

Fig. 2. The normal vector of a point is the average of the vectors pointing form the point itself to its k nearest neighbors. The red arrow shows the direction of the normal vector. (Color figure online)

Algorithm 1 The Boundary Extraction Method used in SCUBI

Input: $D, k, \alpha_{noise}, \alpha_{boundary}$
Output: $Boundary, NV$
1: $scores \leftarrow -\infty$
2: **for** each $x \in D$ **do**
3:     Find $kNN(x)$
4:     $kNN_{average}(x) \leftarrow Average\ distance\ of\ kNN(x)\ to\ x$
5: **end for**
6: $Sort\ kNN_{average}\ in\ descending\ order, Mark\ first\ \alpha_{noise}\ as\ noises$
7: **for** each $x \in D$ **do**
8:     **if** x is not the noise **then**
9:         $Exclude\ the\ noises\ in\ the\ kNN(x)$
10:         $Find\ the\ NV(x)\ of\ x$
11:         $Calculate\ cos(NV(x), u_i(x))$
12:         $scores(x) \leftarrow \sum_{i=1} cos(NV(x), u_i(x))$
13:     **end if**
14: **end for**
15: $Sort\ scores\ in\ descending\ order$
16: $Select\ first\ \alpha_{boundary}\ as\ Boundary$
17: $Return\ Boudanry, NV\ of\ boundary\ points$

After boundary extraction, SCUBI uses clustering algorithms such as DBSCAN and Affinity Propagation to group boundary points into clusters. Finally, interior points are assigned to the same cluster as its nearest boundary point (Algorithm 2).

Algorithm 2 SCUBI

---

(1) *Extract boundary points using the Normal Vector Method*
(2) *Cluster the boundary points using Dbscan*
(3) *Cluster the non − boundary points*
for *each $x \in D$* **do**
    if x is marked "noise" **then**
        $C_{id}(x) = -1$
    else
        Find the $C_{id}(boundary)$ of nearest boundary for x
        Assign the $C_{id}(boundary)$ to x
    end if
end for

---

3 Robust Boundary Extraction

In practice, the quality of the boundary extracted can directly determine the results of subsequent data mining operations. For example, if the boundary points from the same cluster are sparse with large gaps, they may be grouped into multiple clusters incorrectly. Furthermore, in the dynamic situation, one can use the boundary of the origin data as a reference to group new coming data points or detect possible outliers from the data stream. However, the poor quality of the boundary extracted from the origin data may hamper the accuracy of data mining work greatly.

Figure 3(left) shows a set of high quality boundary points, which is effective for detecting outliers as it gives a reasonably good approximation of the data distribution. For example, the new data point shown as a cyan triangle will be regarded as an interior point but the data point indicated by a red square will be recognized as an outlier, according to their relative locations to the boundary. By contrast, Fig. 3(right) shows an imperfect boundary with a large gap in the top-right region. In this case, it is possible that an outlier detection algorithm based on boundary information may incorrectly regard the triangle point as an outlier.

Due to the consideration of efficiency, the boundary extraction procedure in SCUBI is executed only once and only a small percentage of data points are selected. However, data clusters may be quite different in terms of density and the one-off extraction strategy in SCUBI may result in undesired distribution of boundary points, especially

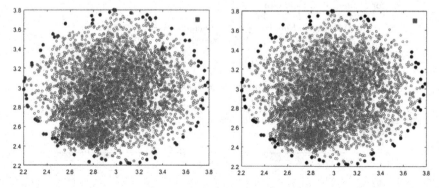

Fig. 3. An effective boundary (left) and an imperfect boundary (right) (Color figure online)

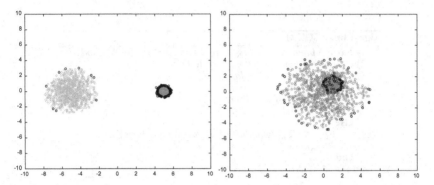

Fig. 4. The challenges of boundary extraction using SCUBI: the two clusters have different levels of density (left) and the density of a cluster is not homogeneous (right).

when the percentage of data points selected is relatively small. Figure 4(left) shows an example with two clusters of different densities. It is clear that most boundary points extracted are concentrated on the denser cluster while the boundary points on the other cluster are very sparse, insufficient as a good representation of the corresponding cluster. Furthermore, this *efficiency-robustness* dilemma cannot always be alleviated by simply increasing the number of boundary points. For example, on inhomogeneous clusters, some interior points may be incorrectly identified as boundary points due to high density level, as shown in Fig. 4(right).

Algorithm 3 Two-stage Boundary Extraction

Input: $D, k_1, k_2, \alpha_1, \alpha_2$
Output: *Boundary*

1: $scores_1 \leftarrow -\infty$
2: **for** each $x \in D$ **do**
3:     $k_1NN(x) \leftarrow$ *Find the k_1 nearest neighbors of x in D*
4:     $u_i \leftarrow (point_i - x) \ \forall \ point \in k_1NN(x)$
5:     $NV_1(x) \leftarrow \sum \frac{u_i}{|u_i|}$
6:     $scores_1(x) \leftarrow \sum cos(NV_1(x), u_i)$
7: **end for**
8: *Sort D in descending order of $scores_1$*
9: $length \leftarrow round(|D| \cdot \alpha_1)$
10: $Boundary_{tmp} \leftarrow D[1 : length]$
11: $Clusters \leftarrow$ *Cluster the $Boundary_{tmp}$ using classical clustering algorithm*
12: $scores_2 \leftarrow -\infty$
13: **for** $cluster_j \in Clusters$ **do**
14:     **for** each $x \in cluster_j$ **do**
15:         $k_2NN(x) \leftarrow$ *Find the k_2 nearest neighbors of x in $cluster_j$*
16:         $v_i \leftarrow (point_i - x) \ \forall \ point \in k_2NN(x)$
17:         $NV_2(x) \leftarrow \sum \frac{v_i}{|v_i|}$
18:         $scores_2(x) \leftarrow \sum cos(NV_2(x), v_i)$
19:     **end for**
20:     *Sort $cluster_j$ in descending order of $scores_2$*
21:     $length_j \leftarrow round(|cluster_j| \cdot \alpha_2)$
22:     $boundary_j \leftarrow cluster_j[1 : length_j]$
23:     $Boundary \leftarrow \{Boundary \cup boundary_j\}$
24: **end for**
25: **return** *Boundary*

This is because SCUBI calculates the scores of all data points according to Eq. 2 and sets a single threshold for distinguishing boundary points from others. If the density of a cluster is low, its data points are sparsely distributed and the neighbors of a data point are likely to spread in a wide range of directions, resulting in relatively low scores. As a result, the selection process in SCUBI tends to favor boundary points from dense clusters. Furthermore, within a cluster, if the density varies significantly, data points with large density gradients are also likely to receive high scores and be regarded as boundary points, even if they are actually interior points.

It should be mentioned that the objective of SCUBI is to conduct clustering as efficiently as possible. Our purpose is different: we want to extract a set of boundary points of high quality from the current dataset for further applications while the computational cost is not our primary concern. In other words, we can afford extracting and clustering more boundary points and incorporating additional processing steps.

The core idea is to replace the one-off extraction strategy in SCUBI by a two-stage procedure. In the first step, SCUBI is applied on the dataset to extract a set of preliminary boundary points and clustering is performed on this point set. The number of boundary points can be set to a value larger than that in normal SCUBI to ensure robustness. In the second step, with the clustering information available, another round of boundary extraction is performed individually on boundary points belonging to the same cluster, to further refine the extraction results (Algorithm 3).

Note that the parameter k in k-nearest neighbors is a key factor in both SCUBI and our boundary extraction technique. For large k values, more data points are involved in the score calculation but noisy points and even data points from other clusters may also be included. In Algorithm 3, a smaller value (k_1) is used in the first round of extraction, which not only reduces the time cost but also enables the boundary details to be preserved. In the second round, a larger value (k_2) is used since the interference from other clusters and noisy points have largely been eliminated, resulting in more accurate normal vectors and boundary information.

4 Experiment

The major purpose of experimental studies is to investigate the effectiveness of the new boundary extraction strategy in comparison to the standard method in SCUBI. Two sets of datasets were used: 2D datasets with highly complex clusters as well as synthetic datasets created by Gaussian distributions with diagonal covariance matrices in which the number of clusters and the dimension were systematically varied from 1 to 9 and 2 to 8 (Table 1), respectively. The number of data points in each cluster was fixed to 10,000 and the variance of each dimension of each cluster was 1.

Table 1. Summary of datasets

Dataset	Dim	Number of cluster	Number of points
Letters	2	19	24,160
Circles	2	30	60,000
U-shapes	2	4	84,600
Spheres	2–8	1–9	10,000–90,000

We extracted the boundary points on datasets *Letters, Circles, U-shapes* with the same parameter settings. The percentage of boundary points was set to 1% in the origin method. For the two-stage boundary extraction, the percentages of data points selected in the two stages were 10% and 10% for the three 2D datasets and 2% and 50% for *Spheres*. For the value of k (number of neighbors), it was fixed to 10 for the original method while k_1 and k_2 were 10 and 20 for our method, respectively.

Circles is a dataset with several clusters and the densities of clusters differ significantly. In Fig. 5, SCUBI tended to extract relatively more boundary points from dense clusters while the boundary points were discontinuous on sparse clusters. Also, SCUBI mistakenly chose some interior points as boundary points. By contrast, our extraction method produced a relatively uniform edge on almost every cluster and the misjudgment of interior points was less likely to happen. *Lines* and *Letters* are two datasets with complex clusters on which SCUBI often regarded interior point as boundary points due to the mutual interference among clusters. With the help of the cluster information, our method effectively reduced the interference of noise and other clusters, ensuring a good boundary on each cluster, as shown in Figs. 6 and 7.

For high-dimensional cases (*Spheres*), the distance between each data point and the corresponding mean was calculated and, for each cluster, the set of outmost data points was regarded as the *ground truth* for boundary points.

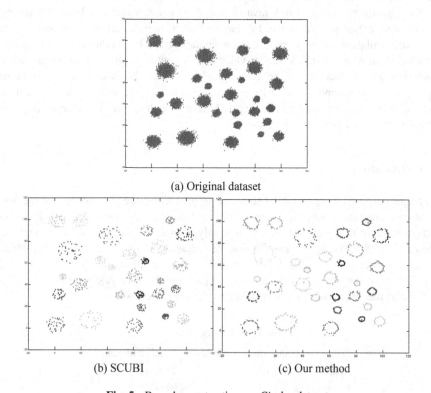

(a) Original dataset

(b) SCUBI   (c) Our method

Fig. 5. Boundary extraction on *Circles* dataset

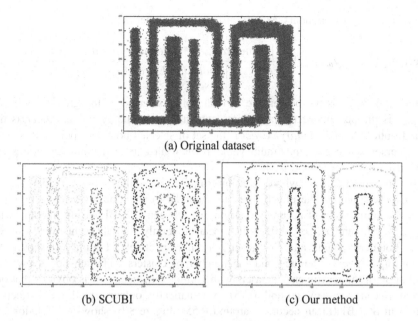

(a) Original dataset

(b) SCUBI   (c) Our method

Fig. 6. Boundary extraction on *U-shapes* dataset

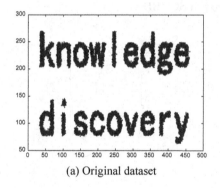

(a) Original dataset

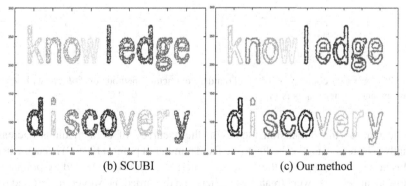

(b) SCUBI   (c) Our method

Fig. 7. Boundary extraction on *Letters* dataset

The accuracy of boundary extraction is defined as below:

$$\text{Accuracy}\left(\alpha_{outer}, \alpha_{boundary}\right) = \frac{\left|\{Outer(\alpha_{outer})\} \cap \{BoundaryExtract(\alpha_{boundary})\}\right|}{\left|\{BoundaryExtract(\alpha_{boundary})\}\right|} \quad (3)$$

In Eq. 3, α_{outer} is the percentage of data points selected as the ground truth and $\alpha_{boundary}$ is the percentage of data points extracted as boundary points. *Outer* is the ground truth set and *BoundaryExtract* is the set of selected boundary points. It is clear that accuracy reaches its maximum value 1 when all extracted boundary points are contained within the ground truth.

We created 63 datasets with dimension from 2 to 8 and the number of clusters from 1 to 9 to demonstrate the effectiveness of boundary extraction algorithms. Figure 8 shows the comparison of accuracy distribution between SCUBI and our method with $\alpha_{boundary} = 0.01$ and $\alpha_{outer} = 0.01$, which means that only 1% data points were selected as the boundary points and the ground truth, respectively. Figure 8(a) shows the relationship between accuracy and the number of clusters and each box shows the distribution of the results on 7 datasets with various dimensions. It is obvious that our method (mean accuracy around 0.75) systematically outperformed the extraction method in SCUBI (mean accuracy around 0.55). Figure 8(b) shows the relationship between accuracy and dimension and each box shows the distribution of the results on 9 datasets with different numbers of clusters. Again, our method was clearly superior to the extraction method in SCUBI.

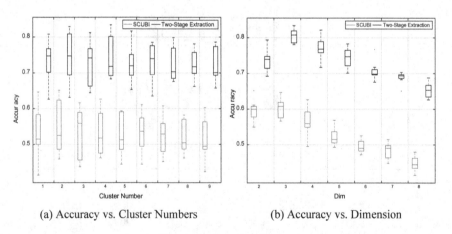

(a) Accuracy vs. Cluster Numbers   (b) Accuracy vs. Dimension

Fig. 8. The accuracy comparison of two boundary extraction methods on *Spheres* with varying dimensions and cluster numbers ($\alpha_{boundary} = 0.01$ and $\alpha_{outer} = 0.01$)

Note that, as the dimension increased, the accuracy of both methods decreased gradually. The reason is largely due to the fixed number of points (10,000) in each cluster: the cluster became sparser as the dimension increased and some boundary points with very low local densities were treated as outliers and discarded. However, the ground truth set was determined by distances only, regardless of the density factor. As a result, some data points in the ground truth set were missing from the boundary points actually extracted.

5 Conclusion

In the era of big data, data mining algorithms are often confronted with unprecedented volume of data. Due to the time complexity of these algorithms, most of them cannot be directly applied to the massive datasets, creating a major gap between academic research and industrial applications. Recently, SCUBI was proposed as a scalable clustering scheme, which strategically employs boundary information as a compact representation of the original dataset to accelerate the clustering procedure. As a complement to mainstream solutions relying on high performance computing infrastructure such as GPU and distributed computing, SCUBI provides a unique perspective for handling massive datasets by exploiting the structural information of datasets while leaving existing clustering algorithms largely unchanged.

In this paper, we argue that boundary information is not only highly valuable for standard clustering tasks but also important for other key applications such as stream clustering [23], outlier/anomaly detection [24, 25] and classification. As the major contribution of our work, a novel two-stage boundary extraction technique was proposed to address the issues that we have identified with SCUBI. More specifically, we found that the current one-off extraction method in SCUBI was ineffective at handling datasets with large inter-cluster or intra-cluster variance in terms of density. Experimental results on a variety of purposefully designed datasets show that our extraction technique, which takes advantage of the clustering information, is clearly superior to the standard extraction technique in SCUBI, especially on cases with large inhomogeneity in terms of cluster density.

As to future work, it is important to conduct formal analysis of existing density-based boundary extraction techniques to provide a rigorous foundation for better understanding their effectiveness and possible limitations. Currently, these methods are mostly heuristic ones and there is still a lack of principled guidance on key issues such as parameter setting and performance evaluation. Meanwhile, with the help of a competent technique that can produce high quality boundary information, it would be very interesting to further extend its application scope to more challenging situations. For example, instead of extracting a fixed set of boundary points, a dynamic boundary set can be maintained in an online manner, which is very useful for processing data steams or when the dataset is too large to be processed as a whole in the memory.

References

1. Jain, K., Murty, N., Flynn, J.: Data clustering: a review. ACM Comput. Surv. **31**(3), 264–323 (1999)
2. Kaufman, L., Rousseeuw, P.J.: Finding Groups in Data: An Introduction to Cluster Analysis. Wiley, Hoboken (2008)

3. MacQueen, J.: Some methods for classification and analysis of multivariate observations. In: Proceedings of the 5th Berkeley Symposium on Mathematical Statistic and Probability, vol. 1, pp. 281–297 (1967)

4. Arthur, D., Manthey, B., Röglin, H.: K-means has polynomial smoothed complexity. In: Foundations of Computer Science, vol. 157, pp. 405–414 (2009)

5. Ester, M., Kriegel, H.P., Xu, X.: A density-based algorithm for discovering clusters a density-based algorithm for discovering clusters in large spatial databases with noise. In: International Conference on Knowledge Discovery and Data Mining, pp. 226–231. AAAI Press, Portland (1996)

6. Frey, B.J., Dueck, D.: Clustering by passing messages between data points. Science 315, 972–976 (2007)

7. Berkhin, P.: A survey of clustering data mining techniques. In: Kogan, J., Nicholas, C., Teboulle, M. (eds.) Grouping Multidimensional Data. Springer, Heidelberg (2006). https://doi.org/10.1007/3-540-28349-8_2

8. Tong, Q.H., Li, X., Yuan, B.: A highly scalable clustering scheme using boundary information. Pattern Recogn. Lett. 89, 1–7 (2017)

9. Edelsbrunner, H., Kirkpatrick, D., Seidel, R.: On the shape of a set of points in the plane. IEEE Trans. Inf. Theory 29(4), 551–559 (1983)

10. Moreira, A.J.C., Santos, M.Y.: Concave hull: a k-nearest neighbors approach for the computation of the region occupied by a set of points. In: Proceedings of the Second International Conference on Computer Graphics Theory and Applications, vol. 3520, pp. 61–68. Springer, Barcelona (2006)

11. López Chau, A., Li, X., Yu, W., Cervantes, J., Mejía-Álvarez, P.: Border samples detection for data mining applications using non convex hulls. In: Batyrshin, I., Sidorov, G. (eds.) MICAI 2011. LNCS (LNAI), vol. 7095, pp. 261–272. Springer, Heidelberg (2011). https://doi.org/10.1007/978-3-642-25330-0_23

12. Hoogs, A., Collins, R.: Object boundary detection in images using a semantic ontology. In: Conference on Computer Vision and Pattern Recognition Workshop, pp. 956–963 (2006)

13. Liu, D., Nosovskiy, G.V., Sourina, O.: Effective clustering and boundary detection algorithm based on delaunay triangulation. Pattern Recogn. Lett. 29, 1261–1273 (2008)

14. Estivill-Castro, V., Lee, I.: AUTOCLUST: automatic clustering via boundary extraction for mining massive point-data sets. In: International Conference on Geocomputation, vol. 26, pp. 23–25 (2000)

15. Yang, J., Estivill-Castro, V., Chalup, S.K.: Support vector clustering through proximity graph modelling. In: International Conference on Neural Information Processing, vol. 2, pp. 898–903. IEEE, Singapore (2002)

16. Chen, X.J., Zhang, G., Hua, X.H.: Point cloud simplification based on the information entropy of normal vector angle. Chin. J. Lasers 42(8), 328–336 (2015)

17. Xia, C., Hsu, W., Lee, M.L.: BORDER: efficient computation of boundary points. IEEE Trans. Knowl. Data Eng. 18(3), 289–303 (2006)

18. Nosovskiy, G.V., Liu, D., Sourina, O.: Automatic clustering and boundary detection algorithm based on adaptive influence function. Pattern Recogn. 41, 2757–2776 (2008)

19. Zhu, F., Ye, N., Yu, W., Xu, S., Li, G.: Boundary detection and sample reduction for one-class support vector machines. Neurocomputing 123, 166–173 (2014)

20. Qiu, B.-Z., Yue, F., Shen, J.-Y.: BRIM: an efficient boundary points detecting algorithm. In: Zhou, Z.-H., Li, H., Yang, Q. (eds.) PAKDD 2007. LNCS (LNAI), vol. 4426, pp. 761–768. Springer, Heidelberg (2007). https://doi.org/10.1007/978-3-540-71701-0_83

21. Li, Y.: Selecting training points for one-class support vector machines. Pattern Recogn. Lett. 32(11), 1517–1522 (2011)

22. He, Y.Z., Wang, C.H., Qiu, B.Z.: Clustering boundary points detection algorithm based on gradient binarization. Appl. Mech. Mater. **266**, 2358–2363 (2013)
23. Silva, J.A., Faria, E.R., Barros, R.C.: Data stream clustering: a survey. ACM Comput. Surv. **46**(1), 13 (2013)
24. Pokrajac, D., Lazarevic, A., Latecki, L.J.: Incremental local outlier detection for data streams. In: IEEE Symposium on Computational Intelligence and Data Mining, pp. 504–515. IEEE, Honolulu (2007)
25. Salehi, M., Leckie, C., Bezdek, J.C.: Fast memory efficient local outlier detection in data streams. IEEE Trans. Knowl. Data Eng. **28**(12), 3246–3260 (2017)

Adaptive Fuzzy Clustering Algorithm with Local Information and Markov Random Field for Image Segmentation

Jialiang Hu and Ying Wen[✉]

Shanghai Key Laboratory of Multidimensional Information Processing,
Department of Computer Science and Technology, East China Normal University,
Shanghai, China
ywen@cs.ecnu.edu.cn

Abstract. Fuzzy c-means (FCM) clustering as one of the clustering method is widely used in image segmentation field, but some methods based on FCM are unable to obtain satisfactory performance for image segmentation under intense noise condition. This paper presents a novel local spatial information based fuzzy c-means clustering and Markov random field method for image segmentation. In the method, a new dissimilarity function is proposed by using the prior relationship degree and local neighbor distances, which enhances its resistance to noise. And a novel prior probability approximation is considered with spatial Euclidean distance and the difference of the mean color level between the center pixel and its neighborhoods. Experiments over synthetic images, real-world images and brain MR images indicate that the proposed method obtains better segmentation performance, compared to the FCM extended methods.

Keywords: Image segmentation · Fuzzy c-means clustering
Markov random field · Local information

1 Introduction

Image segmentation is one of key tasks in image processing and computer vision. The target of image segmentation is to separate source image into several non-overlapping regions which have the same features such as intensity, color, tone, texture, etc. Many clustering-based methods have been proposed for image segmentation [2,6,11,18,23]. Compared with the other clustering methods, Fuzzy C-means (FCM) is one of the simplest and the most popular algorithms in field of image segmentation [2,5,10]. However, the performance of traditional FCM

This work was supported by National Nature Science Foundation of China (No. 61773166), Natural Science Foundation of Shanghai no. 17ZR1408200 and the Science and Technology Commission of Shanghai Municipality under research grant no. 14DZ2260800.

decreases very fast by the impact of noise, outliers and other image artifacts because in the algorithm, all of pixels in the images are regarded as individual points without any relationship [9].

To improve the performance of FCM algorithm, many modified FCM algorithms have been proposed [13, 20, 22]. Spatial relationship of neighborhood pixels, as a significant feature, is widely used in image segmentation [4, 19, 24], because the pixels in the immediate neighborhood usually have similar characteristics and have a high probability of belonging to the same cluster. Based on spatial information, Ahmed et al. [1] presented an FCM with spatial constraints (FCM_S) that the label of a pixel is influenced by labels in its neighborhood. Krinidis et al. [15] proposed a robust fuzzy local information c-means clustering algorithm (FLICM) by introducing a fuzzy factor, which works as a role to control noise tolerance and outlier resistance. Li and Qin [16] added L_p norm into FLICM named fuzzy local information L_p clustering (FLILp) and proposed a method of cluster center estimation, but it is hard to estimate accurately.

Markov random field model (MRF) [7, 17, 21] is a widely used tool to describe the mutual influences between data points, which is able to be extended to the image segmentation for representing the relationship between pixels. In [3], Chatzis applied hidden Markov random field into fuzzy clustering (HMRF-FCM) by taking Kullback-Leibler divergence information into fuzzy objective function. However, the procedure of the prior probability approximation and the potential parameter update in Markov model consume much time. In order to reduce the cost of time, Zhang [25] utilized mean template in distance function and prior probability. In [18], Liu et al. defined the dissimilarity function and prior probability function based on region-level information as well as pixel-level information to develop the robustness of the method.

In this paper, based on the HMRF-FCM algorithm, we propose a novel fuzzy clustering algorithm with local information and Markov random field for image segmentation with intense noise, which studies local information to remove the complex noise and preserve the details and edges. First, we take local spacial information and membership information into a new dissimilarity function to enhance the relationship of neighborhood. Second, in step of prior probability approximation in Markov model, the neighbor Euclidean distances and mean color level of neighborhood constitute a new weight in prior probability function. By using the new dissimilarity function and the prior function, the novel segmentation method considers more impacts of local information to produce a better result. In experiments, we compare our method with five other fuzzy c-means algorithms for image segmentation with intense noise environment to validate the proposed methods effectiveness and robustness.

2   Method

2.1   HMRF-FCM Algorithm

The HMRF-FCM Algorithm [3] was first introduced by Chatzis and Varvarigou. The algorithm is to segment image $X = \{x_1, x_2, ..., x_N\}$ into $C(C \geqslant 2)$ classes.

The HMRF-FCM illustrates the segmentation problem using hidden Markov random field model, and employs the posteriors and prior membership function to indicate the excellent modeling ability of Markov random field and the flexibility of fuzzy clustering [18]. The object function of HMRF-FCM is defined as follows:

$$Q_\lambda = \sum_{i=1}^{N} \sum_{k=1}^{C} r_{i,k} d_{i,k} + \lambda \sum_{i=1}^{N} \sum_{k=1}^{C} r_{i,k} \log \left(\frac{r_{i,k}}{\pi_{i,k}} \right) \tag{1}$$

where $r_{i,k}$, the membership degree in FCM algorithm, is the posteriors probability in HMRF model. $\pi_{i,k}$ is the prior probabilities of the HMRF model, and the parameter λ is the fuzziness degree of the fuzzy membership values. $d_{i,k}$ presents the dissimilarity between a pixel i and cluster centroids k. In HMRF-FCM, all observed data are emitted from multivariate Gaussian form, so we can obtain $d_{i,k}$ from the negative log-posterior of a Gaussian distribution as:

$$d_{i,k} = \frac{1}{2} \left(Blog\,(2\pi) + (x_i - \mu_k)' \Sigma_k^{-1} (x_i - \mu_k) + \log\,(|\Sigma_k|) \right) \tag{2}$$

where μ_k and Σ_k are the mean and covariance matrix of the Gaussian distribution of the kth class respectively. And B is the spectral number of image to be segmented.

In HMRF-FCM algorithm, $\pi_{i,k}$ is the prior probabilities of the HMRF model, which is seen as the actual membership degree according to Kullback-Leibler divergence. The paper [3] gives a function of MRF $\pi_{i,k}$ approximation as:

$$\pi_{i,k} = \frac{\exp\{\beta E_{i,k}\}}{\sum_{l=1}^{C} \exp\{\beta E_{i,l}\}} \tag{3}$$

where β is the parameter of clique potentials and $E_{i,k}$ is the energy function, which counts the number of neighbor pixels whose label is k.

2.2   Proposed Method

In this paper, motivated by HMRF-FCM and FLICM, we proposed a new local spatial information incorporating Markov random field model based fuzzy C-means clustering, which is defined as:

$$Q_\lambda = \sum_{i=1}^{N} \sum_{k=1}^{C} r_{i,k} D_{i,k} + \lambda \sum_{i=1}^{N} \sum_{k=1}^{C} r_{i,k} \log \left(\frac{r_{i,k}}{\pi_{i,k}} \right) \tag{4}$$

where $D_{i,k}$ is a novel dissimilarity function different from the $d_{i,k}$ in Eq. (1). We add a local neighborhood data $d_{N_i,k}$ into the dissimilarity measure to enhance the relationship between neighborhood pixels, which improves the robustness to noise.

$$D_{i,k} = d_{i,k} + (1 - \pi_{i,k})^m d_{N_i,k} \tag{5}$$

where $d_{i,k}$ is the center-pixel distance as Eq. (2), m is the fuzzifier used in FCM, N_i is the set of the neighbors of data x_i, and $d_{N_i,k}$ is the neighbor-pixels distance to class k, given as:

$$d_{N_i,k} = \frac{1}{Z_{N_i}} \sum_{j \in N_i} \frac{1}{1 + L_{ij}} d_{j,k} \tag{6}$$

where Z_{N_i} is a normalized factor as $Z_{N_i} = \sum_{j \in N_i} (1 + L_{ij})^{-1}$. L_{ij} is the spatial Euclidean distance between pixels i and j. By using L_{ij}, the influence of neighbor pixels on the center pixel, $(1 + L_{ij})^{-1}$, can be determined automatically and decreases with Euclidean distance growing.

In the proposed method, the prior probability $\pi_{i,k}$ is defined as a more simple and efficient form which removes time-consuming parameter β and energy term $E_{i,k}$ in the Eq. (3).

$$\pi_{i,k} = \frac{\left(\sum_{j \in N_i} \omega_j r_{j,k}\right)^\theta}{\sum_{l=1}^{C}\left(\sum_{j \in N_i} \omega_j r_{j,l}\right)^\theta} \tag{7}$$

where θ is the strength factor to control the performance, and ω_j is the weighted parameter which determines the impact of the neighbor pixels on the central pixel, defined as

$$\omega_j = \frac{1}{1 + L_{ij} \exp(|\overline{x_i} - \overline{x_j}|/T)} \tag{8}$$

where $\overline{x_i}$ is the local mean value of pixel i neighborhood, T presents the color level of the image. When the value of $\overline{x_j}$ is close to $\overline{x_i}$, it means that pixel i and j are homogeneous, otherwise, heterogeneous. By using $|\overline{x_i} - \overline{x_j}|$ and L_{ij}, the close and homogeneous pixels around would gain more weight and greater contribution to the prior probability approximation, while pixels far away and heterogeneous do less impact on the prior.

Applying Lagrange multiplier method to Eq. (4) with the constrain of $r_{i,k}$ (sum to one), $r_{i,k}$ is given as:

$$r_{i,k} = \frac{\pi_{i,k} \exp(-(1/\lambda)D_{i,k})}{\sum\limits_{l=1}^{C} \pi_{i,l} \exp(-(1/\lambda)D_{i,l})} \tag{9}$$

The derivation of the mean μ_k and the covariance matrix Σ_k is obtained from Lagrange function, given as

$$\mu_k = \frac{\sum_{i=1}^{N} r_{i,k}(x_i + (1 - \pi_{i,k})^m \bar{x}_i)}{\sum_{i=1}^{N} r_{i,k}(1 + (1 - \pi_{i,k})^m)} \tag{10}$$

$$\Sigma_k = \frac{\sum_{i=1}^{N} r_{i,k}((x_i - \mu_k)(x_i - \mu_k)^T + (1 - \pi_{i,k})^m((\overline{x_i} - \mu_k)(\overline{x_i} - \mu_k)^T + \Sigma_{\overline{x_i}})}{\sum_{i=1}^{N} r_{i,k}(1 + (1 - \pi_{i,k})^m)} \tag{11}$$

where $\Sigma_{\overline{x_i}}$ is local covariance matrix around pixel i.

The procedure of the proposed method can be summarized as Algorithm 1.

Algorithm 1. Proposed Method

Input: the data $x_i, i = 1, ..., N$; the number of clusters C.
Output: the membership degrees $r_{i,k}^t$.
1: Set parameters: the degree of fuzziness λ; the stopping condition ξ; the max iteration time Max$T$.
2: Initialize the local mean $\overline{x_i}$ and local covariance matrix $\Sigma_{\overline{x_i}}$ from input image.
3: Initialize the fuzzy membership $r_{i,k}^t$ from the original FCM algorithm.
4: **repeat**
5:     Calculate each of $\pi_{i,k}^t$ by using $r_{i,k}^t$ according to Eqs. (7) and (8).
6:     Based on Eqs. (10) and (11), derive the means μ_k and the convariance matrixes Σ_k, respectively. Next, compute the dissimilarity function $D_{i,k}$ between each pixel and cluster center with the Eq. (5).
7:     Update $r_{i,k}^{t+1}$ by $D_{i,k}$ and $\pi_{i,k}^t$ as Eq. (9).
8:     Update $t = t + 1$.
9: **until** $(|Q_\lambda^{t+1} - Q_\lambda^t|/Q_\lambda^t) \leqslant \xi$ or $t >$ Max$T$.

---

3   Experiment

To validate the effectiveness of the proposed algorithm, we compare it with five fuzzy clustering algorithms, i.e., FCM_S [1], HMRF-FCM [3], FLICM [15], FLILp [16] and Zhang's method [25] which are extensively used in image segmentation. For all of the algorithms, we generally choose the parameters $m = 2$, $\lambda = 2$, $\xi = 0.001$, $\theta = 2$, $N_R = 9$ (a 3×3 window centered around each pixel) in experiments. All parameters explained above are decided by reports of [16, 25] to perform better results.

We take the segmentation accuracy (SA) [12] to evaluate the six algorithms. It is defined as the sum of the correctly classified pixels divided by the sum of the total number of pixels.

$$\text{SA} = \sum_{i=1}^{C} \frac{A_i \cap G_i}{\sum_{j=1}^{C} G_j} \qquad (12)$$

where A_i represents the set of pixels belonging to the ith class by segmentation algorithm, while G_i represents the set of pixels belonging to the ith class in ground truth.

3.1   Comparison Experiments on Synthetic Images

We apply the algorithms on a classification with four classes. Figure 1(a) is an image with four gray levels with different shapes. Figure 1(b) is an image of Fig. 1(a) contaminated with Gaussian noise ($\delta = 0.1$). Figures 1(c–h) are the results of the six algorithms on Fig. 1(b). From these figures, we can see that FCM_S is sensitive to noise. The result of HMRF-FCM shows a rough segmentation image, even if the boundary is not clear enough. Due to improved spatial information employed, FLICM and FLILp are capable of removing a proportion of the noise, but still contain noise inside the area and fail to obtain good

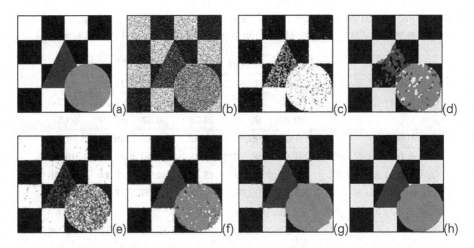

Fig. 1. Clustering of a synthetic image with four classes. (a) Original image. (b) The original image with Gaussian ($\delta = 0.1$). (c) Result of FCM_S. (d) Result of HMRF-FCM. (e) Result of FLICM. (f) Result of FLILp. (g) Result of Zhang's method. (h) Result of proposed method

results. Since Zhang's method and the proposed method combine the advantages of HMRF and FCM, their segmentation results are satisfactory. However, the edge of Zhang's result is not as clear as that of the proposed.

In order to further verify the robustness of the algorithm, all fuzzy clustering algorithms are implemented on Fig. 1(a) corrupted by 4 types of noise, i.e., Gaussian noise, 'salt & pepper' noise and two types of mixed noise. For the two types of mixed noise, one is that the variance of Gaussian noise is a constant while the noise intensity of 'salt & pepper' increases, while the other is the opposite.

Table 1 gives the segmentation accuracies of the six algorithms on Fig. 1(a) corrupted by 4 types of noise with different levels. It can be seen that the mean segmentation accuracy of our method (above 97%) is higher than that of the others. Besides, the accuracy of our method shows an expectable result in each item. Compared with Zhang's algorithm, our method is able to obtain stable and robust performances on mixed noise, while the Zhang's drops down very fast.

3.2   Comparison Experiments on Natural Image

We also apply the six algorithms to a natural image. Figure 2(a) [8] is an image composed of flowers, leaves and background. Figure 2(b) is an image of Fig. 2(a) contaminated with 10% 'salt & pepper' noise and Gaussian noise ($\delta = 0.2$). We set the number of clusters as 3 in the case. Figures 2(c–h) show the segmentation results obtained by the six algorithms, respectively. It is clearly indicated that FCM_S, FLICM and FLILp algorithms are all affected by the noise to different extent, which illustrates that these algorithms is not robust enough to image corrupted by mixed noise. From Figs. 2(d)(g)(h), HMRF shows its advantages

Table 1. Segmentation accuracy (%) of each algorithm on Fig. 1(a) corrupted by different noise with different levels. (Gaussian noise(G) and 'Salt & Pepper' noise(S))

	FCM_S	HMRF	FLICM	FLILp	Zhang's	Ours
$G(\delta = 0.05)$	87.49	98.17	96.17	88.28	94.23	**99.63**
$G(\delta = 0.1)$	75.59	88.62	84.76	88.87	97.75	**99.50**
S = 5%	**99.80**	99.75	99.78	90.37	99.79	99.57
S = 10%	99.60	99.65	99.52	86.49	**99.70**	96.74
$G(\delta = 0.05)$ & S = 5%	86.17	97.76	93.38	87.00	98.84	**99.22**
$G(\delta = 0.1)$ & S = 5%	76.91	92.36	83.11	84.75	87.34	**97.06**
$G(\delta = 0.05)$ & S = 10%	80.37	98.08	87.38	86.36	97.98	**99.46**
$G(\delta = 0.1)$ & S = 10%	75.46	90.47	82.25	83.43	80.73	**92.63**
Mean	85.17	95.60	90.79	86.94	95.54	**97.96**

of spatial modelling capabilities in natural image, which is more smooth than the original FCM with spatial information. Compared Fig. 2(h) with Fig. 2(g), we can see that our method performs a remarkable result of petals and stamens than Zhang's method.

Figure 3 plots the curves of segmentation accuracy for different methods on Fig. 2(a). Experiments are implemented on the image corrupted by four types of noise. With the increase of noise level, the performances of all methods slide down but the proposed method deceases more slowly than the others. Besides,

Fig. 2. Clustering of a natural image. (a) Original image. (b) The original image with 10% 'salt & pepper' noise and Gaussian ($\delta = 0.2$). (c) Result of FCM_S. (d) Result of HMRF-FCM. (e) Result of FLICM. (f) Result of FLILp. (g) Result of Zhang's method. (h) Result of proposed method

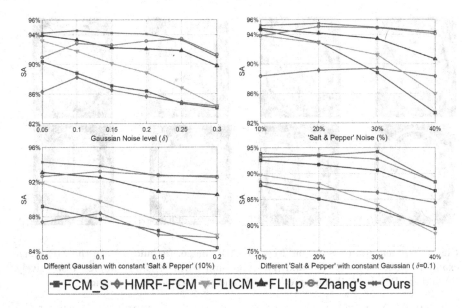

Fig. 3. The segmentation accuracy (SA) vs. different noise with different level on Fig. 2(a)

compared with Zhang's method, our method is visibly robust in all cases. It can be seen that our method suppresses the influence of the noise.

3.3 Comparison Experiments on Brain MR Images

In this subsection, our method are evaluated on a synthetic brain MR image database-BrainWeb [14], which is a 3D simulated brain database that contains a set of realistic MRI data. The segmentation object is to divide the brain MR image into four part: gray matter (GM), white matter (WM), CSF and background. We select sixty brain slice images produced by the MRI simulator with different level noise. We choose two brain MR images randomly for presentation and carry out the numerical analysis on the whole set.

Figure 4 shows the results on two 3% noised MR image examples. As the part in red rectangle in Figs. 4(c)(d) and (g)(h) shown, our result preserves the details and local information, while Zhang's method is hard to distinguish between noise and details so that the details are smoothed. It reflects that the mean template function in distance measurement and prior probability approximation might lose too much details and lead an unsatisfied result. Furthermore, we calculate the segmentation accuracy of the six algorithms on BrainWeb dataset with different level noise. In Table 2, our method obtains satisfied results compared with the other algorithms under 0%, 3% and 9% noise.

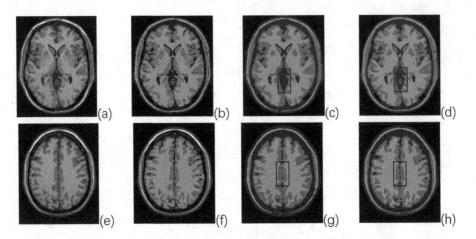

Fig. 4. Segmentation results on T1 brain MR image. (a) and (e) are the MR image examples with 3% noise. (b) and (f) are ground truth of two examples, respectively. (c) and (g) Result of Zhang's method. (d) and (h) Result of our method

Table 2. Segmentation accuracy (%) of each algorithm on MRI dataset with different level noise

	FCM_S	HMRF	FLICM	FLILp	Zhang's	Ours
0% noise	97.29	96.25	97.10	97.37	93.73	**97.70**
3% noise	95.49	95.28	96.77	96.35	90.33	**97.03**
9% noise	81.97	90.57	**94.43**	76.63	88.31	93.08

4   Conclusion

We proposed an adaptive fuzzy clustering algorithm with local information and Markov random field for image segmentation under intense noise. In this method, the new dissimilarity function created by the local information with distances of neighbors greatly enhances noise robustness and the novel prior probability function improves the performance on denoising and reduces computing time. The experiments validate the excellent clustering performance of our algorithm, and show that the proposed distance function with MRF model provides a better applicable way to address image segmentation under intense noise.

References

1. Ahmed, M.N., Yamany, S.M., Mohamed, N., Farag, A.A., Moriarty, T.: A modified fuzzy c-means algorithm for bias field estimation and segmentation of MRI data. IEEE Trans. Med. Imaging **21**(3), 193–199 (2002)
2. Bezdek, J.C.: Pattern Recognition with Fuzzy Objective Function Algorithms. Springer, New York (2013)

3. Chatzis, S.P., Varvarigou, T.A.: A fuzzy clustering approach toward hidden markov random field models for enhanced spatially constrained image segmentation. IEEE Trans. Fuzzy Syst. **16**(5), 1351–1361 (2008)
4. Chen, Y., Zhang, J., Wang, S., Zheng, Y.: Brain magnetic resonance image segmentation based on an adapted non-local fuzzy c-means method. IET Comput. Vis. **6**(6), 610–625 (2012)
5. Choy, S.K., Lam, S.Y., Yu, K.W., Lee, W.Y., Leung, K.T.: Fuzzy model-based clustering and its application in image segmentation. Patt. Recogn. **68**, 141–157 (2017)
6. Fazendeiro, P., de Oliveira, J.V.: Observer-biased fuzzy clustering. IEEE Trans. Fuzzy Syst. **23**(1), 85–97 (2015)
7. Gharieb, R., Gendy, G., Abdelfattah, A., Selim, H.: Adaptive local data and membership based KL divergence incorporating c-means algorithm for fuzzy image segmentation. Appl. Soft Comput. **59**, 143–152 (2017)
8. Gong, M., Liang, Y., Shi, J., Ma, W., Ma, J.: Fuzzy c-means clustering with local information and kernel metric for image segmentation. IEEE Trans. Image Process. **22**(2), 573–584 (2013)
9. Gong, M., Zhou, Z., Ma, J.: Change detection in synthetic aperture radar images based on image fusion and fuzzy clustering. IEEE Trans. Image Process. **21**(4), 2141–2151 (2012)
10. Guo, F.F., Wang, X.X., Shen, J.: Adaptive fuzzy c-means algorithm based on local noise detecting for image segmentation. IET Image Process. **10**(4), 272–279 (2016)
11. Havens, T.C., Bezdek, J.C., Leckie, C., Hall, L.O., Palaniswami, M.: Fuzzy c-means algorithms for very large data. IEEE Trans. Fuzzy Syst. **20**(6), 1130–1146 (2012)
12. Hosotani, F., Inuzuka, Y., Hasegawa, M., Hirobayashi, S., Misawa, T.: Image denoising with edge-preserving and segmentation based on mask NHA. IEEE Trans. Image Process. **24**(12), 6025–6033 (2015)
13. Huang, H.C., Chuang, Y.Y., Chen, C.S.: Multiple kernel fuzzy clustering. IEEE Trans. Fuzzy Syst. **20**(1), 120–134 (2012)
14. Ji, Z, Sun, Q.: A fuzzy clustering with bounded spatial probability for image segmentation. In: 2017 IEEE International Conference on Fuzzy Systems (FUZZ-IEEE), pp. 1–6. IEEE (2017)
15. Krinidis, S., Chatzis, V.: A robust fuzzy local information c-means clustering algorithm. IEEE Trans. Image Process. **19**(5), 1328–1337 (2010)
16. Li, F., Qin, J.: Robust fuzzy local information and LP-norm distance-based image segmentation method. IET Image Process. **11**(4), 217–226 (2017)
17. Li, X., Cui, G., Dong, Y.: Graph regularized non-negative low-rank matrix factorization for image clustering. IEEE Trans. Cybern. **47**, 3840–3853 (2017)
18. Liu, G., Zhang, Y., Wang, A.: Incorporating adaptive local information into fuzzy clustering for image segmentation. IEEE Trans. Image Process. **24**(11), 3990–4000 (2015)
19. Mei, J.P., Chen, L.: LinkFCM: relation integrated fuzzy c-means. Patt. Recogn. **46**(1), 272–283 (2013)
20. Nongmeikapam, K., Kumar, W., Singh, A.D.: A fast and automatically adjustable GRBF kernel based fuzzy c-means for cluster-wise coloured feature extraction and segmentation of MR images. IET Image Process. (2017)
21. Ren, Y., Tang, H., Wei, H.: A markov random field model for image segmentation based on gestalt laws. In: Lu, B.-L., Zhang, L., Kwok, J. (eds.) ICONIP 2011. LNCS, vol. 7064, pp. 582–591. Springer, Heidelberg (2011). https://doi.org/10.1007/978-3-642-24965-5_66

22. Tran, D.C., Wu, Z., Tran, V.H.: Fast generalized fuzzy c-means using particle swarm optimization for image segmentation. In: Loo, C.K., Yap, K.S., Wong, K.W., Teoh, A., Huang, K. (eds.) ICONIP 2014. LNCS, vol. 8835, pp. 263–270. Springer, Cham (2014). https://doi.org/10.1007/978-3-319-12640-1_32
23. Wu, C.H., Ouyang, C.S., Chen, L.W., Lu, L.W.: A new fuzzy clustering validity index with a median factor for centroid-based clustering. IEEE Trans. Fuzzy Syst. **23**(3), 701–718 (2015)
24. Zaixin, Z., Lizhi, C., Guangquan, C.: Neighbourhood weighted fuzzy c-means clustering algorithm for image segmentation. IET Image Process. **8**(3), 150–161 (2013)
25. Zhang, H., Wu, Q.M.J., Zheng, Y., Nguyen, T.M., Wang, D.: Effective fuzzy clustering algorithm with Bayesian model and mean template for image segmentation. IET Image Process. **8**(10), 571–581 (2014)

Efficient Direct Structured Subspace Clustering

Wen-ming Cao[1], Rui Li[1], Sheng Qian[1], Si Wu[2], and Hau-San Wong[1(✉)]

[1] Department of Computer Science, City University of Hong Kong,
Hong Kong, China
{wenmincao2-c,ruili52-c,sqian9-c}@my.cityu.edu.hk, cshswong@cityu.edu.hk
[2] School of Computer Science and Engineering,
South China University of Technology, Guangzhou, China
cswusi@scut.edu.cn

Abstract. Subspace clustering splits data instances that are drawn from special low-dimensional subspaces via utilizing similarities between them. Traditional methods contain two steps: (1) learning the affinity matrix and (2) clustering on the affinity matrix. Although these two steps can alternatively contribute to each other, there exist heavy dependencies between the performance and the initial quality of affinity matrix. In this paper, we propose an efficient direct structured subspace clustering approach to reduce the quality effects of the affinity matrix on performances. We first analyze the connection between the affinity and partition matrices, and then fuse the computation of affinity and partition matrices. This fusion allows better preserving the subspace structures which help strengthen connections between data points in the same subspaces. In addition, we introduce an algorithm to optimize our proposed method. We conduct comparative experiments on multiple data sets with state-of-the-art methods. Our method achieves better or comparable performances.

Keywords: Subspace clustering · Unsupervised learning

1 Introduction

Subspace clustering has been investigated and applied to various field including image content compress and feature representation learning [1], image segmentation [2] and gene expression profile clustering [3]. The high-dimensional characteristics of data potentially produce adverse effects on performances due to the noise disturbance [4]. Rather than being uniformly distributed across the high-dimensional space, these data often lie a union of low-dimensional subspace. For example, when clustering face images of a person under various illumination conditions or tracking a moving object in a continuously temporal sequence, we can observe that data points from a cluster are distributed in a low-dimensional space. Searching a union of low-dimensional subspaces to explore structures in

© Springer Nature Switzerland AG 2018
L. Cheng et al. (Eds.): ICONIP 2018, LNCS 11304, pp. 181–190, 2018.
https://doi.org/10.1007/978-3-030-04212-7_16

data helps alleviate negative effects of high-dimensional noise. Traditional clustering methods exploit proximity distances of data point belonging to each cluster. However, the above data distribute arbitrarily across the high-dimensional ambient space rather than around a centroid, which means traditional methods are not applicable to subspace clustering.

Subspace clustering approaches are grouped into [14]: algebraic-based methods, factorization-based methods, statistical-based methods and self expressive-based methods. The first type of methods converts subspace clustering problems by fitting and differentiating polynomials concerning the input matrix $\mathbf{X}$. The second type of methods obtains segmentations through searching a proper low-rank approximate factorization concerning $\mathbf{X}$. The third type of methods addresses subspace clustering issues by introducing a probabilistic assumption that data are sampled from a certain mixture of Gaussian distributions. The fourth type of methods contains two stages: learning the self-expressive matrix and performing clustering on the learnt self-expressive matrix. Among these methods, sparse subspace clustering (SSC) [10] and low-rank representations (LRR) [11] are representative. The difference between them lies in that SSC imposes ℓ_1-norm regularization on the affinity matrix, while LRR employs ℓ_*-norm regularization on the affinity matrix. Structured sparse subspace clustering (S^3C) [3,14] performs the affinity matrix learning and clustering by introducing a regularization to quantify the disagreement between the affinity and partition matrices.

Unlikely S^3C, we first analyze the connection between the affinity and partition matrices. This connection makes it possible the fusion of computing the affinity and partition matrices, where the transformation on the partition matrix is used as a substitute for the affinity matrix. Thus, the subspace clustering problem is cast as finding a proper partition matrix without explicitly computing the affinity matrix. Besides, since both the affinity and partition matrices are dedicated to capturing the segmentation of data, the above fusion allows better preserving the subspace structural information, which strengthens connections between data instances that are drawn from the same subspace. The contributions of this paper are summarized as follows:

1. We introduce a subspace clustering framework based on the connection between the affinity and partition matrices to avoid learning the affinity matrix.
2. We design an algorithm to optimize the objective function of our method, and conduct comparative experiments with state-of-the-art subspace clustering methods. Our method achieves better or at least comparable performances.

2   Related Work

We review existing related literatures on various kinds of subspace clustering methods. SSC [10] learns sparse representations by imposing ℓ_1-norm regularization on the affinity matrix. LRR [11] finds low-rank representations by employing

ℓ_*-norm regularization on the affinity matrix. Although both SSC and LRR construct the self-expressive matrix in the iteration manner, they perform badly in dealing with noisy data. Low rank subspace clustering (LSRC) derives a closed-form solution by recovering a clean dictionary. Efficient Dense Subspace Clustering (EDSC) [25] employs ℓ_2-norm regularization to handle noise and outliers. Correlation adaptive subspace segmentation (CASS) [18] employs the TraceLasso norm [21] on the representation matrix. Dohyung et al. [19] propose the nearest subspace neighbor algorithm to automatically identify neighbor data instances that are most likely drawn from the same subspace. Yang et al. [20] propose ℓ_0-induced sparse subspace clustering to discover the subspace-sparse representation for arbitrary distinct underlying subspaces. In [27], Ren et al. propose a weighted adaptive mean shift algorithm to reduce effects of the sparsity caused by high-dimensional data and the noise features. Friedman et al. [28] assign proper weights of features based on importance of features on clustering.

The combination of deep learning and subspace clustering has emerged recently. In [5,6], autoencoder is utilized to learn latent representations for clustering. The difference between them is that the former encodes representations from raw data, while the latter treats the predefined affinity matrix as the input of neural network. In [7], deep autoencoder is adopted to provide initialization for deep embedding for clustering. Deep subspace clustering with sparsity prior [8] clusters deep representations, which incorporate the locality for reconstructing input and structured global prior. Improved Deep Embedded Clustering [9] integrates the clustering loss and autoencoder reconstruction loss into a unified framework, and considers the structural preservation when performing clustering and feature learning jointly. Additionally, a self-expressive matrix is learnt by utilizing a fully connected layer in [26] with latent representations as input.

3   Methodology

In this section, we present our proposed subspace clustering approach. Let $\mathbf{X} = [\mathbf{x}_1, \mathbf{x}_2, \cdots, \mathbf{x}_N] \in \mathbb{R}^{d \times N}$ be a set of N data instances $\{\mathbf{x}_i \in \mathbb{R}^d\}_{i=1}^N$ that are sampled from a special union of k subspaces $\{S_j\}_{j=1}^k$, i.e., $S_1 \bigcup S_2 \bigcup \cdots \bigcup S_k$, of dimensions $\{d_j\}_{j=1}^k$ in $\mathbb{R}^d$. Let $\mathbf{X}_j \in \mathbb{R}^{d \times N_j}$ be a sub-matrix of X of rank d_j with $N_j > d_j$ data points that lie in S_j with $\sum_{j=1}^k N_j = N$. The subspace clustering aims at partitioning data instances in the $\mathbf{X}$ into their corresponding subspaces.

3.1   Efficient Direct Structured Subspace Clustering

The objective function of a traditional clustering method is given by:

$$\min \sum_{i=1}^N \sum_{j=1}^k \mathbf{P}_{ji} \left\| \mathbf{X}_{\cdot i} - \mathbf{C}_{\cdot j} \right\|^2 = \min \left\| \mathbf{X} - \mathbf{CP} \right\|_F^2, \tag{1}$$

where $\mathbf{X}_{\cdot i}$ and $\mathbf{C}_{\cdot j}$ denote the ith and j^{th} columns of $\mathbf{X}$ and $\mathbf{C}$, respectively. $\mathbf{C} \in \mathbb{R}^{d \times k}$ and $\mathbf{P} \in \{0, 1\}^{k \times N}$ denote the cluster center and partition matrices.

The solution to Eq. (1) can be obtained in a closed form [23] which is $\mathbf{C} = \mathbf{X}\mathbf{P}^T(\mathbf{P}\mathbf{P}^T)^{-1}$. Thus, the above optimization problem is equivalent to:

$$\min \left\| \mathbf{X} - \mathbf{X}\mathbf{P}^T(\mathbf{P}\mathbf{P}^T)^{-1}\mathbf{P} \right\|_F^2, \ s.t., \ \mathbf{P}^T\mathbf{1}_k = \mathbf{1}_N, \tag{2}$$

where $\mathbf{1}_k \in \mathbb{R}^k$ and $\mathbf{1}_N \in \mathbb{R}^N$ are column vectors whose entries are 1.

Spectral clustering-based subspace clustering learns the affinity matrix under the assumption that any data point can be approximated by a linear combination of other data points, $i.e.$, $\mathbf{X} = \mathbf{X}\mathbf{Z}$, where $\mathbf{Z}$ is the coefficient matrix whose entity $\mathbf{Z}_{ij}$ measures the similarity between the ith and j^{th} data points. When data points are corrupted by noise, the subspace clustering problem is formulated by

$$\min_{\mathbf{Z},\mathbf{E}} \|\mathbf{Z}\|_1 + \lambda \|\mathbf{E}\|_\ell, s.t., \mathbf{X} = \mathbf{X}\mathbf{Z} + \mathbf{E}, \ diag(\mathbf{Z}) = 0, \tag{3}$$

where $\mathbf{E} = [\epsilon_1, \cdots, \epsilon_N]$ is a noise matrix with $\{\|\epsilon_i\|_2 \le \xi\}_{i=1}^N$, and $\xi > 0$. λ is a trade-off parameter. The constraint $diag(\mathbf{Z}) = 0$ is introduced to avoid the solution of $\mathbf{Z}$ deteriorating into an identity matrix.

From Eqs. (2) and (3), the coefficient matrix $\mathbf{Z}$ is obtained by

$$Z = (\mathbf{P}^T(\mathbf{P}\mathbf{P}^T)^{-1}\mathbf{P}) \odot (\mathbf{1}_N\mathbf{1}_N^T - \mathbf{I}) \tag{4}$$

where $\odot$ is an elementwise product. $\mathbf{I}$ is an identity matrix with a proper size.

Structured sparse subspace clustering [3] holds that both $\mathbf{P}$ and $\mathbf{Z}$ attempt to capture the segmentation of data. In other words, $\mathbf{Z}$ and $\mathbf{P}$ have related zero patterns, $i.e.$, for each $\mathbf{Z}_{ij} \ne 0$, the ith and j^{th} data instances should be split into the same subspace, and $\mathbf{P}_{\cdot i} = \mathbf{P}_{\cdot j}$, where $\mathbf{P}_{\cdot i}$ and $\mathbf{P}_{\cdot j}$ denote the ith and j^{th} columns of matrix $\mathbf{P}$. The subspace structured measure [14] is introduced to evaluate the disagreement between $\mathbf{Z}$ and $\mathbf{P}$ as follows

$$\|\Upsilon \odot \mathbf{Z}\|_0 = \sum_{i,j:\mathbf{P}_{\cdot i} \ne \mathbf{P}_{\cdot j}} \mathbb{I}_{\{|\mathbf{Z}_{ij} \ne 0|\}}, \tag{5}$$

where $\Upsilon_{ij} = \frac{1}{2}\|\mathbf{P}_{\cdot i} - \mathbf{P}_{\cdot j}\|^2$, and $\mathbb{I}$ is an indicator function. Since there are two possible values for Υ_{ij}, $i.e.$, $\Upsilon_{ij} \in \{0, 1\}$, we can utilize the number of nonzero entries in $\mathbf{Z}$ to quantify the cost of Eq. (5) when the ith and j^{th} data instances are split into different subspaces. Since the above measure results in a NP-hard problem, a subspace structured norm [14] of $\mathbf{Z}$ concerning $\mathbf{P}$ is defined by

$$\|\mathbf{Z}\|_P = \beta \|\Upsilon \odot \mathbf{Z}\|_1 = \beta \sum_{ij} |\mathbf{Z}_{ij}| (\frac{1}{2}\|\mathbf{P}_{\cdot i} - \mathbf{P}_{\cdot j}\|^2). \tag{6}$$

Combining Eqs. (3), (4) and (6), we can obtain the expression by

$$\min_{\mathbf{P},\mathbf{E}} \|\mathbf{Z}\|_{1,P} + \lambda \|\mathbf{E}\|_\ell, s.t., \mathbf{X} = \mathbf{X}((\mathbf{P}^T(\mathbf{P}\mathbf{P}^T)^{-1}\mathbf{P}) \odot (\mathbf{1}_N\mathbf{1}_N^T - \mathbf{I})) + \mathbf{E}, \tag{7}$$

where

$$\|\mathbf{Z}\|_{1,P} = \sum_{ij} \left| ((\mathbf{P}^T(\mathbf{P}\mathbf{P}^T)^{-1}\mathbf{P}) \odot (\mathbf{1}_N\mathbf{1}_N^T - \mathbf{I}))_{ij} \right| (1 + \frac{\beta}{2}\|\mathbf{P}_{\cdot i} - \mathbf{P}_{\cdot j}\|^2) \tag{8}$$

3.2   Optimization

Given the objective function in Eq. (7), we solve it by introducing a Lagrange multiplier Y^1, and obtain the augmented Lagrange as follows:

$$L(\mathbf{P}, \mathbf{E}, Y^1) = \|\mathbf{Z}\|_{1,P} + \lambda \|\mathbf{E}\|_{\ell} + <Y^1, \mathbf{X} - \mathbf{XZ} - \mathbf{E}> + \frac{\mu}{2}(\|\mathbf{X} - \mathbf{XZ} - \mathbf{E}\|_F^2). \tag{9}$$

When $\mathbf{E}$ and Y^1 are fixed, $\mathbf{P}$ is updated via:

$$\mathbf{P}_{t+1} \Leftarrow \mathbf{P}_t[(\mathbf{X}^T\mathbf{X} + \mathbf{I})(\mathbf{X}^T(\mathbf{X} - \mathbf{E}_t - \frac{1}{\mu_t}Y_t^1)) \odot (\mathbf{1}_N\mathbf{1}_N^T - \mathbf{I})]. \tag{10}$$

Given $\mathbf{P}$, the update of $\mathbf{Z}_{t+1}$ can be obtained from Eq. (8) naturally. When $\mathbf{P}$ and Y^1 are fixed, the update of $\mathbf{E}$ can be obtained by:

$$\mathbf{E}_{t+1} = \arg\min_{\mathbf{E}} \frac{\lambda}{\mu_t} \|\mathbf{E}\|_{\ell} + \frac{1}{2} \left\| \mathbf{E} - \mathbf{X} - \mathbf{XZ}_{t+1} + \frac{1}{\mu_t}Y_t^1 \right\|_F^2. \tag{11}$$

When $\mathbf{E}$ and $\mathbf{P}$ are fixed, Y^1 is updated as follows:

$$Y_{t+1}^1 \Leftarrow Y_t^1 + \mu_{t+1}(\mathbf{X} - \mathbf{XZ}_{t+1} - \mathbf{E}_{t+1}) \tag{12}$$

The optimization algorithm is shown in the Algorithm 1 below.

Algorithm 1. Efficient Direct Structured Subspace Clustering

Input: Data matrix $\mathbf{X}$, $MAXITE$, λ, β and ρ;
Output: Partition Matrix $\mathbf{P}$.
 1: Initialize: $\mathbf{P}$ with k-means clustering, $c_{nt} = 1$, $\mathbf{Z} = \mathbf{0}$;
 2: **while** $c_{nt} \leq MAXITE$ **do**
 3:     Update the noise matrix $\mathbf{E}$ via Eq. (11);
 4:     Update the partition matrix $\mathbf{P}$ via Eq. (10);
 5:     Update Y^1 via Eq. (12);
 6:     Update $\mu_{t+1} \leftarrow \rho\mu_{t+1}$
 7:     $c_{nt} \leftarrow c_{nt} + 1$;
 8: **end while**

---

4   Experiment

We conduct comparative experiments on multiple data sets with the following methods: LRR [13], LatLRR [12], BDLRR [16], CASS [18], LSR1 [17], LSR2 [17], LRSC [15], SSC [4], KSSC [22], BDSSC [16], AE+SSC [26], SSCOPM [24], EDSC [25], AE+EDSC [26], S^3C [14] and DSC [26]. Data sets contain the Extended Yale B, ORL, COIL20 and COIL100. We quantify performances by exploiting clustering error (Err), normalized mutual information (NMI) and subspace-preserving rate (SPR). We repeat 10 times for each algorithm and record the averages.

4.1 Evaluation Metrics

ERR is defined as below:

$$ERR(r, \hat{r}) = 1 - \max_{\pi} \frac{1}{N} \sum_{i=1}^{N} 1_{\{\pi(\hat{r}_i)=r_i\}} \tag{13}$$

where r and $\hat{r}$ denote the ground-truth and the estimated labels, respectively.
NMI is defined as follows:

$$NMI(r, \hat{r}) = \frac{\sum_{i=1}^{c} \sum_{j=1}^{c} r_{ij} \log(\frac{n \times r_{ij}}{r_i \hat{r}_j})}{\sqrt{(\sum_{i=1}^{c} r_i \log(\frac{r_i}{n}))(\sum_{j=1}^{c} \hat{r}_j \log(\frac{\hat{r}_j}{n}))}} \tag{14}$$

where r_{ij} denotes the number of data points shared by the ith and j^{th} classes.
SPR quantifies the extent to which a clustering solution satisfies the subspace-preserving property, and is defined as follows:

$$SPR = \frac{1}{N} \sum_{j=1} (\sum_{i=1} (G_{ji} \cdot |Z_{ji}|)/(\|Z_{j\cdot}\|_1)) \tag{15}$$

where $G_{ji} \in \{0, 1\}$ is the ground-truth affinity. $Z_{j\cdot}$ is the jth row of the learnt affinity matrix Z, and $\|\cdot\|_1$ is the ℓ_1-norm.

4.2 Experimental Analysis

In this part, we present comparative experimental results. First, we show trends of cost function values of our proposed method concerning iterations in Fig. 1. Cost values of our method decrease quick in the first 20 iterations, and remain steady after 30 iterations. It means that our proposed method can converge fast. In particular, the number of iterations is less than 15 for COIL20 and COIL100. Second, we present the comparisons between our method and other subspace clustering methods on the Extended Yale B in Table 1 in terms of Err and NMI. From this table, we have the following observations:

1. Among traditional methods, S^3 achieves the best performances, which is followed by EDSC and LRSC. The advantage of S^3 attributes to the special structured-norm that allows learning structured sparse representations for each data point. EDSC benefits from Frobenius-norm to obtain denser connections between data points, while LRSC benefits from the learning of a clean self-expressive dictionary that reduce negative effects caused by noise.
2. Deep subspace clustering methods achieve better performances than traditional methods. For example, AE+SSC and AE+EDSC outperform SSC and EDSC, respectively. It demonstrates that representations learnt by deep networks can capture the abstract structural information to contribute to performance improvement. Besides, DSC achieves better performances than AE+SSC and AE+EDSC. The reason is that DSC utilizes a self-expressive layer to learn pairwise affinity between all the data points effectively.
3. Our proposed method achieves better performances than traditional methods including S^3, EDSC and LRSC. It verifies that the combination of computing

the affinity and partition matrices produces positive effects on performance improvement. In addition, our method achieves better or comparable performances, compared with DSC. It indicates that the above combination can conduce to the learning of informative structural representations in the coefficient matrix, which helps improve performances.

Third, we show the effects of computing the affinity and partition matrices in a unified framework concerning SPR by varying the trade-off parameter β obtained in the above data sets in Fig. 2. From this figure, we notice that when the value of β increases in the range between 0 and 10, SPR rises quickly before remaining steady or showing slight decrease trends. It is worthy mentioning that the horizontal axis in Fig. 2 is in the log scale rather than in the linear scale. Besides, it is noted that the SPR obtained by our proposed method is higher than that by S^3 when β changes between 0.3 and 10. It means that our method can achieve better subspace-preserving property, which is brought about by the combination of computing the affinity and partition matrices simultaneously.

Table 1. Performance comparison on extended Yale B w.r.t. (Err %) and (NMI %)

No. Sub.	5		15		25		35	
Criterion	Err	NMI	Err	NMI	Err	NMI	Err	NMI
LRR	13.94	84.14	23.22	71.65	27.92	70.58	41.85	58.45
LatLRR	6.90	88.90	32.47	62.47	32.76	62.76	38.75	60.75
BDLRR	12.97	85.38	31.58	64.86	34.67	61.25	35.76	63.68
CASS	21.25	68.89	33.65	62.38	36.45	62.29	36.57	62.58
LSR1	13.87	75.22	37.64	59.87	40.23	63.12	40.42	62.94
LSR2	13.91	76.91	37.01	61.21	40.89	63.04	39.77	64.03
LRSC	7.59	95.86	15.84	85.99	11.46	89.57	14.79	92.05
SSC	4.32	96.38	13.13	88.69	26.22	68.28	28.55	65.65
KSSC	7.58	91.64	14.49	86.75	16.55	82.75	20.48	80.37
BDSSC	27.5	63.74	38.46	48.85	42.47	41.95	38.47	47.35
AE+SSC	11.76	86.76	18.65	79.26	18.72	83.48	22.13	78.01
SSC-OPM	9.68	93.45	16.22	86.65	18.89	84.40	20.29	83.46
EDSC	5.11	94.22	7.63	89.68	10.67	90.82	13.10	89.66
AE+EDSC	4.45	96.68	6.70	90.25	10.27	90.75	13.28	88.26
S^3C	3.41	95.89	7.54	90.86	9.59	91.89	14.67	88.47
DSC	1.80	97.38	**2.30**	94.45	**2.38**	**94.95**	2.85	93.75
Proposed	**1.62**	**98.55**	**2.29**	**95.26**	2.44	94.44	**2.65**	**94.03**

Fourth, we show the comparative clusterings between our proposed method and other methods obtained in ORL, COIL20 and COIL100 in terms of Err and NMI in Figs. 3 and 4, respectively. From these figures, we observe our proposed method can still achieve best performance than other clustering methods

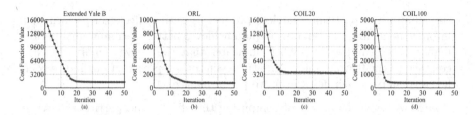

Fig. 1. Trends of cost function values with respect to iterations on Extended Yale B, ORL, COIL20 and COIL100

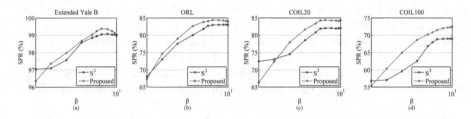

Fig. 2. Trends of SPR concerning β on Extended Yale B, ORL, COIL20 and COIL100

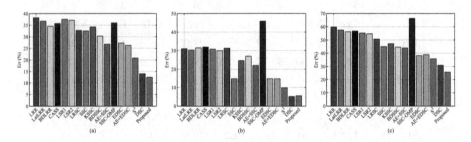

Fig. 3. Clustering Error (%) in ORL (a), COIL20 (b) and COIL100 (c)

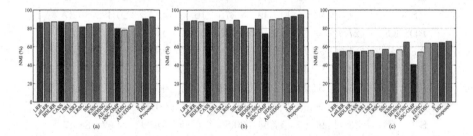

Fig. 4. NMI (%) in ORL (a), COIL20 (b) and COIL100 (c)

including traditional methods and deep subspace clustering methods except on COIL20, where DSC outperforms our method. The reason behind is that our method may consider the affinity between data points from different subspaces when attempting to preserve subspace structural property.

5 Conclusion

In this paper, we propose a direct subspace clustering method without learning the affinity matrix. We analyze the connection between the affinity matrix and partition matrix, and incorporate the computation of the affinity and partition matrices into a unified framework. Besides, we design an algorithm to optimize the objective function of our proposed method. Finally, we conduct extensive comparative experiments on multiple data sets with other representative subspace clustering methods. Our proposed method achieves better or comparable performances, which demonstrates the effectiveness of our method.

Acknowledgment. The work described in this paper was partially supported by a grant from the Research Grants Council of the Hong Kong Special Administrative Region, China [Project No. CityU 11300715), and a grant from City University of Hong Kong [Project No. 7004884].

References

1. Zhao, C., Zhang, J., Ma, S., Fan, X., Zhang, Y., Gao, W.: Reducing image compression artifacts by structural sparse representation and quantization constraint prior. IEEE Trans. Circuits Syst. Video Technol. **27**(10), 2057–2071 (2017)
2. Lai, T., Wang, H., Yan, Y., Chin, T., Zhao, W.: Motion segmentation via a sparsity constraint. IEEE Trans. Intell. Transp. Syst. **18**(4), 973–983 (2017)
3. Li, C., You, C., Vidal, R.: Structured sparse subspace clustering: a joint affinity learning and subspace clustering framework. IEEE Trans. Image Process. **26**(6), 2988–3001 (2017)
4. Elhamifar, E., Vidal, R.: Sparse subspace clustering: algorithm, theory and applications. IEEE Trans. Patt. Anal. Mach. Intell. **35**(11), 2765–2781 (2013)
5. Ma, Y., Shang, C., Yang, F., Huang, D.: Latent subspace clustering based on deep neural networks. In: 6th International Symposium on Advanced Control of Industrial Processes, pp. 502–507. IEEE Press, Hiroshima (2014)
6. Tian, F., Gao, B., Cui, Q., Chen, E., Liu, T.: Learning deep representations for graph clustering. In: 28th AAAI Conference on Artificial Intelligence, pp. 1293–1299. AAAI Press, Québec (2014)
7. Xie, J., Girshick, R., Farhadi, A.: Unsupervised deep embedding for clustering analysis. In: 33rd International Conference on Machine Learning, New York, pp. 478–487 (2016)
8. Peng, X., Xiao, S., Feng, J., Yau, W., Yi, Z.: Deep subspace clustering with sparsity prior. In: 25th International Joint Conference on Artificial Intelligence, New York, pp. 1925–1931 (2016)
9. Guo, X., Gao, L., Liu, X., Yin, J.: Improved deep embedded clustering with local structure preservation. In: 26th International Joint Conference on Artificial Intelligence, Melbourne, pp. 1753–1759 (2017)
10. Elhamifar, E., Vidal, R.: Sparse subspace clustering. In: 22nd IEEE Conference on Computer Vision and Pattern Recognition, pp. 2790–2797. IEEE Press, Florida (2009)
11. Liu, G., Lin, Z., Yu, Y.: Robust subspace segmentation by low-rank representation. In: 27th International Conference on Machine Learning, Haifa, pp. 663–670 (2010)

12. Liu, G., Yan, S.: Latent low-rank representation for subspace segmentation and feature extraction. In: 13th International Conference on Computer Vision, pp. 1615–1122. IEEE Press, Barcelona (2011)
13. Liu, G., Lin, Z., Yan, S., Sun, J., Yu, Y., Ma, Y.: Robust recovery of subspace structures by low-rank representation. IEEE Trans. Patt. Anal. Mach. Intell. **35**(1), 171–184 (2013)
14. Li, C., Vidal, R.: Structured sparse subspace clustering: a unified optimization framework. In: 28th IEEE Conference on Computer Vision and Pattern Recognition, pp. 277–286. IEEE Press, Boston (2015)
15. Vidal, R., Favaro, P.: Low rank subspace clustering (LRSC). Patt. Recogn. Lett. **43**(1), 47–61 (2014)
16. Feng, J., Lin, Z., Xu, H., Yan, S.: Robust subspace segmentation with block-diagonal prior. In: 27th IEEE Conference on Computer Vision and Pattern Recognition, pp. 3818–3825. IEEE Press, Ohio (2014)
17. Lu, C.-Y., Min, H., Zhao, Z.-Q., Zhu, L., Huang, D.-S., Yan, S.: Robust and efficient subspace segmentation via least squares regression. In: Fitzgibbon, A., Lazebnik, S., Perona, P., Sato, Y., Schmid, C. (eds.) ECCV 2012. LNCS, vol. 7578, pp. 347–360. Springer, Heidelberg (2012). https://doi.org/10.1007/978-3-642-33786-4_26
18. Lu, C., Lin, Z., Yan, S.: Correlation adaptive subspace segmentation by trace lasso. In: 14th International Conference on Computer Vision, pp. 1345–1352. IEEE Press, Sydney (2013)
19. Park, D., Caramanis, C, Sanghavi, S.: Greedy subspace clustering. In: 28th Conference on Neural Information Processing Systems, Montréal, pp. 2753–2761 (2014)
20. Yang, Y., Feng, J., Jojic, N., Yang, J., Huang, T.S.: ℓ^0-sparse subspace clustering. In: Leibe, B., Matas, J., Sebe, N., Welling, M. (eds.) ECCV 2016. LNCS, vol. 9906, pp. 731–747. Springer, Cham (2016). https://doi.org/10.1007/978-3-319-46475-6_45
21. Grave, E., Obozinski, G., Bach, F.: Trace lasso: a trace norm regularization for correlated designs. In: 25th Conference on Neural Information Processing Systems, Granada, pp. 2187–2195 (2011)
22. Patel, V., Vidal, R.: Kernel sparse subspace clustering. In: 22nd International Conference on Pattern Recognition, pp. 2849–2853. IEEE Press, Paris (2014)
23. De la Torre, F., Kanade, T.: Discriminative cluster analysis. In: 23rd International Conference on Machine Learning, Pittsburgh, pp. 241–248 (2006)
24. You, C., Robinson, D., Vidal, R.: Scalable sparse subspace clustering by orthogonal matching pursuit. In: 29th IEEE Conference on Computer Vision and Pattern Recognition, pp. 3918–3927. IEEE Press, Las Vegas (2016)
25. Ji, P., Salzmann, M., Li, H.: Efficient dense subspace clustering. In: IEEE Winter Conference on Application of Computer Vision, pp. 461–468. IEEE Press, Steamboat Springs (2014)
26. Ji, P., Zhang, T., Li, H., Salzmann, M., Reid, I.: Deep subspace clustering networks. In: 32nd Conference on Neural Information Processing Systems, Montréal, pp. 24–33 (2017)
27. Ren, Y., Domeniconi, C., Zhang, G., Yu, G.: A weighted adaptive mean shift clustering algorithm. In: SIAM International Conference on Data Mining, pp. 794–802. SIAM Press, Pennsylvania (2014)
28. Friedman, J.H., Meulman, J.J.: Clustering objects on subsets of attributes. J. Royal Stat. Soc. Ser. B (Stat. Methodol.) **66**, 815–849 (2004)

Privacy-Preserving K-Means Clustering Upon Negative Databases

Xiaoyi Hu[1], Liping Lu[1], Dongdong Zhao[1]([✉]), Jianwen Xiang[1]([✉]), Xing Liu[1],
Haiying Zhou[2], Shengwu Xiong[1], and Jing Tian[1]

[1] Hubei Key Laboratory of Transportation of Internet of Things, School of Computer
Science and Technology, Wuhan University of Technology, Wuhan, China
{huxiaoyi,luliping,zdd,jwxiang,liu.xing,swxiong,jtian}@whut.edu.cn
[2] Institution of Automotive Engineering,
Hubei University of Automotive Technology, Shiyan, China
zhouhy_dy@huat.edu.cn

Abstract. Data mining has become very popular with the arrival of
big data era, but it also raises privacy issues. Negative database (*NDB*)
is a new type of data representation which stores the negative image
of data and can protect privacy while supporting some basic data mining operations such as classification and clustering. However, the existing clustering algorithm upon *NDB*s is based on Hamming distance,
when facing datasets which have many categories for each attribute, the
encoded data will become very long and resulting in low computational
efficiency. In this paper, we propose a privacy-preserving k-means clustering algorithm based on Euclidean distance upon *NDB*s. The main step
of k-means algorithm is to calculate the distance between each record
and cluster centers, in order to solve the problem of privacy disclosure in
this step, we transform each record in database into an *NDB* and propose a method to estimate Euclidean distance from a binary string and
an *NDB*. Our work opens up new ideas for data mining upon negative
database.

Keywords: Privacy protection · Data mining · Negative database
k-means clustering

1 Introduction

With the development of information technology in modern society, large amount
of data is generated. This brings about many data mining algorithms, such as
k-means clustering algorithm, k-nearest neighbor algorithm, but it also raises privacy issues since user's personal information may be revealed during the process
of data mining. In this paper, we focus on privacy-preserving k-means clustering
algorithm, which can effectively cluster data while protecting data privacy.

Negative database (*NDB*) · is inspired by Artificial Immunes System and
stores the information which is not contained in traditional databases [10]. Traditional database (*DB*) stores user's original data, once it is disclosed, user's

© Springer Nature Switzerland AG 2018
L. Cheng et al. (Eds.): ICONIP 2018, LNCS 11304, pp. 191–204, 2018.
https://doi.org/10.1007/978-3-030-04212-7_17

privacy will be leaked, unlike *DB*, *NDB* stores information in the complementary set of *DB*, it is *NP*-hard to reverse an *NDB* to get its hidden string, so user's information can be hidden by using *NDB*.

Like the traditional *DB*, *NDB* supports some basic database operations such as intersection, delete, update, select [13,15]. It also supports distance estimation [22]. Due to these characteristics of *NDB*, many applications apply *NDB*s to protect privacy, for example, information hiding [12,14], negative survey [1,11], negative authentication [4–7], negative biometric recognition [2,31], but there are few works that use *NDB* for privacy-preserving data mining. Liu et al. [22] demonstrated that classifying and clustering operations can be effectively carried out upon negative databases, but his algorithm is based on Hamming distance and the encoding method is one-hot encoding, in this case, when the categories of attribute is very large, the encoded string should become very long and will then increase computation cost. In this paper we propose a privacy-preserving *k*-means clustering algorithm based on Euclidean distance upon *NDB*s, which convert decimal numbers to binary numbers directly and has no limitation on the number of categories of attribute. Also, we use *K*-hidden algorithm [30] to generate *NDB* that is more hard-to-reverse than *q*-hidden algorithm in Liu's method. The main step of *k*-means algorithm is to cluster each record to cluster centers and in this step the distance between each record and cluster centers is required to be calculated, if data are not protected, they will be disclosed when calculating the distance. In order to solve the privacy disclosure issues in the step, we transform each record into an *NDB* and propose a method to estimate the Euclidean distance from *NDB*s.

In the following part of this paper, Sect. 2 introduces related work, Sect. 3 introduces the method for estimating Euclidean distance upon *NDB*s. We give the privacy-preserving *k*-means clustering algorithm in Sect. 4 and in Sect. 5, we use experiments to demonstrate the efficiency of our algorithm. We conclude the whole paper in Sect. 6.

2   Related Work

2.1   Privacy Preserving K-Means Clustering

Presently, there are many privacy-preserving methods for *k*-means algorithms. Typically, a privacy-preserving clustering method based on rotation transformation was proposed in [25], which transforms the original data but retains the global distance between data points to obtain accurate clustering results. Dhiraj et al. [9] proposed a method based on clustering rotation, in which each cluster rotates to make data disturbed. Chen et al. [3] proposed a Geometric Data Perturbation (GDP) method that uses three elements: rotation perturbation, translation perturbation, and distance perturbation to carry out random geometric perturbation on data to protect privacy. Lin et al. [20] applied random linear transformation and random perturbation to clustering to protect privacy. Data perturbation distorts the original data to achieve the purpose of privacy protection, however, when the clustering situation is complex, it is difficult to retain important features of original data using data perturbation.

Some privacy-preserving methods are designed for distributed data. Christopher Clifton et al. [29] proposed a privacy-preserving k-means clustering method for vertically distributed data. Jagannathan et al. [18] proposed the concept of arbitrarily partitioned data, which is a generalization of horizontally and vertically partitioned data, and provided an efficient privacy protection protocol for k-means clustering. A method based on zero knowledge identification scheme was proposed in [26] for horizontally partitioned data. These methods can protect user's privacy and achieve a certain clustering accuracy, but they will increase the computation cost of the clustering algorithms and are not suitable for large-scale data.

2.2 Negative Database

The specific definition of NDB is given in [14]: U is a universal set that contains all m-bit binary strings, DB is a dataset contains numbers of m-bit binary strings that require to be protected or hidden. U-DB is the complementary set of DB. Usually the size of U-DB is much larger than DB, so we need to compress the set of U-DB to get NDB by introducing a "don't-care" symbol, written as '*', it means the value at a certain bit where this symbol appear can be either '0' or '1'. Generally, we call '0' and '1' as specified bit and the '*' as unspecified bit. An example of NDB is shown in Table 1. Several NDB generation algorithms have been proposed. Generally, we need NDBs to be hard-to-reverse to protect privacy. The prefix algorithm [14] is known as the first NDB generation algorithm which can generate NDBs from a set of strings, but it is proved to be easy-to-reverse. The $RNDB$ algorithm [14] makes some improvement on the prefix algorithm and is expected to be hard-to-reverse, but it cannot control its reversibility. Several works applied SAT formula generation algorithms to the generation of hard-to-reverse NDBs. In these NDBs, usually each record has K (K is a certain number) specified bits and type i ($i \in [1,K]$) entry has i specified bits different from the hidden string. For example, the q-hidden algorithm proposed in [19] uses a parameter q to control the proportions of different types of records in negative database. The q-hidden algorithm is

Table 1. An example of negative database

DB	U-DB	NDB
110	001	1*1
011	010	0*0
	000	001
	100	100
	101	
	111	

combined with the prefix algorithm to generate hard-to-reverse NDBs in [21]. Liu [23] proposed the p-hidden algorithm, which uses two parameters p_1 and p_2 to generate 3-NDBs (i.e. each entry of the NDB has 3 specified bits), p_i refers to the probability of generating type i entry. Experiments showed that the p-hidden-NDB could be more hard-to-reverse than q-hidden-NDB. Zhao [30] proposed K-hidden algorithm which uses K–1 parameters to control the distribution of different types of entries. The K-hidden algorithm is a fine-grained algorithm which can control NDBs generation process more flexibly. When $K = 3$, the K-hidden algorithm can be regarded as p-hidden algorithm and when K is larger than 3, K-hidden-NDB could be more hard-to-reverse than p-hidden-NDB. The p-hidden algorithm and K-hidden algorithm are proved to be hard-to-reverse against typical SAT solver, (e.g., WalkSAT [16,28] and zChaff [24,27]). In our work, we use the fine-grained K-hidden algorithm to generate hard-to-reverse NDBs.

2.3   The K-Hidden Algorithm

The proposed privacy-preserving k-means algorithm is based on hard-to-reverse K-hidden-NDBs. The K-hidden algorithm is a fine-grained algorithm proposed by Zhao et al. [30]. It uses K-1 parameters (i.e., $p_1 \cdots p_K$, where $p_1 + \cdots + p_K = 1$) to control the distribution of different types of entries in NDBs. In these NDBs, each entry has exactly K specified bits and the type i entry has i bits different from the hidden string. The probability parameters $\{p_1 \cdots p_K\}$ refer to the probabilities of generating each type of entries. Algorithm 1 is the pseudo code of the K-hidden algorithm.

Algorithm 1. The K-hidden algorithm

Input: an m-bit string s; a constant r; the probability parameters $\{p_1 \cdots p_K\}$
Output: NDB_s

1: $NDB_s \leftarrow \phi$
2: $N \leftarrow m \times r$
3: Initialize $\{Q_0, Q_1, \cdots, Q_k\} : Q_0 \leftarrow 0, Q_i \leftarrow p_1 + \cdots + p_i$
4: **while** $(|NDB_s| < N)$ **do**
5:     $rnd \leftarrow random(0, 1)$
6:     Find the i satisfies: $Q_{i-1} \leq rnd < Q_i$
7:     Generate a type i record v
8:     $NDB_s \leftarrow NDB_s \cup v$
9: **return** NDB_s

In the K-hidden algorithm, s is the string required to protect or hide, NDB_s is the corresponding NDB of s. r is a constant parameter which controls the size of NDB_s, usually we have $N = m \times r$ (N is the number of entries in NDB, m is the string length of hidden string), when the size of NDB_s is smaller than N, type i entry is generated with probability p_i and added to NDB_s.

Not all *NDBs* are hard-to-reverse, according to Theorem 1 in [30], the parameters $\{p_1 \cdots p_K\}$ should satisfy the following restriction to ensure *NDBs* to be hard-to-reverse.

$$\sum_{i=1}^{K}(K - 2i)p_i > 0 \tag{1}$$

3   Euclidean Distance Estimation for Negative Databases

To apply *NDBs* to *k*-means clustering, the main process is the Euclidean distance estimation for *NDBs*. The theoretical analyses of our method are given in Sect. 3.1 and two groups of experiments are given in Sect. 3.2 to demonstrate the effectiveness of our method.

3.1   Theoretical Analyses

In this part, the theoretical analyses on how to estimate the Euclidean distance from a binary string and an *NDB* are given. We mainly use three theorems to estimate the Euclidean distance and we will get an estimation formula in the end.

Theorem 1. *For a given m-bit binary string s, in the process of generating NDB_s, the probability that each entry in NDB_s is different from s at a specified bit can be calculated by the following formula.*

$$P_{diff} = \frac{\sum_{i=1}^{K} p_i \times i}{K} \tag{2}$$

Proof. The proof is similar to formula (6) in [31]. In *K*-hidden-*NDB*, type i ($i \in [1, K]$) entry has i specified bits different from s, p_i is the probability that type i entry appears in *NDB*, there are $m \times r$ entries in the *K*-*NDB*, so the total number of different specified bits is

$$N_{diff} = \sum_{i=1}^{K} p_i \times i \times m \times r$$

Since each entry has K specified bits, and total number of specified bits is $K \times m \times r$, so $P_{diff} = \frac{\sum_{i=1}^{k} p_i \times i}{K}$.

Theorem 2. *For an m-bit binary string s which is hidden by NDB_s and unknown to us, suppose at the ith bit, there are n_0 numbers of '0' and n_1 numbers of '1' in NDB_s, the probability that each entry in NDB_s is different from s at the ith bit is P_{diff}, the probability that each entry in NDB_s is the same as s at the ith bit is P_{same} ($P_{same} = 1 - P_{diff}$), then the probability that the value of s_i is '0' can be estimated as:*

$$Pr(s_i = 0) = \frac{(P_{diff})^{n_1} \times (P_{same})^{n_0}}{(P_{diff})^{n_1} \times (P_{same})^{n_0} + (P_{diff})^{n_0} \times (P_{same})^{n_1}} \tag{3}$$

Proof. Suppose there are three events:

event A: the value of s_i is '0'.

event B: the value of s_i is '1'.

event C: at ith bit of the entries in NDB_s, there are n_0 numbers of '0' and n_1 numbers of '1'.

Since NDB_s is known, n_0 and n_1 can be counted, so event C is known to us. $P(A|C)$ is the probability that $s_i = $ '0' when NDB_s is known. According to the Bayes' theorem, we have:

$$P(A|C) = \frac{P(C|A) \times P(A)}{P(C|A) \times P(A) + P(C|B) \times P(B)}$$

s_i is unknown and it can be either '0' or '1', according to the Bayesian hypothesis, it is reasonable to assume $P(A) = P(B) = 1/2$.

Based on Theorem 1, we can get P_{diff} and P_{same}, so if event A happens, the probability that event C happens is:

$$P(C|A) = \binom{n_0}{n_0 + n_1} (P_{diff})^{n_1} \times (P_{same})^{n_0}$$

Likewise, if the event B happens, we have

$$P(C|B) = \binom{n_0}{n_0 + n_1} (P_{diff})^{n_0} \times (P_{same})^{n_1}$$

So $P(A|C)$ can be calculated and we have formula (3).

Theorem 3. *For two m-bit binary strings s and t, if s is hidden by NDB_s, t is an unprotected binary string. The Euclidean distance between s and t can be estimated from NDB_s and t by:*

$$E(s,t) = \sqrt{\sum_{i=0}^{2^m-1} (i - t^d)^2 \times Q_i} \tag{4}$$

where Q_i is the probability that the decimal value of s is i, which can be calculated through Theorem 2, t^d is the value of t.

Proof. Through Theorem 2, we can calculate the probability that each bit of s is '0' or '1', so we can get the probability that the decimal value of s is i, called Q_i.

Because s has m bits, so there are totally 2^m possible values, which are $0, 1, 2, \cdots, 2^m - 1$ respectively. We can calculate the Euclidean distance for each possible value and multiply by the possibility of each value to get the expectation of Euclidean distance between s and t.

For the n-dimensional Euclidean distance, assume s and t have n attributes, which occupies $m_1, m_2, \cdots m_n$ bits, respectively. The Euclidean distance between s and t can be estimated as:

$$E(\mathbf{s}, \mathbf{t}) = \sqrt{\sum_{i=0}^{2^{m_1}-1} \left(i - t_1^d\right)^2 \times Q_{1i} + ... + \sum_{i=0}^{2^{m_n}-1} \left(i - t_n^d\right)^2 \times Q_{ni}} \qquad (5)$$

where Q_{ni} is the probability that the decimal value of the nth attribute of $\mathbf{s}$ is i, t_n^d is the decimal value of the nth attribute of $\mathbf{t}$.

3.2 Experiments on Euclidean Distance Estimation Error

In this part, we use experiments to demonstrate the effectiveness of the proposed estimation method. First, we generate 100 pairs of binary strings and calculate their actual Euclidean distance $H(s,t)$, then we get their estimated Euclidean distance $\bar{H}(s,t)$ through formula (5), we use the formula $error = |H(s,t) - \bar{H}(s,t)|$ to get the error between the estimated Euclidean distance and the actual Euclidean distance.

The Impact of the Similarity of Two Strings. This subsection discusses the influence of the similarity of two strings on the Euclidean estimation results, we achieve it by changing the number of different attributes. The specific steps are as follows: first, a 5000-bits binary string s_1 is generated randomly, suppose every 5 bits is an attribute, so there are 1000 attributes totally. s_2 is generated to have numbers of attributes different from s_1, we set the number of different attributes as 100, 200, ..., 1000, respectively. The parameter K is 4 and because we need NDB to be hard-to-reverse for security issues, formula (1) should be satisfied and we set $P_1 = 0.9, P_2 = 0.05, P_3 = 0.03$ and $r = 8$. The average error and the maximum error are recorded for each parameter settings. Experimental result below demonstrate the impact of the number of different attributes on the error between actual Euclidean distance and estimated Euclidean distance.

Figure 1 shows the average error and maximum error between the actual Euclidean distance and the estimated Euclidean distance of 100 pairs of 5000-bit binary strings. It shows that the less similarity between the two strings, the better estimation accuracy is. When the number of different attributes is 1000, i.e., s_1 and s_2 is totally different, the accuracy of the Euclidean distance estimation method is at its highest level.

The Impact of p. In this part, experiments are conducted to demonstrate the impact of p. In K-hidden algorithm, p is a very important parameter since it controls the distribution of different types of entries in $NDBs$. We randomly generate 100 pairs of binary strings and the length of each string is 5000, each attribute have 5 bits so there are totally 1000 attributes for each string. $r = 15$ and according to [30], the selection of p must follow formula (1) to reach the security requirements, we set four sets of p: (1) $p_1 = 0.5, p_2 = 0.2, p_3 = 0.2$; (2) $p_1 = 0.5, p_2 = 0.4, p_3 = 0.05$; (3) $p_1 = 0.75, p_2 = 0.1, p_3 = 0.1$; (4) $p_1 = 0.9, p_2 = 0.05, p_3 = 0.03$; The results are shown in Table 2:

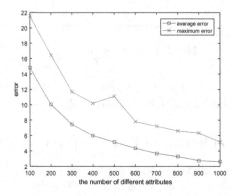

Fig. 1. The average error and maximum error of Euclidean distance estimation method when the number of different attributes changes

Table 2. The values of error when p changes

p	$error_{avg}$	$error_{max}$
$p_1 = 0.5$, $p_2 = 0.2$, $p_3 = 0.2$	75.996104	94.156512
$p_1 = 0.5$, $p_2 = 0.4$, $p_3 = 0.05$	24.156995	34.501286
$p_1 = 0.75$, $p_2 = 0.1$, $p_3 = 0.1$	4.709376	10.250686
$p_1 = 0.9$, $p_2 = 0.05$, $p_3 = 0.03$	0.255267	1.140664

$error_{avg}$ is the average error of estimated Euclidean distance from an NDB and a binary string. $error_{max}$ is the maximum error. As p_1 increases, the error decreases significantly, when p_1 is the same, increasing p_2 also makes estimation results more accurate. A simple understanding is when p_1 gets a larger value, it means the entries in NDB are more likely to the hidden string, so after the probability calculation, larger p_1 can get a better result.

4   Privacy-Preserving K-Means Algorithm upon $NDBs$

In this part, the Euclidean distance estimation method in Sect. 3 is applied to the k-means clustering algorithm to protect privacy, the algorithm description is given below:

Algorithm 2. Privacy-preserving k-means algorithm

Input: The set $NX = \{NDB_{x_i} | x_i \in X\}$, the number of cluster k
Output: The k clusters set

1: Randomly initialize k different binary string $c_i (i = 1, 2, \cdots, k)$
2: **repeat**
3:     **for** each $NDB_{x_i} (x_i \in X)$ **do**
4:         Calculate the distance between $c_i (i = 1, \cdots, k)$ and NDB_{x_i} using formula (5)
5:         Each NDB is clustered to its nearest cluster
6:     Update k cluster centers $c_i (i = 1, \cdots, k)$
7: **until** each $c_i (i = 1, \cdots, k)$ not change
8: **return** k clusters set

The main process of the privacy-preserving k-means algorithm is similar to the traditional k-means algorithm. The difference is that the inputs of this algorithm are NDBs, and in step 4, the Euclidean distance between c_i and NDB is calculated by formula (5).

The above algorithm protects the privacy of the original DB, since every record in DB is hidden by its corresponding NDB and these NDBs are hard-to-reverse, we consider algorithm 2 is a privacy-preserving clustering algorithm.

5   Experiments

Some experiments are conducted in this part to show the effectiveness of the proposed method. We run experiments on the iris dataset in the UCI Machine Learning Repository [8]. The iris dataset contains 150 instances, each instance has four attributes and it is widely used for classification and clustering. The data in iris dataset are labeled into 3 classes, each class has 50 instances.

Since the original data in iris dataset is floating point number, and K-hidden algorithm is based on binary string, each original record should be encoded into binary string to satisfy the requirements of our algorithm.

Our algorithm is based on Euclidean distance, so we convert each floating point number to decimal number and then transform it to binary string. In our algorithm each attribute occupies the same numbers of bits, so we find the largest floating point number and convert it to binary string, the number of bits in this binary string is set as the length (i.e., the number of bits) of attribute. If a floating-point number is converted to a binary string and its number of bits is less than the length of attributes, the remaining bits of the binary string will be set to 0. For example, the maximum value of the first attribute is 7.9, we multiply it by ten and then convert it to binary string, that is 1001111, so the first attribute occupies 7 bits. When encoding the floating point number 5.0, the binary string we get is 110010 and it occupies 6 bits, we will add one bit to the binary string and the final encoded string we get is 0110010. After all data in the iris dataset are encoded, each record in the dataset is transformed into a 25-bit binary string, the four attributes have 7, 6, 7, 5 bits, respectively.

Because the original data has class labels, we use these class labels to evaluate the accuracy of our algorithm.

Define $a_i(i = 1, 2, 3)$ as the number of instances that have the same label i as the original dataset after clustering. We have:

$$Accuracy = \frac{a_1 + a_2 + a_3}{n_{total}} \tag{6}$$

[22] where n_{total} is the number of instances in DB, that is 150. Two sets of experimental results are given below to demonstrate the impact of r and the impact of p on the clustering accuracy.

5.1   Experiments on the Impact of r

r is an important parameter which controls the size of NDB since $N = m \times r$. In this part, we evaluate the accuracy of our algorithm with the transformed data and we compare the accuracy with traditional k-means algorithm and Liu's algorithm [22]. When running experiments on Liu's algorithm, we use the encoding method in [22] and each record is changed to a 119-bit binary string.

Comparison with Liu's Algorithm When $K = 3$: The NDB generation algorithm used in Liu's algorithm is q-hidden algorithm. According to [22], the clustering accuracy is better when given smaller q, so we set $q = 0.3$ and $K = 3, p_1 = 0.752, p_2 = 0.226$ (q-hidden algorithm is a special case of K-hidden algorithm, under this parameter setting, K-hidden algorithm can be regarded as q-hidden algorithm), r ranges from 5 to 25. Each experiment is conducted by 100 times, the average accuracy and the maximum accuracy when r changes are shown in Figs. 2 and 3.

From Figs. 2 and 3, we notice that with the increase of r, the clustering accuracy of both our algorithm and Liu's algorithm increases, but the clustering

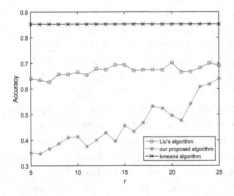

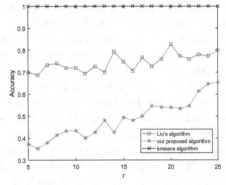

Fig. 2. The average accuracy of clustering when $K = 3$

Fig. 3. The maximum accuracy of clustering when $K = 3$

accuracy of our algorithm is not as good as that of Liu's algorithm. The possible reason is the *NDB* generation algorithm we use treats each bit of binary strings equally but Euclidean distance treats different bits in an attribute differently. This issue does not exist in Liu's algorithm because Hamming distance treats each bit equally.

Comparison with Liu's Algorithm When $K = 4$: In this part, we set $K = 4$ and $p_1 = 0.75, p_2 = 0.1, p_3 = 0.1$ in our algorithm and compare the experimental results with Liu's algorithm with $q = 0.3$. r ranges from 5 to 25. Each experiment is conducted 100 times, the average accuracy and the maximum accuracy when r changes are shown in Figs. 4 and 5.

Figure 4 shows the average accuracy of 100 experiments, Fig. 5 shows the maximum accuracy of 100 experiments. With the increase of r, the accuracy of our algorithm is increased. The accuracy of traditional k-means algorithm is around 0.85. When r is greater than 20, the average accuracy of our algorithm is close to the accuracy of the traditional k-means algorithm. When $r > 9$, our algorithm shows better accuracy than Liu's algorithm. Larger r increases the size of *NDB* and can make the Euclidean distance estimation more accurate to get better clustering result, but it might also increase the probability of revealing information.

Our algorithm also has lower computation cost than Liu's algorithm. The encoding method of Liu's method is ont-hot encoding, when the categories of attribute is very large, the encoded string will become very long and will then increase the computation cost. For example, if a dataset has 10 attribute and each attribute has 1,000,000 categories, the encoded string of Liu's algorithm should be 10,000,000-bit and computation cost will become very large.

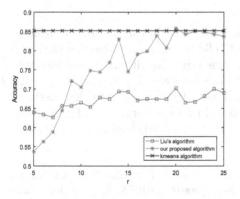
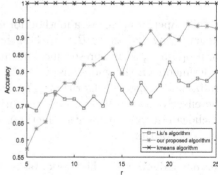

Fig. 4. The average accuracy of clustering when $K = 4$

Fig. 5. The maximum accuracy of clustering when $K = 4$

5.2   Experiments on the Impact of p

p is an important parameter in K-hidden algorithm since it controls the distribution of different types of entries in NDBs. This part shows the impact of p on the clustering accuracy. According to [30], the selection of p must follow formula (1) to reach the security requirements, so we set $r = 15$ and four sets of p: (1) $p_1 = 0.5$, $p_2 = 0.2$, $p_3 = 0.2$; (2) $p_1 = 0.5$, $p_2 = 0.4$, $p_3 = 0.05$; (3) $p_1 = 0.75$, $p_2 = 0.1$, $p_3 = 0.1$; (4) $p_1 = 0.9$, $p_2 = 0.05$, $p_3 = 0.03$;

Table 3. The clustering accuracy when p changes

p	acc_{avg}	acc_{max}
$p_1 = 0.5$, $p_2 = 0.2$, $p_3 = 0.2$	0.355600	0.406667
$p_1 = 0.5$, $p_2 = 0.4$, $p_3 = 0.05$	0.654733	0.713333
$p_1 = 0.75$, $p_2 = 0.1$, $p_3 = 0.1$	0.742200	0.793333
$p_1 = 0.9$, $p_2 = 0.05$, $p_3 = 0.03$	0.860467	0.933333

Table 3 shows the average accuracy and maximum accuracy for 100 times experiments. As p_1 increases, the clustering accuracy increases significantly, when p_1 is the same, increasing p_2 also makes the clustering results more accurate. p_1 means the probability of generating type 1 entry, so larger p_1 means the entries in NDB are more likely to the hidden string, so the estimation result is more accurate. When p_1 is the same, increasing p_2 also makes entries in NDB more likely to the hidden string to have a better result.

6   Conclusions

In this paper, we propose a method for estimating the Euclidean distance from a binary string and an NDB; Based on the Euclidean distance estimation method, we propose a privacy-preserving k-means clustering algorithm. We conduct the experiments on the widely-used iris dataset [8] and compare the accuracy with traditional k-means algorithm [17] and Liu's algorithm [22] to demonstrate the effectiveness of our proposed algorithm. In future work, we will apply our Euclidean distance estimation method to other advanced clustering and classification algorithms.

Acknowledgement. This work was partially supported by the National Natural Science Foundation of China (Grant No. 61806151, 61672398, 61702387), the Hubei Provincial Natural Science Foundation of China (Grant No. 2017CFA012, 2017CFB302), the Key Technical Innovation Project of Hubei (Grant No. 2017AAA122), Provincial Science & Technology International Cooperation Program of Hubei (Grant No. 2017AHB048), the Applied Fundamental Research of Wuhan (Grant No. 20160101010004), and the Open Fund of Hubei Key Lab. of Transportation of IoT (Grant No. 2017III28-004).

References

1. Bao, Y., Luo, W., Zhang, X.: Estimating positive surveys from negative surveys. Stat. Prob. Lett. **83**(2), 551–558 (2013)
2. Bringer, J., Chabanne, H.: Negative databases for biometric data. In: Proceedings of the 12th ACM Workshop on Multimedia and Security, pp. 55–62. ACM (2010)
3. Chen, K., Liu, L.: Geometric data perturbation for privacy preserving outsourced data mining. Knowl. Inf. Syst. **29**(3), 657–695 (2011)
4. Dasgupta, D., Azeem, R.: An investigation of negative authentication systems. In: Proceedings of 3rd International Conference on Information Warfare and Security, pp. 117–126 (2008)
5. Dasgupta, D., Roy, A., Nag, A.: Negative authentication systems. Advances in User Authentication. ISFS, pp. 85–145. Springer, Cham (2017). https://doi.org/10.1007/978-3-319-58808-7_3
6. Dasgupta, D., Saha, S.: A biologically inspired password authentication system. In: Proceedings of the 5th Annual Workshop on Cyber Security and Information Intelligence Research: Cyber Security and Information Intelligence Challenges and Strategies, p. 41. ACM (2009)
7. Dasgupta, D., Saha, S.: Password security through negative filtering. In: 2010 International Conference on Emerging Security Technologies (EST), pp. 83–89. IEEE (2010)
8. Dheeru, D., Karra Taniskidou, E.: UCI machine learning repository (2017). http://archive.ics.uci.edu/ml. Accessed 27 Aug 2018
9. Dhiraj, S.S., Khan, A.M.A., Khan, W., Challagalla, A.: Privacy preservation in k-means clustering by cluster rotation. In: TENCON 2009–2009 IEEE Region 10 Conference, pp. 1–7. IEEE (2009)
10. Esponda, F.: Everything that is not important: negative databases [research frontier]. IEEE Comput. Intell. Mag. **3**(2), 60–63 (2008)
11. Esponda, F.: Negative surveys. arXiv preprint. arXiv: math/0608176 (2006)
12. Esponda, F.: Hiding a needle in a haystack using negative databases. In: Solanki, K., Sullivan, K., Madhow, U. (eds.) IH 2008. LNCS, vol. 5284, pp. 15–29. Springer, Heidelberg (2008). https://doi.org/10.1007/978-3-540-88961-8_2
13. Esponda, F., Ackley, E.S., Helman, P., Jia, H., Forrest, S.: Protecting data privacy through hard-to-reverse negative databases. Int. J. Inf. Secur. **6**(6), 403–415 (2007)
14. Esponda, F., Forrest, S., Helman, P.: Enhancing privacy through negative representations of data. Technical report, Department of Computer Science, University of New Mexico (2004)
15. Esponda, F., Trias, E.D., Ackley, E.S., Forrest, S.: A relational algebra for negative databases. University of New Mexico, Technical report (2007)
16. Ferris, B., Froehlich, J.: WalkSAT as an informed heuristic to DPLL in sat solving. Technical report, CSE 573: Artificial Intelligence (2004)
17. Hartigan, J.A., Wong, M.A.: Algorithm as 136: a k-means clustering algorithm. J. R. Stat. Soc. Ser. C (Appl. Stat.) **28**(1), 100–108 (1979)
18. Jagannathan, G., Wright, R.N.: Privacy-preserving distributed k-means clustering over arbitrarily partitioned data. In: Proceedings of the Eleventh ACM SIGKDD International Conference on Knowledge Discovery in Data Mining, pp. 593–599. ACM (2005)
19. Jia, H., Moore, C., Strain, D.: Generating hard satisfiable formulas by hiding solutions deceptiveily. In: National Conference on Artificial Intelligence, pp. 384–389 (2005)

20. Lin, K.P.: Privacy-preserving kernel k-means clustering outsourcing with random transformation. Knowl. Inf. Syst. **49**(3), 885–908 (2016)
21. Liu, R., Luo, W., Wang, X.: A hybrid of the prefix algorithm and the q-hidden algorithm for generating single negative databases. In: 2011 IEEE Symposium on Computational Intelligence in Cyber Security (CICS), pp. 31–38. IEEE (2011)
22. Liu, R., Luo, W., Yue, L.: Classifying and clustering in negative databases. Front. Comput. Sci. **7**(6), 864–874 (2013)
23. Liu, R., Luo, W., Yue, L.: The p-hidden algorithm: hiding single databases more deeply. Immune Comput. **2**(1), 43–55 (2014)
24. Mahajan, Y.S., Fu, Z., Malik, S.: Zchaff2004: an efficient SAT solver. In: Hoos, H.H., Mitchell, D.G. (eds.) SAT 2004. LNCS, vol. 3542, pp. 360–375. Springer, Heidelberg (2005). https://doi.org/10.1007/11527695_27
25. Oliveira, S., Zaiane, O.: Data perturbation by rotation for privacy-preserving clustering. Technical report TR04-17 (2004)
26. Patel, S., Patel, V., Jinwala, D.: Privacy preserving distributed k-means clustering in malicious model using zero knowledge proof. In: Hota, C., Srimani, P.K. (eds.) ICDCIT 2013. LNCS, vol. 7753, pp. 420–431. Springer, Heidelberg (2013). https://doi.org/10.1007/978-3-642-36071-8_33
27. Pipatsrisawat, K., Darwiche, A.: On the power of clause-learning SAT solvers with restarts. In: Gent, I.P. (ed.) CP 2009. LNCS, vol. 5732, pp. 654–668. Springer, Heidelberg (2009). https://doi.org/10.1007/978-3-642-04244-7_51
28. Selman, B., Kautz, H.A., Cohen, B.: Noise strategies for improving local search. In: AAAI, vol. 94, pp. 337–343 (1994)
29. Vaidya, J., Clifton, C.: Privacy-preserving k-means clustering over vertically partitioned data. In: Proceedings of the Ninth ACM SIGKDD International Conference on Knowledge Discovery and Data Mining, pp. 206–215. ACM (2003)
30. Zhao, D., Luo, W., Liu, R., Yue, L.: A fine-grained algorithm for generating hard-toreverse negative databases. In: 2015 International Workshop on Artificial Immune Systems (AIS), pp. 1–8 (2015)
31. Zhao, D., Luo, W., Liu, R., Yue, L.: Negative iris recognition. IEEE Trans. Dependable Secure Comput. **15**(1), 112–125 (2018)

Self-Paced Multi-Task Multi-View Capped-norm Clustering

Yazhou Ren[✉], Xin Yan, Zechuan Hu, and Zenglin Xu

SMILE Lab, School of Computer Science and Engineering, University of Electronic Science and Technology of China, Chengdu 611731, China
yazhou.ren@uestc.edu.cn

Abstract. Recently, multi-task multi-view clustering (MTMVC) which is able to utilize the relation of different tasks and the information from multiple views under each task to improve the clustering performance has attracted more and more attentions. However, MTMVC typically solves a non-convex optimization problem and thus is easy to stuck into bad local optima. In addition, noises and outliers generally have negative effects on the clustering performance. To alleviate these problems, we propose a novel self-paced multi-task multi-view capped-norm clustering (SPMT-MVCaC) method, which progressively selects data samples to train the MTMVC model from simplicity to complexity. A novel capped-norm term is embedded into the objective of SPMTMVCaC model to reduce the negative influence of noises and outliers, and to further enhance the clustering performance. An efficient alternating optimization method is developed to solve the proposed model. Experimental results on real data sets demonstrate the effectiveness and robustness of the proposed method.

Keywords: Multi-Task Muti-View Clustering · Self-paced learning Capped-norm

1 Introduction

Clustering is to group samples into different groups in an unsupervised manner. Over the past several decades, a lot of single-task single-view clustering methods have be developed, such as partitional clustering [17, 22, 26, 28, 34], density-based clustering [1, 5, 29], mean shift clustering [2, 27, 30], spectral clustering [11, 15, 16], etc. Clustering has been applied in various fields, such as document analysis [18], time series data clustering [40], computer vision [37], regional science [25], etc. However, these traditional methods can only tackle clustering problems with single task and single view.

Different from single-task clustering methods, multi-task clustering learns the information shared by multiple related tasks to improve individual clustering performance on the ground that it is a common issue that tasks are closely related to each other in real clustering problems [8]. Generally, learning multiple tasks simultaneously can achieve higher performance than learning different

© Springer Nature Switzerland AG 2018
L. Cheng et al. (Eds.): ICONIP 2018, LNCS 11304, pp. 205–217, 2018.
https://doi.org/10.1007/978-3-030-04212-7_18

tasks independently [6,21]. In contrast with single-view clustering which tackles only one individual view, multi-view clustering exploits the connections among multiple views to improve clustering performance [9,10,12].

Both multi-view clustering and multi-task clustering have been prone to achieve great performance in many real machine learning problems. However, we often confront with complex problems that involve both multi-view clustering and multi-view clustering in the real data sets, that is to say, the tasks are closely related to each other and each task can be analyzed from multiple views [39]. For instance, the tasks for clustering the leaves data set with 100 species of leaves can be considered as 100 related tasks. For each species, there are several common views, such as margin, shape, and texture. Utilizing both the multi-task and multi-view information can obtain better performance when addressing the clustering issue of such data sets.

A number of multi-task multi-view clustering methods have been proposed in the past few years. The co-clustering based multi-task multi-view clustering framework bridges multi-task learning method and multi-view learning method together to make full advantages of both worlds, which consists of three parts: within-view-task clustering, multi-view relationship learning, and multi-task relationship learning [38]. The bipartite graph based multi-task multi-view clustering (BMTMVC) algorithm applies the bipartite graph co-clustering method in within-view-task clustering to deal with the multi-task multi-view clustering of the data with nonnegative feature values [38]. After that, the semi-nonnegative matrix tri-factorization based multi-task multi-view clustering (SMTMVC) algorithm is proposed by using semi-nonnegative matrix tri-factorization to co-cluster the data in each view of each task in within-task-view clustering to deal with negative data values [39].

Despite their usefulness in clustering multi-task multi-view data, the existing MTMVC methods typically solve a non-convex problem and are easy to find bad local optima. To address this problem, in this work we propose a novel MTMVC model based on self-paced learning (SPL). SPL is a training strategy which is inspired by the human learning process [20]. Instead of training the model with all data samples, SPL paradigm first trains the model on 'easy' samples and then gradually takes 'complex' samples into account. It has been proved that it is beneficial in avoiding bad local minima and can achieve a better generalization ability [14]. Recently, SPL has attracted people's increasing attentions and has been applied in a variety of machine learning fields, such as computer vision [13,35], feature corruption [33], boosting learning [23], diagnosis of Alzheimer's Disease [24], multi-class classification [32], clustering [31], etc.

Furthermore, outliers and noisy data in multi-task multi-view clustering can negatively affect the clustering performance. In order to tackle this issue, we design a novel capped-norm term in the model. In this way, our model is robust to noises and outliers. The resulting model is called self-paced multi-task multi-view capped-norm clustering (SPMTMVCaC).

Overall, the main contributions of this paper are stated as follows:

- We propose a novel multi-task multi-view clustering method based on SPL to avoid bad local optimum and find better clustering solutions.
- The proposed method reduces the negative influence of noises and outliers by utilizing a novel capped-norm loss term. Thus, it can achieve stable clustering results in the presence of noises and outliers.
- The proposed SPMTMVCaC model can be easily solved by an efficient alternating optimization method.
- Time complexity is given and experiments on real multi-task multi-view data sets show that our SPMTMVCaC model is efficient and robust.

2 Preliminaries

Suppose T clustering tasks are given, each associates with V_t views, $\mathbf{X}_t^{(v)} = \{\mathbf{x}_1^{(v)}, \mathbf{x}_2^{(v)}, \ldots, \mathbf{x}_{n_t}^{(v)}\} \in \mathbb{R}^{d_t^{(v)} \times n_t}$, $t = 1, 2, \ldots, T$, $v = 1, \ldots, V_t$, where $d_t^{(v)}$ is the dimension of feature vectors for the v-th view in each task t and n_t is the number of samples in the t-th task. Clustering approaches seek to group data samples of each task t into $c^{(t)}$ disjoint clusters. It can be observed that the related tasks share a lot of features in some views. We view such views as common views, and the other views as specific views [39]. In the multi-task learning literature [8,39], it is usually assumed that $c^{(1)} = c^{(2)} = \cdots = c^{(T)} = c$.

2.1 Multi-Task Multi-View Clustering

A semi-nonnegative matrix tri-factorization based multi-task multi-view clustering (SMTMVC) framework has been developed, which consists of three components: within-view-task clustering, multi-view relationship learning, and multi-task relationship clustering. Then, its common optimization model is defined as [39]:

$$
\min_{\mathbf{R}_t^{(v)}, \tilde{\mathbf{R}}_t^{(v)}, \mathbf{F}_t^{(v)}, \mathbf{F}^{(v)}, \mathbf{G}_t^{(v)}} \sum_{t=1}^{T} (\sum_{v=1}^{V_t} K_1 + \xi \sum_{v=1}^{V_t} \sum_{q \neq v}^{V_t} K_2) + \mu \sum_{v \in S} \sum_{t \in T_v} K3 \tag{1}
$$

$$
\text{s.t.} \quad \mathbf{F}_t^{(v)\mathrm{T}} \mathbf{F}_t^{(v)} = \mathbf{I} \ (t = 1, \ldots, T, \ v = 1, \ldots, V_t)
$$

$$
\mathbf{F}^{(v)\mathrm{T}} \mathbf{F}^{(v)} = \mathbf{I} \ (v \in S)
$$

$$
\mathbf{G}_t^{(v)\mathrm{T}} \mathbf{G}_t^{(v)} = \mathbf{I} \ (t = 1, \ldots, T, \ v = 1, \ldots, V_t)
$$

$$
\mathbf{F}_t^{(v)}, \mathbf{F}^{(v)}, \mathbf{G}_t^{(v)} \geq 0,
$$

where $K_1 = ||\mathbf{X}_t^{(v)} - \mathbf{F}_t^{(v)} \mathbf{R}_t^{(v)} \mathbf{G}_t^{(v)\mathrm{T}}||_F^2$, $K_2 = -tr(\mathbf{G}_t^{(v)} \mathbf{G}_t^{(v)\mathrm{T}} \mathbf{G}_t^{(q)} \mathbf{G}_t^{(q)\mathrm{T}})$, and $K_3 = ||\tilde{\mathbf{X}}_t^{(v)} - \mathbf{F}^{(v)} \tilde{\mathbf{R}}_t^{(v)} \mathbf{G}_t^{(v)\mathrm{T}}||_F^2$. In Eq. (1), S is the index set of common views. T_v is the index set of tasks under the common view v. $\mathbf{R}_t^{(v)}$ is the correlation matrix between the feature clusters and the sample clusters under view v in each task t, $\tilde{\mathbf{R}}_t^{(v)}$ is the correlation matrix between the sample clusters and the shared subspace which is learned by the related tasks with the common view.

$\mathbf{F}_t^{(v)}$ is the feature partition matrix under view v in each task t, $\mathbf{F}^{(v)}$ is the shared subspace. $\mathbf{G}_t^{(v)}$ can be seen as the sample partition matrix under view v in each task t. $\tilde{\mathbf{X}}_t^{(v)}$ is the data matrix between the samples of t task and the shared subspace in the task. Note that when considering a data set $\mathbf{X} \in \mathbb{R}^{d \times n}$, the corresponding matrices $\mathbf{F} \in \mathbb{R}^{d \times l}$, $\mathbf{G} \in \mathbb{R}^{n \times c}$, and $\mathbf{R} \in \mathbb{R}^{l \times c}$, where l and c are the numbers of feature clusters and sample clusters, respectively. Usually, l is equal to c [39].

In Eq. (1), $\sum_{v=1}^{V_t} K_1$ is to co-cluster the feature and sample under each task t, $\sum_{v=1}^{V_t} \sum_{q \neq v}^{V_t} K_2$ is to minimize the disagreement between the cluster assignments of any two different views in each task t, and $\sum_{v \in S} \sum_{t \in T_v} K_3$ is to learn the subspace shared by the related tasks under each common view. These three parts actually correspond to within-view-task clustering, multi-view relationship learning and multi-task relationship clustering, respectively. Please refer to [39] for more details.

2.2 Self-Paced Learning

Self-paced learning seeks to jointly learn the model parameter θ and the self-paced learning weight variable $\mathbf{v}$ by solving [20]:

$$\min_{\theta, \mathbf{v}} \sum_{i=1}^{n} v_i L(\mathbf{x}_i, \theta) + f(\lambda, \mathbf{v}), \tag{2}$$

where $\lambda > 0$, $\mathbf{v} \in [0,1]^n$ (v_i is the corresponding weight of $\mathbf{x}_i$), n is the number of samples, and $L(\mathbf{x}_i, \theta)$ denotes the loss function of sample $\mathbf{x}_i$. $f(\lambda, \mathbf{v})$ denotes the SPL regularization term and is defined as [20]:

$$f(\lambda, \mathbf{v}) = -\lambda \sum_{i=1}^{n} v_i. \tag{3}$$

When fix θ, the optimal values of $\mathbf{v}$ is obtained by:

$$v_i = \begin{cases} 1, & \text{if } L(\mathbf{x}_i, \theta) \leq \lambda, \\ 0, & \text{otherwise,} \end{cases} \tag{4}$$

When λ is initially small, a small number of samples are selected during training. As λ grows gradually, more samples will be selected to train the model until all the samples are added.

3   Self-Paced Multi-Task Multi-View Capped-norm Clustering

3.1   The Objective Function

In this work, we focus on the semi-nonnegative matrix tri-factorization based multi-task multi-view clustering (SMTMVC) model. Based on SMTMVC, we propose self-paced multi-task multi-view capped -norm clustering method (SPMTMVCaC), which can alleviate the non-convex issue and is robust to the noises and outliers, by utilizing SPL training strategy and the capped-norm term. The common optimization model of SPMTMVCaC is defined as:

$$\min_{\mathbf{R}_t^{(v)}, \tilde{\mathbf{R}}_t^{(v)}, \mathbf{F}_t^{(v)}, \mathbf{F}^{(v)}, \mathbf{G}_t^{(v)}, \mathbf{w}_t} \sum_{t=1}^{T} \sum_{i=1}^{n_t} w_t^i g(l_t^i, \theta_t) + \xi \sum_{t=1}^{T} \sum_{v=1}^{V_t} \sum_{q \neq v}^{V_t} \hat{K}_2$$

$$+ \mu \sum_{v \in S} \sum_{t \in T_v} \hat{K}_3 + f(\{\lambda_t\}_{t=1}^{T}, \{\mathbf{w}_t\}_{t=1}^{T}) \qquad (5)$$

$$\text{s.t.} \qquad \mathbf{F}_t^{(v)\mathrm{T}} \mathbf{F}_t^{(v)} = \mathbf{I} \ (t = 1, \dots, T, \ v = 1, \dots, V_t),$$

$$\mathbf{F}^{(v)\mathrm{T}} \mathbf{F}^{(v)} = \mathbf{I} \ (v \in S),$$

$$\mathbf{G}_t^{(v)\mathrm{T}} \mathbf{G}_t^{(v)} = \mathbf{I} \ (t = 1, \dots, T, \ v = 1, \dots, V_t),$$

$$\mathbf{F}_t^{(v)}, \mathbf{F}^{(v)}, \mathbf{G}_t^{(v)} \geq 0,$$

$$\mathbf{w}_t \in [0, 1]^{n_t},$$

where $l_t^i = \sum_{v=1}^{V_t} ||[\mathbf{X}_t^{(v)}]_i - [\mathbf{F}_t^{(v)} \mathbf{R}_t^{(v)} \mathbf{G}_t^{(v)\mathrm{T}}]_i||_2^2$, which can be considered as the reconstruction error among different views of the i-th sample in task t. Since there is no label information in the clustering task, we let such reconstruction error l_t^i be the loss value of the corresponding sample. Here, w_t^i indicates the weight for the i-th sample in the t-th task. Let vector $\mathbf{w}_t = (w_t^1, w_t^2, \cdots, w_t^{n_t})^{\mathrm{T}}$, and matrix $\mathbf{W}_t = diag(w_t^1, w_t^2, \cdots, w_t^{n_t})$.

In Eq. (5), $\hat{K}_2 = -tr([\mathbf{G}_t^{(v)} \mathbf{G}_t^{(v)\mathrm{T}} \mathbf{G}_t^{(q)} \mathbf{G}_t^{(q)\mathrm{T}}]_+)$, where $[\cdot]_+$ indicates that only the selected samples (whose weights are positive) are used to compute $\hat{K}_2$. In contrast, all the data samples in task t are used to compute K_2 in Eq. (1). Similarly, $\hat{K}_3 = ||[\tilde{\mathbf{X}}_t^{(v)} - \mathbf{F}^{(v)} \tilde{\mathbf{R}}_t^{(v)} \mathbf{G}_t^{(v)\mathrm{T}}]_+||_F^2$. The only difference between $\hat{K}_3$ in Eq. (5) and K_3 in Eq. (1) is that, only those samples with positive weights participate in computing $\hat{K}_3$ in our model, while all the data samples in task t are used to compute K_3 in Eq. (1). In the case that the weights of all the data samples are positive, $\hat{K}_2 = K_2$ and $\hat{K}_3 = K_3$ hold, indicating all the data samples are selected to train the model.

Similarly with Eq. (1), the first three parts of Eq. (5) correspond to within-view-task clustering, multi-view relationship learning, and multi-task relationship clustering, respectively. The difference is that, the SPL weighted term is used in the first part of Eq. (5) and only the selected samples are utilized to compute the second and the third parts of Eq. (5). In Eq. (5), the fourth part $f(\{\lambda_t\}_{t=1}^{T}, \{\mathbf{w}_t\}_{t=1}^{T})$ is the SPL regularization term and $g(l_t^i, \theta_t)$ is the capped-norm term. Both will be defined later in this paper.

On the one hand, we gradually train the multi-task multi-view clustering model from 'easy' to 'complex' by increasing the penalty on the regularizer. In this way, our model can alleviate the risk of getting bad local minima. On the other hand, we further design a capped-norm term to remove the effect of noises and outliers.

3.2 Optimization

In this section, we develop an alternative search strategy to optimize Eq. (5). At first, we get the initial values of all the model parameters (i.e., $\mathbf{R}_t^{(v)}$, $\tilde{\mathbf{R}}_t^{(v)}$, $\mathbf{F}_t^{(v)}$, $\mathbf{F}^{(T)}$, and $\mathbf{G}_t^{(v)}$) by running the SMTMVC model for a small number of iterations (which is set to 20 in our experiments). In this way, we can obtain the initial estimation of reconstructing error of each sample. Then, the optimization process can be divided into two steps as follows.

Step 1: Fixed model parameters, update $\mathbf{w}_t$. We first define the SPL regularizer $f(\{\lambda_t\}_{t=1}^T, \{\mathbf{w}_t\}_{t=1}^T)$ as:

$$f(\{\lambda_t\}_{t=1}^T, \{\mathbf{w}_t\}_{t=1}^T) = -\sum_{t=1}^T \lambda_t \sum_{i=1}^{n_t} w_t^i, \tag{6}$$

where a separate SPL controlling parameter λ_t is set for each task, $t = 1, 2, \ldots, T$. Then, we define a novel capped-norm term $g(l_t^i, \theta_t)$ as:

$$g(l_t^i, \theta_t) = \frac{2l_t^i}{sign(\theta_t - l_t^i) + 1}, \tag{7}$$

where $sign(\cdot)$ is a function whose value is equal to 1 when the input is positive and is equal to -1 otherwise.

Given fixed model parameters, the second and the third terms of Eq. (5) are constant. Thus, minimizing Eq. (5) is equivalent to solve:

$$\min_{\{\mathbf{w}_t\}_{t=1}^T} \sum_{t=1}^T \sum_{i=1}^{n_t} w_t^i g(l_t^i, \theta_t) - \sum_{t=1}^T \lambda_t \sum_{i=1}^{n_t} w_t^i. \tag{8}$$

To solve Eq. (8), we consider the following two cases.

Case 1: $l_t^i \geq \theta_t$. When $l_t^i \geq \theta_t$, the value of $g(l_t^i, \theta_t)$ is $+\infty$, and $g(l_t^i, \theta_t) - \lambda_t = +\infty$. Thus, the weight of the corresponding sample w_t^i should be set to 0 to minimize Eq. (8) and this sample will not be selected to train the model. In this way, the noisy data and outliers (whose loss values are typically large) are eliminated in the training process so that they will not affect the final result adversely. As a consequence, the capped-norm term defined by Eq. (7) achieves the same goal with the existing capped-norm methods [7, 10].

Case 2: $l_t^i < \theta_t$. When $l_t^i < \theta_t$, it is easy to show that $g(l_t^i, \theta_t) = l_t^i$. Then, Eq. (8) becomes

$$\min_{\{\mathbf{w}_t\}_{t=1}^T} \sum_{t=1}^T \sum_{i=1}^{n_t} w_t^i (l_t^i - \lambda_t), \tag{9}$$

where the optimal $\mathbf{w}_t$ can be computed as below:

$$w_t^i = \begin{cases} 1, & \text{if } l_t^i \leq \lambda_t, \\ 0, & \text{otherwise.} \end{cases} \tag{10}$$

Step 2: Fixed $\mathbf{w}_t$, update model parameters. For fixed $\mathbf{w}_t$, the last term of Eq. (5) is a constant number. Thus, Eq. (5) can be considered as a weighted version of Eq. (1). That is, only the selected samples (whose weights are 1) are used to update the model parameters. The way of updating model parameters $\mathbf{R}_t^{(v)}$, $\tilde{\mathbf{R}}_t^{(v)}$, $\mathbf{F}^{(v)T}_t$, $\mathbf{F}^{(v)T}$ and $\mathbf{G}_t^{(t)}$ is the same as SMTMVC method [39].

SPMTMVCaC iteratively runs the above two steps. The hyperparameters θ_t and λ_t are automatically determined according to the data. At first, we set the SPL controlling parameter λ_t such that half of data samples with smaller loss values of each task are selected for training. Then, we run SPMTMVCaC by increasing the value of λ_t in each task to add 10% more data each time. In these procedures, θ_t is always equal to λ_t because until now the easy samples are typically chosen to train. After 6 iterations, when all the data are selected, we set θ_t according to the ratio of outliers. For instance, if the ratio of outliers is 0.05, then we set θ_t such that the loss values of 5% of data samples in this task are larger than θ_t. Those samples whose loss values are less than θ_t participate in the final training process. Then, the learned model parameters are the final parameter values and are used to give the final cluster assignments of these samples. For those samples which are not selected in the last iteration, we use the initialized parameter values (i.e., $\mathbf{R}_t^{(v)}$, $\tilde{\mathbf{R}}_t^{(v)}$, $\mathbf{F}_t^{(v)}$, $\mathbf{F}^{(T)}$, and $\mathbf{G}_t^{(v)}$) calculated by the first 20 iterations of SMTMVC model to give their cluster assignments.

3.3 Time Complexity Analysis

Let n be the number of samples in each task and d be the feature dimensionality in each view, the overall computational complexity of SMTMVC model is $O(dn + n^2 + d^2)$ [39]. The proposed SPMTMVCaC needs to solve the weighted version of SMTMVC a few numbers of times (which is 6 according to our setting). Note that SPL can typically speed the convergence [20]. Thus, the time complexity of the proposed SPMTMVCaC is similar to SMTMVC, which is also $O(dn + n^2 + d^2)$.

4   Experimental Setup

To verify the effectiveness of the SPMTMVCaC model, we evaluate the proposed method on three multi-task multi-view data sets, i.e., the Leaves data set, the WebKB data set, and the NUS-WIDE data set, which have been also used in [39].

4.1   Data Sets

The Leaves[1] data set consists of 100 kinds of leaves, each providing sixteen different samples. Each sample is photographed as a color image on a white background. 18 kinds of leaves are selected to form three tasks and there are three common views for each task: shape, margin, and texture.

---

[1] https://archive.ics.uci.edu/ml/datasets/One-hundred+plant+species+leaves+data+set.

The WebKB[2] data set is derived from computer science department websites of various universities. We select only four categories from the seven categories. The data set has four tasks, each is comprised of three views. There is only one common view, which is consisted of the words of the websites of four universities. The specific views are hyperlinks and titles of websites since tasks share lots of words among the main texts and share a few words among hyperlinks and titles.

The NUS-WIDE[3] data set is a real-world image web image data set created by National University of Singapore. This data set includes 81 concepts of web images and associated tags from Flickr. We select 20 concepts to form four tasks, which share seven common views.

The details of the data sets are given in Table 1.

Table 1. Data sets used in the experiments.

	Task	#Class	#Sample	#View	Common view
Leaves	Task1	6	96	3	1–3
	Task2	6	96	3	1–3
	Task3	6	96	3	1–3
WebKB	Task1	4	226	3	1
	Task2	4	252	3	1
	Task3	4	255	3	1
	Task4	4	307	3	1
NUS-WIDE	Task1	5	3900	7	1–7
	Task2	5	3107	7	1–7
	Task3	5	4087	7	1–7
	Task4	5	4709	7	1–7

4.2   Baseline Methods

We compare the proposed method with state-of-the-art methods, which are listed as below:

- k-means [22].
- kk-means: kernel k-means [4].
- BiCo: bipartite graph co-clustering algorithm [3].
- SNMTF: the semi-nonnegative matrix tri-factorization algorithm [36].
- CoRe: co-regularized multi-view spectral clustering algorithm [19].
- LSSMTC: the shared subspace learning multi-task clustering algorithm [8].
- SMTMVC: semi-nonnegative matrix tri-factorization based multi-task multi-view clustering algorithm [39].

---

[2] http://www.cs.cmu.edu/afs/cs.cmu.edu/project/theo-20/www/data/.
[3] http://lms.comp.nus.edu.sg/research/NUS-WIDE.htm.

Since the existing multi-task clustering methods can only apply to a view that includes the features shared by all the tasks, we let LSSMTC work on the common view.

4.3 Evaluation Measure

To evaluate the clustering results, we use two widely-used evaluation measures: clustering accuracy (Acc) and normalized mutual information (NMI). A bigger value of Acc or NMI means a better clustering result.

4.4 Parameter Setting

We set the parameters of the comparing methods in the same way as [39]. Concretely, for kk-means and CoRe, we let δ (the Gaussian kernel width) be the median euclidean distance among samples for each task. And we set the other parameters by searching grid for the optimal values. For LSSMTC, we tune parameter λ and l (dimensionality of the shared subspace) and report the best results, where $\lambda \in \{0.1, 0.2, ..., 0.9\}$ and $l \in \{2, 4, 6, 8, 10\}$. Since BiCo can only work for non-negative data, we perform this algorithm only on the Leaves and the WebKB data sets. The other methods are performed on all data sets.

For SMTMVC and SPMTMVCaC, we set parameters ξ, μ by searching the grid $\{0.1, 0.2, ..., 0.9, 1\}$. For SPMTMVCaC, the threshold parameter θ_t lies on the ratio of outliers. Hence, we set θ_t by searching the values of this ratio with the grid $\{0, 0.01, ..., 0.1\}$. As in [39], we choose the results of SPMTMVCaC from the most informative view, and compute the mean Acc and NMI as well as the standard deviation. Then, we report the Acc and NMI of each task under the best parameters.

All the reported results are the average values of 10 independent runs.

5 Results and Analysis

5.1 Clustering Results on Real Data

In this section, we show the average Acc/NMI values and standard deviations of all the comparing clustering methods for every task of the three data sets. The results are given in Tables 2, 3, and 4, respectively. The results of the first six comparing methods are obtained from [39]. In each column, the best clustering results are highlighted in boldface. From these tables, the following observations can be obtained:

(1) For single-task single-view clustering methods, the co-clustering algorithms BiCo and SNMTF generally perform better than k-means and kk-means.
(2) The multi-task multi-view method SMTMVC improves upon single-task single-view methods (i.e., BiCo and SNMTF), multi-view method (i.e., CoRe), and muti-task clustering method LSSMTC by a large margin. This verifies the usefulness of simultaneously considering the relation among different tasks and the information among different views under each task.

(3) By comparing SPMTMVCaC against SMTMVC, we can find that SPMT-MVCaC always outperforms SMTMVC except on Task2 of Leaves and Task1 of WebKB. SPMTMVCaC also achieves the best performance in most of time. This main reason is that SPMTMVCaC trains the model in a SPL manner and can eliminate the negative influence of noisy data and outliers.

Table 2. Results on leaves.

Methods	Task1		Task2		Task3	
	Acc (%)	NMI (%)	Acc (%)	NMI (%)	Acc (%)	NMI (%)
k-means	68.75 ± 11.34	68.11 ± 6.89	68.02 ± 13.64	78.23 ± 6.07	71.35 ± 3.41	75.60 ± 3.02
kk-means	74.25 ± 4.73	69.24 ± 4.77	76.25 ± 6.49	79.94 ± 5.29	72.60 ± 8.53	69.78 ± 9.86
BiCo	82.40 ± 7.19	75.36 ± 3.93	81.98 ± 2.95	80.78 ± 2.76	76.56 ± 6.57	76.94 ± 1.31
SNMTF	77.08 ± 5.89	73.70 ± 5.49	73.85 ± 10.32	82.82 ± 6.73	76.25 ± 6.15	77.38 ± 3.65
CoRe	86.15 ± 9.19	86.85 ± 4.01	87.81 ± 11.40	91.55 ± 5.79	85.83 ± 11.48	92.16 ± 4.99
LSSMTC	72.81 ± 7.21	82.94 ± 5.79	67.60 ± 4.57	83.24 ± 3.83	76.56 ± 10.87	78.62 ± 7.62
SMTMVC	94.68 ± 1.99	92.01 ± 1.76	97.50 ± 1.91	$\mathbf{99.01 \pm 0.67}$	97.08 ± 1.08	94.27 ± 1.59
SPMTMVCaC	$\mathbf{95.93 \pm 0.33}$	$\mathbf{92.28 \pm 0.69}$	$\mathbf{97.71 \pm 1.12}$	97.57 ± 2.02	$\mathbf{97.50 \pm 0.73}$	$\mathbf{95.17 \pm 1.15}$

Table 3. Results on WebKB.

Methods	Task1		Task2		Task3		Task4	
	Acc(%)	NMI(%)	Acc(%)	NMI(%)	Acc(%)	NMI(%)	Acc(%)	NMI(%)
k-means	61.41 ± 1.81	16.56 ± 3.29	52.07 ± 8.89	13.75 ± 3.91	52.23 ± 3.51	10.82 ± 4.31	61.36 ± 5.35	20.43 ± 8.96
kk-means	48.04 ± 2.17	13.62 ± 2.34	44.72 ± 2.31	19.04 ± 0.73	46.47 ± 1.20	14.64 ± 1.90	56.22 ± 0.97	29.42 ± 0.65
BiCo	65.39 ± 9.36	34.01 ± 4.76	62.29 ± 5.64	28.52 ± 2.36	65.31 ± 5.79	31.75 ± 4.30	72.78 ± 9.22	51.62 ± 6.33
SNMTF	70.08 ± 4.45	42.03 ± 5.71	68.25 ± 4.12	29.32 ± 12.49	75.09 ± 4.86	58.18 ± 6.30	70.00 ± 10.21	44.31 ± 9.59
CoRe	75.66 ± 0.00	48.33 ± 0.00	68.65 ± 0.03	31.84 ± 0.10	67.25 ± 0.06	50.62 ± 0.04	70.75 ± 0.04	41.78 ± 0.05
LSSMTC	62.18 ± 8.22	25.61 ± 4.86	62.03 ± 5.14	29.66 ± 2.77	58.62 ± 6.31	25.66 ± 2.29	66.22 ± 9.84	33.57 ± 5.94
SMTMVC	$\mathbf{70.53\pm6.70}$	$\mathbf{47.79\pm7.27}$	71.55 ± 5.78	36.22 ± 12.8	85.61 ± 3.68	74.72 ± 3.89	75.70 ± 4.18	52.07 ± 8.31
SPMTMVCaC	67.25 ± 1.04	43.51 ± 1.57	$\mathbf{75.60\pm0.50}$	42.04 ± 2.34	$\mathbf{90.27\pm0.99}$	$\mathbf{78.54\pm1.47}$	$\mathbf{78.18\pm0.00}$	$\mathbf{56.66\pm0.47}$

5.2 Sensitivity Analysis

In this section, we analyze the effect of the ratio of outliers (which determines the value of θ_t). We fix parameters ξ and μ, and set θ_t by searching grid $\{0, 0.01, ... 0.1\}$. For instance, a value of 0.05 means that 5% of data samples with large loss values do not participate in the final iteration of SPMTMVCaC. Then, we plot the clustering results with respect to the parameter θ_t on Task3 of Leaves and Task1 of WebKB, as shown in Fig. 1. From Fig. 1 we can observe that: For the Leaves data set, SPMTMVCaC performs worse when the ratio is 0, which means all the data samples are selected to train the model. As the ratio increases, SPMTMVCaC achieves stable better performance when the ratio is in a wide range [0.1, 1]. This demonstrates the effectiveness of removing the effect of outliers by using the capped-norm method. For the WebKB data set,

Table 4. Results on NUS-WIDE.

Methods	Task1		Task2		Task3		Task4	
	Acc(%)	NMI(%)	Acc(%)	NMI(%)	Acc(%)	NMI(%)	Acc(%)	NMI(%)
k-means	59.39±2.43	46.43±2.70	44.74±3.04	24.62±4.84	43.09±2.71	28.43±2.75	59.86±5.44	41.62±4.93
kk-means	59.39±2.43	46.43±2.70	44.74±3.04	24.62±4.84	43.09±2.71	28.43±2.75	59.86±5.44	41.32±4.93
SNMTF	69.84±4.11	48.73±3.92	54.16±1.72	27.37±1.70	55.85±2.77	35.02±1.23	60.07±2.10	41.43±2.89
CoRe	61.11±0.99	45.28±0.44	51.73±2.75	27.19±1.31	52.32±0.70	31.89±0.08	64.55±1.58	46.22±1.33
LSSMTC	68.40±7.07	47.61±6.15	54.59±5.23	27.24±3.38	49.45±2.51	29.47±4.44	64.44±7.83	45.56±5.80
SMTMVC	73.04±7.45	57.06±2.35	58.10±2.17	23.89±5.08	59.74±1.19	37.64±5.18	59.58±6.76	44.49±8.00
SPMTMVCaC	**78.04±2.44**	**58.20±2.89**	**60.29±2.68**	**29.78±2.10**	**60.10±3.26**	**38.95±0.31**	**69.21±3.01**	**49.67±0.89**

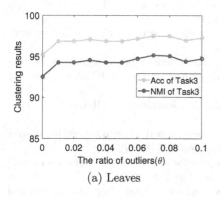
(a) Leaves

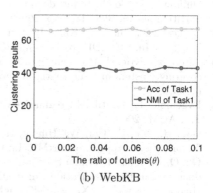
(b) WebKB

Fig. 1. Clustering result of Leaves and WebKB with respect to the setting ratio of outliers (determines the parameter θ_t).

the disparity of the performance in range $[0, 1]$ is small. This may be because that there are few (even no) outliers or noises in Task1 of WebKB data set. Another observation obtained from Fig. 1 is that SPMTMVCaC performs stably in a wide range of the ratio of outliers.

6   Conclusion

In this paper, we propose a novel muti-task muti-view clustering model by integrating self-paced learning strategy and capped-norm. The data samples are gradually selected to train the model from simplicity to complexity, which inherits the advantages of SPL that has been verified to be beneficial in avoiding bad local minima in non-convex problems. A novel capped-norm term is designed to significantly reduces the negative effect of outliers and noises, and to further enhance the clustering performance. We adopt an efficient alternating optimization method to solve the proposed model. Computational complexity analysis is given and experimental results on real benchmark data sets demonstrate the effectiveness and robustness of the proposed method.

Acknowledgments. This paper was in part supported by Grants from the Natural Science Foundation of China (Nos. 61806043, 61572111, and 61872062), a Project funded by China Postdoctoral Science Foundation (No. 2016M602674), a 985 Project of UESTC (No. A1098531023601041), and two Fundamental Research Funds for the Central Universities of China (Nos. ZYGX2016J078 and ZYGX2016Z003).

References

1. Ankerst, M., Breunig, M.M., Kriegel, H.P., Sander, J.: OPTICS: ordering points to identify the clustering structure. In: SIGMOD, pp. 49–60 (1999)
2. Comaniciu, D., Meer, P.: Mean shift: a robust approach toward feature space analysis. PAMI **24**(5), 603–619 (2002)
3. Dhillon, I.S.: Co-clustering documents and words using bipartite spectral graph partitioning. In: SIGKDD, pp. 269–274 (2001)
4. Dhillon, I.S., Guan, Y., Kulis, B.: Kernel k-means: spectral clustering and normalized cuts. In: SIGKDD, pp. 551–556 (2004)
5. Ester, M., Kriegel, H.P., Sander, J., Xu, X.: A density-based algorithm for discovering clusters in large spatial databases with noise. In: SIGKDD, pp. 226–231 (1996)
6. Evgeniou, T., Pontil, M.: regularized multi-task learning. In: SIGKDD, pp. 109–117. ACM (2004)
7. Gao, H., Nie, F., Cai, W., Huang, H.: Robust capped norm nonnegative matrix factorization: capped norm NMF. In: CIKM, pp. 871–880 (2015)
8. Gu, Q., Zhou, J.: Learning the shared subspace for multi-task clustering and transductive transfer classification. In: ICDM, pp. 159–168 (2009)
9. Huang, S., Kang, Z., Xu, Z.: Self-weighted multi-view clustering with soft capped norm. Knowl. Based Syst. **158**, 1–8 (2018)
10. Huang, S., Ren, Y., Xu, Z.: Robust multi-view data clustering with multi-view capped-norm k-means. Neurocomputing **311**, 197–208 (2018)
11. Huang, S., Wang, H., Li, T., Li, T., Xu, Z.: Robust graph regularized nonnegative matrix factorization for clustering. Data Min. Knowl. Discov. **32**(2), 483–503 (2018)
12. Huang, S., Xu, Z., Lv, J.: Adaptive local structure learning for document co-clustering. Knowl. Based Syst. **148**, 74–84 (2018)
13. Jiang, L., Meng, D., Mitamura, T., Hauptmann, A.G.: Easy samples first: self-paced reranking for zero-example multimedia search. In: ACM MM, pp. 547–556 (2014)
14. Jiang, L., Meng, D., Zhao, Q., Shan, S., Hauptmann, A.G.: Self-paced curriculum learning. In: AAAI, pp. 2694–2700 (2015)
15. Kang, Z., Lu, X., Yi, J., Xu, Z.: Self-weighted multiple kernel learning for graph-based clustering and semi-supervised classification. In: IJCAI, pp. 2312–2318 (2018)
16. Kang, Z., Peng, C., Cheng, Q., Xu, Z.: Unified spectral clustering with optimal graph. In: AAAI (2018)
17. Kaufman, L., Rousseeuw, P.: Clustering by means of medoids. Faculty of Mathematics and Informatics (1987). http://books.google.com/books?id=HK-4GwAACAAJ
18. Kim, H.K., Kim, H., Cho, S.: Bag-of-concepts: comprehending document representation through clustering words in distributed representation. Neurocomputing **266**, 336–352 (2017)
19. Kumar, A., Rai, P.: Co-regularized multi-view spectral clustering. In: ICONIP, pp. 1413–1421 (2011)

20. Kumar, M.P., Packer, B., Koller, D.: Self-paced learning for latent variable models. In: NIPS, pp. 1189–1197 (2010)
21. Liu, B., et al.: Learning from semantically dependent multi-tasks. In: IJCNN, pp. 3498–3505 (2017)
22. MacQueen, J.: Some methods for classification and analysis of multivariate observations. In: Proceedings of the 5th Berkeley Symposium on Mathematical Statistics and Probability, pp. 281–297 (1967)
23. Pi, T., et al.: Self-paced boost learning for classification. In: IJCAI, pp. 1932–1938 (2016)
24. Que, X., Ren, Y., Zhou, J., Xu, Z.: Regularized multi-source matrix factorization for diagnosis of Alzheimer's disease. In: Liu, D., Xie, S., Li, Y., Zhao, D., El-Alfy, E.S. (eds.) Neural Information Processing. ICONIP 2017. LNCS, vol. 10634, pp. 463–473. Springer, Cham (2017). https://doi.org/10.1007/978-3-319-70087-8_49
25. Ren, Y.: Big data clustering and its applications in regional science. In: Schintler, L.A., Chen, Z. (eds.) Big Data for Regional Science, pp. 257–264. Routledge, London (2018). chap. 21
26. Ren, Y., Domeniconi, C., Zhang, G., Yu, G.: Weighted-object ensemble clustering. In: ICDM, pp. 627–636 (2013)
27. Ren, Y., Domeniconi, C., Zhang, G., Yu, G.: A weighted adaptive mean shift clustering algorithm. In: SDM, pp. 794–802 (2014)
28. Ren, Y., Domeniconi, C., Zhang, G., Yu, G.: Weighted-object ensemble clustering: methods and analysis. KAIS **51**(2), 661–689 (2017)
29. Ren, Y., Hu, X., Shi, K., Yu, G., Yao, D., Xu, Z.: Semi-supervised DenPeak clustering with pairwise constraints. In: Geng, X., Kang, B.-H. (eds.) PRICAI 2018. LNCS (LNAI), vol. 11012, pp. 837–850. Springer, Cham (2018). https://doi.org/10.1007/978-3-319-97304-3_64
30. Ren, Y., Kamath, U., Domeniconi, C., Zhang, G.: Boosted mean shift clustering. In: Calders, T., Esposito, F., Hüllermeier, E., Meo, R. (eds.) ECML PKDD 2014. LNCS (LNAI), vol. 8725, pp. 646–661. Springer, Heidelberg (2014). https://doi.org/10.1007/978-3-662-44851-9_41
31. Ren, Y., Que, X., Yao, D., Xu, Z.: Self-paced multi-task clustering. arXiv preprint arXiv:1808.08068 (2018)
32. Ren, Y., Zhao, P., Sheng, Y., Yao, D., Xu, Z.: Robust softmax regression for multi-class classification with self-paced learning. In: IJCAI, pp. 2641–2647 (2017)
33. Ren, Y., Zhao, P., Xu, Z., Yao, D.: Balanced self-paced learning with feature corruption. In: IJCNN, pp. 2064–2071 (2017)
34. Strehl, A., Ghosh, J.: Cluster ensembles - a knowledge reuse framework for combining multiple partitions. JMLR **3**, 583–617 (2002)
35. Tang, K., Ramanathan, V., Li, F.F., Koller, D.: Shifting weights: adapting object detectors from image to video. In: NIPS, pp. 647–655 (2012)
36. Wang, H., Nie, F., Huang, H., Makedon, F.: Fast nonnegative matrix tri-factorization for large-scale data co-clustering. In: IJCAI, pp. 1553–1558 (2011)
37. Xie, P., Xing, E.P.: Integrating image clustering and codebook learning. In: AAAI, pp. 1903–1909 (2015)
38. Zhang, X., Zhang, X., Liu, H.: Multi-task multi-view clustering for non-negative data. In: IJCAI, pp. 4055–4061 (2015)
39. Zhang, X., Zhang, X., Liu, H., Liu, X.: Multi-task multi-view clustering. TKDE **28**(12), 3324–3338 (2016)
40. Zhu, H., Pan, X., Xie, Q.: Merging students-t and Rayleigh distributions regression mixture model for clustering time-series. Neurocomputing **266**, 247–262 (2017)

Shape Clustering as a Type of Procrustes Analysis

Kazunori Iwata[✉][iD]

Graduate School of Information Sciences, Hiroshima City University, 3-4-1,
Ozuka-Higashi, Asaminami-ku, Hiroshima 731-3194, Japan
kiwata@hiroshima-cu.ac.jp

Abstract. The ordinary Procrustes sum of squares is one of the most important measures in Procrustes analysis of shape. In this paper, we incorporate a competitive learning scheme into Procrustes analysis. We introduce a measure of distance between the landmarks of shapes for a competitive learning scheme. Thus, we present novel shape clustering as a type of Procrustes analysis. Using datasets of line drawings and outlines, we show that shape clustering performs well for shape classification compared with shape clustering founded on typical vector-based distances.

Keywords: Shape clustering · Procrustes analysis
Competitive learning

1 Introduction

A competitive learning scheme is an effective clustering strategy, and hence has been applied to classification systems, such as learning vector quantization and self-organizing map networks [6,8]. This scheme relies on the distance between a pattern and prototypes that represent cluster centers of patterns, and modifies prototypes stepwise for each pattern. Accordingly, it is important for the scheme to choose an appropriate measure of distance for the respective application domain. Most typical measures focus on patterns as numerical vectors.

In statistical shape analysis, Procrustes analysis has been a practical tool for biological morphology, image analysis, and bioinformatics [1,12]. In Procrustes analysis, the shape of an object is represented by several points called landmarks. A landmark is a point of correspondence that matches between and within classes of similar shapes. Some measures of distance between landmarks on a shape and those on another shape exist to determine whether the two shapes are similar. The ordinary Procrustes sum of squares (OSS) [1] is one of the most important measures in Procrustes analysis. It is obtained by matching the landmarks on a shape with those on another shape as closely as possible over the similarity transformations that consist of translation, rotation, and isotropic scaling. The OSS is a measure of distance between the matrices of landmarks but not numerical vectors.

ⓒ Springer Nature Switzerland AG 2018
L. Cheng et al. (Eds.): ICONIP 2018, LNCS 11304, pp. 218–227, 2018.
https://doi.org/10.1007/978-3-030-04212-7_19

In this paper, we incorporate a competitive learning scheme into Procrustes analysis. Specifically, for the scheme, we use the OSS as a measure of distance between the landmarks of shapes. As a result, we present novel shape clustering as a type of Procrustes analysis. Using datasets of line drawings and outlines, we show that shape clustering performs well for shape classification compared with that founded on typical vector-based distances.

This paper is organized as follows: We briefly describe Procrustes analysis in Sect. 2 and combine it with a competitive learning scheme in Sect. 3. Using several datasets, we compare shape clustering based on the OSS with that based on typical vector-based distances in Sect. 4. In Sect. 5, we provide our summary.

2   Preliminaries

We begin with a representation of the configuration of k landmarks in a vector space of m dimensions.

2.1   Configuration Matrix

Let $\mathbb{R}$ be the set of real numbers and $\mathbb{R}^+$ the set of positive real numbers. Let I_k denote the $k \times k$ identity matrix and 1_k the k-dimensional vector of ones. With the coordinates of k landmarks in m dimensions, $\left(x_1^{(1)}, \ldots, x_m^{(1)}\right), \ldots, \left(x_1^{(k)}, \ldots, x_m^{(k)}\right)$, we compose a $k \times m$ matrix as follows:

$$\begin{pmatrix} x_1^{(1)} \cdots x_m^{(1)} \\ \vdots \quad \cdots \quad \vdots \\ x_1^{(k)} \cdots x_m^{(k)} \end{pmatrix}, \tag{1}$$

and call this a configuration matrix [1]. Henceforth, X denotes a configuration matrix. To align the center of gravity of k landmarks to the origin, we define a $k \times k$ matrix given by

$$C = I_k - \frac{1}{k} 1_k 1_k{}^{\mathrm{T}}, \tag{2}$$

and call this a centering matrix. Accordingly, the rows of CX imply the coordinates of the aligned k landmarks. For any configuration matrix X, its Frobenius norm is expressed as

$$\|X\|_{\mathrm{F}} = \sqrt{\mathrm{Tr}\left(X^{\mathrm{T}}X\right)}, \tag{3}$$

where X^{T} is the transpose of X. Because $X^{\mathrm{T}}X$ becomes a square matrix, $\mathrm{Tr}\left(X^{\mathrm{T}}X\right)$ denotes its trace. Similarly, its max norm is expressed as

$$\|X\|_{\max} = \max_{i,j} \left\{ x_j^{(i)} \right\}, \tag{4}$$

that is, the maximum component of X. Essentially, the square Frobenius norm refers to measuring the distance between configuration matrices with an ordinary square Euclidean norm, regarding a $k \times m$ configuration matrix as a vector in $k \times m$ dimensions. Similarly, the max norm corresponds to the max norm of a vector. Hence, in this paper, they are referred to as vector-based distances.

2.2   Ordinary Procrustes Sum of Squares

In Procrustes analysis, the OSS is one of the most important distances between centered configuration matrices. For any pair of centered configuration matrices $CX_1, CX_2 \in \mathbb{R}^{k \times m}$, the OSS between CX_1 and CX_2 is defined as

$$\text{OSS}\,(CX_1, CX_2) = \inf_{\Gamma \in \text{SO}(m),\, \gamma \in \mathbb{R}^m,\, \beta \in \mathbb{R}^+} \left\| CX_2 - \beta CX_1 \Gamma - 1_k \gamma^{\mathrm{T}} \right\|_{\mathrm{F}}^2, \qquad (5)$$

where $\text{SO}(m)$ denotes the m-dimensional rotation group, β denotes a scaling factor, and γ denotes a transition vector in vector space. Briefly, this is the closest distance between the centered configuration matrices under the action of similarity transformations that consist of translation, rotation, and isotropic scaling. OSS (CX_1, CX_2) can be computed by solving the following problem with respect to (Γ, γ, β):

$$\min \left\| CX_2 - \beta CX_1 \Gamma - 1_k \gamma^{\mathrm{T}} \right\|_{\mathrm{F}}^2$$
$$\text{s.t. } \Gamma \in \text{SO}(m),$$
$$\gamma \in \mathbb{R}^m,$$
$$\beta \in \mathbb{R}^+.$$

Based on [5], the solution to the problem may be analytically expressed as

$$\hat{\gamma} = \mathbf{0} \qquad (6)$$

$$\hat{\Gamma} = UV^{\mathrm{T}} \qquad (7)$$

$$\hat{\beta} = \frac{\text{Tr}\left((CX_2)^{\mathrm{T}} (CX_1) \hat{\Gamma} \right)}{\text{Tr}\left((CX_1)^{\mathrm{T}} (CX_1) \right)}, \qquad (8)$$

where U and V are given by

$$(CX_2)^{\mathrm{T}} (CX_1) = \|CX_1\|\|CX_2\| V \Lambda U^{\mathrm{T}}, \qquad (9)$$

$$U, V \in \text{SO}(m), \qquad (10)$$

where Λ is a diagonal matrix whose diagonal components $\lambda_1, \ldots, \lambda_m$ satisfy $\lambda_1 \geq \lambda_2 \geq \cdots \geq \lambda_{m-1} \geq |\lambda_m|$. In fact, these diagonal components are the square roots of the eigenvalues of $(CX_1)^{\mathrm{T}} (CX_2) (CX_2)^{\mathrm{T}} (CX_1)$, and λ_m becomes the negative square root if and only if $\det\left((CX_1)^{\mathrm{T}} (CX_2) \right) \leq 0$ holds.

2.3   Full Procrustes Mean

The full Procrustes mean [1, Result 7.2] of shapes is an arithmetic mean of configuration matrices of similar shapes. Let $\mathcal{S}$ be a set of configuration matrices of similar shapes. An algorithm to compute the full Procrustes mean for $\mathcal{S}$ is as follows [1, Sect. 7.4.2]:

1. Using centered landmarks, set the full Procrustes mean, written as

$$W \leftarrow \frac{1}{|\mathcal{S}|} \sum_{X \in \mathcal{S}} CX. \tag{11}$$

2. Modify W using

$$W \leftarrow \frac{1}{|\mathcal{S}|} \sum_{X \in \mathcal{S}} \hat{\beta} CX \hat{\Gamma}, \tag{12}$$

where $\hat{\Gamma}$ and $\hat{\beta}$ are defined as

$$\hat{\Gamma} = UV^{\mathrm{T}}, \tag{13}$$

$$\hat{\beta} = \frac{\mathrm{Tr}\left(W^{\mathrm{T}}(CX)\,\hat{\Gamma}\right)}{\mathrm{Tr}\left((CX)^{\mathrm{T}}(CX)\right)}, \tag{14}$$

where U and V are given by

$$W^{\mathrm{T}}(CX) = \|CX\| \|W\| V \Lambda U^{\mathrm{T}}, \tag{15}$$

$$U, V \in \mathrm{SO}(m). \tag{16}$$

3. Repeat step 2 until the difference in the Procrustes sum of squares,

$$\sum_{X \in \mathcal{S}} \left\| W - \hat{\beta} CX \hat{\Gamma} \right\|_{\mathrm{F}}^{2}, \tag{17}$$

cannot be reduced further.

The symbol $\leftarrow$ implies that a variable on the left-hand side is updated by an expression on the right-hand side. Because the full Procrustes mean for $\mathcal{S}$ demonstrates a configuration matrix of the representative shape for $\mathcal{S}$, in a sense, we call this a prototype matrix according to [6].

Thus, we have explained an algorithm to obtain the prototype matrix of labeled data. Before moving to the main part of this paper, we incorporate prototype matrices into a clustering task. Suppose we are given a set of configuration matrices $\mathcal{X}$ that are unlabeled, and want to assign each of them to one of n clusters. Considering n-means clustering, we can readily derive a batch clustering algorithm with n prototype matrices:

1. Choose n configuration matrices $X_1, \ldots, X_n$ from $\mathcal{X}$, and perform $W_i \leftarrow CX_i / \|CX_i\|$ as an initial prototype, for $i = 1, \ldots, n$.
2. Modify W_i using

$$W_i \leftarrow \frac{\widetilde{W}_i}{\left\|\widetilde{W}_i\right\|}, \tag{18}$$

where

$$\widetilde{W}_i = \frac{1}{|\mathcal{X}_i|} \sum_{X \in \mathcal{X}_i} \hat{\beta} CX \hat{\Gamma}, \tag{19}$$

where $\hat{\Gamma}$ and $\hat{\beta}$ are defined similarly according to (13) and (14), respectively, and

$$\mathcal{X}_i = \left\{ X \in \mathcal{X} \,\middle|\, i = \arg\min_{j \in \{1,\dots,n\}} \text{OSS}\,(CX, W_j) \right\}. \tag{20}$$

3. Repeat step 2 until the difference in the OSS,

$$\sum_{i=1}^{n} \sum_{X \in \mathcal{X}_i} \text{OSS}\,(CX, W_i), \tag{21}$$

cannot be reduced further.

We normalized W_i in (18) because elements in different clusters are not always the same size. This n-means-like algorithm is not efficient in terms of space complexity when $|\mathcal{X}|$ is very large. Additionally, when new unlabeled shapes enter, the algorithm requires a large computational effort to recompute the prototype matrices.

3   Main Results

We have described a clustering algorithm to obtain the prototype matrices that correspond to the full Procrustes means of shape clusters. In this section, we develop the algorithm using a competitive learning scheme.

3.1   Competitive Learning Scheme

A competitive learning scheme provides an effective strategy for clustering [6,8]. It is easy to perform clustering online, that is, allow it to treat new unlabeled data sequentially. Then, a large space is not required to store all unlabeled data. More importantly, only a small computational effort is required to modify the prototype matrices for new data.

Now we use the competitive learning scheme for shape clustering. With the OSS and prototype matrices, our clustering algorithm is described as follows:

1. Choose n configuration matrices $X_1, \dots, X_n$ from $\mathcal{X}$, and perform $W_i \leftarrow CX_i / \|CX_i\|$ as an initial prototype, for $i = 1, \dots, n$.
2. Observe configuration matrix X randomly from $\mathcal{X}$.
3. Determine the nearest prototype matrix W_{i^*} to the X, where

$$i^* = \arg\min_{i \in \{1,\dots,n\}} \text{OSS}\,(CX, W_i). \tag{22}$$

4. Modify W_{i^*} using

$$W_{i^*} \leftarrow \frac{\widetilde{W}_{i^*}}{\left\| \widetilde{W}_{i^*} \right\|}, \tag{23}$$

where

$$\widetilde{W}_{i^*} = W_{i^*} - \eta \left. \frac{\partial \text{OSS}\,(CX, W_i)}{\partial W_i} \right|_{W_i = W_{i^*}}, \tag{24}$$

where η is the learning rate between zero and one, and

$$\frac{\partial \mathrm{OSS}\,(CX, W_i)}{\partial W_i} = 2W_i - 2\hat{\beta} C X \hat{\Gamma}. \tag{25}$$

Again, $\hat{\Gamma}$ and $\hat{\beta}$ are defined similarly according to (13) and (14), respectively.
5. Repeat step 4 until the difference in the Procrustes sum of squares in (21) cannot be reduced further.

Equation (25) is derived by differentiating the OSS with respect to a prototype matrix. As space is limited in this paper, we have omitted the details. At step 4, we used a stochastic gradient descent method [11] to move the nearest prototype matrix closer to an observed configuration matrix. This is because performing one modification along with such a stochastic gradient descent method is typically efficient and can also be used to learn online.

3.2 Aim of the Clustering Algorithm

Our clustering algorithm based on the OSS is designed to minimize

$$J\,(W_1, \ldots, W_n) = \frac{1}{|\mathcal{X}|} \sum_{X \in \mathcal{X}} \min_{i \in \{1, \ldots, n\}} \mathrm{OSS}\,(CX, W_i), \tag{26}$$

which, in this paper, is referred to as the free energy. The prototype matrices that minimize the free energy are closely related to the sample full Procrustes mean [1, Definition 6.10], written as

$$\overline{W}_i = \arg\inf_W \frac{1}{|\mathcal{X}_i|} \sum_{X \in \mathcal{X}_i} \mathrm{OSS}\,(CX, W). \tag{27}$$

In fact, assume that each $\mathcal{X}_i$ is invariant during the minimization, that is,

$$\min_{W_1, \ldots, W_n} \sum_{i=1}^{n} \sum_{X \in \mathcal{X}_i} \mathrm{OSS}\,(CX, W_i) = \sum_{i=1}^{n} \min_{W_i} \sum_{X \in \mathcal{X}_i} \mathrm{OSS}\,(CX, W_i). \tag{28}$$

Then, we obtain

$$\min_{W_1, \ldots, W_n} J\,(W_1, \ldots, W_n) = \min_{W_1, \ldots, W_n} \frac{1}{|\mathcal{X}|} \sum_{X \in \mathcal{X}} \min_{i \in \{1, \ldots, n\}} \mathrm{OSS}\,(CX, W_i), \tag{29}$$

$$= \min_{W_1, \ldots, W_n} \frac{1}{|\mathcal{X}|} \sum_{i=1}^{n} \sum_{X \in \mathcal{X}_i} \mathrm{OSS}\,(CX, W_i), \tag{30}$$

$$= \frac{1}{|\mathcal{X}|} \sum_{i=1}^{n} \sum_{X \in \mathcal{X}_i} \mathrm{OSS}\,(CX, \overline{W}_i), \tag{31}$$

and hence energy minimization corresponds to obtaining the sample full Procrustes means of shape clusters. This implies that we can visually check the sample full Procrustes means by plotting the landmarks of their prototype matrices. Such a visual check can be useful in shape analysis.

4   Computational Experiments

We demonstrate that our clustering algorithm is effective in terms of shape classification compared with that for typical vector-based distances.

4.1   Datasets

Line Drawing Dataset. We used a time series dataset of handwritten line drawings [10] that consisted of 90 drawings that were classified into three clusters ($n = 3$). The dataset is available from [9]. Figure 1 shows examples of the line drawings. We located six landmarks manually on each line drawing ($k = 6$, $m = 2$) and then created their 6×2 configuration matrices. In the figure, the landmarks are indicated by dots.

Outline Dataset. To verify our algorithm with various types of shape, we used the ETHZ shape classes dataset [2], which is related to object detection from picture images [3]. This dataset contained the images of ground-truth outlines of 44 apple logos, 55 bottles, 92 giraffes, 48 mugs, and 32 swans processed using the Berkeley natural boundary detector. Accordingly, there were 271 ground-truth outlines that were classified into five clusters ($n = 5$). We binarized and thinned the ground-truth outline images, and then located six landmarks manually on each ground-truth outline ($k = 6$, $m = 2$). Figure 2 shows these examples. We created the images' 6×2 configuration matrices.

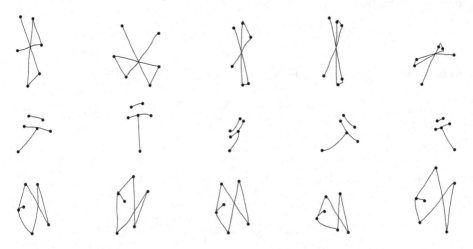

Fig. 1. Example of line drawings: Bluetooth, Japanese postal mark, and fish

4.2   Assessment

We compared our algorithm based on the OSS with the same competitive learning schemes used for shape clustering, based on the square Frobenius

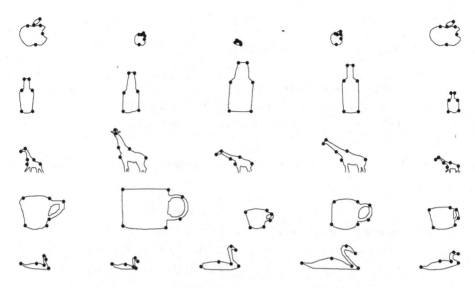

Fig. 2. Example of outlines: apple logo, bottle, giraffe, mug, and swan

norm and max norm, respectively. Specifically, the schemes were obtained using $\|W_i - CX\|_{\mathrm{F}}^2$ and $\|W_i - CX\|_{\mathrm{max}}^2$, respectively, as substitutes for $\mathrm{OSS}(CX, W_i)$ in (24). We ran 1,000 trials, each of which consisted of $450 \, (= 90 \times 5)$ and $542 \, (= 271 \times 2)$ iterations of step 4 of the scheme for the line drawing and outline datasets, respectively. The learning rate was set to $\eta = 0.05$. At each trial, as the initial prototypes, we randomly chose three configuration matrices regarding the line drawing dataset and five regarding the outline dataset. We assessed the resultant clusters using the most popular three measures in the clustering field: entropy, purity, and adjusted rand index (ARI) [4,7]. These measures were calculated using a contingency table of pairs of cluster and answer labels.

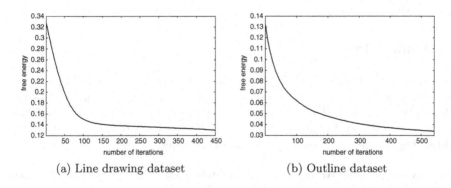

(a) Line drawing dataset   (b) Outline dataset

Fig. 3. Free energy transition averaged over 1,000 trials

Table 1. Clustering assessments using three measures in the line drawing dataset

Norm	Measure		
	Entropy	Purity	ARI
Square Frobenius norm	0.81 (0.09)	0.57 (0.08)	0.17 (0.10)
Max norm	0.89 (0.05)	0.50 (0.05)	0.09 (0.06)
OSS	0.09 (0.18)	0.93 (0.13)	0.89 (0.22)

Table 2. Clustering assessments using three measures in the outline dataset

Norm	Measure		
	Entropy	Purity	ARI
Square Frobenius norm	0.13 (0.11)	0.91 (0.11)	0.83 (0.17)
Max norm	0.31 (0.07)	0.77 (0.07)	0.57 (0.09)
OSS	0.10 (0.14)	0.91 (0.11)	0.85 (0.20)

4.3   Results

Figure 3 shows the free energies of our algorithm based on the OSS that were averaged over 1,000 trials. We see from the figures that the free energy improved steadily as the number of iterations increased. Hence, we infer that the prototypes moved to good positions in terms of the classification of configuration matrices. Next, Tables 1 and 2 show the clustering assessments using the three measures that were also averaged over 1,000 trials. Their standard deviations are shown in parentheses. Briefly, the lower the entropy the better, the higher the purity the better, and the higher the ARI the better. We confirm that our algorithm based on the OSS was more effective regardless of the type of shape dataset. Particularly, the OSS worked well for the line drawing dataset because some line drawings were substantially rotated and uneven in size. Accordingly, we infer that the OSS worked well for clustering the configuration matrices of various shapes.

5   Conclusion

By building the OSS into a competitive learning scheme for clustering strategy, we presented novel shape clustering as a type of Procrustes analysis of shape. With the experimental results on the shape datasets, we demonstrated that shape clustering performed well for shape classification compared with shape clustering founded on typical vector-based distances.

Acknowledgments. We wish to thank Yoshitaka Toda for his support with the experiments. We thank Maxine Garcia, PhD, from Edanz Group (www.edanzediting.com/ac) for editing a draft of this manuscript. This work was supported in part by JSPS KAKENHI Grant Number JP16K00308 and Hiroshima City University TOKUTEI-KENKYUHI Grant Number 1150347.

References

1. Dryden, I.L., Mardia, K.V.: Statistical Shape Analysis with Applications in R. Wiley Series in Probability and Statistics, 2nd edn. Wiley, Chichester (2016)
2. Ferrari, V.: ETHZ Shape Classes, September 2009. http://www.vision.ee.ethz.ch/datasets/index.en.html
3. Ferrari, V., Jurie, F., Schmid, C.: From images to shape models for object detection. Int. J. Comput. Vis. **87**(3), 284–303 (2010)
4. Hubert, L., Arabie, P.: Comparing partitions. J. Classif. **2**(1), 193–218 (1985)
5. Kent, J.T., Mardia, K.V.: Shape, procrustes tangent projections and bilateral symmetry. Biometrika **88**(2), 469–485 (2001)
6. Kohonen, T.: Self-Organizing Maps. Springer Series in Information Sciences, vol. 30. Springer, Heidelberg (2001). https://doi.org/10.1007/978-3-642-56927-2
7. Kuncheva, L.I., Hadjitodorov, S.T.: Using diversity in cluster ensembles. In: Proceedings of the IEEE International Conference on Systems, Man and Cybernetics, pp. 1214–1219. IEEE, The Hague, October 2004
8. Likas, A.: A reinforcement learning approach to online clustering. Neural Comput. **11**(8), 1915–1932 (1999)
9. Okamoto, T., Iwata, K.: Data set of skewed line drawings, September 2017. http://www.prl.info.hiroshima-cu.ac.jp/~kiwata/okamotoiwata/
10. Okamoto, T., Iwata, K., Suematsu, N.: Extending the full procrustes distance to anisotropic scale in shape analysis. In: Proceedings of the 4th Asian Conference on Pattern Recognition, pp. 634–639. IAPR, Nanjing, December 2017. https://doi.org/10.1109/ACPR.2017.139
11. Ruder, S.: An overview of gradient descent optimization algorithms, June 2017. http://ruder.io/optimizing-gradient-descent/
12. Zelditch, M.L., Swiderski, D.L., Sheets, H.D.: Geometric Morphometrics for Biologists: A Primer, 2nd edn. Elsevier Academic Press, Amsterdam (2012)

Classification

Aspect-Level Sentiment Classification with Conv-Attention Mechanism

Qian Yi[1,3], Jie Liu[1,2], Guixuan Zhang[1,2(✉)], and Shuwu Zhang[1,2]

[1] Beijing Engineering Research Center of Digital Content Technology,
Institute of Automation, Chinese Academy of Sciences, Beijing, China
{yiqian2016,jie.liu,guixuan.zhang,shuwu.zhang}@ia.ac.cn
[2] Advanced Innovation Center for Future Visual Entertainment, Beijing, China
[3] University of Chinese Academy of Sciences, Beijing, China

Abstract. The aim of aspect-level sentiment classification is to identify the sentiment polarity of a sentence about a target aspect. Existing methods model the context sequence with recurrent network and
employ attention mechanism to generate aspect-specific representations.
In this paper, we introduce a novel mechanism called Conv-Attention,
which can model the sequential information of context words and generate the aspect-specific attention at the same time via a convolution
operation. Based on the new mechanism, we design a new framework
for aspect-level sentiment classification called Conv-Attention Network
(CAN). Compared to the previous attention-based recurrent models, the
Conv-Attention Network can compute much faster. Extensive experimental results show that our model achieves the state-of-the-art performance
while saving considerable time in model training and inferring.

Keywords: Sentiment analysis · Convolution neural network
Attention

1 Introduction

Aspect-level sentiment classification is a fine-grained task in sentiment analysis
[1]. Given an aspect and its context, the aim of aspect-level sentiment classification is to figure out the sentiment polarity (e.g. positive, negative, neutral) of
the context towards the aspect. For example, for the same context sentence *"So
what if the laptops look chic and cool? The after sales support is terrible."*, when
the target aspect is *"laptops"*, its polarity is positive. But for the aspect *"after
sales support"*, the same sentence has the opposite sentiment polarity.

In this task, the sentiment polarity depends on both context and aspect. In
order to figure out the sentiment of an aspect in its context, we need to obtain an
appropriate aspect-specific context representation. So, abstracting the semantic
information through context words and modeling the correlation between aspect
and context are two critical problems in aspect-level sentiment classification task.
Recurrent Neural Networks (RNNs) perform well in modeling semantic information in a variety of NLP tasks. TD-LSTM separates the sentence into left and

© Springer Nature Switzerland AG 2018
L. Cheng et al. (Eds.): ICONIP 2018, LNCS 11304, pp. 231–243, 2018.
https://doi.org/10.1007/978-3-030-04212-7_20

right part of target, and feeds the two parts into two long short-term memory (LSTM) networks to model the context [2]. Attention mechanism, which can enforce the model to pay more attention to the words related to the target aspect is widely used to model the correlation between aspect and its context. Memory Network adopts attention mechanism to obtain the aspect-specific context representation without considering the sequential information through the context words [3]. Some models combine RNN and attention to obtain better representations. RAM network uses Bidirectional LSTM to generate context memory and combines the multiple attention with Gated Recurrent Unit (GRU) [4].

Although the models based on RNNs and attention have achieved good performance, they still have two main shortcomings. Firstly, RNNs need to maintain the hidden states of the entire past, so they cannot compute in parallel within a sentence, which makes these RNN-based models less efficient. In addition, the existing attention mechanism only considers the target word when computing the attention scores, which may cause errors in some cases. For example, in the sentence *"The taste of the dessert was not delicious enough."*, the word *"delicious"* itself has positive sentiment. But in this case, the polarity for the aspect *"taste"* should be negative. The polarity of *"delicious"* can be reversed only when considering the word *"not"* next to it.

In this paper, we propose a novel mechanism called Conv-Attention to address these two problems. This mechanism can integrate sequential information into the attention scores by convolution operation which can parallel well over a sentence. And in this mechanism, the attention paid on a word depends not only on the word itself but also on the words before and after it. Based on the Conv-Attention, we design a multilayer Conv-Attention network for the aspect-level sentiment classification. The multilayer architecture can capture more fine-grained semantic information.

The contributions of our work are summarized as follows:

- We propose a new mechanism called Conv-Attention. It can integrate the sequential information into the attention scores to generate an aspect-specific context representation by convolution operation,which can parallel well through a sentence.
- We design a model for aspect-level sentiment classification called Conv-Attention Network. The multilayer Conv-Attention Network can capture more fine-grained information for aspect-level sentiment classification.
- Extensive experimental results show that our model achieves the state-of-the-art performance while saving considerable time in model training and computing.

2   Related Work

Before the deep learning techniques are widely used, researchers adopt conventional method for aspect sentiment classification. Jiang et al. [5] and Mohammad et al. [6] extract some sentiment and semantic features like sentiment lexicon and bag of words, and then use SVM classifier to identify the sentiment polarity. Deng and Wiebe [7] use Probabilistic Soft Logic Model to analyze the sentiment of an entity. Ganapathibhotla et al. [8] employ rule-based method to identify the sentiment of an entity in comparative sentences. Although these methods have achieved good performance, they either highly depend on the quality of handcraft features or need massive extra resources.

In recent years, neural network methods are employed for aspect sentiment analysis [2,9,10]. Neural network can automatically generate features that are suitable for the task through back propagation. However, different from sentiment analysis for sentence or document, aspect-level sentiment classification is a fine-grained task, in which the aspect information plays an important role. Dong et al. [9] adjust Recursive Neural Network [10] which has been used for sentence sentiment analysis to predict sentiment for an aspect. TD-LSTM [2] employs two LSTMs to separately model the context words before and after the aspect.

Attention mechanism [11], which has achieved great performance in a variety of fields in NLP, is then widely used in aspect-level sentiment analysis. Wang et al. [12] combine attention with LSTM to generate the context representation. Memory Network [3] employs multilayer attention network to predict the sentiment polarity. IAN [13] uses the attention of aspect words to generate a representation for an aspect and use the aspect vector to help the classification. RAM [4] uses nonlinear unit to combine the results of multiple-attention layers to obtain the features. Ma et al. [14] improve the performance by embedding commonsense knowledge into a LSTM with hierarchical attention mechanism.

Convolutional Neural Network (CNN), which is widely used in computer vision, has been gradually applied to some NLP tasks. Kim [15] applies CNN to text classification. Tang et al. [16] use CNN and LSTM to design a hierarchical architecture for document sentiment classification. The ConvS2S [17], a model based entirely on convolutional neural network, achieves the state-of-the-art performance in machine translation while consuming much less time. Because the filter can attend to several words at once, the features abstracted by CNN are thought to be like the n-gram information. Inspired by this and attention mechanism, we designed the Conv-Attention unit.

3   Model

In this section, we describe the architecture of the Conv-Attention Network and the mechanism of Conv-Attention in detail. The overall architecture of our model is shown in Fig. 1(a).

3.1   The Multilayer Attention Architecture

The input of the model is a context consisting of n words $S = \{s_1, s_2, ..., s_i, ..., s_n\}$ and an aspect s_i. For example, *"So what if the laptops look chic and cool? The after sales support is terrible."* is the context and *"laptops"* and *"after sales support"* are aspects. For sentence with more than one aspect,

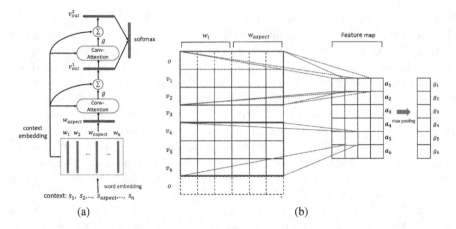

Fig. 1. (a) The overall architecture of a 2-layer CAN; (b) An illustration of Conv-Attention unit. The boxes with thick edges on the left represent the convolution filter.

we input the sentence with only one aspect into the network at a time. In order to represent a word, we map each word s into a low dimensional, continuous and real-valued vector w, which is called word embedding [18]. We lookup the embedding w from the embedding matrix $L \in \mathbb{R}^{d \times |V|}$, in which d is the dimension of word vector and $|V|$ is the vocabulary size. So the context can be represented as $W = \{w_1, w_2, ..., w_i, ..., w_n\}$. The embedding matrix can be trained as parameters of the model or directly adopts the pre-trained vectors generated by unsupervised method like GloVe or Word2Vec [18,19].

To extract the representation of the context for a specific aspect, we use a multilayer attention network. In the first layer, we regard the aspect vector w_{aspect} as the input (we average the word vectors for aspect consisting of more than one word), and use Conv-Attention to generate the attention g_i of each context word.

$$g = CovA(W; w_{aspect}) \tag{1}$$

The symbol $CovA$ represents the Conv-Attention operation. We will elaborate it later in Sect. 3.2. Then, we feed the attention $g = \{g_1, g_2, ..., g_i, ..., g_n\}$ to a softmax function to generate the importance score α_i

$$\alpha_i = \frac{exp(g_i)}{\sum_{j=0}^{n} exp(g_j)}, \tag{2}$$

where n is the length of context. The output vector is computed as a weighted sum of context words

$$v_{out} = \sum_{i=0}^{n} \alpha_i w_i. \tag{3}$$

The output is fed into the next layer like the w_{aspect} in the first layer. We repeat these steps multiple times, and in each layer we obtain an output vector v^k_{out}.

The symbol k is the index of the layer. After obtaining all output vectors, we concatenate them as $V_{out} = \{v^1{}_{out}, v^2{}_{out}, ..., v^i{}_{out}, ..., v^m{}_{out}\}$, where m is the number of layers. The final representation vec, is generated by a fully connected layer

$$vec = tanh(W_l \cdot V_{out} + b_l), \qquad (4)$$

where $W_l \in \mathbb{R}^{d_{out} \times d \cdot m}$, $b_l \in \mathbb{R}^{d_{out} \times 1}$ and d_{out} is the dimension of final features.

In order to train the model in an end-to-end way, we add a *softmax* layer to convert the representation vector to conditional probability distribution $\hat{y}$.

$$\hat{y} = softmax(W_c \cdot vec + b_c), \qquad (5)$$

where $W_c \in \mathbb{R}^{|C| \times d_{out}}$, $b_c \in \mathbb{R}^{|C| \times 1}$ and $|C|$ is the class number.

Because the output vector of each layer can be considered as a representation vector of the context, the attention operation after second layer can be regarded as a kind of self-attention. Self-attention will emphasize more on words corresponding to the semantic of the whole context. So the multilayer architecture enables the model to obtain more fine-grained semantic information.

3.2 Conv-Attention Mechanism

In this part, we will describe the mechanism of Conv-Attention in detail. An illustration of Conv-Attention mechanism is given in Fig. 1(b).

The aspect vector is $w_{aspect} \in \mathbb{R}^d$, and the context is $W = \{w_1, w_2, ..., w_i, ..., w_n\}$. To generate the attention score of each word in context, we first concatenate each word vector with the aspect vector

$$v_i = w_i \oplus w_{aspect}. \qquad (6)$$

The $\oplus$ represents the concatenating operation ,therefore the dimension of v_i is $2d$. The context-aspect matrix can be expressed as $V = \{v_1, v_2, ..., v_i, ..., v_n\}$. For a more concise expression, let $V_{i:i+j}$ refer to the concatenation of $v_i, v_{i+1}, ..., v_{i+j}$. The filter that is applied to a window of h words is $W_f \in \mathbb{R}^{2d \times h}$. Because the attention score generated by one convolution operation corresponds to the center word in the filter, the length of the filter is usually an odd. An attention feature a_i corresponding to the word w_i is generated from a window of context-aspect vectors $V_{i-(h-1)/2:i+(h-1)/2}$ by

$$a_i = relu(W_f \otimes V_{i-(h-1)/2:i+(h-1)/2} + b_f). \qquad (7)$$

Here $\otimes$ refers to the convolution operation and $b_f \in \mathbb{R}$ is a bias term. For each context window $V_{i-(h-1)/2:i+(h-1)/2}$, we use k filters of the same size to generate k attention features, so that we get a feature vector $\boldsymbol{a}_j \in \mathbb{R}^k$. To ensure that the length of feature map is consistent with the context length after convolution operation, we concatenate a zero vector $o \in \mathbb{R}^{2d \times (h-1)/2}$ at the beginning and the end of the context-aspect matrix before each convolution operation likes $\{o, v_1, v_2, ..., v_i, ..., v_n, o\}$. Then, the filter is applied to each possible window of words in the context-aspect matrix $\{V_{1:h}, V_{2:l+1}, ..., V_{n:n+h-1}\}$ to produce a feature map

$$A = \{a_1, a_2, ..., a_j, ..., a_n\} \tag{8}$$

where $A \in \mathbb{R}^{n \times k}$.

At last, we apply a max pooling operation over the feature map to get g_i. For each context word w_i, we take the maximum value of each feature vector as its attention $g_i = max\{a_i\}$.

It is obvious that when $h = 1$ and $k = 1$, the Conv-Attention is very similar to the original attention mechanism. So the original attention mechanism can be seen as a special case of the Conv-Attention. The convolution operation captures the semantic information of the words in the filter to generate the attention of the center word. This can integrate the sequential information into the attention scores. So that the Conv-Attention could capture richer sentiment information.

3.3 Model Training

We train our model in an end-to-end way by back propagation. The loss function is the cross-entropy error of sentiment classification with L_2-regularization. The final objective function to be minimized is

$$loss = -\sum_{i=1}^{|C|} y_i \cdot log(\hat{y}_i) + \lambda ||\theta||^2, \tag{9}$$

where i is the index of class and $|C|$ is the class number, $y \in \mathbb{R}^{|C|}$ denotes the ground truth and λ is the L_2-regularization coefficient. θ is the parameter set of the model. We also use the dropout strategy in each fully connected layer to avoid overfitting.

4   Experiments

4.1   Experiment Setting

We conduct experiments on target-based sentiment classification task of SemEval 2014, SemEval2015 and SemEval2016 datasets to evaluate our model. The SemEval 2014 dataset is composed of reviews of Restaurant and Laptop domains, SemEval2015 and SemEval2016 dataset contain reviews of Restaurant domain. We remove a few examples have the "conflict" label, and the remaining reviews are labeled with three sentiment polarities: positive, neutral and negative. The details of the datasets are shown in Table 1 (the R and L in the brackets refer to the restaurant domain and laptop domain).

To evaluate the performance of the model, we adopt two evaluation metrics. The first metric is the accuracy, which is commonly used in classification task. The macro-F1 score is also adopted, because the classes of the dataset is unbalanced.

Table 1. Details of datasets

Dataset	Positive		Neutral		Negative	
	Train	Test	Train	Test	Train	Test
SemEval2014(R)	2164	728	637	196	807	196
SemEval2014(L)	994	341	464	169	870	128
SemEval2015	963	353	36	37	280	207
SemEval2016	1319	483	72	32	489	135

We use 300-dimension word vectors pre-trained by GloVe as the word embedding for all models [19]. All out-of-vocabulary words and weight matrices are randomly initialized by uniform distribution $U(-0.1, 0.1)$, and all biases are set to zero. We set the L_2-regularization coefficient at 5×10^{-5} and dropout rate at 0.5. The dimension of the output representation is 300. Training stops until convergence or maximum epochs. The maximum epoch is set at 100 for all models. We will discuss the specific settings of CAN in Sect. 4.2.

4.2   Impacts of Attention Layer and Filter Size

The layer number and the filter size are two crucial factors that influence the performance of our model. In this section, we analyze the impacts of them. The filter size here refers to the width of filter. Figure 2 shows how the accuracy changes with the layer number and filter size on the four datasets.

The number of attention layers is one important factor that influences the performance of our model. As we can see in Fig. 2, the model performance leaps when the layer number increases from 1 to 2 or from 2 to 3. It indicates that multilayer attention can capture more sentiment information than 1 layer. Then, the increasing becomes slow and the accuracy even starts to slightly decline when the layer number continues growing. We think this is because the increasing of the complexity makes the model less generalizable.

Another factor of the model is the filter size. The width of the filter stands for the number of words that one convolution operation can attend to. For example, when the filter size is 3, for each word, its Conv-Attention weight is related to one word before and after it. This is why the Conv-Attention mechanism can capture the sequential information. As shown in Fig. 2, the classification accuracy of our model first rises and declines with the filter size increasing. This is because the larger filter size can capture longer sequential information. But when the filter size becomes larger, the number of parameters increases, making the model less generalizable and harder to train.

Based on the experimental results and the analysis above, we pick the layer number 4 and filter size 3 as the settings for later experiments.

4.3   Model Comparisons

We compare the CAN model with following models:

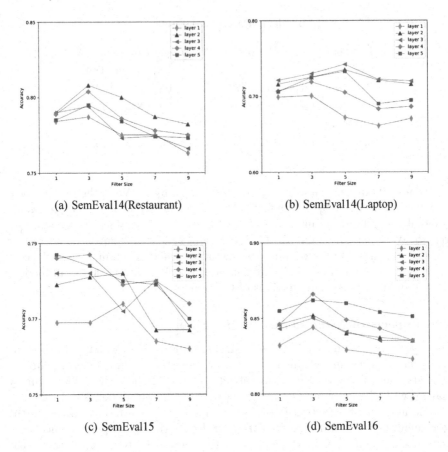

(a) SemEval14(Restaurant)   (b) SemEval14(Laptop)

(c) SemEval15   (d) SemEval16

Fig. 2. Impacts of attention layer and filter size

- Average Context (AC): Using the average of the word vectors of the full context as the representation, which doesn't consider the target aspect.
- LSTM: Feeding the context words into a LSTM network, and using the last hidden states as the representation of the context.
- TD-LSTM [2]: It uses two LSTM to separately model the context words before and after the target aspect. The last hidden states of the two LSTMs are concatenated to predict sentiment polarity.
- Memory Network [3]: Memory Network (Mem) uses multilayer attention network to abstract the information of context words and aspect. It adopts the output of the last attention layer to predict the sentiment polarity.
- IAN [13]: It uses bi-directional LSTM to model both the context and the aspect, and uses attention mechanism to generate aspect-specific context representation and context-specific aspect representation. It further concatenates the two representations as the final representation.
- RAM [4]: It firstly feeds the context words to a bi-directional LSTM, and regards the hidden states as the memory. Then it uses the position weighted

memory to generate multiple attention and a GRU unit to combine the multiple attention results. At last, it uses the output of the GRU for prediction.

Table 2. Model comparisons

Method	SemEval14(R)		SemEval14(L)		SemEval15		SemEval16	
	Acc	M-F1	Acc	M-F1	Acc	M-F1	Acc	M-F1
AC	0.750	0.662	0.676	0.598	0.710	0.459	0.822	0.511
LSTM	0.760	0.654	0.699	0.633	0.752	0.508	0.851	0.542
TD-LSTM	0.773	0.671	0.718	0.667	0.727	0.480	0.789	0.482
Mem	0.785	0.672	0.708	0.663	0.765	0.518	0.835	0.527
IAN	0.786	0.686	**0.727**	0.681	0.770	0.531	0.848	0.537
RAM	0.788	0.683	0.725	0.684	0.772	**0.577**	0.846	0.538
CAN	**0.804**	**0.715**	**0.727**	**0.689**	**0.789**	0.532	**0.868**	**0.555**

Experimental results are given in Table 2. As Table 2 shows, our method outperforms all the contrast models on all datasets. AC has the worst performance. Because it just averages the context words, which doesn't consider neither the correlation between the context and aspect or the sequential information of context words.

Although the LSTM method models the sequential information of the context with long short-term memory network, it doesn't make full use of the target information. So its performance is not as good as other neural network models. The TD-LSTM takes the aspect word into consideration. But it simply thinks that the word besides the aspect is more important. So some important words that are far from the aspect word may be ignored, which makes it doesn't perform well enough.

The memory network adopts the similar multilayer architecture like us and uses attention mechanism. But our model steadily outperforms it on all datasets. It is because it ignored the sequential information. And rather than only use the output of the last layer, we combine the output of every layer to get richer information. Our model also outperforms IAN on three out of the four datasets. This indicates that the Conv-Attention unit can help to generate better features than those attention-based LSTM models.

The RAM network uses complicated architecture to obtain more semantic information. But our model still outperforms it on all datasets. Our method uses Conv-Attention to capture the sequential information and combines the output of each layer with a linear layer, which is more efficient. A more detailed analysis about computing time will be reported in the next part.

Table 3. Computing time comparisons

Method	Train(s)	Test(s)
AC	2.021	0.205
LSTM	31.725	0.489
TD-LSTM	55.815	0.567
Mem	3.168	0.595
IAN	78.172	0.841
RAM	148.103	1.882
CAN	4.362	0.477

4.4 Computing Time

We compare the computing time of all models in the same GPU server. Apart from the architecture of the model, all the factors are set to be same. The computing time of training(one epoch) and testing on the SemEval14(Restaurant) dataset are shown in Table 3. The time is measured in seconds.

From the Table 3, we can observe that the LSTM-based models like LSTM, TD-LSTM, IAN and RAM are much more time consuming than other models. Our model is 10 times faster than most of them in training. This is because the recurrent unit like LSTM and GRU must maintain the hidden state of entire past which prevents parallel computation through a sequence. In contrast to this, our model uses convolution operation which doesn't depend on the results of previous time step and therefore allows parallelism, making the model more efficient. Although the AC method and Memory Network perform efficiently, they haven't modeled the sequential information of the context, which limits their performance.

4.5 The Effect of Conv-Attention

In this section we discuss the effects of Conv-Attention mechanism. In order to confirm the superiority of the mechanism, we compare the performance of Conv-Attention to the original attention mechanism. We compare the classification accuracy of the AC model in Sect. 4.3 with the original attention mechanism and the Conv-Attention mechanism. The results are shown in Table 4.

Table 4. The effect of Conv-Attention mechanism

Method	SemEval14(R)	SemEval14(L)	SemEval15	SemEval16
AC	0.750	0.676	0.710	0.822
Attention	0.761	0.682	0.755	0.833
Conv-Attention	**0.787**	**0.725**	**0.769**	**0.844**

The AC model can be seen as paying equal attention to each context word embedding. The attention model generates attention by $g_i = tanh(w_i \cdot W_a \cdot w_{aspect} + b_a)$, where the $W_a \in \mathbb{R}^{d \times d}$ is the attention weight and $b_a \in \mathbb{R}$ is the bias term. Then the attention g_i is fed to a softmax function to generate the attention scores. Further, the attention scores are used as the weights to compute the weighted sum of the context word embeddings for prediction. The Conv-Attention model uses the same architecture except that the attention scores are generated by Conv-Attention mechanism.

We can observe from the results that Conv-Attention model and the original attention both outperform the AC model, which suggests that adding the aspect information via attention can enhance the performance. Furthermore, the Conv-Attention model consistently outperforms the original attention model on all datasets. This result confirms that the sequential information captured by convolution operation does improve the original attention mechanism, which proves the superiority of the Conv-Attention mechanism.

4.6 Case Study

In this section, we pick an example from the test dataset to visualize the attention computed by the convolutional attention network. Figure 3 shows the attention of context "*I was starving and the small portions were driving me crazy!*" for aspect word "*portions*". It is generated by a two-layers CAN model, and the context's polarity is *negative*. The deeper color indicates the higher attention score.

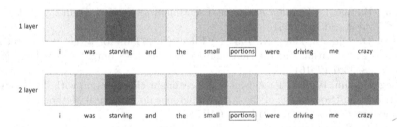

Fig. 3. Case Study: attention weights of a two-layer Conv-Attention network, the deeper color indicates the higher attention weight. (Color figure online)

As shown in Fig. 3, in the first layer, some words like "*starving*" and "*small*", which express a kind of negative sentiment towards the aspect "*portions*", are paid more attention. However, it also emphasizes some meaningless words like "*was*". This is contrary to our intuition that these meaningless words make little contribution to judging the sentiment polarity. In the second layer, the meaningless words are paid less attention. Another important phrase "*driving me crazy*" is now emphasized. As we can see, it cannot be captured completely in the first layer. This illustrates that the higher attention layer can capture more decent information.

The attention visualized in Fig. 3 indicates that our model can correctly emphasize the words which can influence the sentiment for the target aspect. Therefore the final representation can well model the sentiment of a context for the target aspect.

5   Conclusion

In this paper, we propose a novel mechanism called Conv-Attention, which can model the sequential information of context words and generate the aspect-specific attention at the same time through a convolutional operation. Based on Conv-Attention, we design a multilayer framework for aspect-level sentiment classification. Experiment shows that the Conv-Attention can abstract richer information than the original attention and the multilayer architecture can capture more fine-grained semantic information. Our model achieves the state-of-the-art performance on all datasets while using much less time on training and computing.

Acknowledgments. This work has been supported by the National Key R&D Program of China under Grant NO.2017YFB1401000 and the Key Laboratory of Digital Rights Services, which is one of the National Science and Standardization Key Labs for Press and Publication Industry.

References

1. Pang, B., Lee, L.: Opinion mining and sentiment analysis. J. Found. Trends® Inf. Retrieval **2**, 1–135 (2008)
2. Tang, D., Qin, B., Feng, X., Liu, T.: Effective LSTMs for target-dependent sentiment classification. In: International Conference on Computational Linguistics, pp. 3298–3307 (2016)
3. Tang, D., Qin, B., Liu, T.: Aspect level sentiment classification with deep memory network. In: Proceedings of the 2016 Conference on Empirical Methods in Natural Language Processing, pp. 214–224 (2016)
4. Chen, P., Sun, Z., Bing, L., Yang, W.: Recurrent attention network on memory for aspect sentiment analysis. In: Proceedings of the 2017 Conference on Empirical Methods in Natural Language Processing, pp. 452–461 (2017)
5. Jiang, L., Yu, M., Zhou, M., Liu, X., Zhao, T.: Target-dependent twitter sentiment classification. In: Proceedings of the 49th Annual Meeting of the Association for Computational Linguistics, pp. 151–160 (2011)
6. Mohammad, S. M., Kiritchenko, S., Zhu, X.: NRC-Canada: building the state-of-the-art in sentiment analysis of tweets. arXiv preprint arXiv:1308.6242 (2013)
7. Deng, L., Wiebe, J.: Joint prediction for entity/event-level sentiment analysis using probabilistic soft logic models. In: Proceedings of the 2015 Conference on Empirical Methods in Natural Language Processing, pp. 179–189 (2015)
8. Ganapathibhotla, M., Liu, B.: Mining opinions in comparative sentences. In: Proceedings of the 22nd International Conference on Computational Linguistics, pp. 241–248 (2008)

9. Dong, L., Wei, F., Tan, C., Tang, D., Zhou, M., Xu, K.: Adaptive recursive neural network for target-dependent twitter sentiment classification. In: Proceedings of the 52nd Annual Meeting of the Association for Computational Linguistics, pp. 49–54 (2014)
10. Socher, R., et al.: Recursive deep models for semantic compositionality over a sentiment treebank. In: Proceedings of the 2013 Conference on Empirical Methods in Natural Language Processing, pp. 1631–1642 (2013)
11. Bahdanau, D., Cho, K., Bengio, Y.: Neural machine translation by jointly learning to align and translate. arXiv preprint arXiv:1409.0473 (2014)
12. Wang, Y., Huang, M., Zhao, L.: Attention-based LSTM for aspect-level sentiment classification. In: Proceedings of the 2016 Conference on Empirical Methods in Natural Language Processing, pp. 606–615 (2016)
13. Ma, D., Li, S., Zhang, X., Wang, H.: Interactive attention networks for aspect-level sentiment classification. In: International Joint Conference on Artificial Intelligence, pp. 4068–4074 (2017)
14. Ma, Y., Peng, H., Cambria, E.: Targeted aspect-based sentiment analysis via embedding commonsense knowledge into an attentive LSTM. In: AAAI (2018)
15. Kim, Y.: Convolutional neural networks for sentence classification. arXiv preprint arXiv:1408.5882 (2014)
16. Tang, D., Qin, B., Liu, T.: Document modeling with gated recurrent neural network for sentiment classification. In: Proceedings of the 2015 Conference on Empirical Methods in Natural Language Processing, pp. 1422–1432 (2015)
17. Gehring, J., Auli, M., Grangier, D., Yarats, D., Dauphin, Y. N.: Convolutional sequence to sequence learning. In: International Conference on Machine Learning, pp. 1243–1252 (2017)
18. Mikolov, T., Chen, K., Corrado, G., Dean, J.: Efficient estimation of word representations in vector space. arXiv preprint arXiv:1301.3781 (2013)
19. Pennington, J., Socher, R., Manning, C.: Glove: global vectors for word representation. In: Proceedings of the 2014 Conference on Empirical Methods in Natural Language Processing, pp. 1532–1543 (2014)

Attention-Based Combination of CNN and RNN for Relation Classification

Xiaoyu Guo[1], Hui Zhang[1,2], Rui Liu[1(✉)], Xin Ding[1], Runqi Tian[1],
and Bencheng Wang[1]

[1] State Key Laboratory of Software Development Environment, School of Computer
Science and Engineering, Beihang University, Beijing 100191, China
`{guoxiaoyu,hzhang,lr,dasurax,tianrunqi,WangBencheng}@buaa.edu.cn`
[2] Beijing Advanced Innovation Center for Big Data and Brain Computing,
Beijing University, Beijing 100191, China

Abstract. Relation classification is an essential task in natural language
processing (NLP) in order to extract structured data from sentences. In
this paper, we propose a novel model Att-ComNN combining convolutional neural network (CNN) and bidirectional recurrent neural network (RNN) for relation classification. By combining RNN and CNN,
we obtain more accurate context representations of words, which benefits classifying relations. Besides, with both shortest dependency path
(SDP) attention and pooling attention added, this model captures the
most informative context representation for better classification without using other handcrafted features. The results of experiments show
that our model improves the relation classification performance on the
SemEval-2010 Task 8 and outperforms most of previous state-of-the-art
methods, including those depending on much richer forms of handcrafted
features and prior knowledge.

Keywords: Relation classification · Deep neural network
Attention mechanism

1 Introduction

Relation classification is a vital task in natural language processing (NLP) and
aims to identify the semantic relation between two entities in a sentence. Take
the following sentence as an example:

Example 1. Diet fizzy $[drinks]_{e_1}$ cause heart disease and $[diabetes]_{e_2}$.

The marked entities can be denoted as e_1 = "drinks" and e_2 = "diabetes". The
result of relation classification is to recognize that these two entities are of relation Cause-Effect. Because in this sentence, "drinks" is the cause of "diabetes".

In order to improve the performance of relation classification, many traditional methods have been proposed, such as feature-based [7] and kernel-based
[1] approaches. These methods either rely on handcrafted lexical, syntactic and

© Springer Nature Switzerland AG 2018
L. Cheng et al. (Eds.): ICONIP 2018, LNCS 11304, pp. 244–255, 2018.
https://doi.org/10.1007/978-3-030-04212-7_21

semantic features including Part of Speech (POS) tagging, dependency parsing, and paraphrases, or introduce prior knowledge such as WordNet and named entity recognizers (NER). Although they are able to extract representations and structures of sentences, they suffer from poor generalization and error propagation problems [12].

Recently, deep learning methods have shown its feature extraction ability and made considerable improvement in relation classification. By using deep architectures, features of sentences can be automatically extracted without manual intervention, thus a number of neural network (NN) based approaches have achieved better results than traditional methods. Generally, NN-based methods can be divided into three categories: CNN-based [16], RNN-based [14,17] and combination of CNN and RNN [2,8]. Although these models may be effective to exploit features for classifying relations, some of them still need artificial features that take time to elaborately design and suffer from error propagation problems.

In this paper, we propose a novel model Att-ComNN using a combination of convolutional neural network (CNN) and bidirectional recurrent neural network (RNN) with gated recurrent unit (GRU). Different from previous models, we also add two levels of attention to our model. Our chief contributions are as follows:

1. Our model uses a combination of bidirectional RNN and CNN to capture both the sentence level features and word level features. This allows us to take the advantage of the characteristics of these two neural network structures. Apart from these, our model does not need so many extra handcrafted features as other models.
2. We not only introduce a SDP-based attention, but also use a pooling attention to get more accurate classification result. And we observe that both attention mechanisms lead to salient improvements.
3. We find a new way to remove noise in the dataset to some extent. By introducing the SDP information of sentences, we only keep the continuous fragment and remove other parts of sentences that may bring noise into the classification procedures.

2   Related Work

Although there are a few unsupervised methods [3] dealing with relation classification task, the principal methods on relation classification are mainly supervised which can achieve a better result. Over the years, many traditional methods on relation classification have emerged and they can be divided into two categories: feature-based and kernel-based methods. According to feature-based methods, many linguistic features are extracted from sentences and then used to classify the relation between entities [7]. Feature-based methods rely on handcrafted features, thus take time to elaborately design and suffer from error propagation problems. Furthermore, it is hard for such methods to capture structural features of sentences. Kernel-based methods exploit structural information to determine the relation between two entities, such as the dependency tree kernel [1]. Although kernel-based methods can discover the structure between entities, they also suffer from error propagation problems and have limited performance on unseen words.

In order to solve those problems mentioned above, deep neural networks are widely used because of its strong feature extraction ability. Zeng et al. [16] proposed a deep model using CNN with word embeddings and position embeddings as input to extract sentence level features, combined with lexical level features to produce relation classification result by a softmax classifier. To reduce the influence of noise, dos Santos et al. [9] proposed a new convolutional neural network named Classification by Ranking CNN (CR-CNN), which replaced the softmax classifier with a class embedding matrix. Besides, CR-CNN used a pairwise ranking loss function in order to train the model. Also based on CNN, Yu et al. [15] proposed a Combined CNN model with mirror instance (Comb+RMI), while Wang et al. [12] applied entity-specific attention and relation-specific attention to the convolutional layer and pooling layer respectively. The latter one achieved the state-of-art result in SemEval- 2010 Task 8.

Apart from these CNN-based models, recurrent neural network (RNN) models are also popular on relation classification. Zhou et al. [18] proposed a novel neural network Att-BLSTM for relation classification by using Bidirectional Long Short-Term Memory (LSTM) Networks with attention mechanism to capture semantic features of sentences. Like Att-Pooling-CNN model, Att-BLSTM does not utilize any features derived from NLP systems. In addition, Cai et al. [2] combined LSTM with CNN (BRCNN) and introduced the shortest dependency path (SDP) of sentences to keep the key components of sentences. Although BRCNN can capture features more precisely, it still need handcrafted features to train, which will result in error propagation problem. Different from them, our model does not rely on these handcrafted features and achieve a better performance.

3   Proposed Model

In this section, our model will be described in detail. To begin with, we provide an overview of our model in Sect. 3.1. In Sect. 3.2, we present the method of removing data noise, and in Sect. 3.3 we provide reasons and details of using SDP-based attention. Next, Sect. 3.4 deals with how does the pooling attention mechanism works. Finally, Sect. 3.5 describes details of our training objective.

3.1   Structure Overview

Given a sentence, we firstly obtain its key components based on the shortest dependency path (SDP) analysis. After looking up the word embeddings, our model uses bidirectional RNN with gated recurrent unit (GRU) to learn the context representations of these words. The output of RNN is then concatenate

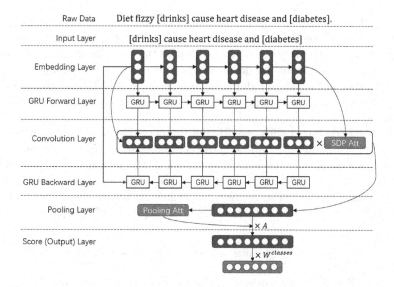

Fig. 1. Overall structure of proposed model

with the original word embeddings to produce the complete word representations. We then apply a convolutional operation to the complete word representations followed by a pooling attention in order to capture the most valuable features in sentences. The output after pooling will be multiplied by a class matrix to calculate scores of each relation class.

The overall structure of our model is shown by Fig. 1. In the remaining part of this section, we will go deep into the model and provide further details about this structure.

3.2 Data Noise Removing Based on SDP

For the purpose of keeping the most important information of sentences, we propose a method to preprocess sentences and remove noise. Given a sentence, we firstly extract its dependency tree. For example, the dependency tree for the example sentence in Sect. 1 is showed in Fig. 2.

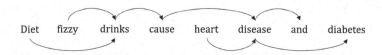

Fig. 2. Dependency tree of the example sentence

According to the dependency tree extracted above, we can move forward a single step further to get the SDP between e_1 = "drinks" and e_2 = "diabetes". Figure 3 shows the SDP of this sentence.

The SDP between two entities contains most informative and representative words for relation classification. It is because (1) if entities are arguments of the same predicate, the SDP between them will pass through the predicate; (2) if two entities belong to different predicate-argument structures that share a common argument, SDP will pass through this argument [2]. Figure 3 shows that the word "cause" is kept, which is the predicate and plays an important role in relation classification task.

But, when compared with the original sentence, the SDP only contains separated words that can not represent the complete meaning of a sentence. As a result, it loses some essential information such as the relation between "drinks" and "disease", which also indicates the relation between entities and helps to classify. Hence, we keep the continuous fragments based on the SDP of sentences. With noise removed, the final result is displayed in Fig. 4.

drinks ⟶ cause ⟶ disease ⟶ diabetes

Fig. 3. Shortest dependency path of the example sentence

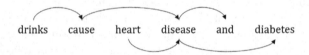

Fig. 4. Final sentence fragment of the example sentence

In this way, we get a smoothed and continuous fragment of sentence whose meaning remains unchanged based on the SDP. Although this kind of extracting key words seems to keep more words than the SDP, the average length of sentences is still shortened by around sixty percent compared with original sentences.

Given a final sentence fragment $F = \{w_1, w_2, ..., w_T\}$, where T is the sentence length, we first get its one-hot vector representation v_t and then transform each word in the fragment into a real-valued vector e_t via looking up the embedding matrix $W^{word} \in \mathbb{R}^{d_w \times V}$. That is:

$$e_t = W^{word} v_t \tag{1}$$

where d_w is the size of word embeddings (a hyper-parameter), and V is the size of vocabulary. Moreover, the embedding matrix W^{word} needs to be learned.

3.3 SDP-Based Attention

For the output e_t of embedding layer, we use bidirectional RNN with GRU [4] to get the context representation of each word. We define the left context of word

w_t as c_t^l, and the right context as c_t^r. Both context representations are vectors and have the same dimension as e_t. (2) to (4) show the calculation of left context representation of word w_t.

$$r_t^l = \text{sigmoid}(W_r^l e_t + U_r^l c_{(t-1)}^l + b_r^l) \tag{2}$$

$$z_t^l = \text{sigmoid}(W_z^l e_t + U_z^l c_{(t-1)}^l + b_z^l) \tag{3}$$

$$c_t^l = (1 - z_t^l) \circ c_{(t-1)}^l + z_t^l \circ \tanh(W_h^l e_t + r_t^l \circ (U_h^l \circ c_{(t-1)}^l) + b_h^l) \tag{4}$$

where W_r, U_r, W_z, U_z, W_h and U_h are weight matrices. r_t and z_t are gates of reset and update respectively. The operator $\circ$ donates the Hadamard product. Similarly, we compute the right context representation c_t^r in the same way as depicted from (5) to (7).

$$r_t^r = \text{sigmoid}(W_r^r e_t + U_r^r c_{(t+1)}^r + b_r^r) \tag{5}$$

$$z_t^r = \text{sigmoid}(W_z^r e_t + U_z^r c_{(t+1)}^r + b_z^r) \tag{6}$$

$$c_t^r = (1 - z_t^r) \circ c_{(t+1)}^r + z_t^r \circ \tanh(W_h^r e_t + r_t^r \circ (U_h^r \circ c_{(t+1)}^r) + b_h^r) \tag{7}$$

Then we define the complete word representation of w_t in (8).

$$w_t^{word} = [(c_t^l)^T, (e_t)^T, (c_t^r)^T]^T \tag{8}$$

where the dimension of c_t^l and c_t^r is same as the dimension of e_t, thus $w_t^{word} \in \mathbb{R}^{3 \cdot d_w}$ Since the significance of words is different from each other, we introduce a SDP-based attention to get more precise word representation. Specifically, we multiply the words in SDP by a higher weight (α_h) and the words not in SDP by a lower weight (α_l). Thus the final word representation is showed by (9) ($S_{indices}$ is used to denote the set containing word index of SDP).

$$w_t^{final} = \begin{cases} \alpha_h \times [(c_t^l)^T, (e_t)^T, (c_t^r)^T]^T & \text{if } t \in S_{indices} \\ \alpha_l \times [(c_t^l)^T, (e_t)^T, (c_t^r)^T]^T & \text{if } t \notin S_{indices} \end{cases} \tag{9}$$

By applying a slide window of size k, we make the complete word representation go through a convolutional neural network in order to get the contextual information of word. Formally, we have

$$R^{final} = [(w_1^{final})^T, (w_2^{final})^T, ..., (w_T^{final})^T]^T \tag{10}$$

$$R^* = \tanh(W_{conv} R^{final} + B_{conv}) \tag{11}$$

where $R^{final} \in \mathbb{R}^{3 \cdot d_w \times T}$ is the representation of the full sentence, and $R^* \in \mathbb{R}^{d_c \times T}$ is the output of convolution layer. In addition, $W_{conv} \in \mathbb{R}^{d_c \times k(3 \cdot d_w)}$ is a weight matrix with a filter size of d_c.

3.4   Attention-Based Pooling

Instead of applying regular max pooling, we introduce an attention-based pooling method to decide the importance of features in R^* (inspired by Wang et al. [12]).

To begin with, we compute a correlation matrix between each window and relation class. Hence, we introduce two matrices U and $W^{classes}$. Learnt by network, matrix U acts as a mapping function, which transforms feature vectors of words into class representations. Besides, for each relation $y \in \mathcal{Y}$, we presume that $W_y^{classes}$ denotes the relation representation of class y, which is trained along with other parameters of network. Combining all the relation representations, we can get an embedding matrix $W^{classes}$ whose columns represent different relation classes. And the correlation matrix G is computed by (12).

$$G = R^{*T}UW^{classes} \tag{12}$$

Then we use a softmax function to to gain the attention pooling matrix A as[1]

$$A_{i,j} = \frac{\exp(G_{i,j})}{\sum_{j^*=1}^{n} \exp(G_{i,j^*})} \tag{13}$$

where $A_{i,j}$ is the (i,j)-th entry of A and $G_{i,j}$ is the (i,j)-th entry of G. This softmax function can amplify the differences between features and stress the most important ones. Then we multiply the attention pooling matrix A with the output of convolution layer R^* and adopt a max operation to select the most prominent features as the output vector. The output vector is calculated as follows in (14).

$$output_i = \max_j (R^*A)_{i,j} \tag{14}$$

where $output_i$ is the i-th entry of $output$ vector and $(R^*A)_{i,j}$ is the (i,j)-th entry of (R^*A).

3.5   Training Objective

We design our loss function based on CR-CNN [9] to reduce the influence of noise. Given a network output, we firstly compute the score for a relation class label $y \in \mathcal{Y}$ by (15).

$$s_\theta(F, y) = output^T [W^{classes}]_y \tag{15}$$

where θ denotes the set of our model parameters. Based on this score function, we can get our pairwise logistic loss function to train our model:

$$\begin{aligned} \mathcal{L} = &\log(1 + \exp(\gamma(m^+ - s_\theta(F, y^+)))) \\ &+ \log(1 + \exp(\gamma(m^- + s_\theta(F, y^-)))) + \beta\|\theta\|^2 \end{aligned} \tag{16}$$

Given a sentence fragment F and its relation class y^+, $s_\theta(F, y^+)$ is the score for its ground truth class, while $s_\theta(F, y^-)$ is the score for a negative class, which is computed by (17).

$$s_\theta(F, y^-) = \arg\max_{y \in \mathcal{Y}, y \neq y^+} s_\theta(F, y) \tag{17}$$

---

[1] The dimension we apply softmax function is different from Wang et al. [12].

As it showed in (17), we choose the highest score among all other classes as the negative score. Back to (16), m^+ and m^- are margins determining scores of both correct and incorrect classes. And γ is a factor to adjust the difference between the margin and score. In order to prevent overfitting, we use $L2$ penalty with the regularization coefficient β. As the training procedure goes, we increase the score of ground truth label. Meanwhile, we decrease the score of negative class label. Finally, the loss function will be convergent.

4 Experiments

4.1 Dataset

To evaluate the performance of our model, we choose the SemEval-2010 Task 8 dataset. The dataset contains 8000 sentences for training and 2717 sentences for testing. And there are ten distinguished relations showed in Table 1.

Table 1. Distinguished relations in dataset

Relation	Train distribution	Test distribution
Cause-Effect	1003 (12.54%)	328 (12.07%)
Component-Whole	941 (11.76%)	312 (11.48%)
Entity-Destination	845 (10.56%)	292 (10.75%)
Product-Producer	717 (8.96%)	231 (8.50%)
Entity-Origin	716 (8.95%)	258 (9.50%)
Member-Collection	690 (8.63%)	233 (8.58%)
Message-Topic	634 (7.92%)	261 (9.61%)
Content-Container	540 (6.75%)	192 (7.07%)
Instrument-Agency	504 (6.30%)	156 (5.74%)
Other	1410 (17.63%)	454 (16.71%)
Total	8000 (100%)	2717 (100.00%)

The first nine relations are directed, while the last "Other" type is undirected. Thus, there are $9 \times 2 + 1 = 19$ different classes for 10 relations. We use the official macro-averaged F_1-score to evaluate the performance of all baseline models and our model (excluding Other).

4.2 Settings

We use stochastic gradient descent (SGD) to update model parameters. To get word embeddings, we train the skip-gram model [6] on Wikipedia. Parameters and matrices are initialized according to the methods proposed by Xavier and Yoshua [5]. Furthermore, the hyper-parameters we select are given in Table 2.

Table 2. Hyper-parameters

Parameter	Parameter meaning	Value
d_w	Dimension of word embeddings	300
d_c	Convolutional output channel size	1000
k	Context window size	1
γ	Scalar factor	2
m^+	Margin to ground truth	2.5
m^-	Margin to negative class	0.5
β	Normalization coefficient	0.0001
λ	Initial learning rate	0.01

There are some hyper-parameters we do not list in Table 2. These parameters are important and need detailed explanation. Firstly, we apply a dropout to the embedding layer in Fig. 1. The dropout rate is 0.2. And we use no dropout on GRU. Secondly, for the SDP-based attention, we assign the value of α_h and α_l according to a proportion of 2:1. Finally, we introduce a hyper-parameter as learning rate decay λ^*. In other words, after some epochs of training approaches, we reduce our initial learning rate by (18).

$$\lambda = \lambda^* \times \lambda (0.1 \leq \lambda^* \leq 0.5) \tag{18}$$

4.3   Results

Table 3 provides a comparison between our model and other models.

The first model [7] of Table 3 is the feature-based model achieving highest performance with $F_1 = 82.8\%$, which is fed with various artificial features. Sorted by the public year, the rest of models in Table 3 are all based on neural networks.

Our model outperforms the basic RNN and CNN models proposed earlier. For example, compared with the well-known CR-CNN model [9], we improve the performance by 2.5%. Since our model can be considered as a combination of RNN and CNN, it is similar to BRCNN model [2]. BRCNN makes use of three types of information to improve the performance, including POS tags, NER features and WordNet, while our model only uses word embeddings to exploit features for relation classification. Although BRCNN model introduces many other features, we get 0.3% higher in F_1 score.

4.4   Analysis

To prove that SDP-based attention and pooling attention can result in improving performance. We conduct an experiment to compare the main model Att-ComNN and other two simplified models. One of them does not contain the SDP-based attention, while the other does not rely on both attention mechanisms. And the results are showed in Table 4.

Table 3. Comparison with public relation classification models

Classifier	Features used	F_1
SVM [7]	POS, prefixes, morphological, WordNet, dependency parse, Levin classed, ProBank, FramNet, NomLex-Plus, Google n-gram, paraphrases, TextRunner	82.2
RNN [11]	Word embeddings	74.8
	+ POS, NER, WordNet	77.6
MVRNN [10]	Word embeddings	79.1
	+ POS, NER, WordNet	82.4
CNN+softmax [16]	Word embeddings	69.7
	+ word position embeddings, WordNet	82.7
SDP-LSTM [14]	Word embeddings	82.4
	+ POS, GR, WordNet	83.7
depLCNN [13]	Word embeddings, WordNet, word around nominals	83.7
	+ negative sampling from NYT dataset	85.6
CR-CNN [9]	Word embeddings	82.8
	+ word position embeddings	84.1
Att-BLSTM [18]	Word embeddings, word position indicators	84.0
Comb+RMI [15]	Word embeddings, WordNet, NER, Additional features	85.7
BRCNN [2]	Word embeddings	85.4
	+ POS, NER, WordNet	86.3
Our model Att-ComNN	Word embeddings	**86.6**

Table 4. Comparison between the main model and simplified models

Model	F_1
Att-ComNN (origin model)	86.6
-w/o SDP-based attention (simplified model-1)	85.4
-w/o any attention (simplified model-2)	85.1

As we can see from the Table 4, when we remove the SDP-based attention mechanism, F_1-score of our simplified model-1 is 85.4%. And when we remove all attention mechanisms from our origin model, F_1-score continues to decrease. We observe that all of our two level attention mechanisms lead to noticeable improvements over these baselines.

5 Conclusion

In this paper, we presented a novel combination of CNN and bidirectional RNN with attention mechanism named Att-ComNN to fulfill the relation classification task. Our Att-ComNN model consists mainly of one layer of CNN, one layer of bidirectional RNN with GRU and two important attention mechanisms. We evaluate the effectiveness of Att-ComNN on SemEval-2010 task 8. The result shows that Att-ComNN achieves better performance at learning features and relation classification, when compared with earlier proposed models based on neural networks. We will keep exploring on NN-based models, and expect to do further researches on relation classification and extraction.

Acknowledgements. This work is supported by National Key R&D Program of China (No. 2017YFB1400200).

References

1. Bunescu, R.C., Mooney, R.J.: A shortest path dependency kernel for relation extraction. In: Proceedings of the Conference Human Language Technology Conference and Conference on Empirical Methods in Natural Language Processing, HLT/EMNLP 2005, Vancouver, British Columbia, Canada, 6–8 October 2005, pp. 724–731 (2005)
2. Cai, R., Zhang, X., Wang, H.: Bidirectional recurrent convolutional neural network for relation classification. In: Proceedings of the 54th Annual Meeting of the Association for Computational Linguistics, ACL 2016, Volume 1: Long Papers, Berlin, Germany, 7–12 August 2016 (2016)
3. Chen, J., Ji, D., Tan, C.L., Niu, Z.: Unsupervised feature selection for relation extraction. In: Second International Joint Conference on Natural Language Processing - IJCNLP 2005 - Companion Volume to the Proceedings of Conference including Posters/Demos and tutorial abstracts, Jeju Island, Republic of Korea, 11–13 October 2005 (2005)
4. Cho, K., van Merrienboer, B., Gülçehre, Ç., Bougares, F., Schwenk, H., Bengio, Y.: Learning phrase representations using RNN encoder-decoder for statistical machine translation. CoRR abs/1406.1078 (2014)
5. Glorot, X., Bengio, Y.: Understanding the difficulty of training deep feedforward neural networks. In: Proceedings of the Thirteenth International Conference on Artificial Intelligence and Statistics, AISTATS 2010, Chia Laguna Resort, Sardinia, Italy, 13–15 May 2010, pp. 249–256 (2010)
6. Mikolov, T., Chen, K., Corrado, G., Dean, J.: Efficient estimation of word representations in vector space. CoRR abs/1301.3781 (2013)
7. Rink, B., Harabagiu, S.M.: UTD: classifying semantic relations by combining lexical and semantic resources. In: Proceedings of the 5th International Workshop on Semantic Evaluation, SemEval@ACL 2010, Uppsala University, Uppsala, Sweden, 15–16 July 2010, pp. 256–259 (2010)
8. Rotsztejn, J., Hollenstein, N., Zhang, C.: Eth-ds3lab at semeval-2018 task 7: effectively combining recurrent and convolutional neural networks for relation classification and extraction. CoRR abs/1804.02042 (2018)

9. dos Santos, C.N., Xiang, B., Zhou, B.: Classifying relations by ranking with convolutional neural networks. CoRR abs/1504.06580 (2015)
10. Socher, R., Huval, B., Manning, C.D., Ng, A.Y.: Semantic compositionality through recursive matrix-vector spaces. In: Proceedings of the 2012 Joint Conference on Empirical Methods in Natural Language Processing and Computational Natural Language Learning, EMNLP-CoNLL 2012, Jeju Island, Korea, 12–14 July 2012, pp. 1201–1211 (2012)
11. Socher, R., Pennington, J., Huang, E.H., Ng, A.Y., Manning, C.D.: Semi-supervised recursive autoencoders for predicting sentiment distributions. In: Proceedings of the 2011 Conference on Empirical Methods in Natural Language Processing, EMNLP 2011, A meeting of SIGDAT, a Special Interest Group of the ACL, John McIntyre Conference Centre, Edinburgh, UK, 27–31 July 2011, pp. 151–161 (2011)
12. Wang, L., Cao, Z., de Melo, G., Liu, Z.: Relation classification via multi-level attention CNNs. In: Proceedings of the 54th Annual Meeting of the Association for Computational Linguistics, ACL 2016, Volume 1: Long Papers, Berlin, Germany, 7–12 August 2016 (2016)
13. Xu, K., Feng, Y., Huang, S., Zhao, D.: Semantic relation classification via convolutional neural networks with simple negative sampling. CoRR abs/1506.07650 (2015)
14. Xu, Y., Mou, L., Li, G., Chen, Y., Peng, H., Jin, Z.: Classifying relations via long short term memory networks along shortest dependency paths. In: Proceedings of the 2015 Conference on Empirical Methods in Natural Language Processing, EMNLP 2015, Lisbon, Portugal, 17–21 September 2015, pp. 1785–1794 (2015)
15. Yu, J., Jiang, J.: Pairwise relation classification with mirror instances and a combined convolutional neural network. In: COLING 2016, 26th International Conference on Computational Linguistics, Proceedings of the Conference: Technical Papers, Osaka, Japan, 11–16 December 2016, pp. 2366–2377 (2016)
16. Zeng, D., Liu, K., Lai, S., Zhou, G., Zhao, J.: Relation classification via convolutional deep neural network. In: COLING 2014, 25th International Conference on Computational Linguistics, Proceedings of the Conference: Technical Papers, Dublin, Ireland, 23–29 August 2014, pp. 2335–2344 (2014)
17. Zheng, S., Wang, F., Bao, H., Hao, Y., Zhou, P., Xu, B.: Joint extraction of entities and relations based on a novel tagging scheme. CoRR abs/1706.05075 (2017)
18. Zhou, P., Shi, W., Tian, J., Qi, Z., Li, B., Hao, H., Xu, B.: Attention-based bidirectional long short-term memory networks for relation classification. In: Proceedings of the 54th Annual Meeting of the Association for Computational Linguistics, ACL 2016, Volume 2: Short Papers, Berlin, Germany, 7–12 August 2016

Discrete Sparse Hashing for Cross-Modal Similarity Search

Lu Wang[1], Chao Ma[1], Enmei Tu[1], Jie Yang[1(✉)], and Nikola Kasabov[2]

[1] Institute of Image Processing and Pattern Recognition,
Shanghai Jiao Tong University, Shanghai, China
{luwang_16,sjtu_machao,tuen,jieyang}@sjtu.edu.cn
[2] Knowledge Engineering and Discovery Research Institute,
Auckland University of Technology, Auckland, New Zealand
nkasabov@aut.ac.nz

Abstract. Cross-modal hashing approaches have achieved great success on cross-modal similarity search. However, most existing cross-modal hashing methods relax the discrete constraints to solve the hashing model and determine the weights of different modalities manually, which can significantly degrade the performance of retrieval. Besides, they are sensitive to noises because of the widely-utilized l_2-norm loss function. To address above problems, in this paper, a novel hashing method is proposed to efficiently learn unified binary codes, namely Discrete Sparse Hashing (DSH). In DSH model, unified hash codes are directly learned by discrete sparse coding in sharing low-dimensional latent space for different modalities, where the large quantization error is avoided and the learned codes are robust owing to the sparsity of binary codes. Moreover, the weights of different modalities are adaptively adjusted for training data. Extensive experiments on three databases demonstrate superior performance of DSH over most state-of-the-art methods.

Keywords: Sharing low-dimensional space · Discrete sparse coding
Linear classification framework · Unsupervised hashing
Cross-modal retrieval

1 Introduction

In these years, with the explosive growth of multimedia information, e.g. image, text, video and so on, it is very interest to develop effective cross-modal hashing algorithms for scalable similarity search. Cross-modal hashing can transform high-dimensional data from different modalities into compact binary representation. After getting the hash codes, similarity search can be made fast by finding the entities that have codes with a small Hamming distance (XOR operation) from the query. Additionally, hash codes need less storage cost for compact binary representation.

Generally, cross-modal hashing methods can be typed into two categories, i.e. supervised cross-modal hashing methods and unsupervised ones. Despite

L. Cheng et al. (Eds.): ICONIP 2018, LNCS 11304, pp. 256–267, 2018.
https://doi.org/10.1007/978-3-030-04212-7_22

supervised cross-modal hashing methods could get promising performance, they usually require label information of the entire data, that is difficult when dealing with a large scale data. Unsupervised cross-modal hashing methods would like to learn hash functions from data relevance by way of preserving the semantic similarity of training data. As unsupervised methods do not require extra supervised label information at all, they are more widespread than the supervised ones, which have wide applications for many tasks in the real world. Besides, unsupervised ones always can train faster when compared with supervised methods.

Supervised cross-modal hashing preserves semantic similarities given by available supervised information, such as multi latent binary embedding (MLBE) [23], semantics-preserving hashing (SePH) [9], semantic correlation maximization (SCM) [22], discrete cross-modal hashing (DCH) [21], and linear subspace ranking hashing (LSRH) [7]. Unsupervised cross-modal hashing methods learn hash functions from data distribution when preserving the similarities of data, such as cross-view hashing (CVH) [6], inter-media hashing (IMH) [16], linear cross-modal hashing (LCMH) [25], collective matrix factorization hashing (CMFH) [2], latent semantic sparse hashing (LSSH) [24], predictable dual-view hashing (PDH) [13], composite correlation quantization (CCQ) [10], and robust and flexible discrete hashing (RFDH) [17]. Recently, deep learning based cross-modal hashing methods have an increasing attention for their great performance improvements, which give binary codes for different modalities in an end-to-end deep learning architecture, capturing the intrinsic cross-modal relevance [4,8]. Besides deep learning based cross-modal hashing methods, multi-view dictionary learning (DL) technique based cross-modal hashing methods also has received much attention. In this field, multi-view discriminant dictionary learning (MDSVD) [20] and multi-view low-rank dictionary learning (MLDL) [19] are the representative works, which learn multiple view-specific structured discriminant dictionaries with each corresponding to a specific view to explore and utilize both the diversity and relevance information of different modalities.

While promising performance has been made, there still are some challenges for cross-modal hashing methods. Firstly, most cross-modal hashing methods discard the discrete constraints to solve the hashing model, which can cause large quantization error. Secondly, they are often sensitive to noises since the l_2-norm are widely used in loss function [17]. Thirdly, the weight factors of different modalities are manually determined [5,17], which may not fully suitable for the specific data.

To solve these problems, we propose a novel unified unsupervised hashing framework in cross-modal scenario, namely Discrete Sparse Hashing (DSH). Specifically, one contribution of DSH is that it finds the sharing low-dimensional latent space for different modalities by different orthogonal transformations. In this latent space, discrete sparse coding model with corresponding discrete optimization algorithm is developed to directly learn unified binary codes for heterogeneous data instead of relaxing the discrete constraints. The learned binary codes are robust to noises because as the advantages of sparse representations.

Furthermore, the weights of different modalities are learned according to training data [17].

2   Method

Assume that the training set has n image-text pairs $\mathcal{O} = \{o_i\}_{i=1}^n$, $o_i = (x_1^i, x_2^i)$, where $x_1^i \in \mathbb{R}^{d_1}$ is the image feature vector, $x_2^i \in \mathbb{R}^{d_2}$ is the text feature vector, and d_1, d_2 are the dimensionality of image feature space and text feature space respectively, usually, $d_1 \neq d_2$. Therefore, we have two data matrices $\mathbf{X}_1 = [x_1^1, x_1^2, \ldots, x_1^n]$ and $\mathbf{X}_2 = [x_2^1, x_2^2, \ldots, x_2^n]$. Besides, $sign(x)$ is the signum function, which outputs 1 for $x > 0$, and outputs 0 for $x \leq 0$.

2.1   Discrete Sparse Hashing

The main idea of DSH is to jointly learn the sharing low-dimensional latent space and a discrete similarity-preserving sparse representation as binary codes in a unified optimization framework. To achieve this mission, we firstly map different modalities to the sharing low-dimensional latent space by the orthogonal transformation matrix $\mathbf{R}_1, \mathbf{R}_2$ respectively, whose dimension is $d(d \leq min(d_1, d_2))$. As a result, we transform $\mathbf{X}_1, \mathbf{X}_2$ into $\mathbf{Y}_1, \mathbf{Y}_2$ respectively. To preserve that the interlinked, data should have the same latent representation, i.e. the inter-modality similarity, we have the constraint $\mathbf{Y}_1 = \mathbf{Y}_2 = \mathbf{Y}$. Therefore, we have the following objective function:

$$\min_{\mathbf{R}_1, \mathbf{R}_2, \mathbf{Y}, \alpha_1, \alpha_2} (\alpha_1)^\gamma \|\mathbf{X}_1 - \mathbf{R}_1\mathbf{Y}\|_{\mathbf{F}}^2 + (\alpha_2)^\gamma \|\mathbf{X}_2 - \mathbf{R}_2\mathbf{Y}\|_{\mathbf{F}}^2 \tag{1}$$

$$s.t. \quad \mathbf{R}_1^T\mathbf{R}_1 = \mathbf{I}_{d \times d}, \mathbf{R}_2^T\mathbf{R}_2 = \mathbf{I}_{d \times d},$$
$$\mathbf{R}_t \in \mathbb{R}^{d_t \times d}, t = 1, 2$$
$$\alpha_1 + \alpha_2 = 1, \alpha_1 > 0, \alpha_2 > 0,$$

where α_1, α_2 is the weight of image and text respectively, $\gamma > 1$ is a parameter, which controls the weight distribution.

As we all know, sparse coding [11] can learn an effective data representation in many applications. The learned sparse representation can capture the salient structure, which is robust to noises. In DSH, we use the discrete sparse coding to capture the salient structures of data for different modalities in the sharing low-dimensional latent space. Therefore, we can write the objective function below:

$$\min_{\mathbf{D}, \mathbf{B}} \|\mathbf{Y} - \mathbf{DB}\|_{\mathbf{F}}^2 + \sum_{i=1}^n \lambda \|b_i\|_1 \tag{2}$$

$$s.t. \quad \mathbf{B} \in \{0, 1\}^{k \times n},$$

where $\mathbf{D} \in \mathbb{R}^{d \times k}$ is the overcomplete basis set, i.e. $k > d$, and $\lambda > 0$ is the parameter to balance the reconstruction error and sparsity. Here integrated binary codes $\mathbf{B} = [b_1, b_2, \ldots, b_n] \in \{0,1\}^{k \times n}$ are learned by above discrete sparse coding, where the ith column $b_i \in \{0,1\}^{k \times 1}$ is the k-bits hash codes for the ith image-text pair o_i in image-text pairs $\mathcal{O} = \{o_i\}_{i=1}^n$.

Combining the aforementioned description, furthermore, adding the constraint $\mathbf{Y} = \mathbf{DB}$ for better performance, the objective function of DSH is written below:

$$\min_{\mathbf{R}_1, \mathbf{R}_2, \mathbf{D}, \mathbf{B}, \alpha_1, \alpha_2} (\alpha_1)^\gamma \|\mathbf{X}_1 - \mathbf{R}_1 \mathbf{DB}\|_{\mathbf{F}}^2 + (\alpha_2)^\gamma \|\mathbf{X}_2 - \mathbf{R}_2 \mathbf{DB}\|_{\mathbf{F}}^2 + \sum_{i=1}^n \lambda \|b_i\|_1$$

$$(3)$$

$$s.t. \quad \mathbf{R}_1^T \mathbf{R}_1 = \mathbf{I}_{d \times d}, \mathbf{R}_2^T \mathbf{R}_2 = \mathbf{I}_{d \times d},$$
$$\mathbf{B} \in \{0,1\}^{k \times n}, \mathbf{R}_t \in \mathbb{R}^{d_t \times d} \quad, t = 1, 2$$
$$\alpha_1 + \alpha_2 = 1, \alpha_1 > 0, \alpha_2 > 0.$$

By minimizing (3), the unified hash binary codes will be obtained directly.

2.2 Optimization Algorithm

The optimization problem in (3) can be solved by an iterative framework with the following listed steps until convergency.

(1) Update $\mathbf{R}_t(t = 1, 2)$ by fixing $\mathbf{D}, \mathbf{B}, \alpha_1, \alpha_2$

Equation (3) with $\mathbf{R}_t(t = 1, 2)$ as unknown variables is equivalent to the Orthogonal Procrustes Problem [14], which can be solved exactly using SVD. More specifically, we perform SVD as $\mathbf{X}_t [\mathbf{DB}]^T = USV^T, U \in \mathbb{R}^{d_t \times d}, V \in \mathbb{R}^{d \times d}$, then we achieve

$$\mathbf{R}_t = UV^T. \tag{4}$$

(2) Update $\mathbf{D}$ by fixing $\mathbf{R}_1, \mathbf{R}_2, \mathbf{B}, \alpha_1, \alpha_2$

With all variables but $\mathbf{D}$ fixed, we get an unconstrained quadratic problem, which has an analytic close form solution

$$\mathbf{D} = \left[\sum_{t=1}^2 (\alpha_t)^\gamma \mathbf{R}_t^T \mathbf{X}_t \mathbf{B}^T \right] \left[(\sum_{t=1}^2 (\alpha_t)^\gamma) \mathbf{B} \mathbf{B}^T \right]^{-1}. \tag{5}$$

(3) Update $\mathbf{B}$ by fixing $\mathbf{R}_1, \mathbf{R}_2, \mathbf{D}, \alpha_1, \alpha_2$

With all variables but $\mathbf{B}$ fixed, we can rewrite the Eq. (3) as:

$$\cdot \min_{\mathbf{B}} \|\mathbf{F}^T \mathbf{B}\|_{\mathbf{F}}^2 - 2Tr(\mathbf{B}^T \mathbf{Q}) \tag{6}$$

$$s.t. \quad \mathbf{B} \in \{0,1\}^{k \times n},$$

where $\mathbf{F} = \left[\sqrt{\sum_{t=1}^{2}(\alpha_t)^\gamma}\mathbf{D}^T, \sqrt{\lambda}\mathbf{I}_{k\times k} \right]$, $\mathbf{Q} = \mathbf{D}^T(\sum_{t=1}^{2}(\alpha_t)^\gamma \mathbf{R}_t^T \mathbf{X}_t)$, and $Tr(\cdot)$ is the trace norm. In fact, we find that there is a analytical solution to each row of $\mathbf{B}$ when we fix all other rows. Therefore, we iteratively update the binary bit one by one in discrete cyclic coordinate descent [15]. In particular, we denote $\mathbf{z}_l^T$ as the l-th row of $\mathbf{B}$ and $\mathbf{B}_l'$ as the matrix of $\mathbf{B}$ excluding $\mathbf{z}_l^T$. Similarly, let $\mathbf{q}_l^T$ be the l-th row of $\mathbf{Q}$, $\mathbf{u}_l^T$ be the l-th row of $\mathbf{F}$, and $\mathbf{F}_l'$ be the matrix of $\mathbf{F}$ excluding $\mathbf{u}_l^T$. The optimal solution of $\mathbf{z}_l$ can be achieved as:

$$\mathbf{z}_l = sign(\mathbf{q}_l - \mathbf{B}_l'^T \mathbf{F}_l' \mathbf{u}_l). \tag{7}$$

(4) Update $\alpha_t(t = 1, 2)$ by fixing $\mathbf{R}_1, \mathbf{R}_2, \mathbf{D}, \mathbf{B}$

With all variables but $\alpha_t(t = 1, 2)$ fixed, we can get a linear optimization problem, which has the optimal solution:

$$\alpha_1 = \frac{(\gamma \mathbf{E}_2)^{\frac{1}{\gamma-1}}}{\sum_{t=1}^{2}(\gamma \mathbf{E}_t)^{\frac{1}{\gamma-1}}}, \quad \alpha_2 = \frac{(\gamma \mathbf{E}_1)^{\frac{1}{\gamma-1}}}{\sum_{t=1}^{2}(\gamma \mathbf{E}_t)^{\frac{1}{\gamma-1}}}, \tag{8}$$

where $\mathbf{E}_t = \|\mathbf{X}_t - \mathbf{R}_t\mathbf{D}\mathbf{B}\|_F^2, t = 1, 2$. According to (8), we can get the following explanation for γ to control $\alpha_t(t = 1, 2)$. When $\gamma \rightarrow \infty$, the weights of different modalities are equal. Furthermore, when $\gamma \rightarrow 1$, the weight factor for the modality whose loss $\mathbf{E}_t$ is the smallest can be 1 and 0 otherwise. From above analysis of γ, we should select γ according to the complementary property, which all modalities have. A poor complementary should prefer small γ, otherwise, γ should be large. We will discuss the effect of performance for the parameter γ in the experiments in Subsect. 3.4.

By these four steps, we can alternatively update $\mathbf{R}_1, \mathbf{R}_2, \mathbf{D}, \mathbf{B}, \alpha_1$ and α_2 and iterate the procedure above until the objective function get a stable minimum value. The process of DSH can be outlined in Algorithm 1.

2.3   Learning Hash Functions

Actually, after getting the unified codes, learning the corresponding hash functions can be modeled as classification problem. As linear classifier can be learned efficiently, in this paper, we choose linear classifier to learn hash functions like RFDH did [17]. Specifically, we learn linear mappings $\mathbf{P}_1 \in \mathbb{R}^{k \times d_1}$ and $\mathbf{P}_2 \in \mathbb{R}^{k \times d_2}$, which transforms images features $\mathbf{X}_1$ and text features $\mathbf{X}_2$ as a unified representation matrix $\mathbf{V} \in \mathbb{R}^{k \times n}$ respectively. After that, we can train a linear multiclass classifier to learn the parameter of the classifier $\mathbf{W} \in \mathbb{R}^{k \times k}$, treating hash codes $\mathbf{B}$ as labels and the representation $\mathbf{V}$ as input. As [17], the overall objective function is written below:

$$\min_{\mathbf{P}_t, \mathbf{a}_t, \mathbf{W}, \mathbf{V}} \beta \|\mathbf{B} - \mathbf{W}\mathbf{V}\|_F^2 + \sum_{t=1}^{2} \|\mathbf{V} - \mathbf{P}_t\mathbf{X}_t + \mathbf{a}_t\mathbf{1}_k^T\|_F^2 + \eta R(\mathbf{P}_1, \mathbf{P}_2, \mathbf{W}, \mathbf{V}),$$

$$\tag{9}$$

Algorithm 1. Discrete Sparse Hashing.

Input: feature matrices $\mathbf{X}_1$ and $\mathbf{X}_2$, code length k, latent dimension d, parameters γ and λ.

Output: hash codes $\mathbf{B}$.

1: Initialize $\mathbf{R}_1, \mathbf{R}_2$ by identity.
2: Initialize $\mathbf{D}$ randomly.
3: Initialize binary codes $\mathbf{B}$ randomly, such that 0 and 1 are balanced in the codes.
4: Initialize the weight $\alpha_1 = 0.5, \alpha_2 = 0.5$ for each modality.
5: **repeat**
6:     Update $\mathbf{R}_t(t = 1, 2)$ by Orthogonal Procrustes as Eqn. (1).
7:     Update $\mathbf{D}$ by Quadratic Optimization as Eqn. Eqn. (5).
8:     Update binary codes $\mathbf{B}$ for each bit one by one by Eqn.(7).
9:     Update the weight $\alpha_t(t = 1, 2)$ by Eqn. (8).
10: **until** Objective function of Eqn. (3) converges.

---

where $\beta > 0$ is a balance parameter, $\eta > 0$ is the parameter to avoid overfitting, $\mathbf{a}_t \in \mathbb{R}^{k \times 1}$ is the intercept vector, $\mathbf{1_k} \in \mathbb{R}^{k \times 1}$ is the vector of all ones, and $R(\cdot) = \|\cdot\|_\mathbf{F}^2$ is a regularization term. The model in (9) can be solved following [17]. When hash functions learned, the hash code $\mathbf{b}^q$ of a query data $\mathbf{x}_t^q(t = 1, 2)$ could be obtained by

$$\mathbf{b}^q = sign(\mathbf{W}(\mathbf{P}_t\mathbf{x}_t^q - \mathbf{a}_t\mathbf{1}_k^T) - 0.51_k). \tag{10}$$

Complexity Analysis: In the training phase, the complexity of DSH in the training phase is $O(m^2Tn + r^2T_1n)$, where $r = max\{d_1, d_2\}$, $m = max\{r, d, k\}$, T is the number of iterations for learning hash codes, and T_1 is the number of iterations for learning hash functions. In the query phase, the time complexity for generating hash code of a query is $O(k^2 + kr)$, which is extremely efficient.

3 Experiments

3.1 Datasets Experimental Settings

Three multimodal datasets are utilized to show the efficiency and effectiveness of the proposed DSH model. **Wiki** [12] has $2,866$ image-text pairs which can be partitioned into 10 semantic categories. In **Wiki**, 128-dimensional SIFT feature is utilized for representing each image and we make use of 10-dimensional topics features for each text. We randomly select $2,173$ image-text pairs as training set and 693 image-text pairs as testing set form the **Wiki**. The true semantic neighbors can be find by the ground-truth labels. $269,648$ images and over $5,000$ user-provided tags are contained in **NUS-WIDE** [1]. Following prior works [18,25], we utilize the original NUS-WIDE to compose a new dataset having $195,834$ image-tag pairs by keeping the 21 most frequent concepts. We choose 500-dimensional SIFT feature to represent image, and text is represented by index vector of the most frequent $1,000$ tags. We randomly select $2,000$

Table 1. Comparison of mAP with Four Multimodal Retrieval Tasks on Three Benchmark. We report the results of DBRC (whose code are not public) in its corresponding paper and we distinguish those results by parenthesis (). "–" is used in the place where the result under that specific setting is not reported in their papers.

Tasks	Method	Wiki				UCI Handwritten Digit				NUS-WIDE			
		16 bits	32 bits	64 bits	128 bits	16 bits	32 bits	64 bits	128 bits	16 bits	32 bits	64 bits	128 bits
I→I	IMH	0.1340	0.1289	0.1258	0.1250	0.3498	0.3317	0.3466	0.3464	0.3746	0.3828	0.3859	0.3838
	LCMH	0.1121	0.1127	0.1146	0.1155	0.4760	0.3846	0.3379	0.3011	0.3374	0.3370	0.3371	0.3373
	PDH	0.1326	0.1375	0.1409	0.1500	0.4842	0.5153	0.5173	0.5781	0.3437	0.3529	0.3479	0.3588
	CMFH	0.1316	0.1325	0.1315	0.1304	0.5135	0.5384	**0.5830**	**0.5838**	**0.4337**	**0.4449**	**0.4426**	**0.4384**
	RFDH	**0.1484**	0.1461	0.1410	0.1383	**0.5341**	**0.5432**	0.5320	0.5478	0.4189	0.4249	0.4207	0.4170
	DBRC	–	–	–	–	–	–	–	–	–	–	–	–
	DSH	0.1467	**0.1488**	**0.1506**	**0.1533**	0.5119	0.5279	0.5472	0.5672	0.4287	0.4287	0.4292	0.4363
T→T	IMH	0.4239	0.4470	0.4727	0.4807	0.3578	0.3618	0.3681	0.3404	0.4439	0.4771	0.5114	0.5304
	LCMH	0.3715	0.3541	0.3405	0.3309	0.5962	0.5004	0.4203	0.3613	0.3284	0.3288	0.3286	0.3282
	PDH	0.4756	0.5078	0.5034	0.5138	0.4995	0.5099	0.5304	0.5850	0.3685	0.3946	0.4356	0.4821
	CMFH	0.4764	0.5162	0.5267	0.5270	0.5271	0.5665	0.6210	0.6350	0.4764	0.5122	0.5562	0.5938
	RFDH	**0.5115**	0.5143	0.5116	0.5123	**0.6353**	0.6502	0.6570	0.6437	0.4396	0.4843	0.5332	0.5689
	DBRC	–	–	–	–	–	–	–	–	–	–	–	–
	DSH	0.5007	**0.5241**	**0.5357**	**0.5386**	0.6074	**0.6556**	**0.6857**	**0.7043**	**0.5391**	**0.5894**	**0.6143**	**0.6270**
I→T	IMH	0.1807	0.1637	0.1510	0.1362	0.2319	0.1978	0.1740	0.1501	0.3769	0.3821	0.3816	0.3751
	LCMH	0.1380	0.1349	0.1197	0.1173	0.0756	0.0842	0.1095	0.1254	0.3310	0.3305	0.3302	0.3300
	PDH	0.1228	0.1342	0.1419	0.1458	0.2116	0.1627	0.1801	0.1637	0.3073	0.3203	0.3170	0.3161
	CMFH	0.1936	0.1945	0.1999	0.2053	0.4710	0.5063	0.5600	0.5580	0.4293	0.4533	0.4581	0.4610
	RFDH	0.2338	0.2315	0.2164	0.2182	**0.5502**	0.5577	0.5587	0.5778	0.4192	0.4263	0.4267	0.4215
	DBRC	**0.2534**	**0.2648**	**0.2686**	**0.2878**	–	–	–	–	0.3939	0.4087	0.4166	0.4165
	DSH	0.2275	0.2341	0.2375	0.2515	0.5459	**0.5774**	**0.5966**	0.6119	**0.4533**	**0.4579**	**0.4643**	**0.4732**
T→I	IMH	0.1701	0.1582	0.1442	0.1307	0.2281	0.1903	0.1702	0.1457	0.3740	0.3757	0.3723	0.3661
	LCMH	0.1120	0.1141	0.114	7 0.1152	0.0681	0.0801	0.1051	0.1181	0.3284	0.3288	0.3281	0.3277
	PDH	0.1056	0.1099	0.1205	0.1373	0.2122	0.1517	0.1676	0.1531	0.3064	0.3160	0.3144	0.3117
	CMFH	0.1790	0.1855	0.1904	0.1949	0.4730	0.5009	0.5595	0.5728	**0.4429**	**0.4573**	**0.4553**	0.4495
	RFDH	0.2383	0.2406	0.2294	0.2295	**0.5303**	**0.5529**	0.5551	0.5733	0.4023	0.4123	0.4118	0.4060
	DBRC	**0.5439**	**0.5377**	**0.5476**	**0.5488**	–	–	–	–	0.4249	0.4294	0.4381	0.4427
	DSH	0.2309	0.2477	0.2586	0.2640	0.5218	0.5439	**0.5726**	**0.6019**	0.4314	0.4336	0.4438	**0.4506**

image-text pairs as query set. The remaining $193,834$ pairs are used as the database, and the training set containing $5,000$ pairs is randomly sampled from the database [16,25]. The true semantic neighbors are defined as those sharing at least one label with it. **UCI Handwritten Digit** [3] consists of multimodal features for handwritten numerals (0–9). It has 10 classes, each of which contains 200 patterns. Following the setting of [3], we select 76 Fourier coefficients for the character shapes as one modal feature, and 64 Karhunen-Love coefficients as the another modal feature. we randomly choose $1,500$ images as the training set, and the remaining is seen as query set. The true semantic neighbors are defined by the associated labels. Besides, we perform four types of cross-modal retrieval schemes: (1) I→I: use image query to retrieve relevant image; (2) T→T: use text query to retrieve relevant text; (3) I→T: use image query to retrieve relevant text; (4) T→I: use text query to retrieve relevant image. The first two missions are intra-modal retrieval and the last two missions are cross-modal retrieval.

DSH takes a primary parameter: the latent space dimension d; and two hyper-parameters λ and γ. We select those parameters by using five-fold cross-validation. Specifically, linear search is used in log scale for the latent space dimension d, and set it to 10 for all the experiments. As for λ and γ, we use grid search over both hyper-parameter spaces $\lambda \in \{10^{-2}, 10^{-3}, 10^{-4}, 10^{-5}, 10^{-6}, 10^{-7}\}$ and $\gamma \in \{2, 4, 6, 8, 10\}$. As a result, the parameter λ is set to 10^{-5}, and the parameter γ is set to 7 for all the experiments. The proposed DSH model is compared with below six state-of-the-art unsupervised cross-modal hashing methods: IMH [16], LCMH [25], PDH [13], CMFH [2], RFDH [17], DBRC [8]. The method DBRC is a deep learning work in unsupervised cross-modal hashing methods. In addition, the performance of DBRC (without public code) is presented partially using the results in its paper. The parameters in above methods are set according to the corresponding papers.

3.2   Results and Discussions

In this section, we perform the comparison of mAP results on all three datasets in Table 1 respectively. From mAP results above, we can collect the following observations. On UCI Handwritten Digit and NUS-WIDE, the proposed DSH significantly outperforms all comparison methods with relatively long hash code length on I→T and T→T tasks. Superiority of DSH can be attributed to its capability to capture the salient structure, revise the weights adaptively, and achieve the small quantization error. However, on Wiki and UCI Handwritten Digit, we can observe that RFDH performs a little better than DSH with relatively short code code length. As the proposed DSH utilizes discrete sparse coding to obtain the unified binary codes, the performance of DSH is low when the number of basis in the overcomplete basis set, i.e. the code length k of binary codes, is less than the dimension d of the sharing latent space. Besides, on Wiki dataset, DBRC performs a little better than DSH on I→T task. Since DBRC is a deep learning work for cross-modal hashing methods, DBRC can capture more nonlinear relevance information of different modalities than the proposed DSH. But for the superiority of DSH, DSH can get comparable performance to DBRC on I→T task.

On I→I and T→I tasks, DSH reaches much higher mAP scores than other comparison methods except DBRC on Wiki and UCI Handwritten Digit with relatively long hash code length, for the reason that DSH can capture the salient structure, although RFDH performs a little better than DSH with relatively short code code length. However, DBRC get better performance than DSH on T→I task, which can be attributed to its capability to explore the nonlinear relevance information of different modalities. Besides, on NUS-WIDE, although DSH does not better than CMFH, performance of DSH can achieve comparable performance to CMFH which has the best performance, and DSH performs better than DBRC on T→I task. From the above discussions, we can confirm the effectiveness of the proposed DSH.

3.3   Convergence Analysis

Since DSH is solved by iterative procedure, we empirically study its convergence property. Figure 1 shows that the value of the objective function (the value is averaged by the number of training data) can go down steadily with more iterations. From Fig. 1, we can know the value of the objective function always converge with 15 iterations on all three datasets at 64 bit, that the effectiveness of Algorithm 1 is validated. The results for other code length can be similar to 64 bits.

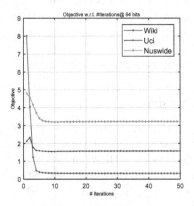

Fig. 1. Convergence analysis.

3.4 Parameter Sensitivity Analysis

In this section, we conduct an empirical analysis for the parameter sensitivity of different settings on all datasets to explore the effects about the algorithm performance, which can validates the proposed DSH can achieve stable and superior performance under a large range of parameter values, verifying that DSH can be robust to parameters. While analyzing one parameter, we fix other parameters to the setting which is mentioned in Subsect. 3.1. Here, we use mAP results at 64 bits on all the evaluated datasets for reflecting the variation of performance with respect to parameter values in the experiment. Our DSH has two hyper-parameters for learning binary codes, which include weight factors hyper-parameter γ and sparsity regularization hyper-parameter λ.

The parameter γ is a hyper-parameter, which controls the weight factor of each modality distribution. As analyzed in Subsect. 2.2, a poor complementary should prefer small γ, otherwise, γ should be large. From Fig. 2, we can see that Wiki, UCI Handwritten Digit and NUS-WIDE achieve the best around $\gamma = 7$. Besides, we can observe that DSH achieves stable and superior performance under a large range of γ.

The parameter λ balances the reconstruction error and sparsity in the DSH model. It can be observed from Fig. 3 that the performance of DSH goes down slightly when λ increasing. We find DSH can achieve best performance around $\lambda = 10^{-5}$ on all three datasets. Fortunately, when we select λ form the range $\left[10^{-8}, 10^{-4}\right]$, the robust performance of the proposed DSH can be guaranteed.

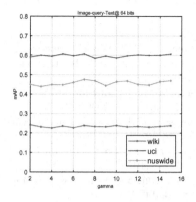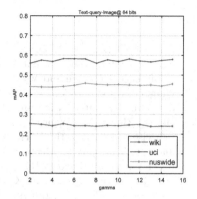

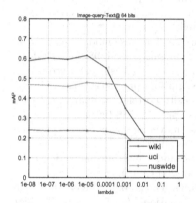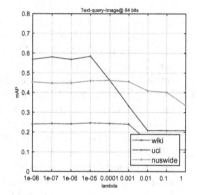

Fig. 2. mAP values versus parameter γ.

Fig. 3. mAP values versus parameter λ.

4   Conclusion

In this paper, we propose a discrete sparse hashing model for cross-modal similarity search. The proposed model uses discrete sparse coding to directly generate unified hash codes in a sharing low-dimensional latent space, which can be effectively solved in the low-dimensional space. By DSH, we can avoid large quantization error, and achieve robust sparse binary representation. Besides, we adaptively adjust the weights of different modalities for better performances. Extensive experiments on three benchmark datasets demonstrate that the proposed model outperforms several state-of-the-art methods.

Acknowledgement. This research is partly supported by NSFC, China (No: 61572315, 6151101179) and 973 Plan, China (No. 2015CB856004).

References

1. Chua, T.S., Tang, J., Hong, R., Li, H., Luo, Z., Zheng, Y.: NUS-WIDE: a real-world web image database from national university of Singapore. In: ACM International Conference on Image and Video Retrieval, pp. 1–9 (2009)
2. Ding, G., Guo, Y., Zhou, J.: Collective matrix factorization hashing for multimodal data. In: Computer Vision and Pattern Recognition, pp. 2083–2090 (2014)
3. He, R., Zhang, M., Wang, L., Ji, Y., Yin, Q.: Cross-modal subspace learning via pairwise constraints. IEEE Trans. Image Process. **24**(12), 5543–5556 (2015)
4. Jiang, Q.Y., Li, W.J.: Deep cross-modal hashing. In: IEEE Conference on Computer Vision and Pattern Recognition, pp. 3270–3278 (2017)
5. Jin, Q., Grama, I., Kervrann, C., Liu, Q.: Nonlocal means and optimal weights for noise removal. SIAM J. Imaging Sci. **10**(4), 1878–1920 (2017)
6. Kumar, S., Udupa, R.: Learning hash functions for cross-view similarity search. In: International Joint Conference on Artificial Intelligence, pp. 1360–1365 (2011)
7. Li, K., Qi, G., Ye, J., Hua, K.: Linear subspace ranking hashing for cross-modal retrieval. IEEE Trans. Pattern Anal. Mach. Intell. **39**(9), 1825–1838 (2017)
8. Li, X., Hu, D., Nie, F.: Deep binary reconstruction for cross-modal hashing. In: Proceedings of the 2017 ACM on Multimedia Conference, pp. 1398–1406 (2017)
9. Lin, Z., Ding, G., Hu, M., Wang, J.: Semantics-preserving hashing for cross-view retrieval. In: Computer Vision and Pattern Recognition, pp. 3864–3872 (2015)
10. Long, M., Cao, Y., Wang, J., Yu, P.S.: Composite correlation quantization for efficient multimodal retrieval. Computer Science, pp. 579–588 (2016)
11. Olshausen, B.A., Field, D.J.: Sparse coding with an overcomplete basis set: a strategy employed by V1. Vis. Res. **37**(23), 3311–3325 (1997)
12. Rasiwasia, N., et al.: A new approach to cross-modal multimedia retrieval. In: International Conference on Multimedia, pp. 251–260 (2010)
13. Rastegari, M., Choi, J., Fakhraei, S., Daume Iii, H., Davis, L.S.: Predictable dual-view hashing. In: International Conference on International Conference on Machine Learning, pp. 1328–1336 (2013)
14. Schnemann, P.H.: A generalized solution of the orthogonal procrustes problem. Psychometrika **31**(1), 1–10 (1966)
15. Shen, F., Shen, C., Liu, W., Shen, H.T.: Supervised discrete hashing. In: Computer Vision and Pattern Recognition, pp. 37–45 (2015)
16. Song, J., Yang, Y., Yang, Y., Huang, Z., Shen, H.T.: Inter-media hashing for large-scale retrieval from heterogeneous data sources. In: ACM SIGMOD International Conference on Management of Data, pp. 785–796 (2013)
17. Wang, D., Wang, Q., Gao, X.: Robust and flexible discrete hashing for cross-modal similarity search. IEEE Trans. Circuits Syst. Video Technol. **99**, 1–1 (2017)
18. Wang, W., Ooi, B.C., Yang, X., Zhang, D., Zhuang, Y.: Effective multi-modal retrieval based on stacked auto-encoders. Proc. VLDB Endow. **7**(8), 649–660 (2014)
19. Wu, F., Jing, X.Y., You, X., Yue, D., Hu, R., Yang, J.Y.: Multi-view low-rank dictionary learning for image classification. Pattern Recognit. **50**(C), 143–154 (2016)
20. Wu, F., Jing, X.Y., Yue, D.: Multi-view Discriminant Dictionary Learning via Learning View-specific and Shared Structured Dictionaries for Image Classification. Kluwer Academic Publishers (2017)
21. Xu, X., Shen, F., Yang, Y., Shen, H.T., Li, X.: Learning discriminative binary codes for large-scale cross-modal retrieval. IEEE Trans. Image Process. **26**(5), 2494–2507 (2017)

22. Zhang, D., Li, W.J.: Large-scale supervised multimodal hashing with semantic correlation maximization. In: Twenty-Eighth AAAI Conference on Artificial Intelligence, pp. 2177–2183 (2014)
23. Zhen, Y., Yeung, D.Y.: A probabilistic model for multimodal hash function learning. In: ACM SIGKDD International Conference on Knowledge Discovery and Data Mining, pp. 940–948 (2012)
24. Zhou, J., Ding, G., Guo, Y.: Latent semantic sparse hashing for cross-modal similarity search. In: Proceedings of the 37th International ACM SIGIR Conference on Research and Development in Information Retrieval, pp. 415–424 (2014)
25. Zhu, X., Huang, Z., Shen, H.T., Zhao, X.: Linear cross-modal hashing for efficient multimedia search. In: ACM Multimedia Conference, pp. 143–152 (2013)

Solving the Double Dummy Bridge Problem with Shallow Autoencoders

Jacek Mańdziuk[(✉)] and Jakub Suchan

Faculty of Mathematics and Information Science, Warsaw University of Technology,
Koszykowa 75, 00-662 Warsaw, Poland
j.mandziuk@mini.pw.edu.pl, kubasuchan@hotmail.com

Abstract. This paper presents a new approach to solving the Double
Dummy Bridge Problem (DDBP). The DDBP is a hard classification
task utilized by bridge playing programs which rely on Monte Carlo
simulations. The proposed method employs shallow autoencoders (AEs)
during an unsupervised pretraining phase and Multilayer Perceptron networks (MLPs) with three hidden layers, built on top of these trained AEs,
in the final fine-tuning training. The results are compared with our previous study in which MLPs with similar architectures, but with no use
of AEs and pretraining, were employed to solve this task. Several conclusions concerning efficient weight topologies and fine-tuning schemes
of the proposed model, as well as interesting weight patterns discovered
in the trained networks are presented and explained.

Keywords: Autoencoder · Double Dummy Bridge Problem
Classification

1 Introduction

Game AI is a popular and fast growing field of Artificial Intelligence (AI) research
which concerns various aspects of machine game playing. While historically this
research domain was mainly focused on perfect-information games such as chess
or checkers, in recent years games in which only partial game-related information
is available to each player (i.e. certain significant game aspects are concealed from
them) gained momentum. Imperfect-information games include, besides others,
most of card games - with the game of bridge being one of the most popular
examples.

Rules of Bridge. Bridge is a popular trick-taking card game played by four
contestants (referred to as *North*, *East*, *South* and *West* or *N*, *E*, *S*, *W*, for
short) in teams of two (*NS* vs *EW*), using a standard 52 card deck. The teams
serve as adversaries during the whole game and firstly participate in a *bidding
phase*.

Afterwards the team which has proposed a higher contract is bound to attain
their prior declaration. Their competitors are expected to spare no efforts in

© Springer Nature Switzerland AG 2018
L. Cheng et al. (Eds.): ICONIP 2018, LNCS 11304, pp. 268–280, 2018.
https://doi.org/10.1007/978-3-030-04212-7_23

order to hinder their success. This constitutes a *play phase*. In this phase, the teammate of the player who won the bidding puts their cards face-up down on the table and ceases henceforth to actively participate in the game. The player left to the bidding winner starts the play phase (makes the initial lead). The participants try to match the suit of the card that was played in the current trick. After everyone has done so the player with the highest-ranked card - including trumps - claims ownership of the trick, furthermore he becomes the one to play first in the next trick.

The goal of the game is to take as many tricks as possible, and certainly the highest number of tricks x that can be collected by a pair equals 13, in which case the opponent pair scores $13 - x$ tricks. Please consult [8] for a detailed explanation of the game rules.

Double Dummy Bridge Problem. Various attempts were made in AI literature to mimic the strategy used by humans during the bidding phase, yet typically a computer program remains inferior to its human counterpart [2,19] in that matter. In Machine Learning (ML) framework, the problem of bidding the appropriate contract can be transformed into a specific learning problem. Please observe that, if, disregarding the rules of Bridge, we assume that all information is available to the players (i.e. the game is a perfect-information one), the problem of bidding the contract becomes deterministic under the assumption that all players follow the optimal strategy in the play phase. In other way, assuming that all information is available to the players the number of tricks to be taken by each of the two playing pairs can be unambiguously assessed for a specified game (with the above-mentioned perfect rationality of the players). This situation is an illustration of the Double Dummy Bridge Problem (DDBP). More precisely, the DDBP consists in answering the question about the number of tricks that will be taken in a given game by the pair NS. The DDBP has been deemed a real challenge by many AI researches, not only because of its complexity, but also due to its high sensitivity to even minute changes in the distribution of cards among the players. For instance, exchanging positions of just two cards in a deal may significantly affect the expected results (i.e. the DDBP solution).

The practical value of fast DDBP solvers comes into play in simulation-based bridge programs in which the outcome of the game is assessed based on massive simulations of possible game scenarios, each of which requires solving a certain DDBP instance related to assumed distribution of cards. Such an approach has been utilized in *partition search* algorithm [3,9] or *cost-sensitive classifiers* and *upper-confidence-bound* algorithms [11].

Related Work. In the literature, there have been several attempts to solve the DDBP using example based learning. In particular, previous approach consisted in applying various Multilayer Perceptron (MLP) architectures with several coding schemes in the input layer and a Resilient Backpropagation (RProp) learning algorithm [14–16]. The best one among the tested MLP architectures $(208 - 52 - 13 - 1)$ accomplished an accuracy of 53.11% for the so-called *suit contracts* and 37.80% for the *no trump contracts*. In the case of suit contracts these results appeared to be superior to those accomplished by the human bridge grandmasters solving exactly the same sets of DDBP instances [13].

Our approach was researched further by others [5], this time with the focus on optimization of the training method - confirming the superiority of the RProp algorithm. Another related work [6] compared the impact of various activation functions on the output error. Yet other neural network approaches to solve the DDBP were proposed in [7] and [17], where respectively the Elman network and the Cascade Correlation network are employed.

Motivation and Research Goals. Encouraged by the promising results, in particular for suit contracts, in this paper we revisit the DDBP, but this time with shallow autoencoders (AEs) as the neural network architecture. In order to make the comparison fair the same best-performing input coding from our previous experiment is used in the current approach. Also the architecture is similar, i.e. shallow AEs with one hidden (intermediate) layer are utilized. This way, we attempt to make a direct comparison of the efficacy of AE training with an unsupervised pretraining phase and the MLP architecture trained in a supervised manner.

In summary, the main contribution of this work is threefold:

– verification of the suitability of the AE-based approach to a very sensitive classification problem, such as the DDBP;
– comparison of the AE's efficacy in solving the DDBP with the classification outcomes obtained previously using the MLP [16];
– shedding light on the intrinsic differences between supervised MLP training and AE training in the considered classification task.

The remainder of the paper is structured as follows. The next section introduces the proposed AE-based approach with a detailed presentation of the coding scheme and several variants of AE network architectures tested in the experimental evaluation of the method. Section 3 presents the experimental setup and analysis of results. Conclusions and directions for further research are summarized in the last section.

2   Autoencoder-Based Architectures

This section describes the AE architectures used in our attempts to solve the DDBP. On a general note, training of an AE consists of two phases. In the first one, the goal is to reconstruct the given input, i.e. to learn the identity function on the training set of examples. Yet, due to the hidden layers being less abundant in neurons this process leads to feature extraction. With every layer being smaller the feature extraction becomes gradually more extensive. In the process of extraction the data is compressed with every consecutive layer, however at a certain depth the data compression may become increasingly lossy and hamper the overall model performance in the recognition phase. *This was actually the case in our experiments, as our initial approaches involving deep autoencoders resulted in highly increasing training error in subsequent AE layers. For this reason we decided to restrict the AE architecture to one hidden layer, which appeared to be most effective in the preliminary experiments.*

After the above-described pretraining phase (consisting in unsupervised feature extraction) the *encoding AE layers* form the final classification network with an additional output layer and are trained in a supervised manner in the same way as MLP networks. The *decoding AE layers* used for input reconstruction have no application for solving this problem and are therefore discarded in the final supervised training phase [18].

2.1 A Deal Representation in the Input Layer

In the input AE layer the deal representation (cards distribution) is fed to the network in the form of 208 binary inputs. This input was divided into 4 sections, each representing the cards of one player, provided in a fixed order, i.e. first W (52 inputs) then N (52 inputs), followed by E and S. Each such section corresponds to the whole deck of cards aligned in the following order: $A\spadesuit, \ldots, 2\spadesuit$, $A\heartsuit, \ldots, 2\heartsuit$, $A\diamondsuit, \ldots, 2\diamondsuit$, $A\clubsuit, \ldots, 2\clubsuit$. Consequently, each card in a deal is represented by 4 inputs (one per player): the input representing the player who actually possesses the card is equal to 1 and the other 3 input neurons associated with that card, representing the three remaining players, are assigned input values equal to 0. This way the player who possesses a given card in a deal is pointed out. In other words, for each player there are exactly 13 inputs equal to 1 - on positions corresponding to the cards he/she possesses, and 39 inputs equal to 0 - on the remaining positions. The above coding proved to be highly efficient in our previous experiments [16] and in subsequent works by other authors [6]. Please consult [16] for a comprehensive discussion on the advantages of this deal coding over the alternative, more concise representations.

In order to encode the remaining deal-related information, i.e. the leading hand (the player who makes the opening lead) and the trump suit we initially extended the input representation by 9 neurons, first 5 of which indicated the game type (No trump, Spades, Hearts, Diamonds, Clubs) and the remaining 4 denoted the player making the opening lead. However, in a set of initial experiments this method proved inefficient as the network could not fully conceive the significance of these dedicated inputs. For this reason, in the final experimental setup we followed the idea presented in [12,16] in which by default the fourth player (S) makes the lead. If the play should start with a lead from N the cards of N and S players are swapped as well as the cards of W and E. As a consequence, the leading player (the fourth one in the input representation) becomes N. As for the representation of the game type, separate networks are trained for trump and no trump (NT) games. In the latter case it is always assumed that the trump color is the first one in each player hand's representation, i.e. Spades. In order to make the another one (say Diamonds) the trump suit, the cards representing Diamonds and Spades are swapped in each hand (please notice that names of suits are just labels and can be exchanged with no consequences). So effectively, the first suit remains the trump one.

2.2   Hidden (Feature) Layers

Hidden layers of the AE network serve as feature extractors that focus on the most relevant aspects of the processed input data and pass it forward for further processing and compression. The size of each subsequent hidden layer is reduced compared to the preceding one, which leads to building gradually more general, high-level feature-based representations in subsequent layers. At the same time, this compression process is prone to certain degree of information loss which, to a large extent, depends on the *compression rate* (CR) between the subsequent layers.

Compression Rate. In order to find efficient AE architectures a bunch of preliminary experiments were conducted, aimed at finding the most suitable CR between layers. First of all, it turned out that for the DDBP **the information loss is high already when the second hidden layer is added**. For this reason we decided to use shallow AE architectures with one hidden layer. Furthermore, **in order to make a direct comparison** with our previous MLP-based approach [16] two more "standard" hidden layers of sizes 52 and 13, respectively were added in the final architectures, which were not trained in AE manner.

In order to find the efficient CR value we applied a grid search procedure within the interval $CR \in [1.1, 3.0]$. As could be expected, the most efficient CRs were found in the middle area of the tested ranges. Higher values of CR resulted in too strong compression and information loss, while low CR values did not offer a relevant advantage compared to the baseline input representation. Based on the outcomes we decided to use two CRs in the final experiments: $CR = 4/3$ and $CR = 2$, which led to 156 and 104 neurons in the first hidden layer (AE compression layer), respectively.

2.3   Output Layer

The above described AE architecture, i.e. $208 - 156/104 - 208$ was used in the pretraining phase in order to build a meaningful feature-based representation in the AE compression layer. Once this unsupervised training process was completed the decoding layer was discarded and replaced by two standard hidden layers (with 52 and 13 units, resp.) and an output layer leading to the **final architecture used for the classification task**: $208 - 156/104 - 52 - 13 - 14$

Initially, two cases were considered for the output layer. Firstly, a classifying *softmax* layer composed of 14 output neurons - one per class (possible DDBP outcomes are integers from $[0, 13]$). Secondly, a layer consisting of one *sigmoid* neuron whose output corresponds to the number of tricks scaled into the range $[0, 1]$. In the latter case the $[0, 1]$ range was divided into 14 segments of pairwise equal lengths and the training (goal) signal for each class was set in the middle of the respective subinterval. In the preliminary tests both approaches yielded comparable results, so we arbitrarily decided to stick to the first option, i.e. one output neuron per class.

2.4   Topology of Connections

One of the main research goals of this paper is to experimentally compare the MLP-based DDBP solution presented in our earlier works and an application of AE architectures solving the same task. In order to make this comparison straightforward we used the same input and output representations and similar network architecture in terms of the number of layers as well as their sizes. In our previous experiments [16] the topologies of the most effective architectures were the ones where the connections between the input and the first hidden layer were restricted to individual hand (player's cards) representations. More precisely each 52-neuron representation of a given hand in the input layer was fully connected with 1/4 of the 1hl neurons (26 or 39 depending on the particular setup), without any connections to other 1hl units. This way the 1hl served as a compression layer where the initial 52-neuron hand representation was transformed to a 26 or 39 unit one, respectively.

For each of the fully connected AE architecture selected for the final experiments a corresponding model with the above described connection topology.

2.5   Network Architectures

In summary of the above description the following network structures were devised as the base models for experimental evaluation:

- Input layer composed of 208 units coding a deal as described in Sect. 2.1. Output layer composed of 14 sigmoid neuron denoting the number of tricks to be taken by the NS pair (cf. Sect. 2.3).
- First hidden layer composed of either 104 or 156 units.
- Topology between input and 1hl being either full connection (F) or dedicated connection (D) pattern (see Sect. 2.4).
- 2hl and 3hl composed of 52 and 13 neurons, respectively.

In all tested models sigmoid neurons are utilized. All together 4 AE models were tested in experiments varying by size of the 1hl (2 options) and topology of connections (2 options). Fully connected AE models are presented in Fig. 1a and those with dedicated connections between input and 1hl in Fig. 1b.

3   Experimental Setup and Results

One of the main observations reported in our previous work was the very distinct quality of the results for suit (trump) contracts and NT ones. More interestingly this observation is true not only with respect to our tests with the MLPs [16], but also among professional human players solving DDBP instances [13]. This property, whose nature is still to be fully discovered, has also been confirmed in our preliminary tests. This is why separate AE networks (of the same architecture) were used for training and testing on suit and NT contracts, respectively. Hence, the number of experimental setups discussed in the previous section was

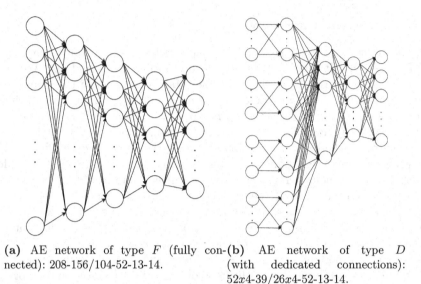

(a) AE network of type F (fully connected): 208-156/104-52-13-14.

(b) AE network of type D
(with dedicated connections):
$52x4$-39/$26x4$-52-13-14.

Fig. 1. Two types of considered network architectures. In each case, the size of the 1hl
is equal to 104 or 156, depending on the experimental setup.

doubled to 8 different scenarios. The source code and the experimental results
presented in this paper are available in our project's Github repository [1].

Training and Testing Data. The data is taken from the Ginsberg's GIB
Library [10] in the form of a text file. Each sample represents one deal, i.e. cards
assigned to each of the 4 hands/players. Each such deal is accompanied by 20
integers (from the set $\{0, 1, \ldots, 13\}$ denoting the number of tricks to be taken
by the NS pair in 20 possible game configurations: 4 trump suits or NT game as
well as 4 possible leading (opening) hands. This way, for a given card distribution
among the players, any possible game contract can be considered.

The GIB Library consists of over 700 000 deals represented in the above-
described way. Following our previous experiments with MLPs, for suit/trump
games we randomly selected 100 000 deals for training and another 100 000 for
testing. These 100 000 training deals were used in 8 game scenarios related to
the choice of a trump suit and leading hand (N or S). In effect, the training
process in suit AE models was performed on a set of 800 000 deals. In the case of
NT models, 400 000 deals was randomly selected (and multiplied by two possible
leading hands) for training and another 100 000 for testing.

Training phase 1 - Training of an Autoencoder. Training is performed in
two phases. In the first one a shallow AE with one compression layer (of size 104
or 156 depending on the model) is trained using the data described above. Once
the training is completed the hidden layer provides a compressed representation
of the 208-dimensional input data.

Since the input and output have the same cardinality and the input vectors are binary and sparse, *crossentropy* is used as an error function:

$$H(x, z) = -\sum_{k=1}^{d} [x_k \log z_k + (1 - x_k) \log (1 - z_k)]. \tag{1}$$

where x_k denotes a desired output, z_k is the actual output and d is the size of a training sample. The Root Mean Square Propagation (RMSProp) algorithm with a learning rate $\eta_1 = 0.001$ is used in this phase. The value of η_1 was selected based on a set of preliminary experiments. Training is terminated when the change in the error function values falls below 0.01.

Training phase 2 - Final Training of a Classifier. In the second training phase, two new hidden layers and an output layer are added to the encoding part of the shallow AE described in the above point comprising the following architecture: $208 - 104/156 - 52 - 13 - 14$, which is trained again based on the whole set of examples in order to serve as a classifier for unknown DDBP deals.

This training phase is performed using the RMSProp algorithm, according to one of the two possible scenarios. Either the whole network is trained or the weights developed in the pretraining phase are frozen and only the remaining part of the network $(104/156 - 52 - 13 - 14)$ undergoes weight modification. While the former is the standard approach, the motivation behind the latter is the following: since the AE pretraining error was close to zero the trained compression layer of the AE can actually serve as an alternative input representation and therefore the weights incoming to this layer need not be modified at this stage.

In either case training is performed with the learning rate $\eta_2 = 0.004$ for the trump deals and $\eta_2 = 0.002$ for the no trump ones. These learning rates were set based on preliminary experiments performed with various values of $\eta_2 \in [0.001, 0.1]$. Training is considered completed when the error function on the validation set increases 25 times. The state of the network corresponding to the overall highest performance, i.e. the lowest error value is considered as the training outcome.

3.1   Results

The results are presented for 16 experiment setups and divided into two tables. Table 1 provides the outcomes for suit contracts and Table 2 for no trump ones. In each case 8 network configurations which correspond to the above-described design decisions are tested.

Conclusions. Several interesting observations can be made based on presented results. **(1)**: they confirm a crucial role of the selected compression rate on the error value in the pretraining phase. On the other hand, quite surprisingly, this initial pretraining error does not have much influence on the final system error, after the fine-tuning phase. **(2)**: in suit contracts the results are close to those

Table 1. Best results for **trump (suit) deals** in 8 possible system configurations. The results in columns F and D denote the *full* and *dedicated* connection topology between the input and 1hl, resp. In either case there are two possible sizes of the 1hl (104 and 156 units). In each of them the weights between the input and 1hl may or may not be frozen during the fine tuning phase. *Compression error* and *fine tuning error* report the *cross-entropy* errors after completion of the training procedure in *phase 1* and *phase 2*, resp. The best reference results for the MLP presented in [16] are equal to $53.11\%, 96.48\%, 99.88\%$ for the exact, one trick off and two tricks off, resp.

Model topology	F				D			
1hl size	104		156		104		156	
Weights partly frozen	No	Yes	No	Yes	No	Yes	No	Yes
Test set								
Suit (♠) - exact	43.66%	39.75%	47.93%	29.06%	51.28%	23.93%	48.97%	32.19%
Suit (♠) - off by 1	90.46%	87.33%	93.83%	72.75%	95.33%	63.50%	94.37%	77.76%
Suit (♠) - off by 2	99.07%	98.45%	99.63%	92.41%	99.72%	86.40%	99.69%	95.07%
Compression error	0.400	0.24	9.560	9.599	46.98	45,90	14.23	13.21
Fine tuning error	0.048	0.050	0.043	0.056	0.043	0.059	0.044	0.054

Table 2. Best results for **no trump deals** in 8 possible system configurations. See the caption of Table 1 for a detailed specification of the table's layout. The best reference results for the MLP presented in [16] are equal to $37.80\%, 84.31\%, 97.34\%$ for the exact, one trick off and two tricks off, resp.

Model topology	F				D			
1hl size	104		156		104		156	
Weights partly frozen	No	Yes	No	Yes	No	Yes	No	Yes
Test set								
NT - exact	36.6%	35.13%	37.30%	28,40%	41.73%	22.46%	39.04%	29.48%
NT - off by 1	82.15%	80.12%	82.04%	70.47%	86.18%	59.73%	84.45%	72.51%
NT - off by 2	95.96%	95.29%	95.61%	90.17%	96.63%	82.90%	96.43%	91.66%
Compression error	0.293	0.325	9.580	9.597	45.13	44.56	13.99	14.49
Fine tuning error	0.052	0.053	0.051	0.057	0.049	0.060	0.051	0.056

obtained in our previous study [16] in the case of perfect accuracy (i.e. 51.28% vs 53.11%), and are on par in the two remaining cases (one/two tricks off). At the same time we managed to slightly surpass the aforementioned results in the case of NT contracts raising its value from 37.80% to 41.73% for perfect classification case and from 84.31% to 86.07% with a one-trick error margin. In both experiments the same deal representation and the same numbers of deals were used for training and testing. Also the resulting architectures were very close with the main difference being a two phase training procedure including an AE pretraining phase in the current studies. (**3**): our initial plans to employ deep autoencoders turned out to be ineffective as any attempt of further compression, extending beyond one hidden layer, resulted in significant raise of the cross-entropy error function. We believe that these three above observations prove that the DDBP is a hard classification task for both MLPs and AEs. (**4**): we observed that networks with dedicated sets of weights between the input and the first hidden layer are more effective than their counterparts with traditional, fully connected topology of weights. (**5**): despite close-to-zero loss value in the pretraining phase of 104 models the idea of freezing the weights between the input and 1hl in the second (fine-tuning) phase was not successful. This came to us as a bit of surprise since the 104 representation (due to the meaningless loss values) seemed to be a good candidate for an alternative deal representation. The reason of failure is subject of further investigation since at the moment we do not possess a convincing explanation. (**6**): similarly to the results of our previous study [16], NT contracts appeared to be much more demanding than suit ones. The reason for that is attributed to different nature of both types of deals: in trump ones the advantage stemming from possession of trump cards can be easier translated to the number of taken tricks, while playing NT contracts is generally more demanding and requires subtle maneuvers - more frequent use of *a finesse* is a typical example.

Weight Patterns. Following our previous study with pure MLP networks (without pretrained autoencoders) [16] we explored the weight spaces of the trained networks in the quest for meaningful patterns, explainable by human bridge players. Figure 2 visualizes the weights of a randomly selected player in one of the randomly selected experiments ($D/104/trump$). Due to space limits we are not able to delve more deeply into this topic, but two general conclusions may be easily drawn. Firstly, after the pretraining phase there are some number of strong (positive or negative) weights, most of which *are not altered* in subsequent training (c.f. left and right subfigures). Secondly, after the final training one can easily spot quite many neurons in the 1hl which are *dedicated to particular suits*, i.e. are "focused" on 13 consecutive inputs representing a given suit. This specific attention is visible in the form of "stripes" in the right subfigure covering the weights from A to 2 of a given suit. Both of the above observations are in line with the MLP-related study [16]. Furthermore, the pertinence of suit lengths (which is the crux of the second observation) is critical in human assessment of hand strength in *trump* deals.

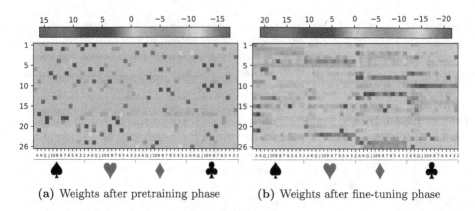

(a) Weights after pretraining phase   (b) Weights after fine-tuning phase

Fig. 2. Visualization of weights representing one player's cards for D/104 network and trump suits data. Each point refers to the respective weight between one of the 52 inputs assigned to that player's cards and one of the dedicated 26 neurons in the 1hl. For instance the weight between $Q\spadesuit$ and 14th 1hl neuron is strongly positive (red color) while the one between $Q\clubsuit$ and the 4th 1 h neuron is strongly negative (orange color). (Color figure online)

4   Summary and Future Work

The main goal of this paper is verification of suitability of AEs for solving the DDBP - a hard classification task with sensitive input-output relation. To this end, several configurations of different network architectures and detailed differences in the training algorithm are proposed and experimentally evaluated. The detailed conclusions are presented and discussed in the previous section.

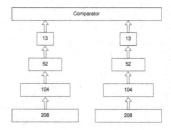

Fig. 3. Schematic presentation of a comparator-based approach to solving the DDBP planned as our next research step.

On a general note, the results of applying shallow AEs in the unsupervised pretraining phase and MLPs in the final fine-tuning phase turned out to be comparable to those of training MLPs directly in a supervised manner. Specific weight patterns observed in weight spaces of the trained MLPs are also visible in the current setup of the system, albeit their detection is not as evident.

Currently we plan to adopt an approach utilized in the DeepChess project [4]: instead of directly answering the question about the number of NS tricks, the system will be trained to compare two deals and answer an easier, qualitative question: *which of the two deals promises higher number of tricks for NS?* (see Fig. 3 for a possible implementation). With the ability of efficient qualitative prediction one could compare the unknown deal with a few reference deals with known numbers of NS tricks to answer the initial, quantitative question.

Acknowledgments. This work was supported by the Polish National Science Centre grant 2017/25/B/ST6/02061.

References

1. DDBP Github repository. https://github.com/holgus103/DDBP/
2. Amit, A., Markovitch, S.: Learning to bid in bridge. Mach. Learn. **63**(3), 287–327 (2006)
3. Beling, P.: Partition search revisited. IEEE Trans. Comput. Intell. AI Games **9**(1), 76–87 (2017)
4. David, O.E., Netanyahu, N.S., Wolf, L.: DeepChess: end-to-end deep neural network for automatic learning in chess. In: Villa, A.E.P., Masulli, P., Pons Rivero, A.J. (eds.) ICANN 2016. LNCS, vol. 9887, pp. 88–96. Springer, Cham (2016). https://doi.org/10.1007/978-3-319-44781-0_11
5. Dharmalingam, M., Amalraj, R.: Articifial neural network architecture for solving the double dummy bridge problem in contract bridge. Int. J. Adv. Res. Comput. Commun. Eng. **2**(12), 4683–4691 (2013)
6. Dharmalingam, M., Amalraj, R.: A solution to the double dummy contract bridge problem influenced by supervised learning module adapted by artificial neural network. ICTACT J. Soft Comput. **5**, 836–843 (2014)
7. Dharmalingam, M., Amalraj, R.: Supervised Elman neural network architecture for solving double dummy bridge problem in contract bridge. Int. J. Sci. Res. (IJSR) **3**(6), 2745–2750 (2014)
8. Francis, H., Truscott, A., Francis, D. (eds.): The Official Encyclopedia of Bridge, 5th edn. American Contract Bridge League Inc., Memphis (1994)
9. Ginsberg, M.L.: http://www.gibware.com
10. Ginsberg, M.L.: Library of double-dummy results. http://www.cirl.uoregon.edu/ginsberg/gibresearch.html
11. Ho, C.Y., Lin, H.T.: Contract bridge bidding by learning. In: AAAI Workshop: Computer Poker and Imperfect Information (2015)
12. Mańdziuk, J., Mossakowski, K.: Example-based estimation of hand's strength in the game of bridge with or without using explicit human knowledge. In: IEEE Symposium on Computational Intelligence in Data Mining, Honolulu, Hawaii, USA, pp. 413–420 (2007)
13. Mańdziuk, J., Mossakowski, K.: Neural networks compete with expert human players in solving the double dummy bridge problem. In: 2009 IEEE Symposium on Computational Intelligence and Games, pp. 117–124, September 2009
14. Mossakowski, K., Mańdziuk, J.: Artificial neural networks for solving double dummy bridge problems. In: Rutkowski, L., Siekmann, J.H., Tadeusiewicz, R., Zadeh, L.A. (eds.) ICAISC 2004. LNCS (LNAI), vol. 3070, pp. 915–921. Springer, Heidelberg (2004). https://doi.org/10.1007/978-3-540-24844-6_142

15. Mossakowski, K., Mańdziuk, J.: Neural networks and the estimation of hands' strength in contract bridge. In: Rutkowski, L., Tadeusiewicz, R., Zadeh, L.A., Żurada, J.M. (eds.) ICAISC 2006. LNCS (LNAI), vol. 4029, pp. 1189–1198. Springer, Heidelberg (2006). https://doi.org/10.1007/11785231_124
16. Mossakowski, K., Mańdziuk, J.: Learning without human expertise: a case study of the double dummy bridge problem. IEEE Trans. Neural Netw. **20**(2), 278–299 (2009)
17. Muthusamy, D.: Double dummy bridge problem in contract bridge: an overview. Artif. Intell. Syst. Mach. Learn. **10**(1), 1–7 (2018)
18. Ng, A., Ngiam, J., Foo, C.Y., Mai, Y., Suen, C.: UFLDL tutorial. http://ufldl.stanford.edu/wiki/index.php/UFLDL_Tutorial
19. Yegnanarayana, B., Khemani, D., Sarkar, M.: Neural networks for contract bridge bidding. Sadhana **21**(3), 395–413 (1996)

Cross-Project Issue Classification Based on Ensemble Modeling in a Social Coding World

Yarong Zeng[✉], Yue Yu, Qiang Fan, Xunhui Zhang, Tao Wang, Gang Yin, and Huaimin Wang

National Laboratory for Parallel and Distributed Processing,
National University of Defence Technology, Changsha, China
{zengyarong16,yuyue,fanqiang09,zhangxunhui,
taowang2005,yingang,hmwang}@nudt.edu.cn

Abstract. The simplified and deformalized contribution mechanisms in social coding are attracting more and more contributors involved in the collaborative software development. To reduce the burden on the side of project core team, various kinds of automated and intelligent approaches have been proposed based on machine learning and data mining technologies, which would be restricted by the lack of training data. In this paper, we conduct an extensive empirical study of transferring and aggregating reusable models across projects in the context of issue classification, based on a large-scale dataset including 799 open source projects and more than 795,000 issues. We propose a novel *cross-project* approach which integrate multiple models learned from various source projects to classify target project. We evaluate our approach through conducting comparative experiments with the *within-project* classification and a typical *cross-project* method called *Bellwether*. The results show that our cross-project approach based on ensemble modeling can obtain great performance, which comparable to the *within-project* classification and performs better than *Bellwether*.

Keywords: Cross-project · Issue classification · Transfer learning
Ensemble modeling · Modeling

1 Introduction

With the popularity of open source, more and more individual developers, research institutes or industrial companies prefer to host their software projects on social coding platforms, *e.g.*, BitBucket[1] and GitHub[2], to organize and manage the whole development life-cycle, which can catch more attentions from a large number of external developers and users. In order to simplify the collaborative process for both experienced and inexperienced contributors, a set of

---

[1] https://bitbucket.org.
[2] https://github.com.

© Springer Nature Switzerland AG 2018
L. Cheng et al. (Eds.): ICONIP 2018, LNCS 11304, pp. 281–292, 2018.
https://doi.org/10.1007/978-3-030-04212-7_24

lightweight tools are widely adopted, *e.g.*, pull request for code-patch submission [6] and Issue Tracking System (ITS) for development task management [4]. For example, by using the ITS on GitHub[3], a contributor only requires a short textual abstract containing a title and an optional description, when s/he wants to report a new found bug, ask an usage question or request an original feature to a software project [5]. On the one hand, those deformalized contribution mechanisms can make a collaborative project easy to collect more contributions from a wider range of the community. On the other hand, the increasing number of arbitrary, half-baked or undesirable contributions flow into the project along with high-quality ones, which poses a serious challenge for the core team in further maintenance, *e.g.*, project integrators are overburdened with evaluating excessive pull requests in time [27,28].

To reduce the burden on the side of project core team, various kinds of automated and intelligent approaches have been proposed based on machine learning and data mining, such as prioritisation tools for recommending the high-priority bug reports [25] and pull requests [26]. In theory, the performances of the training-based approaches are highly correlated with the scale of training data [2,17]. However, within a specific project, a comprehensive historical dataset which can cover the whole development life-cycle is often not available in practice. In this case, building a good training model is a challenging issue as collecting new data and labeling them is cost-expensive. Inspired by the theory and technique of transfer learning [20], *cross-project* approaches have been introduced and brought into focus to solve this problem. However, the existing cross-project work in software engineering, as discussed in Sect. 2.2, faces the challenges in terms of complicated training process and high computational cost. In this paper, we conduct an extensive empirical study of transferring and integrating reusable models across projects in the context of issue classification. We propose a novel approach to classify new issues in target projects by integrating multiple models learned from various source projects. And we evaluate our approach on a collection of 799 projects and more than 795,000 issue reports in GitHub. To the best of our knowledge, this is a first study of exploring cross-project approaches that categorizes issue reports, and also evaluating the cross-project learning topic in software engineering on such a large-scale dataset.

The rest of the paper is organized as follows. In Sect. 2, we discuss the background and related work of issue classification and transfer learning applied in cross-project defect prediction. In Sect. 3, we present the approach employed to address cross-project issue classification. In Sect. 4, we show the dataset used to evaluate our approach. Experiments and results are described in Sects. 5 and 6. In the end, we draw our conclusions in Sect. 7.

---

[3] https://github.com/blog/831-issues-2-0-the-next-generation.

2   Background and Related Work

2.1   Issue Reports Classification

In project development process, both developers and end-users can submit issue reports to ITS when performance does not meet their expectations [5]. According to our statistics to investigate the issue's first response time from core team, for nearly 800 open source projects (see Sect. 4.1) in GitHub. We found that it takes a long time period for core team members to start addressing issues (the average of median response time is 5.2 days, while the average of mean response time is 44 days). Therefore, to reduce the issue analysis time and improve the overall management process, an automated mechanism to classify issue reports is urgently needed.

Categorization of issue reports using techniques from text mining and machine learning has been receiving increasing attention from the research community. Antoniol et al. [1] used three machine learning methods to distinguish bugs from other kinds of issues. They found the information contained in issues posted in ITS can be indeed used to classify such issues, distinguish bugs from other activities, with a correct decision rate as high as 82% (experiment on 3 case projects). There are also some ways to increase the information extracted from ITS to improve the performance of the model. Merten et al. [16] found metadata from ITS can improve classifier performance through conducting empirical research on 4 open-source projects. Fan et al. [5] concluded that semantic perplexity of issue reports is a crucial factor that affects the classification performance. They experimented on 80 projects in GitHub and verified that model which considered semantic complexity do improve the issue classification performance.

Above classify methods rely on the data set within the project to train the classification model. However, Peters et al. [21] found within-projects predictors are weak for small data-sets. And Kitchenham et al. [9] discovered the problems with relying on within-project data-sets. They point out that the time required to collect enough project data can be prohibitive. In these situations, we propose a cross-project approach to address issues categorization problem of small dataset projects. In addition, those related works are based on a set of standard projects, there is no way to prove whether they are suitable for large-scale projects in actual. In this paper, we conduct a large-scale experiment to verify the effectiveness of our cross-project approach on issue classification.

2.2   Cross-Project Prediction

In practice, for new projects or projects with limited training data, It's a time-consuming and effort-intensive task to collect new data and label them. A possible way to solve this problem is using data collected from other projects, called cross-project prediction. Cross-project prediction has been receiving significant attention, and many studies have been proposed in research community.

Initial attempts for cross-project prediction always resulted in pessimistic performance. Zimmermann et al. [33] ran 622 cross-project predictions and found that only 3.4% actually worked. Menzies et al. [15] concluded that the local model were superior to the global model with experiment on four big datasets. Other researchers offer similar disappointing results [3,22,23]. The main reason that leads to the poor performance of cross-project prediction is the data distribution differences between the source and target project [18]. To reduce this difference, many optimized cross-project methods have been proposed. Zhang et al. [30] proposed context-aware rank transformations for predictors, built a universal model on the transformed data of 1398 projects, and found this model obtains prediction performance comparable to the within-project models. Then they proposed connectivity-based unsupervised classifier (via spectral clustering) and evaluated the feasibility for cross-project prediction by experimenting on three publicly available datasets [31].

Transfer learning techniques are often adopted for cross-project prediction to transfer lessons learned from source project S to the target project T in software engineering community. The specific application methods under cross-project scenarios can be divided into two types: data-level and feature-level transform. The data-level approaches directly transfer some subsets of data from source projects to the target project. Turhan et al. [24] collected similar source instances for target instances to train a prediction model using the nearest neighbour filter (NN filter) method. Ma et al. proposed Transfer Naive Bayes (TNB) [14] to predict target project. The data-level approaches use data in raw form, but they suffer from instability issues [12]. This prompted research on feature-level approaches. These approaches use projection to place the source and target data in a common latent feature space. Nam et al. [18] adapted the state-of-the-art transfer learning technique called Transfer Component Analysis (TCA) [19] and proposed TCA+ by adding decision rules to select proper normalization options.

To reduce the complexity of cross-project prediction, Krishna et al. [12] proposed a simple but effective cross-project approach. They chose the prior project which offered most accuracy predictions that generalize across other projects as "bellwether", then used the "bellwether" to generate quality predictors on new project data. This research is very instructive for our work. It proved that directly transferring model learned from the source project to predict the target project is feasible. However we think that the cross-project approach will be more effective [13] when taking the integration of models into consideration. In this work, we propose a new cross-project approach by taking into account the cost of computation and the reusability of cross project resources.

3   Approach

Cross-project prediction may suffer from two limitations. First, conclusion instability proposed by Krishna [12], that is when learning from all available data, prediction model may undergo constant changes whenever new data arrives. Second, the learner is opaque and the cost of calculation is expensive. When encountering new test data, the model needs to be retrained. To ease these two restrictions, we propose a new cross project prediction approach based on **Ensemble Modeling**, which consider the cross-project problem from the perspective of integration with reference to *Ensenmble Learning*. The key idea of our approach is using multiple models of source projects to make a classification decision for the target project. The source project is a project which have sufficient training samples to train a local model. And the target project is a project with limited training data. Comparing with the existing cross-project approaches, doing transfer at model level in our approach can greatly reduce the computational cost, because it can directly reuse the existing trained models. In addition, compared to *Bellwether* mentioned in Sect. 2, which transfer only one model to predict target predict, our approach can achieve better generalization performance than it by combining multiple models. We apply following two operators to classify the new target project P_T. The source project set is denoted as C.

1. *CLASSIFY:*
   - Randomly choose K models of source projects.
   - Integrate these K models to predict the target project P_T using the majority voting rule.
2. *MONITOR:*
   - For source projects, update classification model on time to add the newly arrived training data.
   - For target projects, if enough training data have been accumulated over time, the local model will be used for classification.

For the selection of K value, we test its impact on cross-project classification in Sect. 5, we will elaborate how to choose K value in Sect. 6.

Pay attention to the simplicity of this approach, each model of project in C can be trained in advance. For a new target project, we only need to randomly select models and integrate voting. We can directly reuse the knowledge of other projects to reduce the cost of calculation. And the *MONITOR* operator can relieve conclusion instability.

4   Dataset

4.1   Data Collection

We compose a sample of GitHub projects that each project contains a sufficient number of labeled issues. Using the 05/2017 GHTorrent[4] dump, we get projects that have at least 100 issues in total, including 12,797 projects and 6,414,872 issues. Then, we remove the non-existent projects online and collect issues data for each project through GitHub public API in consideration of data missed in Ghtorrent. The ITS in GitHub uses a labeling system to help organize and prioritize issues. To get a pre-labeled training set, we need to extract

---

[4] http://ghtorrent.org/downloads.html.

category information from the user-defined label system. In this work, we use category extraction method [5] based on the big issue datasets to extract bug-prone labels, such as "[Type] Bug", "Defect", and non-bug-prone label, such as "type: enhancement", "feature request". Through this process, we extract 4,222 labels and about 2,436,308 issues have these labels. Via the statistics of these labels, we found that some labels appeared in only one project. Considering the universality of the labels, we choose the first 40 labels which cover most of the issues(90.88%) and manually determine the issue type "bug" or "non-bug" according to the tendencies of the issue labels. Finally, we get 10,564 projects and 2,214,117 labeled issues.

4.2   Filtering

Repositorys hosted on GitHub have a variety of uses [8], some projects that use GitHub for issue tracking but not for code hosting or reversely. In order to avoid analyzing non-software projects and ensure the effect of classification model, we filter the dataset using the following rules:

- Project must not be forks of existing GitHub projects.
- Project must not be a pure documentation project. This criterion guarantees project have enough contextual information exploring cross project relevance.
- The issue text of project must be in English. This criterion avoids language deviation when classified the issue of other projects.
- Project must have at least 500 labeled issues. This criterion guarantees that the classification models have enough training and testing data.

After the above filterings, the obtained dataset contains 779 projects, and 795,284 issues. The smallest project has 502 issues and the largest project has 13,793 issues in total. The mean number of issues per project is 1020 and the median is 672.

4.3   Preprocessing

Each issue, which have be labeled as "bug" or "non-bug", has two main textual information sources – title and description. We concat the two parts of information together as issue text data. In order to train an automatic classifier, we create a feature vector for each text information through the following preprocessing steps. The first step is data preprocessing, a series of operations (lowercasing, tokenizing, stop-word removing, and stemming) is used to remove noise data and standardize text vocabulary. The issue text is segmented into different terms through this step. The second step is text quantizing. Using *TF-IDF* to calculate term weight, and then each issue can be represented as a weighted vector. For each project, a collection of raw issues text data can be converted to a matrix of *TF-IDF* vectors, which can be used as training data to get a classification model.

5   Experiment

5.1   Model Training

In this paper, we address the issue classification problem using binary classification. For each issue of the target project, we aim to determine the issue type as "bug" or "non-bug". We train a classification model based on the past data that takes the temporal information of issues into account. The temporal information reflects how such classification models can be used in practice. We split the issue data into two clusters at October 1th, 2016 (According to statistics, this division can get more training and testing projects.). All issues before that time point are for training and issues after that time point are for testing. This divides 673,490 issues into training set and 121,794 issues into testing set.

We use the state-of-the-art two-stage classifier [5] to deal with issue classification problem. The first-stage uses SVM to predict the probability of bug-prone and extracts perplex information of sentences from free issue text. The second-stage combines the output of the first-stage and the developer information as training input, then uses logistic regression to predict issue type. In order to make supervised text-based classification performs the best and avoid the influence of unbalanced dataset, we train the classification model using projects with more than 500 labeled issues and the bug rate of which are between 20% and 80%. When validating the performance of the model, we use the projects with more than 50 issues as test projects so as to ensure the authenticity of the classification results.

5.2   Experimental Methodology

To verify our approach, we compare our cross-project approach against *Within-project* and *Bellwethers* approach.

Within-project:

- Using local labeled data of the project to train a classification model.
- Using the model to classify new issue data of project.

Bellwether:

- Traversing cross-project pairs that are independent of the target project and choose the most prior project which offers the highest accuracy value of prediction as "bellwether".
- Using the "bellwether", generate classifier on new issue data of target project.

We use the f_{avg} (See Eq. 1) to evaluate classification performance of the above methods. Statistical analysis is used to verify our conclusions about the difference between the three methods. For the comparison between multiple groups, we use multiple contrast test procedure $\tilde{T}$ [10] in this study. We implement the procedure $\tilde{T}$ by nparcomp[5] package in R to evaluate the classification performance of all the approaches by using our dataset. We set the contrast type

---

[5] https://cran.r-project.org/web/packages/nparcomp/index.html.

to *Tukey* (all-pairs) to compare all groups pairwise. For each pair of groups, the 95% confidence interval is analyzed to test whether the corresponding null hypothesis can be rejected. The null hypothesis of $\hat{T}$ test is that there is no difference between classification result of the two approaches. If $p.Value < 0.05$, it means that the null hypothesis is significantly rejected at 5% level of significance. The *Estimator* denotes the relative effect between two sets of data. If the lower boundary of the interval is greater than 0.5 for groups A and B, it's means that B tends to be larger than A. Similarly, if the upper boundary of the interval is less than 0.5 for groups A and B, it's means that A tends to be larger than B. Finally, if the lower boundary of the interval is less than 0.5 and the upper boundary is greater than 0.5, it's means that the data does not provide enough evidence to reject the null hypothesis [11].

5.3 Experimental Design

According to the model training methods mentioned in the 5.1, we get a projects set S ($|S| = 465$) which have sufficient issues data (at least 500 issues which before October 1*th*, 2016) to train a good classification model within the project, and a projects set T ($|T| = 500$) which have a certain amount of data (at least 50 issues which after October 1*th*, 2016) to verify the performance of the cross classifier. We conduct experiments with the setting below.

We extract 50 projects from the project set S as test projects $Test_{prj}$, which have enough issue data to train a local model, and a certain amount of test issue data to verify the performance (make sure can compare with the *Within-project* method). For the rest of S and the project set T, we construct a set of cross-project pairs as training samples which used to choose prior project "bellwether" in *Bellwethers* approach. The cross-project pair is defined as, for project $A \in S$ and project $B \in T$, we consider a cross-project pair (A, B) to train a model from A to predict B if and only if $A \neq B$. Next, Referring to the steps mentioned in Sects. 3 and 5.2, we use the three approaches (our method, *Bellwether*, *Within-project*) to classify issue data for each project in $Test_{prj}$. We repeated the above process 10 times to enhance the robustness of the experiment and make the conclusions more convincing. Moreover, In order to verify the effect of K value on classification performance, we test the classification performance of our approach with different K values in experiment, within 100 intervals, the step is 1.

5.4 Evaluation Metrics

Using *Precision, Recall, F-measure* metrics as evaluation criteria is a common procedure in related work [7, 14, 18, 29, 33]. *Precision* is used to measure the exactness of the prediction set, while *Recall* evaluates the completeness. *F-measure* represents the harmonic mean of *Precision* and *Recall*. In [5, 32], in order to give an overall performance evaluation, the average *F-measure* is used to evaluate the classification model. This metric uses the weighted average value of *F-measure* for both categories by the proportions of instances in that category [32].

In this paper, considering the performance of both categories (bug or non-bug), we use average *F-measure* to assess the accuracy of the classification. Equation 1 describes the formula to derive the average *F-measure*. The average *F-measure* is denoted as f_{avg} , *F-measure* of bug (nonbug) as f_{bug} (f_{nonbug}), and number of bug (nonbug) as n_{bug} (n_{nonbug}).

$$f_{avg} = \frac{n_{bug} * f_{bug} + n_{nonbug} * f_{nonbug}}{n_{bug} + n_{nonbug}} \tag{1}$$

6   Results

Figure 1 provides a summary of the comparison steps. It shows the median values of f_avg with three approach based on 500 projects. The horizontal axis is the K value, which is mainly used to show the classification performance of our approach with different K values in the experiment. The *Within-project* and the *Bellwether* method are not affected by the K value. From the figure, we can find our approach obtains classification performance comparable to the *Within-project* method when K is large enough. And our approach out-performs *Bellwether* when integrating a few models. The preliminary result provide a evidence that the idea of model integration would achieve ideal performance in cross-project classification. In addition, it can be seen from the figure that there are some fluctuations between adjacent K values when K is greater than 25, which may be related to the randomness of the model selection.

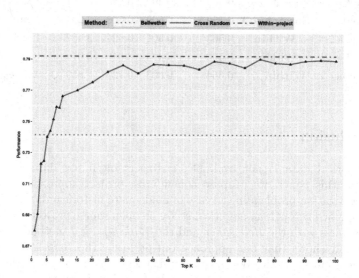

Fig. 1. The median values of f_avg with three approach based on 500 projects

The experimental results in Fig. 1 give us a intuitive view of how well our cross-project approach performs. In addition, we use multiple contrast test to verify whether the conclusion is correct, through comparing the overall f_avg of 500 projects (10 tests, 50 projects at a time) for different approaches(with different K value). Table 1 shows the result of contrast test. And we reveal some of the significant turning points with different K value in performance comparison. The $p.Values$ are over 0.05 in row 2, 5 (gray background in the table), and there are flips between lower boundaries ($Lower < 0.5$) and upper boundaries ($Upper > 0.5$) in these rows. It means there is no statistical significant difference between each group pair. A significant difference at 5% level occurs in row 1, 3, 4, 6. The lower boundaries and upper boundaries are both greater than 0.5 or less than 0.5, which means that there is a significant difference between the two sets of comparison data, one of which is better than the other. And we bold the name of the better method in the figure.

Therefore, we can draw the following conclusions. First, our approach outperform the *Bellwether* method when integrating greater or equal to 9 models of source projects. Second, our cross-project approach can achieve the performance of the *Within-project* method when integrating greater or equal to 25 models.

Table 1. Results of multiple contrast test procedure

Comparison	Estimator	Lower	Upper	Statistic	p.Value
Ensemble_4 vs.**Bellwether**	0.552	0.509	0.594	2.823	0.133e−01
Bellwether vs. Ensemble_5	0.503	0.460	0.545	0.151	9.893e−01
Bellwether vs.**Ensemble_9**	0.559	0.509	0.608	3.234	1.197e−02
Ensemble_24 vs. **Within**	0.563	0.501	0.624	3.462	4.488e−02
Ensemble_25 vs. Within	0.556	0.493	0.618	3.070	1.490e−01
Bellwether vs. **Within**	0.664	0.603	0.720	8.859	0.000e+00

7   Conclusion

When historical data is not available, engineers often use data from other projects. In this paper, we propose a new cross-project approach to address issue categorization problem in the case of inadequate historical data. We evaluate our approach through an empirical study on open source projects in GitHub, which compares with the *Bellwether* and the *Within-project* method. The comparison result shows that our approach out-performs the *Bellwether* and can achieve the accuracy of the *Within-project* method. This means that the idea of integration is encouraging in cross-project classification. In future work, we plan to investigate the effectiveness of our cross-project approach in different software engineering scenarios, such as defect prediction or effort estimation.

References

1. Antoniol, G., Ayari, K., Di Penta, M., Khomh, F., Guéhéneuc, Y.G.: Is it a bug or an enhancement?: A text-based approach to classify change requests. In: Proceedings of the 2008 Conference of the Center for Advanced Studies on Collaborative Research: Meeting of Minds, p. 23. ACM (2008)
2. Batista, G.E., Prati, R.C., Monard, M.C.: A study of the behavior of several methods for balancing machine learning training data. ACM SIGKDD Explor. Newslett. **6**(1), 20–29 (2004)
3. Bettenburg, N., Nagappan, M., Hassan, A.E.: Think locally, act globally: improving defect and effort prediction models. In: 2012 9th IEEE Working Conference on Mining Software Repositories (MSR), pp. 60–69. IEEE (2012)
4. Bissyandé, T.F., Lo, D., Jiang, L., Réveillere, L., Klein, J., Le Traon, Y.: Got issues? Who cares about it? A large scale investigation of issue trackers from GitHub. In: 2013 IEEE 24th International Symposium on Software Reliability Engineering (ISSRE), pp. 188–197. IEEE (2013)
5. Fan, Q., Yu, Y., Yin, G., Wang, T., Wang, H.: Where is the road for issue reports classification based on text mining? In: 2017 ACM/IEEE International Symposium on Empirical Software Engineering and Measurement (ESEM), pp. 121–130. IEEE (2017)
6. Gousios, G., Pinzger, M., Deursen, A.V.: An exploratory study of the pull-based software development model. In: Proceedings of the 36th International Conference on Software Engineering, pp. 345–355. ACM (2014)
7. He, P., Li, B., Ma, Y.: Towards cross-project defect prediction with imbalanced feature sets. arXiv preprint arXiv:1411.4228 (2014)
8. Kalliamvakou, E., Gousios, G., Blincoe, K., Singer, L., German, D.M., Damian, D.: The promises and perils of mining GitHub. In: Proceedings of the 11th Working Conference on Mining Software Repositories, pp. 92–101. ACM (2014)
9. Kitchenham, B.A., Mendes, E., Travassos, G.H.: Cross versus within-company cost estimation studies: a systematic review. IEEE Trans. Softw. Eng. **33**(5), 316–329 (2007)
10. Konietschke, F., Hothorn, L.A., Brunner, E., et al.: Rank-based multiple test procedures and simultaneous confidence intervals. Electron. J. Stat. **6**, 738–759 (2012)
11. Konietschke, F., Placzek, M., Schaarschmidt, F., Hothorn, L.A.: nparcomp: An R software package for nonparametric multiple comparisons and simultaneous confidence intervals (2015)
12. Krishna, R., Menzies, T., Fu, W.: Too much automation? The bellwether effect and its implications for transfer learning. In: Proceedings of the 31st IEEE/ACM International Conference on Automated Software Engineering, pp. 122–131. ACM (2016)
13. Lan, L., Tao, D., Gong, C., Guan, N., Luo, Z.: Online multi-object tracking by quadratic pseudo-boolean optimization. In: IJCAI, pp. 3396–3402 (2016)
14. Ma, Y., Luo, G., Zeng, X., Chen, A.: Transfer learning for cross-company software defect prediction. Inf. Softw. Technol. **54**(3), 248–256 (2012)
15. Menzies, T., Butcher, A., Marcus, A., Zimmermann, T., Cok, D.: Local vs. global models for effort estimation and defect prediction. In: Automated Software Engineering, pp. 343–351. IEEE (2011)
16. Merten, T., Falis, M., Hübner, P., Quirchmayr, T., Bürsner, S., Paech, B.: Software feature request detection in issue tracking systems. In: 2016 IEEE 24th International Requirements Engineering Conference (RE), pp. 166–175. IEEE (2016)

17. Nagappan, N., Ball, T., Zeller, A.: Mining metrics to predict component failures. In: Proceedings of the 28th International Conference on Software Engineering, pp. 452–461. ACM (2006)
18. Nam, J., Pan, S.J., Kim, S.: Transfer defect learning. In: Proceedings of the 2013 International Conference on Software Engineering, pp. 382–391. IEEE Press (2013)
19. Pan, S.J., Tsang, I.W., Kwok, J.T., Yang, Q.: Domain adaptation via transfer component analysis. IEEE Trans. Neural Netw. **22**(2), 199–210 (2011)
20. Pan, S.J., Yang, Q.: A survey on transfer learning. IEEE Trans. Knowl. Data Eng. **22**(10), 1345–1359 (2010)
21. Peters, F., Menzies, T., Marcus, A.: Better cross company defect prediction. In: Mining Software Repositories, pp. 409–418 (2013)
22. Posnett, D., Filkov, V., Devanbu, P.: Ecological inference in empirical software engineering. In: Proceedings of the 2011 26th IEEE/ACM International Conference on Automated Software Engineering, pp. 362–371. IEEE Computer Society (2011)
23. Premraj, R., Herzig, K.: Network versus code metrics to predict defects: a replication study. In: 2011 International Symposium on Empirical Software Engineering and Measurement (ESEM), pp. 215–224. IEEE (2011)
24. Turhan, B., Menzies, T., Bener, A.B., Di Stefano, J.: On the relative value of cross-company and within-company data for defect prediction. Empirical Softw. Eng. **14**(5), 540–578 (2009)
25. Uddin, J., Ghazali, R., Deris, M.M., Naseem, R., Shah, H.: A survey on bug prioritization. Artif. Intell. Rev. **47**(2), 145–180 (2017)
26. Van Der Veen, E., Gousios, G., Zaidman, A.: Automatically prioritizing pull requests. In: Proceedings of the 12th Working Conference on Mining Software Repositories, pp. 357–361. IEEE Press (2015)
27. Yu, Y., Wang, H., Filkov, V., Devanbu, P., Vasilescu, B.: Wait for it: determinants of pull request evaluation latency on GitHub. In: 2015 IEEE/ACM 12th Working Conference on Mining Software Repositories (MSR), pp. 367–371. IEEE (2015)
28. Yu, Y., Wang, H., Yin, G., Wang, T.: Reviewer recommendation for pull-requests in github: what can we learn from code review and bug assignment? Inf. Softw. Technol. **74**, 204–218 (2016)
29. Zanetti, M.S., Scholtes, I., Tessone, C.J., Schweitzer, F.: Categorizing bugs with social networks: a case study on four open source software communities. In: Proceedings of the 35th International Conference on Software Engineering, pp. 1032–1041. IEEE (2013)
30. Zhang, F., Mockus, A., Keivanloo, I., Zou, Y.: Towards building a universal defect prediction model. In: Proceedings of the 11th Working Conference on Mining Software Repositories, pp. 182–191. ACM (2014)
31. Zhang, F., Zheng, Q., Zou, Y., Hassan, A.E.: Cross-project defect prediction using a connectivity-based unsupervised classifier. In: Proceedings of the 38th International Conference on Software Engineering, pp. 309–320. ACM (2016)
32. Zhou, Y., Tong, Y., Gu, R., Gall, H.: Combining text mining and data mining for bug report classification. J. Softw. Evol. Process **28**(3), 150–176 (2016)
33. Zimmermann, T., Nagappan, N., Gall, H., Giger, E., Murphy, B.: Cross-project defect prediction: a large scale experiment on data vs. domain vs. process. In: Proceedings of the 7th Joint Meeting of the European Software Engineering Conference and the ACM SIGSOFT Symposium on the Foundations of Software Engineering, pp. 91–100. ACM (2009)

Decision Tree Twin Support Vector Machine Based on Kernel Clustering for Multi-class Classification

Qingyun Dou and Li Zhang[✉]

School of Computer Science and Technology and Joint International Research
Laboratory of Machine Learning and Neuromorphi Computing,
Soochow University, Suzhou 215006, China
zhangliml@suda.edu.cn

Abstract. To deal with multi-class classification problems using twin support
vector machines (TWSVMs), this paper proposes a novel multi-class classifier,
decision tree twin support vector machine based on kernel clustering
(DT^{2}SVM-KC). We employ the kernel clustering algorithm to generate a binary
tree, and for each non-leaf node, we obtain a pair of non-parallel hyperplanes by
using TWSVM. Simulation results show that the proposed method can keep the
strength of decision tree in computation time and has better performance on
most used datasets compared with other multi-class classification methods based
on TWSVMs.

Keywords: Twin support vector machine · Decision tree
Multi-class classification · Kernel function

1 Introduction

Support vector machine (SVM) is a powerful pattern recognition algorithm for
supervised learning [1]. Twin support vector machine (TWSVM) is one of modified
and improved versions, which has been studied widely in recent years [2]. The aim of
TWSVM is to implement classification by generating two non-parallel planes in which
each plane is closer to one of two classes and is away from the other as far as possible.
Since TWSVMs solve a pair of smaller quadratic programming problems (QPPs)
instead of a single large QPP solved in SVMs, TWSVMs have a lower computational
complexity and work faster than the standard SVMs.

The study of TWSVMs can be divided into two main groups: variant-based and
multi-class-based methods [3]. The binary TWSVMs show excellent performance on
some certain probability model data, e.g. "Cross Planes" data, and therefore, some
variant-based TWSVMs have been studied widely [4–8]. On the other hand, some
extensions of TWSVMs are aimed at solving multi-category classification [9–14]. In
[13], Zhen et al. extended the binary TWSVM to processing multi-class classification
problems and proposed the multi-TWSVM method, which uses the way of one against
all. Another multi-class method proposed in [14] is termed as decision tree twin support
vector machine (DTTSVM), which constructs a decision tree based on maximizing the

© Springer Nature Switzerland AG 2018
L. Cheng et al. (Eds.): ICONIP 2018, LNCS 11304, pp. 293–303, 2018.
https://doi.org/10.1007/978-3-030-04212-7_25

distance between classes and conducts binary classification method on the non-leaf node. However, the way of constructing decision tree may be not fit to the case of applying kernel functions. Thus, the simple formula used in DTTSVM to generate a decision tree leads it hard to obtain the best decision tree when the data distribution is complicated and disordered.

To remedy it, we propose a novel method based on TWSVMs for multi-class classification, called decision tree twin support vector machine based on kernel clustering (DT^{2}SVM-KC). Similar to DTTSVM, our proposed method also decomposes multi-class problems by establishing a binary tree based on kernel clustering, and finds the partition on non-leaf nodes by using TWSVMs. In kernel clustering or TWSVM with kernel functions, we need to map data in the original input space to a high-dimensional feature space in which patterns are more regular and some helpful features can be enlarged. Compared with another multi-class method using the kernel clustering algorithm [15] in [16], the proposed method has faster speed in both theory and experiments. Simulation results confirm that the proposed method has a better generalization performance as well as retains the speed advantage of TWSVM.

The rest of this paper is organized as follows. In Sect. 2, we briefly describe the linear and non-linear TWSVMs. In Sect. 3, we present the novel decision tree TWSVMs based on kernel clustering. In Sect. 4, we compare our proposed method with several multi-class classification methods. In Sect. 5, we give conclusions.

2   Twin Support Vector Machine

In this section, we give a brief review of TWSVM [2]. Given the following training data

$$T = \{(\mathbf{x}_1, y_1), \ldots, (\mathbf{x}_m, y_m)\} \tag{1}$$

where $(\mathbf{x}_i, y_i)$ is the i-th data point, the vector $\mathbf{x}_i$ is in the n-dimensional real space $\mathbf{R}^n$, $y_i \in \{+1, -1\}$ denotes the class label to which $\mathbf{x}_i$ belongs, and m represents the total number of samples in both class +1 and class -1. Let matrices $\mathbf{A} \in \mathbf{R}^{m_1 \times n}$ and $\mathbf{B} \in \mathbf{R}^{m_2 \times n}$ represent data points belonging to classes +1 and −1, respectively, where m_1 and m_2 are the number of samples in classes +1 and classes −1, respectively.

Consider the linearly separable case, the aim of TWSVM is to seek two nonparallel hyperplanes

$$\mathbf{x}^T \mathbf{w}_1 + b_1 = 0 \text{ and } \mathbf{x}^T \mathbf{w}_2 + b_2 = 0 \tag{2}$$

so as to make the data points of one class be closest to one hyperplane and the data points of the other class be away from it as far as possible. In (2), $\mathbf{w}_1$ and $\mathbf{w}_2$ are weight vectors, and b_1 and b_2 are biases for the two hyperplanes, respectively.

Instead of solving a large single quadratic programming problem (QPP) in the standard SVM, two smaller QPPs are solved in TWSVM:

$$\min_{\mathbf{w}_1, b_1, q} \quad \tfrac{1}{2}(\mathbf{Aw}_1 + \mathbf{e}_1 b_1)^T (\mathbf{Aw}_1 + \mathbf{e}_1 b_1) + C_1 \mathbf{e}_2^T \mathbf{q}$$
$$subject\ to \quad -(\mathbf{Bw}_1 + \mathbf{e}_2 b_1) + \mathbf{q} \geq \mathbf{e}_2, \mathbf{q} \geq 0 \tag{3}$$

and

$$\min_{\mathbf{w}_2, b_2, q} \quad \tfrac{1}{2}(\mathbf{Bw}_2 + \mathbf{e}_2 b_2)^T (\mathbf{Bw}_2 + \mathbf{e}_2 b_2) + C_2 \mathbf{e}_1^T \mathbf{q}$$
$$subject\ to \quad -(\mathbf{Aw}_2 + \mathbf{e}_1 b_2) + \mathbf{q} \geq \mathbf{e}_1, \mathbf{q} \geq 0 \tag{4}$$

where C_1 and C_2 are positive regularization parameters, $\mathbf{e}_1$ and $\mathbf{e}_2$ are vectors of all ones, $\mathbf{q} = [q_1, \ldots, q_m]^T$ is the error vector associated with samples. In (3), minimizing the objective function ensures the hyperplane proximal to data points of one class and the constraints keep the hyperplane at least 1 away from the other pattern.

Using the Lagrange multiplier technique and Karush-Kuhn-Tucker (KKT) conditions, one can obtain the following Wolfe dual of (3):

$$\max_{\alpha} \quad \mathbf{e}_2^T \alpha - \tfrac{1}{2} \alpha^T \mathbf{G}(\mathbf{H}^T \mathbf{H})^{-1} \mathbf{G}^T \alpha$$
$$subject\ to \quad 0 \leq \alpha \leq C_1 \tag{5}$$

where $\mathbf{H} = [\mathbf{A}, \mathbf{e}_1]$ and $\mathbf{G} = [\mathbf{B}, \mathbf{e}_2]$. Similarly, the dual of (4) has the following form:

$$\max_{\gamma} \quad \mathbf{e}_1^T \gamma - \tfrac{1}{2} \gamma^T \mathbf{P}(\mathbf{Q}^T \mathbf{Q})^{-1} \mathbf{P}^T \gamma$$
$$subject\ to \quad 0 \leq \gamma \leq C_2 \tag{6}$$

where $\mathbf{P} = [\mathbf{A}, \mathbf{e}_1]$ and $\mathbf{Q} = [\mathbf{B}, \mathbf{e}_2]$.

Once the augmented vector $\mathbf{u} = [\mathbf{w}_1, \mathbf{b}_1]^T$ and $\mathbf{v} = [\mathbf{w}_2, \mathbf{b}_2]^T$ are given by

$$\mathbf{u} = -(\mathbf{H}^T \mathbf{H})^{-1} \mathbf{G}^T \alpha \tag{7}$$

$$\mathbf{v} = (\mathbf{Q}^T \mathbf{Q})^{-1} \mathbf{P}^T \gamma \tag{8}$$

the separating hyperplanes (2) can be obtained. Then a new data point $\mathbf{x} \in \mathbf{R}^n$ can be assigned to class +1 or class −1, depending on which of the pair of nonparallel hyperplanes lies closer to it, i.e.,

$$\mathbf{x}^T \mathbf{w} + b = \min_{l=1,2} |\mathbf{x}^T \mathbf{w}_l + b_l| \tag{9}$$

where $|\cdot|$ denotes the perpendicular distance between a point and a plane.

For the linearly inseparable case, the data in the original space would be implicitly mapped into a high-dimensional feature space using kernel functions. The detail content about it refers to [2].

3 Decision Tree Twin Support Vector Machine Based on Kernel Clustering

Since the original TWSVM is a binary classification algorithm, it is necessary to find a way to implement multi-class classification tasks. Decision tree is an efficient and practicable way for transforming a multi-class classification problem from a large and complex one into binary and multi-layer classification problems. Thus, decision tree has been applied to SVM and TWSVM for solving multi-class classification tasks [14, 16]. However, DTTSVM proposed in [14] possibly does not build the best decision tree, especially under the condition that the sample distribution is disordered or that the pattern is linearly inseparable. The main reason is that this method constructs the decision tree using the original data instead of the mapped data in the case of applying kernel function.

This paper develops a novel decision tree twin support vector machine based on kernel clustering (DT^{2}SVM-KC). It is well known, a decision tree is made up of three kinds of nodes: one root node, non-leaf nodes and leaf nodes. In DT^{2}SVM-KC, the decision tree is the binary one, the simplest form in decision trees, in which each non-leaf node has no more than 2 branches. Using the binary tree for multi-class classification problems can also solve the problem about ambiguous areas.

3.1 Construction of Decision Tree

Different from DTTSVM, we construct the binary decision tree using the kernel clustering algorithm [15] instead of the maximum distance rule. The kernel clustering algorithm is also used in [16], which combines the classical clustering algorithm available with the kernel technique. Thus, it can deal with nonlinear data. Before going to the details of our method, we take a look at the kernel clustering algorithm.

Assume that we have a multi-class training data set $\{(\mathbf{x}_1, v_1), \ldots, (\mathbf{x}_m, v_m)\}$, where $(\mathbf{x}_i, v_i)$ is the i-th data point, the vector $\mathbf{x}_i$ is in the n-dimensional space $\mathbf{R}^n$, $v_i \in \{1, 2, \ldots, c\}$ denotes the class label of $\mathbf{x}_i$, c is the number of classes, and m represents the total number of training samples. To divide whole data set into two subsets, we could use a clustering algorithm. However, points with the same label may be divided in different clusters. Thus, we only cluster the class centers of samples for reducing computational complexity and avoiding mess.

By using kernel functions, all samples $\mathbf{x}_i$ are implicitly mapped into a high-dimensional feature space. Without loss of generality, let $\Phi(\mathbf{x}_i)$ be the image of $\mathbf{x}_i$ in the feature space.

Let X_p be the set of samples in the p-th class. Then, the class center of the p-th class in the feature space is

$$\mathbf{z}_p = \frac{1}{|X_p|} \sum_{\mathbf{x}_i \in X_p} \Phi(\mathbf{x}_i), \ p = 1, \ldots, c \tag{10}$$

where $|X_p|$ denotes the number of samples in the p-th class. Thus, the kernel clustering algorithm deals with only the set $\{\mathbf{z}_p\}_{p=1}^c$, which will be partitioned into two groups.

In the feature space, the distance between the feature sample $\mathbf{z}_p$ and a cluster center $\mathbf{m}_q$ can be computed as:

$$d\left(\mathbf{z}_p, \mathbf{m}_q\right) = \left\| \frac{1}{|X_p|} \sum_{\mathbf{x}_i \in X_p} \Phi(\mathbf{x}_i) - \Phi(\bar{\mathbf{x}}_q) \right\| \tag{11}$$

where $\bar{\mathbf{x}}_q$ is the original image of $\mathbf{m}_q$ in the input space, and $q = 1, 2$.

Actually, $\mathbf{m}_q$ and $\mathbf{z}_p$ are unknown. Thus, (11) could be not computed directly. However, we can introduce kernel tricks to solve this problem. Let $k\left(\mathbf{x}_i, \mathbf{x}_j\right) = \Phi^T(\mathbf{x}_i)\Phi(\mathbf{x}_j)$ be the kernel function. Therefore, (11) can be rewritten as:

$$d\left(\mathbf{z}_p, \mathbf{m}_q\right) = $$
$$\sqrt{k\left(\bar{\mathbf{x}}_q, \bar{\mathbf{x}}_q\right) - \frac{2}{|X_p|} \sum_{\mathbf{x}_i \in X_p} k\left(\mathbf{x}_i, \bar{\mathbf{x}}_q\right) + \frac{1}{|X_p|^2} \sum_{\mathbf{x}_i \in X_p} \sum_{\mathbf{x}_i \in X_p} k\left(\mathbf{x}_i, \mathbf{x}_j\right)}$$

Now, we obtain the distances between class centers and cluster centers. Then the partition matrix $\mathbf{D}$ can be generated by

$$D_{pj} = \begin{cases} 1, & \text{if } j = \underset{q=1,2}{argmin}\left\{d\left(\mathbf{z}_p, \mathbf{m}_q\right)\right\} \\ 0, & \text{otherwise} \end{cases} \tag{12}$$

The main step of kernel clustering is to update $\mathbf{m}_q$. For simplicity, we first define three kernel Gram matrices, $\mathbf{Kxx}$, $\mathbf{Kxm}$ and $\mathbf{Kmm}$:

$$Kxx_{pp'} = \frac{1}{|X_p||X_{p'}|} \sum_{\mathbf{x}_i \in X_p} \sum_{\mathbf{x}_j \in X_{p'}} k\left(\mathbf{x}_i, \mathbf{x}_j\right) \tag{13}$$

$$Kxm_{pq} = \frac{1}{|X_p|} \sum_{\mathbf{x}_i \in X_p} k\left(\mathbf{x}_i, \bar{\mathbf{x}}_q\right) \tag{14}$$

$$Kmm_{qq'} = k\left(\bar{\mathbf{x}}_q, \bar{\mathbf{x}}_{q'}\right) \tag{15}$$

Since $\mathbf{Kxx}$ is fixed, the update $\mathbf{m}_q$ depends on the update of kernel matrixes $\mathbf{Kxm}$ and $\mathbf{Kmm}$. The kernel matrices can be recomputed as:

$$Kxm_{pq} = \frac{\sum_{i=1}^{c} D_{iq} Kxx_{ip}}{\sum_{i=1}^{c} D_{iq}} \tag{16}$$

$$Kmm_{qq'} = \frac{\sum_{i=1}^{c}\sum_{j=1}^{c} D_{ip}D_{jj}Kxx_{ij}}{\left(\sum_{i=1}^{c} D_{cq}\right)\left(\sum_{i=1}^{c} D_{iq'}\right)} \tag{17}$$

We update **Kxm** and **Kmm** until they are convergent.

3.2   Algorithm Description

In our proposed method, the most important thing is to generate a suitable decision tree using kernel clustering. Briefly, we spilt the whole training data points into two groups by performing the kernel clustering algorithm on root node and non-leaf nodes. In doing so, we can get a binary decision tree.

On all non-leaf nodes, we use the binary TWSVM to find the partition of two groups of classes. Each leaf node represents a class, so if we do a c-class problem, our decision tree will totally have $(2c - 1)$ nodes, including $(c - 1)$ non-leaf nodes and c leaf-nodes. Thus, there are $(c - 1)$ TWSVMs in our proposed method.

Subsequently, we describe the proposed method in details. The training algorithm and test algorithm will be given in Algorithms 1 and 2, respectively. From the algorithm description, we can know that in the training process, the proposed method conducts $(c - 1)$ kernel clustering processes and $(c - 1)$ binary TWSVMs on the non-leaf nodes. As the increasing layers, the number of training data and classes on one node decreases. Since the multi-TWSVM needs c binary TWSVMs and each time it uses all training points, our proposed method has the smaller number of training data. In order to obtain better decision tree and better results of generalization, the computational complexity of kernel clustering is a litter higher than that of the distance formula used by DTTSVM. Generally, the proposed method achieves a great balance between efficiency and accuracy.

Algorithm 1 Training of DT^{2}SVM-KC.

Input: Training set $\{(\mathbf{x}_i, y_i)\}_{i=1}^{m}$, parameters C_1 and C_2, kernel function and its kernel parameter, and the number of classes c.

---

1. Check the current nodes: If the node has more than 2 classes, go to step2; if the node has 2 classes, go to step 3; If the node has only 1 class, go to step 4.
2. Split the data points into two groups by using kernel clustering.
3. Train TWSVM using the current training set to obtain two non-parallel hyperplanes.
4. Go to step 1 until all non-nodes has been trained.

Output: A decision tree with trained parameters

Algorithm 2 Test of DT²SVM-KC.

Input: Decision tree with trained TWSVM models and a test sample **x**

1. Search the decision tree. If the node is a leaf node, go to Step 2; if the node is a non-leaf node, go to Step 3.
2. Assign the data point to the class represented by this node and return.
3. Calculate the perpendicular distances $\left| \mathbf{x}^T \mathbf{w}_1 + b_1 \right|$ and $\left| \mathbf{x}^T \mathbf{w}_2 + b_2 \right|$, where $(\mathbf{w}_1, b_1)$ and $(\mathbf{w}_2, b_2)$ are trained TWSVM models on the current node;
4. Move to one of the child nodes based on the minimum distance;
5. Go to Step 1.

Output: The label of sample **x**.

---

4 Experimental Results

In order to verify the effectiveness and feasibleness of the proposed method, we conduct experiments on some benchmark data sets available at the UCI machine learning repository (https://archive.ics.uci.edu/ml/datasets.html): Iris, Glass, Vehicle and Segment. In Table 1, we describe these data sets.

Table 1. Description of data set.

Data Set	Samples	Attributes	Classes
Iris	150	4	3
Glass	214	9	6
Vehicle	846	18	4
Segment	2310	19	7

For the sake of comparison, we also give the classification results of multi-TWSVM [13], DTTSVM [14] and DTSVM [16]. All these data classification methods were implemented by using MATLAB R2015b running on a PC with Intel i5 processor (2.40 GHz) with 4 GB RAM.

The generalization accuracy is determined by the standard ten-fold cross-validation methodology, which is also used to select penalty parameters. The Gaussian radial basis function (RBF) kernel parameter is determined by using the median method in [17].

4.1 Vehicle Dataset

Regularization Parameter Analysis. We first analyze the effect of parameters C_1 and C_2 on the classification performance of DT²SVM-KC on the Vehicle dataset. C_1 and C_2 vary from 2^{-5} to 2^4. The bar graphs of accuracy vs. parameters are shown in Fig. 1, where Fig. 1(a) and (b) show the results of the linear kernel and the RBF kernel, respectively. From Fig. 1, we can know that the accuracy is more impressionable using the linear kernel and relatively stable using the RBF kernel when parameters change between 2^{-5}–2^4. In general, the accuracy decreases with the parameters increasing when using linear function. Comparing two bar graphs, the results of using the RBF kernel are obviously better than those of using the linear kernel. Since to some extent parameter selection is significant to the final results, we apply the ten-fold cross-validation methodology to determine C_1 and C_2 in our experiments.

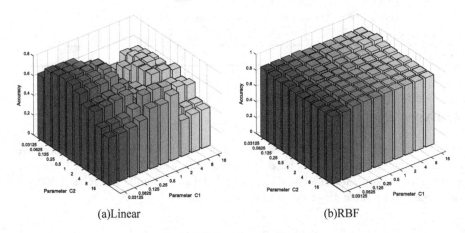

(a)Linear   (b)RBF

Fig. 1. Accuracy vs. regularization parameters on Vehicle using different kernels, (a) linear and (b) RBF.

Decision Tree Analysis. In the algorithm, an important step is to generate a proper binary tree. We present binary trees generated by DTTSVM and our proposed method in Fig. 2. The left decision tree is built by DT^{2}SVM-KC and the right one is constructed by DTTSVM. From Fig. 2, the decision tree generated by our method has less layers and is more balanced. The experimental results also verify that our tree shows better generalization ability than the other decision tree.

4.2   Experimental Results

We compare the classification results of four methods for multi-class classification tasks. During experiments, we choose 80% data for training and the rest for test. The parameters C_1 and C_2 are both obtained through searching in the range of 2^{-5}–2^4 using ten-fold cross-validation. Once parameters are determined on training set, we apply them to the test set and obtain the classification accuracy. We repeat this process 10 runs. The average accuracies on 10 runs obtained by four methods are summarized in Table 2 and 3 for the linear kernel and RBF kernel, respectively, where the best accuracy is in bold.

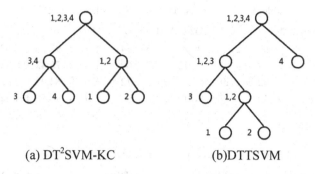

(a) DT²SVM-KC   (b)DTTSVM

Fig. 2. The binary trees generated by DT²SVM-KC (a) and DTTSVM (b).

Table 2. Average test accuracy (%) with linear kernel.

Data Set	Multi-TWSVM	DTTSVM	DTSVM	DT²SVM-KC
Iris	94.00 ± 4.10	96.00 ± 2.63	94.67 ± 3.58	**98.00 ± 2.81**
Glass	40.43 ± 7.82	52.51 ± 5.79	**62.72 ± 5.01**	57.44 ± 2.36
Vehicle	71.05 ± 2.37	73.86 ± 6.75	76.15 ± 3.34	**76.55 ± 2.07**
Segment	80.91 ± 3.05	80.95 ± 0.88	62.08 ± 20.5	**83.07 ± 6.72**

According to Table 2, our proposed method shows higher accuracy on three of four UCI data sets when compared with multi-TWSVM, DTTSVM and DTSVM. Specially, on the Vehicle dataset, we have shown the difference of decision trees constructed by DTTSVM and DT²SVM-KC in Fig. 2. Here, DT²SVM-KC is higher about 3% than DTTSVM. Since both DTSVM and DT²SVM-KC use kernel clustering, the two methods have the same decision tree. The difference between the two methods on the accuracy lies on classifiers themselves. Therefore, we can infer that our algorithm of establishing the decision tree is better than that given in DTTSVM.

Observation on Table 3, DT²SVM-KC performs the best on the four datasets among four methods. Generally, the classifier with the RBF kernel has a better performance than that with the linear kernel if the kernel parameter is determined properly. The results in Table 3 further validate the effectiveness of DT²SVM-KC when dealing with multi-class classification tasks.

Table 4 shows the running time of the proposed method with those of other three classification methods. Because the proposed method does more calculations when it generates the decision tree than DTTSVM does, it is reasonable that the proposed method is slower than DTTSVM in the training procedure. However, since TWSVM has a speed advantage over SVM, DT²SVM-KC keeps the advantage over DTSVM. In addition, DT²SVM-KC is much faster than Multi-TWSVM. DT²SVM-KC has a less test time then other three methods.

Table 3. Average test accuracy (%) with rbf kernel.

Data Set	Multi-TWSVM	DTTSVM	DTSVM	DT²SVM-KC
Iris	96.67 ± 2.22	96.00 ± 2.63	94.67 ± 1.72	**97.33 ± 2.63**
Glass	64.59 ± 4.80	71.13 ± 4.70	69.30 ± 7.82	**75.42 ± 5.48**
Vehicle	74.92 ± 2.59	83.07 ± 4.04	72.44 ± 4.50	**84.03 ± 2.88**
Segment	95.45 ± 1.49	96.15 ± 1.42	94.11 ± 1.24	**96.58 ± 1.02**

Table 4. Comparison of training and test time (sec.).

Data Set	Multi-TWSVM		DTTSVM		DTSVM		DT²SVM-KC	
	Training Time	Test Time	Training Time	Test Time	Training Time	Test Time	Training Time	Test Time
Iris	0.607	0.024	0.326	0.012	0.632	0.15	0.413	0.008
Glass	1.36	0.007	1.07	0.03	2.46	0.343	1.34	0.014
Vehicle	17.38	0.022	5.44	0.078	53.49	1.13	12.52	0.046
Segment	433.84	0.188	71.47	0.215	519.29	4.54	163.96	0.185

Experimental results demonstrate that the proposed method retains a low computational complexity of multi-class classification based on TWSVM and has a better performance with other methods.

5 Conclusions

In this paper, we propose a kernel clustering-based decision tree twin support vector machine for multi-class classification. The method adopts the structure of decision tree in which the leaf nodes represent classes. For non-leaf nodes, the kernel clustering algorithm is used to split classes into two groups before TWSVM is employed to train data points. The proposed method has better generalization performance and keeps the speed advantage of TWSVM, which is supported by the experimental results. In the future, we will consider ways to reduce the training time with keeping the accuracy.

Acknowledgements. We thanks to all those who contribute to the data sets of UCI machine learning repository that are used in this paper. This work was supported in part by the National Natural Science Foundation of China under Grant Nos. 61373093, 61402310, 61672364 and 61672365, by the Soochow Scholar Project of Soochow University, by the Six Talent Peak Project of Jiangsu Province of China, and by the Graduate Innovation and Practice Program of colleges and universities in Jiangsu Province.

References

1. Cortes, C., Vapnik, V.: Support-vector networks. Mach. Learn. **10**(3), 273–297 (1995)
2. Khemchandani, R.J., Chandra, S., et al: Twin support vector machines for pattern classification. IEEE Trans. Pattern Anal. Mach. Intell. **29**(5), 905–910 (2007)
3. Tomar, D., Agarwal, S.: Twin support vector machine: a review from 2007 to 2014. Egypt. Inf. J. **16**(1), 55–69 (2015)
4. Shao, Y., Zhang, C., Wang, X., et al.: Improvements on twin support vector machines. IEEE Trans. Neural Networks **22**(6), 962–968 (2011)
5. Kumar, M.A., Gopal, M.: Least squares twin support vector machines for pattern classification. Expert Syst. Appl. **36**(4), 7535–7543 (2009)
6. Sartakhti, J.S., Ghadiri, N., Afrabandpey, H., et al: Fuzzy least squares twin support vector machines. Artificial Intelligence (2015)
7. Ding, S., Huang, H., Xu, X., et al.: Polynomial smooth twin support vector machines. Appl. Math. Inf. Sci. **8**(4), 2063–2071 (2014)
8. Xu, Y., Yang, Z., Pan, X., et al.: A novel twin support-vector machine with pinball loss. IEEE Trans. Neural Networks **28**(2), 359–370 (2017)
9. Xu, Y., Guo, R., Wang, L.: A twin multi-class classification support vector machine. Cogn. Comput. **5**(4), 580–588 (2013)
10. Tomar, D., Agarwal, S.: An effective weighted multi-class least squares twin support vector machine for imbalanced data classification. Int. J. Comput. Intell. Syst. **8**(4), 761–778 (2015)
11. Xie, J., Hone, K.S., Xie, W., et al.: Extending twin support vector machine classifier for multi-category classification problems. Intell. Data Analysis **17**(4), 649–664 (2013)
12. Yang, Z.X., Shao, Y.H., Zhang, X.S.: Multiple birth support vector machine for multi-class classification. Neural Comput. Appl. **22**(1), 153–161 (2013)
13. Zhen, W., Jin, C., Ming, Q.: Non-parallel planes support vector machine for multi-class classification. In: International Conference on Logistics Systems and Intelligent Management, pp. 581–585. IEEE, Harbin (2010)
14. Shao, Y.H., Chen, W.J., Huang, W.B., et al.: The best separating decision tree twin support vector machine for multi-class classification. Procedia Comput. Sci. **17**, 1032–1038 (2013)
15. Li, Z., Da, Z.W., Cheng, J.L., et al.: Kernel clustering algorithm. Chin. J. Comput. **25**(6), 587–590 (2002)
16. Zhang, L., Zhou, W., Su, T.T., et al.: Decision tree support vector machine. Int. J. Artif. Intell. Tools **16**(01), 1–16 (2007)
17. Zhang, L., Zhou, W., Chang, P., et al.: Kernel sparse representation-based classifier. IEEE Trans. Signal Process. **60**, 1684–1695 (2012)

Machine Learning Techniques for Classification of Livestock Behavior

Natasa Kleanthous[1]([⊠]), Abir Hussain[1]([⊠]), Alex Mason[2,4],
Jennifer Sneddon[3], Andy Shaw[4], Paul Fergus[1], Carl Chalmers[1],
and Dhiya Al-Jumeily[1]

[1] Department of Computer Science, Liverpool John Moores University,
Liverpool, UK
N.K.Orphanidou@2015.ljmu.ac.uk, {A.Hussain,P.Fergus,
C.Chalmers,D.Aljumeily}@ljmu.ac.uk
[2] Animalia AS, Norwegian Meat and Poultry Research Institute, Oslo, Norway
alex.mason@animalia.no
[3] Department of Natural Sciences and Psychology, Liverpool John Moores
University, Liverpool, UK
J.C.Sneddon@ljmu.ac.uk
[4] Department of Built Environment, Liverpool John Moores University,
Liverpool, UK
A.Shaw@ljmu.ac.uk

Abstract. Animal activity recognition is in the interest of agricultural community, animal behaviorists, and conservationists since it acts as an indicator of the animal's health in addition to their nutrition intake when the observation is performed during the circadian circle. Machine learning techniques and tools are used to help identify the activities of livestock. These techniques are helpful to discriminate between complex patterns for classifying animal behaviors during the day; human observation alone is labor intensive and time consuming. This research proposes a robust machine learning method to classify five activities of livestock. To prove the concept, a dataset was utilized based on the observation of two sheep and four goats. A feature selection technique, namely Boruta, was tested with *multilayer perceptron*, *random forests*, *extreme gradient boosting*, and *k-Nearest neighbors* algorithms. The best results were obtained with random forests achieving accuracy of 96.47% and kappa value of 95.41%. The results showed that the method can classify grazing, lying, scratching or biting, standing, and walking with high sensitivity and specificity.

Keywords: Machine learning · Feature extraction · Feature selection
Animal behavior · Signal processing · Accelerometer · Gyroscope
Magnetometer

1 Introduction

Animal activity recognition is of immense importance. It helps with understanding what one can gain from such knowledge, since the animal's wellbeing can be classified and predicted based on daily activities. Evidence suggests that the animal's daily

L. Cheng et al. (Eds.): ICONIP 2018, LNCS 11304, pp. 304–315, 2018.
https://doi.org/10.1007/978-3-030-04212-7_26

activity can be used as an indicator of its health status [1]. Monitoring the animal's daily activity using sensors helps the identification of stress and diseases, such as lameness [2], as well as their daily nutritional intake. Such information is time consuming and labor intensive when it is obtained solely through human observation.

There are a large volume of scientific publications describing the role identification of animal activity plays using inertial measurement units (IMUs) and classification techniques. Animal disease, food intake and overall daily activity or inactivity can be predicted and identified. They are considered as valuable tools in remote monitoring and diagnosis of animal welfare. Positional and animal activity information can be combined and used to qualify patterns of pasture use and animal distribution for targeted animal behavior [3]. It is noted in the literature that animal's decreased activity or hyper activity could be an indicator of disease and distress [4].

Accelerometers are one of the most commonly used sensors, due to their ability to provide information on animal gait patterns. The use of such sensors is convenient since they are light, small, and low power. Other studies have demonstrated computer vision and image analysis for studying such behavior [5, 6]. Furthermore, several studies gathered tracking data from livestock using GPS collars [7–9]. Each of those studies highlight the importance of computer systems for identifying animal behavior and the contribution of such knowledge in supporting the research community as well as the industry.

Different machine learning (ML) techniques are available to classify the activities of animals. Global positioning system (GPS) data was gathered from cattle to identify and understand different activities using various methods including linear discriminant analysis [7], classification tree analysis [8], and k-means [9]. Discriminant analysis was also used by incorporating accelerometer measurements on sheep to study three different behaviors: grazing, ruminating, and resting [10]. Additionally, a combination of GPS and accelerometer measurements were used to detect other behaviors including, foraging, ruminating, traveling, resting and 'other active behaviors' of cattle. This was done with the use of a threshold-based decision trees method [11]. An embedded multilayer perceptron method was used with an accelerometer, gyroscope and magnetometer to predict horse behavior when motionless, trotting, and walking [12]. Furthermore, this method was also applied on sheep accelerometer measurements to predict some behaviors including, grazing, lying down, walking, standing and others [13]. Kamminga et al. [14] used deep neural networks on data collected from sheep and goats to predict activities such as stationary, foraging, walking, trotting, and running. Decision trees have been used with a tilt sensor on sheep to classify active and inactive phases [15]. Additionally, the cows' feeding and standing activities were predicted from accelerometer data using threshold-based statistical analysis [16]. Linear discriminant analysis was also used to identify five activities on sheep using accelerometer data [17]. Discriminant analysis and embedded statistical classifiers were used to study measurements from accelerometer data collected from sheep [18, 19]. All of the abovementioned studies managed to predict the behaviors of the animals with good accuracy. However, more research is needed in the identification of features used and method optimization to improve system performance (i.e. sensitivity, specificity, and accuracy).

An intelligent system is needed to provide information of where the animals are and what they are doing, where they mostly graze, and what their nutritional habits are during the course of a day [20]. Having this vast amount of data, decisions about animal health, animal position monitoring, distribution control, and efficient land utilization [21] can help prevent soil erosion, soil contamination, water pollution and spread of animal diseases [22].

This paper reports the findings from using different features and ML methods. An online open source dataset [23] was used as a testbed to identify the most promising features and ML methods, with the aim of implementing the findings in a larger body of work concerned with virtual fencing. The main novelty of this work is the different combination of feature extraction, feature selection and ML techniques, which allows the identification of an optimal method to improve sensitivity, specificity, kappa value, and accuracy of the classifier. To the best of our knowledge, no previous work combined those techniques before.

The reminder of this paper is organized as follows: Sect. 2 describes the dataset and the methodology used for conducting the animal activity recognition analysis. Section 3 presents the results obtained from the classifiers while Sect. 4 encompasses the discussion and conclusion part of the study. Finally, the future work is included in Sect. 5.

2 Methodology

This section comprises the methods and techniques used for detection of animal behavior using intelligent systems. It also contains a description of the dataset, data preparation steps, feature extraction and feature selection techniques. Additionally, it defines the machine learning algorithms and the evaluation metrics used to assess the performance of the classifiers. Figure 1 shows the stages used during the analysis process; this forms the basis of the methodology.

2.1 Dataset

The dataset used for this study is available online from Data Archiving and Networked Services (DANS) website [14, 23] and contains labeled behavioral data from four goats and two sheep. Each data file contains a record of daily activity for each individual animal. The file holds nine different behaviors including, grazing, scratching or biting, standing, walking, fighting, running, trotting, shaking, and lying.

The measurements are obtained from sensors placed on the collars of the animals. The parameters used in this study are described in Table 1, however, the original dataset included other measurements (e.g. temperature and pressure), which are not included in this study. Additionally, four other activities are omitted, namely, fighting, running, trotting, and shaking, because these behaviors occurred too infrequently to draw statistically significant conclusions. Consequently, five activities are included; grazing, lying, scratching or biting, standing, and walking.

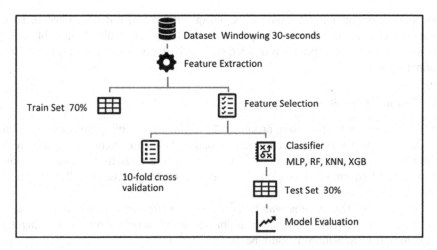

Fig. 1. A breakdown of the stages used during the analysis process

Table 1. Annotation and description of the parameters used from each dataset

Parameters	Description of raw data
ax	Accelerometer x-axis. Sampling Rate: 200 Hz
ay	Accelerometer y-axis. Sampling Rate: 200 Hz
az	Accelerometer z-axis. Sampling Rate: 200 Hz
axhg	High G accelerometer x-axis. Sampling Rate: 200 Hz
ayhg	High G accelerometer y-axis. Sampling Rate: 200 Hz
azhg	High G accelerometer z-axis. Sampling Rate: 200 Hz
cx	Magnetometer x-axis. Sampling Rate: 100 Hz
cy	Magnetometer y-axis. Sampling Rate: 100 Hz
cz	Magnetometer z-axis. Sampling Rate: 100 Hz
gx	Gyroscope x-axis. Sampling Rate: 200 Hz
gy	Gyroscope y-axis. Sampling Rate: 200 Hz
gz	Gyroscope z-axis. Sampling Rate: 200 Hz

The final dataset used for this research study consists of time series measurements from an accelerometer (acc), a high-g accelerometer (hgacc), a magnetometer (magn), and a gyroscope (gyr). Additionally, all sheep and goat datasets were used to examine the performance of different algorithms to learn their activities and identify if the algorithms can generalize well between different animals and activities.

2.2 Data Preparation

The dataset was examined to identify any inconsistencies. Missing values were present in the magnetometer measurements and were replaced using linear interpolation. The sensor measurements were segmented into 30-second data blocks. Each data block comprised individual animal behaviors. Consequently, the final data resulted in 1610 randomly selected data blocks comprising a relatively even distribution of the 5 selected activities. The dataset was divided such as 70% was used for training and 30% for final testing.

2.3   Feature Extraction

A key phase of the methodology process is to identify a suitable pattern mapping for the ML. To achieve that, and as a first step, a total of 23 features were extracted from all parameters of the dataset (refer to Table 1). Therefore, features from various domains (i.e. time and frequency) encompassed a total of 276 (23 features and 12 parameters) newly created features.

The choice of the features was based on their performance in previous research concerned with activity recognition problems using acceleration forces, angular momentum and orientation measurements.

Features from the time domain are calculated using the raw dataset. However, for frequency features, the raw data had to be translated into the frequency domain using Fourier transform [24, 25]. The final selected features are defined in Table 2 providing additional information regarding the formulas of each and the parameters used.

2.4   Exploratory Analysis

To acquire a better understanding of the dataset distribution, a principal component analysis (PCA) mapping was applied and plotted. From Fig. 2, it can be observed that there is a clear distinction between the 5 classes. Additionally, the PCA plot indicates that the walking and scratching or biting classes display substantial disconnection and may establish a simple decision boundary. Conversely, a considerable overlap is demonstrated between the standing and lying classes, thus, constitution of the class predictions might be more difficult.

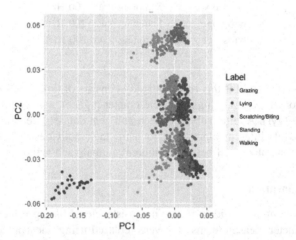

Fig. 2. PCA plot: A 2-dimensional illustration of the extracted behavioral data

Table 2. Definitions of the final features

Feature	Formula	Parameters
Mean	The average value of the signal $\bar{x} = \frac{1}{n}\sum_{i=1}^{n} x_i$	cx
Standard Deviation	Standard deviation of the signal (square root of the variance) $s = \sqrt{\frac{\sum_{i=1}^{n}(x_i - \bar{x})^2}{n-1}}$	gz
Root Mean Square (RMS) velocity	The quadratic mean of the speed of the signal in the time domain $RMS_{velocity} = \sqrt{\frac{1}{n} \cdot \sum_{i=1}^{n} diffinv(x_i)^2}$	gz
Root Mean Square (RMS)	The square root of the mean of the square of the signal $RMS = \sqrt{\frac{1}{n}\sum_{i=1}^{n} x_i^2}$	gz
Energy	The sum of squared magnitudes of the time window $Energy = \sum_{i=1}^{n} x_i^2$	gz
Sum of Changes	The sum of the consecutive differences of the window $\sum_{i=1}^{n} diff(x_i)^1$	az
Mean Absolute Change	The absolute mean of the consecutive differences of the window $\overline{xc} = \frac{1}{n}\sum_{i=1}^{n} |diff(x_i)|$	az
Absolute Integrals	The sum of the integrals of x, y, z accelerometer dimensions where T is the number of data points in each window, and x, y, z are the accelerometer coordinates (n is the number of samples in each window) $I = \int_{t=0}^{T} |x|dt + \int_{t=0}^{T} |y|dt + \int_{t=0}^{T} |z|dt$	gz
Squared Integrals	The sum of squared integrals of x, y, z accelerometer dimensions (T is the time frame, T = n, as n is the number of samples in each window) $I^2 = \left[\int_{t=0}^{T} |x|dt\right]^2 + \left[\int_{t=0}^{T} |y|dt\right]^2 + \left[\int_{t=0}^{T} |z|dt\right]^2$	gz, cx
Madogram	The measure of roughness or smoothness with power index of 1 (p = 1) $\gamma_p(t) = \frac{1}{2}E|X_u - X_{i+u}|^P, p = 1$	az, ax, gy
Peak Frequency	The frequency of maximum power in the Power Spectral Density where Fs is the sample frequency $f_{max} = arg\left(\frac{f_s}{n} max_{i=0}^{n-1} P(i)\right)$	gz, gx

2.5   Feature Selection

The feature selection method is used to reduce the number of features extracted from the raw data. The objective is to have a better understanding of the dataset and to provide a faster predictor [26]. In the feature selection process, the features are chosen based on their ranking and best suitability to the classifiers' performance. The main methods of feature selection are (1) filter, (2) wrapper, and (3) embedded [26].

Filter methods apply statistical measures to allocate ranking scores to each feature. Thus, according to their score, they are either included or excluded from the dataset. Wrapper methods select combinations feature subsets and evaluate their usability on a given machine learning algorithm. Accordingly, the subsets are scored based on their predictive power. On the other hand, embedded methods learn the best features according to the correctness (accuracy) of the learning model.

For this study, the Boruta algorithm was used as the feature selection technique [27]. The Boruta algorithm is a wrapper method which uses a random combination of feature sets and evaluates their importance according to random probes. This method uses the random forest classifier for the selection of all features which are relevant. Consequently, 276 features were used by the Boruta algorithm to identify and accept the most relevant. This process resulted in a set of features which are ranked in decreasing order. Therefore, the top 15 most important features are selected in this study to train the ML algorithms. Table 2 provides the selected feature definitions and their mathematical formulas. From the sensor measurements, only 5 parameters are finally used. The x and z axis of the acceleration, the x and z axis of the gyroscope, and the x axis of the magnetometer.

2.6   Classification Algorithms

A supervised learning approach is used as it is more suitable for the nature of the investigated dataset. In supervised learning, there exists a set of independent variables {x1, x2...xn} and a set of dependent variables {y1, y2...yn}. The aim is to fit a model based on the observations of x, and the response y, and predict y based on a new set of x observations. To achieve that, the model must be trained based on the already known data and then test its ability to predict or classify unknown data. The four classifiers selected for the detection of the five activities of the animals in the pasture are Multilayer Perceptron (MLP) [28], Random Forest (RF) [29], Extreme Gradient Boosting (XGB) [30], and k-Nearest Neighbors (KNN) [31].

2.7   Evaluation Metrics

Sensitivity, specificity, Kappa value, and accuracy are used to evaluate the performance of the models. The classification problem in this study is concerned with a multiclass problem and the classes are evaluated, using the true prediction of the class against the total false prediction of the other classes [32]. The calculations use the true positive (TP), true negative (TN), false positive (FP), and false negative (FN) metrics. The formulas for each metric are illustrated in Table 3.

Table 3. List of the evaluation metrics

Metric	Formula
Sensitivity	$Se = TP/(TP + FN)$
Specificity	$Sp = TN/(TN + FP)$
Accuracy	$Ac = (TP + TN)/(TP + TN + FP + FN)$
Kappa value	$K = (Po - Pe)/(1 - Pe)$, where Po is the proportion of observed agreement and Pe is the proportion of agreement expected by chance.

2.8   Model Training Using 10-Fold Cross Validation

The classifiers are trained using 70% of the original dataset (training set) and tested its performance on 30%. The model training used 10-fold cross validation technique with 1 repetition on the training set to better tune the model parameters as shown in Table 4. The overall results by means of accuracy, kappa value, sensitivity, and specificity indicated a high model performance for all with the best results obtained by RF and XGB. For the RF classifier, the final value used for the model was 8 variables per level and was selected based on the highest accuracy. The XGB's maximum depth of the tree was 3 and collected half of the data instances to grow trees in order to prevent over-fitting and decrease the computation time. Using the same criteria as the RF, the optimal model for KNN selected k = 5 neighbors, and for MLP the size = 5 hidden layers and the regularization parameter decay = 0.1.

Table 4. Summary of performance of the models (10-fold cross validation)

Model	Accuracy	Kappa	Sensitivity	Specificity
MLP	95.21%	93.79%	95.36%	98.76%
RF	96.73%	95.74%	96.40%	99.14%
XGB	97.34%	96.54%	96.96%	99.13%
KNN	92.63%	90.45%	93.19%	98.07%

3   Results

Following the 10-fold cross validation to train the classifiers, the performance of the models was evaluated on the unseen testing set. Table 5 illustrates the results obtained using 30 s windows and 15 features on the test set. It can be observed that the best performance was achieved using RF yielding an accuracy and kappa value of 96.47% and 95.41%, respectively. The second-best results obtained by XGB with 95.85% accuracy and 94.60% kappa value. The grazing, scratching or biting and standing activities identified by RF are higher than the other 3 models. On the other hand, MLP and XGB obtained the highest sensitivity for walking activity with 98.11%.

Table 5. Model performance – test set

ML Algorithms	MLP	RF	XGB	KNN
	Accuracy	Accuracy	Accuracy	Accuracy
	94.40%	**96.47%**	95.85%	93.57%
	Kappa Value	Kappa Value	Kappa Value	Kappa Value
Classes	92.73%	**95.41%**	94.60%	91.66%
Grazing				
Sensitivity	96.09%	97.66%	96.09%	96.88%
Specificity	98.02%	97.74%	98.31%	97.74%
Lying				
Sensitivity	91.53%	93.22%	93.22%	93.22%
Specificity	97.87%	99.76%	99.53%	97.16%
Scratching or Biting				
Sensitivity	94.62%	95.70%	94.62%	92.47%
Specificity	99.74%	99.74%	99.49%	99.74%
Standing				
Sensitivity	92.62%	97.32%	96.64%	90.60%
Specificity	97.30%	98.50%	97.60%	97.60%
Walking				
Sensitivity	98.11%	96.23%	98.11%	96.23%
Specificity	99.77%	99.53%	99.53%	99.53%

All models obtained sensitivity for the grazing activity in the range of 96.09%-97.66%, and specificity in the range of 97.74%-98.31%. Overall, all activities are correctly classified with high results between 91.53% and 99.77%. The KNN classifier obtained the lowest results by means of accuracy and kappa value with 93.57% and 91.66% respectively. The results are promising for the future implementation a system to predict the behavior of the animals.

4   Discussion and Conclusion

Identifying animal behavior and calculation of circadian rhythms is of immense importance since it can act as an indicator of animal wellbeing [33]. Additionally, grazing activity information offers insights on the food intake and preference of the animals. Knowing where the animals mostly graze allows the farmers to efficiently manage animal distribution which can prevent overgrazing and consequently soil erosion and soil contamination.

In this study, an online dataset was used from a recent research study featuring 5 different activities of 4 goats and 2 sheep, including, grazing, lying, scratching or biting, standing, and walking. The aim was to identify the most significant features and to select the machine learning algorithm that could be used to classify the 5 activities with the highest accuracy and kappa value. Various features were extracted from an accelerometer, a gyroscope, and a magnetometer data resulting in a total of 276 feature mappings. Due to the high dimensionality of the dataset, feature selection was applied using the Boruta wrapper method. The top 15 most important features were then used as the main predictors to train random forest, extreme gradient boosting, multilayer perceptron, and k-nearest neighbors. All 4 algorithms achieved high accuracy and Kappa values, indicating that the features could discriminate correctly between the classes. The random forest classifier obtained the best results having an accuracy of 96.47% and Kappa value of 95.41% for 30 s mutually exclusive behaviors. To the best of our knowledge, those results are higher than previous research concerned with the classification of sheep behaviors as shown in Table 6.

Table 6. Comparison with previous studies

Ref	Animals	Classes	Signal	Window	Method	Accuracy
[15]	Sheep	2	Tilt	30-s	LDA	94.4%
[34]	Sheep	5	acc	10-s	DT	91.3%
[17]	Sheep	5	acc	5.3-s	LDA	85.7%
[13]	Sheep	5	acc	–	MLP	76.2%
[14]	Sheep, Goats	5	acc	1-s	DNN	94.00%
[10]	Sheep	3	acc	60-s	DA	93.00%
This work	Sheep, Goats	5	acc, magn, gyr	30-s	RF	96.47%

5 Future Work

In the future, proposed technique will be tested for behavior prediction in real time using embedded sensors on the animals' collars. Additionally, the findings will be implemented in a larger body of on-going work concerned with virtual fencing.

Acknowledgements. The authors would like to acknowledge and thank the Douglas Bomford Trust for the financial and moral support during the project. Additionally, we thank the authors who made their dataset publicly available for use by the community [23].

References

1. McLennan, K.M., et al.: Technical note: validation of an automatic recording system to assess behavioural activity level in sheep (Ovis aries). Small Rumin. Res. **127**, 92–96 (2015)
2. Barwick, J., Lamb, D., Dobos, R., Schneider, D., Welch, M., Trotter, M.: Predicting lameness in sheep activity using tri-axial acceleration signals. Animals **8** (2018)
3. Shepard, E.L.C., et al.: Identification of animal movement patterns using tri-axial accelerometry. Endanger. Species Res. **10**, 47–60 (2008)
4. Gougoulis, D.A., Kyriazakis, I., Fthenakis, G.C.: Diagnostic significance of behaviour changes of sheep: a selected review. Small Rumin. Res. **92**, 52–56 (2010)

5. Krahnstoever, N., Rittscher, J., Tu, P., Chean, K., Tomlinson, T.: Activity recognition using visual tracking and RFID. In: Seventh IEEE Workshops on Application of Computer Vision, WACV/MOTIONS 2005, vol. 1, pp. 494–500 (2005)

6. Cangar, Ö., et al.: Automatic real-time monitoring of locomotion and posture behaviour of pregnant cows prior to calving using online image analysis. Comput. Electron. Agric. **64**, 53–60 (2008)

7. Schlecht, E., Hülsebusch, C., Mahler, F., Becker, K.: The use of differentially corrected global positioning system to monitor activities of cattle at pasture. Appl. Anim. Behav. Sci. **85**, 185–202 (2004)

8. Ungar, E.D., Henkin, Z., Gutman, M., Dolev, A., Genizi, A., Ganskopp, D.: Inference of animal activity from gps collar data on free-ranging cattle. Rangel. Ecol. Manag. **58**, 256–266 (2005)

9. Schwager, M., Anderson, D.M., Butler, Z., Rus, D.: Robust classification of animal tracking data. Comput. Electron. Agric. **56**, 46–59 (2007)

10. Giovanetti, V., et al.: Automatic classification system for grazing, ruminating and resting behaviour of dairy sheep using a tri-axial accelerometer. Livest. Sci. **196**, 42–48 (2017)

11. González, L.A., Bishop-Hurley, G.J., Handcock, R.N., Crossman, C.: Behavioral classification of data from collars containing motion sensors in grazing cattle. Comput. Electron. Agric. **110**, 91–102 (2015)

12. Gutierrez-Galan, D., et al.: Embedded neural network for real-time animal behavior classification. Neurocomputing **272**, 17–26 (2018)

13. Nadimi, E.S., Jørgensen, R.N., Blanes-Vidal, V., Christensen, S.: Monitoring and classifying animal behavior using ZigBee-based mobile ad hoc wireless sensor networks and artificial neural networks. Comput. Electron. Agric. **82**, 44–54 (2012)

14. Kamminga, J.W., Bisby, H.C., Le, D.V., Meratnia, N., Havinga, P.J.M.: Generic online animal activity recognition on collar tags. In: Proceedings of the 2017 ACM International Joint Conference on Pervasive and Ubiquitous Computing and Proceedings of the 2017 ACM International Symposium on Wearable Computers on - UbiComp 2017, pp. 597–606. ACM, New York (2017)

15. Umstätter, C., Waterhouse, A., Holland, J.P.: An automated sensor-based method of simple behavioural classification of sheep in extensive systems. Comput. Electron. Agric. **64**, 19–26 (2008)

16. Arcidiacono, C., Porto, S.M.C.C., Mancino, M., Cascone, G.: Development of a threshold-based classifier for real-time recognition of cow feeding and standing behavioural activities from accelerometer data. Comput. Electron. Agric. **134**, 124–134 (2017)

17. le Roux, S.P., Marias, J., Wolhuter, R., Niesler, T.: Animal-borne behaviour classification for sheep (Dohne Merino) and Rhinoceros (Ceratotherium simum and Diceros bicornis). Anim. Biotelemetry. **5**, 25 (2017)

18. Radeski, M., Ilieski, V.: Gait and posture discrimination in sheep using a tri-axial accelerometer. Animal. **11**, 1249–1257 (2017)

19. Le Roux, S., Wolhuter, R., Niesler, T.: An overview of automatic behaviour classification for animal-borne sensor applications in South Africa (2017)

20. Anderson, D.M., Estell, R.E., Holechek, J.L., Ivey, S., Smith, G.B.: Virtual herding for flexible livestock management - a review. Rangel. J. **36**, 205–221 (2014)

21. Norton, B.E., Barnes, M., Teague, R.: Grazing management can improve livestock distribution: increasing accessible forage and effective grazing capacity. Rangelands **35**, 45–51 (2013)

22. Rutter, S.M.: 13 - Advanced livestock management solutions. In: Ferguson, D.M., Lee, C., Fisher, A. (eds.) Advances in Sheep Welfare, pp. 245–261. Woodhead Publishing (2017)

23. Kamminga, J.W.: Generic online animal activity recognition on collar tags (2017)

24. Mitra, S.K.: Digital Signal Processing: A Computer-Based Approach. McGraw-Hill School Education Group (2001)
25. Rabiner, L.R., Gold, B.: Theory and Application of Digital Signal Processing (1975)
26. Guyon, I., Elisseeff, A.: An introduction to variable and feature selection. J. Mach. Learn. Res. **3**, 1157–1182 (2003)
27. Kursa, M.B., Rudnicki, W.: Feature Selection with Boruta Package (2010)
28. Bishop, C.M.: Neural Networks for Pattern Recognition. Oxford University Press, New York (1995)
29. Breiman, L.: Random forests. Mach. Learn. **45**, 5–32 (2001)
30. Chen, T., Guestrin, C.: XGBoost: A Scalable Tree Boosting System. arXiv1603.02754 [cs], pp. 785–794 (2016)
31. Kramer, O.: K-nearest neighbors. In: Dimensionality Reduction with Unsupervised Nearest Neighbors, pp. 13–23. Springer, Heidelberg (2013). https://doi.org/10.1007/978-3-642-38652-7_2
32. Rifkin, R., Klautau, A.: In defense of one-vs-all classification. J. Mach. Learn. Res. **5**, 101–141 (2004)
33. Scheibe, K.M., et al.: ETHOSYS (R)—new system for recording and analysis of behaviour of free-ranging domestic animals and wildlife. Appl. Anim. Behav. Sci. **55**, 195–211 (1998)
34. Alvarenga, F.A.P., Borges, I., Palkovič, L., Rodina, J., Oddy, V.H., Dobos, R.C.: Using a three-axis accelerometer to identify and classify sheep behaviour at pasture. Appl. Anim. Behav. Sci. **181**, 91–99 (2018)

Two-Stage Attention Network
for Aspect-Level Sentiment Classification

Kai Gao[1,2], Hua Xu[1(✉)], Chengliang Gao[1,2], Xiaomin Sun[1], Junhui Deng[1],
and Xiaoming Zhang[2]

[1] Department of Computer Science and Technology, Tsinghua University,
Beijing 100084, China
xuhua@tsinghua.edu.cn, chengliang_gao@126.com
[2] School of Information Science and Engineering,
Hebei University of Science and Technology, Shijiazhuang 050018, China
gaokai@hebust.edu.cn

Abstract. Currently, most of attention-based works adopt single-stage attention processes during generating context representations toward aspect, but their work lacks the deliberation process: A generated and aspect-related representation is directly used as final output without further polishing. In this work, we introduce the deliberation process to model context for further polishing of attention weights, and then propose a two-stage attention network for aspect-level sentiment classification. The network uses of a two-level attention model with LSTM, where the first-stage attention generates a raw aspect-related representation and the second-stage attention polishes and refines the raw representation by deliberation process. Since the deliberation component has global information what the representation to be generated might be, it has the potential to generate a better aspect-related representation by secondly looking into hidden state produced by LSTM. Experimental results on the dataset of SemEval-2016 task 5 about Laptop indicates that our model achieved the state-of-the-art accuracy of 76.56%.

Keywords: Attention mechanism · LSTM · Text representation
Aspect-level sentiment classification

1 Introduction

The attention mechanism [1,2], which dynamically attends to different parts of source input x while generating each target-side word, is often integrated into neural networks, such as recurrent neural networks(RNN) [4], memory networks [5], and so on [6], to improve the quality of generated context representation. It has adopted for natural language processing (NLP) tasks widely, which includes of machine translation [1], reading comprehension [7], sentiment analysis [8], etc.

Although such works of attention-based neural network has achieved great performances, there is one concern that while modeling the relation between

L. Cheng et al. (Eds.): ICONIP 2018, LNCS 11304, pp. 316–325, 2018.
https://doi.org/10.1007/978-3-030-04212-7_27

context and the special aspect as a real-value vector, it can only leverage words of context and aspect without considering the generated relation representation. That is, when modeling the aspect-related context representation γ, only both tokens w in a given context and tokens v of the special aspect can be used, while γ are not explicitly considered for improving the quality of attention weights. In contrast, in real-world human cognitive processes for sentiment analysis, the global information, including part of the context and the aspect, is leveraged in an iterative polishing process. For example, "It's more expensive but well worth it in the long run." For aspects *price* and *general* of Laptop, the former is negative and the latter is positive, while they are not obvious in context, people first need to understand the raw meaning of this sentence with each aspect respectively. Then use of the raw meaning to refine attention weighs α and to polish γ for identifying the polarity of aspect-related context.

When polishing parts of sentence toward aspect, we take the whole sentence of the meaning into consideration to evaluate how well the local element fits into the global environment rather than only puts the attention weight produced by first-stage attention mechanism as output for aspect-level sentiment classification, which call such the polishing process of the α and the γ as deliberation in the classification.

Motivated by neural machine translation [9], we propose a two-stage attention networks, which leverages the global relation between context and aspect to improve quantities of attention weighs. It is different from the deliberation process in original paper for looking backward and forward in sequence decoding. Concretely, to integrate such process into the context modeling, we carefully design our architecture, which consists of RNN and two stages attention mechanism: a first-stage attention $A1$ and a second-stage/deliberation attention $A2$. Given a source input x, the RNN and $A1$ jointly works like the standard attention-based LSTM model to generate a coarse context information towards aspect $\hat{\gamma}$ as a draft. Afterwards, the deliberation attention $A2$ takes hidden state H and $\hat{\gamma}$ as inputs and outputs the refined attention weighs α and the polished representation γ. In this way, the global information about the aspect-related context can be utilized to refine the generation process.

In order to verify the validity of our model, we conducted some experiments in aspect-level sentiment classification. The main contributions of our work can be summarized as follows: Firstly, for aspect-level sentiment classification, we propose a two-stage attention network, which includes a RNN and a two-stage attention mechanism for context modeling. Secondly, conducting experiment on the SemEval-2016 task 5, we have achieved the state-of-the-art accuracy against baseline methods about the Laptop, and we even outperformed previously reported best result from official website by 0.65%. Thirdly, we further analyzed the gain of the model over single layer attention.

The rest of our paper is organized as follows. We review some related work in Sect. 2, and introduce the proposed deliberation network in Sect. 3, including the model structure. Section 4 details the experimental setup and conducts experiments to justify the effectiveness of our proposals and discusses its results. Section 5 concludes the paper and discusses possible future orientation.

2   Related Works

Aspect-level sentiment classification, a subtask of sentiment analysis, aims to infer the sentiment polarity of the sentence toward target, such as entities or aspect/feature of entities [10], which considers the sentiment information about target. For example, the instance in the *Introduction* has two aspects: *price* and *general* of Laptop, which their sentiment polarity is the opposite.

Existing works about attention have been widely produced for various tasks of sentiment analysis. Reference [8] proposes two models of Target-dependent LSTM and Target-connection LSTM, which regarded the given target as a feature and concatenated it with the context features. Reference [11] takes a real-value vector to represent target feature and further proposes an attention-based LSTM method, which has two ways to interact with target-related context for learning relation between target and context. For aspect-level sentiment classification, reference [12] introduces a deep memory network, a new end-to-end networks model, which combines an attention mechanism and an external memory. Although these works improve performance using the attention-based neural networks framework for context modeling in aspect-level sentiment classification, which take the aspect as additional feature into such networks or the modeling objective, they do not pay much attention to the hierarchical structure of the attention mechanism for deliberation processes. Our work change the structure of the attention mechanism by setting the second-stage attention process on it.

3   Backgrounds

3.1   Long Short-Term Memory (LSTM)

Long Short Term Memory, a special of Recurrent Neural Networks (RNNs) [13], can learn long-term dependency information better than RNNs. Its architecture has an advantage that components of the gradient vector can decay exponentially over long sequences of learning long-term dependencies by adding a memory cell that is able to preserve state over long periods of time. LSTM-based models have become increasingly popular which can be applied to learn hidden task-specific features from sequential data due to its capability of modeling the prefix or suffix context [14]. More formally, each cell in LSTM can be computed as the following.

$$x = \begin{bmatrix} x_t \\ h_{t-1} \end{bmatrix} \tag{1}$$

$$\begin{bmatrix} f_t \\ i_t \\ o_t \end{bmatrix} = \sigma \left(\begin{bmatrix} W_f \\ W_i \\ W_o \end{bmatrix} x + \begin{bmatrix} b_f \\ b_i \\ b_o \end{bmatrix} \right) \tag{2}$$

$$c_t = f_t \odot c_{t?1} + i_t \odot \tanh(W_c \cdot x + b_c) \tag{3}$$

$$h_t = o_t \odot \tanh(c_t) \tag{4}$$

where W_i, W_f and $W_o \in R^{d \times 2d}$ and $W_c \in R^{d \times d}$ are the weighted matrices, and b_i, b_f, b_o and $b_c \in R^d$ are biases of standard LSTM to be learned during training. σ represents the *sigmoid* function and $tanh()$ are the activation functions, and $\odot$ stands for element-wise multiplication. x_t, h_t and c_t is the inputs, hidden state, and cell state of t-th unit of LSTM respectively.

3.2   Attention Mechanism

Moreover, the standard LSTM cannot detect the important part for sentiment classification. Recently, the attention mechanism [3] has become an effective strategy to obtain superior results, because it can be employed to capture the key parts of text [2]. The attention mechanism will produce an attention weights vector α and a weighted representation γ.

$$M = \tanh(W_a X) \tag{5}$$

$$\alpha = softmax(w^T M) \tag{6}$$

$$\gamma = H\alpha^T \tag{7}$$

where $M \in R^{d \times m}, \alpha \in R^m$, and $\gamma \in R^d$. $W_a \in R^{d \times d}$ and $w \in R^d$ are projection parameters.

4   Two-Stage Attention-Based LSTM

In this section, we introduce the overall architecture of two-stage attention networks in detail, and then describe the details of individual components.

4.1   Structure of Our Model

Our proposed two-stage attention network consists of preprocess layer: an RNN P, which usually encodes text, and two-stage attention layer: a first-stage attention $A1$ and a second-stage attention $A2$, as shown in Fig. 1. Deliberation happens at the second-stage attention, so it is also called deliberation attention alternatively.

Briefly speaking, the encode layer P acts as a preprocessing role in the network for source sentence $x = w_1, w_2, w_3, \ldots, w_m$, where m is the length of sentence, which is used to encode source context into a sequence of hidden state H. $A1$ takes the H and target information v_t as inputs and then generates a first-stage the raw global representations towards target $\hat{\gamma}$ as a draft. The $\hat{\gamma}$ is further provided as one of input into the deliberation attention $A2$ for the computing of second-stage attention. In Fig. 1, $\{a_{1,1}, a_{1,2}, a_{1,3}, \ldots, a_{1,m}\}$ and $\{a_{2,1}, a_{2,2}, a_{2,3}, \ldots, a_{2,m}\}$ are attention weights produced by $A1$ and $A2$ respectively, which defines how much of input state should be weighted for global information. For example, if $a_{2,m}$ has a big value, it means that the global information pays a lot of attention to the concatenation of m-th state in the H and v_t. The weights of $a_{1,m}$ the same as $a_{2,m}$ sum to 1 normally.

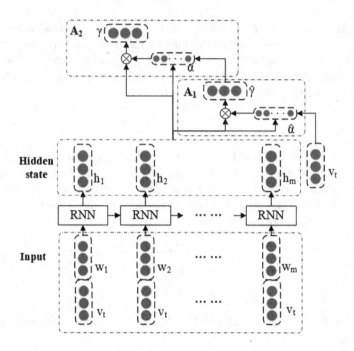

Fig. 1. Framework of deliberation networks. The ⊕ represents a weighed-sum operates. The attention mechanism is omitted for readability.

4.2   Preprocess Layer

Aspect information is vital when classifying the polarity of one sentence toward aspect. For the instance in section *Introduction*, we get opposite polarities about the example towards different aspects. Inspired by in [11], to make the best use of aspect information, we take an embedding vector for each aspect.

Vector $v_{t_i} \in R^{da}$ is represented for the embedding of aspect i, where da is the dimension of aspect embedding. $A \in R^{da \times |A|}$ consists of all aspect embeddings. We introduce the LSTM with aspect embeddings into the preprocess layer due to its performance in [11], which the structure of the model is shown in Fig. 1. Mathematically speaking, in the refinement process, the sampling h_t from the hidden states H is computed as:

$$h_t = LSTM(w_t, v_t, h_{t-1}); t \in [0, m], h_0 = 0 \tag{8}$$

After the input sequence x is fed into the preprocess layer, it is encoded into m hidden states $H = h_1; h_2; \ldots; h_m$. Specifically, $h_i = LSTM(x_i, h_{i?1})$, where x_i acts as the input representation (i.e., word embeddings in this paper) for the i-th word in x and h_0 is a zero vector.

4.3 Two-Stage Attention Layer

In this section, we show details about how the two-stage attention layer generates and polishes the context representation toward a given aspect. The first-stage attention layer $A1$ will generate a series of weighs $a_{1_j}, j \in [1, m]$ for hidden states and a raw attention-aware representation $\hat{\gamma}$, which is defined as follows:

$$\hat{\gamma} = \sum_{i=1}^{m} \alpha_{1,i} h_i; \forall_1 \in [1, m], \sum_{i=1}^{m} \alpha_{1,i} = 1$$

$$\alpha_{1,i} = \exp v_{\alpha_1}^T \tanh W_{h_1} h_i + b_{h_1} \tag{9}$$

Once the raw representation $\hat{\gamma}$ is generated in $A1$, it is fed into the second-stage attention $A2$ for further refining. Based on the representation $\hat{\gamma}$ and the hidden states H provided by recurrent neural network with LSTM, $A2$ eventually outputs the second-stage representation γ via the deliberation process.

Specifically, $A2$ takes the previous representation $\hat{\gamma}$ and the hidden states H as inputs. A detailed point is that the computation of γ is similar to that of $\hat{\gamma}$ as shown in Eq. 9 with two differences: The previous $\hat{\gamma}$ is added as a new input in $A2$ and the model parameters are different. Mathematically, in the refinement process, the second-stage vector γ is computed as:

$$\gamma = \sum_{j=1}^{m} \alpha_{2,j} h_j; \forall_j \in [1, m], \sum_{j=1}^{m} \alpha_{2,j} = 1$$

$$\alpha_{2,j} = \exp \left(v_{\alpha_2}^T \tanh \left(W_{h_2} h_j + W_{pre} \hat{\gamma} + b_{h_2} \right) \right) \tag{10}$$

As it can be seen from the above computation, the deliberation process in the second-stage attention uses the raw aspect-related representation generated by the first-stage attention. Similar to $A1$, $[\hat{\alpha}, \hat{\gamma}]$ will be transformed to generate α and γ further.

Finally, a *softmax* layer is followed to transform γ and h_m into conditional probability distribution.

$$y = softmax(W_r \gamma + W_h h_m + b) \tag{11}$$

where W_r, W_h and b are the parameters for *softmax* layer.

5 Experiment

We conduct experiments on dataset from SemEval-2016 task 5 about laptop domain. Each aspect in its reviews is labeled as three types of sentiment: *positive*, *neutral*, and *negative*. The training/testing part of each sentiment polarity dataset contains about positive of 1637/481, neutral of 188/46 and negative of 1084/274 English samples respectively.

In our experiments, word vectors are pre-trained by *glove* on an unlabeled corpus whose size is about 840 billion.[1] The aspect vectors are initialized by

---

[1] www.glove.com.

sampling from a uniform distribution of $U(-0.01, 0.01)$. We trained the parameters of model by means of initial learning rate of 0.01 for optimizer $AdaGrad$. This paper sets the following hyperparameters as shown in Table 1.

Table 1. The hyperparameters of our model

Hyperparameter	Value
Batch size	25
Word vectors dim.	300
Hidden vectors dim.	300
Aspect vectors dim.	300
Dropout rate	0.5
L2-regularization weight	0.001

5.1   Comparison to Other Methods

In order to evaluate the performance of deliberation networks comprehensively, the compared baseline models are: (1) *Vanilla LSTM* cannot capture any information in a given sentence toward aspect, so it must get the same sentiment polarity although given different aspects; (2) *AE-LSTM* treats an aspect as a target vector and appends it into each word vector as the input of LSTM; (3) *AT-LSTM* puts it into each hidden state vector produced by LSTM as input of the attention mechanism; (4) *ATAE-LSTM* puts a given aspect embedding append into each word input vector and each word hidden vector respectively; and (5) *Leehu* and *INSIG* are the first and second place respectively in evaluation metrics from SemEval-2016 Task 5.[2] To our knowledge, the performance is still best results.

From Table 2, it is clear that our proposed model has achieved the sentiment classification performance over all baselines. For the LSTM model, it cannot recognize the specific parts of text toward a given aspect. It is not surprising that the model has the worst performance due to lacking attention mechanism in network. For three models: AE-LSTM, AT-LSTM, and ATAE-LSTM, they take attention mechanism into LSTM network to identify the key information of context with aspect. From the structural view, these models could be regarded as one-stage attention neural networks, which have a main difference on how to learn aspect embedding. As for our model, although its architecture is similar with that of ATAE-LSTM, it models context representation toward aspect beyond one-stage attention mechanism and gets the state-of-the-art performance

---

[2] Notes that this paper doesn't describe the detail of these models because we don't find liberal about them. The performance of these model is from http://alt.qcri.org/semeval2016/task5/index.php?id=data-and-tools.

Table 2. The Accuracy on Laptops Reviews. Best performance is in **bold**. Results marked with * are re-printed from the references and evaluation results released by SemEval-2016, while those with # are obtained from our own implementation or the codes shared by original authors.

Model	Laptop
LSTM	0.7028#
AE-LSTM	0.7391#
AT-LSTM	0.7278#
ATAE-LSTM	0.7428#
Leehu	0.7591*
INSIG	0.7428*
Our model	**0.7656**

on *Laptop*. Comparing ATAE-LSTM with our model, it is clear that the two-stage attention model improves performance over standard/one-stage attention model. Note that in order to avoid some random results during training, we take the average of the accuracy on multiple experiments as the final result.

5.2 Case Study

To further understand the advantages brought by two-stage attention networks, we show an sentences in Table 3 from aspect-level sentiment classification. We visualize the effect of each stage attention, i.e., first-stage attention weights and deliberation attention weights. We show the advantages through the example that was misclassified by a single-layer attention but correctly identified by the deliberative attention network. For instance, "The processor and GPU are fantastic at running games and multiple apps." for aspect *CPU#miscellaneous*, its original sentiment polarity is positive. The aspect *CPU* and its attribute *miscellaneous* isn't obviously in original sentence. According to common sense, we know that the word *processor* with a underline in a sentence is the same entity as the *CPU*.

In practice, the wrong prediction of sentiment polarity is neutral while the $\hat{\gamma}$ generated by first-stage attention acts as the output of deliberation networks. Two middle results in deliberation network are shown as in Table 3, which means attention weights distribution generated by each stage attention.

From Table 3, we can observe that the common words (e.g., "the", "at", and "and") are paid little attention using deliberation networks in the context. From first-stage attention view, the difference of the attention weight of words in sentence is very small, where the words "fantastic", "at", "running", and "games" play an important role due to higher weight value. There are two shortcomings: (1) Words with higher weights are more like describing *CPU#performance* than *CPU#miscellaneous*, and (2) the common word "at" is given a very high weight. It may be a reason that sentiment polarity is misclassified for the instance. In

Table 3. The Attention Weights Distribution on Deliberation network. "Sentence" is an instance, "F-stage" represents the attention weights generated by first-stage attention, and "S-stage" means the attention weights generated by second-stage attention (deliberation attention)

Sentence	The	processor	and	GPU	are	fantastic
F-stage	0.0833	0.0836	0.0833	0.0816	0.0819	0.0851
S-stage	7.0261e−11	7.5119e−15	4.0854e−08	1.5061e−04	6.4830e−05	9.0657e−01
	at	running	games	and	multiple	apps
F-stage	0.0848	0.0850	0.0841	0.0827	0.0818	0.0827
S-stage	3.0272e−10	3.6631e−15	1.4391e−13	5.7054e−05	9.2843e−02	3.1509e−04

deliberation networks, the second-stage attention acts as a deliberation process role to polish the attention weighs generated by first-stage attention.

From Table 3, we can see that the gap of weights distribution is increased, and the second-stage attention pays more attention to the words "fantastic" and "multiple" than to the other words whereas that of word "at" is reduced. Since the second-stage deliberation component has global information about what the representation to be generated might be, it has the potential to generate a better target-related attention weights by second looking into original words in context. Thus, through the deliberation networks, we can polish attention weights and generate the aspect-related context representation that is helpful for the aspect-level sentiment classification.

6   Conclusions and Future Works

In this paper, we have proposed a deliberation process, which uses two-stage attention mechanism, and take it into RNN to improve aspect-level sentiment classification performance. The key idea of the process is to learn aspect-related embeddings and take aspects into computing attention weights. Our proposed model can concentrate on different parts of a sentence toward the corresponding aspect, so that they are more competitive for aspect-level classification. Through the process there presents potentials for aspect-level sentiment analysis, different aspects are input separately. As for future work, an interesting and possible method would be modelling more than one aspect simultaneously deliberation with the attention mechanism.

Acknowledgments. This paper is sponsored by National Science Foundation of China (61673235, 61772075) and National Science Foundation of Hebei Province (F2017208012). It is also sponsored by the Key Research Project for Hebei University of Science & Technology (2016ZDYY03) and Graduated Student Innovation Project of Hebei Province (CXZZSS2017095).

References

1. Luong, M.T., Pham, H., Manning, C.D.: Effective approaches to attention-based neural machine translation. Computer Science (2015)
2. Yang, Z., Yang, D., Dyer, C., He, X., Smola, A., Hovy, E.: Hierarchical attention networks for document classification. In: 15th Annual Conference of the North American Chapter of the Association for Computational Linguistics: Human Language Technologies, pp. 1480–1489 (2016)
3. Bahdanau, D., Cho, K., Bengio, Y.: Neural machine translation by jointly learning to align and translate. arXiv preprint arXiv:1409.0473 (2014)
4. Ruder, S., Ghaffari, P., Breslin, J.: A hierarchical model of reviews for aspect-based sentiment analysis. arXiv preprint arXiv:1609.02745 (2016)
5. Tay, Y., Tuan, L. A., Hui, S.: Dyadic memory networks for aspect-based sentiment analysis. In: 2017 International Conference on Information and Knowledge Management, pp. 107–116 (2017)
6. Wang, W., Pan, S.J., Dahlmeier, D., Xiao, X.: Recursive neural conditional random fields for aspect-based sentiment analysis. arXiv preprint arXiv:1603.06679 (2016)
7. Wang, S., Jiang, J.: Machine comprehension using match-LSTM and answer pointer. arXiv preprint arXiv:1608.07905 (2016)
8. Tang, D., Qin, B., Feng, X., Liu, T.: Effective LSTMs for target-dependent sentiment classification. Computer Science (2015)
9. Xia, Y., et al.: Deliberation networks: sequence generation beyond one-pass decoding. In: Advances in Neural Information Processing Systems, pp. 1782–1792 (2017)
10. Zhang, L., Wang, S., Liu, B.: Deep learning for sentiment analysis: a survey. Wiley Interdisciplinary Reviews Data Mining & Knowledge Discovery (2018)
11. Wang, Y., Huang, M., Zhao, L., Zhu, X.: Attention-based LSTM for aspect-level sentiment classification. In: 2017 Conference on Empirical Methods on Natural Language Processing, pp. 606–615 (2016)
12. Tang, D., Qin, B., Liu, T.: Aspect level sentiment classification with deep memory network. arXiv preprint arXiv:1605.08900 (2016)
13. Lecun, Y., Bengio, Y., Hinton, G.: Deep learning. Nature **521**, 436 (2015)
14. Qian, Q., Huang, M., Lei, J., Zhu, X.: Linguistically regularized LSTM for sentiment classification. arXiv preprint arXiv:1611.03949 (2016)

The Fuzzy Misclassification Analysis with Deep Neural Network for Handling Class Noise Problem

Anupiya Nugaliyadde[1]([✉]), Ratchakoon Pruengkarn[2],
and Kok Wai Wong[1]

[1] Murdoch University, Perth, Australia
{a.nugaliyadde, k.wong}@murdoch.edu.au
[2] Dhurakij Pundit University, Bangkok, Thailand
ratchakoon.prn@dpu.ac.th

Abstract. Most of the real world data is embedded with noise, and noise can negatively affect the classification learning models which are used to analyse data. Therefore, noisy data should be handled in order to avoid any negative effect on the learning algorithm used to build the analysis model. Deep learning algorithm has shown to outperform general classification algorithms. However, it has undermined by noisy data. This paper proposes a Fuzzy misclassification the analysis with deep neural networks (FAD) to handle the noise in classification ion data. By combining the fuzzy misclassification analysis with the deep neural network, it can improve the classification confidence by better handling the noisy data. The FAD has tested on Ionosphere, Pima, German and Yeast3 datasets by randomly adding 40% of noise to the data. The FAD has shown to consistently provide good results when compared to other noise removal techniques. FAD has outperformed CMTF-SVM by an average of 3.88% in the testing datasets.

Keywords: Class noise · Fuzzy misclassification analysis
Deep neural networks · Noise removal technique

1 Introduction

Deep neural networks have been prominently used for supervised learning for the past few years [1]. The capability of learning from hidden patterns has been the key reason for the success of deep neural networks [2]. The deep neural network has been applied for many classification tasks from image classifications [3], text classification [4] and numeric data classification. However, the noise in data can sometimes affect the performance of deep neural network since the training data pattern can be disrupted [5]. Many techniques have been used to support deep neural networks on noisy data [5]. An instance which is apparently inconsistent with the remaining instances in the dataset

---

A. Nugaliyadde and R. Pruengkarn—Equal contributions to the research.

L. Cheng et al. (Eds.): ICONIP 2018, LNCS 11304, pp. 326–335, 2018.
https://doi.org/10.1007/978-3-030-04212-7_28

can be defined as noise [6]. Class noise consists of inconsistent errors or misclassification errors. This effects on classifier performance [6, 9]. Class noise exists when the instances are labelled in the wrong class or classified into two or more different classes [8]. Fuzzy misclassification analysis is one of the techniques used to deal with class noise by identifying and removing mislabeled data using the prediction results from both truth and falsity classified data [7]. Recently, deep neural networks outperform other learning algorithms in classification accuracy [1] and perform well on numerical data [30]. Nevertheless, to the best of our knowledge, there are not many research in using fuzzy misclassification analysis with deep neural networks as well as to identify and remove the mislabeled instances. In most experiments, in order to handle noise in numerical data, the capability of deep neural networks are used [8]. Furthermore, backpropagation algorithm is also used in pre-training to handle noise in deep neural network [9]. From the literature, most deep neural networks work directly with noisy data without applying any noise handling method first. Deep neural networks has shown to achieve better classification when compared to other machine learning algorithms using large datasets [10]. This implies that the quality of the available data could impact much on the deep neural network. Therefore, this paper proposes Fuzzy Misclassification Analysis with Deep Neural Network (FAD) to improve the quality of the training data and so as to improve the classification accuracy.

2 Related Work

Noisy data can confuse a learning algorithm, blurring the decision boundaries that separate the classes or causing the model to overfit in order to accommodate incorrect data points [11]. Noise can be categorised into two categories: attribute noise and class noise [6, 12, 13]. Erroneous attribute values, missing or unknown attribute values and incomplete attributes are examples of attribute noise [13]. On the other hand, class noise consists of inconsistent errors or misclassification errors. Class noise has more effect on misclassification as there is only one label for each instance [6, 14]. For this reason, this paper focuses on improving the quality of the training data by identifying and eliminating mislabeled instances prior to applying the chosen learning algorithm. There are various techniques to handles class noise data such as filtering [13, 15–18], misclassification analysis [19–21], similarity-based [14, 22], decomposition [23] and ensemble techniques [18, 24, 25]. The deep neural network has shown to perform well in classification compared to other machine learning methods [10]. Noisy data is mostly handled through the deep neural network itself [8]. Bootstrap is used to handle noisy labels in the deep neural network [26]. However, using dropout has also shown capabilities of handling noise [26]. Backpropagation and using deep bidirectional pre-training has also shown to improve the results for noisy data [9]. Furthermore, some models use noise as a generalization method in order to prevent overfitting to the training data [8]. This paper takes the approach of handling the noise in the data used for training the deep neural network.

3 Proposed Method

This section organized as follows: Sect. 3.1 describes the Fuzzy Misclassification Analysis. Section 3.2 presents the methodology to assign fuzzy membership values in fuzzy misclassification analysis process. Finally, the Dense Neural Network is explained in Sect. 3.3.

3.1 Fuzzy Misclassification Analysis

Fuzzy misclassification analysis has shown in past literature as a robust technique which uses to eliminate outlier and class noise in numerical data [7]. Using the complementary technique with Fuzzy Support Vector Machine (FSVM) can help to identify uncertainty data. The fuzzy misclassification analysis is performed by using a truth model and a falsity model. The outputs from the truth model are the actual target output, and the outputs generated from the falsity model are complementary of the truth model [7]. In binary classification problem, if the truth target output is zero, then the falsity target output will be the compliment, which is one. The truth and falsity models are trained to predict with the exponential decaying truth memberships and exponential decaying falsity memberships respectively. The value of memberships is between zero and one, based on the distance from the actual hyperplane [27]. The optimised fuzzy memberships of both truth and falsity model are chosen automatically. The prediction outputs of training data from the truth model (P_{Truth}) is compared with the actual outputs of training data (A_{Truth}). If the prediction outputs and actual outputs are different, then those instances are labelled as misclassification instances. After that, the misclassification instances in the truth model are kept in the misclassification patterns list as M_{Truth}, and the misclassification patterns list of the falsity model as $M_{Falsity}$. The difference between the two misclassification patterns can indicate the uncertainty instances or misclassification instances which are noisy instances and effect to the classification accuracy. Fuzzy misclassification analysis eliminates the misclassification patterns appeared in truth and falsity models $(M_{Truth} \cap M_{Falsity})$. The pseudocode of fuzzy misclassification analysis presents in Fig. 1.

3.2 Assigning Fuzzy Membership Values

By representing the importance of each training examples class, the membership values or weights are assigned to the training examples in order to suppress the effect of outliers and noise [27, 28].

The membership function is defined as follows:

$$m_i = f(x_i) \tag{1}$$

where m_i is the membership value of the data point x_i and $f(x_i)$ represents the importance of x_i in its own class which generates a value between 0 and 1.

The exponential decaying function based on the distance from the actual hyperplane $f(x_i)$ is defined as follows.

a. The Truth and Falsity models are trained by true and false membership values.

b. The prediction outputs *(P)* on the training data *(T)* of both models are compared with the actual outputs *(A)*.

c. The uncertainty information or misclassification patterns of the Truth model *(M_{Truth})* and Falsity model *(M_{Falsity})* are detected.

- For Truth Model: if $P_{Truth(i)} \neq A_{Truth(i)}$ then $M_{Truth} \leftarrow M_{Truth} \cap \{T_i\}$
- For Falsity Model: if $P_{Falsity(i)} \neq A_{Falsity(i)}$ then $M_{Falsity} \leftarrow M_{Falsity} \cap \{T_i\}$

d. The new training data *(N)* is created.

- $N \leftarrow T - (M_{Truth} \cap M_{Falsity})$

Fig. 1. Pseudocode of the fuzzy misclassification analysis technique

$$f_{exp}^{hyp}(x_i) = \frac{2}{1 + exp\left(\beta d_i^{hyp}\right)} \tag{2}$$

where β is the steepness of the decay which steps from 0 to 1, d_i^{hyp} is the functional margin for each example x_i and it is defined in (3).

$$d_i^{hyp} = y_i(\omega \cdot \Phi(x_i) + b) \tag{3}$$

The lower membership value represents that the example is far away from the actual separating hyperplane whereas the example which is close to the actual separating hyperplane has the higher membership value.

3.3 Deep Neural Network (DNN)

Deep learning models have outperformed other classification models as reported in the literature [1]. There are many deep learning models used for classification [29]. However, DNN has been popular in many numeric data classification learning models [30]. Therefore, results from fuzzy misclassification analysis are passed through the DNN for binary classification. The DNN comprises 6 layers, with relu activation function (see Fig. 2). The final layer has an activation of softmax in order to support the binary classification [31].

To remove overfitting, dropouts were introduced. In order to find the best dropout rate, dropout rates were increased and tested. The best dropout rate of 60% was introduced for each layer. The loss function is sparse categorical cross-entropy [32]. The optimiser is Adam technique to optimize the results. For each dataset, the input dimension is changed. However, the same DNN architecture is used for all the datasets.

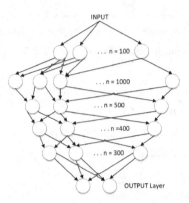

Fig. 2. The DNNs structure. n represents the number of neurons in each layer. Input layer size differs from task to task depending on the input parameters.

4 Experimental Design

This section describes the dataset characteristics and the experimental steps used in this study.

4.1 The Characteristic of the Datasets

In order to illustrate the handling of the class noise problem with the FAD approaches, the four benchmark datasets from the UCI machine learning repository [33] and KEEL repository [34] were used in the experiments. Due to the preprocessing, the attributes of the dataset are normalized from discrete data to numerical data type. The binary imbalance classification problem considered in the experiments consists of German credit data, Johns Hopkins Ionosphere, Pima Indians Diabetes and Yeast3. The summary of the characteristic of each dataset is given in Table 1.

Table 1. The summary of the dataset characteristics used in the experiments

Characteristics	Datasets			
	Ionosphere	Pima	German	Yeast3
Source	UCI	UCI	UCI	KEEL
#instances	351	768	1000	1484
#attributes	33	8	20	8
#majority	225	500	700	1321
#minority	126	268	300	163
#IR	1.79	1.87	2.33	8.10
#training	281	614	800	1187
#testing	70	154	200	297

4.2 Experimental Processes

Regarding class noise elimination with the fuzzy misclassification analysis, it can be applied with any classifier with both the truth and falsity models. DNN has shown capabilities of learning complex patterns [1, 2]. Therefore, this study focuses on implementing fuzzy misclassification analysis with DNN. There are 3 main processes: (1) data cleaning with fuzzy misclassification analysis, (2) classification with DNN and (3) evaluation of classification accuracy.

The experiments were conducted using randomly selected training and testing data. For each dataset, the experiments were conducted 10 times each time randomly selecting the training and testing datasets. The dataset was divided into 80% training set and 20% testing set. The training data was represented into the truth and falsify data. The complementary technique was implemented for the falsity data by complementing the target output of the truth data. After that, both truth and falsity data were performed with fuzzy misclassification method. The fuzzy misclassification analysis applied the radial basis function as a kernel and the exponentially decaying as the membership function. The optimal decay value was chosen from a range of 0 to 1 with a step of 0.1, which presented the highest classification accuracy for each truth and falsify data. The misclassification instances which appeared in both the truth and falsity models were considered as class noise and removed from the training data. In addition, the majority vote and consensus vote are other techniques which are widely used for identifying and removing mislabelled instances [15, 16]. In order to further assess the proposed techniques, Majority Vote [15], Consensus Vote [16] and CMTFSVM [35] noise removal techniques were used. They were compared with 40% of the noise added into the original datasets. The class noise was generated randomly by selecting 40% of the training instances and changing their class labels. For example, the positive class labels were converted to negative class labels and vice versa. It was important to note that noise was added only in the training set, not in the testing set. With this method, classification accuracy could measure a defected prediction model, which has been trained from the noisy dataset [17, 36, 37]. In the classification process, the cleaned data was classified by DNN. To achieve the optimal result, DNN ran for 600 epochs. In order to avoid overfitting of the DNN, a 60% dropout rate was added to each layer separately. Finally, during the evaluation process, classification accuracy was calculated by comparing the predicted results with the actual target results on the testing data.

4.3 Experimental Results

The classification accuracies, after injecting 40% of noise into the original datasets, are shown in Table 2. By comparing the classification accuracies of FSVM and DNN classifiers without applying any noise removal techniques, FSVM outperformed DNN for all datasets on both the original datasets and 40% of noise datasets. It could mean that FSVM classifier could reduce the effect of noise by assigning fuzzy membership value whereas the original DNN classifier is more sensitive to noisy data. Regarding noise removal techniques, the classification accuracies of the FAD algorithm for the original data were higher than the other techniques on the most datasets, while

Consensus Vote presented better results on the Yeast3. In addition, noise removal techniques gave similar classification accuracies for almost all datasets, except the Ionosphere dataset. It could mean that Majority Vote and Consensus Vote removed useful instances for building the classification model from the Ionosphere training dataset, whereas CMTFSVM and FAD could maintain the useful instances in the training dataset. However, the classification accuracies were not significantly different when comparing FSVM with CMTFSVM on the original datasets for the most dataset, while the FAD algorithm could improve the accuracies approximately by 14%, 14%, 12%, and 19% on the Ionosphere, Pima, German and Yeast3 respectively when compared with DNN.

Table 2. The percentage of classification accuracies compared with other techniques. FSVM and DNN are used without any noise removal in order to compare the improvements when noise removal method is added to the classification.

Datasets	IR		FSVM	DNN	Noise removal techniques			
					Majority Vote	Consensus Vote	CMTFSVM	FAD
Ionosphere	1.79	Original	91.17	81.43	89.48	91.04	90.57	**95.71**
		40% of noise	88.71	67.14	64.29	42.34	**88.14**	87.14
Pima	1.87	Original	78.05	62.99	76.21	76.27	75.97	**77.27**
		40% of noise	76.23	62.34	72.08	70.25	**76.36**	73.38
German	2.33	Original	79.40	71.00	77.95	78.00	78.60	**83.00**
		40% of noise	77.70	67.50	71.23	72.86	79.10	**82.00**
Yeast3	8.10	Original	94.68	73.00	94.37	**94.43**	94.34	92.51
		40% of noise	89.50	83.20	90.66	89.19	91.62	**93.94**
Average		Original	85.83	72.11	84.50	84.97	84.87	**88.50**
		40% of noise	83.04	70.05	74.57	68.66	83.81	**87.69**

By adding 40% of noise to the original datasets, FAD performed better in most cases as compared to the other techniques. By implementing 60% of dropout rate, the FAD algorithm could improve DNN classification accuracies approximately by 20%, 11%, 14%, and 10% on Ionosphere, Pima, German and Yeast3 respectively. This could imply that the fuzzy misclassification analysis could implement well on the DNN and enhance the classification accuracy of DNN classifier. Furthermore, the FAD can remove noisy instances and retain most of the cleaned original data better than the other techniques, when the data is highly noisy data, and in this case, with 40% of noise injected.

5   Conclusion

This paper presents the data cleaning and classification technique using fuzzy misclassification analysis with DNN classifier by identifying and removing mislabeled data using the prediction results from both truth and falsity classified data. The benchmark datasets used in the experiments are obtained from UCI and KEEL repositories. The experiment results showed that the FAD with 60% of dropout rate gave the best results in term of classification accuracy as compared to the Majority Vote, Consensus Vote, and CMTFSVM noise removal techniques. It could imply that the FAD is robust to noisy data. In addition, the FAD technique can avoid the overfitting problem and eliminate noise with confidence, by removing high potential misclassification patterns. Finally, the FAD technique presents that in most cases it outperforms the traditional approaches and Fuzzy classification approaches.

Acknowledgement. This work was partially supported by a Murdoch University internal grant on the high-end computer.

References

1. LeCun, Y., Bengio, Y., Hinton, G.: Deep learning. Nature **521**(7553), 436–444 (2015)
2. Nugaliyadde, A., Wong, K.W., Sohel, F., Xie, H.: Reinforced memory network for question answering. In: Liu, D., Xie, S., Li, Y., Zhao, D., El-Alfy, E.S. (eds.) ICONIP 2017. LNCS, vol. 10635, pp. 482–490. Springer, Cham (2017). https://doi.org/10.1007/978-3-319-70096-0_50
3. He, K., Zhang, X., Ren, S., Sun, J.: Deep residual learning for image recognition. In: Proceedings of the IEEE Conference on Computer Vision and Pattern Recognition, pp. 770–778. IEEE Press (2016)
4. Glorot, X., Bordes, A., Bengio, Y.: Domain adaptation for large-scale sentiment classification: a deep learning approach. In: Proceedings of the 28th International Conference on Machine Learning, pp. 513–520 (2011)
5. Xiao, T., Xia, T., Yang, Y., Huang, C., Wang, X.: Learning from massive noisy labeled data for image classification. In: IEEE Conference on Computer Vision and Pattern Recognition (CVPR), Boston, MA, pp. 2691–2699 (2015)
6. Frenay, B., Verleysen, M.: Classification in the presence of label noise: a survey. IEEE Trans. Neural Netw. Learn. Syst. **25**(5), 845–869 (2014)
7. Pruengkarn, R., Wong, K.W., Fung, C.C.: Data cleaning using complementary fuzzy support vector machine technique. In: Hirose, A., Ozawa, S., Doya, K., Ikeda, K., Lee, M., Liu, D. (eds.) ICONIP 2016, Part II. LNCS, vol. 9948, pp. 160–167. Springer, Cham (2016). https://doi.org/10.1007/978-3-319-46672-9_19
8. Gupta, S., Agrawal, A., Gopalakrishnan, K., Narayanan, P.: Deep learning with limited numerical precision. In: International Conference on Machine Learning, pp. 1737–1746 (2015)
9. Audhkhasi, K., Osoba, O., Kosko, B.: Noise benefits in backpropagation and deep bidirectional pre-training. In: International Joint Conference on Neural Networks, pp. 1–8. IEEE (2013)

10. Mahjoubfar, A., Chen, C.L., Jalali, B.: Deep learning and classification. Artificial Intelligence in Label-free Microscopy, pp. 73–85. Springer, Cham (2017). https://doi.org/10.1007/978-3-319-51448-2_8
11. Khoshgoftaar, T.M., Hulse, J.V., Napolitano, A.: Comparing boosting and bagging techniques with noisy and imbalanced data. IEEE Trans. Syst. Man Cybern. Part A Syst. Hum. 41(3), 552–568 (2011)
12. Guan, D., Yuan, W., Lee, Y.K., Lee, S.: Identifying mislabeled training data with the aid of unlabeled data. Appl. Intell. 35(3), 345–358 (2011)
13. Zerhari, B., Lahcen, A.A., Mouline, S.: Detection and elimination of class noise in large datasets using partitioning filter technique. In: 4th IEEE International Colloquium on Information Science and Technology (CiSt), Tangier, pp. 194–199 (2016)
14. Krawczyk, B., Sáez, J.A., Woźniak, M.: Tackling label noise with multi-class decomposition using fuzzy one-class support vector machines. In: IEEE International Conference on Fuzzy Systems (FUZZ-IEEE), Vancouver, BC, pp. 915–922 (2016)
15. Yuan, W., Guan, D., Ma, T., Khattak, A.M.: Classification with class noises through probabilistic sampling. Inf. Fusion 41(C), 57–67 (2018)
16. Zhang, J., Sheng, V.S., Li, Q., Wu, J., Wu, X.: Consensus algorithms for biased labeling in crowdsourcing. Inf. Sci. 382(C), 254–273 (2017)
17. Verbaeten, S., Van Assche, A.: Ensemble methods for noise elimination in classification problems. In: Windeatt, T., Roli, F. (eds.) MCS 2003. LNCS, vol. 2709, pp. 317–325. Springer, Heidelberg (2003). https://doi.org/10.1007/3-540-44938-8_32
18. Luengo, J., Shim, S.O., Alshomrani, S., Altalhi, A., Herrera, F.: CNC-NOS: class noise cleaning by ensemble filtering and noise scoring. Knowl. Based Syst. 140(C), 27–49 (2017)
19. Jeatrakul, P., Wong, K.W., Fung, C.C.: Data cleaning for classification using misclassification analysis. J. Adv. Comput. Intell. Intell. Inform. 14(3), 297–302 (2010)
20. Pendharkar, P.C.: Bayesian posterior misclassification error risk distributions for ensemble classifiers. Eng. Appl. Artif. Intell. 65(C), 484–492 (2017)
21. Ekambaram, R., et al.: Active cleaning of label noise. Pattern Recogn. 51(C), 463–480 (2016)
22. Tomašev, N., Buza, K.: Hubness-aware kNN classification of high-dimensional data in presence of label noise. Neurocomputing 160(C), 157–172 (2015)
23. Lee, G.H., Taur, J.S., Tao, C.W.: A robust fuzzy support vector machine for two-class pattern classification. Int. J. Fuzzy Syst. 8(2), 76–86 (2006)
24. Çatak, F.Ö.: Robust ensemble classifier combination based on noise removal with one-class SVM. In: Arik, S., Huang, T., Lai, W.K., Liu, Q. (eds.) ICONIP 2015, Part II. LNCS, vol. 9490, pp. 10–17. Springer, Cham (2015). https://doi.org/10.1007/978-3-319-26535-3_2
25. Sabzevari, M., Martínez-Muñoz, G., Suárez, A.: A two-stage ensemble method for the detection of class-label noise. Neurocomputing 275(C), 2374–2383 (2017)
26. Reed, S., Lee, H., Anguelov, D., Szegedy, C., Erhan, D., Rabinovich, A.: Training deep neural networks on noisy labels with bootstrapping (2014). arXiv preprint: arXiv:1412.6596
27. Batuwita, R., Palade, V.: FSVM-CIL: fuzzy support vector machines for class imbalance learning. IEEE Trans. Fuzzy Syst. 18(3), 558–571 (2010)
28. Abe, S., Inoue, T.: Fuzzy support vector machines for multiclass problems. In: European Symposium on Artificial Neural Networks, Bruges, Belgium, pp. 113–118 (2002)
29. Chen, Y., Lin, Z., Zhao, X., Wang, G., Gu, Y.: Deep learning-based classification of hyperspectral data. IEEE J. Sel. Top. Appl. Earth Obs. Remote Sens. 7(6), 2094–2107 (2014)

30. Boukoros, S., Nugaliyadde, A., Marnerides, A., Vassilakis, C., Koutsakis, P., Wong, K.W.: Modeling server workloads for campus email traffic using recurrent neural networks. In: Liu, D., Xie, S., Li, Y., Zhao, D., El-Alfy, E.-S.M. (eds.) ICONIP 2017, Part V. LNCS, vol. 10638, pp. 57–66. Springer, Cham (2017). https://doi.org/10.1007/978-3-319-70139-4_6

31. Krizhevsky, A., Sutskever, I., Hinton, G.E.: Imagenet classification with deep convolutional neural networks. In: Advances in Neural Information Processing Systems, pp. 1097–1105 (2012)

32. Zhang, W., Du, T., Wang, J.: Deep learning over multi-field categorical data. In: Ferro, N., et al. (eds.) ECIR 2016. LNCS, vol. 9626, pp. 45–57. Springer, Cham (2016). https://doi.org/10.1007/978-3-319-30671-1_4

33. Lichman, M.: UCI Machine Learning Repository. University of California, Irvine, School of Information and Computer Sciences (2013)

34. Alcalá-Fdez, J., et al.: KEEL data-mining software tool: data set repository, integration of algorithms and experimental analysis framework. J. Multiple-Valued Logic Soft Comput. **17** (2–3), 255–287 (2011)

35. Pruengkarn, R., Wong, K.W., Fung, C.C.: Imbalanced data classification using complementary fuzzy support vector machine techniques and SMOTE. In: IEEE International Conference on Systems, Man, and Cybernetics, Banff, Canada, pp. 978–983 (2017)

36. Daza, L., Acuna, E.: An algorithm for detecting noise on supervised classification. In: The World Congress on Engineering and Computer Science 2007, San Francisco, USA, pp. 1–6 (2007)

37. Kim, S., Zhang, H., Wu, R., Gong, L.: Dealing with noise in defect prediction. In: 33rd International Conference on Software Engineering, Honolulu, HI, pp. 481–490 (2011)

A Neuronal Morphology Classification Approach Based on Deep Residual Neural Networks

Xianghong Lin[✉], Jianyang Zheng, Xiangwen Wang,
and Huifang Ma

College of Computer Science and Engineering, Northwest Normal University,
Lanzhou 730070, China
linxh@nwnu.edu.cn

Abstract. The neuron classification problem is significant for understanding structure-function relationships in computational neuroscience. Advances in recent years have accelerated the speed of data collection, resulting in a large amount of data on the geometric, morphological, physiological, and molecular characteristics of neurons. These data encourage researchers to strive for automated neuron classification through powerful machine learning techniques. This paper extracts a statistical dataset of 43 geometrical features obtained from 116 human neurons, and proposes a neuronal morphology classification approach based on deep residual neural networks with feature scaling. The approach is applied to classify 18 types of human neurons and compares the accuracy of different number of residual block. Then, we also compare the accuracy between the proposed approach and other mainstream machine learning approaches, the classification accuracy of our approach is 100% in the training set and the testing set accuracy is 76.96%. The experimental results show that the deep residual neural network model has better classification accuracy for human neurons.

Keywords: Neuron classification · Geometric features
Deep residual neural network · Feature scaling

1 Introduction

Understanding the morphology and functioning of the brain from a neuroscience point of view, requires adequate computational descriptions of biological neurons, especially an emphasis on their morphologies [1]. The mechanism of the structure-function relationships and brain information processing of the brain is very complex, and the traditional mechanism of cell biology alone in the study of advanced cognitive functions by independent scientists or laboratories is very limited [2]. The purpose of the Human Brain Project is to develop modern information tools, establish common standards for neuronal databases worldwide, and develop a variety of neural informatics tools for integrating, searching and modeling of multidisciplinary neural informatics data [3]. It enables us to make better use of neuroscience data, conduct research on the brain, accelerate human understanding of the brain, and improve human health.

L. Cheng et al. (Eds.): ICONIP 2018, LNCS 11304, pp. 336–348, 2018.
https://doi.org/10.1007/978-3-030-04212-7_29

Neurons are the basic units that make up the structure of the brain. The neuronal structure and function contains many factors, and the morphological and electrical properties of neurons are two of the most important factors [4, 5]. The electrical properties of neurons include different patterns of potential firing. Potential firing can be established in response to the neuron puncture by microelectrodes with the different stimuli. Morphological characteristics can be obtained by staining techniques, mainly including the spatial structure of neurons. However, distinguishing the types of neurons using the potential firing characteristics of neuronal cells is complex. In contrast, the data on the geometry of neuronal cells is easy to obtain. Therefore, it is more convenient and intuitive to study the neuronal cell geometry.

Alavi et al. [6] separately analyzed the two-dimensional and three-dimensional (3D) microscopic images of neurons and obtained neuronal morphological information, and then classified the dopaminergic neurons of rodents. They have compared the performance of three popular classification methods in bioimage classification, such as support vector machines (SVMs), back propagation neural networks (BPNNs) and multinomial logistic regression. Han et al. [7] used SVMs to classify neurons, treating neurons as irregular fragments, and using fractal geometry to describe the spatial structure of neurons. Zhang et al. [8] selected and analyzed the data for 3D coordinate of 49 neurons, then performed factor analysis on 16 features. Three major characteristics, including dimension, degree of divergence and growths, were obtained and chosen for neuron classification. Li [9] proposed the neuronal morphology classification method based on ensemble extreme learning machine, in which 20 geometric features of neurons are used.

This paper extracts the dataset for 3D coordinate of 116 human neurons that are relatively abundant in the samples collect in the current Neuromorpho.org [10]. Using L-Measure tool [11], the 43 morphological features of human neurons are obtained. We present a neuronal morphology classification approach based on deep residual neural networks (DRNNs). In this approach, the original morphological features are firstly transformed into new data by the feature scaling, and then the DRNN is used to discover the intrinsic relationship of the data. Finally, compared with the other machine learning methods, the experimental results show the effectiveness of the proposed method.

2 Human Neuron Data Sets

The morphologies of human neurons are very complex and varied. Different parameters describe the morphological characteristics of neurons from various views. Simple features cannot describe the morphological characteristics of neurons completely. Therefore, we must obtain enough features to describe the morphological characteristics of neurons.

Neuromorpho.org is a neuron database containing 3D reconstruction data and related metadata for neurons [10]. It collects data from over 200 laboratories around the world and continuously collects and publishes new data. So far, Neuromorpho.org is the world's largest publicly accessible source database for neurological sources. In this paper, we select 3D human neuron category data for classification. Some human neuronal type, such as Golgi, Lugaro, granule and ganglion, contains only 5 neurons in Neuromorpho.org. In order to obtain balanced samples, we select 6 or 7 3D neurons from other neuron type. Table 1 lists the 18 human neuronal types and the number of neurons for each type.

Table 1. 18 Types of human neuron from Neuromorpho.org

No.	Human neuronal type	Number of neurons
1	Aspiny	7
2	amacrine	7
3	basket	6
4	bipolar	7
5	Cajal-Retzius	6
6	GABAergic	7
7	glutamatergic	7
8	Golgi	5
9	von Economo	7
10	stellate	7
11	spiny	7
12	sensory receptor	7
13	pyramidal 7 Allen cell type	7
14	lung terminal	7
15	Lugaro	5
16	granule	5
17	induced pluripotent stem cell-(iPSC)-derived	7
18	ganglion	5

Because the morphological data files of human neurons obtain from Neuromorpho. org are 3D data in the SWC format, they cannot directly reflect the geometric features of neurons. Therefore, the original data needs to be preprocessed. L-Measure is a branch of the computational neuroanatomy group subproject of the Human Brain Project [12]. L-Measure is mainly used to quantitatively characterize the morphologies of neurons. Numerous neuronal anatomical parameters are calculated from 3D digital reconstruction files of neurons. This includes the characteristics of cell body surface area, number of stems, number of branches, and mathematical modeling of a series of features of neurons. We use L-Measure to provide the maximum, minimum, mean, and sum values for each neuronal feature parameter. The 43 specific feature parameter names and simple descriptions are shown in Table 2.

Table 2. Neuronal feature parameters extracted by L-Measure

No.	Feature name	Simple description of feature
1	Soma_surface	Surface of the soma
2	N_stems	Number of stems attached to the soma
3	N_bifs	Number of bifurcations
4	N_branch	Number of branches in the given input neuron
5	N_tips	Number of terminal tips for the given input neuron
6	Width	Width is computed on the X-coordinates
7	Height	Height is computed on the Y-coordinates
8	Depth	Depth is computed on the Z-coordinates
9	Type	Type of compartment
10	Diameter	Diameter of each compartment the neuron
11	Diameter_pow	Compute the diameter raised to the power 1.5 for each compartment
12	Length	Length of compartment
13	Surface	Surface of the compartment
14	SectionArea	Section area of the compartment
15	Volume	Volume of the compartment
16	EucDistance	Euclidean distance of a compartment with respect to soma
17	PathDistance	Path distance of a compartment from the soma point
18	Branch_Order	Order of the branch with respect to soma
19	Terminal_degree	The total number of tips that each compartment will terminate into
20	TerminalSegment	Branch that ends as a terminal branch
21	Taper_1	The Burke Taper which is measured between two bifurcation points
22	Taper_2	The Hillman Taper which is measured between two bifurcation points
23	Branch_pathlength	Sum of the length of all compartments
24	Contraction	Ratio between Euclidean distance of a branch and its path length
25	Fragmentation	Total number of compartments
26	Daughter_Ratio	Ratio between the bigger daughter and the other one
27	Parent_Daughter_Ratio	Ratio between the diameter of a daughter and its father
28	Partition_asymmetry	This is computed only on bifurcation
29	Rall_Power	Rall value is computed as best value that fits the equationRall
30	Pk	Average value for Rall Power
31	Pk_classic	Same value as Pk, but with Rall Power sets to 1.5
32	Pk_2	Same value as Pk, but with Rall Power sets to 2
33	Bif_ampl_local	Angle between the first two compartments
34	Bif_ampl_remote	Angle between that bifurcation and the end of the two growing segments

(*continued*)

Table 2. (continued)

No.	Feature name	Simple description of feature
35	Bif_tilt_local	Angle between the previous compartment of bifurcating father and the two daughter branches of the same bifurcation
36	Bif_tilt_remote	Angle between the previous father compartment of the current bifurcating father and its two daughter compartments
37	Bif_torque_local	Angle between the plane of previous bifurcation and the current bifurcation
38	Bif_torque_remote	Angle between current plane of bifurcation and previous plane of bifurcation
39	Last_parent_diam	Diameter of last bifurcation before the terminal tips
40	Diam_threshold	Diameter of first compartment after the last bifurcation leading to a terminal tip
41	HillmanThreshold	Compute the weighted average between 50% of father and 25% of daughter diameters of the terminal bifurcation
42	Helix	Compute the helix by choosing the 3 segments at a time
43	Fractal_Dim	Calculate for branches of the dendrite trees

3 Neuronal Morphology Classification Approach

3.1 Deep Residual Neural Networks

The traditional shallow neural networks usually have only a few hidden layers as opposed to deep neural networks which contain a large number of hidden layers. The researchers found that deep neural networks with the right architectures achieve better results than shallow ones that have the same computational power (e.g. number of neurons or connections) [13]. Deep neural networks enhance the feature representation ability forme from the input data by stacking a number of hidden layers. The main explanation is that the deep neural networks are able to extract better features than shallow models [14]. The model used in this paper is a feedforward deep neural network structure with one input layer, multiple hidden layers, and one output layer. The data is input from the input layer and is subjected to high-dimensional nonlinear transformation through the hidden layer. Finally, the output layer is used for classification prediction. The input layer is generally a data feature, and the output layer is a data category. Its network structure is shown in Fig. 1.

Although the deep neural networks have better feature representation ability, with the increase of the number of network layers, it is inevitable that the gradient disappears [15], and the residual network can further solve this problem. Assume that $H(x)$ represents the corresponding output of the multilayer neural network after the input the sample feature vector x. According to neural network theory, $H(x)$ can fit any function. Assuming that the input and output dimensions are the same, it can be verified that the fitting $F(x)$ is equivalent to the fitting $H(x)$. Under this condition, the original function can be expressed as:

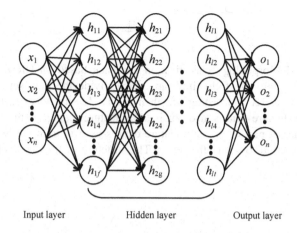

Fig. 1. The deep neural network structure

$$F(x) = H(x) + x \tag{1}$$

In addition, the activation function use in each layer of the residual block is ReLU activation function [16], and the structure of the residual block is shown in Fig. 2.

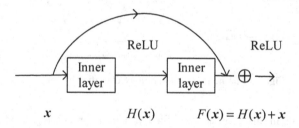

Fig. 2. Shortcut links for residual blocks in the residual network

In the above connection method, the number of parameters and the computational complexity of the network is not increased. In other words, the residual structure can be applied to various existing models without changing the existing architecture of the network.

3.2 Classification Approach of 3D Neurons

In order to improve the classification accuracy of Fisher Linear Discrimination, a scaling algorithm is given [17]. For the feature f, finding the maximum f_{max} and minimum f_{min} found from all training images, the scaling value $\tilde{f}$ can be expressed as:

$$\tilde{f} = \frac{f - f_{\min}}{f_{\max} - f_{\min}} \tag{2}$$

The Neuronal Morphology classification approach based on DRNN with feature scaling is divided into the following parts: First divide the data set of the second chapter into training set and testing set; then use L-Measure to extract feature data; Afterwards, new feature data is also obtained by feature scaling of these feature data. Then the training set part of the new feature data is used to train the DRNN, and the testing set part is used to verify the result. The structure of the approach is shown in Fig. 3.

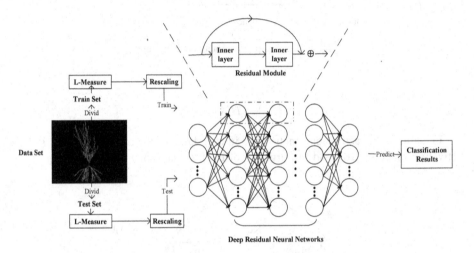

Fig. 3. The neuronal morphology classification process

The core steps in this approach are: Firstly, data preprocessing. That is, traversal is used to perform feature scaling on feature data; Secondly DRNN model training. Before training, use the parameters of the next chapter to initialize the network. In the training process, first feature-scaling features and original labels are shuffled according to the original correspondence; Then the scaled features are input from the input layer, and then these data through DRNN to get the prediction results; After comparing the predicted results with the actual results, the input conditions of the optimizer can be obtained to optimize the error function and update the DRNN parameters until the stop conditions are satisfied (the error is small enough and the number of set iterations reached) [15]. Last output the trained model. For the core process of DRNN is described as Algorithm 1.

Algorithm 1: The deep residual neural networks with feature scaling

---

Input: Human neuron features $X = \{x_1, \ldots, x_i, \ldots, x_N\}$, human neuron labels $Y = \{y_1, \ldots, y_i, \ldots, y_N\}$

Output: Human neuron classification model

---

1.   Calculate $X_{max} = \max\{x_1, \ldots, x_i, \ldots, x_N\}$, $X_{min} = \min\{x_1, \ldots, x_i, \ldots, x_N\}$

2.   **for** $i \leftarrow 1$ **to** N **do**

3.   Use X_{max} and X_{min} and x_i to calculate $xscale_i$ according to formula (2)

4.   **end for**

5.   Initialize basic parameters of DRNN

6.   **repeat**

7.   **for** $i \leftarrow 1$ **to** N **do**

8.   Use $xscale_i$ and y_i **as** input and output, according to BP algorithm to update DRNN's variable parameters

9.   **end for**

10.   **until** the cessation condition is satisfied

---

4   Experimental Results and Analysis

4.1   Experimental Parameter Settings

The experiments in this paper are mainly divided into two parts: the analysis of effects of the relevant parameters of DRNN on neuronal morphology classification, and comparing the classification accuracy between the DRNN model and other machine learning classifier. The accuracy of the experimental results is achieved by randomly shuffling training data and test data each time. The average and standard deviation of the statistical results after 10 repetitions of the experiment are determined as the final result of the experiment.

In order to evaluate the performance of this prediction model, the correct prediction score based on SKLEARN is used as a classification evaluation standard. If $\hat{y}$ is the prediction of the i th sample and y_i is the corresponding true value, then the correctly predicted score on n_{sample} is defined as:

$$accuracy(y, \hat{y}) = \frac{1}{n_{sample}} \sum_{i=0}^{n_{sample}-1} 1(\hat{y}_i = y_i) \tag{3}$$

The DRNN structure is composed of 4-layer residual modules. Each residual block consists of 4 neural networks with the same number of input layer. The last layer of activation function is softmax activation function, and the other layer activation function is ReLU activation function. The initialization weights are uniform, The Adam [18] optimization algorithm is an extension to stochastic gradient descent that has recently seen broader adoption for deep learning, the loss function is binary cross-entropy function, the batchsize is 23, and the epoch is 300. We also use technology of early stopping with a patience value of 50 to speed up the training process.

In the experiments, the training set and testing set use in each training model are generated by randomly sampling the data set of Table 1 by 0.8 and 0.2 ratios. That is, the number of experimental training sets is 93, and the number of experimental testing sets is 23.

4.2   Neuron Classification Process Analysis

DRNN training accuracy and loss function curves are shown in Fig. 4. It can be seen from the figure that when the number of iterations is from 0 to 20, the accuracy of the model increases rapidly, and the model error decreases rapidly; when the number of iterations is from 20 to 60, the accuracy of the model increases slowly and then stabilizes, and the model error also decreases slowly and then stabilizes; after training, the accuracy and loss functions are basically stable, and the model has stability.

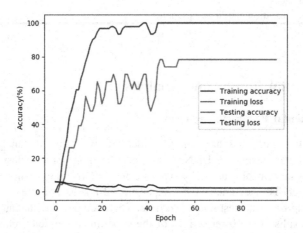

Fig. 4. The change curve of the accuracy and loss rate during training

Figure 5 shows the variation of DRNN accuracy when the number of residual blocks is different. When the residual blocks are 2, 4, 8, and 12, the accuracy of the training set is 97.80%, 100%, 100%, and 100%. The accuracy of the testing set is 60.87%, 73.91%, 73.91%, and 65.22%. It can be seen that when the number of residual blocks increases from 2 to 4, the training set and testing set have the highest classification accuracy; when the number of residual blocks increases from 4 to 8, DRNN performance does not change; when the number of residual blocks increases from 8 to 12, the accuracy of the testing set may begin to decline due to overfitting. Therefore, when the number of residual blocks is 4, the model structure parameters are relatively reasonable.

The main idea of normalization is to calculate its p-norm for each sample, then each element in the sample is divided by the norm, which also has function on feature

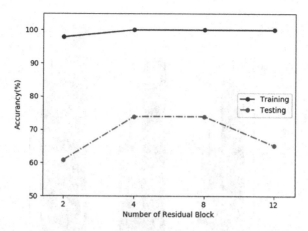

Fig. 5. Comparison of accuracy of different number of residual block

scaling [19]. Commonly used norms is L_1 and L_2. Assume that for a vector of length n, its norm L_p is calculated as follows:

$$L_p = (\sum_{i=1}^{n} |x_i|^P)^{\frac{1}{p}}$$ (4)

Based on this, we compare the model accuracy when using L_1 and L_2 for data preprocessing. When using L_1 for data preprocessing, the accuracy of the training set and testing set is 75.27% and 43.48%. When using L_2 for data preprocessing, the accuracy of the training set and the testing set are 84.34% and 57.61%. As can be seen from Fig. 6, compared with the L_1 and L_2 normalization methods, use feature scaling for data preprocessing performance is better.

4.3   Classification Performance Comparison

To verify the effectiveness of DRNN model in morphological classification of human neurons, we compare the following mainstream machine learning methods: k-nearest neighbors (KNN) classifier [20], ridge regression (RR) [21], SVM [22], BPNN and recurrent neural networks of long short-term memory (LSTM) [23]. The results are shown in Table 3. KNN, RR, and SVM are implemented by SKLEARN, in which the parameters are all default. BPNN and LSTM are implemented by Keras. The main network structures are three layers, each layer with 100 neurons, and the other parameters are the same as DRNN. It can be seen that the accuracy and stability of the approach provides in this paper are all optimal.

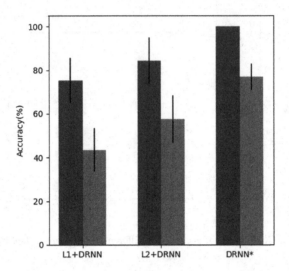

Fig. 6. Comparison of accuracy of pre-processing

Table 3. Comparison of accuracy of different classifiers

Classification approach	Accuracy of different approach	
	Training set (%)	Testing set (%)
KNN	77.91 ± 2.34	62.61 ± 7.99
RR	97.47 ± 1.56	69.13 ± 10.13
SVM	99.34 ± 0.57	72.17 ± 8.74
BPNN	97.91 ± 1.59	55.65 ± 9.78
LTSM	99.89 + 0.34	44.34 ± 8.15
DRNN	100	76.96 ± 6.49

5   Conclusion

This paper proposes an approach based on deep residual neural networks with feature scaling, which can learn effective features from complex human neuron data and classify human neurons. At the same time, we experimentally analyze the influence of the relevant parameters to obtain a relatively better model performance. We also use different machine learning classifiers to compare neuronal classification performance, the effectiveness of the approach in less samples is proved, and it provides a theoretical basis for neuroscience research. However, it may be difficult to achieve perfection due to many problems with structured features, such as lack of specialized extraction standards, poor data enhancement capabilities, and poor interpretability. Future research will be based on this article, extract features using convolution or other processing methods from the perspective of 3D graphics, to make full use of the three-dimensional neuron features (even features such as the three-dimensional angle between the synapses), to further improve the accuracy and stability of neuronal morphology classification problems.

Acknowledgment. The work is supported by the National Natural Science Foundation of China under Grant No. 61762080, and the Medium and Small Scale Enterprises Technology Innovation Foundation of Gansu Province under Grant No. 17CX2JA038.

References

1. Buckmaster, P.S., Alonso, A., Canfiled, D.R., et al.: Dendritic morphology, local circuitry, and intrinsic electrophysiology of principal neurons in the entorhinal cortex of macaque monkeys. J. Comp. Neurol. **470**(3), 317–329 (2004)
2. D'Angelo, E.: The human brain project. Funct. Neurol. **306**(6), 50–55 (2012)
3. Shepherd, G.M., Mirsky, J.S., Healy, M.D., et al.: The human brain project: neuroinformatics tools for integrating, searching and modeling multidisciplinary neuroscience data. Trends Neurosci. **21**(11), 460–468 (1998)
4. Lu, W., Bushong, E.A., Shih, T.P., et al.: The cell-autonomous role of excitatory synaptic transmission in the regulation of neuronal structure and function. Neuron **78**(3), 433–439 (2013)
5. Lin, X., Li, Z., Ma, H., et al.: An evolutionary developmental approach for generation of 3D neuronal morphologies using gene regulatory networks. Neurocomputing **273**, 346–356 (2017)
6. Alavi, A., Cavanagh, B., Tuxworth, G., et al.: Automated classification of dopaminergic neurons in the rodent brain. In: International Joint Conference on Neural Networks, Atlanta, GA, United States, pp. 81–88. IEEE (2009)
7. Han, F., Zeng, J.: Research for neuron classification based on support vector machine. In: Third International Conference on Digital Manufacturing and Automation, Guilin, China, pp. 646–649. IEEE (2012)
8. Zhang, J., Deng, S., Guo, H., et al.: Application of cluster analysis in morphological characteristics of neurons. J. Zhejiang Univ. (Agric. Life Sci.) **37**(5), 493–500 (2011)
9. Li, J.: Research on neuron classification based on ensemble of extreme learning machine. Master Thesis, Donghua University, China (2017)
10. Ascoli, G.A., Donohue, D.E., Halavi, M.: NeuroMorpho.Org: a central resource for neuronal morphologies. J. Neurosci. **27**(35), 9247–9251 (2007)
11. Scorcioni, R., Polavaram, S., Ascoli, G.A.: L-Measure: a web-accessible tool for the analysis, comparison and search of digital reconstructions of neuronal morphologies. Nat. Protoc. **3**(5), 866–876 (2008)
12. Ascoli, G.A.: Computational Neuroanatomy: Principles and Methods. Humana Press, Totowa (2002)
13. Schmidhuber, J.: Deep learning in neural networks: an overview. Neural Netw. **61**, 85–117 (2015)
14. LeCun, Y., Bengio, Y., Hinton, G.: Deep learning. Nature **521**(7553), 436–444 (2015)
15. Erb, R.J.: Introduction to backpropagation neural network computation. Pharm. Res. **10**(2), 165–170 (1993)
16. He, K., Zhang, X., Ren, S., et al.: Deep residual learning for image recognition. In: IEEE Conference on Computer Vision and Pattern Recognition, Las Vegas, NV, United States, pp. 770–778. IEEE (2016)

17. Wang, Y., Moulin, P.: Optimized feature extraction for learning-based image steganalysis. IEEE Trans. Inf. Forensics Secur. **2**(1), 31–45 (2007)
18. Diederik P, Kingma., Jimmy, Ba.: Adam: A Method for Stochastic Optimization. Computer Science (2014)
19. Salton, G.: A vector space model for automatic indexing. Commun. ACM **18**(11), 613–620 (1974)
20. Fukunaga, K., Hostetler, L.: Optimization of k nearest neighbor density estimates. IEEE Trans. Inf. Theory **19**(3), 320–326 (1973)
21. Mcdonald, G.C.: Ridge regression. Wiley Interdisc. Rev. Comput. Stat. **1**(1), 93–100 (2010)
22. Ukil, A.: Support vector machine. Comput. Sci. **1**(4), 1–28 (2002)
23. Greff, K., Srivastava, R.K., Koutník, J., et al.: LSTM: a search space odyssey. IEEE Trans. Neural Netw. Learn. Syst. **28**(10), 2222–2232 (2017)

Privacy-Preserving Naive Bayes Classification Using Fully Homomorphic Encryption

Sangwook Kim[1]($\boxtimes$), Masahiro Omori[1], Takuya Hayashi[2], Toshiaki Omori[1], Lihua Wang[1,2], and Seiichi Ozawa[1]

[1] Kobe University, Kobe, Japan
{kim,omori}@eedeptkobe-u.ac.jp, 165t210t@stukobe-u.ac.jp,
h-wang@peoplekobe-u.ac.jp, ozawasei@kobe-u.ac.jp
[2] National Institute of Information and Communications Technology, Koganei, Japan
t-hayashi@nict.go.jp

Abstract. Many services for data analysis require customer's data to be exposed and privacy issues are critical in related fields. To address this problem, we propose a Privacy-Preserving Naive Bayes classifier (PP-NBC) model which provides classification results without leaking privacy information in data sources. Through classification process in PP-NBC, the operations are evaluated using encrypted data by applying fully homomorphic encryption scheme so that service providers are able to handle customer's data without knowing their actual values. The proposed method is implemented with a homomorphic encryption library called *HElib* and we carry out a primitive performance evaluation for the proposed PP-NBC.

Keywords: Privacy-preserving data mining · Machine learning
Naive Bayes · Fully homomorphic encryption · Classification

1 Introduction

Recently, it is not difficult to find remote services for data analysis available on cloud computing environments which are provided by big companies like Amazon, Google, Microsoft and so on [1]. Big data gathered from companies and individuals provide sufficient amount of data for analysis and governments also open their data in public to promote researches and developments of business applications [2]. Also, cutting-edge machine learning techniques for data analysis improve quality of analyses. In these circumstances, by using machine learning models on the cloud server, users are able to outsource heavy computations and to secure better prediction accuracies by virtue of data obtained from other sources.

However, concerns related to data privacy are also rapidly increasing [3]. Companies are used to collect customers' data without explicit notion and data

© Springer Nature Switzerland AG 2018
L. Cheng et al. (Eds.): ICONIP 2018, LNCS 11304, pp. 349–358, 2018.
https://doi.org/10.1007/978-3-030-04212-7_30

repositories storing private information are not free from hackers. Sometimes, applications of data analysis use very sensitive privacy such as health information or economical situations. To address this problem, several methods were proposed. Federated learning is a machine learning setting to improve the model by aggregating update signals of multiple users instead of their original data [4]. But federated learning cannot prevent privacy leakage at the prediction phase which should be executed on the centralized model. Cryptography is about theories to seal contents against anonymous attackers. In general, cryptography is used for private communications on untrusted channels, i.e., the encrypted message is decrypted by the purposed receiver at the end. When the data should be input to data analysis model, it should be decrypted and there may be security threats. Homomorphic encryption provide chance to outsource computation without loss of privacy [5]. In homomorphic encryption systems, several computations can be evaluated using encrypted message and decrypted results are valid. In this paper, we propose privacy-preserving Naive Bayes classifier on fully homomorphic encryption (PP-NBC) system. Following sections are organized as following. Related works are presented in Sect. 2 and the proposed method is described in Sect. 3. Section 4 provides experimental settings and corresponding results. Finally, in Sect. 5, conclusion and future works are given.

2   Related Works

When analyzing big data, leaking privacy from such data could be a critical issue because recent development of machine learning and data mining technologies allows us to extract and infer personal information from data, even though direct personal information such as name, physical address, and gender is masked. To protect from leaking personal information, data could be encrypted with a typical encryption method like RSA or AES encryption. However, traditional public key encryption scheme does not allow us to perform data processing over encrypted data. To solve this problem, homomorphic encryption can be applied to a machine learning scheme. Li et. al. [5] proposed an outsourcing computation scheme where a Privacy-Preserving Naive Bayes classifier (PP-NBC) model is implemented with an additive homomorphic encryption called Paillier encryption [6].

Before explaining the Li's PP-NBC model, let us briefly mention about the homomorphic encryption and Naive Bayes classifier model.

2.1   Preparation

In public key encryption, secret key sk and public key pk are a pair. Secret key sk must be kept secret so that others cannot decrypt ciphertexts, while the corresponding public key pk is open to every participant. Let plaintext m_1, m_2 are in the plaintext space $\mathcal{M}$, $E_{pk}(m)$ is encrypted ciphertext by public key pk and $D_{sk}(E_{pk}(m))$ is decryption using secret key sk. In homomorphic cryptography, there is an operation '$\odot$' in ciphertext corresponding to operation '$\cdot$' in plaintext.

$$m_1 \cdot m_2 = D_{sk}(E_{pk}(m_1) \odot E_{pk}(m_2)), \tag{1}$$

For a long time, homomorphic encryption enabled only one of addition and multiplication, but in 2009, Gentry realized fully homomorphic encryption [7]. Fully homomorphic cryptosystem has no restrictions on addition and multiplication, and arbitrary arithmetic circuits can be computed on ciphertext. In this paper we adopt this fully homomorphic encryption as the encryption method.

Naive Bayes classifier is a machine learning method based on Bayesian inference, which assumes conditional independence of features. d-dimensional feature vector is represented as $\mathbf{x} = (x_1, x_2, \cdots, x_d)$. Assume that there are λ mutually exclusive classes assignable for sample $\mathbf{x}$, i.e., $C_{\mathbf{x}} \in \{1, 2, \cdots, \lambda\}$. Then a classifier determines the label which has the maximum posterior probability for $\mathbf{x}$

$$C_{\mathbf{x}} = \arg\max_{1 \leq i \leq \lambda} P(C = i|\mathbf{x}). \tag{2}$$

Bayes' theorem used for estimation is expressed by the following equation: posterior probability $P(C = i|\mathbf{x})$ is equal to the product of the prior probability $P(C = i)$ and the likelihood $P(\mathbf{x}|C = i)$ divided by the probability $P(\mathbf{x})$. Because $P(\mathbf{x})$ is common for all class labels $i \in \{1, 2, \cdots, \lambda\}$, $P(C = i|\mathbf{x})$ does not affect magnitude relation, so we can ignore $P(\mathbf{x})$. Therefore, it can be reduced to the following equation:

$$C_{\mathbf{x}} = \arg\max_{1 \leq i \leq \lambda} P(\mathbf{x}|C = i)P(C = i). \tag{3}$$

In the naive Bayes classifier, the conditional probability of test data $\mathbf{x}$ given class label i is obtained by assuming that $P(\mathbf{x}|C = i)$ is independent to each elements $x_1, x_2, \cdots, x_d)$. Therefore, it can be expressed as the product of the probability of them.

$$P(\mathbf{x}|C = i) = \prod_{t=1}^{d} P(x_t|C = i) \tag{4}$$

Finally, the classification result is represented as following equation.

$$C_{\mathbf{x}} = \arg\max_{1 \leq i \leq \lambda} P(C = i) \prod_{t=1}^{d} P(x_t|C = i) \tag{5}$$

From training data, $P(C = i)$ and $P(x_t|C = i)$ are estimated as following equations:

$$P(C = i) = \frac{m_i}{m}, \tag{6}$$

$$P(x_t|C = i) = \frac{m_{it}}{m_i}, \tag{7}$$

where m is the total number of samples in the training data, m_i represents the total number of samples in class i, and m_{it} is the number of occurrence of x_t in class i.

2.2   Li's PP-NBC Model

Participants of the system are the data owner (Alice) who has the training data, the user (Bob) who wants to know the correct answer label for his data (test data) without its own class label, and two cloud servers (CS_1 and CS_2). We will describe a model (Fig. 1) that performs data analysis while maintaining the confidentiality of data in the situation where the above four participants interact.

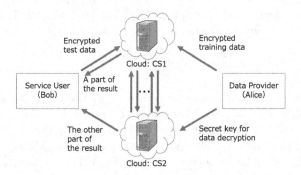

Fig. 1. System configuration for PPDM

Alice has a data set $\mathbf{X}$ and corresponding labels $\mathbf{C}$ for constructing a classi-fier. $\mathbf{X}$ consists of N samples $\mathbf{x}_1, \mathbf{x}_2, \ldots, \mathbf{x}_N$, and each sample is a d-dimensional vector, $\mathbf{x}_i = (x_{i1}, x_{i2}, \ldots, x_{id})$ where $1 \leq i \leq N$. Bob has his test sample $\mathbf{s}$ and $\mathbf{s}$ is also a d-dimensional vector $(s_1, s_2, \ldots, s_d)$. Alice sends $E_{pk}(\mathbf{X})$ to CS_1 and the secret key to CS_2. Bob sends $E_{pk}(\mathbf{x})$ to CS_1. Then CS_1 performs calcula-tions on ciphertexts without observing their actual values. And if there are some operations requiring decryption, CS_1 outsource those operations to CS_2. In this procedure, CS_1 adds some noise to data or randomly permute them to prevent CS_2 gathering meaningful information during operations. For example, CS_1 per-forms classification for Bob's encrypted data. Since obtained classification result is also encrypted, decryption function of CS_2 is needed to notify classification result to Bob. However, if encrypted data is send to CS_2, CS_2 can recognize Bob's label. Therefore, CS_1 randomly permutes element of classification result vector and sends permuted result vector to CS_2 and permutation vector to Bob. Now CS_2 decrypts the result vector but cannot know the label. Now CS_2 sends index of maximum element of permuted result vector to Bob, and then finally Bob can recover his class label by performing inverse-permutation over the receipt index from CS_2.

In this configuration, all participants are assumed to *honest-but-curious*. They follow predefined protocols correctly but they'll try to gather informa-tion about data if there is a chance. Also, it is assumed that CS_1 and CS_2 do not collude with each other to obtain information of the data. This setting is widely accepted by prior researches [5] and it is a plausible assumption. For example, if companies that provide cloud services such as Amazon and Google collaborate and misuse data, they will lost the credibility and take catastrophic disadvantages for their services. In another example, one of them can be government department. Then Alice and bob don't need to trust the company if a reliable government department participates in.

Paillier cryptosystem which is an additive homomorphic encryption adopted by Li et al. can calculate only addition operation. There is a problem in finding class labels for test data in Naive Bayes classifier because division is included in the equation of the probability estimation. Therefore, it is necessary to transform the formula. The posterior probability can be expressed as follows.

$$P(\mathbf{x}|C=i)P(C=i) = P(C=i)\prod_{t=1}^{d} P(x_t|C=i) \tag{8}$$

Applying Eqs. (6) and (7), Eq. (8) can be expressed as

$$P(\mathbf{x}|C=i)P(C=i) = \frac{m_i}{m}\prod_{t=1}^{d}\frac{m_{it}}{m_i} = \frac{\prod_{t=1}^{d} m_{it}}{m_i^{d-1}m}. \tag{9}$$

Let i be the class label for the test data for all $i, j \in \{1, \ldots, \lambda\}$ satisfying the following condition:

$$\frac{\prod_{t=1}^{d} m_{it}}{m_i^{d-1}m} \geq \frac{\prod_{t=1}^{d} m_{jt}}{m_j^{d-1}m}. \tag{10}$$

Multiplying both sides of Eq. (10) by $mm_i^{d-1}m_j^{d-1}$, it yields

$$m_j^{d-1}\prod_{t=1}^{d} m_{it} \geq m_i^{d-1}\prod_{t=1}^{d} m_{jt} \tag{11}$$

In this paper, the cloud server determines the class label for the test data based on Eq. (11). These values are all encrypted because they are calculated on a cloud server.

3   Proposed Privacy-Preserving Naive Bayes Classifier

In this section, we describe the data structure, then explain the protocol of a Privacy-Preserving Naive Bayes Classifier (PP-NBC) using fully homomorphic cryptography. In this paper, we implement the protocols in PP-NBC with a homomorphic encryption library called *HElib*.

3.1   Data Representation

Alice and Bob convert all the attribute values of training and test data to one-hot vectors and send them to cloud server CS_1. One-hot vector means a vector whose attribute value is one and the rest are all zeros. To transform real-valued data, the range of data is divided into p bins and the bin in which the target data is located is set as value one. In the experiments, p, the number of bins, are set to 5.

Data: Alice's encrypted class labels $E_{pk}(\mathbf{c}_1), E_{pk}(\mathbf{c}_2), \cdots, E_{pk}(\mathbf{c}_N)$
Result: $E_{pk}(m_i^{d-1})$ for all $i \in \{1, 2, \cdots, \lambda\}$
CS_1 accumulates $\{E_{pk}(\mathbf{c}_1), E_{pk}(\mathbf{c}_2), \cdots, E_{pk}(\mathbf{c}_N)\}$ to obtain frequencies in encrypted form.

$$E_{pk}((m_1, m_2, \cdots, m_\lambda)) = E_{pk}(\mathbf{c}_1) + E_{pk}(\mathbf{c}_2) + \cdots + E_{pk}(\mathbf{c}_N)$$

CS_1 calculates element-wise exponentials by repeating element-wise multiplications.

$$(E_{pk}(m_i^{d-1})) = (E_{pk}(m_1^{d-1}), E_{pk}(m_2^{d-1}), \cdots, E_{pk}(m_\lambda^{d-1}))$$

Algorithm 1. Computation of $(E_{pk}(m_i^{d-1}))$

3.2 Protocol 1

In this protocol, $E_{pk}(m_i^{d-1})$, $i \in \{1, 2, \cdots, \lambda\}$ is obtained. Alice converts her data into one-hot vectors, encrypts them using her public key pk, and sends them to CS_1. CS_1 receives data from Alice, adds them all together, and take its $d-1$-th power. Here, d is assumed to be shared between participants (Alice, Bob, CS_1, CS_2).

3.3 Protocol 2

To evaluate Eq. (11), the following $E_{pk}(\mathbf{H})$ should be computed:

$$E_{pk}(\mathbf{H}) = E_{pk}\left(\begin{bmatrix} m_1^{d-1} \prod_{t=1}^d m_{1t} & m_1^{d-1} \prod_{t=1}^d m_{2t} & \cdots & m_1^{d-1} \prod_{t=1}^d m_{\lambda t} \\ m_2^{d-1} \prod_{t=1}^d m_{1t} & m_2^{d-1} \prod_{t=1}^d m_{2t} & \cdots & m_2^{d-1} \prod_{t=1}^d m_{\lambda t} \\ \vdots & \vdots & \ddots & \vdots \\ m_\lambda^{d-1} \prod_{t=1}^d m_{1t} & m_\lambda^{d-1} \prod_{t=1}^d m_{2t} & \cdots & m_\lambda^{d-1} \prod_{t=1}^d m_{\lambda t} \end{bmatrix}\right) \tag{12}$$

In Protocol 2, each element $E_{pk}(m_i^{d-1})E_{pk}(\prod_{t=1}^d m_{jt})$, for $i, j \in \{1, 2, \cdots, \lambda\}$ is computed. To do this, the following calculation is performed (see also Algorithm 2).

First, CS_1 takes a tensor product of attribute values and one-hot vector of class labels, and creates a matrix $\mathbf{A}$ containing attribute value information in the row direction and class label information in the column direction. By accumulating the matrix $E_{pk}(\mathbf{A})$, the number of attribute values belonging to each class is counted for all the attribute values in the training data. The result of counting is expressed as a matrix $E_{pk}(\mathbf{B})$. Here, the element of $\mathbf{B}$ is m_{it} which is the result of counting the number of attribute values belonging to each category for all attribute values in training data. Subsequently, CS_1 receives test data from Bob. Since the data is represented by a one-hot vector, by taking the product with the matrix $E_{pk}(\mathbf{B})$, the number of attribute values of the test data identical to the attribute values in the training data is calculated. Then, CS_1 obtains the vector $E_{pk}(\mathbf{v_t})$ and multiplies these vectors to get $E_{pk}(\mathbf{v}')$.

Data: $E_{pk}(\mathbf{x}_1), E_{pk}(\mathbf{x}_2), \cdots, E_{pk}(\mathbf{x}_N), E_{pk}(\mathbf{c}_1), E_{pk}(\mathbf{c}_2), \cdots E_{pk}(\mathbf{c}_N),$
$(E_{pk}(m_i^{d-1}))$, Bob's encrypted sample s

Result: $E_{pk}(\mathbf{H})$

CS_1 calculates $E_{pk}(\mathbf{A}_{ktl})$, where $1 \le l \le p$ and p indicates the dimensionality of one-hot encoding for each feature.

for $k = 1$ *to* N **do**
    for $t = 1$ *to* d **do**
        for $l = 1$ *to* p **do**

$$E_{pk}(\mathbf{A}_{ktl}) = E_{pk}(\mathbf{x}_{ktl})E_{pk}(\mathbf{c}_k)$$

        end
    end
end

CS_1 calculates $E_{pk}(\mathbf{B}_{tl})$ for all t and l.

$$E_{pk}(\mathbf{B}_{tl}) = E_{pk}(\mathbf{A}_{1tl}) + E_{pk}(\mathbf{A}_{2tl}) + \cdots + E_{pk}(\mathbf{A}_{Ntl})$$

CS_1 calculates the vectors $E_{pk}(\mathbf{v}_t)$.

for $t = 1$ *to* d **do**

$$E_{pk}(\mathbf{v}_t) = \sum_{l=1}^{p} E_{pk}(\mathbf{s}_{tl})E_{pk}(\mathbf{B}_{tl})$$

end

CS_1 calculates $E_{pk}(\mathbf{v}')$. Operator $\circ$ represents element-wise multiplication of two vectors.

for $t = 1$ *to* d **do**

$$E_{pk}(\mathbf{v}') = E_{pk}(\mathbf{v}_1) \circ E_{pk}(\mathbf{v}_2) \circ \cdots \circ E_{pk}(\mathbf{v}_d)$$

end

CS_1 calculates $E_{pk}(\mathbf{h}_i) = E_{pk}(m_i^{d-1})E_{pk}(\mathbf{v}')$ and constructs $E_{pk}(\mathbf{H})$.

$$E_{pk}(\mathbf{H}) = [E_{pk}(\mathbf{h}_1)^T, E_{pk}(\mathbf{h}_2)^T, \cdots, E_{pk}(\mathbf{h}_\lambda)^T]$$

Algorithm 2. Computation of $E_{pk}(m_i^{d-1})E_{pk}(\prod_{t=1}^{d} m_{jt})$

3.4   Protocol 3

In this protocol, the class label for the test data is acquired. First, CS_1 performs $rotate()$ on $E_{pk}(\mathbf{H})$ to prevent the CS_2 from obtaining classification result directly. CS_1 sends two rotated matrices $rotate(E_{pk}(\mathbf{H}), k)$, $rotate(E_{pk}(\mathbf{H}^T), k)$ to CS_2 and k to Bob. CS_2 creates a matrix $\mathbf{Y}$ based on the two received matrices. Elements of this matrix $\mathbf{Y}$ are 1 if the elements on the side of $rotate(E_{pk}(\mathbf{H}), k)$ are larger than $rotate(E_{pk}(\mathbf{H}^T), k)$ and 0 if they are smaller, respectively. CS_2 selects an index of matrix in which all 1's are arranged in the column direction in the matrix $\mathbf{Y}$ sends a vector $\mathbf{u}$ such that 1 is on the index and 0 otherwise.

Data: CS_1 has matrices $E_{pk}(\mathbf{H})$ and $E_{pk}(\mathbf{H}^T)$

Result: Classification result $\mathbf{v}$ for Bob's data

CS_1 randomly choose k. CS_1 sends $rotate(E_{pk}(\mathbf{H}), k)$ and $rotate(E_{pk}(\mathbf{H}^T), k)$ to CS_2.

CS_2 compares the elements in the same component of $D_{sk}(rotate(E_{pk}(\mathbf{H}, k)))$ and $D_{sk}(rotate(E_{pk}(\mathbf{H}^T), k))$ by decrypting using secret key sk.

$$\mathbf{Y}_{ij} = \begin{cases} 1 \text{ if } D_{sk}(rotate(E_{pk}(H_{ij}), k)) \leq D_{sk}(rotate(E_{pk}(H_{ij}^T), k)) \\ 0 \text{ if } D_{sk}(rotate(E_{pk}(H_{ij}), k)) > D_{sk}(rotate(E_{pk}(H_{ij}^T), k)) \end{cases}$$

CS_1 sends k to Bob. CS_2 constructs $\mathbf{u}$ and sends $\mathbf{u}$ to Bob. Bob recovers original order of elements by performing $rotate(\mathbf{u}, -k)$ and takes index of 1 to obtain classification result.

Algorithm 3. Computation of the class label for Bob's data

4   Experiments

In the experiment, implementation of privacy-preserving Naive Bayes classifier using fully homomorphic encryption described in Sect. 3 was implemented using *HElib* and C++ language, and the prediction accuracy and calculation time were evaluated. The experiment was conducted on the system which has Ubuntu 18.04 OS, Intel i7-8700K CPU and 64 GB RAM.

4.1   Data Set

In this experiment Iris data set [8] of UCI Machine Learning Repository [9] was used. In this dataset, 150 samples are classified into three categories, Setona, Versicolor, and Virginica. Attributes of samples are Sepal Length, Sepal Width, Petal Length, and Petal Width, i.e., feature dimensionality is four. Each attribute is encoded as one-hot vector with five bins. Bins equality divides the range between the minimum and the maximum value of an attribute. 80% of samples are used for training and remaining 20% of data are used to test.

4.2   Prediction Accuracy

The accuracy of a Naive Bayes classifier which performs the proposed protocols was evaluated. The obtained classification accuracy is 90% for Iris data set using the fully homomorphic encryption. Because Naive Bayes classification multiplies conditional probabilities, if there are a zero probability, posterior probability also goes to zero. The problem occurs when the test data has an attribute value not included in the training data at the stage of protocol 2 and referred to as the zero frequency problem, which causes degradation of the classification accuracy. To avoid this problem, we added 1 to every m_{it}. Following table summarizes accuracies of original privacy-preserving Naive Bayes classifier and regularized version (Tables 1 and 2).

Table 1. Prediction Accuracy of Proposed PP-NBC

	Without smoothing	With smoothing
Accuracy (%)	76.7	90

Table 2. Execution time

Procedure	Serial processing	4-core processing
Protocol 1	1 m	(1 m)
Protocol 2	14 h37 m	3 h22 m
Protocol 3	3 h02 m	1 h40 m
Total	17 h40 m	5 h03 m

4.3   Execution Time

Execution times for proposed protocols was measured. Since many operations calculating each attribute can be executed in parallel, we implemented multi-thread version also. In this experiment four CPU cores were used to perform operations in parallel because there are four attributes in IRIS data set. However, we omitted to implement multi-threaded version of protocol 1 as it consumes relatively little time.

Although we used multi-thread to process parallel computation in the experiment, it is possible to implement the system on distributed environment by adopting operations like MapReduce [10].

5   Conclusions

In this work, we proposed a method to construct a naive Bayes classifier that enables privacy protection using fully homomorphic encryption. The training data provider (Alice) sends the data represented by one-hot vector to CS_1 and sends the secret key to CS_2. CS_1 constitutes a classifier from training data. The class label with a maximum posteriori probability for the test data received from Bob is calculated by CS_1 and CS_2, and is returned back to Bob. The communication cost between CS_1 and CS_2 in the proposed method is $\mathcal{O}(\lambda)$, while the cost of the prior work [5] is $\mathcal{O}(m\lambda d)$. This is because multiplication in the encrypted state becomes possible by using fully homomorphic encryption and it is no longer necessary to transmit data from CS_1 to CS_2. In the experiment, these protocols were implemented using HElib and the operation was verified using Iris data of UCI data repository. Although operations of fully homomorphic encryption are heavy, the method would secure privacies of data owner and service customers. In the aspect of implementation, there are many chances for further optimization and enhancement. In protocol 3, although contents of matrix **H** are randomly rotated to prevent extraction of meaningful information, CS2 still is able to observe a sort of information from the matrix and this can be regarded as information leakage in some sense. Therefore, to hide rotated contents, additional methods such as Yao's garbled circuit [11] can be considered. And to enhance processing time for the classification, it is natural to assume their distributed processing abilities like MapReduce operation, as two participants are cloud service provider. If privacy preserving data classification schemes are integrated into big data platforms, the execution time will be dramatically reduced. On the other hand, the implementations of PP-NBC using fully homomorphic encryption has weak point also. Since only addition and multiplication are used on encrypted ciphertexts, values always increase via iterations of operations. Therefore, to ensure proper representation of values, many bits are required and the speed of operations gets slow. We would address this limitation in future works, to make the method more practical.

References

1. Amazon Web Services (AWS). https://aws.amazon.com/
2. Data.gov. https://www.data.gov/
3. EU GDPR Information Portal. http://eugdpr.org/eugdpr.org-1.html
4. Konen, J., McMahan, H.B., Yu, F.X., Richtarik, P., Suresh, A.T., Bacon, D.: Federated learning: strategies for improving communication efficiency. In: NIPS Workshop on Private Multi-Party Machine Learning (2016)
5. Li, X., Zhu, Y., Wang, J.: Secure Naïve Bayesian classification over encrypted data in cloud. In: Chen, L., Han, J. (eds.) ProvSec 2016. LNCS, vol. 10005, pp. 130–150. Springer, Cham (2016). https://doi.org/10.1007/978-3-319-47422-9_8
6. Paillier, P.: Public-key cryptosystems based on composite degree residuosity classes. In: Stern, J. (ed.) EUROCRYPT 1999. LNCS, vol. 1592, pp. 223–238. Springer, Heidelberg (1999). https://doi.org/10.1007/3-540-48910-X_16
7. Gentry, C.: A Fully Homomorphic Encryption Scheme. Dissertation for Ph.D. degree, Stanford University, United States - California (2009)
8. Fisher, R.A.: The use of multiple measurements in taxonomic problems. Ann. Eugenics **7**(2), 179–188 (1936)
9. UCI Machine Learning Repository. http://archive.ics.uci.edu/ml
10. Dean, J., Ghemawat, S.: MapReduce: simplified data processing on large clusters. Commun. ACM **51**, 107–113 (2008)
11. Yao, A.C.: How to generate and exchange secrets. In: 27th Annual Symposium on Foundations of Computer Science, pp. 162–167 (1986)

Classification of Calligraphy Style Based on Convolutional Neural Network

Fengrui Dai, Chenwei Tang, and Jiancheng Lv[(⊠)]

Machine Intelligence Laboratory, College of Computer Science, Sichuan University,
Chengdu 610065, Sichuan, People's Republic of China
lvjiancheng@scu.edu.cn

Abstract. Calligraphy is the cultural treasure of the Chinese nation for five millenniums, which has always been loved by the Chinese people. This paper collects a large number of characters of Chinese calligraphy and builds a Chinese characters calligraphy data set. By establishing three different Convolution Neural Network (CNN) models, the features of calligraphy handwriting are extracted. In addition, we use some techniques to improve the robustness and generalization ability of the CNN model, so that the model can adapt to more classification tasks. The experimental results prove that the proposed method can not only well identify the style of different calligraphers, but also have good performance in the classification of the font format of soft pen calligraphy and hard pen calligraphy.

Keywords: Calligraphy classification · Characters set
Convolution neural network · Robustness and generalization

1 Introduction

Calligraphy is the cultural treasure of the Chinese nation and has always been followed and loved by the Chinese people [1].

As the Fig. 1 shows, in line with the difference of writing media, traditional calligraphy is mainly divided into hard pen calligraphy and soft pen calligraphy. Generally, according to the difference of calligraphy style, the hard pen calligraphy can be mainly categorized into 4 major styles: regular script, running script, cursive script, and official script [2]; The soft pen calligraphy is mainly divided into 6 major types: regular script, style between regular script and running script, cursive script, official script and seal script. Calligraphers, as a kind of people who push calligraphy to the height of art, divide calligraphy into several styles which are recognized by the world. The Fig. 2 shows several different styles of calligraphers.

In recent years, the computer image processing technology has achieved great success, the development of calligraphy has been furthered. Currently, in the field of calligraphy art, the combination of Artificial Intelligence (AI) technology and calligraphy art is a hot spot of research [3,4]. This paper mainly studies

© Springer Nature Switzerland AG 2018
L. Cheng et al. (Eds.): ICONIP 2018, LNCS 11304, pp. 359–370, 2018.
https://doi.org/10.1007/978-3-030-04212-7_31

Fig. 1. The characters in the left box are the 4 major styles of the hard pen. The characters in the right box are the 6 major styles of the soft pen.

Fig. 2. The eight characters in the figure are from 8 famous calligraphers, including Mao Tse-tung, Wang Xizhi etc.

how to use related technologies of machine learning, in order to extract the features of calligraphy handwriting, categorize different calligraphy styles, assist the identification of calligraphy works and help with calligraphy learning and creation, with the final purpose being to drive forward the digitalization process of calligraphy art and provide reference for the appraisal of art works, as well as assisting in the learning and creation of Chinese calligraphy.

However, the classification of Chinese calligraphy style is still in its initial stage. There is no universal features extraction method at present. Currently, the most widely used calligraphy style classification method is computer-image-processing-based method [5], which has big drawbacks in the generalization and precision of such technologies [6]. In addition, the complicated preprocessing link, low classification precision, low robustness and other limiting factors make the calligraphy style classification method based on computer image processing has not been popularized [7]. Furthermore, the feature engineer [8] in traditional machine learning methods that professionals in the field of the data are required to manually design the data features [9]. Therefore, the traditional machine learning method has not been promoted, neither.

In this paper, an end-to-end [10] calligraphy style classification model based on CNN [11] is proposed. This method effectively avoids the complex steps of

features extraction and data reconstruction, and skips the complicated preprocessing. We summarize our contributions as follows:

- We collected and sorted three calligraphy databases based on different style
  categories and calligraphers, namely, hard pen calligraphy, soft pen calligraphy and calligrapher database. A total of more than 120 thousand character
  images were collated.
- The proposed method is an end-to-end neural network model. This method
  skips the complicated preprocessing, which effectively avoids the complex
  steps of features extraction and data reconstruction [12].
- Local Convolution Neural Network (LCNN), Global Convolution Neural Network (GCNN) and Two Pathway Convolution Neural Network (TPCNN)
  three Convolution Neural Network (CNN) network models are built. The
  experimental results prove that the three models have good classification
  results.
- Through a series of skills [13], the proposed method has high robustness [14]
  and generalization ability. The proposed method has good performance in
  the classification of the font format of soft pen, hard pen calligraphy and
  calligrapher styles.

The rest of this paper is organized as follows: Sect. 2 presents the related work
and the details of our method are described in Sect. 3. In Sect. 4, several experiments are conducted and analyzed. Conclusion and future work are discussed in
Sect. 5.

2 Related Work

There is no public dataset available for Chinese characters in different typefaces.
The Chinese characters before were usually limited to the single category of the
hard pen or soft pen. What's more, there is no corresponding data set for the
classification of Chinese characters in calligraphers.

In addition, the most widely, used calligraphy style classification method is
based on computer graphics and image processing technology [15] and Support
Vector Machine (SVM) [16]. First, a single calligraphy character will be input
in the form of pictures into the Coherent Point Drift (CPD) model [17]. Then
the CPD model can extract the two features of the Chinese character single
stroke and stroke topology. Finally, the features extracted by CPD input to the
SVM model to complete the classification task [18], PCA and other commonly
methods are often used for image feature extraction.

As the Fig. 3 shows, the font classification method based on CPD algorithm
is mainly divided into two steps: character decomposition, feature extraction
and difference calculation. The first step is to decompose the main features of
the characters into every single stroke topology and stroke shape feature. In the
Fig. 3, the red pentagram is marked as the central point of the whole font, the
blue dot marks the center point of each stroke, the green line represents the shape
of a single stroke; The second step compares the weights of character topology

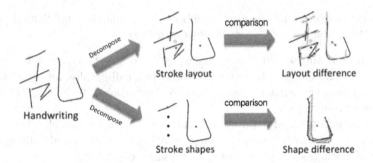

Fig. 3. The red pentagram is marked as the central point of the whole font, the blue dot marks the center point of each stroke, the green line represents the shape of a single stroke. (Color figure online)

and every single handwriting shape between the original font and standard fonts or other fonts. Usually the feature weight of the font structure and the stroke shape is set to 1:1. Finally, SVM is utilized to complete the classification task.

At present, there is a similar method of Chinese character features extraction and normalization based on the recognition method of handwritten Chinese character stroke extraction and reconstruction [19], on handwritten Chinese character stroke width normalization technology [18]. This method has improved greatly, compared with traditional machine handwriting and manual calligraphy handwriting features extraction method [20]. While, due to the complicated preprocessing link, low classification precision, low robustness and other limiting factors, the calligraphy style classification method based on computer image processing has not been fully popularized [21].

Since the convolution neural network was first applied to the ILSVRC [22] image classification competition in 2012 and made remarkable achievements, the convolution neural network has been widely used in image recognition in the field of classification. At the same time, with the emergence of large-scale data sets such as ImageNet [23] and MSCOCO [24], the training intensity of the convolution neural network has been improved continuously, which makes the model more generalized and improves the application effect in the actual image classification problem. Therefore, a deep convolution neural network can extract the features of the calligraphic font in picture format and classify it well.

In a large neural network, if the dataset is small, local convergence and over-fitting may occur in the system. What's more, the change of the data distribution will make the network training difficult and hard to convergence. The two techniques of dropout [25] and batch normalization [26] can well solve the above-mentioned problem. We will also describe them in detail in the following chapters.

3   Methods

In this section, the proposed models for calligraphy recognition tasks are presented. Considering that each font of Chinese character has its own characteristics, which are not only reflected in the overall structure, but also in the details of stroke, three CNN models are designed. One model is called Local Convolution Neural Network (LCNN), which is used to extract stroke features of Chinese characters. One model is called Global Convolution Neural Network (GCNN), which is used to extract the whole frame structure of Chinese characters. The other model is called Two Pathway Convolution Neural Network (TPCNN). We will describe these three models in detail, respectively.

3.1   Local Convolution Neural Network

When Chinese learn to write Chinese characters, they often use the Field-Character Shape (just like a 2×2 table with the same height and width) as a template to standardize the writing format. The Field-Character Shape also plays an auxiliary role in structural arrangement. The field character includes four borders, one horizontal midline, and one vertical middle line. The four lattices are called left upper case, left lower case, right upper case and right lower case.

In order to better extract the stroke information and edge features of the character pictures, the character picture is divided into four squares according to the style of the Field-Character Shape. As the Fig. 4 shows, four small pieces of pictures are put into the LCNN. A series of convolution and pooling operations are used to extract the features of each picture. Then average of the four image's feature vector is input to the softmax layer to get the classification results. It is worth noting that the convolution neural network model is invariable when four sub images are input in turn, and the parameters of the network are shared during the training process.

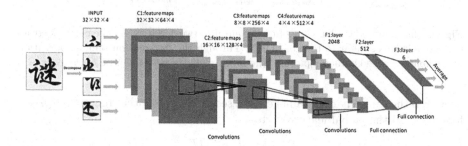

Fig. 4. The proposed Local Convolution Neural Network (LCNN). The input image of a Chinese character is cut into four pieces. Then four feature vectors will be obtained by the LCNN. Finally, by average computing, we can get the local features of the input data.

Given a character image I, splitting it into left upper case I_1, left lower case I_2, right upper case I_3, right lower case I_4, functions f_{LCNN} are sought to learn for extracting the local feature vectors:

$$P_{local} = \frac{\sum_{i=1}^{4} f_{LCNN}(I_i)}{4},$$ (1)

where P_{local} denotes the local feature vector of the input I. The P_{local} will be the input of the softmax layer, and its dimension is related to the number of categories of the dataset. The classification loss of the LCNN is written as:

$$L_{local} = Softmax(P_{local}).$$ (2)

3.2   Global Convolution Neural Network

As mentioned earlier, the characteristics of calligraphy are embodied not only in the meticulous features of each stroke, but also in the overall structure. LCNN can perform well in extracting stroke features, but it will also lead to loss of overall structural information. In order to improve the robustness of the model and make it able to classify soft pen calligraphy, hard pen calligraphy and calligraphers, we have designed GCNN for feature extraction of the whole structure (see Fig. 5).

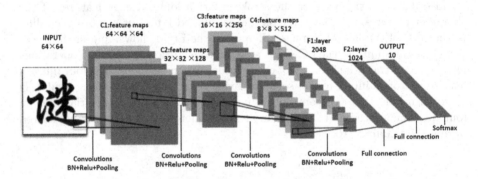

Fig. 5. The proposed Global Convolution Neural Network (GCNN). The network structure of GCNN is similar to that of LCNN. Through the convolution pooling operation of a whole Chinese character picture, the input global features can be obtained.

Given the same character image I, we seek to learn another functions f_{GCNN} for extracted the global feature vector:

$$P_{global} = f_{GCNN}(I),$$ (3)

where P_{global} denotes the global feature vector of the input I. The P_{global} will be the input of the softmax layer. The classification loss of the GCNN is written as:

$$L_{global} = Softmax(P_{global}).$$ (4)

3.3   Two Pathway Convolution Neural Network

In order to take account of the local features and the overall features, we also design the TPCNN. The original input font image is resized to a quarter, which is consistent with the size of the input of LCNN. As shown in the Fig. 6, the network structure of TPCNN is the same as that of LCNN, but there are some differences between input and output processing.

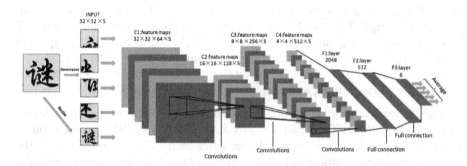

Fig. 6. The proposed Two Pathway Convolution Neural Network (TPCNN). The network structure of TPCNN is same to that of LCNN. The local feature and global feature of input image can be obtained.

Given a character image I, splitting it into left upper case I_1, left lower case I_2, right upper case I_3, right lower case I_4 and resizing it to I_5, we seek to learn functions f_{TPCNN} for extracted the local feature vectors:

$$P_{tp} = \frac{\sum_{i=1}^{5} f_{TPCNN}(I_i)}{5},$$ (5)

where P_{tp} denotes the local feature and global feature vector of the input I. The P_{tp} will also be the input of the softmax layer. The classification loss of the TPCNN is written as:

$$L_{tp} = Softmax(P_{tp}),$$ (6)

Compared with natural pictures, the color of Chinese character pictures is single and the content is simple. Therefore, when designing the deep convolution neural network, we add the two techniques of dropout and batch normalization behind each layer of convolutional layer to alleviate the local convergence, overfitting and the slow training speed. Through the adjustment of the parameters of three convolution neural network models, our method can deal with the calligraphy style classification of hard pen, soft pen and calligraphers at the same time.

4   Experiments

Dataset. We have collected three datasets on calligraphy style: Hard-Pen dataset, Soft-Pen dataset and Calligrapher dataset. More than 6,000 high frequency chinese characters on the Internet are crawled first. Then the existing Chinese character patterns are used to generate different styles of Chinese characters. Finally, each calligraphy character image is made into a 64 * 64 gray value image separately. The Hard-Pen dataste contains 24,000 character images of 4 different categories, including regular script, cursive script, running hand and official script. The Soft-Pen dataste contains 36,000 character images of 6 different categories, including regular script, style between regular script and cursive script, cursive script, running hand and seal script. The Calligrapher dataste contains 65,000 character images of 10 famous calligraphers. We split these dataset into training and test sets by the proportion of 8 : 2.

Network Setup. This paper adopts classical convolution neural network structure to build LCNN, GCNN and TPCNN models. Every model contains 4 convolution layers and 2 fully-connected layer. After each convolution operation, the batch normalization processing, ReLU non-linear activation, max-pooling and dropout are carried out in turn. For each convolution layer, we use the kernels with the size of 4×4. In the process of network training, we use adam with base learning rate 0.00001 and minibatch size 80.

Task. We mainly evaluate the three proposed methods through the task of hard pen style classification, soft pen style classification and calligrapher style classification.

4.1   Classification of Calligraphy Styles

In this section we describe the protocol and results for our classification of calligraphy styles tasks. We use three proposed convolutional neural network models to classify three datasets separately.

Table 1. Classification of calligraphy styles on Hard-Pen dataset, Soft-Pen dataset and Calligrapher dataset. The average value of the last column is the average of the classification accuracy of each model for three data sets.

Accuracy(%)	Hard-Pen	Soft-Pen	Calligrapher	Average
LCNN	92.65	93.71	83.93	90.11
GCNN	92.13	93.13	83.02	89.42
TPCNN	**94.98**	**95.16**	**86.54**	**92.22**

Table 1 summarizes our test results. All in the Hard-Pen, Soft-Pen and Calligrapher dataset, the TPCNN model obtained the best classification results. Furthermore, the classification effect of LCNN on three data is better than that of GCNN. Note that the recognition accuracies on Calligrapher dataset are not so well. A detailed analysis of this situation will be made in the next sub section. Experiments prove that compared with the overall framework, the extraction of local features is more conducive to the style classification of calligraphic fonts. But judging from the TPCNN's best results, the style of a Chinese character is not only influenced by strokes, but also closely related to the overall framework.

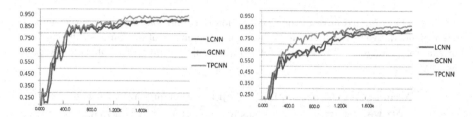

Fig. 7. The graph on the left is the training situation of the three models on the Soft-Pen dataset. The right one is the training situation of the three models on the Calligrapher dataset.

Figure 8 shows the training situation of the three models on the Hard-Pen dataset. We observe that the TPCNN model achieves the highest classification accuracy on the training dataset. What's more, from the growth rate and stability of the accuracy rate, the TPCNN model can converge faster. Similarly, as the Fig. 7 shown, TPCNN always achieves the best recognition accuracy on other two training dataset, and the convergence speed is the fastest.

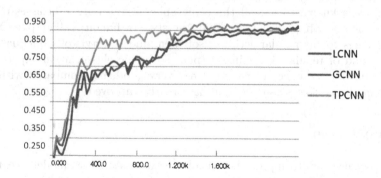

Fig. 8. The training situation of the three models on the Hard-Pen dataset. The horizontal axis represents the number of iterations, and the longitudinal axis represents the classification accuracy of the training set. The blue line represents the training of the LCNN model, the red indicates GCNN, and the green represents TPCNN. (Color figure online)

4.2   Comparison to Related Work

In this section, the proposed method and the related work are compared. Because of the special of Chinese calligraphy style classification, there is only a few relevant research. Comparisons are made only between the proposed method and the method combined CPD with SVM [18], which are the best results of calligraphy classification (Table 2).

Table 2. The comparison result of classification of calligraphy styles on server famous calligraphers dataset. The first line of the form is the names of the famous calligraphers. For example, Gongquan Liu is a top four great regular script calligrapher in late Tang Dynasty.

Accuracy(%)	Gongquan Liu	Shaoji He	Zhenqing Yan	Average
CPD & SVM	77.2	68.6	81.1	75.6
TPCNN	**89.4**	**84.7**	**91.7**	**88.6**

In the calligraphy setting, there are notable improvements. For the Gongquan Liu's style dataset, the proposed method TPCNN (89.4%) outperforms the CPD & SVM method (77.2%) for Chinese calligraphy style classification. Notably for both Shaoji He and Zhenqing Yan, the proposed method TPCNN (84.7% and 91.7%) are higher than the CPD & SVM method (68.6% and 81.1%) with 16.1% and 10.6% accuracy.

In general, our method is much better than the existing calligraphy style classification method. However, it is worth mentioning that the recognition accuracy of the calligrapher style is generally lower than the Hard-Pen dataset, Soft-Pen dataset. As is known to all, all the calligraphers' styles are subclasses of the hard pen or the soft pen. For example, the style of famous calligrapher Xizhi Wang belongs to the regular script of soft pen, and the style of Gongquan Liu is also a kind of regular script in soft pen. So in the classification experiment of calligraphers, the effect is generally not very good, but compared with the existing methods, this paper' method has greatly improved.

5   Conclusion

This paper successfully applies the convolution neural network model to extract and analyze the style of calligraphy and classify different styles. On the one hand, the collection of a large number of calligraphy works, dozens of styles of calligraphy data sets are provided; on the other hand, an end to end of the calligraphy style classification method is proposed, not only the classification precision is high, but also the calculation cost is low. It also has good generalization ability. This method is an effective attempt to combine art with artificial intelligence, the development of the digitalization of calligraphy art can be promoted.

With the help of the method of extracting calligraphy handwriting features, this paper will provide reference for the identification work of calligraphy art works, and can assist in the learning and creation of calligraphy in combination with the classification results of calligraphy. Next move will be taken to study how to use this method to extract a large number of handwriting features in the process of calligraphy style classification, and conduct personal handwriting identification research.

Acknowledgments. This work was supported by the National Science Foundation of China (Grant No. 61625204), partially supported by the State Key Program of National Science Foundation of China (Grant No. 61432012 and 61432014).

References

1. Xu, S., Lau, F.C.M., Cheung, W.K., Pan, Y.: Automatic generation of artistic Chinese calligraphy. IEEE Intell. Syst. **20**(3), 32–39 (2005)
2. Zhang, C.X.: On Chinese calligraphy and hard-pen calligraphy. J. Nanyang Norm. Univ. **10**, 018 (2010)
3. Li, M., Lv, J., Li, X., Yin, J.: Computer-generated abstract paintings oriented by the color composition of images. Information **8**(2), 68 (2017)
4. Chang, C.Y., Lin, I.C.: An efficient progressive image transmission scheme for Chinese calligraphy. Int. J. Pattern Recognit. Artif. Intell. **20**(7), 1077–1092 (2014)
5. Wang, F.Y., Lu, R., Zeng, D.: Artificial intelligence in China. IEEE Intell. Syst. **23**(6), 24–25 (2007)
6. Xiao, J.: Automatic generation of large-scale handwriting fonts via style learning. In: SIGGRAPH ASIA 2016 Technical Briefs, Association for Computing Machinery (2016)
7. Phan, Q.H., Fu, H., Chan, A.B.: FlexyFont: learning transferring rules for flexible typeface synthesis. Comput. Graph. Forum **34**, 245–256 (2015)
8. Scott, S., Matwin, S.: Feature engineering for text classification. In: International Conference on ICML, pp. 379–388 (1999)
9. Seide, F., Li, G., Chen, X., Yu, D.: Feature engineering in context-dependent deep neural networks for conversational speech transcription. In: Automatic Speech Recognition and Understanding, pp. 24–29 (2011)
10. Zhang, Y., Pezeshki, M., Brakel, P., Zhang, S., Bengio, C.L.Y., Courville, A.: Towards end-to-end speech recognition with deep convolutional neural networks, pp. 410–414 (2017)
11. Sahiner, B., et al.: Classification of mass and normal breast tissue: a convolution neural network classifier with spatial domain and texture images. IEEE Trans. Med. Imaging **15**(5), 598–610 (1996)
12. Lv, J., Yi, Z.: An improved backpropagation algorithm using absolute error function. In: Wang, J., Liao, X., Yi, Z. (eds.) ISNN 2005. LNCS, vol. 3496, pp. 585–590. Springer, Heidelberg (2005). https://doi.org/10.1007/11427391_93
13. Lv, J.C., Yi, Z., Zhou, J.: Subspace Learning of Neural Networks. Chemical Rubber Company Press, Incorporated, Boca Raton (2010)
14. Lv, J.C., Tan, K.K., Yi, Z., Huang, S.: A family of fuzzy learning algorithms for robust principal component analysis neural networks. IEEE Trans. Fuzzy Syst. **18**(1), 217–226 (2010)

15. Mcelhaney, R.M: Algorithms for Graphics and Image Processing, pp. 1116–1117. Computer Science Press, Rockville (1982)
16. Ukil, A.: Support vector machine. Comput. Sci. **1**(4), 1–28 (2002)
17. Myronenko, A., Song, X.: Point set registration: coherent point drift. IEEE Trans. Pattern Anal. Mach. Intell. **32**(12), 2262–2275 (2010)
18. Min, W., Baoying, Z., Chenhong, Y., et al.: Chinese calligraphy feature extraction and recognition. Inf. Commun. **7**, 19–20 (2015)
19. Wang, J., Lin, F., Chen, J.: Recognition method based on handwritten Chinese characters stroke extraction extraction recombined. Comput. Eng. **33**(10), 230–232 (2007)
20. Guo, C., Hu, X.: Feature extraction for calligraphy stroke based on computer image processing. J. Tianjin Univ. Sci. Technol. **25**(5), 68–72 (2010)
21. Wen, P.Z., Yao, H., Shen, J.W.: Recognition method of stone inscription font based on convolution neural network. In: Computer Engineering & Design (2018)
22. Yu, W., Yang, K., Bai, Y., Yao, H., Rui, Y.: Visualizing and comparing convolutional neural networks. In: Computer Science (2014)
23. Deng, J., Dong, W., Socher, R., Li, L.J., Li, K., Li, F.F.: ImageNet: a large-scale hierarchical image database. In: Computer Vision and Pattern Recognition, pp. 248–255 (2009)
24. Havard, W., Besacier, L., Rosec, O.: SPEECH-COCO: 600k visually grounded spoken captions aligned to MSCOCO data set, pp. 42–46 (2017)
25. Srivastava, N., Hinton, G., Krizhevsky, A., Sutskever, I., Salakhutdinov, R.: Dropout: a simple way to prevent neural networks from overfitting. J. Mach. Learn. Res. **15**(1), 1929–1958 (2014)
26. Ioffe, S., Szegedy, C.: Batch normalization: accelerating deep network training by reducing internal covariate shift, pp. 448–456 (2015)

Tropical Fruits Classification Using an AlexNet-Type Convolutional Neural Network and Image Augmentation

Alberto Patino-Saucedo[1], Horacio Rostro-Gonzalez[1]($\boxtimes$), and Jorg Conradt[2,3]

[1] Department of Electronics, University of Guanajuato, 36885 Salamanca, Mexico
{alberto.patino,hrostrog}@ugto.mx
[2] Department of Electrical and Computer Engineering,
Neuroscientific System Theory, Technical University of Munich,
80333 Munich, Germany
conradt@tum.de
[3] School of Electrical Engineering and Computer Science, KTH,
114 28 Stockholm, Sweden

Abstract. AlexNet is a Convolutional Neural Network (CNN) and reference in the field of Machine Learning for Deep Learning. It has been successfully applied to image classification, especially in large sets such as ImageNet. Here, we have successfully applied a smaller version of the AlexNet CNN to classify tropical fruits from the Supermarket Produce dataset. This database contains 2633 images of fruits divided into 15 categories with high variability and complexity, i.e. shadows, pose, occlusion, reflection (fruits inside a bag), etc. Since few training samples are required for fruit classification and to prevent overfitting, the modified AlexNet CNN has fewer feature maps and fully connected neurons than the original one, and data augmentation of the training set is used. Numerical results show a top-1 classification accuracy of 99.56 %, and a top-2 accuracy of 100 % for the 15 classes, which outperforms previous works on the same dataset.

Keywords: Convolutional neural networks · AlexNet CNN
Fruit classification · Image augmentation

1 Introduction

The ability to categorize information from the visual world is an essential task in human beings. During the past decades, with the advent of increasingly more complex computational systems, the possibility of building artificial systems able to reach and even surpass human rates of recognition of digital images has become a closer reality. Nonetheless, some artificial vision tasks remain to be solved. One example is that of automatic fruit and vegetable classification. Supermarkets need a fast and reliable way to automatically classify their produce. It has to be robust to fruit variability, highly precise and with few training

© Springer Nature Switzerland AG 2018
L. Cheng et al. (Eds.): ICONIP 2018, LNCS 11304, pp. 371–379, 2018.
https://doi.org/10.1007/978-3-030-04212-7_32

samples needed in the training phase (e.g. 30 pictures per class). Ideally, given an image of fruits or vegetables from only one variety, the system will return a list of possible candidates of the product species and kind, from a wide variety of training classes. This classification system can be used by retail stock managers for further labeling, storage and delivery of the produce, and by customers for automatic generation of prices [5].

In order to create a robust fruit and vegetable classification system, a series of challenges appear, including variability in ripeness state (which affects the shape, color and texture information), shadows, camera occlusions, differences in illumination and quantity of fruits in the picture. Sometimes, the object can be inside a plastic bag which adds specular reflections and hue shifts. Another challenge is how to classify among different varieties of the same fruit or vegetable. Additionally, training a precise image recognition system with few examples in the training set is deemed a complex challenge in computer vision.

These considerations have been taken into account by Rocha et al. [5], while gathering the Supermarket Produce dataset. This dataset contains 2633 images of fruits divided into 15 categories, which were captured on a clear background with a resolution of 1024×768 pixels. All images were stored in a 8 bits RGB color-space. The images were gathered at various times and days for the same category. These features increase the dataset variability and represent a quite realistic scenario, where all the required pictures can be taken rapidly by non-professional operators. The Supermarket Produce dataset also comprises differences in pose and number of elements within an image. Furthermore, the presence of shadows and cropping/occlusions makes the dataset more complicated to analyze. In Fig. 1, we show some samples from the dataset.

This paper presents a system that performs automatic classification of fruits and vegetables in the Supermarket Produce Dataset with a classification error significantly lower to that of previous works, considering a range from 8 to 64 training samples per class, with the use of an AlexNet-type [3] convolutional neural network (CNN) and the data augmentation technique.

Section 2 is an overview of previous methods for automatic fruit classification and presents a justification for the use of CNN's on this specific task. Section 3 presents the topology of the network and delves into the data augmentation technique used in this work. Section 4 offers experimental results and their analysis. Finally, Sect. 5 presents conclusions and future directions.

2   Literature Review

The classical approach for image recognition has been to extract the main features from the digital images and feed those features into a classifier for discrimination. By following this paradigm, complex feature extractors and classifiers have been developed and impressive rates of recognition have been achieved. However, among the main drawbacks of this approach, there is the relation between the robustness of the systems and their computational complexity and the lack of a general method to solve different tasks. Usually, visual recognition

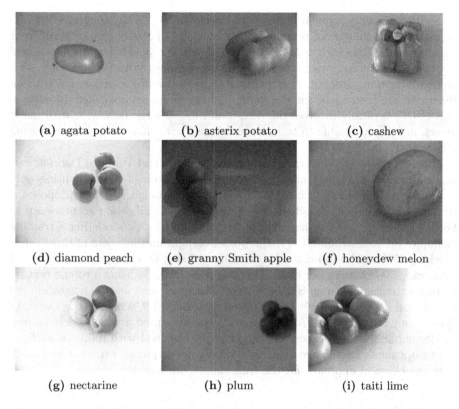

(a) agata potato   (b) asterix potato   (c) cashew

(d) diamond peach   (e) granny Smith apple   (f) honeydew melon

(g) nectarine   (h) plum   (i) taiti lime

Fig. 1. Samples from the Supermarket Produce dataset

systems rely on processing and analyzing the input images, with stages such as segmentation, saliency detection, colorspace change, texture analysis and so on, each stage with different algorithms whose performance depends on the specific problem.

An alternative to the aforementioned approach is to unify all the different subsystems in a macro system, which must be able to extract relevant information and perform the classification task. This is the main goal of several machine learning techniques, from which the most successful has been historically that of neural networks [6]. These networks, inspired by the biological neural system, consist of a set of units or nodes, called neurons or perceptrons, organized in layers, with connections among them, where the weights of the connections transform the information through matrix operations. These weights are adjusted following a learning rule, generally a back-propagation algorithm, whose aim is to reduce the classification error.

To the best of our knowledge, perceptron based neural networks have not an outstanding performance on image classification for large datasets, for they require one-dimensional inputs, dramatically reducing the information contained in spatial rapports. For this reason, and exploiting the fact of increased computing capacity, convolutional neural networks (CNN's), which operate in 2 or more dimensions, were proposed [4], achieving impressive results in different benchmark datasets such as MNIST [6] and ImageNet [3]. This new generation of neural networks, along with other innovative techniques to avoid overfitting, such as regularization, dropout, max pooling and the use of bio-inspired activation functions, are the core of a new paradigm of machine learning called deep learning. CNN's typically involves the design and use of large networks, with many layers, making it feasible to train large datasets with little to no pre-processing whatsoever.

Regarding the classification task on the Supermarket Produce Dataset, two main attempts have been made. Rocha et al. [5], considered that using only a feature descriptor and classifier would not be enough, so they proposed a fusion of features and classifiers, reaching a top-1 classification accuracy up to 98%, considering 64 training samples per class and 85% considering 8 training examples per class. Ram-Duvey [1] reported a slight improvement of these rates, reaching 99% in the top one responses, considering 60 examples per class and 90% using 20 examples per class. This was performed by using a robust texture feature extraction with an SVM classifier. Zhang [8] used feature extraction fed to a neural network for fruit classification, reaching 81.9 % on a dataset with 18 classes and 1653 samples. All the mentioned works used a wide variety of pre-processing and feature extraction algorithms, combined with robust classifiers. This work shows that image augmentation of a set of pictures trained on a CNN beat the classical approach for the fruit classification task.

3   Methodology

3.1   Fruit-AlexNet CNN

An adaptation of the AlexNet Convolutional Neural Network (AlexNet CNN) was used to train the Supermarket Produce Dataset. This network was originally created for image recognition on the very large ImageNet dataset (1000 classes, 1.2 million images), obtaining considerably better results than the previous state of the art [3].

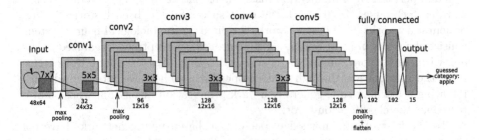

Fig. 2. Proposed AlexNet variation for fruit classification.

As the fruit classification problem addressed in this work requires considerably fewer training examples and classes, using the originally proposed AlexNet CNN leads to overfitting, yielding poor classification accuracy [9]. Ways for reducing overfitting in CNN's have been proposed, such as weight regularization and dropout [7]. Furthermore, according to [2], a cause of overfitting is the use of models that are more complex than necessary, i.e., models containing more parameters than those justified by the data.

Taking the previous considerations into account, we propose to reduce model complexity without losing the robustness that AlexNet CNN offers, allowing an accurate representation of features for every class without overfitting. Hence, starting from the original AlexNet architecture, a fine-tuning of the network parameter such as the number of feature maps, dense neurons and dropout percentage for the layers was performed. The results reported in this work were the best that we could find by experimentally adjusting the previously mentioned parameters.

Briefly depicted, the network uses 5 convolutional layers and 3 fully connected layers, with max-pooling layers after the first, second and fifth convolutional layer, and dropouts after the first and second fully connected layer. A scheme of the architecture of this network is shown in Fig. 2. A comparison of parameters respect to the original AlexNet can be seen in Table 1.

Table 1. Parameter comparison between the original AlexNet and the proposed AlexNet for fruit classification. For convolutional layers, refer to the number of feature maps, and for fully connected layers, refer to the number of neurons.

Parameter/Network	AlexNet	Fruit-AlexNet
Input	$224 \times 224 \times 3$	$48 \times 64 \times 3$
Conv1	48	32
Conv2	128	96
Conv3	192	128
Conv4	192	128
Conv5	128	128
FC1	2048	192
FC2	2048	192
Output	1000	15
Dropout	0.5	0.7

The weight initialization for the convolutional layer was an orthogonal matrix, and for the fully connected layer, there was a Glorot normalized initialization. For all hidden layers, the activation function was a relu and for the output layer, a softmax. Additionally, according to the original AlexNet model, a max pooling layer was added after the first, second and fifth convolutional layer. This was taking into account a downsampling of the input images from the Supermarket Produce dataset by a factor of 16, yielding 48×64 input size images.

3.2 Image Augmentation

We considered pertinent the use of data augmentation of the training set for the fruit classification task. This method artificially enlarges the training data by adding label-preserving transformations. In the case of pictures they are flipped, shifted and rotated, increasing the variability of training examples from a reduced dataset. By performing image augmentation the network is more likely to learn a better, more general representation of the data.

In this work, data augmentation consisted in random horizontal and vertical flips of the image, random rotations up to 45 degrees, random horizontal and vertical shifts up to 20% of the image width and height, or a combination of the preceding techniques. This augmentation was performed with the aid of Keras ImageGenerator class. Examples of augmented images are shown in Fig. 3.

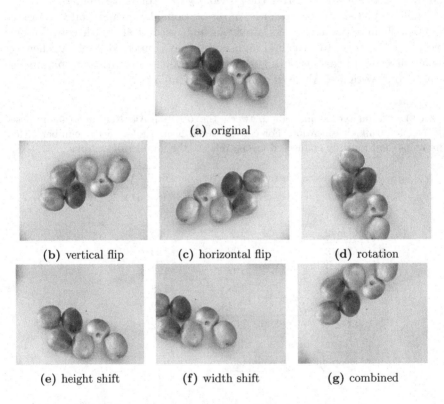

(a) original

(b) vertical flip   (c) horizontal flip   (d) rotation

(e) height shift   (f) width shift   (g) combined

Fig. 3. Image augmentation of a single training example.

4   Results and Discussion

Two experiments were performed to assess the performance of the proposed fruit classification network. The first, baseline experiment was performed without image augmentation and the second experiment was performed with image augmentation. For both cases, the dataset was split into a training and a test set. The training set contained from 8 to 64 randomly selected samples per class, leaving the rest of the dataset (2633 samples, 15 classes) for testing. Both experiments were performed directly feeding the selected training samples into the network, with no preprocessing. In the second experiment, 256 samples per class were artificially generated from the baseline training sets, by the use of image augmentation.

The training phase for both experiments was done by running 300 epochs and using the stochastic gradient descent optimizer with a learning rate of 0.001. This network was implemented in Python, using the Keras deep learning module, and the Theano tensor operation backend, optimized for a GPU implementation in a Nvidia GeForce GTX 775M graphic card using CUDA 8.0.53.

For both experiments, we selected the training images using sampling without replacement from the pool of each image class. We repeated this procedure 10 times, and report the top-1 average classification accuracy (μ), from which the average error is 1-μ. Strict k-fold cross-validation is not used, given that we are interested in different sizes of training sets. Additionally, we consider the situation when the system is asked to provide the top two responses. The top-2 classification error for a specific input is zero if the correct answer is among the two most probable classes provided by the network. We compare the classification error of the Fruit-AlexNet CNN with and without image augmentation, and the state-of-the-art SVM Fusion reported by Rocha et al. [5]. Figure 4 shows the error per class of the top-1 responses, and Fig. 5 shows the same for the top-2 responses.

Experimental results show that the use of image augmentation significantly improve the top-1 and top-2 recognition rates for every training set size considered, with respect to the networks trained without data augmentation. The improvement in performance is more evident when fewer training examples are considered. This was expected, as an augmented training set increases variability in rotations, orientation, and occlusions, inducing the network to learn a better representation of the data. Furthermore, the classification accuracy of the trained network outperformed previous works, by reaching a maximum top-1 accuracy of 99.46 % for top-1 and of 100% for top-2 responses. This work provides the best-known accuracy on the Supermarket Produce Dataset, considering few (8 to 64) training samples per class. Remarkably, for as few as 8 training examples per class, the system is able to perform classification with a precision of 93% for the top-1 response and 97.5% for the top-2 responses. As few samples are required, custom fruit recognition systems based on this work can be developed in a fast way, as the sample gathering process is less tedious, and the training phase takes minutes. This can prove useful in further implementations of fast, robust, low-resource, customizable fruit recognition systems for local markets.

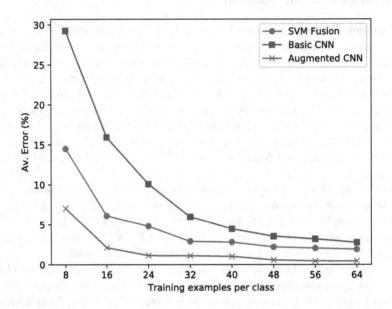

Fig. 4. Average top-1 classification error results.

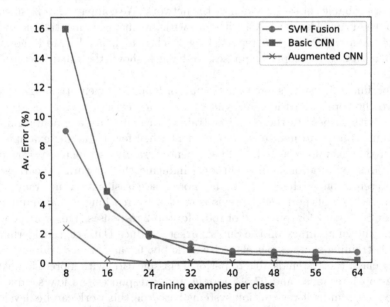

Fig. 5. Average top-2 classification error results.

5   Conclusion

This paper approached the fruit classification task on the Supermarket Produce dataset, considering few training examples per class. An adaptation of the AlexNet CNN architecture was used to train 8 to 64 training examples, with and without augmentation of the training set. The use of data augmentation allowed a considerably better learning when few examples are used, which is desirable for an ideal low resources fruit recognition system. Overall, the proposed system outperforms previous fruit classification systems which use classical feature extraction or neural networks without data augmentation.

Next directions of this work include to apply transfer learning techniques for adaptive classification, to analyze the feasibility of deploying this fruit classification system on embedded electronic devices, e.g., implementing the trained CNN offline and deploying it into a smartphone gathering input data from its integrated camera. Another direction would be to train a system that extracts more complex information from the fruit images such as quantity and maturation state of the fruits.

References

1. Dubey, S.R., Jalal, A.: Robust approach for fruit and vegetable classification. Procedia Eng. **38**, 3449–3453 (2012)
2. Hawkins, D.M.: The problem of overfitting. J. Chem. Inf. Comput. Sci. **44**(1), 1–12 (2004)
3. Krizhevsky, A., Sutskever, I., Hinton, G.E.: Imagenet classification with deep convolutional neural networks. In: Proceedings of the 25th International Conference on Neural Information Processing Systems, NIPS 2012, vol. 1, pp. 1097–1105. Curran Associates Inc., USA (2012)
4. Lecun, Y., Bottou, L., Bengio, Y., Haffner, P.: Gradient-based learning applied to document recognition. Proc. IEEE **86**(11), 2278–2324 (1998)
5. Rocha, A., Hauagge, D.C., Wainer, J., Goldenstein, S.: Automatic fruit and vegetable classification from images. Comput. Electron. Agric. **70**(1), 96–104 (2010)
6. Schmidhuber, J.: Deep learning in neural networks: an overview. Neural Netw. **61**, 85–117 (2015)
7. Srivastava, N., Hinton, G., Krizhevsky, A., Sutskever, I., Salakhutdinov, R.: Dropout: a simple way to prevent neural networks from overfitting. J. Mach. Learn. Res. **15**, 1929–1958 (2014)
8. Zhang, Y., Wang, S., Ji, G., Phillips, P.: Fruit classification using computer vision and feedforward neural network. J. Food Eng. **143**, 167–177 (2014)
9. Zhao, W.: Research on the deep learning of the small sample data based on transfer learning. AIP Conf. Proc. **1864**(1), 020018 (2017)

Supervised and Semi-supervised Multi-task Binary Classification

Rakesh Kumar Sanodiya[1(✉)], Sriparna Saha[1], Jimson Mathew[1],
and Arpita Raj[2]

[1] Indian Institute of Technology Patna, Patna, India
rakesh.pcs16@iitp.ac.in
[2] Indian Institute of Engineering Science and Technology, Shibpur, India

Abstract. In this paper, we interrogate multi-task learning in the background of Gaussian Processes(GP) for constructing different models dealing with the issue of binary classification. At first, we propose a new supervised multi-task classification approach (SMBGC) based on Gaussian processes where kernel parameters for all tasks share a common prior. In recent years great advancement in the field of machine learning domain is being done by exploitation and extraction of information from unlabeled data. Machine learning models require labeled data for training but the amount of labeled data available is quite low since labeling them is expensive. To overcome this problem we came up with a semi-supervised multi-task binary Gaussian process classification (SSMBGC). In this approach, even small amount of labeled data can contribute to our model training and hence they enhance the generalization performance of a model on a learning task with the help of some other related tasks.

Keywords: Supervised · Semi-supervised · Gaussian process
Classification

1 Introduction

In real-world problems, a large amount of data can be easily made available. Supervised learning methods based on completely labeled training datasets can lead to design of well functioning classification systems. Generally, classification is a supervised learning approach in which model first gets trained and then the model uses the learning to classify new observations [1]. Here, we focus on binary classification approach.

Labeling the data for training a model is both tedious and expensive process involving continuous human participation. This motivated the need to incorporate unlabeled data along with the labeled data and it is also possible that there are some related tasks with sufficient labeled data and while learning what is learned for each task can help other tasks to get learned in more better way.

© Springer Nature Switzerland AG 2018
L. Cheng et al. (Eds.): ICONIP 2018, LNCS 11304, pp. 380–391, 2018.
https://doi.org/10.1007/978-3-030-04212-7_33

This combination brings out a better performing classifier in comparison to the case where labeled data is limited.

There have been many different methods introduced recently to improve the generalized performance of classification problems based on utilizing information beyond given labeled data. These studies include integration of semi-supervised learning and multi-task learning (MTL) into a single framework, and combining the idea of semi-supervised learning with transfer metric learning (STML). In Classification problems, MTL is responsible in improvement of the performance of multiple classification tasks by learning them jointly while STML involves improvement of learning performance of a particular new task (Target task) with the support of similar related tasks (Source tasks) [2–4].

In semi-supervised learning, to bring out the usefulness of unlabeled data, we must assume some geometric intuition for the distribution of data. Cluster assumptions and Manifold assumptions [5] are two widely used intuitions. According to cluster assumption, labels of data points in given cluster share same labels while in manifold assumption if two points have a high correlation with respect to some metric on the manifold, they remain similar when performing some projection. Today, graph-based methods are based upon the principle of manifold assumptions. Our proposed semi-supervised approach is also graph-based, in which the geometry of data is modeled in the form of a graph where labeled and unlabeled data form vertices of the graph and the edges denote appropriate neighborhood relationships. The issue of binary classification shown in Fig. 1(a) can be approached by using a Gaussian kernel (RBF), $K = \frac{exp(x-z)^2}{2\sigma^2}$.

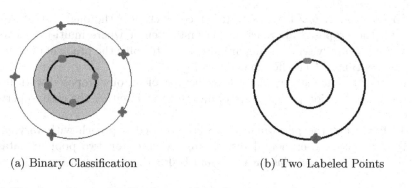

(a) Binary Classification   (b) Two Labeled Points

Fig. 1. Showing a binary classification problem: distribution of two types of data points on two different concentric circles.

Figure 1 represents distribution of data points in two different classes where decision boundary separating two different classes is non-linear. This kernel K has a unique Reproducing Kernel Hilbert Space (RKHS) of functions associated with it which is denoted by $\mathscr{H}$. Suppose we have N_l number of labeled examples (x, y) where $x \in \mathbb{R}^2$ and $y \in \{+1, -1\}$. The best function f for the respective classifier is the one which minimizes the given regularization problem where $\|f\|$ represents the norm of the function in RKHS and E is the error function.

$$E(f^i, X^l, Y^l) + \beta \|f\|_{\mathscr{H}}^2 \tag{1}$$

E can either be square error for regularized least squares (RLS) or can be hinge loss (error function for classification in terms of maximum margin) in case of SVM. As per representer theorem the function f that minimizes the above problem can be expressed as linear combination of Gaussian kernels situated at all the data points $f(\mathbf{x}) = \Sigma_{i=1}^l \alpha_i K(\mathbf{x}, \mathbf{x}_i)$. Thus, the problem gets lessen to compute an appropriate value of coefficient α_i so that $f(\mathbf{x})$ fits the data well.

There exists some earlier works in case of semi-supervised learning for classification problem and multi task learning for regression problem under Gaussian process framework. Sindhwani et al. [6] earlier proposed a semi supervised model that acts as a binary Classifier and incorporates the geometry of the underlying data in its kernel function. Jebara [7] proposed common feature selection for multiple SVM trained on different but interrelated tasks. Evigenou et al. [8] extended the standard single task kernel methods to the case of Multi Task Learning and established that this can outperform single task learning specially when we have related tasks but less data per task. Durichen et al. [9] proposed a model for multivariate physiological time-series analysis using multi-task Gaussian process model. Xue et al. proposed a statistical model of multi-task learning based on Dirichlet Process Priors [10] for classification problem. Gao et al. [11] combined the idea of transfer learning with Gaussian processes regression for visual tracking. As such there exists no method which uses Multi Task Learning and Semi Supervised learning for Gaussian Process Binary Classification.

Contributions of the current work are as follows:

- To the best of our knowledge (after conducting a thorough literature survey), the proposed approach is the first attempt where multi-task learning and semi-supervised learning are integrated to solve the problem of Gaussian Process Binary Classification.
- In order to alleviate the data deficiency problem, our proposed work boosts the generalization performance of all the tasks in either the case, supervised or unsupervised.
- To figure out the performance of our proposed approach with varying the labeled data, experimental results are presented on two popular datasets, Handwritten Letter classification, and USPS Digit classification

2 Supervised Multi-task Binary Gaussian Process Classification

Majority of the machine learning algorithms are supervised learning where we have a set of labeled training data and the aim is to use an algorithm to determine a mapping function from the input to output so that when we have a new data point we can predict the class label for that same. Here, this concept is being used for multi-task binary classification problem where the idea of Gaussian Process serves as a base for this model.

We consider n classification tasks $N_1, N_2......N_n$ related to each other. The training set Q_i of task N_i comprises of l_i labeled data points $(x_j^i, y_j^i)_j^{l_i}$ where input $x_j^i \in \mathbb{R}^d$ and corresponding label y_j^i in $\{+1, -1\}$. For each task N_i we adopt a Gaussian process for supervised learning and proceed by selecting a covariance function $K:X*X \to \mathbb{R}$. We associate a latent variable f_j^i with each input data x_j^i. These latent variables are assumed to be Gaussian distributed with zero mean and covariance K_{θ_i}. Correlation between $f^i(x)$ and $f^i(z)$ is given by $K(x,z)$ where x and z are any two data points, i.e.,

$$f^i|X^i \sim \mathcal{N}(I_{m_i}, K_{\theta_i}) \tag{2}$$

where $\mathbf{f}^i$ denotes the column matrix of all latent variables associated with task N_i, $\mathbf{X}^i = (\mathbf{x}_1^i, \mathbf{x}_2^i, ...\mathbf{x}_{m_i}^i,)$, $\mathcal{N}(\mathbf{m}, \sigma)$ represents multivariate Gaussian distribution with mean $\mathbf{m}$ and covariance σ, I_{m_i} represents m_i x 1 dimension zero vector, and $\mathbf{K}_{\theta_i}$ represents the kernel matrix on $\mathbf{X}^i$ where θ_i is a parameter of kernel function.

The class labels for the above latent function f^i for the ith task are assumed to be Bernoulli distributed. Thus, the likelihood for training dataset is defined as follows:

$$p\left(y^i|f^i\right) = \prod_{j=1}^{m} Ber\left(y_j^i|s\left(f_j^i\right)\right) \tag{3}$$

where $y^i = (y_1^i, y_2^i, ..., y_m^i)$.

$$p\left(y^i|f^i\right) = \prod_{j=1}^{m} s\left(f_j^i\right)^{y_j^i} \left(1 - s\left(f_j^i\right)\right)^{1-y_j^i} \neq \mathcal{N}\left(f^i|.,.\right) \tag{4}$$

Since all the tasks are considered to be related, we take common prior on all the kernel parameters of the tasks, i.e., $(\theta_i)_i^n$.

$$\theta_i \sim \mathcal{N}(m_\theta, \sigma_\theta) \tag{5}$$

Inferences for binary Gaussian process classification [12] of N^i tasks are naturally divided into two steps: in first step, we compute the latent variable corresponding to a test case over the training set latent variables.

$$p(f_*^i|X^i, y^i, x_*^i, \theta_i) = \int p(f_*^i|X^i, x_*^i, f^i)p(f^i|X^i, y^i)df \tag{6}$$

where $p(f^i|X^i, y^i, \theta_i) = p(y^i|f^i)p(f^i|X^i, \theta_i)/p(y^i|X^i)$ is the posterior distribution over the latent variable and in second step, for making probabilistic predictions, we use this distribution over the test set latent variables, f_*^i.

$$\bar{\pi}_* = p(y_*^i = +1|X^i, y^i, x_*^i, \theta_i) = \int s(f_*^i)p(f_*^i|X^i, y^i, x_*^i)df_*^i. \tag{7}$$

where $s(f_*^i) = p(y_*^i|f_*^i)$

The posterior distribution $p(f^i|X^i, y^i, \theta_i)$ is non-Gaussian. In order to preserve computational tractability for this posterior distribution, a number of inference methods can be used to approximate it. Some popular techniques include expectation propagation (EP), KL-Divergence Minimization (KL), Variational Bounds (VB), Factorial Variational Method (FV) and Laplace Approximation (LA). In this paper, we use Laplace's method approximation to the posterior distribution $p(f^i|X^i, y^i)$. To obtain a Gaussian approximation, we need to apply second order Taylor expansion of log $p(f^i|X^i, y^i)$ around the posterior approach $\hat{f}^i$. The mode $\hat{f}^i$ is founded by Newton's method.

$$q\left(f^i|X^i, y^i\right) = \mathcal{N}(f^i|\hat{f}^i, K_i^{-1}) \propto exp(-\frac{1}{2}(f^i - \hat{f}^i)^T K_i(f^i - \hat{f}^i)) \qquad (8)$$

where $\hat{f}^i = \text{argmax}_{f^i} p(f^i|X^i, y^i)$ and $K_i = -\bigtriangledown \bigtriangledown \log p(f^i|X^i, y^i)|_{f^i=\hat{f}^i}$ or $(K_{\theta_i}^{-1} + W_i)$ is the Hessian of the negative log posterior at that point.

Then, the log marginal likelihood can be approximated as follows:

$$\ln p(y^i|X^i, \theta_i) = \ln p(y^i|\hat{f}^i) - \frac{1}{2}\hat{f}^{i^T} K_{\theta_i}^{-1} \hat{f}^i + \frac{1}{2}\ln |I + K_{\theta_i} W_i| \qquad (9)$$

Where $W_i = \bigtriangledown \bigtriangledown \log p(y^i|f^i)$
Now the log-likelihood of all tasks can be calculated as follows:

$$L = -\frac{1}{2}\sum_{i=1}^{N}\left[(y^i)^T (K_\theta)^{-1} y^i + \ln |I + K_{\theta_i} W_i|\right]$$
$$-\frac{1}{2}\sum_{i=1}^{N}\left[(\theta_i - m_\theta)^T \sigma_\theta^{-1}(\theta_i - m_\theta)\ln |\sigma_\theta^{-1}|\right] + \text{Constant} \qquad (10)$$

where $|T|$ is the determinant of a square matrix T. In order to find out the values of $\theta_i, m_\theta, \sigma_\theta$, we need to maximize the above function L. The Kernel function used by us is $k(x_1, x_2) = \theta_1 x_1^T x_2 + \theta_2 exp(-\frac{\|x_1-x_2\|^2}{2\theta_3^2})$. This kernel function and look-likelihood are similar to the supervised kernel function and look-likelihood defined in [13], but they are for regression. For estimating the parameter θ_i, value in t^{th} repetition, we use gradient descent to maximize the log-likelihood as follows.

$$\frac{\partial L}{\partial ln\theta_i} = \frac{1}{2}\text{diag}(\theta_i)\left[tr(C\frac{\partial K_{\theta_i}}{\partial \theta_j}) - 2(\sigma_\theta^t)^{-1}(\theta_i - m_\theta^{(t)})\right] \qquad (11)$$

Where $C = (K_{\theta_i})^{-1}(y^i)^T y^i (K_{\theta_i})^{-1} - (I + K_{\theta_i} W_i)^{-1} W_i$
After finding out the value of θ_i in the tth iteration, now we find out the value of σ_θ and m_θ in the $(i + 1)^{th}$ iteration as follows:

$$m_\theta^{(t+1)} = \frac{1}{n}\sum_{i=1}^{n}\theta_i \qquad (12)$$

$$\sigma_\theta^{(t+1)} = \frac{1}{n} \sum_{i=1}^{n} (\theta_i - m_\theta^{(t+1)})(\theta_i - m_\theta^{(t+1)})^T \tag{13}$$

Here n is the total number of tasks.

The posterior mean of f_*^i for a given test data point x_*^i of task N_i under the Laplace approximation can be expressed by the following equation

$$E_q \left[f_*^i | X^i, y^i, x_*^i \right] = k(x_*^i)^T K_{\theta_i}^{-1} \hat{f}^i = k(x_*)^T \nabla \, logp(y^i | \hat{f}^i) \tag{14}$$

where $\hat{f}^i = q(f^i | X^i, y^i)$ which is taken from the Eq. 8.

Similarly covariance of f_*^i can be defined as follows:

$$V_q \left[f_*^i | X^i, y^i, x_*^i \right] = k_{\theta_i}(x_*^i, x_*^i) - k_*^T (K_{\theta_i} + W_i^{-1})^{-1} k_* \tag{15}$$

Where $k_{\theta_i}(,)$ is a kernel function which is parameterized by θ_i, and $k_*^i = (k_{\theta_i}(x_*^i, x_1^i), k_{\theta_i}(x_*^i, x_2^i), k_{\theta_i}(x_*^i, x_3^i), ..., k_{\theta_i}(x_*^i, x_n^i))^T$

Finally for getting labeled data y_*^i for test data point x_*^i of task N_i, we use sigmoid function s(f_*^i) for posterior mean f_*^i.

3   Semi-supervised Multi-task Binary Gaussian Process Classification

After a brief discussion about multi-task supervised Gaussian process binary classification, we now extend this model to multi-task semi-supervised Gaussian binary classification. We broaden our model to semi-supervised one to incorporate the information from unlabeled data as well. The training set Q_i of task N_i now comprises of l_i labeled data points $\{(\mathbf{x}_m^i, \mathbf{y}_m^i)\}_{m=1}^{l_i}$ and u_i $\{\mathbf{x}_m^i\}_{m=l_i+1}^{u_i}$ unlabeled data points. The amount of unlabeled data is far greater than the amount of labeled data points, i.e., $u_i >> l_i$.

For the Gaussian process, we choose a function K(.,.) that is symmetric positive and semidefinite as our kernel function. This kernel function has Reproducing Kernel Hilbert Space(H) of functions associated with it. Result of classification after applying this RKHS function [6] is shown in Fig. 2 (a). This Fig. 2 (a) shows that the decision boundary doesn't separate the binary class according to data distribution(i.e. decision boundary is linear) . The norm of the function in Hilbert space $\|f\|_{\mathcal{H}}$ is a measure of smoothness of the function, i.e., how well the function fits the given data points.In semi-supervised learning, the unlabeled data may provide additional precision and regularity corresponding to data manifolds or clusters. In order to incorporate the smoothness from the unlabeled data as in semi supervised case we reconstruct the RKHS by redefining the norm as

$$\langle f, g \rangle_{\widetilde{\mathcal{H}}} = \langle f, g \rangle_{\mathcal{H}} + f^T M \mathbf{g} \tag{16}$$

where $\mathbf{f}$ is $\{f(\mathbf{x})\}_{(\mathbf{x}) \epsilon X_Q}$ and $\mathbf{g}$ is $\{g(\mathbf{x})\}_{(\mathbf{x}) \epsilon X_Q}$. After reconstruction of this new RKHS with modified inner product, it results to a new data dependent kernel k associated to it which assimilates the geometry of the data in elemental classes as shown in Fig. 2(b).

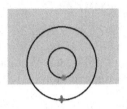
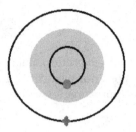

(a) Classifier learnt in the RKHS

(b) Classifier learnt in deformed RKHS

Fig. 2. Showing a binary classification problem approach: decision boundary is linear and non-linear respectively.

We use this kernel function for semi-supervised binary Gaussian process classification. Therefore the covariance between $f_{\mathbf{x}_i}$ and $f_{\mathbf{x}_j}$ now depends on the coordinate as well as the geometry of data in the dataset. For each task N_i we construct a similarity graph using the following given equation :

$$S^i_{mn} = \begin{cases} \exp(\frac{-\|\mathbf{x}^i_m - \mathbf{x}^i_n\|^2}{L^i_m L^i_n}) & \text{if } \mathbf{x}^i_m \epsilon N_K(\mathbf{x}^i_n) \text{ or } \mathbf{x}^i_n \epsilon N_K(\mathbf{x}^i_m), \\ 0 & \text{otherwise.} \end{cases}$$

where set of K nearest points for the given data point $\mathbf{x}^i_m$ in task N_i are represented by $N_K(\mathbf{x}^i_n)$, L^i_m is the separation between $\mathbf{x}^i_n$ and its Kth nearest neighbor, and L^i_n is the separation between $\mathbf{x}^i_n$ and its Kth nearest neighbor. A random variable G^i is proposed for analyzing the data geometry of task N_i. Through Bayesian learning we can estimate the posterior. We define the likelihood of the random variable G_i as follows:

$$P(G^i|\mathbf{f}^i_Q) = Z_i \exp(-\frac{\mathbf{p}^T_Q M^i \mathbf{p}_Q}{2}) \tag{17}$$

$\mathbf{p}_Q = [P(y = 1|f_{\mathbf{x}_1}), ..., P(y = 1|f_{\mathbf{x}_{l+u}})]^T = [\Phi(f_{x_1}), ..., \Phi(f_{\mathbf{x}_{l+u}})]^T$ denotes the conditional probability of the labels with respect to the given latent variables. $\mathbf{M}$ is the matrix for Laplacian graph and Z_i is a hyperparameter which is required to be evaluated. If the value of G_i is large, it is clear that the probability of graph containing manifold structure is high.

Thus, the join prior of f^i_Q conditioned on G_i can be computed as follows

$$f^i_Q|G^i \sim \mathcal{N}(0_n, (K^{-1}_{\theta_i} + Z_i M^i)^{-1}) \tag{18}$$

This data-dependent new inner product assimilates the original smoothness and regularity with the underlying smoothness provided by the Matrix M^i for task N_i. The kernel function for this new RKHS can be derived by using reproducing property of the RKHS and is stated by

$$\widetilde{K}(\mathbf{x}, \mathbf{z}) = K_{\theta_i}(\mathbf{x}, \mathbf{z}) - k(x^i_x)^T (Z^{-1}_i I + M^i K_{\theta_i})^{-1} M^i k(x^i_z) \tag{19}$$

Where $k_z^i = (k_{\theta_i}(x_z^i, x_1^i), k_{\theta_i}(x_z^i, x_2^i), k_{\theta_i}(x_z^i, x_3^i), ..., k_{\theta_i}(x_z^i, x_n^i))^T$

We use k_{θ_i} to represent the kernel matrix whose elements are determined by modifying the kernel function in Sect. 2 on the labeled data of task N_i. Here also, we calculate the values of $\theta_i, m_\theta, \sigma_\theta$ same as calculated for multi-task supervised learning in Sect. 2. But to calculate Z_i, the gradient descent method can be used calculate as:

$$\frac{\partial L}{\partial ln Z_i} = \frac{Z_i}{2} \left\{ tr \left[(K_{\theta_i})^{-1} y^i (y^i)^T (K_{\theta_i})^{-1} \frac{\partial K_{\theta_i}}{\partial Z_i} \right] \right\} - tr \left[(K_{\theta_i} W_i + I)^{-1} \frac{\partial K_{\theta_i}}{\partial Z_i} \right] \quad (20)$$

In this section also, we calculate labeled data for new task as we calculated for multi-task supervised learning in Sect. 2.

4   Experiments

In this section, we have shown experimental results for the above proposed supervised multi-task binary Gaussian process classification (SMBGC) and semi-supervised multi-task binary Gaussian process classification (SSMBGC) on the standard datasets like USPS dataset[1] and English alphabet dataset[2]. In order to show the performance of our proposed approaches SMBGC and SSMBGC, we have compared them with binary Gaussian process classification (BGC) [12] and semi-supervised binary Gaussian process classification (SBGC) [6].

4.1   English Alphabets Binary Classification

For the binary classification problem, we divided this dataset into 6 different tasks such as a/b, b/c, c/a, p/q, q/r and p/r respectively, where each data point of the task has 16 features. In order to improve the generalized learning performance for each task, we grouped these six tasks into two different sets according to their relation. The first set comprises of three tasks task-1, task-2 and task-3 and the second one also comprises of tasks task-4, task-5 and task-6. The generalized learning performance for each set is enhanced with respect to each task present in the set. Each task has approximately 1500 data points or more (where approx 750 data points are for one alphabet and 750 data points for the second one). Out of 1500 data points, 60% data points are reserved for training our proposed models and the remaining 40 percent data points are utilized for testing our models. Out of the reserved training data points, we vary this data for our SMBGC and SSMBGC to show their performance. Means, for training our SMBGC model, we use 5%, 10%, 20%, and 30% of the training data as labeled while for our SSMBGC model, we use same training data as well as rest of unlabeled data. Accuracies of the SMBGC and SSMBGC models on varying the training data are calculated by evaluating the models on the test data. Comparative results shown in Fig. 3 for various tasks show that the performance of all the tasks are improved.

---

[1] https://archive.ics.uci.edu/ml/support/Optical+Recognition+of+Handwritten +Digits.

[2] https://archive.ics.uci.edu/ml/datasets/letter+recognition.

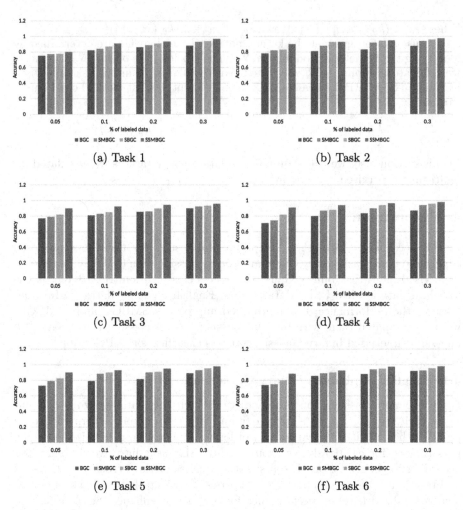

Fig. 3. Comparison of BCG, SMBGC, SBGC and SSMBGC on the Hand-written letter classification

4.2   USPS Digit Binary Classification

We perform our classification Task on the USPS digit dataset consisting of approx 7000 data points and 256 features. With the help of given dataset we examine the performance of our model on 6 classification tasks, namely 3/5, 5/6, 3/6, 2/4, 1/2, and 1/4. Similarly, we group these above six tasks into two different sets where first set constitute tasks task-1, task-2 and task-3 and second set constitutes tasks task-4, task-5 and task-6. keeping all the experimental procedure same as in alphabet binary classification, we observed the accuracies of our two proposed model SMBGC and SSMBGC as well as for BGC and SBGC classification. Results in the Fig. 4 show that our approaches improves the performance of all the tasks together.

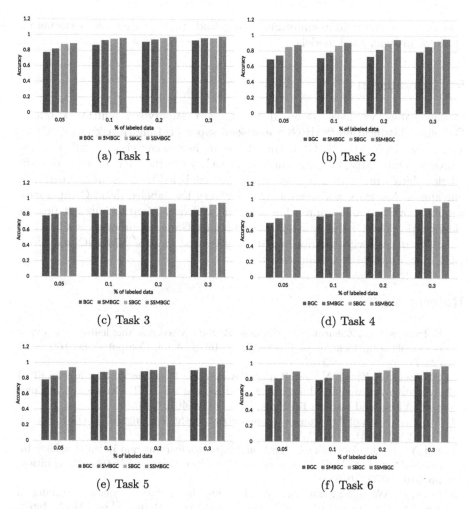

Fig. 4. Comparison of BCG, SMBGC, SBGC and SSMBGC on the USPS Digit classification.

4.3   Evaluation Measures

We have evaluated our proposed methods with binary Gaussian process classification (BGC) [12] and semi-supervised binary Gaussian process classification (SBGC) [6]. BGC algorithm for any task simply develops the predictive model based on its own labeled data, but SBGC algorithm takes helps from the unlabeled data as well. Both of the algorithms (BGC and SBGC) don't take help from other related tasks. Therefore, their performance shown in Figs. 3 and 4 is not as good as our proposed approaches. Our both the proposed approaches (SMBGC and SSMBGC) while developing predictive model take help from other related tasks as well. Their performance is increasing because we have used common Gaussian prior on all the kernel parameters of all tasks. If the unlabeled data doesn't follow manifold assumption, then it leads to decreasing the performance of our semi-supervised models.

5 Conclusion

In this paper, we have proposed two models, supervised multi-task binary Gaussian classification (SMBGC) and semi-supervised multi-task binary Gaussian Classification (SSMBGC), and observed their accuracies for different test datasets. Our SMBGC algorithm learns well with the help of related tasks labeled data while our SSMBGC from the unlabeled data as well. Therefore, our approaches work well even as we have very few labeled data. Comparative results show that our approaches perform well over binary Gaussian process classification (BGC) and semi-supervised binary Gaussian process classification.

In future, we will enhance our above-proposed models (SMBGC and SSM-BGC) for Multi-class classification too.

References

1. Kotsiantis, S.B., Zaharakis, I., Pintelas, P.: Supervised machine learning: a review of classification techniques. Emerg. Artif. Intell. Appl. Comput. Eng. **160**, 3–24 (2007)
2. Oh Song, H., Xiang, Y., Jegelka, S., Savarese, S.: Deep metric learning via lifted structured feature embedding. In: Proceedings of the IEEE Conference on Computer Vision and Pattern Recognition, pp. 4004–4012 (2016)
3. Xu, Y., et al.: A unified framework for metric transfer learning. IEEE Trans. Knowl. Data Eng. **29**(6), 1158–1171 (2017)
4. Wu, Y., Ji, Q.: Constrained deep transfer feature learning and its applications. In: Proceedings of the IEEE Conference on Computer Vision and Pattern Recognition, pp. 5101–5109 (2016)
5. Chen, K., Wang, S.: Semi-supervised learning via regularized boosting working on multiple semi-supervised assumptions. IEEE Trans. Pattern Anal. Mach. Intell. **33**(1), 129–143 (2011)
6. Sindhwani, V., Chu, W., Keerthi, S.S.: Semi-supervised Gaussian process classifiers. In: IJCAI, pp. 1059–1064 (2007)
7. Jebara, T.: Multi-task feature and kernel selection for SVMs. In: Proceedings of the Twenty-First International Conference on Machine learning, p. 55. ACM (2004)
8. Evgeniou, T., Micchelli, C.A., Pontil, M.: Learning multiple tasks with kernel methods. J. Mach. Learn. Res. **6**, 615–637 (2005)
9. Durichen, R., Pimentel, M.A., Clifton, L., Schweikard, A., Clifton, D.A.: Multitask Gaussian processes for multivariate physiological time-series analysis. IEEE Trans. Biomed. Eng. **62**(1), 314–322 (2015)
10. Xue, Y., Liao, X., Carin, L., Krishnapuram, B.: Multi-task learning for classification with dirichlet process priors. J. Mach. Learn. Res. **8**, 35–63 (2007)
11. Gao, J., Ling, H., Hu, W., Xing, J.: Transfer learning based visual tracking with Gaussian processes regression. In: Fleet, D., Pajdla, T., Schiele, B., Tuytelaars, T. (eds.) ECCV 2014. LNCS, vol. 8691, pp. 188–203. Springer, Cham (2014). https://doi.org/10.1007/978-3-319-10578-9_13

12. Rasmussen, C.E., Williams, C.K.: Gaussian Process for Machine Learning. MIT Press, Cambridge (2006)
13. Zhang, Y., Yeung, D.-Y.: Semi-supervised multi-task regression. In: Buntine, W., Grobelnik, M., Mladenić, D., Shawe-Taylor, J. (eds.) ECML PKDD 2009. LNCS (LNAI), vol. 5782, pp. 617–631. Springer, Heidelberg (2009). https://doi.org/10.1007/978-3-642-04174-7_40

Employ Decision Values for Soft-Classifier Evaluation with Crispy References

Lei Zhu$^{(\boxtimes)}$, Tao Ban, Takeshi Takahashi, and Daisuke Inoue

National Institute of Information and Communications Technology, Tokyo, Japan
lzhu@nict.go.jp

Abstract. Evaluation of classification performance has been comprehensively studied for both crispy and fuzzy classification tasks. In this paper, we address the hybrid case: evaluating fuzzy prediction results against crispy references. The proposal is motivated by the following facts: (1) most datasets in practice are produced with crispy labels due to the excessive cost of fuzzy labelling; and (2) many state-of-the-art classifiers can yield fuzzy decision values even if they are trained from data with crispy labels. We derive our fuzzy-crispy evaluation criterion based on a widely adopted fuzzy-set-based evaluation method. By exploiting the distribution of decision values, the proposed criterion bears more comprehensive information than conventional crispy classification evaluation criteria. The advantages of the proposed criterion are demonstrated in artificial and real-world classification case studies.

Keywords: Classification evaluation · Decision value
Crispy reference

1 Introduction

Based on the different form of data labeling, classification tasks can be categorized into *hard classification* or *soft classification*. Hard classification, in which each instance belongs to one and only one of the classes in training and testing, is more conventional. The label representation for hard classification is also known as crispy labeling [10]. On the other hand, data can be labeled in a fuzzy manner, where each instance is associated with degrees of membership to several classes at the same time. Generally, a classifier adopted in a hard/soft classification task is assumed to produce prediction results in a crispy/fuzzy manner.

The evaluation of pure hard/soft classification has been studied for decades. There are numerous measurements for either evaluating hard prediction with crispy references or evaluating soft prediction with fuzzy references [11]. As an evaluation criterion for soft classification, Jager and Benz proposed to measure classification accuracy using fuzzy similarity [6]. In the work of Binaghi *et al.*, fuzzy prediction and ground truth are taken as fuzzy sets, where each element in the confusion matrix is calculated by fuzzy-set intersection [1]. Fuzzy similarity

© Springer Nature Switzerland AG 2018
L. Cheng et al. (Eds.): ICONIP 2018, LNCS 11304, pp. 392–402, 2018.
https://doi.org/10.1007/978-3-030-04212-7_34

and fuzzy-set intersection are also employed to compute fuzzy Kappa for classification map evaluation, which has impacted remote sensing and geographic informatics [3,5]. Refer to [11] for a comprehensive review on hard classification evaluation.

In this paper, we study the hybrid case between hard and soft classification evaluation which is rarely addressed in the literature. We consider the following scenario: First, the dataset is produced with crispy label for both training and testing. Then, for an unseen sample, the classifier outputs not only a crispy prediction about its class label, but decision values which indicate to what extent does it belong to the all the classes as well. The proposal is motivated by the following facts: (1) labeling data in a fuzzy manner is costly in many environments: e.g., most datasets produced in cyber-security field are generated with crispy labels; (2) many classifiers, e.g., SVM and Naive Bayes, output soft decision values along with the crispy prediction even if they are trained upon data with crispy labels; and (3) the soft decision values, commonly normalized to $[0, 1]$, contain extra information than class labels, may be useful for classifier evaluation.

The rest of this paper is organized as follows. Section 2 introduces the fuzzy-set based evaluation criterion proposed by Binaghi et al. [1]. Section 3 presents the proposed fuzzy-crispy evaluation method, including the confusion matrix calculation for both multi-class and binary cases. Then we demonstrate the advantages of proposed method in Sect. 4 with case studies on both artificial and real-world classification task, and finally we conclude the work in Sect. 5.

2 Preliminary

A fuzzy-set-based approach has been proposed by Binaghi et al. [1] to evaluate soft classifiers on multi-class classification. In their setup, both classifier prediction and ground truth (reference) are represented in form of fuzzy-sets. All elements in the confusion matrix are calculated via fuzzy-set intersection.

Given test dataset X with Q classes, let R_n be the set of data assigned to class n in the reference, and C_m be the set of data that are classified into class m by the classifier. Then $\{R_n|1 \leq n \leq Q\}$ and $\{C_m|1 \leq m \leq Q\}$ form two partitions of the test set X according to reference and prediction, respectively.

For the conventional hard classification whose reference and prediction are both crispy, R_n and C_m are taken as crispy sets. Any sample x in X can be either member or non-member of classes n and m according to discrimination (membership) function of R_n and C_m

$$\bullet \quad \mu_{R_n} : X \rightarrow \{0,1\} \qquad \qquad \mu_{C_m} : X \rightarrow \{0,1\}$$

$$\mu_{R_n}(x) = \begin{cases} 1 & \text{if } x \in R_n \\ 0 & \text{otherwise} \end{cases} \qquad \mu_{C_m}(x) = \begin{cases} 1 & \text{if } x \in C_m \\ 0 & \text{otherwise} \end{cases} . \qquad (1)$$

Let M be the confusion matrix, then its element $M(m, n)$ denotes the cardinality of the intersection between R_n and C_m

$$M(m,n) = |R_n \cap C_m| = \sum_{x \in X} \mu_{R_n \cap C_m}(x), \qquad (2)$$

where

$$\mu_{R_n \cap C_m}(x) = \begin{cases} 1 & \text{if } x \in C_m \wedge x \in R_n \\ 0 & \text{otherwise} \end{cases}. \qquad (3)$$

In context of soft classification, one sample is no longer explicitly classified into one specific class. Instead, the prediction is presented as a set of membership grades $\alpha_i(x)$ representing to what extend x belongs to class i, where $\alpha_i(x) \in [0, 1]$ and $i \in \{1, \ldots, Q\}$. Similarly, the reference is also represented as a set of multiple and partial class memberships $\beta_j(x)$.

In such fuzzy scenario, Binaghi *et al.* propose to consider above R_n and C_m as fuzzy sets, which are denoted as $\tilde{R}_n$ and $\tilde{C}_m$, respectively. $\tilde{R}_n$ and $\tilde{C}_m$ are determined by their membership functions

$$\mu_{\tilde{R}_n} : X \to [0, 1] \quad \mu_{\tilde{C}_m} : X \to [0, 1]. \qquad (4)$$

$\mu_{\tilde{R}_n}(x)$ and $\mu_{\tilde{C}_m}(x)$ refer to the gradual membership of the sample x in classes n and m indicated in the reference and prediction, respectively. Then $\{\tilde{R}_n\}$ and $\{\tilde{C}_m\}$ perform two partitioning of test set X in the fuzzy manner. The orthogonality condition (or sum-normalization) is sometimes applied, which force the sum of all membership functions to be one for any sample in the test set

$$\sum_{n=1}^{Q} \mu_{\tilde{R}_n}(x) = 1 \quad \sum_{m=1}^{Q} \mu_{\tilde{C}_m}(x) = 1 \quad \forall x \in X. \qquad (5)$$

In the rest of this paper, we assume the orthogonality condition holds at all time.

To obtain the fuzzy confusion matrix $\widetilde{M}$ for soft classification, each element $\widetilde{M}(m, n)$ requires calculating the degree of membership in the fuzzy intersection set $\tilde{R}_n \cap \tilde{C}_m$. Among various options for fuzzy set intersection, the standard operation [8] in fuzzy set theory is adopted in the work of Binaghi *et al.* The intersection of two membership function is valued as

$$\mu_{\tilde{R}_n \cap \tilde{C}_m}(x) = \min(\mu_{\tilde{R}_n}(x), \mu_{\tilde{C}_m}(x)), \qquad (6)$$

the minimal of two membership functions. Then the cardinality of the overall fuzzy intersection $\tilde{R}_n \cap \tilde{C}_m$ representing $\widetilde{M}(m, n)$ in the fuzzy confusion matrix is computed as

$$\widetilde{M}(m,n) = \left| \tilde{R}_n \cap \tilde{C}_m \right| = \sum_{x \in X} \mu_{\tilde{R}_n \cap \tilde{C}_m}(x). \qquad (7)$$

3   Evaluating Soft Classifiers with Crispy References

In this section, we consider the situation that the references are crispy but the predictions are fuzzy. This can be treated as a special case in above fuzzy-set-based method, where the fuzzy membership function of reference $\mu_{\tilde{R}_n}(x)$ has only two possible values, either 0 or 1.

3.1   The Multi-class Case

Let $\bar{R}_n$ and $\bar{C}_m$ be the crispy reference and fuzzy prediction sets, respectively. Then their membership functions are in form of

$$\mu_{\bar{R}_n} : X \to \{0,1\} \quad \mu_{\bar{C}_m} : X \to [0,1]. \tag{8}$$

We also know that any test sample x can only belongs to one class n, such that

$$\begin{aligned} \mu_{\bar{R}_n}(x) &= 1 \\ \mu_{\bar{R}_m}(x) &= 0 \quad m \in \{1 \dots Q\} \quad m \neq n. \end{aligned} \tag{9}$$

Then, the intersection between two membership function (above Eq. (6)) can be written as

$$\mu_{\bar{R}_n \cap \bar{C}_m}(x) = \min(\mu_{\bar{R}_n}(x), \mu_{\bar{C}_m}(x)) = \begin{cases} \mu_{\bar{C}_m}(x) & \text{if } x \in R_n \\ 0 & \text{otherwise} \end{cases}. \tag{10}$$

Consequently, for the confusion matrix on the complete test set, $\bar{M}(m,n)$ is computed as

$$\bar{M}(m,n) = \left| \bar{R}_n \cap \bar{C}_m \right| = \sum_{x \in X} \mu_{\bar{R}_n \cap \bar{C}_m}(x) = \sum_{x \in R_n} \mu_{\bar{C}_m}(x). \tag{11}$$

3.2   The Binary Case

The binary fuzzy-crispy evaluation is simply a special case of above multi-class case that both m and n can only take value 1 or 2. Since (9), then we have

$$\begin{aligned} \mu_{\bar{R}_1}(x) + \mu_{\bar{R}_2}(x) &= 1 \quad \forall x \in X \\ \mu_{\bar{R}_1}(x) &\in \{0,1\} \quad \mu_{\bar{R}_2}(x) \in \{0,1\}. \end{aligned} \tag{12}$$

Because we assume the orthogonality (5) holds, then we have

$$\sum_{m=1}^{2} \mu_{\bar{C}_m}(x) = 1 \quad \forall x \in X \tag{13}$$

$$\mu_{\bar{C}_1}(x) \in [0,1] \quad \mu_{\bar{C}_2}(x) \in [0,1].$$

Let class 1 be the positive class, and 2 be the negative class. Use class number 0 instead of 2 to have better convention with the binary classification notation.

Also let $D \in [0,1]$ be the decision value of being positive class, i.e., $D(x) = \mu_{\tilde{C}_1}(x)$, then $\mu_{\tilde{C}_0}(x) = 1 - D(x)$. Let $T(x) \in \{0,1\}$ be the class label in reference, then we know $\mu_{\tilde{R}_1}(x) = T(x)$ and $\mu_{\tilde{R}_0}(x) = 1 - T(x)$.

According to (10) and (11), the confusion matrix for a single sample x in the binary fuzzy-crispy case can be written as

$$
\begin{array}{lcc}
 & \text{Truth (R)} & \\
\text{Pred (C)} & + & - \\
+ & min(D(x), T(x)) & min(D(x), 1 - T(x)) \\
- & min(1 - D(x), T(x)) & min(1 - D(x), 1 - T(x)),
\end{array}
\tag{14}
$$

and for the complete test set, the confusion matrix is computed as

$$
\begin{aligned}
TP &= \sum_{x \in X} min(D(x), T(x)) = \sum_{T(x)=1} D(x) \\
FP &= \sum_{x \in X} min(D(x), 1 - T(x)) = \sum_{T(x)=0} D(x) \\
FN &= \sum_{x \in X} min(1 - D(x), T(x)) = \sum_{T(x)=1} (1 - D(x)) \\
TN &= \sum_{x \in X} min(1 - D(x), 1 - T(x)) = \sum_{T(x)=0} (1 - D(x)),
\end{aligned}
\tag{15}
$$

where TP, FP, FN and TN denotes True Positive, False Positive, False Negative and True Negative, respectively.

Having the confusion matrix computed, we can easily obtain those frequently used measurements for binary classification. Let n_+/n_- be the number of positive/negative samples in the reference. For over all accuracy (ACC), we have

$$
ACC = \frac{\sum_{T(x)=1} D(x) + \sum_{T(x)=0} (1 - D(x))}{n_+ + n_-},
\tag{16}
$$

which indicated how well the decision value of positive/negative samples are close to 1/0. Sensitivity, true positive rate (TPR) or recall is computed as

$$
TPR = \frac{\sum_{T(x)=1} D(x)}{\sum_{T(x)=1} D(x) + \sum_{T(x)=1} (1 - D(x))} = \frac{\sum_{T(x)=1} D(x)}{n_+},
\tag{17}
$$

which represents the average D for all positive samples. Specificity or true negative rate (TNR) is given by

$$
SPC = \frac{\sum_{T(x)=0} (1 - D(x))}{\sum_{T(x)=0} D(x) + \sum_{T(x)=0} (1 - D(x))} = \frac{\sum_{T(x)=0} (1 - D(x))}{n_-},
\tag{18}
$$

which shows the average $1 - D$ for all negative samples.

Similarly, we can compute precision or positive predictive value (PPV), negative predictive value (NPV), fall-out or false positive rate (FPR), miss rate or false negative rate (FNR) as

$$PPV = \frac{\sum_{T(x)=1} D}{\sum_{x \in X} D(x)} \qquad NPV = \frac{\sum_{T(x)=0}(1 - D(x))}{\sum_{x \in X}(1 - D(x))}$$

$$FPR = \frac{\sum_{T(x)=0} D(x)}{n_-} \qquad FNR = \frac{\sum_{T(x)=1}(1 - D(x))}{n_+}. \tag{19}$$

4   Experiment

In this section, we demonstrate the advantage of proposed fuzzy-crispy evaluation via case studies on artificial chessboard data and real-world security related
data. All classifiers employed in our experiments are capable of output both
fuzzy decision value and crispy predict label (by applying default threshold 0.5
on decision value), including SVM [2], Fuzzy K-NN [7], Naive Bayes, Random
Forest and Decision Tree.

4.1   On Artificial Chessboard Data

We generate a two-dimensional dataset for training, as shown in Fig. 1, the two
class samples (red and blue dots denote positive and negative class, respectively)
form a 5×5 chessboard. All five classifiers are trained by the same training
dataset. The magenta solid line in Fig. 1 shows the decision boundary for an
ideal classifier.

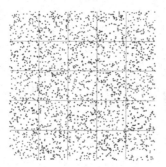

Fig. 1. The chess board training data (Color figure online)

For testing, we randomly generate another testing set that follows the same
5×5 chessboard distribution. We have five trained classifiers conduct both fuzzy
and crispy prediction and evaluate the prediction with crispy ground truth. The
prediction and evaluation results are reported in Fig. 2 and Table 1.

The first column in Fig. 2 shows the prediction results of different classifiers.
The red/blue color of dots represents the crispy class label prediction and the
contour shows the fuzzy prediction result where dark green and bright yellow are
the area has decision value close to 0 and 1, respectively. The magenta solid line

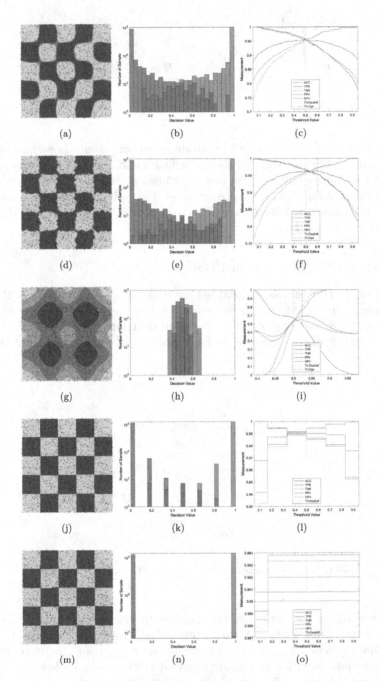

Fig. 2. Prediction results, histogram of decision value and performance variation against threshold shifting. (a–c) SVM, (d–f) Fuzzy K-NN, (g–i) Naive Bayes, (j–l) Random Forest, (m–o) Decision Tree (Color figure online)

Table 1. Evaluation of fuzzy and crispy prediction

Algorithm	Fuzzy					Crispy				
	ACC	TPR	TNR	FPR	FNR	ACC	TPR	TNR	FPR	FNR
SVM	92.33	92.52	92.13	7.87	7.48	95.44	95.62	95.25	4.75	4.38
Fuzzy Knn	93.68	94.42	92.88	7.12	5.58	96.20	96.77	95.58	4.42	3.23
Naive Bayes	51.41	52.26	50.48	49.52	47.74	63.32	66.38	60.00	40.00	33.62
Random Forest	98.32	98.45	98.18	1.82	1.55	99.00	99.15	98.83	1.17	0.85
Tree	99.15	99.08	99.24	0.76	0.92	99.20	99.08	99.33	0.67	0.92

gives the decision boundary of each classifier. As we can see, Random Forest and Decision Tree perform similarly and give the best prediction. Naive Bayes clearly has the worst prediction among all algorithms, and has the largest uncertain area (where the decision value away from either 0 or 1). The performance of SVM and Fuzzy K-NN lie between the best and the worst, and it is hard to say which is better as they are making mistakes in different pattern.

The second column in Fig. 2 gives the histogram of decision value for two classes, brown for positive and blue for negative class. It worth noting that the Y axis of these histograms is in logarithmic scale. The results here mostly agree with the basic judgement we have made from column one. The only new conclusion we can make from this column is that Decision Tree gives better fuzzy prediction than Random Forest, as it gives less uncertain results.

Since we have the decision value of every single test sample, then we are able to adjust the classifier by shifting the decision threshold used in hardening. The plots in third column of Fig. 2 show the variation of performance (in solid line) against any possible decision threshold. The vertical dash yellow line represents the default threshold, and the dash green line shows the optimal threshold to achieve the highest accuracy. Regarding random forest and decision tree, since the decision values are concentrated in a few numbers, the optimal threshold is in form of relatively large interval. Specifically, $[.3333, .6667)$ for random forest and $[.1667, 1)$ for decision tree.

We calculate the accuracy (ACC), true positive rate (TPR), true negative rate (TMR), false positive rate (FPR) and false negative rate (FNR) using fuzzy and crispy prediction respectively and report them in Table 1. As we can see, the result from both method agree with the general interpretation from first column of Fig. 2. If we look at the difference between Fuzzy K-NN and SVM, we find that this difference is enlarged using proposed fuzzy-crispy evaluation. This is because SVM has worse decision value distribution, two class are more 'mixed', as seen in Fig. 2b and e. This phenomenon can be seen more clearly when we compare random forest and decision tree, the worse decision value distribution for random forest leads to harder separation and smaller optimal threshold interval. This drawback is better captured by proposed fuzzy-crispy evaluation in form of enlarged numerical accuracy difference compare with decision tree. For naive bayes, the accuracy computed using fuzzy prediction is much lower than the one calculated from crispy input, this also better agrees our intuition from Fig. 2g and h, a completely screwed classifier has almost none separability.

4.2 On Real-World Data

CIC Tor/NonTor [9] is a dataset generated by Canadian Institute for Cybersecurity (CIC), University of New Brunswick (UNB), Canada. The raw pcap data is captured in a controlled environment as shown in Fig. 3, the workstation is connected to Tor network via a gateway, then the packages between workstation and gateway are taken as regular (NonTor) traffic and the traffic between gateway and Tor network are labeled as Tor. This Tor/NonTor differentiation is the task for CIC Tor/NonTor scenario A (we remark as CICTor-Sa). In scenario B, the workstation executes one application at a time and the traffic captured is labeled according to eight types of application. In our experiment, we take only two classes (audio-streaming and browsing in Tor traffic) to form a binary classification, and remark it as CICTor-Sb.

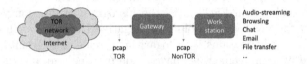

Fig. 3. Capture scenario of CIC Tor/NonTor data

ISCX VPN-nonVPN [4] is a dataset collected also by CIC. It consists traffic of seven types of application over or not over VPN. In its scenario A, the task is to identify VPN/NonVPN traffic (ISCX-Sa in our paper). The scenario B is to differentiate application types in NonVPN traffic. In our experiment, we let Browsing be the positive class and all other type as negative, and refer this task as ISCX-Sb. Table 2 shows the characteristic of all four datasets.

Table 2. Data description

Data set	Pos/Neg class	# feature	# train sample	# test sample
CICTor-Sa	Tor/NonTor	26	805/5979	7239/53805
CICTor-Sb	Audio/Browsing	26	361/802	360/802
ISCX-Sa	NonVPN/VPN	23	3586/3918	5379/5875
ISCX-Sb	Browsing/Other	23	1000/2586	1500/3879

We calculate the accuracy from both fuzzy and crispy prediction, the results on different datasets are reported in Table 3. We also plot the accuracy variation against threshold shifting for all algorithms in each dataset, as shown in Fig. 4.

Table 3. Accuracy computed from fuzzy and crispy prediction

Dataset	Fuzzy					Crispy				
	F. K-NN	N. Bayes	R. Forest	SVM	D. Tree	F. K-NN	N. Bayes	R. Forest	SVM	D. Tree
CICTor-Sa	95.53	84.73	96.44	90.47	97.58	96.14	85.49	98.13	93.71	97.76
CICTor-Sb	83.32	69.04	89.53	64.34	99.46	85.97	70.05	93.63	75.56	99.48
ISCX-Sa	79.32	65.97	83.01	52.05	85.56	81.33	66.80	87.52	57.09	86.51
ISCX-Sb	90.45	81.52	92.31	73.72	93.25	91.99	84.55	95.07	82.10	93.81

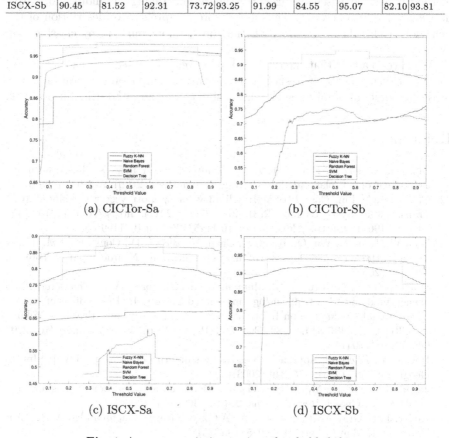

(a) CICTor-Sa   (b) CICTor-Sb

(c) ISCX-Sa   (d) ISCX-Sb

Fig. 4. Accuracy variation against threshold shifting

As we can see, the crispy result only considers the accuracy at the default threshold, however, the fuzzy results take into account that the overall difficulty of finding a good threshold. In the other word, the fuzzy accuracy is more determined by the area below the accuracy-threshold curve. This can be observed more clearly if we look at the Naive Bayes vs. SVM on data CICTor-Sb, as well as the Random Forest vs. Decision Tree on data ISCX-Sa and ISCX-Sb.

5   Conclusion

In this paper, we address the hybrid evaluation task that evaluating fuzzy classification prediction using crispy reference data. This task has practical value especially considering most datasets are crispy and many classifier can make fuzzy prediction along with the crispy label after hardening. Our fuzzy-crispy method is derived based on fuzzy-set theory and taking the more informative decision value into account. Thus gives more comprehensive description of the behavior of classifier compare with the conventional pure crispy classification evaluation.

We have to admit that the relationship between proposed fuzzy accuracy and the area below accuracy-threshold curve is only an empirical observation. Some further theoretical proof for such relationship can be an interesting future work.

References

1. Binaghi, E., Brivio, P.A., Ghezzi, P., Rampini, A.: A fuzzy set-based accuracy assessment of soft classification. Pattern Recogn. Lett. **20**(9), 935–948 (1999)
2. Chang, C.C., Lin, C.J.: LIBSVM: a library for support vector machines. ACM Trans. Intell. Syst. Technol. **2**(3), 27:1–27:27 (2011). https://doi.org/10.1145/1961189.1961199. http://doi.acm.org/10.1145/1961189.1961199
3. Dou, W., Ren, Y., Wu, Q., Ruan, S., Chen, Y., Bloyet, D., Constans, J.M.: Fuzzy kappa for the agreement measure of fuzzy classifications. Neurocomputing **70**(4–6), 726–734 (2007)
4. Draper-Gil, G., Lashkari, A.H., Mamun, M.S.I., Ghorbani, A.A.: Characterization of encrypted and VPN traffic using time-related features. In: Proceedings of the 2nd International Conference on Information Systems Security and Privacy - Volume 1: ICISSP, pp. 407–414. INSTICC, SciTePress (2016). https://doi.org/10.5220/0005740704070414
5. Hagen, A.: Fuzzy set approach to assessing similarity of categorical maps. Int. J. Geogr. Inf. Sci. **17**(3), 235–249 (2003)
6. Jager, G., Benz, U.: Measures of classification accuracy based on fuzzy similarity. IEEE Trans. Geosci. Remote. Sens. **38**(3), 1462–1467 (2000)
7. Keller, J.M., Gray, M.R., Givens, J.A.: A fuzzy k-nearest neighbor algorithm. IEEE Trans. Syst. Man Cybern. **4**, 580–585 (1985)
8. Klir, G.J., Folger, T.A.: Fuzzy Sets, Uncertainty, and Information. Prentice-Hall, Inc., Upper Saddle River (1987)
9. Lashkari, A.H., Gil, G.D., Mamun, M.S.I., Ghorbani, A.A.: Characterization of tor traffic using time based features. In: Proceedings of the 3rd International Conference on Information Systems Security and Privacy - Volume 1: ICISSP, pp. 253–262. INSTICC, SciTePress (2017). https://doi.org/10.5220/0006105602530262
10. Liu, Y., Zhang, H.H., Wu, Y.: Hard or soft classification large-margin unified machines. J. Am. Stat. Assoc. **106**(493), 166–177 (2011)
11. Sokolova, M., Lapalme, G.: A systematic analysis of performance measures for classification tasks. Inf. Process. Manag. **45**(4), 427–437 (2009)

Detection

Guide-Wire Detecting Based on Speeded up Robust Features for Percutaneous Coronary Intervention

Prasong Pusit[1,2(✉)], Xiaoliang Xie[2(✉)], and Zengguang Hou[1,2,3(✉)]

[1] University of Chinese Academy of Sciences, Beijing 100049, China
pusitprasong@gmail.com

[2] State Key Laboratory of Management and Control for Complex Systems, Institute of Automation, CAS, Beijing 100190, China
{xiaoliang.xie,zengguang.hou}@ia.ac.cn

[3] CAS Center for Excellence in Brain Science and Intelligence Technology, Beijing 100190, China

Abstract. Percutaneous coronary intervention (PCI) is a type of endovascular surgery. In the PCI procedure, guide-wire threading under the monitoring of X-ray videos is a vital step widely used to treat narrowing stenosis of a coronary artery. Detection of guide-wire in X-ray videos is not a trivial task because guide-wire has various shapes, and the signal to noise rate is pretty low. Besides, some anatomical skeleton contours are similar to guide-wires. Therefore, it urgently needs accuracy and robust method. In this research, we present a fast and robust guide-wire detection method we offer a fast and robust guide-wire detection method, speeded up robust features (SURF) is applied to locate the tip of guide-wire in various shapes and situations. Our approach was evaluated by testing on 18 X-ray sequence images, total 1073 frames (50 frames for training and 1023 frames for testing). The detection accuracy is 92.7% with 20 fps speed that shows a promising result for guide-wires detection.

Keywords: Guide-wire · Signal-to-noise rate
Cardiovascular diseases · Percutaneous coronary intervention
Guide-wire detection

1 Introduction

Endovascular surgery is a kind of intervention surgery widely used to treat cardiovascular diseases (CVDs). PCI surgery under the monitoring of X-ray videos

This work is partially supported by the National Natural Science Foundation of China (Grant #61533016, #U1613210), European Commission Marie Skodowska-Curie SMOOTH project (H2020-MSCA-RISE-2016-734875), and the Royal Thai Government.

© Springer Nature Switzerland AG 2018
L. Cheng et al. (Eds.): ICONIP 2018, LNCS 11304, pp. 405–415, 2018.
https://doi.org/10.1007/978-3-030-04212-7_35

is a type of endovascular surgery that uses coronary catheterization and guide-wires is inserted into a patient's vascular system [1]. During the intervention surgery, the instant visual feedback offered by the X-ray images and the physician's knowledge are needed in the operation. It has become a popular method presently. In the last few years, PCI X-ray guide-wires detection has become very popular among navigations, because many applications such as robot-assisted surgery, guide-wire motion tracking reconstruct guide-wire model in 3-D, rely upon it. However, a guide-wire detecting is a troublesome task. Due to lacking automatic tracking feedback, to improve the safety and reduce the surgery time are very challenging. Therefore, any navigation which related to a guide-wire detection becomes a challenging and important task. There are many challenging problems with a guide-wire detection. Firstly, guide-wires can be changed into various shapes, thus it hard to set up a set of training template that covers all of the guide-wire shapes. Secondly, it's very hard to distinguish a guide-wire from the cluttered backgrounds because of the influence of some other adjacents such as bones and organs. Thirdly, there is a low signal-to-noise ratio (SNR) [2] during the interventions that cause the guide-wire look similar to backgrounds. Furthermore, a patient's organ is always moving, so the background image is always changing. Furthermore, a patient's organ is always moving, so the background image is always changing.

Previously, several methods have been offered to solve guide-wire detecting problems. The first step of the methods aims at improving X-ray image's quality, then use other means such as B-spline model method [3,4], Hessian method [5], steerable and enhancing diffusion method [6] to achieve the proposed. The main idea of these methods always present in the following steps: (i) use multi-filter methods to enhance the quality of guide-wire X-ray images; (ii) find some important points which may contain a guide-wire image or interest key pixels by computing a probabilistic map; (iii) detect global curve segments by finding a relation between the key points from the previous step, next connect the key points together. However, all of the possible adjacent pixels are needed to take into consideration. Thus they require a complex computation [7,8]. Afterward, machine learning with convolutional neural networks (ConvNet) methods are presented [9,10], the methods are highly efficient and can solve many problems of previous methods. However, the system is hard to train and setup. For example, ConvNet system requires considerable training data.

Consider the problem of the conventional methods. We present a new method based on SURF. It is accuracy and robust method for guide-wire detection. SURF has many significant advantages as follows: First of all, it does not require any manual indication to automatically detect the position of guide-wires. Secondly, our proposed method is easier to train compared with the ConvNet method. We only need moderate training templates to train the system. Furthermore, we can use a feedback information from a previous frame to improve the detection accuracy and speed.

The arrangement of this paper is arranged as follows: Sect. 2 gives an introduction. Section 3 performs our experiments, and Sect. 4 presents the conclusions.

2 Method

SURF is one of potential feature detector and descriptor. It is often used in many tasks such as object recognition, 3D reconstruction, image registration, image classification and so on. SURF is partly inspired by the scale-invariant feature transform (SIFT). However, the efficient of SURF is several times faster than SIFT, furthermore it more robust against different image transformations than SIFT [11,12]. In this research, we present the efficiently guide-wires detection method base on SURF. In the first step, the selected X-ray images for training template datasets are set to train the system. Then, we demonstrate our method of testing datasets.

2.1 Preprocessing

In the first step, we need to do a preprocessing because sometimes guide-wire is very similar to a background image, and there are too many unnecessary objects and noise in a PCI X-ray image. In this research, we prefer to use the binarize 2-D grayscale image by thresholding method [13,14] in the preprocessing process. The method is used to convert some parts of a PCI X-ray image to black or white color based on a set threshold.

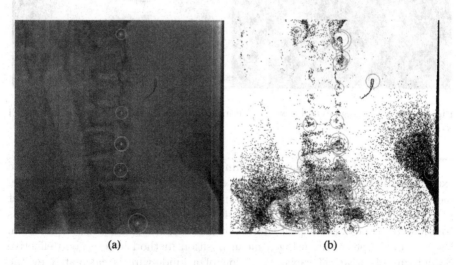

(a)   (b)

Fig. 1. The strong point detecting result between preprocessing the original frame and the frame after preprocessing: (a) is the original X-ray frame; (b) is the X-ray frame after preprocessing.

Figure 1 has shown that the guide-wire on the original frame image is easy to be classified as an unimportant object. On the other hand, almost all the guide-wire on the frame image, which has passed the preprocessing method, is classified as an important object.

2.2 SURF

SURF technique is finding a correspondence point between the reference image and the target image. It is mainly used in many object detection works because it is very efficient to detect objects despite a scale change or in-plane rotation. Furthermore, It can be used in the case of a small amount of out-of-plane rotation and occlusion [15]. Recently, there are many pieces of research which use SURF on determining feature searching area [16,17] and the feature extraction algorithm [18]. In this research, we apply SURF to find the matching points between training templates and image from testing PCI X-ray sequences.

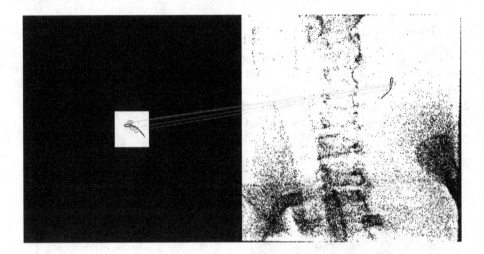

Fig. 2. The matching point between training templates and image from testing PCI X-ray sequences

From Fig. 2, there are many matching points, thus we need to calculate an exact position before marking the position of the guide-wire. In the conventional method [11] normally use estimate geometric transform to calculate the exact position. However, it requires a lot of information and a lot of matching points to calculate. We can not expect that all of the cases can provide that much data. Base on our experiment, using a mean position method have a good effective enough in this case. The exact position of a guide-wire is calculated by the follows:

$$P(x, y) = \frac{\sum_{i=1}^{N} P_i(x_i, y_i)}{N} \tag{1}$$

While P(x, y) is the exact position of a guide-wire, $P_i(x_i, y_i)$ is the position of matching points, and N is the total number of matching points.

In some cases, there is more than one template, which has a matching point with the testing image. Therefore, it requires a method to decide which template is correct. Local binary pattern (LBP) is a common method used for comparing an image texture and can be used to compare a rotation image [19]. Thus, we can use LBP to decide the correct result as follows:

$$LBP_{diff} = \sum LBP_d[x_i, ..., x_n] \tag{2}$$

while:

$$LBP_d[x_i, ..., x_n] = k_t(LBP_t[x_i, ..., x_n] - LBP_T[x_i, ..., x_n])^2 \tag{3}$$

While $LBP_t[x_i, ..., x_n]$ is the LBP set value of template, $LBP_T[x_i, ..., x_n]$ is the LBP set value of testing image, $LBP_d[x_i, ..., x_n]$ is different set value, k_t is a constant value for each training image, and LBP_{diff} is the total sum of $LBP_d[x_i, ..., x_n]$ values.

Then, we can compare each LBP_{diff} from each template. The lowest one can be descried as the correct answer.

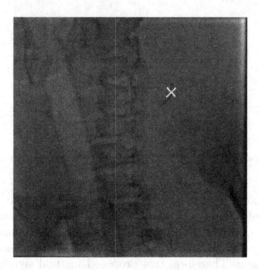

Fig. 3. The mean position of detected PCI guide wire X-ray image.

Figure 3 has shown that the system has marked the position of the lowest one. We can see that the position of the mark is around the middle of the guide-wire. Thus, the position can be described as the correct position of the guide-wire.

2.3   Detection Method

From Fig. 4, the detection method follows these steps: (i) preprocessing; (ii) pre-detection If, there is a feedback data, then go to (iii) detection. The steps (i) and (ii) have been already described in Sects. 2.1 and 2.2 respectively. In step (iii), we mainly apply the method from step (ii). However, feedback data from the previous image frame is also included. The system will create a new testing image and a temporary template base on the feedback data.

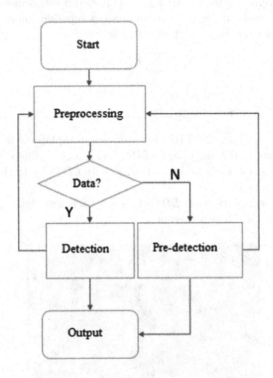

Fig. 4. The flow chart of our SURF guide-wire detection method.

Figure 5 has shown that the strong points around the guide-wire on (b) are many more amounts than (a). In our experiment, we have found out that if in the guide-wire area has more strong points, the detecting accuracy will improve because there is more possibility that strong points between test images and the template is matching. Therefore, this proposed method significantly increases the accuracy of guide-wire detection.

3   Results

We perform our proposed method on a Windows 10, 64-bit operating system which has the following specs: CPU Intel(R), Core(TM) i7-4720HQ, 2.60 GHz, and 8.00 GB RAM. The platform is MATLAB. Our method is tested on 1073 frames from 18 sequences collected during the animal trials of the medical robot [20], and each sequence frame size is 512 by 512, including the variety of guide-wire shape and motion conditions. The training image represents a variety of

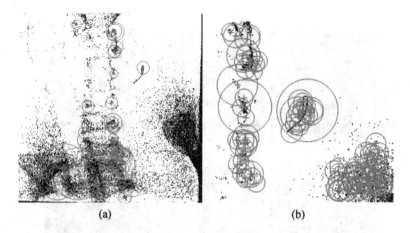

(a)   (b)

Fig. 5. Comparing between original size image and the new testing image: (a) the original size image; (b) a new testing image.

guide-wire shape and motions. We split our experiments into three phases as follows: Phase 1, we just use the original images with the pre-detection method. Phase 2, the images with pre-processing with the pre-detection method are used. Phase 3 is using the feedback data from Phase 2 to improve the overall performant by considering the value from Phase 2's pre-detection method and the detection method. In our proposed method the guide-wire detection in each sequence does not require any position of a guide-wire from the initial frames. The system has the ability to locate the initial position.

The experiment in Phase 1, there is almost none correct detection result. Due to the very low SNR and the other unnecessary part of the images, the guide-wire is easy to be described as an unimportant object. therefore SURF could not detect any useful feature around the guide-wire. The result is shown in Fig. 6.

The experiment in Phase 2, there is around a 50% correct detection result. Due to the preprocessing method, some of the unnecessary parts such as noise and the part of the adjacent organ were cut off. Therefore, SURF could detect a useful feature around the guide-wire. However, there are still some parts have a similar feature with the guide-wire which, lead to miss detecting. The result is shown in Fig. 7.

The experiment in Phase 3, The accuracy can reach 90% with an average speed is around 20 fps. Due to the data from a pre-detecting method, we can create a new frame which is used for the detecting methods. The important feature around a guide-wire area that is detected by SURF is many more compare with Phase 2 method. Therefore, comparing between the guide-wire and the trained template is more accurate. The result is shown in Fig. 8 and Table 1.

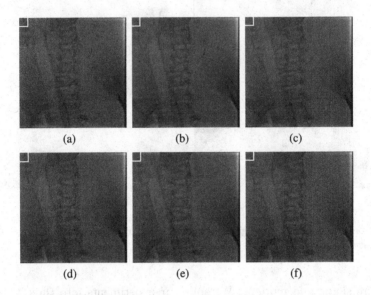

Fig. 6. The detection result of Phase 1: (a) is the 3th frame; (b) is 14th frame; (c) is 31th frame; (d) is 35th frame; (e) is 39th frame; (f) is 59th frame. The white rectangles represent the detection results.

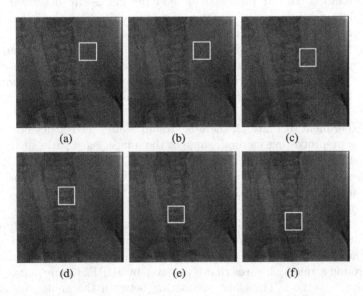

Fig. 7. The detection result of Phase 2: (a) is the 3th frame; (b) is 14th frame; (c) is 31th frame; (d) is 35th frame; (e) is 39th frame; (f) is 59th frame. The white rectangles represent the detection results.

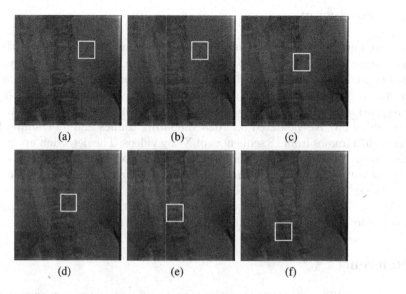

Fig. 8. The detection result of Phase 3: (a) is the 3th frame; (b) is 14th frame; (c) is 31th frame; (d) is 35th frame; (e) is 39th frame; (f) is 59th frame. The white rectangles represent the detection results.

Table 1. The result of experiment in Phase 3

Sequence	Average time (sec.)	Accuracy (%)
1	0.0492	80.0
2	0.0504	96.8
3	0.0449	90.9
4	0.0466	92.6
5	0.0470	94.7
6	0.0442	95.1
7	0.0443	95.4
8	0.0453	96.1
9	0.0466	95.4
10	0.0445	97.0
11	0.0443	95.0
12	0.0463	94.1
13	0.0469	90.7
14	0.0521	95.4
15	0.0496	87.5
16	0.0513	96.0
17	0.0523	93.4
18	0.0556	83.4
Average	0.0478 (20 fps)	**92.7**

4   Conclusion

We presented an automatic guide-wire detection method to detect a tip of the guide-wire on X-ray videos base on SURF. Our approach can automatically detect guide-wire under changing motion and shape, and it can ignore an unimportance object that looks similar to guide-wire by using the data from the pre-processing method and take the data from the pre-detecting method into consideration. We validated on 1023 validating frames and 50 training frames, total 1073 frames from 18 sequences of X-ray videos. The detection average accuracy is 92.7% with 4.56% standard deviation proves that the method is robust and able to provide the accurate detection. The average speed of the process is around 20 fps while the speeds of the 18 X-ray video sequences are between 7–15 fps, shows that the speed of this proposed method can perform in real-time guide-wire detection. Thus, the experiment shows a reliable result.

References

1. Mohr, F.W., et al.: Coronary artery bypass graft surgery versus percutaneous coronary intervention in patients with three-vessel disease and left main coronary disease: 5-year follow-up of the randomised, clinical SYNTAX trial. Lancet **381**(9867), 629–638 (2013)
2. Heibel, H., Glocker, B., Groher, M., Pfister, M., Navab, N.: Interventional tool tracking using discrete optimization. IEEE Trans. Med. Imaging **32**(3), 544–555 (2013)
3. Chen, B.J., Wu, Z., Sun, S., Zhang, D., Chen, T.: Guidewire tracking using a novel sequential segment optimization method in interventional X-ray videos. In: 2016 IEEE 13th International Symposium on Biomedical Imaging (ISBI), pp. 103–106. IEEE (2016)
4. Chang, P.L., et al.: Robust catheter and guidewire tracking using B-spline tube model and pixel-wise posteriors. IEEE Robot. Autom. Lett. **1**(1), 303–308 (2016)
5. Fazlali, H., et al.: Vessel region detection in coronary X-ray angiograms. In: 2015 IEEE International Conference on Image Processing (ICIP), pp. 1493–1497. IEEE (2015)
6. Hernandez-Vela, A., et al.: Accurate coronary centerline extraction, caliber estimation, and catheter detection in angiographies. IEEE Trans. Inf. Technol. Biomed. **16**(6), 1332–1340 (2012)
7. Heibela, T.H., Glockera, B., Grohera, M., Paragios, N., Komodakis, N., Navaba, N.: Discrete tracking of parametrized curves. In: IEEE Conference on Computer Vision and Pattern Recognition, CVPR 2009, pp. 1754–1761. IEEE (2009)
8. Honnorat, N., Vaillant, R., Paragios, N.: Graph-based geometric-iconic guide-wire tracking. In: Fichtinger, G., Martel, A., Peters, T. (eds.) MICCAI 2011. LNCS, vol. 6891, pp. 9–16. Springer, Heidelberg (2011). https://doi.org/10.1007/978-3-642-23623-5_2
9. Wang, L., Xie, X.L., Gao, Z.J., Bian, G.B., Hou, Z.G.: Guide-wire detecting using a modified cascade classifier in interventional radiology. In: 2016 IEEE 38th Annual International Conference of the Engineering in Medicine and Biology Society (EMBC), pp. 1240–1243. IEEE (2016)

10. Wang, L., Xie, X.L., Bian, G.B., Hou, Z.G., Cheng, X.R., Prasong, P.: Guide-wire detection using region proposal network for X-ray image-guided navigation. In: 2017 International Joint Conference on Neural Networks (IJCNN), pp. 3169–3175. IEEE (2017)
11. Bay, H., Ess, A., Tuytelaars, T., Van Gool, L.: Speeded-up robust features (SURF). Comput. Vis. Image Underst. **110**(3), 346–359 (2008)
12. Hamid, N., Yahya, A., Ahmad, R.B., Al-Qershi, O.M.: A comparison between using SIFT and SURF for characteristic region based image steganography. Int. J. Comput. Sci. Issues **9**(3), 110–116 (2012)
13. Otsu, N.: A threshold selection method from gray-level histograms. IEEE Trans. Syst. Man Cybern. **9**(1), 62–66 (1979)
14. Bradley, D., Roth, G.: Adaptive thresholding using the integral image. J. Graph. Tools **12**(2), 13–21 (2007)
15. Bay, H., Tuytelaars, T., Van Gool, L.: SURF: speeded up robust features. In: Leonardis, A., Bischof, H., Pinz, A. (eds.) ECCV 2006. LNCS, vol. 3951, pp. 404–417. Springer, Heidelberg (2006). https://doi.org/10.1007/11744023_32
16. Wasson, V., et al.: An efficient content based image retrieval based on speeded up robust features (SURF) with optimization technique. In: 2017 2nd IEEE International Conference on Recent Trends in Electronics, Information and Communication Technology (RTEICT), pp. 730–735. IEEE (2017)
17. Wu, L., Liu, B., Zhao, B.: Unsupervised change detection of remote sensing images based on SURF and SVM. In: 2017 International Conference on Computing Intelligence and Information System (CIIS), pp. 214–218. IEEE (2017)
18. Kim, Y., Jung, H.: Reconfigurable hardware architecture for faster descriptor extraction in SURF. Electron. Lett. **54**, 210–212 (2018)
19. Ojala, T., Pietikainen, M., Maenpaa, T.: Multiresolution gray-scale and rotation invariant texture classification with local binary patterns. IEEE Trans. Pattern Anal. Mach. Intell. **24**(7), 971–987 (2002)
20. Feng, Z.Q., Bian, G.B., Xie, X.L., Hou, Z.G., Hao, J.L.: Design and evaluation of a bio-inspired robotic hand for percutaneous coronary intervention. In: 2015 IEEE International Conference on Robotics and Automation (ICRA), pp. 5338–5343. IEEE (2015)

New Default Box Strategy of SSD
for Small Target Detection

Yuyao He, Baoqi Li[✉], and Yaohua Zhao

School of Marine Science and Technology, Northwestern Polytechnical University,
Beilin District, Xi'an 710072, Shaanxi, People's Republic of China
heyyao@nwpu.edu.cn, {bqli,2016200388}@mail.nwpu.edu.cn

Abstract. SSD, which combines the advantages of Faster-RCNN and
YOLO, has excellent performance in both detection speed and precision
by merging the default boxes of six different layers. As the original default
box strategy cannot accurately capture the small target information, the
detection precision of SSD for small target images is not as good as
normal size targets. In this paper, a new default box strategy, which
can give the appropriate size and number of default boxes, is proposed
to improve the performance of SSD for small target detection. The new
default box strategy is made up of new scales and new aspect ratios. The
new scales, which provide the basic scales for the six layers, are defined
by the size ratio of the kernel to the convolutional layer. In addition,
the new scale range is reduced from [20, 90] to [20, 60]. The new aspect
ratios, which determine the size and the number of default boxes of
the six layers, are defined as [[1.1], [1.1], [1.1], [1.1], [0.8, 1.2], [1.1]].
Experiment results on the small ground target dataset show that the
detection precision of SSD with the new strategy is 99.5 mAP, which is
4.6 mAP higher than that of the original SSD. More importantly, the
training time of SSD with the new strategy is 963 s or 326 s less than
that of the original SSD.

Keywords: Ground small target detection · SSD · Default boxes
Scales · Aspect ratios

1 Introduction

Ground target detection [1–4] plays an important role in unmanned aerial vehicle
(UAV) navigation, searching, accurate striking, and the evaluation after attack-
ing. Compared with image classification [5], target detection is a more challenging task that requires classifying targets and localizing them within an image [6–
8]. Due to flight mileage and safety, UAVs must shoot the ground targets from a
farther distance. This causes the target pixel ratio in the image to be very small
and increases the difficulty of target detection.

Classical target detection models, like the deformable part model (DPM), are
run at evenly spaced locations over the entire image by using a sliding window

© Springer Nature Switzerland AG 2018
L. Cheng et al. (Eds.): ICONIP 2018, LNCS 11304, pp. 416–425, 2018.
https://doi.org/10.1007/978-3-030-04212-7_36

approach [9]. However, it is difficulty to obtain satisfactory results when detecting complex targets by these models. By embedding the deep learning [10–15] model of convolutional neural networks (CNN) in the target detection model, the precision of the target detection has been continuously refined over the last few years. Regions-CNN (R -CNN) [16], which is the first model to use CNN for target detection, uses a region proposal method to generate 2,000 region proposals from the input image and zoom all the region proposals into a fixed size. Then, a CNN is run on these region proposals for extracting features. Two fully connected layers (support vector machine (SVM) and linear regression layers) are added to the last feature layer of CNN. SVM is used to classify the input feature. Linear regression is used to fine tune the location and size of the border. Because the SVM and the linear regression layers must be trained separately and information duplication exists between the 2,000 region proposals, R-CNN is slow and hard to be optimized. Fast R-CNN [17], which combines classification with regression by a Multi-task function, is an improved model for R-CNN. Faster R-CNN [18,19], which uses nine different anchor boxes of final convolution layer to instead of a large number of region proposals, is an improved model for Fast R-CNN. Despite this pipeline has been continuous improved, the detection speed is still slowly due to the complex network structure. You only Look once (YOLO) [20] simplifies this complex pipeline to an end-to-end network that predicts and classifies multiple bounding boxes simultaneously. The detection speed of YOLO has been greatly improved, but the detection precision of that is lower than that of Faster R-CNN.

Using the anchor mechanism used in Faster R-CNN, single shot multiBox detector (SSD) [21] also improves detection precision and maintains the high speed as YOLO. SSD produces a series of default boxes and scores for the presence of object class instances in those boxes of six layers (conv4_3, fc7, conv6_2, conv7_2, conv8_2 and conv9_2). The six layers are shown in Fig. 1. In order to facilitate the analysis of SSD, conv4_3, fc7, conv6_2 and conv7_2 are defined as the lower layers, and conv7_2, conv8_2 and conv9_2 are defined as the higher layers. SSD has a diverse set of predictions and covers various input object sizes and shapes by a default box strategy [22]. The default box strategy has two components. One is the scale, which is responsible for providing the basic scale for the six different layer default boxes. The other is the aspect ratio, which is responsible for determining the size and number of the those layer default boxes. Because the small targets may not even have any information at the very top layers, SSD has much worse performance on smaller targets than bigger ones [21]. The deeper reason is that the original default boxes in each layer is only suitable for the detection of normal size targets and not for the detection of small targets. In original default box strategy, the original scale range is [20,90], which is suitable for detecting normal size targets. The oversized scales will cause the size of the default ratios to be too large to capture the small target in the image accurately. At the same time, the original scale increment of each layer is fixed, which does not match the changing rules of the six different layers in size. In addition, the original aspect ratios produce 8,532 default boxes. It seems that more default boxes will bring higher detection precision, but more negative samples of default boxes will reduce the convergence of SSD.

To the problem that the small ground targets can not be accurately detected by SSD with the original default box strategy, a new default box strategy is proposed for small target detection in this paper. In the new strategy, the scale range is fit for small targets, and the scale increment is redesigned according to the six layer changing rules in size, and the aspect ratio is redefined according to the characteristics of the small target information in six different layers. SSD with the new default box strategy can detect small target images faster and more accurately.

2   A New Default Box Strategy of SSD for Small Target Detection

SSD with the original default box strategy is not suitable for small target detection. Therefore, we propose a new default box strategy for SSD. The new default box strategy comprises new scales and new aspect ratios. In the new scales, the ratio of kernel to convolution layer in size is used to replace the fixed scale increment, and the new range of the scales is changed to [20, 60]. In the new aspect ratios, the number of aspect ratios of each layer are reduced because of the too many negative samples. Compared to the aspect ratio number in the lower layers, we increase the number of aspect ratios in the higher layers. The advantage of doing this is that it does not introduce too many default boxes and can capture more useful information (as much as possible) from the top layer. In Sect. 2.1, we introduce the new scales. In Sect. 2.2, we introduce the new aspect ratios.

2.1   New Scales Based on the Change in Size of the Six Layers

The framework of SSD with the new default box strategy is shown in Fig. 1. The same six layers are selected for extracting default boxes, but the size and number of default boxes of the six layers are redesigned. The lower layers contain finer details of the targets, and the higher layers include target contour information. In other words, the proportion of the targets in the feature maps (from lower layers to higher layers) increases gradually. Therefore, the small value scales are used to the lower layers and the big value scales are used to the higher layers. The sizes of the six layers are $38*38$, $19*19$, $10*10$, $5*5$, $3*3$ and $1*1$, and the length of the six feature layers are twice reduced. The current convolution layer is calculated from previous kernel and convolution layer. Based on the above two facts, we use the size ratio of the previous kernel to convolution layer to replace the fixed scale increment. The new scale is computed as:

$$s_k = s_{min} + \frac{s_{max} - s_{min}}{m_{(k-1)}} n_{(k-1)}, k \in [1, 6] \tag{1}$$

where $m_{(k-1)}$ represents the length of the $k-1$ layer, and $n_{(k-1)}$ is the length of the $k-1$ convolution layer kernel. When k is equal to 1, m_0 is the size of conv3_3 of VGG, and n_0 is the size of convolution kernel of conv3_3. Due to the sizes of the convolution kernels of six layers are all $3 * 3$, the new scale increment is only related to the size of the six layers.

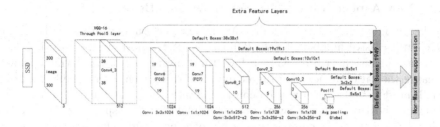

Fig. 1. Simplified network architecture of SSD

The original scale range [21] is used to cover all sizes of the target. At the lower layers, small scale increments are used to ensure that small target detail information can be well captured in spite of less number aspect ratios in lower layers. At the higher layers, big scale increments are used to capture the small target contour information completely. For small ground target detection, the proportion of the targets in the feature maps (the six layers) is relatively small. That is, the proportion of small targets will not be as high as normal size targets in the top layer (conv9_2). Accordingly, the top layer scales should be smaller than the original scale. In this paper, the new scale range is defined as $s_{min} = 20$ and $s_{max} = 60$.

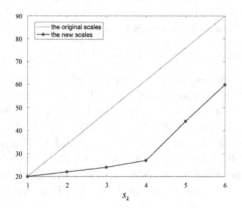

Fig. 2. New and original scales

The new scales and the original scales are shown in Fig. 2. The original scale values of six layers are 20, 34, 48, 62, 76 and 90. The original scale increment of the six layers is fixed and equal to 0.14. The new scale values of six layers are

20, 22, 24, 27, 32, 44 and 60. The new scale line is approximated to a broken line. The value of the new scale increment is small at the lower layers and large at the higher layers.

2.2 New Aspect Ratios with Fewer Default Boxes

In theory, more default boxes will bring higher performance of the SSD. However, an excessive number of default boxes leads to another problems; i.e., the detection precision of SSD will be low; the convergence of SSD will be slow. Because the ratio of the small target information in each layer default boxes is small, most default boxes will be considered as negative samples. The number of default boxes of the six layers are determined by the aspect ratios directly. Selecting a reasonable number and value of aspect ratios for six layers is an effective way to improve the performance of SSD.

We find that a series different number of default boxes are produced by adding one aspect ratio to the six layers. For example, $38*38$ default boxes are increased by adding one aspect ratio to conv4_3, but $1*1$ default box is only increased by adding one aspect ratio to conv9_2. The scales provide a basic size for the design of default boxes. If the scales are well designed, each layer only needs one aspect ratio to detect the small target accurately. Based on this hypothesis, the number of the six layer aspect ratios can be reduced greatly. However, SSD have serious information loss problems in the higher layers. Therefore, more number of aspect ratios are added to higher layers. By increasing the number of higher layer aspect ratios, the number of default boxes are not significantly increased, and the small target information in the higher layers can be better captured. In one word, we control the number of default boxes by reducing the number of lower-layer aspect ratios and increasing the number of higher-layer aspect ratios. The new aspect ratios of the six layers are defined as $a_k^r = [[1.1], [1.1], [1.1], [1.1], [0.8, 1.2], [1.1]]$. The total number of the default boxes is calculated as:

$$N = \sum d_k^2 * l_k, k \in [1, m] \tag{2}$$

where d_k represents the length of k-th layer and l_k is the aspect ratio number of k-th layer. For the six layers, the original aspect ratio numbers of the six layers are 4, 6, 6, 6, 4 and 4, and the total default box number is $8532 = 38*38*4+19*19*6+10*10*6+5*5*6+3*3*4+1*1*4$. The new aspect ratio numbers of the six layers are 1, 1, 1, 1, 2 and 1. The total default boxes number only are $1,949 = 38*38*1+19*19*1+10*10*1+5*5*1+3*3*2+1*1*4$, which is much smaller than the number of the original default boxes. It is help to improve the detection speed and precision of SSD by using less and effective default boxes.

3   Experiment and Analysis

SSD is based on VGG16, which is pre-trained on the ILSVRC CLS-LOC dataset. The resulting model is fine-tuned by using a fixed learning rate of 0.001, momentum of 0.9, weight decay of 0.0005, and batch size of 32. To verify the performance of the proposed strategy, the experiments are performed on a small ground target dataset which contains four categories. In Sect. 3.1, the collection and composition of the small ground target dataset are introduced. In Sect. 3.2, the original default box strategy and the proposed strategy are compared for detecting small ground targets. In Sects. 3.3 and 3.4, the influence of the scales and the aspect ratios are studied.

3.1   Small Ground Target Dataset

The small ground target is a relative concept, where 'small' refers to the small size of the target in the image. We built a dataset for small ground target detection: SGT-det. SGT-det comprises four types of ground targets: tanks, missiles, trucks and helicopters. To get the small ground target image dataset, a UAV collects 20,000 small ground target images from different angles of which 15,000 are used for model training and the remaining 5,000 for model testing. The pixel ratio of the four small ground targets in the image is less than 0.05.

3.2   Performance Comparison of the Proposed Strategy with the Original Default Box Strategy

To analyze the new default box strategy better, the scales and the aspect ratios are considered as two separate factors. "$\_\bullet\_$" indicates the adoption of new factor, and "$\_\_$" indicates the adoption of the original factor; e.g., two "$\_\_$" represent the original default box strategy, and two "$\_\bullet\_$" represent the new default box strategy. When the SSD model is iterated 10,000 times, the mAP of the test dataset is used as the quantitative evaluation index.

Table 1. Performance comparison of the proposed strategy with the original default box strategy

Factor	SSD			
	The original strategy	Only aspect ratios	Only scales	The new strategy
s_k			$\bullet$	$\bullet$
a_k^r		$\bullet$		$\bullet$
mAP	94.9	96.2	97.3	**99.5**

From Table 1, we can find that both s_k and a_k^r can improve the detection precision of the ground small target. With the s_k, the detection precision of SSD is 96.2 mAP and 1.3 mAP higher than that of the original default box strategy. With the a_k^r, the detection precision of SSD is 97.3 mAP and 2.4 mAP higher than that of the original default box strategy. When s_k and a_k^r work at the same time, the detection precision of SSD is 99.5 mAP and 4.6 mAP higher than that of the original default box strategy.

(a) tank: mAP=1.00   (b) missile: mAP=1.00

(c) truck: mAP=1.00   (d) helicopter: mAP=0.99

Fig. 3. Detection result of the four small ground targets

By using jupyter notebook, the trained SSD model is used to detect four small ground target images. The detection results are shown in Fig. 3. The SSD model with the new default box strategy can detect small ground targets with high precision. Even the small target is consistent with the background color or the small target is at the edge of the image.

3.3   Influence of the Scale Range on the Detection Precision of SSD

To better analyze the impact of the scale range on the default box strategy, seven different scale ranges are used to detect small target images. The mAP of the test dataset is used as an evaluation index.

Table 2. The influence of the scale range on the detection precision of SSD

Scale range	20–30	20–40	20–50	20–60	20–70	20–80	20–90
mAP	95.2	97.20	98.6	**99.5**	97.4	96.5	95.3

The results presented in Table 2 show that the detection precision of the scales in [20, 90] is 95.3 mAP, in [20 30] is 95.2 mAP, and in [20 60] is 99.5 mAP.

The scales in large or small ranges are not suitable for small target detection. In fact, the scale in the top layer is directly related to the proportion of the targets in the image. In this paper, the scales in [20 60] are the best choice for small ground target detection.

3.4   Influence of Different Aspect Ratios on the Performance of SSD

In this experiment, the original a_k^r is used as a reference. The positive triangle and inverted triangle are defined according to the number shape of the aspect ratios. These two kinds of a_k^r are designed for finding which layer will affect the detection precision of the SSD. The performance of the aspect ratios is evaluated based on two measures: mAP and training time.

Table 3. The influence of different number of aspect ratios on the performance of SSD

	Original	Positive triangle	Inverted triangle	New
$conv4_3$	[1,1/2,2]	[1,1/2,2]	[2]	[1.1]
$fc7$	[1,1/2,2,1/3,3]	[1,1/2,2,1/3,3]	[1/2,2]	[1.1]
$conv6_2$	[1,1/2,2,1/3,3]	[1,1/2,2,1/3]	[1,1/2,2]	[1.1]
$conv7_2$	[1,1/2,2,1/3,3]	[1,1/2,2]	[1,1/2,2,1/3]	[1.1]
$conv8_2$	[1,1/2,2]	[1/2,2]	[1,1/2,2,1/3,3]	[0.8,1.2]
$conv9_2$	[1,1/2,2]	[2]	[1,1/2,2]	[1.1]
Boxes	8932	8571	4554	1949
Time	1289 s	1185 s	1023	963
mAP	94.9	95.9	97.8	**99.5**

From Table 3 we find that SSD with the original aspect ratios have the lowest detection precision at 94.9 mAP and a maximum training time of 1,289 s. SSD with the new aspect ratios have the best detection precision at 99.5 mAP and a minimum training time of 963 s. Comparing and analyzing the results of the positive triangle and inverted triangle aspect ratios, each layer contains negative samples, and the number of negative samples in the lower layers are higher than that in the higher layers. For the original and inverted triangle aspect ratios, the detection accuracy and speed of SSD are obviously improved by reducing the number of lower layer aspect ratios. For the results of the original and the new aspect ratios, the performance of SSD can be improved by a small number and effective aspect ratios. By using the new aspect ratios, the detection speed of the SSD is faster, and the detection precision of the SSD is higher.

4   Conclusion

SSD with the original default box strategy can not detect small ground targets accurately, a new default box strategy is proposed by redesigning the scale values and the number of aspect ratios of the six layers. The new scales can better reflect the essential relationship of the six feature layers, as well as capture the small target information. Based on the new designed scales, the new aspect ratios greatly reduce the number of default boxes and improved the detection speed of SSD. On the four small ground target dataset, SSD with the new strategy demonstrates an improvement by more than 4.6 mAP compared with the original strategy. What is more valuable is that the time consumption of SSD with the new strategy is also reduced. SSD with the new default box strategy is more suitable for detection of small ground targets.

References

1. Abdullah, R.S.A.R., Salah, A.A., Ismail, A., Hashim, F., Rashid, N.E.A., Aziz, N.H.A.: LTE-based passive bistatic radar system for detection of ground moving targets. ETRI J. **38**, 302–313 (2016)
2. Kim, S., Song, W.J., Kim, S.H.: Robust ground target detection by sar and IR sensor fusion using adaboost-feature selection. Sensors **16**, 1117–1134 (2016)
3. Xu, H., Yang, Z., Chen, G., Liao, G., Tian, M.: A ground moving target detection approach based on shadow feature with multichannel high-resolution synthetic aperture radar. IEEE Geosci. Remote. Sens. Lett. **13**, 1572–1576 (2016)
4. Breitenstein, M.D., Reichlin, F., Leibe, B., Kollermeier, E., Gool, L.V.: Online multiperson tracking-by-detection from a single, uncalibrated camera. IEEE Trans. Pattern Anal. Mach. Intell. **33**, 1820–1833 (2011)
5. He, K., Zhang, X., Ren, S., Sun, J.: Deep residual learning for image recognition. In: 34th IEEE Conference on Computer Vision and Pattern Recognition, pp. 770–778. IEEE Press, Las Vegas (2016)
6. Shrivastava, A., Gupta, A., Girshick, R.: Training region-based object detectors with online hard example mining. In: 34th IEEE Conference on Computer Vision and Pattern Recognition, pp. 761–769. IEEE Press, Las Vegas (2016)
7. Felzenszwalb, P.F., Girshick, R.B., Mcallester, D., Ramanan, D.: Object detection with discriminatively trained part-based models. IEEE Trans. Pattern Anal. Mach. Intell. **32**, 1627–1630 (2010)
8. Najibi, M., Rastegari, M., Davis, L.S.: G-CNN: an iterative grid based object detector. IEEE Access **5**, 24023–24031 (2017)
9. Erhan, D., Szegedy, C., Toshev, A., Anguelov, D.: Scalable object detection using deep neural networks. In: 32th IEEE Conference on Computer Vision and Pattern Recognition, pp. 2155–2162. IEEE Press, Columbus (2014)
10. Hinton, G.: Where do features come from? Cogn. Sci. **38**, 1078–101 (2014)
11. Krizhevsky, A., Sutskever, I., Hinton, G. E.: ImageNet classification with deep convolutional neural networks. In: 26th International Conference on Neural Information Processing Systems, pp. 1097–1105. Curran Associates Inc. Press (2012)
12. Szegedy, C., Toshev, A., Erhan, D.: Deep neural networks for object detection. Adv. Neural Inf. Process. Syst. **26**, 2553–2561 (2013)
13. Hinton, G.E., Osindero, S., Teh, Y.W.: A fast learning algorithm for deep belief nets. Neural Netw. **18**, 1527 (2006)
14. Bengio, Y.: Learning deep architectures for AI. Found. Trends Mach. Learn. **2**, 1–127 (2009)

15. Cao, X., Zhang, X., Yu, Y., Niu, L.: Deep learning-based recognition of underwater target. In: 20th IEEE International Conference on Digital Signal Processing, pp. 89–93. IEEE Press, Beijing (2017)
16. Girshick R., Donahue, J., Darrell, T., Malik, J.: Rich feature hierarchies for accurate object detection and semantic segmentation. In: 32th IEEE Conference on Computer Vision and Pattern Recognition, pp. 580–587. IEEE Press, Columbus (2014)
17. Girshick, R.: Fast R-CNN. In: 33th IEEE International Conference on Computer Vision, pp. 1440–1448. IEEE Press, Santiago (2015)
18. Ren, S., He, K., Girshick, R., Sun, J.: Faster R-CNN: towards real-time object detection with region proposal networks. IEEE Trans. Pattern Anal. Mach. Intell. **39**, 1137–1149 (2017)
19. He, K., Zhang, X., Ren, S., Sun, J.: Spatial pyramid pooling in deep convolutional networks for visual recognition. IEEE Trans. Pattern Anal. Mach. Intell. **37**, 1904–1916 (2015)
20. Redmon, J., Divvala, S., Girshick, R., Farhadi, A.: You only look once: unified, real-time object detection. In: 34th IEEE Conference on Computer Vision and Pattern Recognition, pp. 779–788. IEEE Press, Las Vegas (2016)
21. Liu, W., Anguelov, D., Erhan, D., Szegedy, C., Reed, S., Fu, C.Y.: SSD: single shot multibox detector. In: 34th IEEE Conference on Computer Vision and Pattern Recognition, pp. 21–37. IEEE Press, Las Vegas (2016)
22. Cai, Z., Fan, Q., Feris, R.S., Vasconcelos, N.: A unified multi-scale deep convolutional neural network for fast object detection. In: Leibe, B., Matas, J., Sebe, N., Welling, M. (eds.) ECCV 2016. LNCS, vol. 9908, pp. 354–370. Springer, Cham (2016). https://doi.org/10.1007/978-3-319-46493-0_22

Weakly Supervised Temporal Action Detection with Shot-Based Temporal Pooling Network

Haisheng Su[1], Xu Zhao[1(✉)], Tianwei Lin[1], and Haiping Fei[2]

[1] Department of Automation, Shanghai Jiao Tong University, Shanghai, China
[2] Industrial Internet Innovation Center (Shanghai) Co., Ltd., Shanghai, China
{suhaisheng,zhaoxu,wzmsltw}@sjtu.edu.cn, feihaiping@3in.org

Abstract. Weakly supervised temporal action detection in untrimmed videos is an important yet challenging task, where only video-level class labels are available for temporally locating actions in the videos during training. In this paper, we propose a novel architecture for this task. Specifically, we put forward an effective shot-based sampling method aiming at generating a more simplified but representative feature sequence for action detection, instead of using uniform sampling which causes extremely irrelevant frames retained. Furthermore, in order to distinguish action instances existing in the videos, we design a multi-stage Temporal Pooling Network (TPN) for the purposes of predicting video categories and localizing class-specific action instances respectively. Experiments conducted on THUMOS14 dataset confirm that our method outperforms other state-of-the-art weakly supervised approaches.

Keywords: Temporal action detection · Weak supervision
Shot-based sampling · Temporal pooling network · Class-specific

1 Introduction

Recently, impressive progress has been achieved on video analysis, which motivates two important tasks: action recognition and temporal action detection. Action recognition [1–4] is a crucial problem for video understanding which aims to classify manually trimmed videos. However, temporal action detection is more challenging since it not only deals with the classification of action categories in long untrimmed videos, but also localizes the boundaries of action instances. It can be applied in many areas such as smart surveillance and security system.

There are many deep learning based works focusing on the task of temporal action detection [5–9]. Most of them are performed with full supervision which relies on the temporal annotations of action instances greatly. However, it is time-consuming to annotate the temporal boundaries of each action instance for

---

This research has been supported by NSFC Program (61673269, 61273285).

L. Cheng et al. (Eds.): ICONIP 2018, LNCS 11304, pp. 426–436, 2018.
https://doi.org/10.1007/978-3-030-04212-7_37

a large-scale dataset. Furthermore, due to the subjective judgement, the annotating results can vary from person to person. Therefore, weakly supervised action detection draws attention of many researchers. In order to effectively locate the action instances in the videos using only video-level class labels during training, there are two crucial points: (1) remove considerable amount of irrelevant frames existing in the untrimmed videos; (2) generate high-quality detections especially in the videos containing multiple action instances of various classes.

A long and untrimmed video usually comes up with extremely irrelevant information, where action instances only occupy small parts, thus a sampling measure would contribute to removing the irrelevant frames and accelerating the speed of action detection. Uniform sampling method is widely used [10,11], however, it fails to utilize action structure information. Besides, traditional snippet-based classifiers rely on discriminative parts of actions greatly. Based on these two issues, we propose a shot-based sampling method to sample the input visual feature sequence generated by two-stream network [12], which can generate a more simplified and representative feature sequence for action detection.

How to generate proposals from a video sequence is another difficult problem for action detection under weak supervision. Bottom-up mechanism adopted in [13] first selects a set of clips as proposals, which are further classified, then the classification results are merged to match the video-level class labels. However, it fails to distinguish multiple action instances from each other existing in the video. We propose temporal pooling network to detect the class-specific discriminative frames of a given video in a top-down way, by means of computing one dimensional weighted Temporal Class Activation Maps (T-CAM). Besides, a novel attention module is designed to modulate the video representations, which can highlight the salient frames while suppressing the irrelevant counter-parts. To the best of our knowledge, we are the first to extend the class activation mapping [14] used for discriminative localization to the temporal domain, and the attention module designed for temporal attention weights learning is unique to our work. In sum, our main contributions include: (i) video representative parts selection using an effective shot-based sampling method; (ii) a top-down temporal pooling network for weakly supervised temporal action detection; (iii) extensive experiments reveal the good performance of our method.

2 Our Approach

In this section, we introduce the technical details of our approach. The framework is illustrated in Fig. 1.

2.1 Two-Stream Network for Feature Extraction

We adopt pretrained multiple two-stream network [13] to extract video feature representations, since this kind of architecture has shown great performance in action recognition task. The architecture of each two-stream network contains two branches: spatial network handles a single RGB frame and temporal network

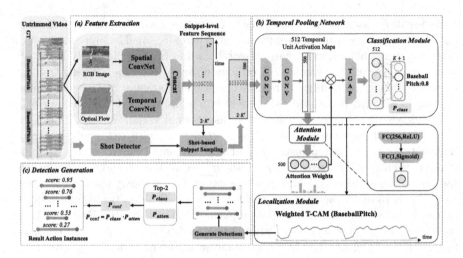

Fig. 1. Framework of our approach. (a) Two-stream network is used to extract visual features in snippet-level, then we adopt shot-based sampling strategy to sample the input feature sequence. (b) The architecture of temporal pooling network: *classification module* handles the sampled feature sequence for action classification; *attention module* learns attention weights for each sampled snippet; *localization module* generates weighted temporal class activation maps (T-CAM) to locate actions in temporal domain. (c) Detection generation: during prediction, we choose top-2 prediction results and group consecutive snippets with high activations to form the final detections

takes optical flow stacking of 5 frames as input. We compute optical flow using GPU implementation of [15] from the OpenCV toolbox.

Denote a given video $V = \{x_n\}_{n=1}^{T_v}$ consists of T_v frames, to extract the video features, we divide the video into $T_s = T_v/\sigma$ consecutive video snippets, where σ is the frame number of a snippet. A snippet is represented as $s = \{x_n\}_{n=x_f}^{x_f+\sigma}$, where x_f is the starting frame, $x_f + \sigma$ is the ending frame. Each snippet is independent without overlapping with each other and is processed by pretrained two-stream network respectively. Then we concatenate output scores in fc-action layer of two-stream network including both RGB and flow modalities of all snippets to form the feature sequence $F = \{f_{s_i} = \{f_{S,s_i}, f_{T,s_i}\}\}_{i=1}^{T_s}$, where f_{S,s_i} and f_{T,s_i} are output score vector of spatial and temporal network respectively with length K', where K' includes K action categories and one background category. This feature sequence is then fed to the shot-based temporal pooling network.

2.2 Shot-Based Sampling Method

We claim that the speed of temporal action detection task can be accelerated using a subset of discriminative snippets in a video. In order to remove the irrelevant frames and reduce the computational cost, we sample the input feature sequence instead of using all snippets for video classification. However, different sampling methods are bound to result in various classification performance.

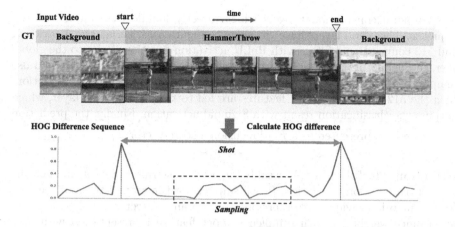

Fig. 2. A qualitative example of shot generated on THUMOS14

Generally, an effective sampling method should possess the ability to generate a simplified sequence which can contain the discriminative parts of action instances within the long untrimmed video. We propose the **shot-based snippet sampling method** that firstly detects the shot changes of the video based on the difference of histogram of oriented gradient (HOG) features [16] between two adjacent frames. An action shot will be detected if the absolute value of this difference is higher than a threshold. Each shot is denoted by (s_i^b, s_i^e), where s_i^b, s_i^e represent the beginning and ending of each action shot respectively. A qualitative example on THUMOS14 dataset is shown in Fig. 2. It should be noted that the shots are generated unsupervisedly.

Since the definition of the action boundaries is usually vague, we sample $\dfrac{N_t}{N_{shot}(V_j)}$ snippets around the middle of each shot as key frames to compose a new snippets sequence $S' = \left\{ s_i' \right\}_{i=1}^{N_t}$, where $N_{shot}(V_j)$ denotes the number of shots generated in the video V_j and N_t is the number of sampled snippets in total used for training phase and we set it as 500 empirically. Finally, the sampled feature sequence can be represented as $F' = \left\{ f_{s_i'} \right\}_{i=1}^{N_t}$.

2.3 Temporal Pooling Network

Inspired by the idea using CAM [14] for discriminative parts localization of images in CNN, we extend it to the temporal domain and propose a novel multi-stage architecture for weakly supervised temporal action detection. We use the sampled feature sequence as the input of Temporal Pooling Network (TPN). The architecture of TPN mainly contains three sub-modules listed as follows.

Classification Module. This module aims to classify the categories of each input untrimmed video based on the sampled feature sequence. It begins with two temporal convolutional layers which directly handle the sampled feature sequence. These two layers have the same configurations: kernel size 3, stride 1, and 512 convolutional filters with ReLU activation function. Refer to the global average pooling proposed in [17], we extend it to the temporal domain and use temporal global average pooling (T-GAP) to encode video-level representation, then the 512 temporal pooled features are fed to the classification layer, which outputs the classification result with Sigmoid activation. Finally, the prediction result can be represented as a $K^{'}$-dimensional score vector $p_{class} = \{p_{class}^c\}_{c=1}^{K^{'}}$.

Attention Module. In this weakly supervised learning system, neither the ground-truth action categories of each snippet nor the ground-truth of the confidence map is provided. We therefore propose an indirect method to learn an importance weight for each sampled snippet feature representation with only weak-labels, thus to further highlight the snippet relevant to the actions as a soft selection. Concretely, we adopt a multilayer perceptron model with one hidden layer to learn the attention weights. Each snippet feature representation in the last convolutional layer is fed to the attention module respectively which consists of two fully connected layers with ReLU and Sigmoid activation separately. Then the temporal feature maps combined with the generated attention weights $\lambda = \{\lambda_i\}_{i=1}^{N_t}$ are followed by T-GAP to aggregate the video-level representation $\overline{R} = \sum_{i=1}^{N_t} \lambda_i r_i$, where r_i denotes the ith snippet feature map in the last convolutional layer. Note that the temporal attention weights are trained in a class-agnostic way.

Localization Module. The goal of this module is to temporally identify the discriminative parts of the video which contribute most to the classification results. Refer to CAM in [14], we derive one-dimensional temporal class activation mapping (T-CAM) combined with temporal attention weights. We denote $w^c(z)$ as the zth element of the weight matrix $W \in \mathbb{R}^{Z \times K^{'}}$ in the classification layer corresponding to class c. The input to the final sigmoid layer for class c is

$$s^c = \sum_{z=1}^{Z} w^c(z)\overline{R}(z) = \sum_{z=1}^{Z} w^c(z) \sum_{i=1}^{N_t} \lambda_i r_i(z) = \sum_{i=1}^{N_t} \lambda_i \sum_{z=1}^{Z} w^c(z)r_i(z). \quad (1)$$

Finally, we define the weighted T-CAM for class c of length N_t as $M_i^c = \lambda_i \sum_{z=1}^{Z} w^c(z)r_i(z)$. The weighted T-CAM describes the probability of each snippet being in a specific action class of the video.

2.4 Training Procedure

Using the simplified feature sequence generated by the shot-based sampling method, we train a classification model for the K action categories as well as background. The training data is constructed as $\Omega_{class} = \left\{(F^{'}(V_j), y_j)\right\}_{j=1}^{M}$, where y_j denotes the ground-truth class label of video V_j. To train the classification module, we combine the multi-label cross-entropy loss and l_2 regularization loss to form the loss function:

$$L_{class} = \sum_{j=1}^{M} y_{label}^j log y_{pred}^j + \lambda \cdot L_2(\Theta_{class}), \tag{2}$$

where y_{pred}^j and y_{label}^j are predicted results and ground-truth video-level class labels of video V_j respectively, M is the number of training videos. λ balances the cross-entropy loss and l_2 regularization loss, and Θ_{class} is the classification module. We set $\lambda = 10^{-4}$ empirically. As for parameters in SGD, we train the model 300 epochs and the learning rate decays from 10^{-3} to 10^{-4} after 50 epochs. The batch size is set to 1.

2.5   Detection Generation and Prediction

During prediction, we use uniform sampling method to sample the input feature sequence and group consecutive snippets with activations higher than threshold θ among the weighted T-CAM into candidate detections with varied durations. Through empirical validation, we set θ as 5% of the maximum score. For a better observation of the localization ability of our model, we use top-2 video-level classification results of [13]. By combining the predicted action class score p_{class} and proposal attention weight p_{atten}, we can get a confidence p_{conf} for each proposal:

$$p_{conf} = p_{class} \cdot p_{atten}, \tag{3}$$

$$p_{atten} = \frac{\sum_{i=t_{start}}^{t_{end}} ReLU(M_i^c)}{t_{end} - t_{start} + 1}, \tag{4}$$

where p_{atten} is weighted mean T-CAM followed by a ReLU of all the snippets within the proposal durations denoted by $[t_{start}, t_{end}]$.

3   Experiments

3.1   Dataset and Setup

We evaluate our STPN on THUMOS14 dataset [18] for temporal action detection task. The THUMOS14 dataset contains 1010 and 1574 untrimmed videos for validation and testing respectively, while only 200 and 213 videos are temporal annotated among 20 action categories. Each video can contain multiple action instances of various classes. We train our model over 200 untrimmed videos on validation set and use 213 temporally annotated videos on the testing set to evaluate action detection performance. It should be noted that we don't use any temporal action instance annotations during training phase.

For temporal action detection task, we use mean Average Precision (mAP) as evaluation metric on this dataset. A predicted result instance is regarded as correct only when it gets the correct category label and its intersection over union (IoU) with the ground-truth instance is higher than θ.

Parameters used in each module have been given before. We extract features from pretrained two-stream network using Caffe [19]. And we implement STPN using TensorFlow [20].

3.2   Evaluation on Visual Feature Encoder

Visual encoders are used to extract snippet-level features. To study the contributions of different visual encoders, we evaluate them individually and coherently with the strictest IoU threshold 0.5 as shown in Table 1. We can observe that two-stream network [12] shows better performance than C3D network [3].

Table 1. Comparison of different visual encoders used in STPN on THUMOS14

Video feature encoders	mAP ($\theta = 0.5$)
C3D Network	7.9
Spatial-Stream Network	7.3
Temporal-Stream Network	11.3
Two-Stream Network	**14.0**

3.3   Evaluation on Attention Module

Attention module serves as a soft selection method to guide the model to explicitly focus on important parts of the input videos. To study the contribution of attention module, we evaluate STPN with and without attention module separately under the strictest IoU threshold 0.5 as shown in Table 2. We can observe that STPN with attention module gives a significant boost in performance.

Table 2. Comparison of different architectures used in STPN on THUMOS14

Networks	mAP ($\theta = 0.5$)
STPN (w/o attention module)	10.9
STPN	**14.0**

3.4 Evaluation on Shot-Based Sampling Method

We claim that the speed of action detection task can be accelerated using a subset of discriminative snippets in a video. To check the effects of our proposed sampling method, we evaluate temporal pooling network without snippet sampling, with uniform sampling and with shot-based sampling method during training phase using one Nvidia 1080 graphic card. We use the strictest threshold 0.5 and compute the mean time cost of per video during evaluation. As shown in Table 3, TPN with snippet sampling performs better than without sampling and our proposed shot-based sampling method leads to the best performance in both mAP and time consumption.

Table 3. Comparison of different sampling methods used in STPN on THUMOS14

Methods	mAP ($\theta = 0.5$)	Time cost
w/o snippet sampling	2.2	2.35 s
Uniform sampling	11.1	0.21 s
Shot-based sampling	**14.0**	**0.19 s**

3.5 Comparison with the State-of-the-Art Methods

Our approach is also compared with some state-of-the-art methods [6–8,11,13, 21,22] of full supervision or weak supervision. In [11], Singh et al. introduce a hide-and-seek strategy to force the network to seek other relevant parts when the most discriminative parts are hidden. However, this method hides the temporal regions randomly and blindly without guidance, thus is inefficient. Wang et al. [13] adopt a bottom-up mechanism for weakly supervised action detection in untrimmed videos, where clip proposals are first generated for classification. However, the use of softmax function across proposals blocks it from distinguishing multiple action instances existing in the videos. Compared with these weakly supervised methods, our STPN can not only highlight the important parts of the input video feature sequence automatically and efficiently, but also detect the class-specific action instances accurately. Comparison results are shown in Table 4. We can observe that our approach outperforms other weakly supervised state-of-the-art methods and even achieves comparable performance to that of fully supervised methods, which demonstrates the effectiveness of our temporal pooling network on learning from long and untrimmed videos. Qualitative examples on THUMOS14 dataset are shown in Fig. 3.

Table 4. Comparison of our method with other state-of-the-art methods on THU-MOS14 for action detection. * indicates using full supervision for training

mAP@$IoU(\theta)$	0.5	0.4	0.3	0.2	0.1
Oneata et al. [21]*	14.4	20.8	27.0	33.6	36.6
Shou et al. [6]*	19.0	28.7	36.3	43.5	47.7
Lin et al. [7]*	24.6	35.0	43.0	47.8	50.1
Zhao et al. [8]*	29.8	41.0	51.9	59.4	66.0
Gao et al. [22]*	**31.0**	41.3	50.1	56.7	60.1
Singh et al. [11]	6.8	12.7	19.5	27.8	36.4
Wang et al. [13]	13.7	21.1	28.2	**37.7**	44.4
Ours	**14.0**	**21.4**	**29.1**	37.3	**44.8**

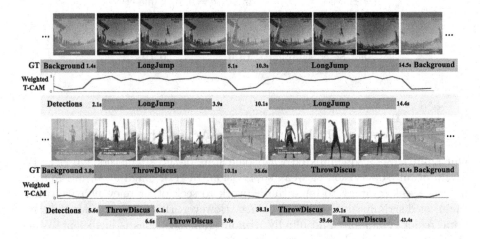

Fig. 3. Qualitative examples of detections generated by STPN on THUMOS14

4   Conclusion

In this paper, we propose a top-down architecture, called STPN, for weakly supervised temporal action detection. In our approach, we first adopt the shot-based sampling method and two-stream network to generate a more representative and simplified feature sequence. Then the classification module can recognize the action categories accurately and the attention module can highlight the relevant frames while suppressing the irrelevant counter-parts simultaneously. Next, the localization module can detect the class-specific discriminative snippets of untrimmed videos by means of generating the weighted T-CAM. Final, we group consecutive snippets with high activations into proposals of variable lengths. Our approach achieves the state-of-the-art performance on the THUMOS14 dataset. In the future, we will try more advanced detection methods and post-processing strategies to generate higher quality proposals with lower redundancy.

References

1. Wang, L., Qiao, Y., Tang, X.: MoFAP: a multi-level representation for action recognition. 2016 Int. J. Comput. Vis. **119**, 254–271 (2016). https://doi.org/10.1007/s11263-015-0859-0
2. Karpathy, A., Toderici, G., Shetty, S., Leung, T., Sukthankar, R., Fei-Fei, L.: Large-scale video classification with convolutional neural networks. In: 2014 IEEE Conference on Computer Vision and Pattern Recognition, pp. 1725–1732. IEEE Press, New York (2014)
3. Tran, D., Bourdev, L.D., Fergus, R., Torresani, L., Paluri, M.: Learning spatiotemporal features with 3D convolutional networks. In: 2015 IEEE International Conference on Computer Vision, pp. 4489–4497. IEEE Press, New York (2015)
4. Wang, L., et al.: Temporal segment networks: towards good practices for deep action recognition. In: Leibe, B., Matas, J., Sebe, N., Welling, M. (eds.) ECCV 2016. LNCS, vol. 9912, pp. 20–36. Springer, Cham (2016). https://doi.org/10.1007/978-3-319-46484-8_2
5. Lin, T., Zhao, X., Fan, Z.: Temporal action localization with two-stream segment-based RNN. In: 2017 IEEE Conference on Image Processing, pp. 1–4. IEEE Press, New York (2017)
6. Shou, Z., Wang, D., Chang, S.: Action temporal localization in untrimmed videos via multi-stage CNNs. In: 2016 IEEE Conference on Computer Vision and Pattern Recognition, pp. 1049–1058. IEEE Press, New York (2016)
7. Lin, T., Zhao, X., Shou, Z.: Single shot temporal action detection. In: 25th ACM International Conference on Multimedia, pp. 988–996. ACM, California (2017)
8. Zhao, Y., Xiong, Y., Wang, L., Wu, Z., Tang, X., Lin, D.: Temporal action detection with structured segment networks. In: 2017 IEEE International Conference on Computer Vision, pp. 6–7. IEEE Press, New York (2017)
9. Lin, T., Zhao, X., Su, H., Wang, C., Yang, M.: BSN: boundary sensitive network for temporal action proposal generation. arXiv preprint arXiv:1806.02964 (2018)
10. Gan, C., Wang, N., Yang, Y., Yeung, D., G.Hauptmann, A.: DevNet: a deep event network for multimedia event detection and evidence recounting. In: 2015 IEEE Conference on Computer Vision and Pattern Recognition, pp. 2568–2577. IEEE Press, New York (2015)
11. Singh, K.K., Lee, Y.J.: Hide-and-Seek: forcing a network to be meticulous for weakly-supervised object and action localization. In: 2017 IEEE International Conference on Computer Vision, pp. 1961–1970. IEEE Press, New York (2017)
12. Simoyan, K., Zisserman, A.: Two-stream convolutional networks for action recognition in videos. In: Advances in Neural Information Processing Systems, pp. 568–576. Curran Associates Inc., New York (2014)
13. Wang, L., Xiong, Y., Lin, D., van Gool, L.: UntrimmedNets for weakly supervised action recognition and detection. In: 2017 IEEE Conference on Computer Vision and Pattern Recognition, pp. 2–6. IEEE Press, New York (2017)
14. Zhou, B., Khosla, A., Lapedriza, A., Oliva, A., Torralba, A.: Learning deep features for discriminative localization. In: 2016 IEEE Conference on Computer Vision and Pattern Recognition, pp. 2921–2929. IEEE Press, New York (2016)
15. Brox, T., Bruhn, A., Papenberg, N., Weickert, J.: High accuracy optical flow estimation based on a theory for warping. In: Pajdla, T., Matas, J. (eds.) ECCV 2004. LNCS, vol. 3024, pp. 25–36. Springer, Heidelberg (2004). https://doi.org/10.1007/978-3-540-24673-2_3

16. Yamasaki, T.: Histogram of oriented gradients. In: Journal of the Institute of Image Information and Television Engineers, pp. 1368–1371 (2010)
17. Lin, M., Chen, Q., Yan, S.: Network in network. In: 2014 IEEE International Conference on Learning Representations, pp. 1–4. IEEE Press, New York (2014)
18. Jiang, Y.G., et al.: THUMOS challenge: action recognition with a large number of classes. In: ECCV Workshop, vol. 5. Springer, Heidelberg (2014)
19. Jia, Y., et al.: Caffe: convolutional architecture for fast feature embedding. In: 22nd ACM International Conference on Multimedia, pp. 675–678 (2014)
20. Abadi, M., Agarwal, A., Barham, P., et al.: Tensorflow: large-scale machine learning on heterogeneous distributed systems. arXiv preprint arXiv:1603.04467 (2016)
21. Oneata, D., Verbeek, J., Schmid, C.: The LEAR submission at thumos2014. In: Thumos14 Action Recognition Challenge, pp. 1–7. Springer, Heidelberg (2014)
22. Gao, J., Yang, Z., Nevatia, R.: Cascaded boundary regression for temporal action detection. arXiv preprint arXiv:1705.01180 (2017)

Drogue Detection for Autonomous Aerial Refueling Based on Adaboost and Convolutional Neural Networks

Yanjie Guo, Yimin Deng[✉], and Haibin Duan

School of Automation Science and Electrical Engineering, Beihang University,
Beijing 100083, China
ymdeng@buaa.edu.cn

Abstract. Autonomous aerial refueling (AAR) is an important capability for the future development of unmanned aerial vehicles (UAVs). A robust and accurate algorithm of detecting the drogue is crucial to such a capability. In this paper, we present an innovative algorithm based on the adaptive boosting algorithm and convolutional neural networks (CNN) classifier with improved focal loss (IFL). The IFL function addresses the sample imbalance during the training stage of the CNN classifier. The pytorch deep learning framework with the graphics processing units (GPUs) is used to implement the system. Real scenario images that contain drogue carried by UAVs are for training and testing. The results show that the algorithm not only accelerates the speed but also improves the accuracy.

Keywords: Autonomous aerial refueling · Adaboost · CNN
Sample imbalance

1 Introduction

Nowadays, unmanned aerial vehicles (UAVs) are widely used in various domains. And the autonomous aerial refueling (AAR) is a crucial technique to expand the range of activity, payload and endurance. Probe-drogue aerial refueling system (PDARS) [1] is one of the current methods and it is the standard for the United States Navy and the air forces of other nations. Because of its economy and flexibility, it is preferred to be used for UAVs. Convolutional neural network (CNN) has been the research hotspot in image recognition problem [2] recently and it is excellent in extract features without manually design. The region-based convolution neural network (R-CNN) [3] reached high object detection accuracy. And next come the one-stage detectors YOLO [4] and SSD [5] with even better accuracy. However, their heavy structure, complex training process and low speed make them redundant to a dichotomous problem. The sample imbalance during the training [6], due to the positive and negative sample generation method, makes the plenty of easy samples overwhelm the loss back propagation stage. As a result, the model will pay less attention to the hard samples and finally lead to a low accuracy. Until the research in focal loss [7], it proposes an algorithm using all the examples without sampling and makes a progress in solving the imbalance problem.

© Springer Nature Switzerland AG 2018
L. Cheng et al. (Eds.): ICONIP 2018, LNCS 11304, pp. 437–443, 2018.
https://doi.org/10.1007/978-3-030-04212-7_38

In this paper, we propose an innovative method combining the Adaboost algorithm with the CNN classifiers to do the drogue detection.

2   Structure and Basic Model

2.1   Framework Overall

The whole system framework is shown as Fig. 1. The top-left is the input of the system including annotated training or testing images at different stages. The bottom-left is the Adaboost algorithm used to make region proposal. Then the top right is the CNN classifier to make further classification. Finally, the result is shown in the bottom-right.

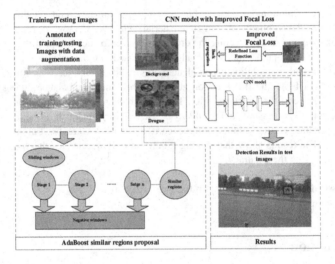

Fig. 1. The framework of drogue detection.

In the training stage, the positive and negative samples are fed into the Adaboost algorithm, and the CNN model uses the calibration images which contain the drogue for training. During the training of the CNN model, the improved focal loss algorithm is used to manage the imbalance of the samples. In the testing stage, large-scale images which contain the drogue are taken as the input. The proposal regions in the original image are resized to the input size of the CNN model. So, the CNN model can do the work of extracting features and making final classification.

2.2   Adaboost Algorithm

The Adaboost algorithm [8] is a machine learning meta-algorithm formulated by Yoav Freund and Robert Schapire that combines plenty of weak classifiers. The final score is defined as the linear combination of the classifiers from every stage. Adaboost has been confirmed to be excellent in two-class classification problems.

In this paper, the cascade Adaboost algorithm is employed to implement the similar regions proposal job [9]. The principle is to cascade the weak classifiers which contain simple feature to boost strong classifiers. It can allow the background regions to be discarded at earlier stages. Thus, the stage number will influence the accuracy and efficiency of the system. HOG is chosen as the feature descriptors for its outstanding feature expression among the plenty of feature descriptors. To meet the demand of getting a high recall rate, the stage number of the Adaboost algorithm is the main factor that matters. And the detail about setting the number of the stage will be discussed in Sect. 4.

2.3 Tiny CNN Model

After the first step of the whole algorithm, the similar regions which contain the real drogue regions and non-drogue regions are proposed. Softmax has a unique advantage in disposing feature vectors and it is widely used in deep learning with the development of computer vision. Unlike SVM, softmax maps the feature vectors to the probability field. During the training, it always expects the Cross-entropy to be as small as possible. In other words, softmax pursues near-perfect results. In consideration of the efficiency, a tiny CNN model is designed to extract the feature which is shown in Fig. 2.

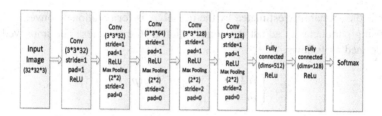

Fig. 2. The architecture of the CNN model.

Similar with the general architecture of a CNN model, the model above contains: convolution and max pooling, nonlinear activation function Rectified Linear Unit (ReLU), full connection and loss function. The input images are entered into the model and after several convolution and max pooling layers and ReLU is used after each layer. Then the feature maps are put into inner product layers and after the softmax layer the model outputs the final probability to predict whether the region is positive or not. The loss function will be discussed in next part.

3 Improved Focal Loss

Faster R-CNN [10] designs a strategy to rebalance the positive and negative samples ratio by fixed the ratio to 1:3, which makes 25% of the mini-batch is positive samples. The online hard example mining makes a further step to pay attention the loss of the sample. Instead of fixing the ratio, it chooses the samples by sorting the loss for which the current model performs worst. Then the focal loss proposes a method to make back propagation for all the examples but add a weight coefficient according to their loss level.

In this section, we introduce the focal loss method because it fully uses all the samples without manually resample relying on experience. And it is improved by adding a damping parameter to further increase the probability of the cross-entropy to meet our need and the result achieves a better performance.

The original softmax loss function for binary classification is given as follows (without consideration of regularization loss):

$$L = \begin{cases} -\log(p) & \text{if } y = 1 \\ -\log(1-p) & \text{otherwise} \end{cases} \tag{1}$$

where y specifies the ground-truth class and p is the probability estimated by the model for the target. The focal loss function is defined as:

$$FL(p_t) = -(1-p_t)^\gamma \log(p_t) \tag{2}$$

$$p_t = \begin{cases} p & \text{if } y = 1 \\ 1-p & \text{otherwise} \end{cases} \tag{3}$$

where p is the probability estimated by the model, p_t is for notational convenience. γ is the focusing parameter to adjust the contribution of the samples. The improved focal loss is defined as follows:

$$\begin{cases} IFL = -(1-p_t)^\omega \log(p_t) \\ \omega = e^{-\frac{k}{N}} \cdot \gamma \end{cases} \tag{4}$$

where k represents the current epoch in the training stage, N is the total epoch designed for the training, and ω is the decayed focus parameter.

The principle is that if the samples are misclassified and the probability of the right class is small, the loss is nearly unaffected because it is the sample that needs to be learned well. On the contrary, the loss for the well-classified examples will be down-weighted. This method adjusts the weights of the samples by their loss, and leads the model paying more attention on the hard samples. However, the goal of the softmax function is to reach a nearly perfect probability. The original focal loss indeed improves the samples imbalance matter, but somehow resists the optimization of the probability. For instance, if we take the γ for 5, and the well-classified target sample has a 0.75 probability. Then the loss of this sample is down-weighted by divide almost 1000. That may resist the optimization of the model. So, we improve the focal loss by adding a decay factor, the factor is an exponential decay function which is controlled by the ratio of the current epoch and total epoch. At the early epoch of the training, the factor ω is decayed little to prevent the overwhelming influence due to the sample imbalance. Along with the training, the factor will be decayed more to let the model learn to output better probability even when the sample is well classified.

4 Experimental Result

4.1 Dataset

Our drogue dataset is collected from our drogue model placed on the UAV at different angles and various illuminations. Horizontal flip and rotation were applied for the data augment. The dataset for training Adaboost contains 39283 negative sample images and 14392 positive images. The CNN classifier is trained by 3501 annotated images, the positive and negative examples are acquired by random cropping and classifying according to the intersection over union of the samples and the annotation bounding box.

4.2 Detailed Settings

The CNN model is trained with stochastic gradient descent (SGD) with 16 images per mini-batch on 2 GPUs. The model is trained for 40 epochs with an initial learning rate of 0.001, which is divided by 10 every 10 epochs. The training loss is the sum of the regularization L2 loss and several loss functions discussed above. Three different loss functions are used, including the general cross entropy loss function, the original focal loss function and the improved loss function. The focus parameter γ is respectively set to 0, 2 and 2 with the decayed parameter.

4.3 Results

The results of the drogue detection in real scenario images are shown in Fig. 3. The box left after the filtration of the CNN classifier is the final detection result of the system. It can be seen that after the classification of the CNN model, the false positive regions are successfully rejected by the CNN classifier.

Fig. 3. The detection results.

The entire system is tested on the dataset that contains about 4200 annotated real scenario images. The precision/recall (PR) curve that describes the performance of the system is shown in Fig. 4. It can be seen that the system with the improved focal loss keeps a high precision against the other two algorithms and reaches a highest mAP at 0.88.

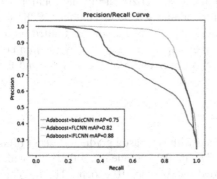

Fig. 4. The PR curve of the systems

The Adaboost cascade algorithm can run at 50FPS. And the tiny CNN model finish one proposal box at 0.0033 s on NVIDIA 1080ti. And the entire time of it depends on the number of the proposal boxes, but generally less than 0.01 s.

5   Conclusions

Autonomous aerial refueling is a challenging task in the future application of UAV, and the detection of the drogue is the crucial link. In view of the requirement of the system, the Adaboost with the CNN classifier trained by improved focal loss is proposed above. The cascade Adaboost algorithm is used to do the similar region proposal job and reach a high recall. The tiny CNN model is applied to make further classification. With the improved focal loss method, the sample imbalance is handled and it outperforms the original focal loss. The final result shows that the whole system has good performance and surpasses the baseline methods of cascade Adaboost with CNN classifier and it with the original focal loss.

References

1. Chen, C.I., Stettner, R.: Drogue tracking using 3d flash lidar for autonomous aerial refueling. In: Laser Radar Technology and Applications XVI, pp. 1–11. SPIE Press, Orlando (2011)
2. Russakovsky, O., Deng, J., Su, H., Krause, J., Satheesh, S., Ma, S.: Imagenet large scale visual recognition challenge. Int. J. Comput. Vis. 115(3), 211–252 (2015)
3. Girshick, R.: Fast r-cnn. In: 17th Proceedings of IEEE International Conference on Computer Vision, pp. 1–9. IEEE Press, Santiago (2015)

4. Redmon, J., Divvala, S., Girshick, R., Farhadi, A.: You only look once: unified, real-time object detection. In: 29th IEEE Conference on Computer Vision and Pattern Recognition, pp. 779–788. IEEE Computer Society, Las Vegas (2016)
5. Liu, W., et al.: Single shot multibox detector. In: The 14th European Conference on Computer Vision, pp. 21–37. ECCV Press, Amsterdam (2016)
6. He, H., Garcia, E.A.: Learning from imbalanced data. IEEE Trans. Knowl. Data Eng. **21**(9), 1263–1284 (2009)
7. Lin, T.Y., Goyal, P., Girshick, R., He, K., Dollar, P.: Focal loss for dense object detection. arXiv preprint arXiv:1708.02002 (2017)
8. Rätsch, G., Onoda, T., Müller, K.R.: Soft margins for adaboost. Mach. Learn. **42**(3), 287–320 (2001)
9. Viola, P., Jones, M.: Fast and robust classification using asymmetric adaboost and a detector cascade. In: 14th Advances Neural Information Processing Systems, pp. 1311–1318. MIT Press, Cambridge (2001)
10. Ren, S.Q., He, K.M., Girshick, R., Sun, J.: Towards real-time object detection with region proposal networks. IEEE Trans. Pattern Anal. Mach. Intell. **39**(6), 1137–1149 (2017)

Deep Neural Network Based Salient Object Detection with Image Enhancement

Lecheng Zhou and Xiaodong Gu[✉]

Department of Electronic Engineering, Fudan University,
Shanghai 200433, China
xdgu@fudan.edu.cn

Abstract. Salient object detection aims to discover the most visually attractive regions from images. It allows more efficient follow-up processing of images without handling redundant information. In this paper, we propose a novel framework based on deep neural network to detect salient objects. The proposed framework introduces feature enhancement to input images to improve the performance of the fully convolutional neural network (FCN). Images are segmented and weighted through superpixel based pulse coupled neural networks. Low-level features including contrast and spatial features are extracted during this procedure by removing background disturbance in images. Subsequent neural network takes the enhanced images in and produces the saliency maps. Finally, some refinements are made afterwards to achieve better saliency results. Experimental results on five representative benchmarks show the superiority of our model than other state-of-the-art methods. Furthermore, comparisons are made to verify the effectiveness of image enhancement part in our model.

Keywords: Salient object detection · Fully convolutional neural network
Image enhancement

1 Introduction

There is a broad space for application of salient object detection in the field of image processing. Automatically separating regions of interest from the background can significantly decrease the amount of data to be processed, especially when large scales of image data are involved. Many computer vision applications such as object recognition, image compression, image retrieval, and semantic segmentation may take salient object detection as a preprocessing.

The first computational visual saliency model [1] was designed by L. Itti, which combines low-level features including color, intensity and orientation features of input images to compute the saliency scores. This work promoted later studies [2–5, 8–10] on mining image features and building computational models for salient object detection. Though great progress was made, handcrafted features are fundamentally based on prior knowledge of images, and mistakes are inevitable when meeting with challenging benchmarks. Learning based approaches [6, 7, 22] were then developed, to

© Springer Nature Switzerland AG 2018
L. Cheng et al. (Eds.): ICONIP 2018, LNCS 11304, pp. 444–453, 2018.
https://doi.org/10.1007/978-3-030-04212-7_39

better combine different image features and improve the generalization capacity of the model. Yet problems remain because of the manual design of fusion details.

Recently, the continuous researches in machine learning lead to a breakthrough in computer vision. A neural network can automatically learn high-level features from a raw image, and gradually form a fusion method according to the task given. The enormous amount of parameters in a deep neural network enables a more accurate mapping between features and results. Among various types of network, convolutional neural network (CNN) is an ideal tool to detect salient objects from images, for its progressively larger receptive field like human visual system. Tremendous progress on performance is achieved by some CNN based salient object detection approaches [11–15]. Fully convolutional neural network (FCN), as an improvement of CNN, produces pixelwise prediction results, making it accessible to get a saliency map directly from the output layer.

Inspired by the outstanding performance made by FCN, meanwhile to make use of the priors, we proposed an end-to-end model for salient object detection. The model enhance low-level features before the neural network extract semantic features. An input image is first segmented and weighted through a superpixel based pulse coupled neural network (PCNN), thus contrast and structural features of the input image, which are critical factors to define salient objects, are represented by the segmentation result. The enhanced image then goes into an FCN to compute its saliency. As the saliency map is a deconvolution of the high-level features, object details and contours may be damaged, so low-level features are combined to refine the detection result.

We next introduce the detail framework of our proposed approach in Sect. 2. In Sect. 3, we provide the experiment settings and comparative results on different benchmarks. Finally, a brief summary of this paper is given in Sect. 4.

2 Proposed Model

2.1 Network Architecture

We can divide the whole architecture of our model into two parts (as shown in Fig. 1): an image enhancement network, and a saliency prediction network. The image enhancement network uses contrast and structural features to enhance the input image. We segment the input image into superpixels. Based on the superpixel segmentation, a multi-level unit-linking PCNN performs weighting to the image and highlight image foreground. We combine priors of salient object to the image by low-level features enhancement in the first part of our model. The saliency prediction network consists of an FCN to compute pixelwise saliency scores, and a fusion with low-level features to help refine the final saliency map.

2.2 Image Enhancement Network

As it is proved that human cortical cells may be hard-wired to preferentially respond to the highest contrast stimulus in their receptive fields, contrast prior becomes a critical factor to determine a salient object. Since low-level features extracted by the FCN

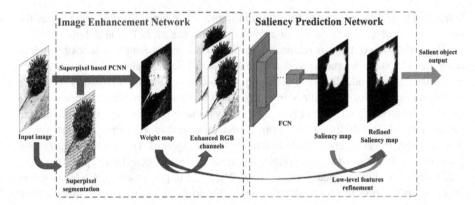

Fig. 1. Illustration of the proposed salient object detection model.

shallower layer may be uncertain, we can enhance the contrast and spatial features before sending the image into the network. Hence the first part of our work is to efficiently extract these low-level features from the image.

Trivial Unit-Linking PCNN. Pulse coupled neural network (PCNN) is a spatio-temporal neural network inspired by the study of cat's visual cortex [16]. A simplified model, unit-linking PCNN [17], is widely used in image processing, for its lower computational cost and simpler network parameters.

Figure 2 illustrates how a neuron works in the network. Each pixel in the image has a neuron corresponding to it. The neuron takes pixel intensity as its input in F channel, meanwhile it links with its neighbor neurons in L channel to transfer ignition signal. We will compare the combined two channel with the ignition threshold. If the value exceeds the threshold, the neuron will generate a spark. The neuron spark will pass through the network until it reaches an obstacle, where there is an obvious intensity difference (like object boundaries). After several iterations, the final firing image of PCNN will segment the image, separating different objects apart. As a result, the ignition process of neuron network can reflect the contrast and spatial features of the image.

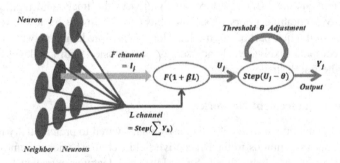

Fig. 2. A single neuron's iteration diagram in unit-linking PCNN.

Multi-level Segmentation. Trivial PCNN only divide objects in the image into two categories by the binary firing image. This is insufficient to describe an image with complex semantics. We introduce multi-level segmentation to enhance the contrast descriptive ability of the network. Since the ignition threshold of the network significantly affects the segmentation result, we can use multi threshold to get different objects segmented separately. For an input image, we select troughs in its CIE-Lab color histogram as the initial thresholds of the network, for we can best distinguish different objects at these troughs. By doing so, we can get a set of segmentation $\{S\}_{i=1}^{N}$.

For each segmentation result S_i of the image I, we compute the mutual information MI_i [18] between the segmentation result and original image by the probability distribution P_{S_i} and P_I, and joint probability distribution P_{S_iI}:

$$MI_i = -\sum_{x \in S_i} P_{S_i}(x) lg\, P_{S_i}(x) - \sum_{y \in I} P_I(y) lg\, P_I(y) + \sum_{x,y \in S_i,I} P_{S_iI}(x,y) lg\, P_{S_iI}(x,y), \quad (1)$$

Since larger MI_i represents stronger structural similarity between S_i and I, we sort the segmentation results according to MI_i, and assign a normalized weight coefficient w_i to each S_i. We can fuse S_i to a multi-level segmentation S_{fuse} as followed:

$$S_{fuse}(x) = \sum_{i=1}^{N} w_i S_i(x). \quad (2)$$

Superpixel Based PCNN. The weighted image created by multi-level PCNN may ignore locality in some cases. To overcome this weakness, we combine superpixel segmentation with unit-linking PCNN. The image is segmented into superpixels by SLIC algorithm [19]. We make neurons in the same superpixel to have a tighter connection by enlarging their receptive fields and linking weights with each other. After the PCNN generates the weighted image, a smoothing process is performed based on superpixel, to make sure that each superpixel has the same weight. Let $w(i)$ denote the weight of a pixel i, $N(p)$ denote the number of pixels in a superpixel p, and the smoothed weight $w_{sm}(p)$ can be computed by

$$w_{sm}(p) = \frac{1}{N(p)} \sum_{i \in p} w(i). \quad (3)$$

Spatial Weighting. As center prior suggest that salient objects are generally closer to image center, we introduce spatial weighting in the image enhancement process. For each superpixel p, its spatial weight $w_{sp}(p)$ is computed by

$$w_{sp}(p) = min\{exp[(D_0 - D(p))/\sigma^2], 1\}, \quad (4)$$

where $D(p)$ is the Euclidean distance between superpixel p and the image center, normalized to [0,1]. D_0 is a modifying factor, to let regions near image center have the largest spatial weight. σ^2 can control the strength of spatial weighting. In our implementation, we choose $D_0 = 0.4$, $\sigma^2 = 0.4$. After that, we perform phase inversion in this step to ensure the image contrast is enhanced after weighted as illustrated in Fig. 3(e).

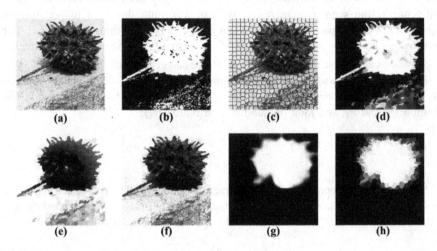

Fig. 3. Examples to illustrate different steps in our salient object detection network. (a) Raw input image. (b) Trivial PCNN segmentation. (c) SLIC superpixel segmentation. (d) Superpixel based PCNN segmentation. (e) Adding spatial weighting. (f) Enhanced RGB channels for FCN input. (g) Saliency map generated by FCN. (h) Refined saliency map.

2.3   Saliency Prediction Network

FCN Salient Object Detection. Long et al. first design fully convolutional neural network (FCN) [20] for semantic segmentation. We slightly modify VGG-16 network into the FCN used in our model, which has 16 convolutional layers and a deconvolution layer. Since we need a grayscale saliency map (or saliency probabilities between [0, 1]) from the network, a sigmoid function is added to normalize the saliency map. Moreover, we use cross entropy as our loss function.

The three RGB input channels of FCN are multiplied by the weighted image generated from the image enhancement network. For each channel, the weighted intensity $I(i)$ of a pixel i is computed as follow

$$I(i) = (1 - \theta) + \theta \cdot w_{sm}(p(i))w_{sp}(p(i)) \cdot I_o(i), \tag{5}$$

where $p(i)$ is the superpixel consisting of pixel i, and $I_o(i)$ is the original intensity of pixel i. θ controls the strength of the image enhancement, and we use $\theta = 0.7$ in our implementation. We manually add contrast and spatial priors to the learning procedure of the network, and make the network concentrate on extracting saliency-oriental features.

Low-Level Features Fusion. The output of FCN is actually a deconvolution of high-level feature maps, and because of its large stride taken, the details and boundaries of salient objects may not be well preserved. On the contrary, low-level features generally concentrate on local details. Hence, we perform a fusion between the enhancement result with the saliency map generated by the neural network. We directly conduct a product with exponent to the original saliency map $S_0(i)$:

$$S_{rf}(i) = S_0(i)^\gamma \left[w_{sm}(p(i)) w_{sp}(p(i)) \right]^{1-\gamma}, \tag{6}$$

where $\gamma = 0.8$ is the control factor. We then normalize the refined saliency map S_{rf} to [0,1], and smooth it based on superpixel. The smoothing process has been described in the image enhancement network.

3 Experimental Results

3.1 Experiment Settings

Experiment Datasets. We conduct comparative experiments on five benchmarks to evaluate our proposed approach, including MSRA10 k [21], THUR [22], SED2 [23], ECSSD [8] and PASCAL-S [24]. MSRA10 k contains 10000 images, we use 8000 images in our test, for the rest 2000 are used as training set. There are 15000 images in THUR altogether, while only 6232 of them have pixel-wise saliency annotation maps. SED2 has 100 images with two separate salient objects in each of them. ECSSD has 1000 images, and PASCAL-S has 850 images.

Implementation Details. The proposed model is implemented on Matlab and Caffe [25]. The training set contains 2000 images randomly picked from MSRA-10 k dataset. We randomly sort and horizontally flip each input image, and resize them into 500×500 before training. We train and test all these datasets on a NVIDIA GeForce GTX1080Ti GPU, with in total 200000 training iterations.

Evaluation Metrics. We use precision and recall (P-R) curve, F-measure and mean absolute error (MAE) to evaluate the performance of these salient object detection approaches.

By using different binary thresholds of the saliency map (from 0 to 255), we can get a set of precision and recall to draw the P-R curve. The performance of saliency detection can be observed from the curve.

F-measure provides a balanced metric between precision and recall, which can be computed by:

$$F_{\alpha^2} = \frac{(1 + \alpha^2) Precision \times Recall}{\alpha^2 \times Precision + Recall}, \tag{7}$$

where α^2 is set to 0.3 in our experiment. From the P-R curve, we can directly get maximum F-measure ($F_{\max}$) with its corresponding precision and recall. Average F-measure (*Favg*) can be typically calculate by the adaptive binary thresholds (twice the average gray value in this paper) of each saliency maps.

MAE measures the dissimilarity between the saliency map and binary ground truth, which is defined as:

$$MAE = \frac{1}{W \times H} \sum_{x=1}^{W} \sum_{y=1}^{H} |S(x,y) - GT(x,y)|. \tag{8}$$

3.2 State-of-the-Art Performance Comparison

Our approach is compared with several state-of-the-art methods in the experiment including FT [2], HC [3], RC [3], GC [10], MC [6], GMR [4], RBD [9], DSR [5] and DRFI [7]. The results of these approaches are generated by their open source code and all the saliency maps are normalized to [0,255].

The comparative P-R curves with different approaches on five datasets are displayed in Fig. 4. The red line represents the P-R curve of our approach. Comparing with other approaches, we have a larger area under the curve on these five benchmarks, which suggests a better performance of our model.

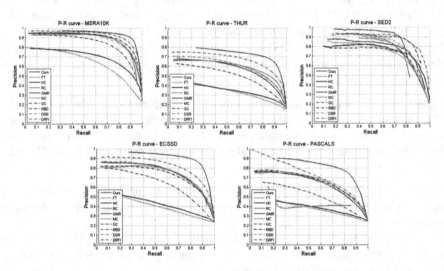

Fig. 4. P-R curve comparison between our approach with state-of-the-art approaches. (Color figure online)

We also quantitatively compare our approach with others on F-measure and MAE as displayed in Figs. 5 and 6, and the leftmost bars represent our approach. We can clearly see that our approach have a higher F-measure and lower MAE in most cases. It demonstrates the superiority of our model.

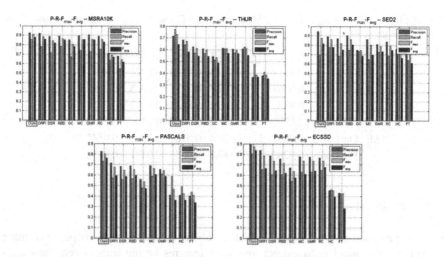

Fig. 5. Bar chart of F_{max} with corresponding Precision-Recall and F_{avg}.

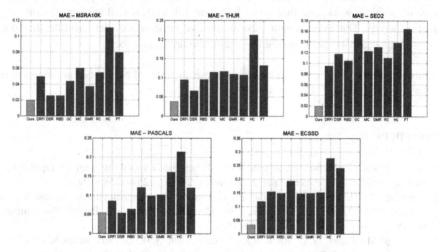

Fig. 6. Comparison for mean absolute error (MAE) of different approaches.

3.3 Contribution of Image Enhancement Network

To further analyze the utility of the image enhancement network in our model, we train a trivial FCN salient object detection network with the same training settings. We test the two models on different benchmarks, and compute F-measure and MAE of their results. From the information given in Table 1, our proposed model obviously have a better performance in all the cases. Hence the contribution of image enhancement is remarkable in our model.

Table 1. Comparison of F_{max}, F_{avg} and MAE between FCN with image enhancement and trivial FCN on five datasets.

		MSRA10 k	THUR	SED2	ECSSD	PASCALS
F_{max}	Ours	0.9179	0.7423	0.8850	0.8926	0.8031
	FCN only	0.9007	0.7267	0.8753	0.8675	0.7899
F_{avg}	Ours	0.8858	0.6792	0.8137	0.8447	0.7634
	FCN only	0.8776	0.6606	0.7668	0.8307	0.7252
MAE	Ours	0.0194	0.0394	0.0187	0.0336	0.0545
	FCN only	0.0290	0.0689	0.0657	0.0581	0.0852

4   Conclusion

We have proposed a novel model for salient object detection in this paper. The main concept of our model is to extract low-level features before images sent into deep neural network. Contrast and spatial features also help refine the details of saliency results. We perform an enhancement to input images with superpixel based pulse coupled neural network, and then use a fully convolutional neural network to compute image saliency. The experimental results on five benchmarks demonstrate that our proposed approach outperforms these state-of-the-art approaches.

Acknowledgments. This work was supported in part by National Natural Science Foundation of China under grant 61771145 and 61371148.

References

1. Itti, L., Koch, C., Niebur, E.: A model of saliency-based visual attention for rapid scene analysis. IEEE Trans. Pattern Anal. Mach. Intell. **20**(11), 1254–1259 (1998)
2. Achanta, R., Hemami, S., Estrada, F., Susstrunk, S.: Frequency-tuned salient region detection. In: 22th IEEE Conference on Computer Vision and Pattern Recognition, pp. 1597–1604. IEEE Press, Miami (2009)
3. Cheng, M.M., Zhang, G.X., Mitra, N.J., Huang, X., Hu, S.M.: Global contrast based salient region detection. IEEE Trans. Pattern Anal. Mach. Intell. **37**(3), 409–416 (2011)
4. Yang, C., Zhang, L., Lu, H., Ruan, X., Yang, M.H.: Saliency detection via graph-based manifold ranking. In: 26th IEEE Conference on Computer Vision and Pattern Recognition, pp. 3166–3173. IEEE Press, Portland (2013)
5. Li, X., Lu, H., Zhang, L., Xiang, R., Yang, M.H.: Saliency detection via dense and sparse reconstruction. In: 26th IEEE Conference on Computer Vision and Pattern Recognition, pp. 2976–2983. IEEE Press, Portland (2013)
6. Jiang, B., Zhang, L., Lu, H., Yang, C., Yang, M.H.: Saliency detection via absorbing markov chain. In: 14th IEEE International Conference on Computer Vision, pp. 1665–1672. IEEE Press, Sydney (2013)
7. Jiang, H., Wang, J., Yuan, Z., Wu, Y., Zheng, N., Li, S.: Salient object detection: a discriminative regional feature integration approach. In: 26th IEEE Conference on Computer Vision and Pattern Recognition, pp. 2083–2090. IEEE Press, Portland (2013)

8. Yan, Q., Xu, L., Shi, J., Jia, J.: Hierarchical saliency detection. In: 26th IEEE Conference on Computer Vision and Pattern Recognition, pp. 1155–1162. IEEE Press, Portland (2013)

9. Zhu, W., Liang, S., Wei, Y., Sun, J.: Saliency optimization from robust background detection. In: 27th IEEE Conference on Computer Vision and Pattern Recognition, pp. 2814–2821. IEEE Press, Columbus (2014)

10. Cheng, M.M., Warrell, J., Lin, W.Y., Zheng, S., Vineet, V., Crook, N.: Efficient salient region detection with soft image abstraction. In: 14th IEEE International Conference on Computer Vision, pp. 1529–1536. IEEE Press, Sydney (2013)

11. Chen, T., Lin, L., Liu, L., Luo, X., Li, X.: DISC: deep image saliency computing via progressive representation learning. IEEE Trans. Neural Networks Learn. Syst. 27(6), 1135–1149 (2016)

12. Li, X., Zhao, L., Wei, L., Yang, M.H.: DeepSaliency: multi-task deep neural network model for salient object detection. IEEE Trans. Image Process. 25(8), 3919–3930 (2016)

13. He, S., Lau, R.H.W., Liu, W., Huang, Z., Yang, Q.: SuperCNN: a superpixelwise convolutional neural network for salient object detection. Int. J. Comput. Vision 115(3), 330–344 (2015)

14. Li, H., Chen, J., Lu, H., Chi, Z.: CNN for saliency detection with low-level feature integration. Neurocomputing 226(C), 212–220 (2017)

15. Hou, Q., Cheng M.M., Hu, X., Borji, A., Tu, Z., Torr, P.H.S.: Deeply supervised salient object detection with short connections. In: 30th IEEE Conference on Computer Vision and Pattern Recognition, pp. 3203–3212. IEEE Press, Honolulu (2017)

16. Eckhorn, R., Reitboeck, H.J., Arndt, M., Dicke, P.: Feature linking via synchronization among distributed assemblies: simulations of results from cat visual cortex. Neural Comput. 2(3), 293–307 (1990)

17. Gu, X.: Feature extraction using unit-linking pulse coupled neural network and its applications. Neural Process. Lett. 27(1), 25–41 (2008)

18. Wei, W., Li, Z.: Automated image segmentation based on modified PCNN and mutual information entropy. Comput. Eng. 36(13), 199 (2010)

19. Anchanta, R., Shaji, A., Smith, K., Luchhi, A., Fua, P., Susstrunk, S.: SLIC superpixels compared to state-of-the-art superpixel methods. IEEE Trans. Pattern Anal. Mach. Intell. 34(11), 2274–2282 (2012)

20. Long, J., Shelhamer, E., Darrell, T.: Fully convolutional networks for semantic segmentation. In: 28th IEEE Conference on Computer Vision and Pattern Recognition, pp. 3431–3440. IEEE Press, Boston (2015)

21. Liu, T., Sun, J., Zheng, N., Tang, X., Shum, H.: Learning to detect a salient object. In: 20th IEEE Conference on Computer Vision and Pattern Recognition, pp. 353–367. IEEE Press, Minneapolis (2007)

22. Cheng, M.M., Mitra, N.J., Huang, X., Hu, S.M.: SalientShape: group saliency in image collections. Visual Comput. Int. J. Comput. Graphics 30(4), 443–453 (2014)

23. Alpert, S., Galun, M., Basri, R., Brandt, A.: Image segmentation by probabilistic bottom-up aggregation and cue integration. IEEE Trans. Pattern Anal. Mach. Intell. 34(2), 315–327 (2012)

24. Li, Y., Hou, X., Koch, C., Rehg, J.M., Yuille, A.L.: The secrets of salient object segmentation. In: 27th IEEE Conference on Computer Vision and Pattern Recognition, pp. 280–287. IEEE Press, Columbus (2014)

25. Jia, Y., Shelhamer, E., Donahue, J., et al.: Caffe: convolutional architecture for fast feature embedding. In: 22nd ACM International Conference on Multimedia, pp. 675–678. ACM Press, Orlando (2014)

Brain Slices Microscopic Detection Using Simplified SSD with Cycle-GAN Data Augmentation

Weizhou Liu[1,2], Long Cheng[1,2(✉)], and Deyuan Meng[3]

[1] State Key Laboratory of Management and Control for Complex Systems,
Institute of Automation, Chinese Academy of Sciences, Beijing 100190, China
long.cheng@ia.ac.cn
[2] School of Artificial Intelligence, University of Chinese Academy of Sciences,
Beijing 100049, China
[3] School of Automation Science and Electrical Engineering,
Beihang University, Beijing 100083, China

Abstract. Orderly automatic collection of brain slices on the silicon substrate is critical for understanding the working principle of the whole-brain neural network. Accurate and real-time brain slices detection with microscopic CCD is crucial for automatic collection of brain slices. To solve this task, an efficient simplified SSD detection model with Cycle-GAN data augmentation is presented in this paper. The proposed simplified SSD streamlines the detection network of the original SSD architecture, leading to a more rapid detection. Moreover, the proposed Cycle-GAN data augmentation method overcomes the limitation of training images. To verify the effectiveness of the proposed method, experiments are conducted with a self-made brain slices dataset. The experiment results suggest that, the proposed method has a good performance of rapidly detecting brain slices with only a small training dataset.

Keywords: Microscopic object detection · Deep learning
Data augmentation

1 Introduction

Three-dimensional (3D) representations of cerebral ultrastructure are essential for fully understanding the structure and working principle of the whole-brain neural network [1]. Currently, 3D imaging with the scanning electron microscopy provides the resolution necessary to reliably reconstruct all neuronal circuits contained within brain tissues [2].

For achieving high quality 3D imaging, brain tissues are sliced to brain slice ribbons, which need to be collected on silicon substrates. The slicing mechanism and the manually collection process are shown in Fig. 1, ribbons of brain slices floating on a knife boat are manually deposited on a silicon substrate. However,

© Springer Nature Switzerland AG 2018
L. Cheng et al. (Eds.): ICONIP 2018, LNCS 11304, pp. 454–463, 2018.
https://doi.org/10.1007/978-3-030-04212-7_40

Fig. 1. Process for manually collection of brain slices. (a) shows the operating environment. (b) shows the slicing mechanism of Leica UC7 ultramicrotome. (c) shows the manually collection process. (d) shows the collected brain slices.

manually collection for huge amounts of brain slices requires the high experienced operation skills, and also consumes intensive work. Studies have been carried out for brain slices automatic collection, e.g., Schalek *et al.* developed an automated tape ultramicrotome (ATUM) which automatically collects thousands of serial sections on a plastic tape, but the charge accumulation of the sample caused by low conductivity of the plastic tape will influence the imaging quality [3].

In order to realize the automatic collection with silicon substrate, a simple automatic collection prototype was proposed, which is shown in Fig. 2. The collection device allows precise position and manipulation of a hydrophilic circular silicon wafer. The position and spin velocity of the silicon wafer could be adjusted by a displacement platform and a rotary motor, respectively. In the collection process, the silicon wafer was inclined dipped into a custom-built knife boat which has a baffle plate to ensure brain slices floating in the collection area, and as sectioning progressed the wafer was moved to the position of slices to be collected and slowly rotated to capture the nascent sections of the brain slice ribbons. This procedure needs an accurate and fast detection of brain slices in the microscopic live video.

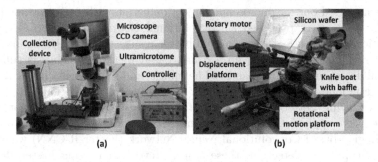

Fig. 2. The automatic collection device for brain slices. (a) shows the collection device and the ultramicrotome. (b) shows the detailed structure of the collection device.

This detection task is challenging due to factors like variations in brain slice shape, color and density. And the existence of reflection light of diamond knife and lighting condition changes in the background will also pose challenges to detection. In addition, unlike natural image object detection task, the difficulty for obtaining large-scale dataset poses another challenge to train deep learning based detection models.

To address these challenges, a simplified SSD detection model was proposed for real-time brain slices detection task. In addition, Cycle-GAN data augmentation was proposed to improve the detection accuracy with limited training dataset. The proposed simplified SSD with Cycle-GAN data augmentation could achieve an accurate and real-time detection in the brain slices microscopic detection task.

2   Related Works

2.1   Data Augmentation

It is a common knowledge that a deep learning based algorithm would be more effective when accessing more training data. Previous studies have demonstrated the effectiveness of data augmentation through minor modifications to the available training data, such as image cropping, rotation, and mirroring [4].

In recent years, generative adversarial network (GAN) has been proposed as a powerful technique to perform unsupervised generation of desired image samples [5]. For improving image classifier performance, several data augmentation methods based on GAN have been proposed and have shown to be extremely good at augmenting training dataset [6, 7].

2.2   Microscopic Object Detection

In recent years, deep learning based methods have achieved a great success in object detection. Unlike traditional handcrafted features, the features learned by deep learning based methods are more effective and general. A series of studies have been carried out for microscopic object detection with deep learning based methods. Mao *et al.* proposed a Convolutional Neural Network (CNN) with automatically learned features achieved better results than traditional methods [8]. Oscar *et al.* proposed a deep learning based method which was sufficient for visual detection of soil-transmitted helminths and schistosoma haematobium through a mobile, digital microscope [9].

For achieving better detection results, the state-of-the-art detection models have been applied to microscopy images. A series of methods based on Faster Region-based Convolutional Neural Network (Faster R-CNN) were used for microscopic cell detection, these studies reveal that using Faster R-CNN model to detect cells in microscopic image is very effective [10–12]. However, the two-stage detection models like Faster R-CNN based on region proposal still need expensive computations, and are difficult to use for real-time applications. To overcome this problem, end-to-end detection methods such as Single Shot Multibox Detector (SSD) are much simple and faster than Faster R-CNN, which could be used for real-time applications [13]. Yi *et al.* proposed an efficient neural cell detection method based on a light-weight SSD neural network [14], the detection speed is 10 FPS by testing on a single Nvidia K40 GPU. However, there is still room for improving the detection precision and speed with SSD model.

3   Data Augmentation

3.1   Data Augmentation with Traditional Techniques

Considering that the training image set is relatively small, data augmentation by some traditional methods was performed. In order to increase the robustness of the model for detecting various input object sizes and shapes, each training image was randomly sampled to patches which have minimum jaccard overlap with the ground truth bounding box of 0.1, 0.3, 0.5, 0.7, or 0.9. In addition, extra patches were randomly sampled from the input training image, and the overlapped part of the ground truth box will only be kept when the center of ground truth box was in the sampled patch. After the above-mentioned sample step, each sample patch was resized to 300 × 300 size, and some random flipping and photo-metric distorting were additionally performed on each sampled patch.

3.2   Data Augmentation with Cycle-GAN

In this paper, Cycle-GAN is used to augment the training data for improving microscopic object detection performance. Unlike traditional GAN, Cycle-GAN captures special characteristics of one image set and learns how these character-istics could be translated into other image sets [15].

For the microscopic images set augmentation task, Cycle-GAN is used to transfer images from training dataset into the artistic styles of Monet, Van Gogh, Cezanne, and Ukiyo-e. Figure 3 gives an augmentation example of a single train-ing image. Through this augmentation process, additional data annotation is avoided due to Cycle-GAN kept the target locations in original images. The generated training images via Cycle-GAN have different color, density, lighting conditions compared to original training images, using these augmented images for training will improve the generalization ability of the detection network.

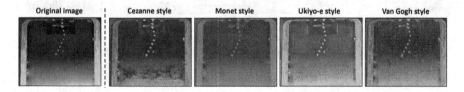

Fig. 3. An augmentation example of a single training image. The brain slices micro-scopic image in the training dataset was transferred into artistic styles of Monet, Van Gogh, Cezanne, and Ukiyo-e.

4   Simplified SSD Detection Model

4.1   Network Architecture

The overall pipeline of the proposed approach is shown in Fig. 4. Firstly, each input image is initially preprocessed into 300 × 300 input size. Then, the simplified SSD is formed by following the main architecture of the original SSD with some improvements, which aims to increase detection speed and accuracy in the brain slices microscopic detection task. The network architecture of the proposed simplified SSD model includes two main parts: a standard base network based on VGG-16 architecture (without any classification layers) and an auxiliary structure consists of some extra feature layers. The base network is used for feature extraction and the auxiliary structure is used to increase detection ability at multiple scales.

Compared to the original SSD model, the proposed simplified SSD model removes several extra feature layers on the top of the original SSD network. Each feature layer in the original SSD model was designed to detect objects within the specific range of scale, the large feature maps aim to detect small objects in the input image. The brain slices in the microscopic images are relatively small and without large scale change. For the brain slices detection task, three large feature layers (Conv4_3, Conv7, Conv8_2) which aim to detect small objects were used as the feature maps.

Similar to the original SSD, each feature map contains a specific number of default boxes, which are of certain shapes. Then object confidences and locations of these default boxes are predicted by two convolutional filters, i.e., localization filter and object filter. Finally, a fixed-size prediction of locations and scores for the presence of brain slices in those default boxes can be obtained, and a followed non-maximum suppression step will produce the final detection results.

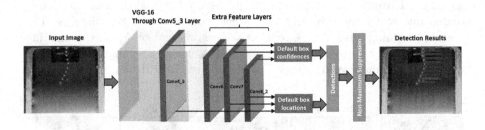

Fig. 4. The network architecture of the proposed simplified SSD.

4.2   Training

Unlike two-stage detection methods such as Faster R-CNN, SSD has a simple end-to-end training process. Next the training details are described in the following three aspects.

Transfer Learning. Several data augmentation methods have been used to solve the problem of the limited training images. However, it is not enough to train a large CNN model. In order to solve this problem, transfer learning were adopt to initialize the proposed simplified SSD model. In particular, the VGG-16 base network pre-trained on ImageNet classification dataset was transferred to the proposed simplified SSD model as the base network. This transfer learning process is an effective paradigm for helping the proposed detection model extract the features of brain slices when training data is scarce.

Matching Strategy. Each default box in feature maps needs to be matched with the corresponding ground truth bounding boxes. In this way, the positive examples and negative examples for training the proposed simplified SSD can be obtained. More specially, the jaccard overlap index of each default box with each ground truth box is calculated. For each ground truth box, the default box with the highest jaccard overlap index will be selected for matching. In addition, for each ground truth box, if it's jaccard overlap index with any default box is higher than a threshold (0.5), then they will be matched.

Training Objective. The proposed simplified SSD training objective contains two objective losses: the localization loss and the confidence loss. The localization loss is defined as a Smooth L1 loss, which can be written as:

$$L_{loc}(x, l, g) = \sum_{i \in Pos}^{N} \sum_{m \in \{cx, cy, w, h\}} x_{ij}^p smooth_{L_1}(l_i^m - \hat{g}_j^m), \tag{1}$$

$$smooth_{L_1}(z) = \begin{cases} 0.5z^2 & if |z| < 1 \\ |z| - 0.5 & otherwise \end{cases}, \tag{2}$$

$$\hat{g}_j^{cx} = (g_j^{cx} - d_i^{cx})/d_i^w \qquad \hat{g}_j^{cy} = (g_j^{cy} - d_i^{cy})/d_i^h,$$
$$\hat{g}_j^w = \log(\frac{g_j^w}{d_i^w}) \qquad\qquad \hat{g}_j^h = \log(\frac{g_j^h}{d_i^h}), \tag{3}$$

where N is the total number of matched default boxes; $x_{ij}^p = \{1, 0\}$ is an indicator, $x_{ij}^p = 1$ only when the i-th default box and the j-th ground truth box of category p were matched; l and g represent the predicted box and the ground truth box, respectively; (cx, cy) represents the center offset of the default bounding box, and (w, h) represents the width offset and height offset of the default bounding box.

The confidence loss is a traditional softmax loss over multiple classes confidences(c), which is defined as:

$$L_{conf}(x, c) = -\sum_{i \in Pos}^{N} x_{ij}^p \log(\hat{c}_i^p) - \sum_{i \in Neg} \log(\hat{c}_i^0), \qquad where \quad \hat{c}_i^p = \frac{exp(c_i^p)}{\sum_p exp(c_i^p)}. \tag{4}$$

The localization loss and confidence loss are combined to obtain the fully training objective:

$$L(x, l, g, c) = \frac{1}{N}(L_{loc}(x, l, g) + L_{conf}(x, c)). \tag{5}$$

5   Experiments

5.1   Data

In these experiments, rat hippocampus tissues are used for serial sectioning. Ribbons of rat hippocampus sections were cut at 30 nm section thickness using an ultramicrotome (Leica UC7, Wetzlar, Germany) equipped with a diamond knife angled at 35° (Ultra Diatome Knife, Biel, Switzerland). Then the CCD camera (Leica IC90E, Wetzlar, Germany) was adopted to collect several microscopy videos, from which 76 microscopic images which size of 1280×1024 were sampled. Among these images, 60 images were used for training, 16 images for testing.

The brain slices in these microscopic images were marked by bounding boxes as ground-truth annotations. Then data augmentation process was performed by Cycle-GAN. After the data augmentation process by Cycle-GAN, the training dataset contains 300 training images and 16 testing images. All the images and annotations were made into VOC format for convenient data reading by Caffe framework.

5.2   Results

In this subsection, the effectiveness of the proposed data augmentation method and simplified SSD are illustrated respectively. Firstly, the proposed Cycle-GAN data augmentation method is evaluated with three recent state-of-the-art detection algorithms: Faster R-CNN, SSD and YOLO-v3. Secondly, several comparison experiments were performed to illustrate the high performance of the proposed simplified SSD. The proposed data augmentation method and object detection model are trained and tested on a workstation with an Intel i7-6850K processor of 3.6 GHz and a single Nvidia Titan-xp GPU.

Cycle-GAN Data Augmentation. Faster R-CNN, SSD and YOLO-v3 are implemented on the brain slices microscopic dataset, respectively, and for comparison, the three state-of-the-art detection model are also implemented on the augmented dataset. For a fair comparison, VGG16 is chosen as the base network for all of the three detection models. The experiment results are given in Table 1. It can be seen that the Cycle-GAN data augmentation process could effectively improve the detection performance of all the three detection models.

Table 1. Effects of Cycle-GAN data augmentation on three state-of-the-art detection models

| Method | mAP | | |
| | Use Cycle-GAN augmentation? | | |
	Yes	No	
SSD	0.950189	0.886907	
YOLOv3	0.929375	0.869549	
Faster R-CNN	0.909146	0.867713	

Simplified SSD. Currently, SSD is an excellent detection model for its high speed detection performance while keep a high detection accuracy. It can be seen that SSD detection model is also very suitable for the brain slices detection task from Table 1. Since the ultimate objective is to detect brain slices with microscopic CCD in real time, the detection speed needs to be further accelerated by using the proposed simplified SSD.

The comparison experiments results are listed in Table 2. To compare with the original SSD detection model, the proposed simplified SSD has a better detection accuracy than the original SSD. Furthermore, the detection speed of the proposed simplified SSD has a distinct improvement than the original SSD detection model, and the GPU memory usage is reduced too. This characteristics enhanced the model portability of the proposed simplified SSD, specifically, it will be convenient to migrate the simplified SSD detection model to other graphics cards with lower computing ability.

Figure 5 demonstrates several examples of the final detection results. It can be seen that the proposed simplified SSD model is able to deal with the color change, background interference and shape deformation of the brain slices.

Table 2. Comparison of the proposed simplified SSD and the original SSD

| Method | mAP | | FPS | GPU memory |
| | Use Cycle-GAN augmentation? | | | |
	Yes	No		
Original SSD	0.950189	0.886907	24	737 MB
Simplified SSD	0.965129	0.890029	29	689 MB

To illustrate the reason for selecting the three feature layers (Conv4_3, Conv7, Conv8_2) as the feature maps, layers of the original SSD are progressively removed for comparing the detection results. Table 3 shows that using Conv4_3, Conv7 and Conv8_2 as the feature maps can achieve the best detection results. From these experiments, some interesting trends can be observed. For example, it increases the detection performance if the small feature layers are progressively removed, the reason might be that the small feature maps aim to detect

Fig. 5. Visualization of the detection results using simplified SSD. The red boxes represent the final detection results. (Color figure online)

large objects which are not present in the microscopic images, so remove these small feature layers on the top of SSD network may reduce the interference of the useless large default boxes. But when only keeping Conv4_3 for prediction, although detection speed is greatly improved, the detection performance is the worst. In order to obtain a better detection accuracy and a comparable detection speed to the original SSD, Conv4_3, Conv7 and Conv8_2 are chosen as the feature maps for prediction.

Table 3. Comparison of using different output layers in original SSD

Prediction source layers from						mAP Use Cycle-GAN?		FPS
Conv4_3	Conv7	Conv8_2	Conv9_2	Conv10_2	Conv11_2	Yes	No	
✓	✓	✓	✓	✓	✓	0.950189	0.886907	24
✓	✓	✓	✓	✓		0.958645	0.890341	26
✓	✓	✓	✓			0.932281	0.890109	27
✓	✓	✓				**0.965129**	**0.890029**	**29**
✓	✓					0.963461	0.897847	33
✓						0.900577	0.873513	42

6   Conclusions

A real-time detection model is proposed in this paper to detect brain slices in microscopic live video. To this end, in order to improve detection accuracy and speed with the relatively small training dataset, firstly, a data augmentation method based on Cycle-GAN is proposed, which significantly increases the detection accuracy. Secondly, a simplified SSD is proposed, which aims to increase the detection speed while reduce the GPU memory usage. The experiment results show that the proposed simplified SSD with Cycle-GAN data augmentation can overcome the problem of limited training dataset, and has a good ability to detect brain slices in microscopic images in real time.

Acknowledgments. This work was supported in part by the National Natural Science Foundation of China under Grants 61873268, 61633016, in part by the Research Fund for Young Top-Notch Talent of National Ten Thousand Talent Program, in part by the Beijing Municipal Natural Science Foundation under Grant 4162066.

References

1. Kubota, Y.: New developments in electron microscopy for serial image acquisition of neuronal profiles. Microscopy **64**, 27–36 (2015)
2. Mikula, S.: Progress towards Mammalian whole-brain cellular connectomics. Front. Neuroanat. **10**, 62–71 (2016)
3. Schalek, R., et al.: Development of high-throughput, high-resolution 3D reconstruction of large-volume biological tissue using automated tape collection ultramicrotomy and scanning electron microscopy. Microsc. Microanal. **17**, 966–967 (2011)
4. Perez, L., Wang, J.: The effectiveness of data augmentation in image classification using deep learning. arXiv preprint arXiv:1712.04621 (2017)
5. Goodfellow, I., et al.: Generative adversarial nets. In: Advances in Neural Information Processing Systems, pp. 2672–2680 (2014)
6. Wei, L., Zhang, S., Gao, W., Tian, Q.: Person transfer GAN to bridge domain gap for person re-identification. arXiv preprint arXiv:1711.08565 (2017)
7. Mariani, G., Scheidegger, F., Istrate, R., Bekas, C., Malossi, C.: BAGAN: data augmentation with balancing GAN. arXiv preprint arXiv:1803.09655 (2018)
8. Mao, Y., Yin, Z., Schober, J.M.: Iteratively training classifiers for circulating tumor cell detection. In: IEEE 12th International Symposium on Biomedical Imaging, pp. 190–194 (2015)
9. Holmström, O., et al.: Point-of-care mobile digital microscopy and deep learning for the detection of soil-transmitted Helminths and Schistosoma haematobium. Glob. Health. Action. **10**, 1337325 (2017)
10. Yang, S., Fang, B., Tang, W., Wu, X., Qian, J., Yang, W.: Faster R-CNN based microscopic cell detection. In: International Conference on Security, Pattern Analysis, and Cybernetics, pp. 345–350 (2017)
11. Hung, J., Carpenter, A.: Applying faster R-CNN for object detection on malaria images. In: IEEE Conference on Computer Vision and Pattern Recognition Workshops, pp. 808–813 (2017)
12. Zhang, J., Hu, H., Chen, S., Huang, Y., Guan, Q.: Cancer cells detection in phase-contrast microscopy images based on Faster R-CNN. In: 9th International Symposium on Computational Intelligence and Design, pp. 363–367 (2016)
13. Liu, W., et al.: SSD: single shot multibox detector. In: Proceedings of the European Conference on Computer Vision, pp. 21–37 (2016)
14. Yi, J., Wu, P., Hoeppner, D.J., Metaxas, D.: Fast neural cell detection using lightweight SSD neural network. In: Proceedings of the IEEE Conference on Computer Vision and Pattern Recognition Workshops, pp. 108–112 (2017)
15. Zhu, J.-Y., Park, T., Isola, P., Efros, A.A.: Unpaired image-to-image translation using cycle-consistent adversarial networks. arXiv preprint arXiv:1703.10593 (2017)

Agglomeration Detection in Gas-Phase Ethylene Polymerization Based on Multi-scale Convolutional Neural Network

Wenqian Zhang, Jing Wang[✉], and Haiyan Wu

College of Information Science and Technology, Beijing University of Chemical
Technology, Beijing, China
jwang@mail.buct.edu.cn

Abstract. Fault will affect product quality, damage the reaction device, and
cause property damage, so fault detection is a crucial part of the industrial
production process. The traditional multivariate statistics method is gradually
limited because the industrial data for inspection is mostly time series with the
characteristics of difficult modeling and large noise interference. Multi-Scale
Convolutional Neural Network (MCNN) has achieved remarkable results in
time series processing and the computational efficiency. This paper applies
MCNN to the fault detection and classification of the industry process. MCNN
incorporates the feature extraction and the classification in a single framework. It
will lead to further feature representations and superior fault detection performance at the industrial process. MCNN is conducted in the TensorFlow
framework, and its fault detection performance is evaluated with existing BP
neural network on a large amount of time series industrial data from a real gas-
phase ethylene polymerization industry.

Keywords: Agglomeration detection · Multi-scale convolution neural network
Multi-branch feature extraction · Time series industrial data

1 Introduction

Industrial big data refers to a large amount of diversified time series generated at a high
speed by industrial equipment, and it is usually more structured, more correlated, more
orderly in time and more ready for analytics. The classification of time series data has
been around for decades within the community of statistical analysis and machine
learning, and found many important applications such as fault detection and clinical
diagnosis. There are many methods of time series classification that can be divided into
three categories: distance-based methods, feature-based methods and model-based
methods [1, 2].

The key of distance-based methods is to define a distance function to measure the
similarity between two time series. K nearest neighbor (KNN) and support vector
machines (SVM) use this principle for time series classification. The second category is
feature-based methods in which the key information of the time series is extracted and
represented by eigenvectors. Based on this, feature-based classifiers are used to complete the classification. This method uses the dimensionality reduction techniques, such

© Springer Nature Switzerland AG 2018
L. Cheng et al. (Eds.): ICONIP 2018, LNCS 11304, pp. 464–475, 2018.
https://doi.org/10.1007/978-3-030-04212-7_41

as spectral method and feature eigenvalue reduction method. The spectral method uses a Fourier transform to transform the time series from the time domain space to the frequency domain space. The eigenvalue reduction method mainly includes Principal Component Analysis (PCA), singular value decomposition (SVD) and Latent Discriminant Analysis (LDA). Model-based methods assume that the time series in each class is generated from a model, and the parameters of the model can be determined through the training of samples. The models of different classification are different, and a new time series is sequentially matched with each model to determine which class it belongs to. The autoregressive (AR) model and the Markov model (MM) are typical methods based on model. However, the model-based methods are greatly limited since most of the time series are difficult or even impossible to generate by models.

Deep learning has made remarkable achievements in recent years. As a type of deep learning method, convolutional neural networks (CNN) have achieved significant progress in speech processing, object recognition, and face verification [3–5]. Convolutional neural network (CNN) has local receptive field, weights sharing and pooling, which are different from other forward networks. These features make CNN automatically extract and generate new image features [6]. Widespread usage of CNN motivates researchers to consider whether CNN can also be used in time series classification. Zheng et al. proposed a model called a multi channels deep convolution neural network that extracts features by placing multiple time series into different CNNs, and then puts the extracted features into a new CNN framework [7]. This type of network requires a large number of multivariate data sets which are not suitable for using in the situations that industrial fault data is less. Cui et al. proposed a multi-scale convolution neural network (MCNN) that is an end-to-end model without requiring any handcrafted features. Besides, MCNN is a network specifically designed for time series, which specially adds times series processing stage before convolution layer. Experimental results indicate that MCNN has outstanding achievement comparing with other traditional methods in time series classification [8]. This paper will apply the MCNN method to the fault detection of the tradition chemical engineering process: polyethylene preparation.

As a significant raw material for chemical reactions, polyethylene is widely used in agriculture, industry, defense and other fields. The gas-solid Fluid Bed Reactor (FBR) has been widely used for the polyethylene preparation because of its advantages of high stability, simple operation, economy, and safety. However, the hydrodynamic properties in the reactor are changed due to the electrostatic impact that results in the failure of removing the reaction heat and other issues. This leads to particle heap, smelting, and finally agglomeration. When the agglomerate is too large, it will block the distribution plate or discharge system and lead to the emergency shutdown of the explosion. It can affect the safe and stable operation of the reactor and cause property damage and even life-threatening in the most serious circumstances. Hence, it is an important task to detect agglomeration in FBRs in time.

Numerous methods of agglomeration detection have been proposed, such as acoustic wave method, pressure fluctuations, electrostatic method and temperature pressure method [9–12]. However, the electrostatic method is easily influenced by the working conditions and the technical level still needs to be broken. The warm pressure method has a large lag and depends on the working experience of the operators. As one of the Nondestructive Testing (NDT) methods, the acoustic wave method is safe, sensitive, and easy to implement, which makes it widely used and has attracted the attention of researchers. Acoustic signals reflect the characteristic of particles in the fluidized bed, such as particle distribution and movement. The concrete information of the fluidized bed and the real-time monitoring of the important variables of the fluidized bed can be obtained through comprehensive analysis of the acoustic signals collected in real time.

The acoustic signal obtained by the sensor equipped on the FBR is a set of time series data, thus the method of time series processing can be used for the agglomeration detection. It should be noted that the sampling acoustic wave signal has a large difference due to the different sensor installation positions. It is difficult for a single sensor to obtain all the information of an industrial reactor. Therefore, four-channel acoustic sensors installed at different locations outside the reactor are used to detect the acoustic signal. The four channel signals are collected and fused into the MCNN for determining whether the agglomerate occurs. MCNN performs three transformations to decompose the acoustic signal into the multiple temporal and spectrum scales. Tunable local convolution is employed to identify and learn features from each decomposed time series. MCNN consists of the multi-branch input layer, convolution layers and the full connected network. It is capable of extracting the salient features that exists in the multiple frequency components and completing agglomeration detection. The result shows that MCNN is feasible in industrial fault detection.

The objective of this paper is to detect the agglomeration fault in FERs based on MCNN. The remaining of this paper is organized as follows. The architecture of MCNN and MCNN classification for time series are described in Sect. 2. The agglomeration data acquisition and the MCNN computational map in the Tensorflow framework are described in Sect. 3. Experimental results and conclusions are separately drawn in Sects. 4 and 5.

2 MCNN Methodology

2.1 MCNN Architecture

MCNN is a feedforward neural network, which can automatically extract features of different positions and different scales and then link them up to continue the feature extraction. This characteristic of MCNN makes it superior in the classification of time series over other methods. The architecture of MCNN in this work is given in Fig. 1. The MCNN architecture has three sequential stages: transformation stage, local convolution stage, and full convolution stage.

Different time scales have different intrinsic characteristics. Long term (also known as low frequency) indicates the overall trends and short term (also known as high frequency) reflects the minor changes in local changes. Both of them may be a key factor in determining the prediction quality for several certain tasks. In the MCNN transformation stage, various conversions are made to the input time series, including identity mapping, down-sampling in the time domain, and spectral transformation in the frequency domain.

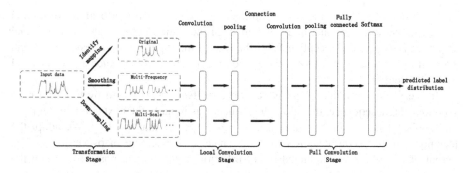

Fig. 1. Architecture of MCNN

In the original branch, the input sequence is identified mapping without any changing.

In the multi-scale branch, down-sampling technique is used to generate multiscale of acoustic signals. Common sampling methods in signal processing include down sampling, up-sampling and over-sampling. The purpose of over-sampling is to change the distribution of noise and obtain an optimal signal-noise-ratio. Up-sampling increases the amount of data while creating new signals. Only down-sampling can reduce the amount of data by reducing the frequency of data sampling. Meanwhile, dawn-sampling retains useful information about original signals. The down-sampled signal at rate of p is given as follows:

$$T_{ds} = \{t(1+p \times i)\}, i = 0, 1, 2, \cdots, \left\lfloor \frac{n-1}{p} \right\rfloor, \tag{1}$$

where T_{ds} is the down-sampled signal, t is the actual signal, and n is the number of samples.

In the multi-frequency branch, multiple low pass filters act on the raw input to make it smooth. As the filter length increases the smoothness of the output increases, whereas the sharp transitions in the data are made increasingly blunt. Low pass filters will decrease the variance of the time series and reduce the effects of the high-frequency disturbances and the random noise. Thus, newly generated time series represent general low frequency information. The decomposed acoustic signal (T_l) is obtained by an average filter with a window size l.

$$T_l = \frac{t_i + t_{i+1} + t_{i+2} + \ldots + t_{i+l-1}}{l}, i = 0, 1, 2, \cdots n - l - 1 \tag{2}$$

Local convolution and max pooling are separately performed on the output of multi-scale and multi-frequency branches in the local convolution stage. First, we obtain multiple time series with different lengths from a single input sequence. Then apply several 1-D local convolutional layers on these newly generated time series to extract the features independently in the local convolution stage. All the local convolution filters have the same size, down sampling the time series rather than increasing the filter size can greatly reduce the number of parameters in the local convolutional layer. It is worth noting that under the effect of the same filter size, the shorter signal will have a larger local receptive field (the size of the input layer corresponding to an element in the output of a layer), while the longer signal will produce a smaller local receptive field [13].

In the full convolution stage, we connect the features extracted by the local convolutions into a sequence, then flow the sequence as input to the next convolution layer that applies 1-D convolution, and finally completes the max pooling. We adopt the technique of deep concatenation to concatenate all the feature maps vertically. The output of MCNN is the predicted the distribution result of each possible label in the input time series. The final classification result is obtained according to the most possibility label.

2.2 Classification Based on MCNN

Training is the process of adjusting weights and deviations using MCNN algorithm and data samples. The training process is usually a time-consuming process. The completion of training may take several hours, days or even weeks due to the number of training samples and the complexity of the algorithm. The MCNN in this paper uses GPU to calculate, which has high calculation efficiency and shortens the training time. In order to evaluate the learning effect during (or after) training, it is important to test the network with some data that does not belong to the training data set. Therefore, part of the training data is separated as validation data before the training begins. Only when the loss of validation data is less than the loss of training data, the neural network is said to have a good learning effect. If the opposite happens, that is, the training error decreases when the verification error increases, the network is considered to be ineffective for learning.

Deep learning algorithms typically use loss function to measure the error between the output produced by the MCNN and the actual target output. The choice of loss function mainly depends on whether the neural network performs classification or regression tasks. In this work, a cross entropy loss function is used for the purpose of classification. Firstly the Softmax function (3) is used to calculate the probability distribution of the input data to each label [14],

$$y_i = \frac{e^{x_i}}{\sum_{j=1}^{n} e^{x_j}} \tag{3}$$

where n is the number of labels within the class, x_i is the output from the neuron corresponding to the ith label, and y_i represents the probability that the classified input belongs to the ith label. The larger the input value is, the greater the probability that the input data belongs to the label. So we have,

$$\sum_{i=1}^{n} y_i = 1 \tag{4}$$

In this paper, each class uses a separate Softmax activation function. The cross entropy CE_i^k function is [14],

$$CE_i^k = -\sum_i \hat{y}_i^k \times \log(y_i^k), \tag{5}$$

where CE_i^k is the cross entropy of the ith label in the kth class, y_i^k is the output of the neural network corresponding to the ith label in the kth class, and $\hat{y}_i^k$ is the actual output of the ith label in the kth class. The cross entropy is used to evaluate the difference between probability distribution and real distribution of current training. The cross entropy will also be smaller, when the output of the model is closer to the desired output. In other words, reducing cross entropy loss will improve the accuracy of the model.

The classification accuracy Acc is defined as follows [13],

$$Acc = \frac{TP + FN}{TP + TN + TN + FP} \times 100\%, \tag{6}$$

where

- TN (True Negative) denotes the case of a negative sample being predicted negative correctly;
- TP (True Positive) refers to the case of a positive sample being predicted positive correctly;
- FN (False Negative) refers to the case that a positive sample being predicted negative incorrectly;
- FP (False Positive) denotes the case that a negative sample being predicted positive incorrectly.

3   Related Experimental Work

3.1   Experimental Device and Experimental Data

The experimental data comes from a real polyethylene fluidized bed of China Sinopec Group. The device diagram in Fig. 2 shows a gas-solid fluidized bed facility for polyethylene production. The production process mainly includes six major processes such as raw material refining, catalyst feeding, polymerization, polymer degassing & tail gas recovery, granulation & product storage, and packaging. The raw materials for the reaction include ethylene, butene and gaseous hydrogen, and the product is a powdered polyethylene. The catalyst is a solid triethyl aluminum powder and the cocatalyst is an aluminum alkyl.

Figure 3 shows a schematic of an acoustic aggregation monitoring system in FBR. The system is divided into two main parts: a pilot plant of the polyethylene fluidized

Fig. 2. Gas phase polyethylene fluidized bed apparatus diagram

bed and an acoustic signal acquisition system. Four acoustic signal sensors are divided into two pairs affixed to the outer bed wall. Sensors 1 and 3 are located near the distribution plate, and sensor 2 and 4 are located near the expansion section. The collected acoustic signal is input to the Microcontroller Unit (MCU) in the explosion-proof box, and is transmitted wirelessly to the industrial control computer to monitor the operating conditions in real time. The sampling period of acoustic signal is 10 s, and 2500 data are continuously sampled. Since 4 acoustic signal sensors are equipped on the fluidized bed wall. Finally, 4×2500 sets data are collected.

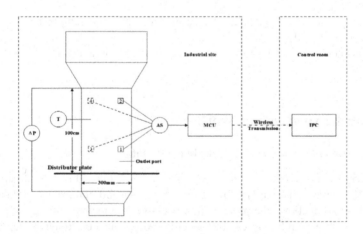

Fig. 3. Schematic diagram of the experimental apparatus

3.2   Implementing the MCNN Computational Graph Using TensorFlow

TensorFlow framework is used to complete the training of MCNN. TensorFlow is the second generation of artificial intelligence learning system developed by Google based on DistBelief. It can transfer complex data structures to the artificial intelligence neural network for data analysis and processing. One of TensorFlow's strengths is to visualize the algorithm's calculation graph after the encoding is completed [15, 16]. The calculation map of the MCNN used for fault detection is shown in Fig. 4. For comparison, the same data set will also be processed using a Back Propagation Neural Network (BPNN) with a single hidden layer.

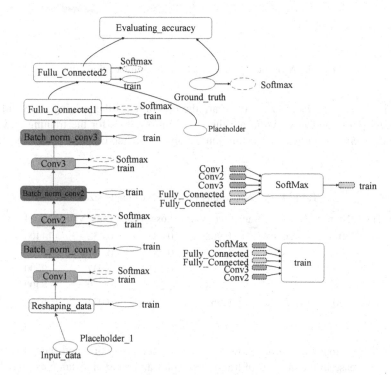

Fig. 4. Computational graph for CNN generated using TensorFlow

4   Experimental Results

4.1   Experimental Data and Experimental Sample Selection

At 18:40 on January 6 2017, the FBR system had an alarm due to pressure drop. Figure 5 shows the waveform of the original sound signal within 3 h of the alarm to detect the time of agglomeration. Here the acoustic signal fluctuates caused by agglomeration fault occurring around 17:20. Therefore, we divide the normal data and fault data according to the fault time 17:20.

Several different experiments are completed in order to select the appropriate number of training samples and validation samples. The training and validating results at different sample size are shown in the Table 1. The symbols' meaning is as follows, *Num_Train* and *Num_Valid* are the number of training and validation set samples,

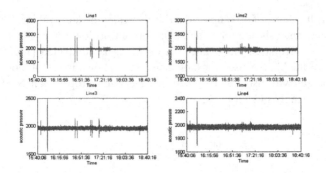

Fig. 5. Acoustic signal example with agglomeration fault

respectively. *Acc_Train* and *Acc_Valid* are the accuracy for the training and validation data. *Loss_Train* and *Loss_Valid* are the cross entropy loss for the training and validation data. *Min_Iteration* is the minimum number of iterations to obtain the stable accuracy.

Table 1. MCNN results for different sample sizes

Num_Train	Num_Valid	Acc_Train	Acc_Valid	Loss_Train	Loss_Valid	Min_Iteration
500	300	0.984	0.962	0.48	0.132	14400
400	100	0.988	0.967	0.28	0.051	11200
200	50	0.99	0.992	0.24	0.047	9000
100	25	0.992	0.998	0.2	0.029	8400
70	30	0.986	0.962	0.2	0.034	6200

It is shown that *Acc_Train* shows a trend of increasing first and then decreasing with the decrease in the number of training samples and validation samples. *Acc_Valid* is consistent with the trend of changes in *Acc_Train*. *Acc_Valid* is higher than *Acc_Train* under *Nun_Train*=200 and *Num_Train*=100 and *Acc_Valid* is always lower than *Acc_Train* in other cases.

The *Loss_Train* gradually decreases and remains stable, but the *Loss_Valid* decreases and then increases. The different trends between *Loss_Train* and *Loss_Valid* indicate that MCNN has been overfitted with the change of sample size. There are many reasons for overfitting, such as wrong sampling method, too much noise data in the sample, and too many parameters. In our work, early stopping technique has been applied for preventing overfitting. Therefore, we can select apposite sample to avoid overfitting.

The minimum number of iterations tends to decrease with the decrease in the number of samples. The reason is that the decrease in the number of samples reduces the number of parameters that the network needs to train, so the network can reach stability more quickly. Consider the trade-off among the accuracy, loss, and the minimum number of iterations, we get the optimal learning condition: *Num_Train*=100 and *Num_Valid* = 25.

4.2   Agglomeration Detection Based on MCNN

The training data of MCNN is obtained from the real polyethylene fluidized bed data of China Sinopec Group. 100 sets of data are selected as training data and 25 sets of data as validation data. The learning rate is 0.00002 for all the epochs. Here we apply a BP with two hidden layers in which the neural neurons take softmax function for transform, and a linear function is used in output layer for transform to get values with a broad range. And there are four neurons in the input layer, 4 neural cells in the first hidden layer and 2 neural cells in the second hidden layer [17].

The improvement in average classification accuracy of training and validation is shown in Fig. 6. Neither the train accuracy nor the test accuracy changes at the beginning of the iteration, then the accuracy rate rises and finally reaches a stable value. The cross entropy losses for the training and validation process are shown in Fig. 7. Both cross entropy loss of validation and training increase first, then decrease and final converge to the stable state. Besides, the cross entropy loss of validation is lower than that of training.

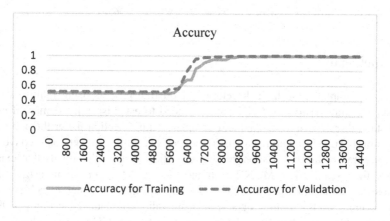

Fig. 6. Average classification accuracy of training and validation for MCNN

Table 2 gives the final classification accuracy. From the results, it is shown that using MCNN for fluidized bed agglomerate fault detection is feasible. For the training and validation data set, the cross entropy loss and the classification accuracy are both improved compared with BPNN methods. This indicates that CNN is still popularizing in memory. When using validation data, the results of BPNN are significantly different from that of MCNN, which indicates that the generalization ability of BPNN under this data is not very effective. The training time is about 1 min for the MCNN and about 3 min for the BPNN.

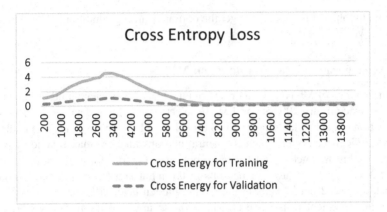

Fig. 7. Cross entropy loss function curve of training and validation for MCNN

Table 2. Comparing between MCNN and BPNN

NN type	Cross entropy loss		Average accuracy	
	Training	Validation	Training	Validation
MCNN	0.20	0.029	99.2%	99.8%
BPNN	0.79	0.098	97.3%	96.6%

5   Conclusion

This paper completes the fault detection in industrial process using multi-scale convolutional neural network. MCNN does not need to establish a mathematical model, and it starts directly from the data and complete fault detection under the appropriate data samples. Transformation stage is introduced before the conventional convolution stage to represent the non-stationary characteristics of acoustic signals that exists in different time scales. The MCNN is trained using data from an industrial site to complete the multi-label classification. Use the Softmax classifier to achieve up to 99.2% accuracy on the training data set and the validation data set accuracy is 99.8%. The results show that it is feasible and accurate to use MCNN for industrial fault detection.

At present, the detection of fault time takes about one minutes, and the accuracy of the initial iteration is only small. The next work attempts to use other time series conversion methods to improve the MCNN, so that the algorithm can complete the barrier detection in fewer iterations, thus shortening the detection time.

Acknowledgements. This work is supported by the National Natural Science Foundation of China (No. 61573050) and the open-project grant funded by the State Key Laboratory of Synthetical Automation for Process Industry at the Northeastern University (No. PAL-N201702).

References

1. Xing, Z., Pei, J., Keogh, E.: A brief survey on sequence classification. ACM SIGKDD Explor. Newsl. **12**, 40–48 (2010)
2. Wang, J., Liu, P., She, M.F., Nahavandi, S., Kouzani, A.: Bag-of-words representation for biomedical time series classification. Biomed. Signal Process. Control **8**, 634–644 (2013)
3. Lee, H., Pham, P., Largman, Y.: Unsupervised feature learning for audio classification using convolutional deep belief networks. In: Advances in Neural Information Processing Systems, pp. 1096–1104. MIT Press, Massachusetts (2009)
4. Krizhevsky, A., Sutskever, I., Hinton, G.E.: Imagenet classification with deep convolutional neural networks. In: Advances in Neural Information Processing Systems, pp. 1097–1105. MIT Press, Massachusetts (2012)
5. Schroff, F., Kalenichenko, D., Philbin, J.: Facenet: a unified embedding for face recognition and clustering. In: Proceedings of the IEEE Conference on Computer Vision and Pattern Recognition, pp. 815–823. IEEE Press, New York (2015)
6. Swietojanski, P., Ghoshal, A., Renals, S.: Convolutional neural networks for distant speech recognition. IEEE Signal Process. Lett. **21**, 1120–1124 (2014)
7. Zheng, Y., Liu, Q., Chen, E., Ge, Y., Zhao, J.L.: Time series classification using multi-channels deep convolutional neural networks. In: Li, F., Li, G., Hwang, S.-w., Yao, B., Zhang, Z. (eds.) WAIM 2014. LNCS, vol. 8485, pp. 298–310. Springer, Cham (2014). https://doi.org/10.1007/978-3-319-08010-9_33
8. Cui, Z., Chen, W., Chen, Y.: Multi-scale convolutional neural networks for time series classification. arXiv preprint arXiv:1603.06995 (2016)
9. Bartels, M., Lin, W., Nijenhuis, J., Kapteijn, F.: Agglomeration in fluidized beds at high temperatures: mechanisms, detection and prevention. Prog. Energy Combust. Sci. **34**, 633–666 (2008)
10. Werther, J.: Measurement techniques in fluidized beds. Powder Technol. **102**, 15–36 (1999)
11. Vervloet, D., Van Nijenhuis, J., Ommen, J.R.: Monitoring a lab-scale fluidized bed dryer: a comparison between pressure transducers, passive acoustic emissions and vibration measurements. Powder Technol. **197**, 36–48 (2010)
12. Chen, A., Bi, H.T., Grace, J.R.: Charge distribution around a rising bubble in a two-dimensional fluidized bed by signal reconstruction. Powder Technol. **177**, 113–124 (2007)
13. Yao, Z., Zhu, Z., Chen, Y.: Atrial fibrillation detection by multi-scale convolutional neural networks. In: International Conference on Information Fusion, pp. 1–6. IEEE Press, Piscataway (2017)
14. Plathottam, S.J., Hossein, S., Prakash, R.: Convolutional Neural Networks (CNNs) for power system big data analysis. In: 2017 North American Power Symposium (NAPS), pp. 1–6. IEEE Press, New York (2017)
15. Abadi, M., Agarwal, A, Barham, P.: Tensorflow: Large-scale machine learning on heterogeneous distributed systems. arXiv preprint arXiv:1603.04467 (2016)
16. Abadi, M., Barham, P., Chen, J.: TensorFlow: A System for Large-Scale Machine Learning. In: OSDI. vol. 16, pp. 265–283. USENIX Association, Berkeley (2016)
17. Xiao, Z., Ye, S.J., Zhong, B., Sun, C.X.: BP neural network with rough set for short term load forecasting. Expert Syst. Appl. **36**, 273–279 (2009)

Intra-class Structure Aware Networks
for Screen Defect Detection

Chengchao Shen, Jie Song, Shuyi Song, Sihui Luo, Li Sun, and Mingli Song[✉]

College of Computer Science, Zhejiang University, Hangzhou, China
{chengchaoshen,sjie,brendasoung,sihuiluo829,lsun,brooksong}@zju.edu.cn

Abstract. Typically, screen defect detection treats different types of defects as a single category and ignores the large variation among them, which may pose large difficulty to the model learning and thus lead to inferior performance. In this paper, we propose a novel network model, called Intra-class Structure Aware Networks (ISANs), to alleviate the difficulty of learning one single concept which exhibits in various forms. The proposed model introduces more neural units for the "defect" category rather than a single one, to accommodate the large variations in this category, which can significantly improve the representation power. Regularized by prior distribution of intra-class variants, our approach can learn intra-class structure of screen defect without extra fine-grained labels. Experimental results demonstrate that ISANs can effectively discriminate intra-class variants and gain significant performance improvement on screen defect detection task as well as the classification task in MNIST.

Keywords: Intra-class Structure Aware
Convolutional Neural Networks · Screen defect detection

1 Introduction

Screen defects are flaws in the screen which may lead to an unpleasant visual experience. Hence, it's important for manufacturers to detect these screen defects during the manufacture of screen and prevent these defective screens from being sold to consumers. Screen defects often exhibit in different forms, such as dot-like defect, line-like defect and blob-like defect, as shown in Fig. 1. These sub-category defects are caused by different production technologies. It's important for screen manufactures to analyse the occurrence of these defect types. So screen manufactures can improve their production technology to produce high-quality screen. However, manually collecting these screen defect data into a fine-grained labeled dataset is a time-consuming job. Hence, we need a classifier that can not only recognize whether there are defects in the image but also discriminate different intra-class variants.

Convolutional Neural Networks (CNNs) have shown great success in visual object recognition [2,4,6,12,14], which is a promising option for the recognition of screen defects. A straightforward solution for screen detection is directly

© Springer Nature Switzerland AG 2018
L. Cheng et al. (Eds.): ICONIP 2018, LNCS 11304, pp. 476–485, 2018.
https://doi.org/10.1007/978-3-030-04212-7_42

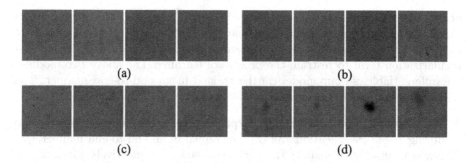

Fig. 1. Screen defect dataset. Row (a) contains screen images without defects. Row (b) is composed of screen images with line-like defects. Images of row (c) have various dot-like defects. Row (d) contains images with blob-like defects.

applying CNNs to recognize whether the screen image contains defects, which can be regarded as a binary classification task. This solution generally focuses on learning the common discriminative features of screen defects and ignores the difference among different screen defects even if some screen defects have a very different visual appearance. CNNs tend to learn the common features from intra-class variants and eliminate the intra-class variation, such as pose, deformation, texture, etc. [11]. Hence, plain CNNs can't effectively discriminate intra-class variants of screen defect without fine-grained labeled data.

Recently, some approaches have been proposed to tackle the above problem. Capsule Networks [11] retain the various properties of a particular entity that are present in the image with an active capsule. Since Capsule Networks obtain instantiation parameters of a single image, they can't directly summarize a mixture of images into several clusters. STNs [5] are proposed to obtain rotation-invariant features by adding a spatial transformer module with affine transformation to the features of the CNNs. TIPooling [7] networks are built by duplicating the plain CNN eight times to capture different augmented rotated version of input images and appending a transform-invariant pooling layer before the output layer. The above two methods are all limited to eliminate the effect of rotated intra-class variants on classification. They can't effectively discriminate general intra-class variants of screen defect without fine-grained labeled data.

To address the above problem, we propose a novel approach, called Intra-class Structure Aware Networks (ISANs), which can learn not only discriminative inter-class features but also intra-class semantic structure from data. We assume that the weight parameters from the final fully connected layer of CNNs can be represented as $W = [w_1, ..., w_n]^T$, where n denotes the number of category, $w_i \in \mathbb{R}^m$. Each weight parameter vector w_i can correspond to a category [10]. This weight parameter vector can be regarded as the embedding vector of the corresponding category. Instead of representing one category with only one embedding vector, ISANs introduce more embedding vectors under one parent class. Each embedding vector of ISANs can represent the general features of the corresponding intra-class variant. However, because of the lack of super-vision among intra-class variants, we can't effectively control the direction of networks' optimization. Hence, we introduce an auxiliary prior distribution as a regularization term to restrain the process of learning. The above two modules are differentiable, so our model can be trained in an end-to-end manner. The major contributions of our work can be summarized as follows:

1. We propose a novel approach that explicitly encodes the intra-class structure during image classification and can be trained in an end-to-end manner.
2. Experimental results show that our approach can effectively discriminate intra-class variants of screen defect and achieve superior or competitive clas-sification performance on screen defect dataset and MNIST.

2   Intra-class Structure Aware Networks

In this section, we first introduce intra-class variant embedded space, in which the intra-class structure is explicitly encoded. Afterward, we propose a regularization term to balance the cluster assignments of intra-class variants. Finally, we present the optimization method of our model.

2.1   Intra-class Variant Embedded Space

Assume that there are N samples, $\mathbf{X} = \{x_1, ..., x_N\}$, where $x_i \in \mathbb{R}^{d_x}$. These samples can be classified into M categories with labels $\mathbf{Y} = \{y_1, ..., y_N\}$, where $y_i \in \{1, ..., M\}$. With a embedding function $\phi_W : \mathbf{X} \to \mathbf{Z}$, we can transform the raw samples x_i into a embedding vector z_i, where $\mathbf{Z} = \{z_1, ..., z_N\}$ is the embedded space of $\mathbf{X}$ and $z_i \in \mathbb{R}^{d_z}$ is a more compact representation of x_i. The dimension of z_i is far smaller than raw sample x_i ($d_z \ll d_x$) to avoid the "curse of dimensionality" [1]. With consideration of the ability of non-linear function approximation [3], we select CNNs to parameterize ϕ_W.

If we use labeled dataset $[\mathbf{X}, \mathbf{Y}]$ to train a plain CNN, we can generally formulate the last fully connected layer and softmax as follow:

$$p_{ij} = P(z_i; \Theta) = \frac{\exp(\theta_j^T z_i)}{\sum\limits_{j'=1}^{M} \exp(\theta_{j'}^T z_i)}, \tag{1}$$

where $\Theta = [\theta_1, ..., \theta_M] \in \mathbb{R}^{d_z \times M}$ are the parameters of the fully connected layer, θ_j denotes the parameters of class j and p_{ij} denotes the probability of sample x_i belonging to class j.

Equation (1) computes inner product between the representation of sample z_i and the parameters θ_j to evaluate the similarity between sample x_i and class j. If there is a sample $x_{i'}$ that is quite different from other samples from the same class, also called intra-class variant, it's difficult for CNN to extract the common class features from this sample. The CNNs trend to remember this case in parameters [15]. However, it may distort the hidden feature space.

We propose a more elegant and effective network to tackle this problem, called Intra-class Structure Aware Networks (ISANs). Figure 2 demonstrates an overview of our proposed ISANs. We introduce more neural units for every class to accommodate different intra-class variants in a natural way. We assume that there are V intra-class variants under each class and formulate the corresponding parameters of class j as:

$$\theta_j = [\theta_{j1}, ..., \theta_{jV}]. \tag{2}$$

We call the extended parameter space Intra-class Variant Embedded Space (IVES). For simplicity of notation, we will adopt the same number of intra-class variants under the different classes for the rest of this paper, but the model with different numbers of intra-class variants under each class is a straightforward extension.

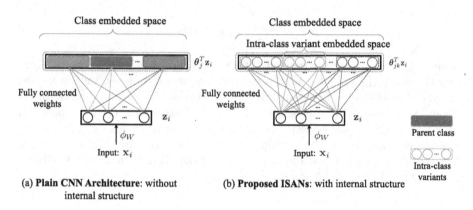

(a) **Plain CNN Architecture**: without internal structure

(b) **Proposed ISANs**: with internal structure

Fig. 2. A comparison of (a) plain CNN architecture and (b) proposed ISANs. Proposed ISANs extend a extra dimension for each class to explicitly encode intra-class structure, which can guide CNN to learn more intra-class diversities.

The probability of sample x_i belonging to variant k of class j can be formulated as follow:

$$p_{ijk} = \frac{\exp(\theta_{jk}^T z_i)}{\sum\limits_{j'=1}^{M} \sum\limits_{k'=1}^{V} \exp(\theta_{j'k'}^T z_i)}. \tag{3}$$

We can naturally obtain the probability of sample x_i belonging to class j as:

$$p_{ij} = \sum_{k=1}^{V} p_{ijk}. \tag{4}$$

Because our task only offers supervision for the parent class of related intra-class variants, we need to introduce a mechanism to supervise the learning process of different variants. We adopt a method that supervises the p_{ij} with parent class labels and indirectly supervises the p_{ijk} at the same time. The detail of this method will be introduced in Sect. 2.2.

2.2 The Regularization for Cluster Assignments of Intra-class Variant

As mentioned in Sect. 2.1, there is no direct supervision for the learning of intra-class structure. This may result in a degenerate solution, which assigns most of the samples to a few clusters (corresponding to intra-class variant) or assigns a cluster to some outlier samples. To address this problem, we introduce a regularization term to supervise the learning of the probability p_{ijk}. For class j, there are $N_j = \sum_{i=1}^{N} \mathbb{1}\{y_i = j\}$ samples, where $\mathbb{1}\{y_i = j\}$ is an indictor function. We can obtain predicted cluster assignments as:

$$p_{jk} = \frac{1}{N_j} \sum_{i=1}^{N} \mathbb{1}\{y_i = j\} p_{ijk}, \tag{5}$$

where p_{jk} denotes the soft frequency that intra-class k is assigned to category j on the dataset. Combined with predicted cluster assignments, we can impose our prior knowledge of data distribution to supervise the learning process of the model. More specifically, we exploit a uniform distribution $\mathbf{Q}$ to support our prior knowledge of the dataset: balancing the cluster assignments of intra-class variants is beneficial to the learning of intra-class structure. We measure the distance between predicted distribution and prior distribution with Kullback-Leibler (KL) divergence as follow:

$$\mathcal{L}_{prior} = \sum_{j=1}^{M} KL(\mathbf{P}_j \| \mathbf{Q}_j) \tag{6}$$

$$= \sum_{j=1}^{M} \sum_{k=1}^{V} p_{jk} \log \frac{p_{jk}}{q_{jk}},$$

where $\mathbf{P}_j$ denotes the distribution of variants under class j, $\mathbf{Q}_j$ denotes the prior distribution of variants under class j ($\mathbf{Q}_j \in \mathbf{Q}$, $q_{jk} \in \mathbf{Q}_j$). This regularization item forces the networks to learn the main variation of intra-class variants from the corresponding category.

2.3 Optimization

Our objective function contains two parts: (1) classification loss to learn feature extractor and discriminate the inter-class features; (2) the regularization term of the balance among cluster assignments. For classification task, we select cross-entropy between target distribution and predicted distribution as loss:

$$\mathcal{L}_{cls} = -\frac{1}{N} \sum_{i=1}^{N} \sum_{j=1}^{M} \mathbb{1}\{y_i = j\} \log p_{ij}. \tag{7}$$

For balance regularization, we simplify Eq. (6) by setting the value of q_{jk} to 1 where the distribution $\mathbf{Q}_j$ is not normalized. The formula can be presented as follow:

$$\mathcal{L}_{prior} = \sum_{j=1}^{M} \sum_{k=1}^{V} p_{jk} \log p_{jk}. \tag{8}$$

We balance the importance of two terms with Lagrange multiplier λ as follow:

$$\mathcal{L}_{total} = \mathcal{L}_{cls} + \lambda \mathcal{L}_{prior}. \tag{9}$$

The gradients of $\mathcal{L}_{total}$ with respect to p_{ijk} can be computed as:

$$\frac{\partial \mathcal{L}_{total}}{\partial p_{ijk}} = \mathbb{1}\{y_i = j\} \left[\frac{1}{N} \left(\sum_{k=1}^{V} p_{ijk} \right)^{-1} + \frac{\lambda}{N_j} \log \left(\frac{1}{N_j} \sum_{i'=1}^{N} \mathbb{1}\{y_{i'} = j\} p_{i'jk} \right) + \frac{\lambda}{N_j} \right]. \tag{10}$$

With consideration of the scalability on the large dataset, we don't directly apply the regularization term of balance to the entire dataset. Instead, we evaluate the total loss and optimize the network with mini-batch gradient descent strategy. The gradients of $\mathcal{L}_{total}$ with respect to p_{jk} in a mini-batch can be formulated as:

$$\frac{\partial \mathcal{L}_{total}}{\partial p_{jk}} = \sum_{i=1}^{B} \frac{\partial \mathcal{L}_{total}}{\partial p_{ijk}}, \tag{11}$$

where B denotes the size of a mini-batch, and the N of $\partial \mathcal{L}_{total} / \partial p_{ijk}$ in Eq. (10) is modified to B for the compatibility with a mini-batch.

Hence, our model is differentiable, which means it can be embedded into modern neural network framework seamlessly for end-to-end training.

3 Experiments

In this section, we evaluate the performance of the proposed method: ISANs on screen defect dataset and MNIST [8]. Firstly, we introduce basic experimental setup, including datasets, implementation and evaluation metrics. Then, we report the experimental results with the above evaluation metrics. Finally, we analyse the results of our experiments.

3.1 Experimental Setup

Datasets. Screen Defect Dataset is collected from automatic screen manufacturing system. As shown in Fig. 1, there are many sub-category defects on defect images in practice. These defects can be roughly classified into dot-like defects, blob-like defects, line-like defects and so on. These defect types have a very imbalanced distribution. Since some defect types are difficult to collect, it's time-consuming to collect a balanced dataset with fine-grained labels. Hence, we collect this dataset by splitting these images into two categories: with defects and without defects. These defect types can be regarded as intra-class variants under the defect category. Overall, this dataset contains $28,485$ 48×48 gray-scale images, which can be randomly split into a training set ($13,625$ images without defects, $12,860$ images with defects) and a testing set ($1,000$ images without defects, $1,000$ images with defects).

MNIST is composed of $70,000$ 28×28 gray-scale handwritten digits images from 10 classes. It can be split into a training set with 60,000 images and a test set with 10,000 images. In order to obtain a dataset with balanced intra-class variant data, we rotate the images of MNIST by 0, 90, 180 and $270\,°$, respectively. So we can get a novel MNIST dataset, in which each class has 4 intra-class variants with different orientations. We call this novel MNIST rotMNIST.

Implementation. We choose LeNet-5 as backbone network for MNIST, rotM-NIST and screen defect dataset. For the stability of optimization, we select large mini-batch size: 256. A large mini-batch can cover the diversity of dataset better. We evaluate our proposed method with two experimental setups: (1) evaluation on the dataset with imbalanced intra-class variants, such as MNIST and screen defect dataset; (2) evaluation on the dataset with balanced intra-class variants, such as rotMNIST. For setup (1), we set a small λ to balance the assignment of cluster but not punish too much. For setup (2), we set a large λ to make data assignment more balanced and the cluster more meaningful.

Evaluation Metrics. We evaluate the performance of our proposed model with different metrics according to different tasks. For classification, we evaluate our method with accuracy as follow:

$$Acc = \frac{1}{N} \sum_{i=1}^{N} \mathbb{1}\{\arg\max_{j}(p_{ij}) = l_i\}, \tag{12}$$

where l_i is the ground truth label.

For clustering, we evaluate our method with unsupervised clustering accuracy as follow:

$$Acc_j^{cluster} = \max_{h} \frac{1}{N_j} \sum_{i=1}^{N} \mathbb{1}\{h(\arg\max_{k}(p_{ijk})) = l_i\}, \tag{13}$$

where $h(\cdot)$ covers all possible one-to-one mapping between clusters and labels. This metric finds the best matching between cluster assignments and labels. The optimal map $h(\cdot)$ can be computed using Kuhn-Munkres algorithm [9].

Another measure for clustering is normalized mutual information (NMI) [13]. It's defined as:

$$\text{NMI}(L, C) = \frac{I(L, C)}{\sqrt{H(L) \cdot H(C)}}, \tag{14}$$

where L denotes the set of ground truth labels, C denotes the set of predicted clusters, $I(\cdot)$ denotes the mutual information between L and C, and $H(\cdot)$ denotes their entropy.

3.2 Experimental Results

Image Recognition. For a fair comparison, we set up the same baseline network for ISANs with different hyper-parameters and counterpart methods. Table 1 demonstrates the results of ISANs compared with other state-of-the-art methods. Our proposed method achieves superior or competitive performance on MNIST and rotMNIST. ISAN-1 with $\lambda = 1.0$ can balance classification loss and cluster assignment loss better. Hence, ISAN-1 obtains better accuracy than other methods on MNIST. ISAN-2 with $\lambda = 2.0$ achieves the best performance on rotMNIST. We believe that it's because that rotMNIST has well-balanced intra-class rotated variants. A larger λ can guide the model to learn a better representation for intra-class variants.

We also evaluate our proposed method on screen defect dataset. The results of classification are shown in Table 2. Our proposed method with setup of ISAN-1 outperforms the baseline CNN with a significant margin. It demonstrates that ISANs can improve the classification performance by accommodating intra-class variants.

Intra-class Variants Clustering. Without the supervision of ground truth labels, we regard the learning of intra-class structure as a clustering problem. Hence, we evaluate the quality of intra-class structure with $Acc_j^{cluster}$ and NMI. Since there is no general definitive standard for deciding whether two images belong to one variant, we choose rotMNIST that has clear and balanced rotated variants as dataset to evaluate our method. Table 3 presents the performance of ISANs on digits from rotMNIST. On the one hand, the performances of digits: "0", "1" and "8" are obviously lower than other digits. We conclude that symmetry property of these digits can confuse the recognition of these intra-class

Table 1. Classification performances on the MNIST and rotMNIST. ISAN-1 denotes ISAN with hyper-parameter: $\lambda = 1.0$ and $V = 4$. ISAN-2 denotes ISAN with hyper-parameter: $\lambda = 2.0$ and $V = 4$. ISAN-3 denotes ISAN with hyper-parameter: $\lambda = 1.0$ and $V = 6$.

Method	Accuracy (MNIST)	Accuracy (rotMNIST)
STN [5]	0.9934	0.9892
TIPooling [7]	0.9903	0.9884
Baseline	0.9894	0.9878
ISAN-1	**0.9937**	0.9899
ISAN-2	0.9928	**0.9902**
ISAN-3	0.9932	0.9891

Table 2. Classification performances on the screen defect dataset. ISAN-1 denotes ISAN with hyper-parameter: $\lambda = 0.25$ and $V = 4$. ISAN-2 denotes ISAN with hyper-parameter: $\lambda = 0.5$ and $V = 4$. ISAN-3 denotes ISAN with hyper-parameter: $\lambda = 0.25$ and $V = 6$.

Method	Baseline	ISAN-1	ISAN-2	ISAN-3
Accuracy	0.904	0.948	0.886	0.935

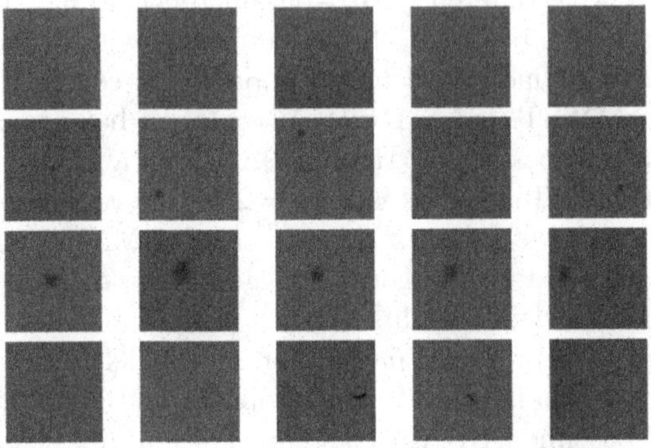

Fig. 3. Some quality results of defect cluster. Each row represents the defect from one intra-class variant.

variants. Hence, the performances of these digits can't reflect the intra-class structure reasonably. On the other hand, ISANs can discriminate intra-class variants of other digits efficiently. It demonstrates that our proposed method has learned a good intra-class structure without the supervision of ground truth labels.

For the dataset with unbalanced intra-class variants, we display some quality results from test set. As shown in Fig. 3, our proposed method can effectively discriminate the defects with different shapes. It illustrates that ISANs can also learn intra-class semantic structure from the dataset with unbalanced intra-class variants.

Table 3. Clustering performances on intra-class variants of the rotMNIST.

rotMNIST digits	0	1	2	3	4	5	6	7	8	9
Clustering Acc	0.699	0.711	0.941	0.922	0.921	0.915	0.935	0.915	0.748	0.927
NMI	0.671	0.702	0.905	0.892	0.894	0.884	0.905	0.885	0.722	0.893

4   Conclusion

In this paper, we propose a novel image classification model: ISANs for screen defect detection, which can effectively encode intra-class semantic structure. On the one hand, we introduce more neural units to accommodate different intra-class variants. On the other hand, we exploit uniform distribution regularization term for the learning of intra-class structure, which can effectively balance the cluster assignments of intra-class variants. Our proposed model can explicitly encode intra-class structure and alleviate the difficulty to learn intra-class variants. Experimental results demonstrate that ISAN is a more reasonable classifier for image recognition, since it achieves superior or competitive results compared to other state-of-the-art methods.

Acknowledgments. This work is supported by National Natural Science Foundation of China (61572428,U1509206), Fundamental Research Funds for the Central Universities (2017FZA5014), National Key Research and Development Program (2016YFB1200203) and Key Research and Development Program of Zhejiang Province (2018C01004).

References

1. Bellman, R.: Adaptive Control Process: A Guided Tour. Princeton University Press, New Jersey (1961)
2. He, K., Zhang, X., Ren, S., Sun, J.: Deep residual learning for image recognition. In: CVPR, pp. 770–778. IEEE (2016)
3. Hornik, K.: Approximation capabilities of multilayer feedforward networks. Neural Netw. **4**(2), 251–257 (1991)
4. Huang, G., Liu, Z., Weinberger, K.Q., van der Maaten, L.: Densely connected convolutional networks. In: CVPR. IEEE (2017)
5. Jaderberg, M., Simonyan, K., Zisserman, A., et al.: Spatial transformer networks. In: NIPS, pp. 2017–2025. NIPS (2015)
6. Krizhevsky, A., Sutskever, I., Hinton, G.E.: Imagenet classification with deep convolutional neural networks. In: NIPS, pp. 1097–1105. NIPS (2012)
7. Laptev, D., Savinov, N., Buhmann, J.M., Pollefeys, M.: Ti-POOLING: transformation-invariant pooling for feature learning in convolutional neural networks. In: CVPR, pp. 289–297. IEEE (2016)
8. LeCun, Y., et al.: Handwritten digit recognition with a back-propagation network. In: NIPS, pp. 396–404. NIPS (1990)
9. Munkres, J.: Algorithms for the assignment and transportation problems. J. Soc. Ind. Appl. Math. **5**(1), 32–38 (1957)
10. Qiao, S., Liu, C., Shen, W., Yuille, A.: Few-shot image recognition by predicting parameters from activations. arXiv preprint arXiv:1706.03466 (2017)
11. Sabour, S., Frosst, N., Hinton, G.E.: Dynamic routing between capsules. In: NIPS, pp. 3859–3869. NIPS (2017)
12. Simonyan, K., Zisserman, A.: Very deep convolutional networks for large-scale image recognition. arXiv preprint arXiv:1409.1556 (2014)
13. Strehl, A., Ghosh, J.: Cluster ensembles: a knowledge reuse framework for combining multiple partitions. J. Mach. Learn. Res. **3**, 583–617 (2002)
14. Szegedy, C., et al.: Going deeper with convolutions. In: CVPR, pp. 1–9. IEEE (2015)
15. Zhang, C., Bengio, S., Hardt, M., Recht, B., Vinyals, O.: Understanding deep learning requires rethinking generalization. arXiv preprint arXiv:1611.03530 (2016)

Mobile Malware Detection - An Analysis of the Impact of Feature Categories

Mahbub E. Khoda[1]([✉]), Joarder Kamruzzaman[1], Iqbal Gondal[1], and Tasadduq Imam[2]

[1] Internet Commerce Security Laboratory, Federation University Australia, Ballarat, Australia
{m.khoda,joarder.kamruzzaman,iqbal.gondal}@federation.edu.au
[2] CQUniversity Australia, Rockhampton, Australia
t.imam@cqu.edu.au

Abstract. The use of smartphones and hand-held devices continues to increase with rapid development in underlying technology and widespread deployment of numerous applications including social network, email and financial transactions. Inevitably, malware attacks are shifting towards these devices. To detect mobile malware, features representing the characteristics of applications play a crucial role. In this work, we systematically studied the impact of all categories of features (i.e., *permission, application programmers interface calls, inter component communication* and *dynamic* features) of android applications in classifying a malware from benign applications. We identified the best combination of feature categories that yield better performance in terms of widely used metrics than blindly using all feature categories. We proposed a new technique to include contextual information in API calls into feature values and the study reveals that embedding such information enhances malware detection capability by a good margin. Information gain analysis shows that a significant number of features in ICC category is not relevant to malware prediction and hence, least effective. This study will be useful in designing better mobile malware detection system.

Keywords: Mobile malware · Feature categories · Classifiers · Context

1 Introduction

Smartphones, tablets and other hand held mobile devices are increasingly used in mobile commerce, on-line transactions and sharing of personal data due to the ease of use, continuous connectivity and convenience of these devices. A recent study predicts that the number of mobile phone users will exceed five billion by 2019 [1]. Moreover, a study by Google in 2016 reveals that there are two billion android devices active monthly and this has allowed android system to capture more than 80% of shares in the mobile operating system market [2,3].

© Springer Nature Switzerland AG 2018
L. Cheng et al. (Eds.): ICONIP 2018, LNCS 11304, pp. 486–498, 2018.
https://doi.org/10.1007/978-3-030-04212-7_43

This widespread use has made mobile devices an inevitable target of cyber-crimes through dissemination of malware [4]. Malware developers target smart devices to steal information and gain privileged access to the device. The stolen information could be used for blackmailing, utilizing the compromised device to call premium numbers and steal money, and for stealing banking information to use in illegal transactions. Hence, detecting malware in mobile platform is becoming increasingly important.

Mobile malware detection is primarily based on extracting features from the applications to train classifiers. In android system, malware detection features of an application can be divided into two broad categories: static and dynamic. Static features are the ones that can be extracted from the application code without installing the application in a real device. Static features have three sub-categories: permission, API calls and Inter Component Communication (ICC) features. On the other hand, dynamic features are the ones that require the application to be installed and run in a real device so that the dynamic behavior of the application can be monitored.

Different works have evolved over time based on the above mentioned feature categories in Android malware detection. In [5] Aiman et al. studied two categories of applications: business and tools. The work extracted the permission of the applications and applied k-means clustering algorithm to classify malware. The approach achieved 71% recall rate on a dataset provided by Frank et al. [6]. In [7] Suleiman et al. used Bayesian classifier on permission features and achieved an overall accuracy of 93% on Malware Genome dataset [8]. These approaches only considered the permission features which were effective mainly for the early generation of malware that mostly tried to gain over-privilege by requesting more permissions than necessary.

Drebin [9] is a static analysis method that extracts static features from an application categorized by one of the eight sets: hardware features, requested permissions, app components, filtered intents, restricted API calls, used permissions, suspicious API calls, and network addresses. The work achieved an overall accuracy of 94% on their dataset which is publicly available. However, since Drebin did not consider dynamic behavior it is vulnerable to code obfuscation and dynamic code loading attacks. Contextual information of security sensitive API calls is a relatively new concept in android malware detection. Since android system relies on call back methods greatly, it is important to distinguish between the activation events of a security sensitive behaviour. For example, a method could be called upon user button click or it could be called based on some system generated event such as, change in signal strength. The user is aware of the first call but not of the second. This is the contextual information associated with an API call. The idea of contextual information and it's effect is discussed in detail in Sect. 3. AppContext [10] is a method to identify the entry point of a security sensitive method call from the application call graph and to extract environmental information to construct a feature vector and make prediction. Narayanan et al. [11] introduced a graph similarity based method based on the graph kernel computed from the contextual dependency graph. The work was further improved in [12] where multiple views of the graph kernel was considered. One major difference of their works from others is that they considered the structural information of the application. However, capturing all the structural information is an overwhelmingly difficult task due to call-back methods and execution jumps of android applications.

ICCDetector [13] designed a parser to extract ICC related features falling into one of the four categories: component, explicit intent, implicit intent and intent filters. Feizollah et al. in [14] considered intents (explicit and implicit) as ICC features along with required permission. Implementing Bayesian Network algorithm for detection, the work achieved accuracy as high as 91% on Drebin dataset. The methods considering ICC features are effective in revealing malicious intentions, information leaks and collusion attacks. However, they also add redundancy in the feature space since all the communicating components are not related to security threats.

Static analysis fails to detect code modification and intrusion of virus through code obfuscation and dynamic code loading. To address these some researchers adopted dynamic approaches for malware detection. For example, Afonso et al. [15] proposed a dynamic malware detection system which logs the frequency of android system calls to detect malicious behavior. The technique fails to detect malware that do not meet a certain API level requirement. Maline was a system developed by Marko et al. [16] that recorded system call patterns of android app achieving an overall accuracy of 96% on Drebin dataset.

Hybrid approaches on the other hand utilize both static and dynamic features. For example, Tong and Yan [17] dynamically collected the data and statically analyzed them. The work classified new samples by matching with precomputed benign and malicious patterns based on the frequency and weight of sequential system calls. The approach largely depends on the number of applications used to build the database. Droid-Sec [18] considered static and dynamic features for classification using a deep belief network however, it considered only a limited number of features.

Based on the studies of current literature we have found that no work has studied the impact of individual feature categories. Since these feature categories represent different characteristics of android applications it is necessary to investigate the behaviors of these feature categories regarding malware detection. Moreover, an attempt to find the best combination of feature categories is also necessary because a blind combination of all the feature categories may not be the best course of action for android malware detection. In this paper we aim to systematically investigate these issues. In this regard our work makes the following contributions:

– The impact of each feature category in detecting android malware is presented.
– The best combination of feature categories that yields the highest detection performance across the most widely used classifiers are determined.
– Defined a method to encode the contextual information into a numeric value so that it can be embedded in the feature space of a classifier. Our experimental

results show that this embedding leads to a better performance than the recent state-of-the-art techniques resulting in an overall accuracy of 99.1%.

2   Characteristics of Feature Categories

Permission (Static Feature): In android operating system every application has to declare the necessary permission before the application can be installed. For instance, if an application sends SMS it needs to declare that it requires SEND_SMS permission in its manifest file.

Application Programmer Interface (API) Calls (Static Feature): API calls are functions of android operating system that are available to android application developers to perform different operations. For example, to get the location of a device the application will call "TelephonyManager.getDeviceID()".

Inter Component Communication (ICC) (Static Feature): Android OS allows the components of an app to communicate with each other through ICC mechanism. This was made available to programmers to reduce the extra burden of app development. But malware developers can exploit this facility to construct collusion attack through communicating components [13]. Hence, the components that can communicate with each other are taken as features for android malware detection. In this work we extracted these components as ICC features including: activities, content providers, broadcast receivers and intents.

System Calls (Dynamic Feature): Android system is built on top of Linux kernel. As a result, almost every operation is done through some Linux system call in this system. The system calls made by an application can be traced during runtime that can reveal dynamic malicious activities of an application.

3   Encoding Contextual Information as Feature Value

In this section, we discuss the contextual information that can reveal malicious use of API calls and we also propose a scheme to encode this information into API category of features before modeling classifiers.

The importance of contextual information of an API call can be illustrated through the working principle of a malware named MoonSMS. The call graph of the application is shown in Fig. 1. This malware is a repackaged application that apparently sends greeting SMSs to get the SEND_SMS permission. As shown in the figure SmsManager.sendTextMessage() is a security sensitive method that can be invoked under three conditions. Firstly, the method is invoked if the user clicks the "send" button from an activity named "SendTextActivity". Secondly, when the signal strength of the device is changed the application sends a premium SMS through "ActionReceiver.OnReceiver()" method. Thirdly, whenever the application is launched its "SplashActivity.OnCreate()" lifecycle method automatically sends another premium SMS based on some database query. A detailed description of how this application works could be found in [10]. It is important to note that only the first case here is the expected behavior of the application where the user wants the SMS to be sent. The other two invocations happen without the user's consent and malicious in nature.

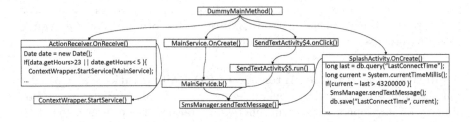

Fig. 1. Call graph of Moon SMS malware [10]

To extract the contextual information we first built the application call graph using flowdroid [19]. Then we tracked back to the entry point of the method that calls a security sensitive API. An entry point of a method could be one of UI callback method, System event callback method or Hardware event callback method.

If a method call is initiated through the UI event (user initiated), the call is less likely to be malicious. Other two type of initialization increases the possibility of the call to be malicious. Simply representing the security sensitive API calls by a binary value as in [9,20] can lead to classifying both benign and malicious calls to the same class. Our initial experiment confirms that this way of representation produces more false negatives, i.e., misses malware. To encode this contextual information into a numeric value we first calculated a maliciousness indicator score M_m using weighted average formula where we assign more weights to the user unaware calls and lesser weights to user aware calls. The reason for assigning weights in such a manner is that a malware is most likely to exhibit malicious behavior by making user-unaware API calls. The score is calculated as

$$M_m = \frac{W_{ui}v_{ui} + W_{se}v_{se} + W_{he}v_{he}}{W_{ui} + W_{se} + W_{he}}. \tag{1}$$

Here, W_{ui}, W_{se} and W_{he} are the weights associated with UI callback, system event callback and hardware event callback respectively. v_{ui}, v_{se} and v_{he} are binary values representing whether the method m can be reached using the respective activation event (1) or not (0). Then we feed this value to the sigmoid function to obtain the feature value V_m:

$$V_m = sig(kM_m). \tag{2}$$

Taking $x = M_m$, k is the steepness of the sigmoid function, defined as

$$sig(kx) = \frac{1}{1 + \exp(-kx)} \tag{3}$$

To determine the weights and k value we conducted empirical experiments where we varied the weights from 0.5 to 5 with the interval 0.5 and varied the value of k from 0.5 to 2 with the interval 0.5. Through experimentation we determined 0.5 as the weight for UI event activation (W_{ui}), 4 as the weight for hardware and system event activation (W_{se} and W_{he}) and 1 as the value of k. Choice of such values results in feature values that produce most clearly distinguishable values of user aware and unaware calls.

4   Experiments and Results

Deep neural network architectures have attracted a lot of focus of the malware researcher in recent times due to its improved performance, as compared to traditional machine learning tools, in various domains such as, pattern recognition, image classification and voice matching. Deep Belief Network (DBN) is the pioneering work that led to the ever gaining popularity of deep learning and this is most common method adopted by malware researchers as discussed in Sect. 1. Hence, in our first experiment we used DBN as our base classifier to study the impact of each feature category. We used a two hidden layer DBN classifier with 600 and 500 nodes respectively and used sigmoid as activation function. For the next experiment we embedded the contextual information of API features. We determined the best combination of feature category with and without contextual information embedding. We took 20 trials of each experiments and presented the average value as results.

4.1   Dataset

For malware samples we used Malgenome [8] and Drebin datasets [9]. These datasets are freely available upon request. Benign applications are downloaded from google play store and we wrote a python crawler to download app from a mirror[1]. We used 5500 malware and 6000 benign applications in our experiment that were developed between 2010 and 2014. From this dataset we extracted approximately 600, 1000, 30000 and 80 "Permission", "API", "ICC" and "Dynamic" features respectively. It is noteworthy that there are numerous APIs called from the apps; however, when these are filtered through the security sensitive API list provided by Pscout [21] we ended up with approximately 1000 APIs. Also, Android system is built on top of Linux kernel hence, it uses the system calls provided by Linux. Depending on the kernel version, there can be about 250 system calls available. When we recorded the system calls during our dynamic analysis, we found 80 distinct system calls used by the apps.

---

[1] https://archive.org/details/android_apps.

4.2 Performance Measure

The following performance metrics were used in our evaluation:

$$Accuracy = \frac{TP+TN}{TP+FP+TN+FN} \times 100\%$$
$$Precision = \frac{TP}{TP+FP} \times 100\%$$

$$Recall = \frac{TP}{TP+FN} \times 100\%$$

$$Specificity = \frac{TN}{TN+FP} \times 100\%$$

$$F1 - score = 2.\frac{Precision \times Recall}{Precision + Recall}$$

$$G - mean = \sqrt{Specificity \times Sensitivity}$$

Where TP is the number of true positives, i.e., malware accurately predicted; FP is the number of false positives, i.e., benign apps that were predicted as malware; TN is the number of true negatives, i.e., benign apps accurately predicted and FN is the number of false negatives, i.e., malware that were predicted as benign apps.

4.3 Impact of Feature Categories and Finding Best Combination of Feature Categories

The four different categories of features can be arranged in 15 different combinations. We used these 15 different combinations and performed classification using our DBN classifier. We used 10-fold cross validation in our experiments. While dividing the data into 10-fold we separated the benign and malware applications, and divided each group of applications into 10 parts. We then joined the two groups to obtain the final set divided into 10-folds. This was done to ensure that the benign and malware ratio is consistent among all the folds.

Table 1 shows the results of our experiments. It can be seen that as an individual feature category both "Permission" and "API" have reasonable strength. On the other hand, "ICC" performs the least in all performance metrics. This indicates that the "ICC" features do not have a significant effect on the classifier's accuracy. Our explanation for such phenomena is that, since the names of the communicating components are taken as "ICC" features and functionally similar attacks can be designed using different component name, ICC features will fail to capture this because the extracted features will be different. We also see that as an individual category, "Dynamic" features performs nearly equal to "Permission" and slightly better than "API", attaining 89.8% accuracy while capturing a completely different type of characteristics of malware (i.e., dynamic runtime behavior). The combination of "Permission", "API" and "Dynamic" yields the best detection performance as altogether it captures diverse and strong characteristics of malware, specially the combination reduces FN and FP considerably. Further, adding "ICC" to this combination adds redundancy to the feature space due to the reason mentioned above. As a result, the classifier performance degrades for all feature category combination. Table 2 shows the result of the experiments that incorporate contextual information with "API" calls. It shows that with contextual information, "API" call outperforms Permission as a singular feature. This is because most of the operations (benign or malicious) are done through some "API" call. Same set of permissions will be required for a malicious or a benign app to perform the same operation. But the maliciousness can be distinguished based on the context of the operation. This table also shows that the combination of "Permission", "API" and Dynamic features results in the best accuracy (99.1%), and further improve the sensitivity and specificity. However, since no. of FP decreases for all feature category combination, to reach a decisive conclusion we use another measure, g-mean, that indicates a balance between false positive and false negative. We found the best g-mean value of 0.991 for the same combination of feature categories that provided best accuracy further confirming our observation. Our achieved performance here is better than other reported performance such as, 94% accuracy by Arp et al. [9], 96% accuracy by Su et al. [20] and 96.5% accuracy by Wang et al. [10].

Table 1. Malware detection performance using different categories of feature and their combination with deep learning not considering contextual information

Metrics	Feature categories						
	Permission	API	ICC	Dynamic	Permission + API + Dynamic	API + ICC + Dynamic	ALL
TN	5651	5834	5436	5843	5862	5871	5963
FN	593	1183	1621	1023	196	411	313
TP	4967	4377	3939	4537	5364	5174	5247
FP	349	166	564	157	138	104	37
Accuracy (%)	91.9	88.4	81	89.8	**97.1**	95.5	97
Recall (%)	89.4	78.8	70	81.7	**96.6**	92.7	94.4
Specificity (%)	94.2	97.2	90.7	97.3	97.7	98.3	**99.3**
F1-score	0.91	0.87	0.78	0.88	**0.97**	0.95	**0.97**
G-mean	0.917	0.875	0.801	0.891	**0.971**	0.954	0.968

The Wilcoxon Signed-Rank test comparing our scheme with best feature category combination without context (Table 1) and with context (Table 2) on all performance metrics at different runs yielded p-value, $p <= 2.5 \times 10^{-3}$, validating performance improvement with contextual information being statistically significant.

4.4   Verification Using Different Classifiers

We further experimented with four other well known classifiers to record their performance for different combination of feature categories. These experiments were done to make sure that the relative performance with different combination of feature categories for other classifiers is consistent with our finding from previous experiments. The classifiers we used for these experiments are: K-Nearest Neighbors, Support Vector Machine, Random Forest and C4.5. The result of these experiments is shown in Fig. 2.

Table 2. Malware detection performance using different categories of feature and their combination with deep learning considering contextual information. [(c) indicates contextual information of API calls was considered]

Metrics	Feature categories			
	API(c)	Permission + API(c) + Dynamic	API(c) + ICC + Dynamic	ALL(c)
TN	5933	5914	5964	5984
FN	203	18	174	153
TP	5357	5542	5426	5407
FP	67	86	36	16
Accuracy (%)	97.7	**99.1**	98.2	98.6
Recall (%)	96.4	**99**	96.8	97.3
Specificity (%)	98.8	98.5	99	**99.5**
F1-score	0.98	**0.99**	0.98	0.98
G-mean	0.976	**0.991**	0.981	0.985

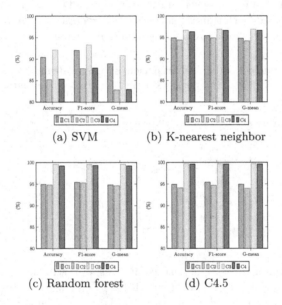

(a) SVM　　　(b) K-nearest neighbor

(c) Random forest　　　(d) C4.5

Fig. 2. Performance comparison of different feature category combinations for other classifiers. Combination - *C1*: Permission, API, Dynamic, *C2*: All, **C3: Permission, API with context, Dynamic** and *C4*: All with context

Figure 3a shows the 1-fold training time of the DBN classifier. It's noteworthy that while the combination of all feature categories and the combination "Permission", "API" and "Dynamic" gives comparable performance in terms of accuracy, the training time for the later is significantly less. Training time will be an issue for an adaptive classification mechanism to tackle concept drift which is an acknowledged problem in mobile malware literature [11]. As new vulnerabilities are exposed, the architecture and behavior of malware change. As a result, features and their values change as well. To devise an adaptive system to deal with new malwares it is required to forget the trend of old malware and retrain the system with new ones. Thus, it is preferable to have training time as less as possible. We also plotted the Receiver Operating Characteristic (ROC) curve from our experiments with DBN in Fig. 3b. As shown in the figure the feature category combination: permission, API with context and dynamic resulted in the highest Area Under ROC Curve (AUC) of 0.998.

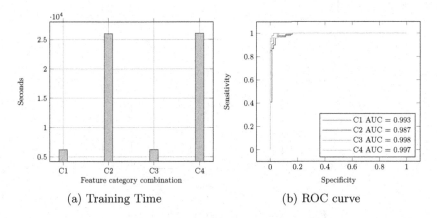

(a) Training Time   (b) ROC curve

Fig. 3. Training time and ROC curve for combination *C1*: Permission, API, Dynamic, *C2*: All, **C3: Permission, API with context, Dynamic** and *C4*: All with context

4.5 Further Analysis of ICC Features

We further extensively studied the "ICC" features to see how much redundancy it adds to the feature space. For this we used a relevancy measure based on entropy and information gain. The entropy of a random variable X is defined as

$$E_n(X) = -\sum_i P(x_i) log_2(P(x_i)) \tag{4}$$

and the entropy of X given another variable Y is

$$E_n(X|Y) = -\sum_j P(y_j) \sum_i P(x_i|y_i) log_2(P(x_i|y_i)) \tag{5}$$

Then the *information gain* of X given Y can be calculated as [22]

$$I_g(X|Y) = E_n(X) - E_n(X|Y). \tag{6}$$

We compute the information gain of each feature given the class label and use it as the feature relevancy indicator i.e., if this value is greater than a threshold ϑ, the feature is considered to be relevant. The value of ϑ is predefined by the user. Figure 4 shows the number of relevant features with respect to ϑ threshold. We see the number of relevant features drastically reduces with slight increase in threshold after $\vartheta = 8 \times 10^{-5}$. We also inspected the features having information gain value less than 8×10^{-5} and found that the number of apps in our dataset that have those features is less than 10 indicating that a large number of ICC features are redundant.

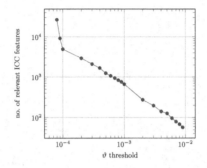

Fig. 4. Number of relevant ICC features

5   Conclusion and Future Work

In this work we systematically studied the impact of all different categories of features in android malware detection. We identified the best combination among the feature categories and showed that blind accumulation of all feature categories is not the best way to address android malware detection problem. We also encoded the contextual information of API calls into numeric values and embedded it in the feature space which enhances malware classifier performance. Using information gain we further examined the relevancy of the ICC features and found that a significant number of those features are not relevant which further bolsters our observation about best feature category combination. As malware continue to develop exploiting new found vulnerabilities of the system (known as concept drift), we plan to extend this work to design an adaptive classifier with incremental training to tackle this.

References

1. Number of Smartphone Users Worldwide. https://www.statista.com/statistics/330695/number-of-smartphone-users-worldwide/. Accessed 16 Nov 2017
2. Number of Android Devices. https://www.macrumors.com/2017/05/17/2-billion-active-android-devices/. Accessed 16 Nov 2017

3. Alzaylaee, M.K., Yerima, S.Y., Sezer, S.: Emulator vs real phone: android malware detection using machine learning. In: Proceedings of the 3rd ACM on International Workshop on Security and Privacy Analytics, pp. 65–72. ACM (2017)
4. Yang, C., Zhang, J., Gu, G.: Understanding the market-level and network-level behaviors of the android malware ecosystem. In: 2017 IEEE 37th International Conference on Distributed Computing Systems (ICDCS), pp. 2452–2457. IEEE (2017)
5. Samra, A.A.A., Yim, K., Ghanem, O.A.: Analysis of clustering technique in android malware detection. In: 2013 Seventh International Conference on Innovative Mobile and Internet Services in Ubiquitous Computing (IMIS), pp. 729–733. IEEE (2013)
6. Frank, M., Dong, B., Felt, A.P., Song, D.: Mining permission request patterns from android and Facebook applications, pp. 870–875, December 2012. https://doi.org/10.1109/ICDM.2012.86
7. Yerima, S.Y., Sezer, S., McWilliams, G.: Analysis of bayesian classification-based approaches for android malware detection. IET Inf. Secur. 8(1), 25–36 (2014)
8. Zhou, Y., Jiang, X.: Dissecting android malware: characterization and evolution. In: 2012 IEEE Symposium on Security and Privacy (SP), pp. 95–109. IEEE (2012)
9. Arp, D., Spreitzenbarth, M., Hubner, M., Gascon, H., Rieck, K., Siemens, C.: DREBIN: effective and explainable detection of android malware in your pocket. In: Ndss, vol. 14, pp. 23–26 (2014)
10. Yang, W., Xiao, X., Andow, B., Li, S., Xie, T., Enck, W.: Appcontext: differentiating malicious and benign mobile app behaviors using context. In: Proceedings of the 37th International Conference on Software Engineering, ICSE 2015, vol. 1, pp. 303–313. IEEE Press, Piscataway (2015). http://dl.acm.org/citation.cfm?id=2818754.2818793
11. Narayanan, A., Chandramohan, M., Chen, L., Liu, Y.: Context-aware, adaptive and scalable android malware detection through online learning (extended version). CoRR abs/1706.00947 (2017). http://arxiv.org/abs/1706.00947
12. Narayanan, A., Chandramohan, M., Chen, L., Liu, Y.: A multi-view context-aware approach to android malware detection and malicious code localization. Empir. Softw. Eng. (2017). https://doi.org/10.1007/s10664-017-9539-8
13. Xu, K., Li, Y., Deng, R.H.: ICCDetector: ICC-based malware detection on android. IEEE Trans. Inf. Forensics Secur. 11(6), 1252–1264 (2016)
14. Feizollah, A., Anuar, N.B., Salleh, R., Suarez-Tangil, G., Furnell, S.: Androdialysis: analysis of android intent effectiveness in malware detection. Comput. Secur. 65, 121–134 (2017)
15. Afonso, V.M., de Amorim, M.F., Grégio, A.R.A., Junquera, G.B., de Geus, P.L.: Identifying android malware using dynamically obtained features. J. Comput. Virol. Hacking Tech. 11(1), 9–17 (2015)
16. Dimjašević, M., Atzeni, S., Ugrina, I., Rakamaric, Z.: Evaluation of android malware detection based on system calls. In: Proceedings of the 2016 ACM on International Workshop on Security And Privacy Analytics, pp. 1–8. ACM (2016)
17. Tong, F., Yan, Z.: A hybrid approach of mobile malware detection in android. J. Parallel Distrib. Comput. 103, 22–31 (2017)
18. Yuan, Z., Lu, Y., Wang, Z., Xue, Y.: Droid-sec: deep learning in android malware detection. In: ACM SIGCOMM Computer Communication Review, vol. 44, pp. 371–372. ACM (2014)
19. Arzt, S., et al.: Flowdroid: Precise context, flow, field, object-sensitive and lifecycle-aware taint analysis for android apps. ACM Sigplan Not. 49(6), 259–269 (2014)

20. Su, X., Zhang, D., Li, W., Zhao, K.: A deep learning approach to android mal-
    ware feature learning and detection. In: Trustcom/BigDataSE/I SPA, pp. 244–251.
    IEEE (2016)
21. Au, K.W.Y., Zhou, Y.F., Huang, Z., Lie, D.: Pscout: analyzing the android per-
    mission specification. In: Proceedings of the 2012 ACM Conference on Computer
    and Communications Security, pp. 217–228. ACM (2012)
22. Quinlan, J.R.: C4.5: Programs for Machine Learning. Morgan Kaufmann Publish-
    ers Inc., San Francisco (1993)

Recurrent RetinaNet: A Video Object Detection Model Based on Focal Loss

Xiaobo Li, Haohua Zhao, and Liqing Zhang[(✉)]

Key Laboratory of Shanghai Education Commission for Intelligent Interaction
and Cognitive Engineering, Department of Computer Science and Engineering,
Shanghai Jiao Tong University, Shanghai 200240, China
{sz3052167,haoh.zhao,lqzhang}@sjtu.edu.cn

Abstract. Object detection in still images has been extensively investigated recent years. But object detection in videos is still a challenging research topic. Directly using those methods for still images in videos would suffer from blurring and low resolution in video images. Some methods utilized the temporal information to boost the detection accuracy, but they are usually expensive in time in estimating optical flow. In this paper, we propose Recurrent RetinaNet, a flexible end-to-end approach for object detection in videos. In this work, a backbone network is leveraged to generate several feature maps, then a feature pyramid network extracts pyramid features from the feature maps. Detection boxes are generated according to the shapes of pyramid features. Two subnets with convolutional layers and Convolutional LSTM layers, are added on the top for box regression and classification. Note that the boxes are generated regardless of the content of images, there may be an extreme foreground-background imbalance. Thus, focal loss, which has been shown effective in object detection on images, is employed as the loss function for the classification subnet. Experiments show that the approach improves the detection accuracy and avoids the detection loss of some objects in some cases compared to RetinaNet and the model complexity is still good enough for real-time applications.

Keywords: Object detection · Focal loss · Convolutional LSTM

1 Introduction

Object detection aims to find the objects appearing in images or videos and their locations. Object detection in still images has been extensively investigated, especially since convolutional neural networks (CNN) [10] were shown powerful in computer vision. A simple way to extend image object detection to videos is to detect objects in each frame. However, in a video, objects may be blur or occluded, and sometimes the light change may degenerate the detection performance to some extent. In these cases, consecutive frames may help improve the performance in videos using the temporal features. The interaction between frames along with the spatial features has been shown effective in several computer vision tasks like action recognition [14,15].

© Springer Nature Switzerland AG 2018
L. Cheng et al. (Eds.): ICONIP 2018, LNCS 11304, pp. 499–508, 2018.
https://doi.org/10.1007/978-3-030-04212-7_44

Kang et al. [9] proposed Tubelet Proposal Networks which detect in two stages: a tubelet proposal network to generate tubelet proposals and a encoder-decoder CNN-LSTM network to extract tubelet features and classifies each proposal boxes into different classes. Hetang et al. [6] proposed Impression Network which needs flow-warp, a warping operation based on the optical flow. Both of these methods need a preprocessing stage, which would influence the detection speed.

In this paper, we propose a one-stage end-to-end video detection model without much preprocessing. The proposal boxes for detections are generated uniformly according to the shape of frames of the input video. The frames are fed into a mature backbone convolutional neural network, like ResNet [5], MobileNet [8]. The results of the last several layers are extracted as feature maps. The feature maps are further put into a feature pyramid network [11] which yields the pyramid features from different levels for classification and regression of the boxes. Two subnets are added for classification and regression respectively. In the two subnets, traditional convolutional layers and Convolutional LSTM [13] layers are used to ultilize both the spatial and temporal information. In the classification subnet, a focal loss [12] is used for training to reduce influence of object-background imbalance. In detection, non maximum suppression is taken.

The rest of this paper is organized as follows. Section 2 reviews two important background methods for this work. Section 3 introduces the details of our model. Experiments are introduced in Sect. 4 and the last section delivers conclusions.

2   Background

This section gives a brief review of two base methods, focal loss and convolutional LSTM network, which are highly related to this work.

Focal Loss. Without much preprocessing, one-stage detection methods usually generate proposal boxes regardless of the values in the three channels of images, so most of these proposals may locate in the background area. If we regard the loss for background and objects equally, the model will tend to classify each box as background. Lin et al. [12] proposed a focal loss for objection detection on images to solve this problem. The α-balanced form of focal loss is defined as

$$\text{FL}(p,y) = \begin{cases} -\alpha(1-p)^\gamma \log p, & y=1 \\ -(1-\alpha)p^\gamma \log(1-p), & y=0 \end{cases} \tag{1}$$

where $y \in \{0,1\}$ specifies the ground-truth class in one-hot and $p \in [0,1]$ is the probability that model estimated for the class with label $y=1$. This loss function reduces the loss compared to cross entropy when $\gamma > 0$, but it makes the loss for well-classified samples relatively smaller than those $p < 0.5$, preventing the model to misclassify boxes with objects as background.

Convolutional LSTM Network. Long short term memory (LSTM) [7] has been widely used for processing temporal information. However, LSTM networks are usually used to processing sequences of low dimensional vector features, and deploying it into images directly would increase the computing cost dramatically. Shi et al. [13] proposed convolutional LSTM network, which takes the techniques like parameter sharing from CNN into LSTM. It is similar to an LSTM layer, but the input transformations and recurrent transformations are both convolutional. Thus, it may extract the temporal information in an efficient way.

3   Recurrent RetinaNet

This section gives the structure and some settings of our training and prediction model.

3.1   Backbone

Many convolution neural networks have been shown powerful on ImageNet dataset [3]. When an image is fed into such a network, the results of its last a few layers would be an effective representation of the input. Thus we extract feature maps with a backbone whose parameters are initialized with the pre-trained model on ImageNet dataset. In our experiments, we use MobileNet [8] with 128 layers and width parameter $\alpha = 0.75$, which is a balance between the accuracy and computation consuming.

3.2   Feature Pyramid Network

We adopt the Feature Pyramid Network [11] to extract further pyramid features. The structure we use is similar as the structure described in RetinaNet [12]. The structure can be seen in Fig. 1, where C3, C4 and C5 are the outputs of 3 layers of the backbone. Note that the number of filters of each layer in FPN is set as 64.

With the FPN, we can get pyramid features in 5 levels, as P3–P7 in Fig. 1. The proposal boxes are generated based on the shapes of these features, as described in the next subsection.

3.3   Anchors

Anchors are the proposed boxes for objects. As mentioned above, the anchors are generated according to the shape of pyramid features. With the settings in former layers, we know that Px is $\frac{1}{2^x}$ of the original input in height and width. Thus for each of the pixels in Px, we generate an anchor whose height and width are both 2^x, and then we shift their centers into the corresponding pixels of the input. Furthermore, as in [11,12], we expand each anchor into 9 with three aspect ratios $\{1 : 2, 1 : 1, 2 : 1\}$ and three anchor resize scales $\{2^0, 2^{\frac{1}{3}}, 2^{\frac{2}{3}}\}$. However, with these expansions, there may be some anchors whose centers are located out of the input, so a filter is deployed to delete these illegal anchors. These operations are operated on all of the pyramid features, P3–P7. As a result, we have many dense anchors, which should be able to cover the objects in videos.

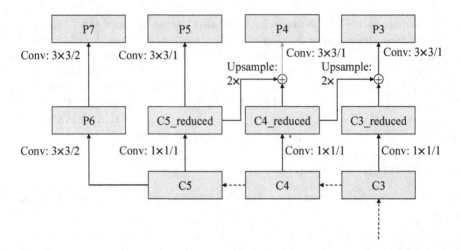

Fig. 1. The structure of FPN. The dashed lines are relationships in the backbone and the solid lines are those in FPN. In the figure, Conv: a × b/c means a 2D convolutional layer with kernel size a × b and stride c, and this annotation remains in the rest of this paper. The boxes marked with P are extracted features and those with C are intermediate results. The number x in the boxes implies that the result is $\frac{1}{2^x}$ of the original input is width and height. All of the units pad the inputs such that the outputs keep the shape of the inputs. Note that with the structure of the backbone, the shapes of the results of upsamplings agree with the layers they added to.

3.4   Regression Subnet

The regression subnet is aimed at regressiong the offset of anchors to nearby ground-truth objects. This subnet is a combination of 2D Convolutional and Convolutional LSTM layers. The structure is shown in Fig. 2.

It consists of 2 Colvolutional LSTM layers and 3 2D Convolutional layers, taking advantage of deep CNNs without making the model too complex. Except the last layer, the number of filters in each layer is 64. The number of filters in the last layer is 9 × 4 for each of the four boundaries of the 9 anchors of a given pixel. The offset of the boundaries is set as the settings in R-CNN [4]. The regression subnet is connected to the regression loss function with L1 regularization in the training model.

3.5   Classification Subnet

The classification subnet is deployed to predict the probability of object presence for each of the anchors and object classes. The structure of classification subnet is the same as regression subnet, as can be seen if Fig. 2.

The number of filters in each layer except the last one is also 64. The last layer, whose number of filters is set as 9× the number of classes, is responsible for predicting the probability for each of the 9 anchors centered at a given pixel in the feature map containing each of the classes, so it is activated by a sigmoid

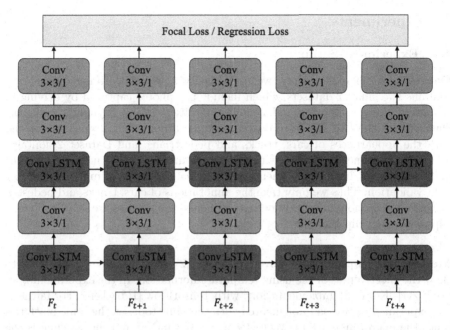

Fig. 2. The structure of two subnets of a 5-frame video clip. F_t is the feature from the former layers at time t.

function. The classification subnet is further connected to the focal loss function in the training model. As described in [12], we take $\gamma = 2.0$ and $\alpha = 0.75$ in the focal loss.

3.6 Detection

When predicting with our model, some additional steps are needed. Firstly, we assume that a frame can only contain no more than 100 objects, so we can just abandon the anchors that is very likely to be located in the background areas. In the experiments, the anchors are sorted desendingly according the maximum predicted probability among all classes, and only the first 100 anchors are left.

In addition, since the anchors were generated in a dense way, there may be several anchors containing the same object. Thus, non maximum suppression is taken to further reduce the number of detected anchors. In our experiments, anchors are selected desendingly according to the predicted probability among classes and the anchors with an IoU greater than 0.5 with at least one of the selected anchors are abandoned.

4 Experiments

4.1 Experiment Setup

Dataset. In our experiments, we use Udacity annotated driving dataset[1]. It contains two smaller datasets, which include the photos captured by driving in Mountain View, California and neighboring cities during daylight conditions. Dataset 1 of the two smaller datasets contains around 9500 frames annotated with the appearances of cars, trucks and pedestrians, and Dataset 2 contains about 15000 frames annotated with the appearances of cars, trucks, pedestrians and traffic lights. In the experiments, we take Dataset 2 for training and Dataset 1 for validation. Thus we only take the annotations of cars, trucks and pedestrians as our detection objects. In our experiments, the framess are split into video clips with each clip 5 frames.

Models and Implementation. RetinaNet [12] is the baseline in this work to show the effectiveness of the using temporal information. In the experiments, we use Fizyr's Keras [2] implementation[2] with TensorFlow [1] backend. For fairness, our experiments are also implemented with the same tools on the same platform. Since the main difference between Recurrent RetinaNet and the baseline is the Convolutional LSTM layers in the two subnets, we further implement two models with only one of the subnets having Convolutional LSTM layers, to find the effectiveness of Convolutional LSTM layers on regression and classification respectively. To show the effect of focal loss, we also evaluated the performance of our model without focal loss. In a word, five models including the baseline, Recurrent RetinaNet, Recurrent RetinaNet with only Convolutional LSTM layers in regression subnet, Recurrent RetinaNet with only Convolution LSTM layers in classification subnet, and Recurrent RetinaNet with cross entropy loss are investageted in our experiments.

4.2 Performance and Analysis

The models are firstly evaluated on the accuracy of detections. In addition, since the Convolutional LSTM layers increase the complexity of model, so it may increase the time cost for detections. However, a fast detection is one of the key advantages of one-stage methods, so it is essential to evaluate the computational performance of our model.

Classification Performance. The classification performance of models is measured by Average Precision (AP) with IoU threshold 0.5, denoted as AP^{50}. After around 100 epochs, all of the five model converged. The AP^{50} of the five models can be found in Table 1.

---

[1] https://github.com/udacity/self-driving-car/tree/master/annotations.
[2] https://github.com/fizyr/keras-retinanet.

Table 1. The AP^{50} of the five models. Baseline is RetinaNet [12], R-only is Recurrent RetinaNet with Convolutional LSTM layers only in the regression subnet, C-only is the model with Convolutional LSTM layers only in the classification subnet, and R&C is Recurrent Retina with Convolutional LSTM layers in both the regression and classification subnets. The last column shows the result of R&C with cross entropy loss instead of focal loss in the classification subnet.

Models	Baseline	R-only	C-only	R&C	R&C(CE loss)
Car	0.6496	0.6616	**0.6959**	0.6769	0.5270
Truck	0.2491	0.2252	0.2284	**0.2706**	0.2362
Pedestrian	0.2828	0.3175	0.3076	**0.3454**	0.2298
mAP	0.3938	0.4014	0.4106	**0.4310**	0.3310

Firstly, the model with cross entropy loss results in the worst result, showing that the focal loss can improve the detection in the object detection. Furthermore, we can find that the three editions of Recurrent RetinaNet outperform the baseline in AP^{50} for cars and pedestrians, but the models R-only and C-only have a slight decrement in AP^{50} for trucks compared to the baseline. As for R-only and C-only models, it proves that the precision improvements of Convolutional LSTM layers in the two submodels are almost the same. But when we compare the first four models together, we can conclude that utilizing Convolutional LSTM layers in both classification subnet and regression subnet improves the pricision compared to either R-only model or C-only model, this may due to the collaboration of the Convolutional LSTM layers in the two subnets.

To show the effectiveness of our models in an intutive way, we give the detection results of the five models for part of three contiguous frames from the validation set in Fig. 3. The baseline model loses the detection of the car in Frame 398, which does not happen in models with Convolutional LSTM layers. This is likely due to the baseline model does not take the temporal information into account.

However, the precision is quite different for different classes. This may partially due to that some vehicles are difficult to be classified as a car or a truck and pedestrians often appear on the roadside with a bad light condition. In addition, this may also due to the extreme imbalance of the three classes. In the training set, the number of appearances of car is 60788, whereas 9866 for pedestrians and 3503 for trucks, resulting in a model that may not be trained enough for trucks and pedestrians.

Computational Performance. We measure the computational performance in detection speed. The time cost for detections of the models can be seen in Table 2. The time is measured on GPU Nvidia GTX 1080 Ti and the time contains the cost of saving the JPEG files for the detection results. Note that the complexity of the model R&C with CE loss is the same as that of R&C, so we do not show the detection time for R&C model with CE loss.

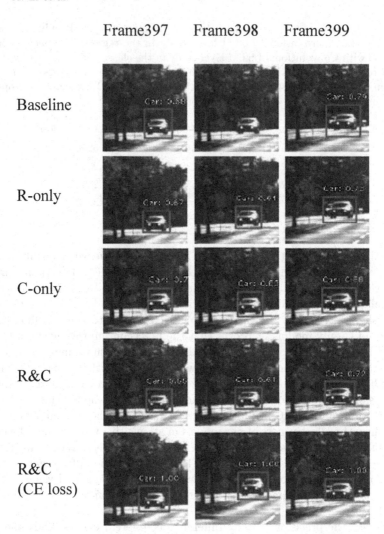

Fig. 3. The detection results of the five models for Frame 397, 398 and 399 from the validation set.

Table 2. The time cost (in second) for detections on validation set.

Models	Baseline	R-only	C-only	R&C
The Whole Set	997	1055	1077	1440
Per Frame	0.106	0.112	0.114	0.153

It can be seen that the Convolutional LSTM layers slow down the detections to some extent. In training, the time needed for detecting each epoch of our model is alomost twice as that of baseline. But, in detection, the time adding is not so high and the prediction time is still good enough for real-time applications.

5   Conclusion

In this paper, we propose Recurrent RetinaNet, a one-stage end-to-end network for video object detection. The network could take advantage of the temporal relationship between frames and avoid object-background imbalance. Three varieties of our model and the model without focal loss are compared with RetinaNet, the baseline, and the experiments show that the focal loss and Convolutional LSTM layers in the model can improve the accuracy of detections, especially when the Convolutional LSTM layers in the two subnets cocollaborate. And the computational performance evaluation shows that our model does not add much computation complexity compared to the baseline and the time for detection is still good enough for real-time applications. In conclusion, Recurrent RetinaNet would be an effective and efficient model for object detection in videos.

However, due to the limited time, there are still many alternatives that can be tried, such as the hyper parameters γ and α in focal loss, the backbone, the number of filters, etc. These problems will be remained as further work.

Acknowledgement. The work was supported by the National Basic Research Program of China (Grant No. 2015CB856004) and the Key Basic Research Program of Shanghai Municipality, China (15JC1400103,16JC1402800).

References

1. Abadi, M., et al.: TensorFlow: Large-Scale Machine Learning on Heterogeneous Systems (2015). http://tensorflow.org/. Software available from tensorflow.org
2. Chollet, F., et al.: Keras (2015). https://github.com/keras-team/keras
3. Deng, J., et al.: ImageNet: a large-scale hierarchical image database. In: CVPR 2009, IEEE (2009)
4. Girshick, R., Donahue, J., Darrell, T., Malik, J.: Rich feature hierarchies for accurate object detection and semantic segmentation. In: CVPR 2014 (2014)
5. He, K., Zhang, X., Ren, S., Sun, J.: Deep Residual Learning for Image Recognition. Cornell University Library (2017)
6. Hetang, C., Qin, H., Liu, S., Yan, J.: Impression network for video object detection. arXiv preprint arXiv:1712.05896 (2017)
7. Hochreiter, S., Schmidhuber, J.: Long short-term memory. Neural Comput. **9**(8), 1735–1780 (1997)
8. Howard, A.G., et al.: MobileNets: efficient convolutional neural networks for mobile vision applications. arXiv preprint arXiv:1704.04861 (2017)
9. Kang, K., et al.: Object detection in videos with tubelet proposal networks. In: Proceedings of CVPR, vol. 2, p. 7 (2017)
10. Krizhevsky, A., Sutskever, I., Hinton, G.E.: ImageNet classification with deep convolutional neural networks. In: Advances in Neural Information Processing Systems, pp. 1097–1105 (2012)
11. Lin, T.Y., Dollár, P., Girshick, R., He, K., Hariharan, B., Belongie, S.: Feature pyramid networks for object detection. In: CVPR, vol. 1, p. 4 (2017)
12. Lin, T.Y., Goyal, P., Girshick, R., He, K., Dollár, P.: Focal loss for dense object detection. arXiv preprint arXiv:1708.02002 (2017)

13. Xingjian, S., Chen, Z., Wang, H., Yeung, D.Y., Wong, W.K., Woo, W.C.: Convolutional LSTM network: a machine learning approach for precipitation nowcasting. In: NIPS 2015 (2015)
14. Xue, W., Zhao, H., Zhang, L.: Encoding multi-resolution two-stream CNNs for action recognition. In: Hirose, A., Ozawa, S., Doya, K., Ikeda, K., Lee, M., Liu, D. (eds.) ICONIP 2016. LNCS, vol. 9949, pp. 564–571. Springer, Cham (2016). https://doi.org/10.1007/978-3-319-46675-0_62
15. Zhang, K., Zhang, L.: Extracting hierarchical spatial and temporal features for human action recognition. Multimedia Tools Appl. **77**(13), 16053–16068 (2018). Springer

Neural Causality Detection
for Multi-dimensional Point Processes

Tianyu Wang$^{(\boxtimes)}$, Christian Walder, and Tom Gedeon

Australia National University College of Engineering and Computer Science,
Canberra, Australia
u6014854@anu.edu.au

Abstract. In the big data era, while correlation detection is relatively straightforward and successfully addressed by many techniques, causality detection does not have a generally-used solution. Causality provides valuable insights into data and guides further studies. With the overall assumption that causal influence can only be from prior history events, time plays an essential part in causality analysis, and this important feature means the data with strict temporal structure needs to be modelled. Traditionally, temporal point processes are employed to model data containing temporal structure information. The heuristic parameterization property of such models makes the task difficult. Domain related knowledge are needed to design proper parameterization. Recently, Recurrent Neural Networks (RNNs) have been used for time-related data modelling. RNN's trainable parameterization considerably reduces the dependency on domain-related knowledge. In this work, we show that combining neural network techniques with Granger causality framework has great potential by presenting an RNN model integrated with a Granger causality framework. The experimental results show that the same network structure can be applied to a variety of datasets and causalities are detected successfully.

Keywords: Granger causality · Recurrent neural network
Temporal point process

1 Introduction

In the big data era, extracting and understanding relations between data has become increasingly important. The goal is to analyse how influence flows between events from different sources or dimensions, i.e., whether one event's behaviour is affected by an earlier event and if so estimating the strength of that influence. Researchers have shown that many fields can benefit from causality detection.

Granger causality has been shown to be a reliable causality analysis method in fields including econometric [6] and neuroscience [8]. However, the number of the application scenarios of traditional Granger causality is significantly limited. The Granger causality utilizes the dependency between historical events and

© Springer Nature Switzerland AG 2018
L. Cheng et al. (Eds.): ICONIP 2018, LNCS 11304, pp. 509–521, 2018.
https://doi.org/10.1007/978-3-030-04212-7_45

present events to detect causal relations. For most sequence data, the dependency is complicated, but in most cases, a simple linear dependency is assumed, i.e. the value of the current event is a weighted sum of the past events. This oversimplified linear parameterization cannot capture complex dependencies, thereby limiting the performance of Granger causality applications and restricts application scenarios to regression problems on regular time grids. Efforts are made to overcome these limitation by employing radial basis function [1,10], kernel method [11] or locally linear neighborhoods [3,5]

Regarding sequence data modelling, point process models allow the training of sophisticated parameterized models with a maximum likelihood approach [2, 12]. Traditional point process models also have limitations. One limitation is that the sophisticated parameterization needs to be heuristically designed to suit the dependencies of the real world data. Thus, domain related knowledge plays an essential role in the model design. Heuristic parameterization also makes designed model not reusable through different domains. Recently, a recurrent neural network based model called Recurrent Marked Temporal Point Process (RMTPP) [4] was proposed targeting the heuristic parameterization problem for marked point process models, and achieved state-of-the-art results. This model solved the heuristic parameterization problem by approximating the dependency using a recurrent neural network. Experiments showed that the RMTPP model well approximated different one-dimensional point process models with minor hyperparameter tweaking and zero prior knowledge of the true parameterization.

In our work, we aimed to show that the advantage of recurrent neural network based models can be combined with Granger causality framework and provide a domain independent Granger causality detection model. The Granger causality is designed to explicitly detect causality between different dimensions, i.e, the model of each dimension should be independent given the input data. Here we can consider dimensions to be different event sources. To solve the dimension-wise modelling problem, we proposed the Recurrent Multi-dimensional Temporal Point Process (RMDTPP) model that can effectively model multi-dimensional data.

We tested our model on both data on regular time grids (the time differences are constant one and not explicitly modelled) and data with real-valued timestamps (the time differences are explicitly modelled as real value). Experimental results showed that RMDTPP can be integrated into the Granger causality framework seamlessly. Causality between dimensions are successfully detected and the RMDTPP model can be applied to different data without parameterization changing.

2   A Review of Multi-dimensional Point Process

2.1   Multi-dimensional Point Process

Point process is commonly used to study sequence data. Normally, one sequence contains multiple event points, each event point has a timestamp describing the time that the event is observed of the time that the event happens. When will the next event happen is affected by all the previous events. If all the events are generated from one source, we use one-dimensional point process to model the data sequence. Combining multiple one-dimensional point processes with extra parameters capturing interactions between them leads to a multi-dimensional point process. In a multi-dimensional point process, more than one dimension can evolve at the same time and influence each other. One multi-dimensional point process can be uniquely defined by it conditional intensity function (CIF) $\lambda^*(t)$. Here * serves as a reminder that the function is conditional on the past, as first introduced by Daley and Vere-Jones [7]. The CIF describes how many events are expected to occur in an arbitrary time interval. Dimensions can self-intervene or cross intervene in a multi-dimension point process, which means that the intensity of one dimension receives the historical influence from all dimensions.

If we take a multi-dimensional Hawkes point process as an example, the CIF of the multi-dimensional Hawkes is defined as follows:

$$\lambda(t)^{m*} = \lambda_0^m + \sum_{n=1}^{M} \sum_{t_i^n < t} \alpha^{mn} e^{-\beta^{mn}(t-t_i^n)}, \tag{1}$$

where λ_0^m is the initial intensity of m dimension, α^{mn} and β^{mn} are employed to capture the influence flowing from n dimension to m dimension [9,13]. The size of matrix α and matrix β is $M \times M$ if there are in total M dimensions.

The heuristic parameterization problem refers to that the parameterization shown in (1) specifies how the point process evolves. Each event from one dimension stimulates the intensity at the same strength. The intensities of all dimensions gradually decay over time. Only the sequence data that roughly follows the specified evolving pattern can be well modelled. Thus, before we define any CIF for a dataset, we need to know how the data evolves, which is not always possible.

If we treat the length of time interval between two consecutive events as a random variable, the distribution of this random variable can be specified and calculated by a conditional density function $f^*(t)$. One CIF is corresponding to one conditional density function via

$$f^{m*}(t) = \lambda^{m*}(t) \exp\left(-\int_{t_j}^{t} \lambda^{m*}(s)ds\right). \tag{2}$$

The likelihood of the point process of dimension m is the product of the conditional density values of each event. One thing to notice, in a multi-dimensional point process, the process of each dimension ends at the same time, i.e. the domain $[0, T]$ of all dimensions are the same. However, the last event of one dimension may not happen at the timestamp T, which means that the process of one single dimension does not end with the last event on that dimension. Normally, we add a pseudo event for each dimension at timestamp T with the intensity $\lambda^{m*}(T) = 1$. We denote the number of events in dimension m including the pseudo event as L^m and t_i^m is the timestamp of ith event in dimension m, $t_{L_m}^m = T$ and $t_0^m = 0$. We have

$$\mathcal{L}(D_m) = \prod_{i=1}^{L^m} f^{m*}(t_i^m) \tag{3}$$

$$= \prod_{i=1}^{L^m} \lambda^{m*}(t_i^m) \exp\left(-\int_{t_{i-1}^m}^{t_i^m} \lambda^{m*}(s)ds\right). \tag{4}$$

When $i = L^m$, we have

$$\lambda^{m*}(t_i^m) \exp\left(-\int_{t_{i-1}^m}^{t_i^m} \lambda^{m*}(s)ds\right) = \exp\left(-\int_{t_{i-1}^m}^{t_i^m} \lambda^{m*}(s)ds\right), \tag{5}$$

which is the likelihood of the process from the last real event of dimension m to the end.

As described above, in a multi-dimensional point process, the intensity of one dimension is under the influence of the events from all other dimensions. During the interval from t_{i-1}^m to t_i^m, events from another dimension may happen and bring sudden changes to the intensity of dimension m. Thus, the integration has to be done piecewise as

$$\mathcal{L}(D_m) = \prod_{i=1}^{L^m} \lambda^{m*}(t_i^m) \exp\left(-\int_{t_{i-1}^m}^{t_i^m} \lambda^{m*}(s)ds\right) \tag{6}$$

$$= \left(\prod_{i=1}^{L^m} \lambda^{m*}(t_i^m)\right) \exp\left(-\int_0^T \lambda^{m*}(s)ds\right) \tag{7}$$

$$= \left(\prod_{i=1}^{L^m} \lambda^{m*}(t_i^m)\right) \exp\left(-\sum_{j=1}^{L} \int_{t_{j-1}}^{t_j} \lambda^{m*}(s)ds\right), \tag{8}$$

where L is the total number of events produced by the multi-dimensional point process plus the pseudo-event at the end of the domain at timestamp T. The likelihood function of a multi-dimensional point process is the product of the likelihood of each single dimension point process as

$$\mathcal{L}(D) = \prod_{m=1}^{M} \mathcal{L}(D_m) \tag{9}$$

$$= \prod_{m=1}^{M} \left(\prod_{i=1}^{L^m} \lambda^{m*}(t_i^m)\right) \exp\left(-\sum_{j=1}^{L} \int_{t_{j-1}}^{t_j} \lambda^{m*}(s)ds\right). \tag{10}$$

3   The Proposed Approach

3.1   Recurrent Multi-dimensional Temporal Point Process

While one hot is a group of bits where only one bit is allowed to be 1 and the rest are 0, many-hot allows as many bits to be 1 as needed. Simultaneous events can be easily represented by setting the corresponding bits to 1. In the RMDTPP model we use the many-hot representation to allow simultaneous events. The input of RMDTPP at step i contains the many-hot representation of the observation y_i and the time difference $d_i = t_i - t_{i-1}$. The hidden units of the RNN now should contain all the historical information needed to predict the conditional intensity of the selected dimension.

We adopted the CIF parameterization from RMTPP model [4] as

$$\lambda^*(t) = \exp(\mathbf{v}^\top \cdot \mathbf{h}_j + w(t - t_j) + b), \tag{11}$$

where $\mathbf{h}_j$ is the output of the RNN and $\mathbf{v}^\top$, w as well as b are learnable parameters. Then according to Eq. (2), the conditional density function is

$$\log f^*(t) = \mathbf{v}^\top \cdot \mathbf{h}_j + w(t - t_j) + b + \frac{1}{w} \exp(\mathbf{v}^\top \cdot \mathbf{h}_j + b) - \frac{1}{w} \exp(\mathbf{v}^\top \cdot \mathbf{h}_j + w(t - t_j) + b). \tag{12}$$

Equation (10) shows that the calculation of likelihood of dimension m only requires the conditional intensity value $\lambda^{m*}(t)$ and the integral of CIF between two events as $\Lambda_{t_{i+1}}^m = \exp\left(-\int_{t_{i-1}}^{t_i} \lambda^{m*}(s) ds\right)$. Thus, at each step, we let the RMDTPP model outputs $\lambda^{m*}(t)$ and $\Lambda_{t_{i+1}}^m$.

In a formal way, given an M dimensional sequence data $D \triangleq \{(t_i, \tilde{y}_i)\}$, $i = 0, ..., L$, where $L - 1$ is the total number of events in the sequence. $t_i \in [0, T]$ is the time stamp of event i, $t_0 = 0$, $t_L = T$ and $\tilde{y}_i \in \{0, 1\}^M$ is the many-hot representation of the dimensions of event i, $(\tilde{y}_i)_m$ is the mth elements of $\tilde{y}_i$. Notice that there should be no $\tilde{y}$ that every element of $\tilde{y}$ is zero. That is for any t_i in D, there is at least one event is observed at that timestamp. Here, we make the start point $(t_0, \tilde{y}_0)$ and the end point $(t_L, \tilde{y}_L)$ of the point process as pseudo events to simplify our work and $\tilde{y}_0 = \tilde{y}_L = (0, 0, ..., 0)^T$. RMDTPP model maximizes the likelihood of point process of dimension m with duration T:

$$\mathcal{L}(D_m) = \prod_{i=1}^{L} [\lambda^{m*}(t_i)]^{(\tilde{y}_i)_m} \exp\left(-\int_{t_{i-1}}^{t_i} \lambda^{m*}(s) ds\right). \tag{13}$$

With the output of the RMDTPP, we can compute the likelihood of the dimension $\mathcal{L}^m$ and use backpropagation to train the whole model in a maximum likelihood fashion. The structure of the RMDTPP model is shown in Fig. 1.

Another advantage of RMDTPP is that at each step after the output of RNN is computed, the likelihood calculation of each dimension can be run in parallel. We call the parallel training process as joint training. The hidden units in a joint training model should compress all the historical information instead of the information needed by a single dimension.

To achieve joint training, we initialize weights $\mathbf{v}$, w and b for each dimension. At each step, the output of the RNN is sent to each dimension, and from here the operations of each dimension are done independently and simultaneously. The likelihood is now the joint likelihood of all the dimensions, i.e. the product of the likelihood of each dimension. During the backpropagation phrase, the hidden layer of the RNN is updated by the gradients from all dimensions and the RNN is forced to learn a history representation that compresses the information needed by every dimension.

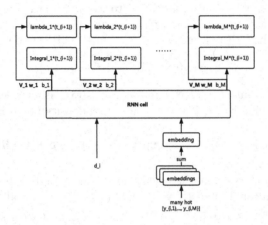

Fig. 1. Structure of RMDTPP model. M is the number of dimensions. As shown in the figure the conditional intensity and integral calculation of each dimension can run in parallel

3.2   Discrete RMDTPP

For data on a regular time grid, the length of each grid cell is fixed, which means that the time difference d between two time steps is a constant. In this case, the temporal structure of the data is simple and does not need to be modelled explicitly.

The input data is now a matrix D of M rows and T columns where M is the number of dimensions, and T is the length of the data. One row in D corresponds to the events record of one dimension. $D_{n,t} = 1$ if a event from dimension n is observed at time step t, else $D_{n,t} = 0$. Notice that in this setup, each column in D is a many-hot representation.

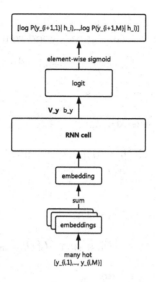

Fig. 2. Discrete RMDTPP structure

As shown in Fig. 2, the input of the discrete RMDTPP at time step t is simply the elements at the corresponding column $D_{:,t}$ and the output is a vector of predicted probabilities $P(D_{:,t+1} = 1|\mathbf{h}_t)$. The likelihood that the discrete RMDTPP tries to maximize is:

$$\log \mathcal{L}(D_n) = \sum_{t=1}^{T} \sum_{n=1}^{M} D_{n,t} P(D_{n,t} = 1|\mathbf{h}_t) + (1 - D_{n,t}) P(D_{n,t} = 0|\mathbf{h}_t). \quad (14)$$

3.3 Granger Causality

In this work we adopted the Granger causality defined by likelihood reduction [8]. The general idea is that for two dimensions X and Y, if the historical events of X contribute to the prediction of feature events of Y, then we say dimension X Granger-causes Y. The reason we use 'Granger-cause' or 'g-cause' in short instead of 'cause' is that there is a gap between the Granger causality and the real causality in terms of philosophy. The prediction contribution is measured by likelihood reduction. First, we train a model to use both historical events from dimension X and dimension Y to predict the feature event of Y. We use the model to calculate a likelihood $\mathcal{L}(Y)$ of the events of dimension Y. Then we train a new model to use only the historical events from dimension Y to predict the feature event of Y. We use the new model to calculate a new likelihood $\hat{\mathcal{L}}(Y)$ of the events of dimension Y. If $\hat{\mathcal{L}}(Y) - \mathcal{L}(Y) < 0$ then we say X contributes to the prediction of Y.

The detailed framework is introduced in Algorithm 1. Then the causality will be predicted according to the resulting likelihood reduction matrix.

Data: The training and testing dataset D_{train} and D_{test}
The number of dimensions M
Result: Likelihood reduction matrix Φ
$\Phi = M \times M$ zeros matrix;
$rmdtpp = $ RMDTPP.initialize(M);
$rmdtpp$.train(D_{train});
$\mathcal{L}(D_{test}) = [\mathcal{L}(D_{test\ 1}), ..., \mathcal{L}(D_{test\ M})] = rmtpp$.predict($D_{test}$);
for $m = 1; m \leq M; m = m + 1$ **do**
    $D'_{train} = D_{train} - D_{train\ m}$;
    $D'_{test} = D_{test} - D_{test\ m}$;
    $rmdtpp = $ RMDTPP.initialize(M-1);
    $rmdtpp$.train(D'_{train});
    $\mathcal{L}(D'_{test}) = [\mathcal{L}(D_{test\ 1}), ..., \mathcal{L}(D_{test\ m-1}), \mathcal{L}(D_{test\ m+1}), ..., \mathcal{L}(D_{test\ M})] = rmdtpp$.predict($D'_{test}$);
    for $n = 1; n \leq M; n = n + 1$ **do**
        if $n < m$ **then**
            | $\Phi_{n,m} = \mathcal{L}(D'_{test})n - \mathcal{L}(D_{test})n$
        end
        else
            | $\Phi_{n,m} = \mathcal{L}(D'_{test})n - \mathcal{L}(D_{test})n+1$
        end
    end
end

Algorithm 1: Granger causality detection process

4 Experiment

4.1 Dataset and Evaluation Metric

To demonstrate the capacity of RMDTPP model, we first test our model on a three-dimensional piecewise homogeneous Poisson process dataset. The homogeneous Poisson point process is one of the most simple point processes. If we treat the time difference between two adjacent events in a homogeneous Poisson point process as a random variable r, then r obeys an exponential distribution. The probability density function of an exponential distribution is $\lambda e^{-\lambda x}$ when $x > 0$ and 0 when $x \leq 0$. The λ is a positive parameter. The intensity of a homogeneous Poisson process will be constant $\frac{1}{\lambda}$. We generate 1,000 sequences, each of them containing 300 events. 300 sequences are randomly chosen to serve as the test set. The rest 700 serve as the training set.

We also adopted the neural activity simulation (NAS) framework specified by [8]. A dataset containing 20 sequences is generated to test the discrete RMDTPP model. Each sequence contains 100,000 events from 5 different dimensions. Half of the dataset is used as training set and the other half is used as testing set.

Then the discrete NAS dataset is converted into a continuous NAS dataset by removing all the time step where there are no events observed and the time difference between events is calculated as integer.

Moreover, a four-dimensional Hawkes point process dataset is generated. Both the four-dimensional Hawkes point process dataset and the continuous NAS dataset are used to test the continuous RMDTPP model.

The causality ground truth is obtained according to the parameters the parameters of the data generation frameworks. If the parameters of the data generation model indicate that events from dimension A increase or decrease the intensity of dimension B then we consider that there is a causal relation from A to B.

We treat the causality detection as a binary classification problem. The receiver operating characteristic curve (ROC) and area under curve (AUC) are used for performance measurement. Higher AUC indicates a better classification performance.

4.2   Experiment Results

To demonstrate the data fitting performance of the RMDTPP model, we first test our model on the three-dimensional piecewise Poisson process dataset.

As shown in the Fig. 3, The RMDTPP model successfully predicts the conditional intensity value, captures the un-natural sudden change of the intensity and predicts intensity to be zero when the process of that dimension finishes. It is worth noticing that the model is trained using maximum likelihood approach. The model is not directly trained to minimize the gap between the predicted conditional intensity value and the true value. The result shows that our RMDTPP model can approximate the true CIF with no priori knowledge about it.

Then we combine the RMDTPP model with the Granger causality framework. The causality detection performance of the discrete RMDTPP model is tested on the discrete NAS dataset. The resulting likelihood reduction matrix and the ground truth causality matrix is shown in Fig. 4(a) and (b). We also test the causality detection performance of the continuous RMDTPP model on the continuous NAS dataset. The resulting likelihood reduction matrix is shown in Fig. 4(c). The ROC curve of both continuous and discrete NAS data is shown in Fig. 4(d).

Considering that the NAS dataset is not a natural continuous time dataset, we then test the causality performance on the four-dimensional Hawkes point process dataset. The results are shown in Fig. 5.

The experiments results are reported in Table 1. The first row is the performance of the NAS model [8]. The NAS model itself is a discrete time point process. NAS dataset can be seen as the results of performing sampling operation on the corresponding NAS model.

From the comparison, we can see that the performance of discrete RMDTPP matches the performance of the NAS model which has full awareness of the true parameterization. Continuous RMDTPP model is outperformed by both NAS

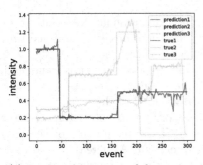
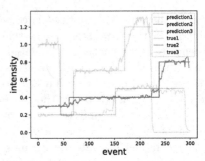

(a) Predicted intensity of dimension 1.   (b) Predicted intensity of dimension 2.

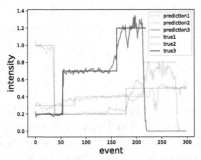

(c) Predicted intensity of dimension 3.

Fig. 3. Predicted intensity of the three-dimension piecewise Poisson process toy dataset.

model and discrete RMDTPP model on the NAS dataset. Part of the reason is that the likelihood reduction value is closely related to the true likelihood that is calculated with the full dataset. Likelihood reductions are calculated based on the true likelihood. The fluctuation of the true likelihood increases after we make the RMDTPP model to learn the real-valued time difference. As a result, it is harder to find a reasonable threshold for all dimension pairs.

No previous application of the Granger causality framework with traditional continuous time point process models can be found. Since that the causality can be easily identified via the learned parameters model.

Table 1. Model AUC value comparison

	NAS	4 dimensional hawkes point process
NAS model	**1.0**	N/A
Discrete RMDTPP	**1.0**	N/A
Continuous RMDTPP	0.83	**0.72**

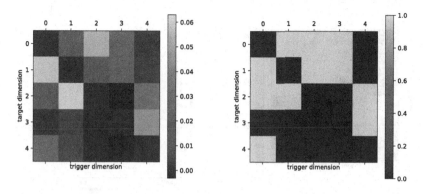

(a) Likelihood reductions of discrete NAS dataset.

(b) The ground truth causality matrix.

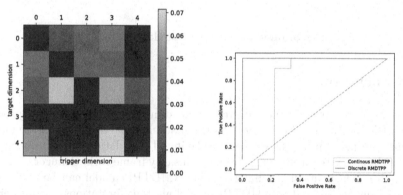

(c) RMDTPP likelihood reduction of NAS.

(d) ROC curve of the causality detection of NAS.

Fig. 4. The likelihood reduction matrices and the ROC curves.

However, only continuous RMDTPP model can work with both the NAS dataset and the multi-dimensional Hawkes point process dataset. The only hyperparameter that has to be manually adjusted is the size of the RNN units, which can be decided by observing the likelihood results and picking the one with the highest likelihood. Thus, we can say that combining RMDTPP model with Granger causality has a high potential. In discrete cases, the discrete RMDTPP model is practically usable. The heuristic parameterization problem of traditional point process model is overcome allowing Granger causality to be applied to various scenarios.

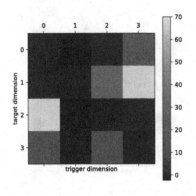
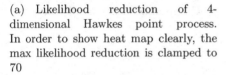
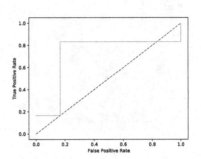

(a) Likelihood reduction of 4-dimensional Hawkes point process. In order to show heat map clearly, the max likelihood reduction is clamped to 70

(b) ROC curve of causality detection on 4-dimensional Hawkes point process.

Fig. 5. Resulting Granger causality matrices.

5   Conclusion

In this work, we propose the recurrent multi-dimension temporal point process model. RMDTPP model is inspired by the mathematical definition of the multi-dimensional point process. A separate likelihood calculation of each dimension allows seamless integration with Granger causality framework. Granger causality detection experiments show that the same RMDTPP model can be applied to various sequence data. The RMDTPP model also has limitations. In this conditional intensity function, w is a learned constant scalar value, which means that the time difference between the last event and current time $(t - t_j)$ can only linearly influence the current conditional intensity in log space. This feature is not true for all temporal point process models. It is possible to give the conditional intensity function a more flexible parameterization. Despite of the limitations, we proved that the combining neural network work techniques with Granger causality leads to a powerful model. The heuristic parameterization problem is overcome and causal relation between dimensions can be detected. Our work enables the application of Granger causality on sequence data from any domain.

References

1. Ancona, N., Marinazzo, D., Stramaglia, S.: Radial basis function approach to nonlinear granger causality of time series. Phys. Rev. E Stat. Nonlinear Soft Matter Phys. **70**(5 Pt 2), 056221 (2004)

2. Bacry, E., Delattre, S., Hoffmann, M., Muzy, J.: Some limit theorems for Hawkes processes and application to financial statistics. Stoch. Process. Their Appl. **123**(7), 2475–2499 (2013). https://doi.org/10.1016/j.spa.2013.04.007. http://www.sciencedirect.com/science/article/pii/S0304414913001026. A Special Issue on the Occasion of the 2013 International Year of Statistics
3. Chen, Y., Rangarajan, G., Feng, J., Ding, M.: Analyzing multiple nonlinear time series with extended Granger causality. Phys. Lett. A **324**, 26–35 (2004). https://doi.org/10.1016/j.physleta.2004.02.032
4. Du, N., Dai, H., Trivedi, R., Upadhyay, U., Gomez-Rodriguez, M., Song, L.: Recurrent marked temporal point processes: embedding event history to vector. In: Proceedings of the 22nd ACM SIGKDD International Conference on Knowledge Discovery and Data Mining, KDD 2016, pp. 1555–1564. ACM, New York (2016). https://doi.org/10.1145/2939672.2939875. http://doi.acm.org/10.1145/2939672.2939875
5. Freiwald, W.A., et al.: Testing non-linearity and directedness of interactions between neural groups in the macaque inferotemporal cortex. J. Neurosci. Methods **94**(1), 105–119 (1999). https://doi.org/10.1016/S0165-0270(99)00129-6. http://www.sciencedirect.com/science/article/pii/S0165027099001296
6. Granger, C.W.J.: Investigating causal relations by econometric models and cross-spectral methods. Econometrica **37**(3), 424–438 (1969). http://www.jstor.org/stable/1912791
7. Guttorp, P.: An introduction to the theory of point processes (D. J. Daley and D. Vere-Jones). SIAM Rev. **32**(1), 175-2 (1990). https://search.proquest.com/docview/926135796?accountid=8330. Copyright © 1990 Society for Industrial and Applied Mathematics. Last updated 05 Mar 2012. CODEN - SIREAD
8. Kim, S., Putrino, D., Ghosh, S., Brown, E.N.: A granger causality measure for point process models of ensemble neural spiking activity. PLOS Comput. Biol. **7**(3), 1–13 (2011). https://doi.org/10.1371/journal.pcbi.1001110
9. Liniger, T.J.: Multivariate Hawkes processes. Ph.D. thesis, ETH Zurich. Diss., Eidgenössische Technische Hochschule ETH Zürich, Nr. 18403 (2009). https://doi.org/10.3929/ethz-a-006037599
10. Marinazzo, D., Pellicoro, M., Stramaglia, S.: Nonlinear parametric model for Granger causality of time series **73**(6), 066216 (2006). https://doi.org/10.1103/PhysRevE.73.066216
11. Marinazzo, D., Pellicoro, M., Stramaglia, S.: Kernel method for nonlinear granger causality. Phys. Rev. Lett. **100**(14), 144103 (2008). https://doi.org/10.1103/PhysRevLett.100.144103
12. Ogata, Y.: Statistical models for earthquake occurrences and residual analysis for point processes. J. Am. Stat. Assoc. **83**(401), 9–27 (1988). https://doi.org/10.1080/01621459.1988.10478560. https://amstat.tandfonline.com/doi/abs/10.1080/01621459.1988.10478560
13. Toke, I.M.: An introduction to Hawkes processes with applications to finance, p. 6, February 2011

Density-Induced Support Vector Data Description for Fault Detection on Tennessee Eastman Process

Yangtao Xue, Li Zhang$^{(\boxtimes)}$, Bangjun Wang, and Baige Tang

School of Computer Science and Technology and Joint International Research Laboratory of Machine Learning and Neuromorphic Computing, Soochow University, Suzhou 215006, China
zhangliml@suda.edu.cn

Abstract. Fault detection can be taken as a behavior of detecting abnormal data in process data. Support vector data description (SVDD) has been successfully used for fault detection. Although density-induced support vector data description (D-SVDD) can give a better description of target data by introducing relative density degrees than SVDD, the problem of an additional parameter selection hinders the application of D-SVDD, which has a great influence on the performance of D-SVDD. This paper bounds this additional parameter for D-SVDD and applies D-SVDD to fault detection on TE process monitoring. Experiment shows D-SVDD is promising.

Keywords: D-SVDD · Fault detection · TE process

1 Introduction

With the rapid development of modern industrial processes, process monitoring has been attracted considerably increasing attention. In the past several decades, the model-based methods for process monitoring have difficulty in building accurate mathematical models, while the data-driven approaches have advantages of extracting process information from huge amounts of the recorded data [12]. Fault detection based on data-driven approaches has found wide applications in various industrial processes, including chemicals, polymers, pharmaceutical process and power distribution network [6].

It is easy to apply the data-driven fault detection methods to real industrial systems [6]. Especially, multivariate statistics methods have obtained great achievement [2]. Classical multivariate statistics methods include principal component analysis (PCA) and partial least squares (PLS). However, the traditional multivariate statistics approaches are under the assumption that the relationship between variables is linear and the process data are Gaussian-distribution [10]. In reality, the characteristics of process data always are dynamical, non-linear, non-Gaussian and larger-scale [13]. The unreasonable assumption could result

© Springer Nature Switzerland AG 2018
L. Cheng et al. (Eds.): ICONIP 2018, LNCS 11304, pp. 522–531, 2018.
https://doi.org/10.1007/978-3-030-04212-7_46

in a poor performance for the traditional multivariate statistics approaches. To solve these problems, many researchers take fault detection as a classification problem [10,13].

Particularly, fault detection can be regarded as a one-class classification problem which is to find the outliers among normal samples. Support vector data description (SVDD), also called one-class support vector machine (SVM), can not only overcome the disadvantages of traditional multivariate statistics approaches, but also avoid the curse of dimensionality and the local minima problem [8]. SVDD takes use of the normal data to construct a hypersphere for detecting whether a sample is normal or outlier [8,9]. This hypersphere is a closed decision boundary which contains most of training (normal) samples with a minimal radius. In [10], an alternative approach was presented, which is to construct a hyperplane to separate the region containing data from the region containing no data. When using the radial basis function (RBF) kernel, these two methods are identical to each other. Both methods are finally formulated as quadratic programming (QP) problems. The method in [7] has been applied to fault detection [5].

Since SVDD does not consider the density distribution of a given dataset, Lee et al. proposed density-induced support vector data description (D-SVDD) by introducing relative density degrees into SVDD in [3,4]. The relative density degrees can reflect the distribution of a given dataset. Specifically, if there are more data points in a denser region, the relative density degree of a sample is large. Therefore, D-SVDD can shift the center of hypersphere to the denser region [3]. In order to obtain better performance, D-SVDD introduces a new parameter which needs setting beforehand. However, the parameter was not discussed in both [3,4].

As we know, D-SVDD has not been applied to fault detection. Because the parameter selection has a great influence on the performance of D-SVDD. Therefore, this paper focuses on analyzing the parameter T in D-SVDD and applied D-SVDD to fault detection.

2   Density-Induced Support Vector Data Description

D-SVDD is a kind of one-class classifier that it can find the optimal hypersphere including most training data by reflecting the relative density degrees of the training data [9].

2.1   Methods for Extracting Relative Density Degrees

In [3,4], it shows that samples with different density degrees may play a different role in constructing the boundary. Compared with the samples located in a low density region, the ones located in a high density region should shrink into the compact description scope as possible as they can. There are several methods for extracting relative density degrees. Here, we consider the method proposed in [4], which uses a nearest neighborhood approach to estimate relative density degrees.

In the nearest neighborhood method, we can extract the relative density degrees of given training target samples $\{x_i\}_{i=1}^N$, where $x_i \in \mathbb{R}^m$, m is the dimensionality of training samples, and N is the number of training samples. For each target sample x_i, we first find its K nearest neighbors. Let x_i^K be its Kth neighbor and $D(x_i, x_i^K)$ represent the distance between x_i and x_i^K. Then, we can extract the relative density degree ρ_i of x_i, which is defined as :

$$\rho_i = \exp\left(w \times \frac{D^K}{D(x_i, x_i^K)}\right)$$

where $0 \leq w \leq 1$ is a weighting factor and $D^K = \frac{1}{N}\sum_{i=1}^N D(x_i, x_i^K)$. The method reports that the lower the distance between the data point and its Kth neighbor is, the higher the value of the relative density degrees is.

2.2 Density-Induced SVDD

Given the set of training sample $\{x_i, \rho_i\}_{i=1}^N$, D-SVDD tries to obtain the optimal hypersphere by solving the following primal programming:

$$\max_{R,a,\xi} \quad R^2 + C\sum_{i=1}^N \xi_i$$

$$s.t. \quad \rho_i\|x_i - a\|^2 \leq R^2 + C\sum_{i=1}^N \xi_i, \xi_i \geq 0, i = 1, ..., N, \tag{1}$$

where $C \geq 0$ is the penalty factor, R is the radius of the optimal hypersphere, a is the center of the optimal hypersphere, and ξ_i is the slack variable. By introducing kernel tricks, the dual programming of (1) can be described as

$$\max_{\alpha} \quad \sum_{i=1}^N \alpha_i\rho_i k(x_i, x_i) - \frac{1}{T}\sum_{i=1}^N\sum_{j=1}^N \alpha_i\alpha_j\rho_i\rho_j k(x_i, x_j)$$

$$s.t. \quad \sum_{i=1}^N \alpha_i = 1, \sum_{i=1}^N \alpha_i\rho_i = T, 0 \leq \alpha \leq C, i = 1, ..., N, \tag{2}$$

where $k(x_i, x_j)$ is the Mercer kernel function, and α_i is the Lagrange multiplier.

3 D-SVDD for Fault Detection

3.1 Analysis on Parameter T

In the optimization problem (2), there has an equality constraint which relates to the new parameter T, namely, $\sum_{i=1}^N \alpha_i\rho_i = T$. Thus, the value of T needs setting beforehand as the penalty factors C and kernel parameters. However, references [3,4] did not discuss the parameter T even in experiments. Since we know nothing about T, it is difficult to determine the value of T. In addition, this parameter also has an important effect on the performance of D-SVDD. If we randomly appoint a value to T, the performance of D-SVDD cannot be guaranteed. Therefore, it is necessary to analyze the parameter T.

In the following, we give a theorem to indicate the range of T.

Theorem 1. Given a set of training samples $\{\mathbf{x}_i, \rho_i\}_{i=1}^{N}$, the domain of T is

$$\rho_{min} \leq T \leq \rho_{max}, \tag{3}$$

where $\rho_{min} = min\{\rho_1, ...\rho_N\}$ and $\rho_{max} = max\{\rho_1, ...\rho_N\}$.

Proof: Given a set of training samples $\{\mathbf{x}_i, \rho_i\}_{i=1}^{N}$, there are two extreme situations for the well-trained D-SVDD. One is that there exists only one support vector $\mathbf{x}_{sv}$, and the other is that all samples are support vectors.

In the first situation, there exists only one support vector $\mathbf{x}_{sv}$. Since the equality constraint $\sum_{i=1}^{N} \alpha_i = 1$, the corresponding Lagrange multiplier $\alpha_{sv} = 1$ and other $\alpha_i (i \neq sv)$ is zero. Thus,

$$T = \sum_{i=1}^{N} \alpha_i \rho_i = \alpha_{sv}\rho_{sv} = \rho_{sv}.$$

Obviously, we have

$$\rho_{min} \leq \rho_{sv} \leq \rho_{max}.$$

Therefore, $\rho_{min} \leq T \leq \rho_{max}$.

In the second extreme situation, all samples are support vectors. Since $\sum_{i=1}^{N} \alpha_i = 1$, $\alpha_i = \frac{1}{N}$ for all samples. Thus,

$$T = \sum_{i=1}^{N} \alpha_i \rho_i = \frac{1}{N} \sum_{i=1}^{N} \rho_i = \rho_{ave}.$$

Since $\rho_{min} \leq \rho_{ave} \leq \rho_{max}$, we also have $\rho_{min} \leq T \leq \rho_{max}$.

Except for the two extreme situations, we also have $\sum_{i=1}^{N} \alpha_i = 1$ and $\sum_{i=1}^{N} \alpha_i \rho_i = T$. T can be considered as a convex combination of $\rho_i, i = 1, ..., N$. Thus, we have

$$T = \sum_{i=1}^{N} \alpha_i \rho_i \leq \sum_{i=1}^{N} \alpha_i \rho_{max} = \rho_{max}, \tag{4}$$

and

$$T = \sum_{i=1}^{N} \alpha_i \rho_i \geq \sum_{i=1}^{N} \alpha_i \rho_{min} = \rho_{min}. \tag{5}$$

By combining (4) and (5), we finally have $\rho_{min} \leq T \leq \rho_{max}$. This completes the proof.

Theorem 1 shows that the range of parameter T is closely related to the relative density degrees of the given dataset. Therefore, the parameter T could take any value in the interval $[\rho_{min}, \rho_{max}]$. However, the optimal value for T is still unknown. Actually, the parameters in most methods depend on the given tasks. Thus, it is difficult to select the optimal parameters. The same situation occurs on the selection of T. However, in this paper, the experiments results indicate that the average of relative density can be as a good choice for parameter T.

3.2   D-SVDD for Fault Detection

Generally, fault detection requires a distance metric and a threshold, such as Hotellings T^2 and SPE (Square Predicted Error) [2,11,12]. For one-class SVM, the perpendicular distance of a given point from the margin is adopted as a measure [5]. Similar to [5], we use the distance of a given point from the center as the distance metric in D-SVDD. The measure of this distance of a given point $\mathbf{x}$ from the center a is

$$d(\mathbf{x}) = (k(\mathbf{x}, \mathbf{x}) - \frac{2}{T} \sum_{i=1}^{N} \alpha_i \rho_i k(\mathbf{x}_i, \mathbf{x})$$

$$+ \frac{1}{T^2} \sum_{i=1}^{N} \sum_{j=1}^{N} \alpha_i \alpha_j \rho_i \rho_j k(\mathbf{x}_i, \mathbf{x}_j))^{1/2}.$$

Moreover, we need a threshold to determine whether a given point is normal or not. The radius of hypersphere could be taken as a threshold. If the distance of a given sample from the center is smaller than or equal to the radius, the sample will be considered as normal data. Namely, the decision rule is

$$J = \frac{d^2(\mathbf{x})}{R^2},$$

where

$$R = (k(\mathbf{x}_{sv}, \mathbf{x}_{sv}) - \frac{2}{T} \sum_{i=1}^{N} \alpha_i \rho_i k(\mathbf{x}_i, \mathbf{x}_{sv})$$

$$+ \frac{1}{T^2} \sum_{i=1}^{N} \sum_{j=1}^{N} \alpha_i \alpha_j \rho_i \rho_j k(\mathbf{x}_i, \mathbf{x}_j))^{1/2},$$

$\mathbf{x}_{sv}$ is a support vector with $0 \leq \alpha \leq C$. If $J \leq 1$, $\mathbf{x}$ is fault-free, otherwise, is fault or an outlier.

4   Experiments

To validate the detection performance of D-SVDD, we perform experiments on the Tennessee Eastman process, which is a benchmark simulation problem in chemical engineering discussed by Downs and Vogel [1]. The dataset can be

A normal process dataset (1460 samples) for training has been collected. 21 kinds of programmed faults (960 samples) are simulated and are collected for test. All faults are introduced into the process on the 161st time point. Each sample consists of 52 variables (features).

In this experiment, we build a model with the normal process dataset. We take 21 kinds of faults as test datasets each of which contains 960 samples, and predict whether a fault occurs or not. Two generally used indices, i.e. fault detection rate (FDR) and false alarm rate (FAR), are mainly considered for evaluating the fault detection performance [11].

$$FDR = \frac{\#Fault\ samples\ with\ J > 1}{\#Fault\ samples}$$

$$FAR = \frac{\#Normal\ samples\ with\ J > 1}{\#Normal\ samples}$$

To balance the two indices, we use the recall index, or

$$Recall = \frac{1 - FAR + FDR}{2}.$$

4.1 Experiments on Parameter T

In reality, there is no absolutely optimal parameters, which could be approximately decided by performing several experiments on given tasks. Here, we take Fault 11 as an example to validate the performance of D-SVDD for different T in the interval $[\rho_{min}, \rho_{max}]$. The recall (4) is used as an indication for finding the suboptimal parameter value.

Since D-SVDD takes into consideration the local distribution of training samples, the value of the parameter T significantly influences its performance. In order to describe the data distribution accurately, we set $w = 1$ and $K = 5$ when calculating the relative density degrees. According to the characteristics of given tasks, we select the linear kernel function in the D-SVDD training model and set $C = 1$.

Table 1. Statistics of relative density degrees on TE process

ρ_{min}	ρ_{ave}	ρ_{max}
2.0937	2.7484	3.6183

The proposed D-SVDD method is applied on the Tennessee Eastman process. The statistics of relative density are given in Table 1. We generate 20 linearly equally spaced values for T between ρ_{min} and ρ_{max}. The result of Fault 11 is shown in Fig. 1. The maximum recall of Fault 11 is achieved when $T = 2.7484$ which indicates the average of density is a good choice for T. That is to say, the parameter T in D-SVDD decided by the average of density is a great method.

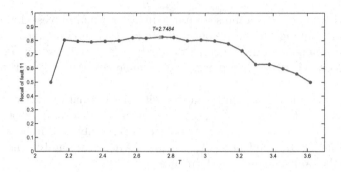

Fig. 1. Recall of Fault 11 for different T using D-SVDD

4.2   Comparison with Other Methods

As both PCA and PLS are typical approaches for fault detection, we list their results of the T^2 statistic of PCA and PLS as comparisons, the number of principal components (PCs) in PCA is 17 and the latent variables (LVs) is 29, which were suggested in [11]. We compare four methods, PCA, PLS, C-SVDD and D-SVDD ($T = 2.7484$) on two faults: Fault 4 and Fault 11.

Fault 4: Fault 4 involves a sudden temperature increase in the reactor and the change is compensated by the control loops. The monitoring results of Fault 4 using PCA, PLS, C-SVDD and D-SVDD are shown in Fig. 2 and Table 2. Note that the first 160 samples are normal, and the other 800 samples are faults. The most fault data can be detected by using D-SVDD while half of fault data can not be identified by PLS and C-SVDD. Although the FDR of PCA for Fault 4 is higher by 4% comparing with D-SVDD, the FAR increases by nearly 34%. In addition, the recall of Fault 4 using D-SVDD is 91.88% which is much higher than that of PCA (76.88%), PLS (73.06%) and C-SVDD (78.19%). That is to say, the fault detection performance of D-SVDD for Fault 4 is more stable and effective.

Table 2. Comparison of four methods for Fault 4

Method	FDR	FAR	Recall
PCA-T^2	**94.38**	40.63	76.88
PLS-T^2	46.13	**0**	73.06
C-SVDD	57.00	0.63	78.19
D-SVDD	90.63	6.88	**91.88**

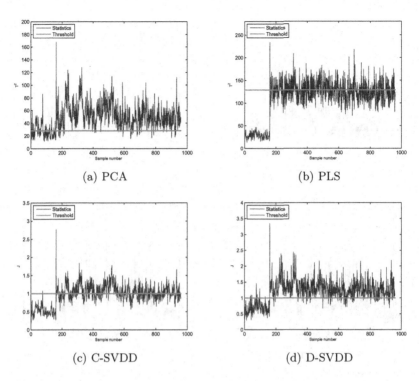

Fig. 2. Monitoring performances for Fault 4 using PCA (a), PLS (b), C-SVDD (c) and D-SVDD (d).

Fault 11: The root cause of Fault 11 is reactor cooling water inlet temperature that results from random variation. From the Fig. 3 and Table 3, we can see that fault monitoring using PLS has the worst performance, where the FDR for Fault 11 merely achieves 37%. The highest FDR is 92.13% using PCA, however, its recall is 65.75% with the highest FAR 60.63%. D-SVDD has significantly reduced the FAR in PCA by 52% and increased the FDR in C-SVDD by nearly 20%. Therefore, the overall monitoring performance of D-SVDD is the best among four methods, since the recall of D-SVDD is 82.63% which is improved much comparing with PCA (65.75%), PLS (68.50%) and C-SVDD (76.94%).

Table 3. Comparison of four methods for Fault 11

Method	FDR	FAR	Recall
PCA-T^2	**92.13**	60.63	65.75
PLS-T^2	37.00	**0**	68.50
C-SVDD	53.88	**0**	76.94
D-SVDD	73.38	8.13	**82.63**

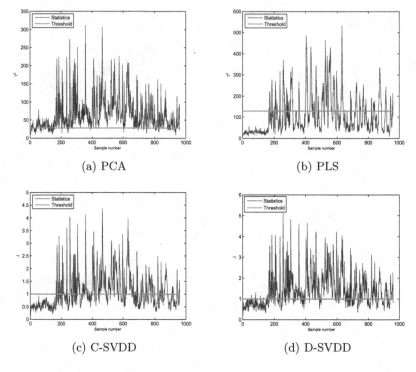

(a) PCA   (b) PLS

(c) C-SVDD   (d) D-SVDD

Fig. 3. Monitoring performances for Fault 11 using PCA (a), PLS (b), C-SVDD (c) and D-SVDD (d).

5   Conclusions

The paper analyzes the parameter T in D-SVDD and indicates its domain. We find the optimal parameter for given tasks and apply it to the fault detection of chemical process and experimental results indicate that the average density is a good choice for T. Compared with C-SVDD, D-SVDD is more stable on the TE process since it makes full use of the information implied in the data. Besides, the D-SVDD also performs much better on the TE process than conventional multivariate statistic approaches such as PCA and PLS for Fault 4 and 11. In conclusion, the D-SVDD is promising in the application of chemical fault detection.

Acknowledgments. This work was supported in part by the National Natural Science Foundation of China under Grant No. 61373093, by the Natural Science Foundation of Jiangsu Province of China under Grant Nos. BK20140008, and by the Soochow Scholar Project.

References

1. Downs, J.J., Vogel, E.F.: A plant-wide industrial process control problem. Comput. Chem. Eng. **17**(3), 245–255 (1993)
2. Ge, Z., Song, Z., Gao, F.: Review of recent research on data-based process monitoring. Ind. Eng. Chem. Res. **52**(10), 3543–3562 (2013)
3. Lee, K., Kim, D.W., Lee, D., Lee, K.H.: Improving support vector data description using local density degree. Pattern Recognit. **38**(10), 1768–1771 (2005)
4. Lee, K., Kim, D.W., Lee, K.H., Lee, D.: Density-induced support vector data description. IEEE Trans. Neural Netw. **18**(1), 284–289 (2007)
5. Mahadevan, S., Shah, S.L.: Fault detection and diagnosis in process data using one-class support vector machines. J. Process. Control. **19**(10), 1627–1639 (2009)
6. Qin, S.J.: Survey on data-driven industrial process monitoring and diagnosis. Annu. Rev. Control **36**(2), 220–234 (2012)
7. Schölkopf, B., Platt, J.C., Shawe-Taylor, J., Smola, A.J., Williamson, R.C.: Estimating the support of a high-dimensional distribution. Neural Comput. **13**(7), 1443–1471 (2001)
8. Tax, D.M., Duin, R.P.: Support vector domain description. Pattern Recognit. Lett. **20**(11), 1191–1199 (1999)
9. Tax, D.M., Duin, R.P.: Support vector data description. Mach. Learn. **54**(1), 45–66 (2004)
10. Xiao, Y., Wang, H., Zhang, L., Xu, W.: Two methods of selecting gaussian kernel parameters for one-class SVM and their application to fault detection. Knowl. Based Syst. **59**, 75–84 (2014)
11. Yin, S., Ding, S.X., Haghani, A., Hao, H., Zhang, P.: A comparison study of basic data-driven fault diagnosis and process monitoring methods on the benchmark tennessee eastman process. J. Process. Control. **22**(9), 1567–1581 (2012)
12. Yin, S., Ding, S.X., Xie, X., Luo, H.: A review on basic data-driven approaches for industrial process monitoring. IEEE Trans. Ind. Electron. **61**(11), 6418–6428 (2014)
13. Zhang, K., Qian, K., Chai, Y., Li, Y., Liu, J.: Research on fault diagnosis of tennessee eastman process based on KPCA and SVM. In: 2014 Seventh International Symposium on Computational Intelligence and Design (ISCID), vol. 1, pp. 490–495 (2014)

ExtTra: Short-Term Traffic Flow Prediction Based on Extremely Randomized Trees

Jiaxing Shang[1,2,3](✉), Xiaofan Yan[1], Linhui Feng[1], Zheng Dong[1],
Haojie Wang[1], and Shangbo Zhou[1,2]

[1] College of Computer Science, Chongiqng University, Chongqing, China
{shangjx,shbzhou}@cqu.edu.cn, xiaofan_yan@foxmail.com,
feng_linhui@163.com, isdongzheng@foxmail.com, wang_hj9766@foxmail.com
[2] Key Laboratory of Dependable Service Computing in Cyber Physical Society,
Ministry of Education, Chongqing University, Chongqing, China
[3] Guangxi Key Laboratory of Trusted Software,
Guilin University of Electronic Technology, Guilin, China

Abstract. Short-term traffic flow prediction is an important task for intelligent transportation systems. Conventional time series based approaches such as ARIMA can hardly reflect the inter-dependence of related roads. Other parametric or nonparametric methods do not take full advantage of the spatial temporal features. Moreover, some machine learning models are still not investigated in solving this problem. To fill this gap, in this paper we propose ExtTra: an extremely randomized trees based approach for short-term traffic flow prediction. To the best of our knowledge, our work is the first effort to apply the extremely randomized trees model on the traffic flow prediction problem. Moreover, our approach incorporates new spatial temporal features which were not considered in previous studies. Experimental results show that our approach significantly outperforms the baselines in prediction accuracy.

Keywords: Traffic flow prediction · Extremely randomized trees
Spatial temporal features · Intelligent transportation · Machine learning

1 Introduction

Intelligent transport system (ITS), as an advanced application system, provides innovative services for governments and traffic regulators. It aims at enabling traffic users to be better informed and making safer and smarter use of transport networks. As a fundamental and important component of the intelligent transportation system, short-term traffic flow prediction aims to predict the traffic flow of the next 5 to 30 min based on historical traffic flows and other available information [1]. It supports proactive dynamic transportation planing, traffic control and serves as a basis for many ITS subsystems. For example, based on the predicted traffic conditions, traffic controllers can intervene in the

© Springer Nature Switzerland AG 2018
L. Cheng et al. (Eds.): ICONIP 2018, LNCS 11304, pp. 532–544, 2018.
https://doi.org/10.1007/978-3-030-04212-7_47

traffic signals to reduce traffic congestion and improve mobility, travelers can have real-time guidance information for a better route choice, and so forth. Due to its many practical values, the study of short-term traffic flow prediction has attracted the attention of many scholars and a wide variety of models and algorithms have been proposed. Roughly, these approaches can be divided into four categories: (1) time series analysis, e.g., ARIMA [1], SARIMA [2], and their extension [3]; (2) parametric methods, e.g., Bayesian model [4], Kalman filtering [5], and stochastic differential equation [6]; (3) nonparametric methods, e.g., SVR [7] and k-NN [8,9]; and (4) neural network-based approaches, e.g., DBN [10], CNN [11], RNN [12,13] and ELM [14].

These technologies solve the traffic flow prediction problem from different aspects, and they have shown their effectiveness through experiments on real world datasets. However, these technologies still have some limitations. The time series based methods, such as ARIMA and SARIMA, predict traffic flows only based on the historical data of the same road and ignored the inter-dependence of related roads. The other approaches, including the parametric, nonparametric, and neural network based methods, do not take full advantage of the spatial temporal features, such as the lane information, time of the day, day of the week, etc., which are available in the traffic data. Moreover, some machine learning models are still not investigated in solving the short-term traffic flow prediction problem. Motivated by the above observations, in this paper we propose **ExtTra**: an **Ext**remely randomized trees based approach for short-term **Tra**ffic flow prediction. Our approach is based on the extremely randomized trees (Extra-Trees) machine learning model proposed by Geurts et al. [15]. To the best of our knowledge, this model has not yet been used before to solve the traffic flow prediction problem. The reason for choosing Extra-Trees is due to its capability of capturing the non-linear features of the short traffic flow prediction problem, which has been shown in previous studies [8,16]. In our model, we not only consider general spatial temporal features such as historical traffic data and inter-dependence of related roads, as many previous works did so, but also involve some spatial temporal features that were not considered before, i.e., the lane information, time of the day, day of the week, etc. Experimental results show that these features can significantly improve the prediction accuracy. Our approach is comprehensively evaluated on the traffic dataset from a city of China. In general, our work has the following contributions:

- We propose ExtTra: an extremely randomized trees (Extra-Trees) based approach for short-term traffic flow prediction, providing the first effort to apply the Extra-Trees model on the traffic flow prediction problem.
- Our approach incorporates some spatial temporal features which were not considered in previous research works, and these features are shown to be helpful in improving the prediction accuracy.
- Experimental results on the dataset from a city of China show that our approach significantly outperforms other baselines in prediction accuracy.

The rest of this paper is organized as follows: Sect. 2 introduces some related works; Sect. 3 presents our ExtTra approach; Sect. 4 gives the experimental evaluation, including the baselines, evaluation metrics, and the results; Sect. 5 concludes this paper.

2   Related Work

A plenty of studies have been proposed to solve the short-term traffic flow prediction problem. We roughly classify these studies into the following four groups: (1) time series; (2) parametric; (3) non-parametric; and (4) neural network.

Most of the time series analysis approaches are based on the ARIMA (Autoregressive Integrated Moving Average) model [1]. Williams et al. [2] extended the ARIMA model and proposed SARIMA (Seasonal ARIMA) to capture the repeatable (seasonal) pattern of the traffic flow. Kumar et al. [3] extended the SARIMA model to handle limited input data.

Parametric methods take a "model-based" approach to solve the problem by imposing prior knowledge on the traffic flow pattern. Pascale et al. [4] constructed a Bayesian network to represent the traffic flow pattern. Wu et al. [5] proposed a STRE (Spatial Temporal Random Effect) model based on Kalman filter to predict traffic flow for a dense urban street network with volatile traffic flows. Rajabzadeh et al. [6] used Hull White model to construct a baseline predictor and employed the time-varying Vasicek model to capture traffic flow fluctuations.

Nonparametric methods take a "data-based" approach to solve the problem, aiming to uncover the correlations between traffic flow and historical information. Jeong et al. [7] combined online SVR with a weighted learning model to predict traffic flow. Dell'Acqua et al. [8] combined time seasonality with NNR (Nearest Neighbor Regression) and proposed Tam-NNR(Time-aware multivariate NNR) model. Xia et al. [9] proposed STW-KNN model with consideration of spatial temporal correlation and implemented their method under a Map-Reduce framework.

Recently, deep learning (DL) has drawn much attention and more and more studies apply the DL models on the traffic flow prediction problem. Huang et al. [10] made the first effort to apply DL model on traffic prediction, in which they trained a DBN (Deep Belief Network) for unsupervised feature learning. Yu et al. [11] converted network-wide traffic speed into a series of static images and used the images to train a CNN (Convolutional Neural Network) for traffic speed prediction. Du et al. [12] applied RNNs (Recurrent Neural Networks) to capture long temporal traffic flow patterns. Zhang et al. [13] proposed a RNN (Residual Neural Network) based approach which considers spatial, temporal, and external influence when predicting the inflow and outflow of traffic crowds. Wang et al. [14] predicted short traffic flow based on a ELM (Extreme Learning Machine) model, which can be incrementally updated with respect to new data.

3   Proposed Approach

3.1   Data Set

The traffic dataset studied in this paper is from the urban road data of Heifei, a city located in Anhui province of China, collected during the time of 11–26th

July, 2–18th August, and 24th August to 5th September, 2016, about 46 days of data in total. The data were collected by the microwave detectors located at the urban road segments along Huangshan road of the city, as shown in Fig. 1. The dataset consists of traffic information from ten road segments, which are marked in Fig. 1, where each road segment is associated with a direction. Unlike most previous studies whose traffic data were obtained through GPS or loop detectors, the data collected by microwave detectors are more accurate and they directly provide traffic speed and volume information at fine grained lane level. The format of collected data is summarized in Table 1.

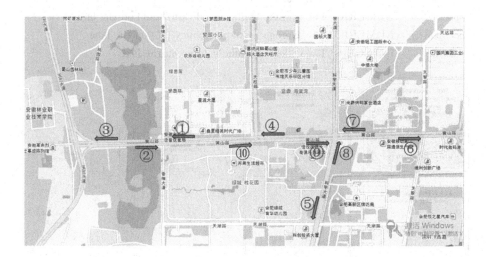

Fig. 1. The Huangshan road where data were collected

3.2 General Framework

Based on the dataset, as most previous studies did, we aggregate the traffic volumes into 20 min intervals and regard the aggregated volumes as the traffic flows to be predicted, i.e., each day is associated with 72 time intervals. Besides, we aim to predict traffic flows at road segment level, so we also aggregate traffic volumes according to *road_id* over different lanes. We give the formal problem definition as follows:

Definition 1. *(Short-term traffic flow prediction): Given a road segment r and a time interval $[t, t + \Delta)$, where $\Delta = 20$ minutes, the short-term traffic flow prediction problem aims to predict the aggregated traffic volume on road r during time $[t, t + \Delta)$ based on historical traffic flow information.*

The general framework of our proposed approach is shown in Fig. 2, which mainly consists of four steps. In the first step, traffic data are collected by microwave detectors, and each record has the format {*time, road_id, lane_id,*

Table 1. Format of the collected traffic data

Field	Type	Description
time	TIME	The time stamp when the record was collected
detector_id	INTEGER	The id of the detector
road_id	INTEGER	The number of the road segment as shown in Fig. 1
lane_id	INTEGER	The number of lane and each road segment has four lanes
vehicle_type	INTEGER	The type of vehicle (0: undefined, 1: pedestrian, 2: bike or motorbike, 3: car, 4: truck)
speed	NUMERIC	The average traffic speed (km/hour) on the lane
occupancy	NUMERIC	The average room occupancy (%) on the lane
volume	INTEGER	The collected traffic volume on the lane

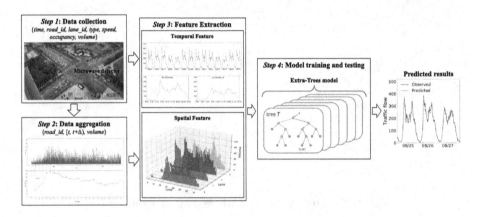

Fig. 2. The general framework of ExtTra

type, speed, occupancy, volume}. Then the data are aggregated into 20-minute intervals, as show in step 2, which provides a "smoother" representation, i.e., {road_id, [t + Δ), volume}. After that we extract spatial temporal features from both original and aggregated traffic data. In step 4 we feed the features to extremely randomized trees to train a learning model, which can be further used to predict the future traffic flows. Our major work lies in step 3, and we will give detailed description in the following subsections.

3.3 Feature Extraction

In our approach we comprehensively consider spatial temporal features, some of which were not considered in previous studies.

Temporal Features: To extract the temporal features, we first analysis the evolution pattern of dynamic traffic flows, as shown in Fig. 3. The blue curve represents the dynamic traffic flow of road 1 during two weeks from 3–16th,

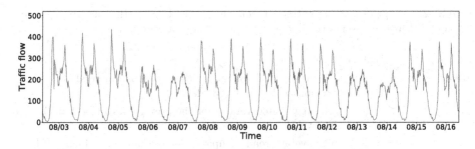

Fig. 3. The dynamic traffic flow during 3–16th August on road 1

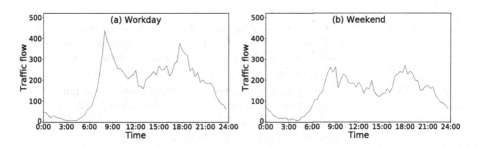

Fig. 4. The dynamic traffic flow for workday and weekend

August 2016. From this figure we can observe both long and short-term periodicity of the traffic follow patterns. The short-term periodicity can be observed within each day, with two peaks corresponding to the rush hours. For the long-term periodicity pattern, two short curves (e.g., 08/06 and 08/07) can be periodically observed in every seven curves. The short curves correspond to the weekends while the tall ones correspond to the workdays.

Figure 4(a) and (b) present detailed dynamic traffic flows of a work day (08/05) and a weekend (08/06) respectively. From both of the sub-figures we see that traffic flow peaks around 8:00 in the morning and 18:00 in the afternoon. Besides, traffic flows of work days are significantly higher than those of the weekends. Through the observations on Fig. 3 and 4, we can conclude that the temporal features, i.e., time of the day and day of the week, play an important role in the traffic flow patterns. However, most existing methods model the temporal features by the traffic flows of the same time in the previous days or the same day of previous weeks, while ignored the time features themselves. In this paper, we directly include the two features in our model and use Extra-Trees model to capture the non-learner correlations between these features and the traffic flows.

Spatial Features: To predict the traffic volume of a road segment during a given time interval, existing studies incorporate spatial features such as the traffic flow data of related roads. However, the lane information were hardly considered in previous studies. Since the lane level traffic data provide fine grained traffic

Table 2. Groups of features extracted from the traffic dataset (the numbers next to the group name represents the number of features in that group).

Group	Feature	Description
Base (12)	vol_i^d	The aggregated volume for the same time interval of the previous i-th ($i = 1, 2, 3$) day
	vol_i^t	The aggregated volume for the previous i-th ($i = 1, 2, 3$) interval
	occ_i	The average road occupancy for the same time interval of the previous i-th ($i = 1, 2, 3$) day
	$speed_i$	The average speed for the same time interval of the previous i-th ($i = 1, 2, 3$) day
Diff. (12)	$diff_{i,j}$	The difference of aggregated volume between current and the next i-th ($i = -1, 1, 2, 3$) time intervals on the previous j-th ($j = 1, 2, 3$) day, $i = -1$ represents the difference between the current and the previous time intervals
Temporal (5)	*time*	Current time of the day, since traffic flow is aggregated over 20 minute intervals, *time* varies form 1 to 72
	day	Current day of the week, which varies from 0 to 6, and 0 denotes Monday while 6 indicates Sunday
	day_i	Day of the week for the previous i-th ($i = 1, 2, 3$) day
Spatial (25)	$l_{i,j}^d$	The aggregated volume for the same time interval of the previous i-th ($i = 1, 2, 3$) day on the j-th ($j = 1, 2, 3, 4$) lane
	$l_{i,j}^t$	The aggregated volume for the previous i-th interval on the j-th ($j = 1, 2, 3, 4$) lane
	road	When this feature is used, all the above features of related roads will be included

flow formation, we also include them as spatial features in our approach by aggregating traffic volume into 20 min intervals at lane level.

In sum, the extracted features can be divided into four groups: (1) base feature group, which includes the historical traffic flow, occupancy, and speed information of the same road; (2) difference feature group, which includes the differences between historical traffic flows; (3) temporal feature group, which includes time of the day and day of the week; (4) spatial feature group, which includes the lane-level and related road traffic flows. The features are summarized in Table 2.

3.4   Extremely Randomized Trees

Extremely randomized trees (Extra-Trees) is a tree-based ensemble machine learning method proposed by Geurts et al. [15]. Compared with the traditional random forest model, Extra-Trees has two major differences: (1) it splits nodes of a decision tree by choosing cut-points fully at random; (2) unlike random forest model which uses a bootstrap resampling strategy to grow the trees, Extra-Trees uses the whole learning sample in order to minimize bias. The Extra-Trees model has the following advantages:

- The node splitting procedure is quite simple, making it more efficient than other ensemble based methods.
- Extra-Trees model supports non-linear prediction, so it is capable to capture our proposed new spatial temporal features very well.
- Due to its high randomness, Extra-Trees model can excellently overcome the overfitting problem.

Since previous studies [8, 16] have shown that the short-term traffic flow prediction problem exhibits strong non-linear properties, and based on the above analysis of the Extra-Trees model, we choose it as our prediction model.

4   Experimental Evaluation

4.1   Experiment Setup

We first introduce the baseline methods, evaluation metrics, and the model training process.

Baseline Methods: We compare ExtTra with the following seven baseline methods: autoregressive integrated moving average (**ARIMA**), random forest (**RF**), gradient boosted decision tree (**GBDT**), artificial neural networks (**ANN**), Bayes ridge regression (**BRR**), linear regression (**LR**), and support vector regression (**SVR**) with linear kernel. These baselines cover a wide spectrum of existing approaches, including the time series method (ARIMA), parametric method (BRR), nonparametric methods (RF, GBDT, LR, SVR), and neural network-based approach (ANN). All the base line methods are implemented using the Scikit-learn package [17].

Evaluation Metrics: We employ three metrics, i.e., mean absolute error (**MAE**), mean absolute percentage error (**MAPE**), and root mean squared error (**RMSE**). These measures are widely used to evaluate the traffic flow prediction performance and are defined as follows:

$$MAE = \frac{1}{n} \sum_{i=1}^{n} |f_i - \hat{f}_i| \tag{1}$$

$$MAPE = \frac{1}{n} \sum_{i=1}^{n} \left| \frac{f_i - \hat{f}_i}{f_i} \right| \tag{2}$$

$$RMSE = \sqrt{\frac{\sum_{i=1}^{n} (f_i - \hat{f}_i)^2}{n}} \tag{3}$$

where $\hat{f}_i$ refers to the predicted traffic flow, f_i is the observed traffic flow.

Training and Testing: As we have introduced in Sect. 3.1, the traffic dataset contains about 46 days' traffic flow data. We use the first 34 days (about 75%) for model training and the rest 12 days (about 25%) for testing. The testing set covers the time from 25th August to 5th September. The training set has 2324 examples while the testing set has 864 examples. Because the traffic flows on most of the road segments (except road 1 and 4) suffer severe data missing problem, we only include road 1 and 4 in our experiments. Specifically, we predict the traffic flow of road 1 based on the features extracted from both road 1 and 4, where road 4 is considered as the related road of road 1, as shown in Fig. 1. The parameters of all the methods are tuned through 10-fold cross-validation and we export the results with the best parameters for all methods.

4.2   Experimental Results

Overall Results: We first give the overall prediction performance comparison of different methods, as shown in Fig. 5, where we used all the features to train the models. From Fig. 5 we see that the ExtTra approach significantly outperforms all the baseline methods in term of MAPE, RMSE, and MAE. Since the ARIMA model only relies on the historical traffic flows of the same road without considerations of related roads or other information, it exhibits the worst performance. From the results of MAPE, we see that non-linear models (ExtTra, RF, GBDT and ANN) performs much better than linear models (LR and SVR), this is because the former models are more capable of capturing the non-linear correlations between the features (time of the day, day of the week) and the traffic flows, which again verifies the effectiveness of our extracted new features. In sum, the results in Fig. 5 exhibit the superiority of our approach over the baselines.

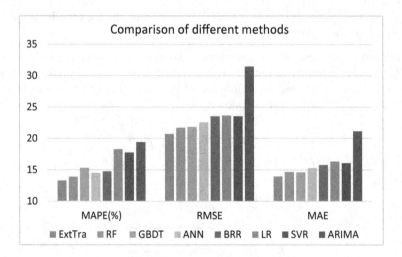

Fig. 5. The overall prediction performance of different methods (all features)

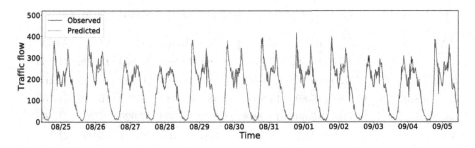

Fig. 6. The observed dynamic traffic flows versus those predicted by ExtTra (Color figure online)

Predicted vs Observed: We plot the dynamic traffic flows predicted by Ext-Tra (with all features) and compare them with the observed (real) traffic flows, as shown in Fig. 6, where the blue curve denotes the observed results while the red dotted curve represents the predicted ones. Overall, the predicted curve is quite close to the observed one, except for some points where the observed curve exhibits strong fluctuations while the predicted curve is more smooth, e.g., the data points around 9/05 and 9/05. From the figure we see the fluctuations are more likely to occur during the midday, from about 11:00 to 15:00. These fluctuations are the major sources of prediction errors and they are very difficult to predict, because some of them are caused by road accidents, which are unpredictable.

Table 3. MAPE comparison of different methods under different feature groups

MAPE(%)	ExtTra	RF	GBDT	ANN	BRR	LR	SVR	ARIMA
Base	**15.274**	15.918	16.050	15.841	16.117	15.862	16.158	19.398
Base+Diff.	**14.375**	15.120	15.490	14.691	15.204	15.119	14.946	19.398
Base+Spatial	**14.677**	14.935	15.924	14.791	16.084	15.346	15.768	19.398
Base+Temporal	**14.511**	15.071	14.810	14.527	21.417	22.041	20.461	19.398
All	**13.291**	13.896	15.311	14.483	14.742	18.250	17.726	19.398

Table 4. RMSE comparison of different methods under different feature groups

RMSE	ExtTra	RF	GBDT	ANN	BRR	LR	SVR	ARIMA
Base	**23.069**	24.202	23.748	24.631	25.748	25.711	25.763	31.452
Base+Diff.	**22.320**	23.944	22.934	23.211	24.642	24.726	24.537	31.452
Base+Spatial	**22.522**	23.416	23.244	23.326	24.379	24.326	24.312	31.452
Base+Temporal	21.814	22.479	**21.784**	24.043	25.462	25.516	25.249	31.452
All	**20.713**	21.733	21.844	22.548	23.536	23.659	23.540	31.452

Table 5. MAE comparison of different methods under different feature groups

MAE	ExtTra	RF	GBDT	ANN	BRR	LR	SVR	ARIMA
Base	**15.917**	16.562	16.276	16.408	17.086	17.053	17.090	21.165
Base+Diff.	**15.020**	16.212	15.621	15.283	16.180	16.259	16.129	21.165
Base+Spatial	**15.547**	15.897	15.999	15.631	16.348	16.280	16.246	21.165
Base+Temporal	14.710	15.394	**14.698**	15.950	17.500	17.626	17.201	21.165
All	**13.959**	14.661	14.597	15.270	15.765	16.311	16.054	21.165

Comparison of Feature Groups: We compare the prediction performance of different methods under different feature groups, and the results of MAPE, RMSE, and MAE are shown in Tables 3, 4 and 5, respectively. In Table 3 we see ExtTra exhibits the best performance in terms of MAPE under all the feature groups while ARIMA gives the worst performance. In general, the non-linear models (ExtTra, RF, GBDT, and ANN) performs better than the linear models (LR and SVR). For most of the methods, including more features can significantly improve the prediction accuracy. Take our ExtTra method for example, its MAPE under the *Base* feature group is 15.274. When we add the *Diff.*, *Spatial*, and *Temporal* feature groups, its MAPEs are reduced to 14.375, 14.677 and 14.511, respectively. If we use all the features, its MAPE can be further reduced to 13.291. However, on three models (LR, SVR, and BRR), we see that when the temporal features are included, the prediction accuracy becomes even worse. This is because the features in the *Temporal* group (time of the day, day of the week) are non-linearly related to the traffic flow and the linear models (actually BRR is also a linear model) are incapable to capture these non-linear characteristics. Table 4 and Table 5 show similar results, where our ExtTra approach exhibits the best overall performance, except for the *Base + Temporal* feature group, where it slightly loses to the GBDT method.

In sum, through the extensive experiments on the traffic dataset, our ExtTra approach exhibits its effectiveness in capturing the non-linear features of traffic flow patterns and generating accurate predictions.

5   Conclusion

In this paper we proposed an extremely randomized trees based approach for short-term traffic flow prediction. Our work provides the first effort to apply the extremely randomized trees model on the traffic flow prediction problem. In our approach we used some spatial temporal features which were not considered in previous studies, and showed that these features can improve the prediction accuracy. We compared our approach with a spectrum of baselines through extensive experiments and the results verified the effectiveness of our approach. Our work provides a new perspective to the research of traffic flow prediction.

Acknowledgements. This work was supported in part by: National Natural Science Foundation of China (No. 61702059), Frontier and Application Foundation Research Program of Chongqing City (No. cstc2018jcyjAX0340), Chongqing Industrial Generic Technology Innovation Program (No. cstc2017zdcy-zdzxX0010), Guangxi Key Laboratory of Trusted Software (No. kx201702).

References

1. Hamed, M.M., Al-Masaeid, H.R., Said, Z.M.B.: Short-term prediction of traffic volume in urban arterials. J. Transp. Eng. **121**, 249–254 (1995)
2. Williams, B.M., Hoel, L.A.: Modeling and forecasting vehicular traffic flow as a seasonal ARIMA process: theoretical basis and empirical results. J. Transp. Eng. **129**, 664–672 (2003)
3. Kumar, S.V., Vanajakshi, L.: Short-term traffic flow prediction using seasonal ARIMA model with limited input data. Eur. Transp. Res. Rev. **7**, 21 (2015)
4. Pascale, A., Nicoli, M.: Adaptive Bayesian network for traffic flow prediction. In: 2011 IEEE Statistical Signal Processing Workshop (SSP), pp. 177–180 (2011)
5. Wu, Y.-J., Chen, F., Lu, C.-T., Yang, S.: Urban traffic flow prediction using a spatio-temporal random effects model. J. Intell. Transp. Syst. **20**, 282–293 (2016)
6. Rajabzadeh, Y., Rezaie, A.H., Amindavar, H.: Short-term traffic flow prediction using time-varying Vasicek model. Transp. Res. Part C Emerg. Technol. **74**, 168–181 (2017)
7. Jeong, Y.S., Byon, Y.J., Castro-Neto, M.M., Easa, S.M.: Supervised weighting-online learning algorithm for short-term traffic flow prediction. IEEE Trans. Intell. Transp. Syst. **14**, 1700–1707 (2013)
8. Dell'Acqua, P., Bellotti, F., Berta, R., Gloria, A.D.: Time-aware multivariate nearest neighbor regression methods for traffic flow prediction. IEEE Trans. Intell. Transp. Syst. **16**, 3393–3402 (2015)
9. Xia, D., Wang, B., Li, H., Li, Y., Zhang, Z.: A distributed spatial-temporal weighted model on MapReduce for short-term traffic flow forecasting. Neurocomputing **179**, 246–263 (2016)
10. Huang, W., Song, G., Hong, H., Xie, K.: Deep architecture for traffic flow prediction: deep belief networks with multitask learning. IEEE Trans. Intell. Transp. Syst. **15**, 2191–2201 (2014)
11. Yu, H., Wu, Z., Wang, S., Wang, Y., Ma, X.: Spatiotemporal Recurrent Convolutional Networks for Traffic Prediction in Transportation Networks. Sensors **17**, 1501 (2017)
12. Du, S., Li, T., Gong, X., Yang, Y., Horng, S.J.: Traffic flow forecasting based on hybrid deep learning framework. In: 12th International Conference on Intelligent Systems and Knowledge Engineering (ISKE), pp. 1–6 (2017)
13. Zhang, J., Zheng, Y., Qi, D.: Deep spatio-temporal residual networks for citywide crowd flows prediction. In: AAAI, pp. 1655–1661 (2017)
14. Wang, D., Xiong, J., Xiao, Z., Li, X.: Short-term traffic flow prediction based on ensemble real-time sequential extreme learning machine under non-stationary condition. In: IEEE 83rd Vehicular Technology Conference (VTC Spring), pp. 1–5 (2016)
15. Geurts, P., Ernst, D., Wehenkel, L.: Extremely randomized trees. Mach. Learn. **63**, 3–42 (2006)

16. Yang, H.F., Dillon, T.S., Chen, Y.P.P.: Optimized structure of the traffic flow forecasting model with a deep learning approach. IEEE Trans. Neural Netw. Learn. Syst. **28**, 2371–2381 (2017)
17. Pedregosa, F., Varoquaux, G., Gramfort, A., Michel, V., Thirion, B., Grisel, O., Blondel, M., Prettenhofer, P., Weiss, R., Dubourg, V.: Scikit-learn: machine learning in Python. J. Mach. Learn. Res. **12**, 2825–2830 (2011)

Actor Model Anomaly Detection Using Kernel Principal Component Analysis

Chunze Wang$^{(\boxtimes)}$, Jing Wang, Chun Wang, and Qiwei Shen

Beijing University of Posts and Telecommunications, Beijing 100876, China
wangchunze@126.com

Abstract. With the increasing complexity of Internet applications, traditional software architectures have been unable to support the pressure of system access brought about by user growth. Distributed systems have gradually become the mainstream architecture, and messaging has become a widely adopted model. Akka is a distributed framework based on the Actor message communication model. At present, the fault and anomaly detection for the Actor system is mainly to capture the anomaly in the code writing, it is difficult to decouple from the program, so an algorithm using kernel principal component analysis algorithm based on message monitoring is proposed to detect anomaly on Actor system. In this paper, we obtain the message of Actor system by using AspectJ's slicing of the byte code injection of Java code, and we can use Kernel Principal Component Analysis algorithm to perform data dimension reduction and feature extraction through nonlinear mapping. Then the k-means algorithm was used for cluster analysis. The LOF (local outlier points factor) algorithm was used to compare the density of each point p and its neighborhood points to determine abnormal points. Finally, we took the spider program based on the Actor model as a case to collect data and do the experiment, which verified the validity and rationality of the method.

Keywords: Anomaly detection · Kernel principal component analysis
Akka actor · Nonlinear process · K-means

1 Introduction

With the continuous development of technology, the complexity of Internet applications is increasing. The traditional software architecture has been difficult to support the pressure of system access brought about by user growth. Distributed systems have gradually become the mainstream architecture, and task parallel design has become the mainstream of program designs [1]. In order to solve the problem of concurrency under distributed conditions, message communication has become a widely used mode [2]. The Actor model is a parallel computing model proposed by Hewitt et al. [3] in the computer science fields in 1973. The model treats the actor as a generic parallel computing primitive: an actor responds to received messages, makes local decisions; creates more actor, or sends more messages; it also prepares to receive the next message. Actors do not depend on each other and communicate with asynchronous messages [4]. Its emerging solves the cumbersome and inefficient problems of Java in the

© Springer Nature Switzerland AG 2018
L. Cheng et al. (Eds.): ICONIP 2018, LNCS 11304, pp. 545–554, 2018.
https://doi.org/10.1007/978-3-030-04212-7_48

use of multithreading and locking mechanisms in concurrent programming. Now the Actor model has become the theoretical basis of many parallel computing systems.

Theoretical research shows that existing technologies cannot completely eliminate software failures [5]. In the system that uses the Actor message communication model, message sending and message consumption are interacted through the network. Network instability can cause the message to be lost or duplicated, which may cause inconsistency in the message. The design and management of the Actor system are complicated. Once the system is abnormal, it is difficult to be found in time, resulting in the system timeliness reduction, hardware resource waste and other problems. There are many causes of system exceptions, such as software design flaws, code errors, network errors, and insufficient hardware resources.

The commonly used methods for anomaly detection include statistical time series analysis [6], supervised support vector machine model, random forest algorithm [7, 8], and unsupervised k-means clustering algorithm [9] and so on.

This topic discusses a Java/Scala-based Actor model that runs on the JVM. At present, the abnormality detection method for the Actor system is mainly based on Java code and JVM level detection. Yang et al. [10] proposed a static analysis method of the Java null pointer abnormality that describes the fault by the fault mode state machine and iterate the fault state according to the result of the data flow analysis. This method is suitable for null pointer anomaly detection. Song et al. [11] proposed a detection method based on Java method call chain, which is suitable for detecting the data competition anomaly of Java multi-thread. Wang et al. [12] implemented an invariant detection tool that uses objects as granularity and is suitable for error detection in Java concurrent processes. Yu et al. [13] acquired the stack information in the Java program running in real time through JDI, and proposed a statistical algorithm for object reachability, which is suitable for dynamic detection of memory leakage in Java programs. These methods are aimed at the static analysis of Java code and the dynamic detection of JVM and can play a certain role in the abnormal detection of the null pointer of the Actor model system, data competition anomaly during concurrency, and memory overflow. However, these algorithms have large limitations, which are difficult to decouple, and it is difficult to discover the abnormal in complex Actor systems in time.

Therefore, an algorithm based on the kernel principal component analysis algorithm for Actor system anomaly detection by message monitoring is proposed. The following three aspects of research were mainly carried out on the subject.

(1) The message passing in the Actor system will be obtained by an annotation-based program slicing method. The main method is to use AspectJ to perform byte code injection on Scala/Java code. AspectJ is an easy-to-use, powerful AOP programming language that extends Java's implantation rules in pointcuts, join points, advice, and aspects [14].

(2) For a large Actor system, there may be hundreds of messages. The method of kernel principal component analysis is used to reduce the dimension, and the main direction vector of the original data in the high-dimensional feature space is obtained. PCA (Principal Component Analysis) is an unsupervised learning algorithm and is often used for data reduction, lossy data compression, feature

extraction, data visualization, exploratory data analysis, pattern recognition, and time series prediction [15]. PCA is linear, and it tends to be powerless for non-linear data. When input data are linearly indivisible, KPCA (Kernel Principal Component Analysis) can be used to map linearly indivisible data to linearly separable feature Spaces by nonlinear mapping [16–18].

(3) Using k-means algorithm for cluster analysis, k-means clustering is one of the most classically used and most widely used clustering methods [19, 20]. Then we can use LOF algorithm to determine whether the point is an abnormal point by comparing the density of each point p and its neighboring points.

2 Introduction

2.1 Feature Extraction

PCA is a classic linear feature extraction method. By projecting data from high-dimensional n to low-dimensional k, the original data information is represented by a small number of k-dimensional features. This k-dimensional feature is a completely reconstructed new orthogonal feature. Instead of simply removing the remaining n-k dimension features from the n-dimensional features. PCA can be defined as an orthogonal projection of data in a low-dimensional linear space called the principal subspace, which maximizes the variance of the projection data (Hotelling 1933), i.e., the maximum variance Theory [21].

Let $x_i \in R^d (i = 1, 2, 3, \ldots, N)$ be N samples in d-dimensional space, that is, $d \times N$ matrix. If d-dimensional spatial species sample X is linear, d-dimensional spatial sample Projected onto the plane $W = (w_1, w_2, w_3, \cdots, w_d)$, w is the eigenvector in space and λ is the eigenvalue.

Then the size of the projected value can be expressed as $w^T \cdot x_j$, and the projected variance is

$$\sigma^2 = \frac{1}{N-1} \sum_{i=1}^{N} \left(w^T x_i - 0\right)^2 = \frac{1}{N-1} \sum_{i=1}^{N} \left(w^T x_i\right)\left(w^T x_i\right) \tag{1}$$

From this, we can get the covariance matrix of the feature space as

$$C = \frac{1}{N-1} \sum_{i=1}^{N} x_i x_i^T \tag{2}$$

Where C is the covariance matrix. And the expression for solving the eigenvalues and eigenvectors of C is

$$Cw = \lambda w \tag{3}$$

Where λ is the eigenvalue and w is the eigenvector. From this we can get

$$w_j = \frac{1}{\lambda} \sum (x_i \cdot x_i^T) w_j = \frac{1}{\lambda} \sum (x_i^T \cdot w_j) x_i \qquad (4)$$

Where $x^T \cdot w$ is scalar, let $\alpha = \frac{1}{\lambda} x_i^T \cdot w_j$, we can get

$$w_j = \sum \alpha x_i \qquad (5)$$

In order to better deal with nonlinear data, introduce a nonlinear mapping function Φ, mapping the nonlinear data in the original space to a linear high-dimensional space, according to formula (5) can be obtained

$$w_j = \sum \alpha \phi(x_i) \qquad (6)$$

By substituting $\phi(x_i)$ into (3), we can get

$$\sum_{i=1}^{d} (\phi(x_i) * \phi(x_i)^T) w_j = \lambda w_j \qquad (7)$$

Then substitutes (6) into (7), we can get

$$\phi(x_i)^T \sum_{i=1}^{d} (\phi(x_i) * \phi(x_i)^T) \sum \alpha \phi(x_i) = \lambda \phi(x_i)^T \sum \alpha \phi(x_i) \qquad (8)$$

We can let

$$k(x_1, x_2) = \phi_{(x_i)}^T \cdot \phi(x_i) \qquad (9)$$

Substitute (9) into (8), we can get

$$K\alpha = \lambda \alpha \qquad (10)$$

The matrix K is called the nucleus matrix, where α is an eigenvector of K. Then the eigenvector w in the high dimensional space can be obtained from α by Eq. (6).

Commonly used kernel functions mainly include:

(1) Linear Kernel: $k(x, y) = x^T y + c$
(2) Polynomial Kernel: $k(x, y) = (ax^T y + c)^4$
(3) Gaussian Kernel: $k(x, y) = \exp\left(-\frac{\|x-y\|^2}{2\sigma^2}\right)$
(4) Sigmoid Kernel: $k(x,y) = \tanh(\alpha x^T + c)$

We can select the appropriate kernel function to solve the eigenvalues and eigenvectors of the kernel matrix K, so as to further obtain the main direction vector. Assuming that we take the feature vector corresponding to the first m eigenvalues, the main direction vector of the original data is

$$w = \lambda_1 w_1 + \lambda_2 w_2 + \cdots + \lambda_m w_m \tag{11}$$

In order to make the main directional vector unity, $\lambda = (\lambda_1, \lambda_2, \lambda_3, \ldots, \lambda_m)$ is the normalized result.

2.2   Data Clustering

K-means clustering algorithm is an unsupervised learning algorithm for classifying a given data set into k clusters, where similar objects are grouped into the same cluster [22, 23]. Assume that the original sample set is $X = \{x^{(1)}, x^{(2)}, \cdots, x^{(m)}\}$, the calculation method is:

(1)   Randomly select k cluster centroids, set as $\mu_1, \mu_2, \ldots, \mu_k \in R^n$
(2)   Traverse all data and calculate the similarity $S(x_i, \mu_k)$,between the sample x_i and all centroids μ_k, and divide each data into the nearest center point.
(3)   Calculate the average of each cluster as a new center point, where the algorithm for calculating the center of the point group can be Minkowski Distance formula:

$$d_{ij} = \lambda \sqrt{\sum_{k=1}^{n} |x_{ik} - x_{jk}|^{\lambda}} \tag{12}$$

Where λ can be arbitrary, can be negative, it can be positive, or infinite.

(4)   Repeat step (2) and (3) until all center points no longer change or reach the set number of iterations.

We can use the mean square error as the evaluation index.

$$J = \sqrt{\sum_{j=1}^{k} \sum_{q=1}^{nj} \sum_{p=1}^{d} \left(x_{jq}^p - u_j^p\right)^2 / (n-1)} \tag{13}$$

d is the sample dimension, n is the number of samples in the i-th cluster in the clustering result, and k is the clustering parameter.

(5)   Calculate the local outlier factor (LOF)

The LOF algorithm determines whether the point is an anomaly by comparing the density of each point p and its neighboring points. The density is calculated by the distance between the points. The farther the distance between points, the lower the density, and the closer the distance, the higher the density. If the density of point p is lower, the more likely it is to be recognized as an abnormal point [24]. The formula of LOF is

$$\text{LOF}_k(p) = \frac{\sum_{o \in N_k(p)} \frac{lrd_k(o)}{lrd_k(p)}}{|N_k(p)|} = \frac{\sum_{o \in N_k(p)} lrd_k(o)}{|N_k(p)|} / lrd_k(p) \tag{14}$$

If the ratio is closer to 1, it shows that the neighborhood density of p is almost the same, and p may be in the same cluster as the neighborhood. If the ratio is less than 1, the density of p is higher than that of its neighborhood point, and p is dense. If the ratio is greater than 1, it means that the density of p is less than the density of its neighborhood points, and p is more likely to be an anomaly.

2.3   Actor Model Anomaly Detection Algorithm

It can be seen that through the processing of KPCA, the non-linear information contained in the data set can be mined to achieve the purpose of data compression, noise separation and correlation removal among samples. The k-means algorithm is used to cluster the data processed by KPCA algorithm, and the abnormal information of the data is judged according to LOF, the local outlier factor.

The process diagram of the proposed KPCA algorithm for Actor model exception detection is as follows (Fig. 1 and Table 1):

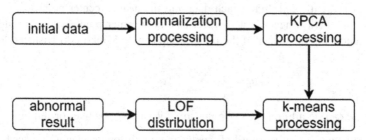

Fig. 1. Actor model exception detection flow diagram.

Table 1. The experimental results

Dimension m	Neighbor number k	Run time (in seconds)	Accuracy
2	4	4.62	0.68
2	6	5.01	0.73
2	8	5.73	0.81
2	10	6.16	0.77

The proposed KPCA algorithm based on the Actor model exception detection algorithm is as follows:

Input: raw Data, main dimension m, nearest neighbor k.

Output: abnormal message data.

(1)   data = FORMAT(rawData); //Standardize raw data
(2)   mainData = KPCA(data);
(3)   res = KMEANS(mainData); //k-means clustering
(4)   size = length(m);
(5)   for i in 1 to size:
(6)   res(i).lof = LOF(res(i));
(7)   return res;

Algorithm complexity analysis: for dimension for p, data of size n, t iteration, divided into k clusters, to standardize data processing time complexity is O (n), the complexity of the general analysis of the KPCA to $O(n \times p^3)$, k-means cluster complexity to $O(ntk)$, comprehensive analysis of the algorithm's time complexity is $O(n \times m)$, m for max (p^3, ntk).

3 The Experimental Results

3.1 Experimental Data Collections

Distributed crawler system developed by Actor system based on Akka framework is adopted to collect message data and verify the algorithm. The original data consists of the consumption number of various messages per second. Experimental data were collected data of program running for a period of time, and the DDOS attacks in the process of the program is running, change the database connection number, modify memory size, limit the CPU usage, and limit the network bandwidth to simulate operation anomalies, and record the corresponding operation time. The final raw data include 32 types of messages, with a total of 1.6 million pieces of data.

3.2 Experimental Results Display

It was found that the cumulative contribution rate of the first two principal components was more than 93%, so the number of principal components m was 2. The LOF is calculated by this algorithm. If the LOF value of p at a certain point is greater than 1, the density of p is less than that of its neighborhood point, and p is more likely to be an abnormal point.

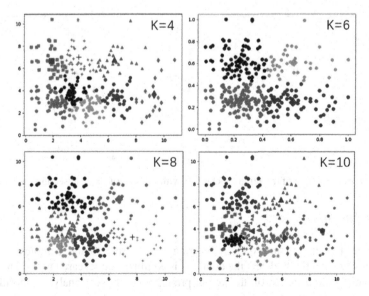

Fig. 2. K-means clustering effect diagram (k = 4,6,8,10).

Figure 2 shows the clustered results when m = 2, k is 4, 6, 8 and 10, respectively. And Fig. 3 shows the distribution of LOF values as a function of message sequence. Data greater than 1 marked in the figure may be an abnormal situation.

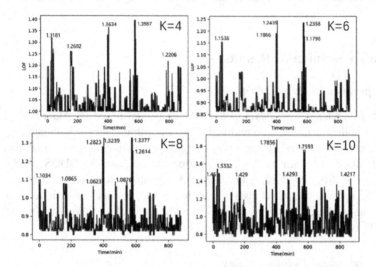

Fig. 3. LOF value distribution diagram of data points (k = 4,6,8,10).

Both the size of the data set and the value of the algorithm may affect the efficiency of the algorithm. It can be seen from Fig. 4 that as the number of neighbors k increases, the algorithm consumption time increases gradually. At the same time, as the size of the data set increases, the time consumed by the algorithm also gradually increases.

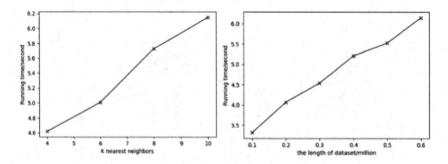

Fig. 4. The running time of the program changes with the size of the data set and the value of k.

4   Conclusions

For systems that apply the Actor messaging model, due to the complexity of the system, how to effectively detect the anomaly is an important research topic. This paper presents an algorithm based on the kernel principal component analysis algorithm for

Actor system anomaly detection by message monitoring, we can use byte code injection through AspectJ to obtain multidimensional message sequence, and use Kernel Principal Component Analysis for data reduction and feature extraction, then we can cluster the data by K-means algorithm, and use the LOF algorithm to compare the density of each point p and its neighborhood points for abnormal point judgments. Finally, we have taken the spider program based on the Actor model as a case to collect data and do an experiment, which verified the validity and rationality of the method. At present, only anomaly detection has been performed for the already running program, but the future state of the program cannot be judged. How to predict the abnormality of the program in the future and classify the anomaly will be further studied in the future.

References

1. Chen, G.L., Zhong, G.: Study on parallel computing. J. Comput. Sci. Technol. **21**(5), 665–673 (2006)
2. Zhao, Y.: On conversion mechanism of asynchronous communication message based on time decoupling in distributed object-oriented middleware. Comput. Appl. Softw. **25**(10), 160–162 (2008)
3. Hewitt, C., Bishop, P., Steiger, R.: A universal modular ACTOR formalism for artificial intelligence. In: International Joint Conference on Artificial Intelligence, pp. 235–245 (1973)
4. Zhang, X.W., Jia, X.Y., Chang, L., Liu, Q.C., Hu, X.H.: Parallel dynamic symbolic execution method based on actor model. J. Chin. Comput. Syst. **39**(1), 12–16 (2018)
5. Wang, Z.B., Li, C.Y.: A survey of the technique for software fault diagnosis. Microcomput. Inf. **26**(34), 161–163 (2010)
6. Sun, T.: Data stream outlier detection study based on time series analysis. University of Electronic Science and Technology of China, Xi'an (2016)
7. Erfani, S.M., Rajasegarar, S., Karunasekera, S., Leckie, C.: High-dimensional and large-scale anomaly detection using a linear one-class SVM with deep learning. Pattern Recogn. **58**(C), 121–134 (2016)
8. Chen, F., Liu, Z., Sun, M.T.: Anomaly detection by using random projection forest. In: IEEE International Conference on Image Processing, pp. 1210–1214 (2015)
9. Lahrache, A., Cocconcelli, M., Rubini, R.: Anomaly detection in a cutting tool by K-means clustering and support vector machines. Pol. Towarz. Diagn. Tech. **18**(3), 21–29 (2017)
10. Yang, R., Jin, D.H., Gong, Y.Z., Ma, Y.: Static analysis method for detecting null pointer deference in Java. J. Tsinghua Univ. (Sci. Technol.) **51**(s1), 1509–1514 (2011)
11. Song, D.H., Chen, E.H.: A call-chain-based static data race detection algorithm for java program. Ship Electron. Eng. **33**(12), 59–63 (2013)
12. Wang, D., Yang, M., Zhou, X.: An object-centric concurrency bug detection framework for Java. J. Chin. Comput. Syst. **34**(06), 97–102 (2013)
13. Yu, Q., Jiang, S.J., Wang, X.Y., Ju, X.L., Dong, Q.J.: Detection and measurement method of memory leaking objects. J. Front. Comput. Sci. Technol. **8**(8), 978–988 (2014)
14. Ma, X., Li, Q.S., Chen, P.G.: Basic information needed in reversing-engineer by Aspect. Comput. Syst. Appl. **20**(02), 63–67 (2011)
15. Tipping, M.E., Bishop, C.M.: Probabilistic principal component analysis. J. R. Stat. Soc. **61**(3), 611–622 (2010)

16. Sheriff, M.Z., Karim, M.N., Nounou, M.N., Nounou, H., Mansouri, M.: Fault detection of nonlinear systems using an improved KPCA method. In: International Conference on Control, pp. 0036–0041 (2017)
17. Li, Q., Zhou, X.S.: Multivariate time series anomaly detection method based on KPCA. Comput. Measur. Control **19**(4), 822–825 (2011)
18. Fezai, R., Jaffel, I., Taouali, O., Harkat, M.F., Bouguil, N.: Online process monitoring based on Kernel method. In: International Conference on Control, pp. 236–241 (2017)
19. Dave, T.: OpenFlow: enabling innovation in campus networks. ACM SIGCOMM Comput. Commun. Rev. **38**(2), 69–74 (2008)
20. Farias, F.N.N., Salvatti, J.J., Cerqueira, E.C., Abelem, A.J.G.: A proposal management of the legacy network environment using OpenFlow control plane. In: Network Operations and Management Symposium, vol. 104, no. 5, pp. 1143–1150 (2012)
21. Hotelling, H.: Analysis of a complex of statistical variables into principal components. J. Educ. Psychol. **24**(6), 417–520 (1933)
22. Zhai, R.: Virtual machine exception detection based on nearest neighbor algorithm. Netw. Secur. Technol. Appl. (6), 55-56 (2016)
23. Meng, H.D., Ren, J.P.: Clustering algorithm based on cloud computing platform. Comput. Eng. Des. 2990–2994 (2015)
24. Breunig, M.M.: LOF: identifying density-based local outliers. In: ACM Sigmod International Conference on Management of Data, vol. 29, no. 2, pp. 93–104 (2000)

Passive Detection of Splicing and Copy-Move Attacks in Image Forgery

Mohammad Manzurul Islam[(⊠)], Joarder Kamruzzaman,
Gour Karmakar, Manzur Murshed, and Gayan Kahandawa

School of Science, Engineering and Information Technology,
Federation University Australia, Churchill, Australia
{mm.islam,joarder,gour,manzur.murshed,
g.appuhamillage}@federation.edu.au

Abstract. Internet of Things (IoT) image sensors for surveillance and moni-toring, digital cameras, smart phones and social media generate huge volume of digital images every day. Image splicing and copy-move attacks are the most common types of image forgery that can be done very easily using modern photo editing software. Recently, digital forensics has drawn much attention to detect such tampering on images. In this paper, we introduce a novel feature extraction technique, namely Sum of Relevant Inter-Cell Values (SRIV) using which we propose a passive (blind) image forgery detection method based on Discrete Cosine Transformation (DCT) and Local Binary Pattern (LBP). First, the input image is divided into non-overlapping blocks and 2D block DCT is applied to capture the changes of a tampered image in the frequency domain. Then LBP operator is applied to enhance the local changes among the neigh-bouring DCT coefficients, magnifying the changes in high frequency compo-nents resulting from splicing and copy-move attacks. The resulting LBP image is again divided into non-overlapping blocks. Finally, SRIV is applied on the LBP image blocks to extract features which are then fed into a Support Vector Machine (SVM) classifier to identify forged images from authentic ones. Extensive experiment on four well-known benchmark datasets of tampered images reveal the superiority of our method over recent state-of-the-art methods.

Keywords: Digital forensics · Splicing attack · Copy-move attack
Discrete Cosine Transformation · Local Binary Pattern
Support Vector Machine

1 Introduction

Today, Internet of Things (IoT) has emerged as an integrated technology in our daily life. According to Business Insider Intelligence [1], there will be more than 24 billion IoT devices by 2020 which results in approximately four devices per person living on earth. Our everyday essential devices such as wearable sensors, visual sensors, home appliances, security devices, etc. are increasingly being connected to the Internet. Among them, visual sensors play a vital role in physical and cyberspace security and surveillance. Digital social media platforms like Facebook and Instagram are being flooded with millions of images each day. For many cutting-edge applications, people

© Springer Nature Switzerland AG 2018
L. Cheng et al. (Eds.): ICONIP 2018, LNCS 11304, pp. 555–567, 2018.
https://doi.org/10.1007/978-3-030-04212-7_49

rely on image data more than any other form of data. However, sophisticated digital image editing tools and software have become available. They are very easy to use and they can generate fake images that appear to be very natural. The forged images generated by these tools do not leave any trace for human visual system. Hiding facts, spreading negative propaganda, disrupting operational and decision-making processes have become very common in today's online media. Among all the possible image tampering operations, splicing and copy-move are the most notorious and commonly used attacks on digital images [2]. Image splicing forgery is done by copying one or more portion of an image and pasting it on another image, while in copy-move forgery, one or more objects of an image is copied and then are pasted on the other part of the same image.

As we know that 'a picture is worth a thousand words', an artificially altered image can have devastating consequences. During the 2017 G-20 summit in Germany, AP photojournalist Markus Schreiber captured the image in Fig. 1a prior to the first working session on the very first day of the summit. Later this picture was most likely edited and uploaded in social media as Fig. 1b by a Russian journalist and Putin loyalist Vladimir Soloviev [3]. Although he soon deleted the post from Facebook, it already spread all over the world and introduced new debate and confusion in world politics. In the same way, an altered image can mislead the world leaders in making business decision, taking political steps or even starting a nuclear war.

(a) Authentic image   (b) Spliced image

Fig. 1. Image splicing example

Modern photomontage does not leave any trace for naked eyes, yet they can be identified through digital forensics. The existing methods for identifying image forgery can be roughly divided into two categories: active and passive. Active methods (e.g., [4]) rely on injecting digital watermark or signature into the original image. To verify the authenticity of an image, the receiver checks if the digital watermark or signature is unchanged or not. Unfortunately, most of the image sensors do not have the capability to integrate complex digital watermarking functionalities because of high cost and resource requirements. As a result, active techniques are not commonly observed and practised in today's data driven IoT network. On the other hand, passive approaches (e.g., [5, 6]) do not need such prior knowledge, require less resources, and hence have drawn much attentions in digital forensics in recent years. The main idea behind passive (blind) detection is that an altered image might not be visibly identifiable as tampered, but tampering obviously introduces disturbance in the structural and statistical characteristics of an image. To be more specific, image tampering introduces new micro-patterns and sharp edges along the boundary of the pasted area. From signal processing's point of view, splicing and copy-move artifacts are the 'noise' inserted into a clear signal.

A major portion of images that are targeted for tampering are security sensitive images captured through security and surveillance cameras installed in factory warehouses, shops, financial institutes, military installations, government vaults, border defence etc. These images are mostly in gray scale due to the nature of their applications, lighting condition and recording time (e.g., night time). Again, color images can also be converted into gray scale images. All these justify the advancement of detecting attacks on gray scale images as the attack detection methods for gray scale images can be used in both gray and color images.

Although many researchers have proposed different approaches to image forgery detection with promising accuracy, there are still scopes for the advancement of these techniques using innovative features that are more discriminative and sensitive to the tampering artifacts produced by splicing and copy-move attacks. To achieve this, in this paper, firstly, we introduce a novel feature extraction technique, namely Sum of Relevant Inter-Cell Values (SRIV) for propagating the effects of splicing and copy-move attacks into all features more explicitly than representing it using typical features such as histogram or higher order statistical moments based features. Secondly, using SRIV features, we then propose a passive (blind) detection method using Discrete Cosine Transformation (DCT) and Local Binary Pattern (LBP) for detecting splicing attacks on image. Since LBP can enhance the local changes among the neighbouring DCT coefficient values, first we identify the micro-patterns introduced by splicing operation applying 2D block DCT transformation on image and then, apply LBP in those DCT coefficients. For propagating the effects of the changes into all features, we then extract the features using our proposed SRIV technique applied to the LBP image 2D array. Finally, we feed these features to support vector machine (SVM) for learning and classification. Improved classification accuracy over recent methods described in [5, 6] using four benchmark datasets substantiate the efficacy of our proposed SRIV technique and image forgery detection approach.

2 Related Works

A number of approaches have been proposed in recent years to detect image tampering. They differ mainly on the techniques they adopt to model the structural and statistical changes in forged images. The works reported below utilized SVM for classification once features have been extracted from an image. Among them, the authors who implemented their work based on gray scale image used Columbia dataset [7] while others used different color datasets [8–10].

In [11], Ng et al. proposed bicoherence features to detect image splicing and suggested several methods to improve the capabilities of bicoherence features for splicing detection. They achieved as high as 72% detection accuracy over their own gray image dataset named Columbia [7]. Later, this dataset turned into one of the most popular benchmark datasets for gray scale image splicing detection. Hilbert-Huang transform (HHT) and moments of characteristics function of wavelet sub-band were used to extract features in [12]. It was the first work to utilize HHT to identify image splicing. The authors reported 80.15% detection accuracy. Chen et al. in [13] adopted statistical moments of characteristics functions of wavelet sub-band and 2D phase congruency to identify splicing artifacts and achieved 82.32% detection accuracy.

A few researchers adopted run-length based approach to identify image splicing. Dong et al. [14] investigated the disturbance of pixel correlation and rationality introduced by image splicing operation. They proposed a run-length and edge statistics based approach to identify spliced images from authentic ones and attained 76.52% accuracy. Later, this method was improved by He et al. [15] in terms of accuracy (80.58%), computational cost and feature dimensionality.

Shi et al. [16] proposed a method based on a natural image model where statistical moment features and Markov features are extracted from a given image as well as from multi block DCT of the same image. He et al. [17] expanded the original Markov features by Shi et al. and modelled the splicing artifacts based on Markov features in DCT and DWT domains. Unlike [16], they considered both intra-block and inter-block correlation among DCT coefficients. Although methods in [16, 17] achieved satisfactory result on Columbia dataset, the detection accuracy was reduced to 84.86% and 89.76%, respectively when applied on CASIA 2 dataset [10] which is a more challenging dataset in nature [17]. In [18], Wang et al. proposed a method to identify splicing attacks by modelling the edge information of image in chroma space as a finite-state Markov chain and considered its stationary distribution as features. This method achieved 95.6% accuracy on CASIA 2 dataset.

Zhang et al. [5] and Alahmadi et al. [6] proposed their methods utilizing both DCT and LBP. They mainly differ based on the order of DCT and LBP application on image blocks and feature extraction technique. Zhang et al. applied LBP operator on the magnitude component of 2D-DCT coefficients of the gray scale input image. They extracted features by calculating the histogram of the resultant LBP 2D array. In contrast, Alahmadi et al. divided the chrominance channels of the input image into blocks. Then LBP is applied and the resultant LBP 2D array of each block is transformed into frequency domain using 2D-DCT. Finally, features were extracted by calculating the standard deviation of the corresponding inter-cell DCT coefficients. Both the methods are promising in terms of detection accuracy. Inspired by the ability of DCT and LBP to generate discriminative features of authentic and spliced images, we propose a new feature extraction technique and using it, an image forgery detection approach, which is described in the following section.

3   Proposed Method

Image splicing and copy-move attacks are very widespread attacks on images. The detection mechanism is a binary decision problem – whether an image is forged or not. These attacks introduce structural and statistical changes in the host image which, in turn, affect features that can be extracted to describe the image. Therefore, a number of techniques need to be applied on the images before final features can be derived to feed into a chosen classifier. Figure 2 depicts the overall mechanism in our proposed method and its key components are described in the following sections.

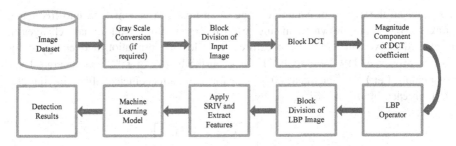

Fig. 2. Proposed image splicing and copy-move detection system

3.1 Converting Images into Gray Scale

We have implemented our system using four benchmark datasets commonly used for image splicing and copy-move detection. Among them, one dataset is already in gray scale and remaining datasets are in color space. As a result, we converted color datasets into gray scale. It is worth noting that many applications in surveillance and security system rely on gray scale images that are collected in night time environment.

3.2 Block Division of Input Image

Splicing and copy-move operation can be applied in different ways on host images. Again, different image fragments may be pasted into different parts of the host image. It is not expected to be able to identify the splicing artifacts by one single block size. Hence, for different types of images, different sized block divisions are essential to identify discriminative features of the forged images. Our proposed method performs block divisions in two phases. In the first phase, we divide an input image into square-sized blocks. The second phase is explained later in Sect. 3.5. We have tested our system with different block sizes: 4×4, 8×8 and 16×16 as well as combining features from all three mentioned blocks. The following procedure divides an image into blocks. Let $I^{wb \times hb}$ be a gray scale image of size $wb \times hb$ pixels. We divide $I^{wb \times hb}$ into $w \times h$ non-overlapping blocks of size $b \times b$ pixels. The resultant image block 2D array is,

$$I^{wb \times hb} = \begin{bmatrix} I_{1,1}^{b \times b} & \cdots & I_{1,w}^{b \times b} \\ \vdots & \ddots & \vdots \\ I_{h,1}^{b \times b} & \cdots & I_{h,w}^{b \times b} \end{bmatrix}. \tag{1}$$

3.3 Block Discrete Cosine Transformation (BDCT)

Image tampering introduces new micro patterns and sharp edges along the affected regions. It changes the local frequency distribution by altering regularity, smoothness, continuity of the tampered image and thus it disturbs the natural correlation between image pixels [16]. It is essential to reduce the diversity of image content and magnify the effects of image splicing and copy-move attack before final feature extraction.

To represent the degree of content change of an image, it is converted into frequency domain. BDCT has shown promising result in representing pixel domain changes in local frequency distribution as it exhibits excellent decorrelation and energy compaction properties [19]. We apply 2D-DCT on the blocks of $I^{wb \times hb}$ to generate DCT coefficients. Let $Y^{wb \times hb}$ be the resultant transform domain coefficient after applying 2D-DCT on each block and it is given by,

$$
Y^{wb \times hb} = \begin{bmatrix} Y_{1,1}^{b \times b} & \cdots & Y_{1,w}^{b \times b} \\ \vdots & \ddots & \vdots \\ Y_{h,1}^{b \times b} & \cdots & Y_{h,w}^{b \times b} \end{bmatrix}, \tag{2}
$$

where $Y_{i,j}^{b \times b} = 2D\text{-}DCT\left(I_{i,j}^{b \times b}\right), 1 \leq i \leq w, 1 \leq j \leq h$. The 2D-DCT of an input block $I_{i,j}^{b \times b}$ produces the output block $Y_{i,j}^{b \times b}$ as,

$$
Y_{i,j}^{b \times b}(p,q) = \alpha_p \alpha_q \sum_{m=0}^{b-1} \sum_{n=0}^{b-1} I_{i,j}^{b \times b}(m,n) \cos\frac{\pi(2m+1)p}{2b} \cos\frac{\pi(2n+1)q}{2b}, \tag{3}
$$

where $0 \leq p \leq b-1, 0 \leq q \leq b-1$ and

$$
\alpha_p = \begin{cases} \sqrt{\frac{1}{b}}, & \text{if } p = 0 \\ \sqrt{\frac{2}{b}}, & \text{otherwise} \end{cases}, \quad \alpha_q = \begin{cases} \sqrt{\frac{1}{b}}, & \text{if } q = 0 \\ \sqrt{\frac{2}{b}}, & \text{otherwise} \end{cases}. \tag{4}
$$

3.4   Local Binary Pattern (LBP) Operator

To identify and enhance different splicing artifacts, we employ LBP operator on the magnitude component of $Y^{wb \times hb}$. LBP is a computationally inexpensive yet robust texture descriptor. The main idea for adopting LBP in our system is to enhance the local changes among the neighbouring DCT coefficient values because of the occurrences of micro-patterns and sharp edges that are introduced by splicing and copy-move attacks. LBP can effectively highlight these tampering artifacts and enhance them in the host images. In LBP, each pixel of a given 2D array is compared with its neighbouring pixels and an LBP code is generated for that pixel. It is computed as below:

Let $L^{wb \times hb}$ be the resultant LBP array generated by applying LBP operator on the magnitude components of $Y^{wb \times hb}$ and is given by,

$$
L^{wb \times hb} = LBP_{N,R}\left(\left|Y^{wb \times hb}\right|\right), \tag{5}
$$

$$
LBP_{N,R} = \sum_{n=0}^{N-1} g(p_n - p_c)2^n. \tag{6}
$$

Here, N is the number of neighbor pixels; R is the radius and p_c is the central pixel which is compared with each neighbouring pixel $p_n (n = 0, 1, \ldots, N - 1)$. In our proposed method, we use $N = 8$ and $R = 1$. The function $g(p_n - p_c)$ is given by:

$$g(p_n - p_c) = \begin{cases} 1, & p_n - p_c \geq 0 \\ 0, & p_n - p_c < 0 \end{cases}. \tag{7}$$

For $N = 8$ and $R = 1$, the central pixel p_c compares its own value with neighbouring 8 pixels. If the neighbor pixel's value is greater than or equal to the central pixel value, then 1 is recorded; otherwise 0. Based on these comparisons, central pixel p_c stores it's LBP code. Figure 3 explains the procedure with an example. Here, the binary values are obtained after comparison between central pixel p_c and the 8 neighboring pixels. Then the 8-bit binary digit is formed starting from Least Significant Bit (LSB) to Most Significant Bit (MSB). Finally, the binary digit is converted into decimal and the LBP code is stored in place of central pixel p_c.

Fig. 3. LBP code generation procedure

3.5   Block Division of LBP Image

In the second phase of block division, we divide the LBP image 2D array $L^{wb \times hb}$ into same size of blocks similar to the block division done in Sect. 3.2. We divide $L^{wb \times hb}$ into $w \times h$ non-overlapping blocks of size $b \times b$ pixels. The resultant LBP image block 2D array is given by,

$$L^{wb \times hb} = \begin{bmatrix} L_{1,1}^{b \times b} & \cdots & L_{1,w}^{b \times b} \\ \vdots & \ddots & \vdots \\ L_{h,1}^{b \times b} & \cdots & L_{h,w}^{b \times b} \end{bmatrix}. \tag{8}$$

3.6   Apply SRIV and Feature Generation

As shown in Fig. 2, in our proposed method, the SRIV features are derived from LBP codes generated using DCT coefficients. The main reason for adopting such approach in a specific order is that DCT coefficients represent the pixel value variations in the spatial domain, while LBP enhances the local changes among the neighboring DCT coefficient values, magnifying the changes of splicing and copy-move attacks in higher frequency components. To make the detection system more accurate, we need to preserve the local changes captured by LBP as much as possible. Since splicing attacks usually make subtle changes in an image, these local changes can be regarded as outliers. Mean is most affected by outliers than other statistical measures. The SRIV features in our proposed method are based on an aggregation operator (sum).

These features are similar to the mean based features as the number of blocks having a particular size (e.g., 8×8) in a specific image remains always the same. Therefore, this vindicates the SRIV features can represent the local changes because of splicing and copy-move attacks more accurately than the standard deviation based features used in [6] and the histogram-based features applied in [5]. We experimented our method with different block sizes as mentioned in Sects. 3.2 and 3.5. Consequently, we have varying dimensionality of features as listed in Table 2. The SRIV features are computed as below:

Let $Z_k^{w \times h}$ be the k-th LBP code values of all blocks in $L^{wb \times hb}$. Therefore,

$$Z_k^{w \times h} = \begin{bmatrix} L_{1,1}^{b \times b}(k) & \cdots & L_{1,w}^{b \times b}(k) \\ \vdots & \ddots & \vdots \\ L_{h,1}^{b \times b}(k) & \cdots & L_{h,w}^{b \times b}(k) \end{bmatrix}, \quad 1 \le k \le b^2, \tag{9}$$

where $L_{u,v}^{b \times b}(k)$ is the k-th LBP code of that block. Then the k-th feature F_k on the whole image is calculated as,

$$F_k = \sum_{u=1}^{w} \sum_{v=1}^{h} L_{u,v}^{b \times b}(k). \tag{10}$$

To justify our argument as mentioned before that the SRIV features are more discriminative and more effective than the standard deviation based features, we extracted features by our approach using both SRIV and standard deviation. For a representative sample, we selected one authentic image (Fig. 4a) and its spliced version (Fig. 4b) from CASIA 2 dataset. We then plotted the extracted features in a graph (Fig. 5) where x-axis represents feature number and y-axis represent the feature values. From Fig. 5, it is clearly visible that the SRIV feature values vary more sharply than those of standard deviation for both the original and its spliced image. This is also evidenced by the fact that the standard deviation of SRIV feature values for both the original and its spliced image (0.18, 0.17) are higher than those of standard deviation based feature values (0.16, 0.15). All of these evidences show the SRIV features are more discriminating and hence more effective than those for standard deviation.

(a)   (b)

Fig. 4. Authentic image (a) and its spliced image (b) from CASIA 2

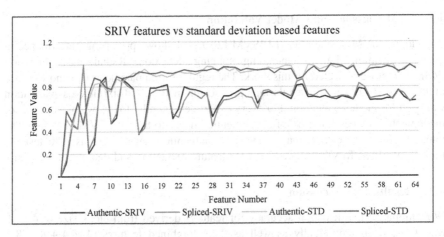

Fig. 5. Comparing the SRIV features with the standard deviation based features

4 Experiments and Results

4.1 Description of Datasets

We have evaluated our proposed system using four publicly available and well recognized benchmark datasets for image splicing detection: (i) Columbia gray [7], (ii) Columbia Uncompressed [8], (iii) CASIA 1 [9] and (iv) CASIA 2 [10]. We have summarized the datasets used to evaluate our method in Table 1.

Table 1. Summary of the datasets

| Dataset | Image size | Image type | No. of images | | | Tampering method |
			Authentic	Tampered	Total	
Columbia	128 × 128	JPG	933	912	1845	Simple crop-and-paste
Columbia Uncomp.	757 × 568 − 1152 × 768	TIF, BMP	183	180	363	Simple crop-and-paste, spliced image from exactly 2 cameras
CASIA 1	384 × 256, 256 × 384	JPG	800	921	1721	Photoshop with pre-processing; No post-processing
CASIA 2	240 × 160 − 900 × 600	JPG, TIF, BMP	7491	5123	12614	Photoshop with pre-processing and/or post - processing

4.2   SVM Classifier and Model Validation

We adopted SVM as classifier (LIBSVM [20]) as it shows promising performance in many application domains including splicing detection. Radial Basis Function (RBF) kernel was selected for this work. The regularisation parameter (C) and variance of RBF kernel (γ) were chosen through grid-search method and sixfold cross-validation was used for model evaluation. For every experiment, similar to [5], we picked 5/6th of the tampered images and 5/6th of the authentic images to train the SVM classifier. The remaining 1/6th tampered images and 1/6th authentic images were used to test the trained classifier. MATLAB was used for feature extraction and data pre-processing.

4.3   Results and Discussion

We summarise the detection accuracy for features derived from block size of 4×4, 8×8, 16×16 individually as well as their combined features $(4 \times 4 + 8 \times 8 + 16 \times 16)$ in Table 2. The effect of different sized block DCT varies from dataset to dataset. Our proposed method achieves detection accuracy of 85.64%, 94.49%, 95.40% and 99.76% over Columbia gray, Columbia Uncompressed, CASIA 1 and CASIA 2 datasets respectively. Additionally, the precision, recall and AUC (Area Under ROC curve) of our system is also reported in Table 2.

Columbia gray dataset is a popular but older dataset with low resolution fixed dimension (128×128) JPG images. Our method performs best (85.64%) for block size of 8×8 on this dataset while block size 4×4 and 16×16 reduces detection accuracy by 7% and 5%, respectively. Combined features from all three blocks provides 84.34% detection accuracy. Similar trend is observed for Columbia Uncompressed dataset where combining features from all blocks does not yield the best result. However, we achieved the best results by combining features for CASIA 1 (95.40%) and CASIA 2 (99.76%) datasets. Our method has produced quite encouraging result in these datasets, which demonstrates the strength of the feature extraction and overall techniques used in our approach.

4.4   Comparison with Recent Methods

Among various methods for detecting splicing and copy-move attacks (Sect. 2), two existing ones adopt both DCT and LBP in their systems and report good detection accuracy. Since they have not reported results with all four datasets, to make a fair comparison, we implemented those two methods to get their detection capability for each dataset. The basic experimental setup remains the same as mentioned in Sect. 4.2.

In [5], Zhang et al. found best accuracy for combined features extracted from block size 4×4, 8×8 and 16×16. They identified best parameters for SVM and RBF kernel through grid-search method. In [6], Alahmadi et al. attained the best accuracy with 16×16 blocks and LBP parameter P(neighbour) = 8, R(radius) = 1, SVM parameter $C = 2^5$ with RBF kernel $\gamma = 2^{-5}$. We implemented their methods using their reported parameters. Table 3 depicts the comparison of detection accuracy among different methods across different datasets. It is clearly visible that our method's overall accuracy is higher (up to 5%) than two existing state-of-the-art methods in all four

Table 2. Overall detection accuracy in our proposed method with varying block size. Note that image in Columbia Uncomp., CASAI 1 and CASIA 2 are converted into gray scale

Block size	Feature dimensionality	Evaluation	Columbia	Columbia Uncomp.	CASIA 1	CASIA 2
4 × 4	16	Accuracy (%)	78.5908	87.6033	72.8438	95.0844
		Precision	0.789	0.857	0.761	0.935
		Recall	0.773	0.900	0.718	0.945
		AUC	0.786	0.876	0.729	0.950
8 × 8	64	Accuracy (%)	**85.6369**	92.2865	93.5897	97.6453
		Precision	0.852	0.942	0.934	0.968
		Recall	0.859	0.900	0.947	0.975
		AUC	0.856	0.923	0.935	0.976
16 × 16	256	Accuracy (%)	80.8672	**94.4904**	87.7622	99.231
		Precision	0.803	0.944	0.875	0.988
		Recall	0.813	0.944	0.900	0.993
		AUC	0.809	0.945	0.876	0.992
4 × 4 + 8 × 8 + 16 × 16	336	Accuracy (%)	84.336	94.2149	**95.3963**	**99.7622**
		Precision	0.830	0.944	0.946	0.996
		Recall	0.860	0.939	0.970	0.998
		AUC	0.844	0.942	0.953	0.998

Table 3. Comparison of detection accuracies of the proposed method with [5, 6]

Dataset	Evaluation	Proposed method	Method in [5]	Method in [6]
Columbia	Accuracy (%)	**85.6369**	81.1924	77.1816
	Precision	0.852	0.806	0.768
	Recall	0.859	0.827	0.772
	AUC	0.856	0.816	0.772
Columbia Uncomp.	Accuracy (%)	**94.4904**	92.8375	93.3884
	Precision	0.944	0.910	0.994
	Recall	0.944	0.950	0.872
	AUC	0.945	0.929	0.933
CASIA 1	Accuracy (%)	**95.3963**	92.5991	78.0886
	Precision	0.946	0.951	0.821
	Recall	0.970	0.909	0.755
	AUC	0.953	0.927	0.783
CASIA 2	Accuracy (%)	**99.7622**	84.1433	94.1965
	Precision	0.996	0.812	0.918
	Recall	0.998	0.793	0.935
	AUC	0.998	0.834	0.939

benchmark datasets. To the best of our knowledge, detection accuracy of 99.76% is the highest among all other methods available in the literature that deal with gray scale images. Our method outperforms others in terms of precision, recall and AUC in all cases except for recall in Columbia Uncomp. and precision in CASIA 1. Specially, our method attains better AUC, which is a more accepted performance metric, for all four benchmark datasets.

5 Conclusion

In this paper, we introduced SRIV, a novel feature extraction technique using which we proposed a robust model for detecting splicing and copy-move attacks on image data adopting both DCT and LBP in the mentioned order. These attacks change the pixel values in the spatial domain by introducing sharp edges, alien micro-patterns and so on. DCT shows excellent image pixel decorrelation and energy compaction properties which is used to capture the change in the spatial domain. Then, LBP is applied on the magnitude component of the 2D array returned by DCT to enhance the local changes among the neighbouring DCT coefficient values. Finally, SRIV is applied on the LBP image blocks to extract features. These features are used to train an SVM with RBF kernel to detect the tampered images. Experimental results confirm that our method outperforms other methods across four benchmark image forgery detection datasets. Future work will target detection of splicing and copy-move attacks on color images.

Acknowledgement. This work is supported by the Research Priority Area (RPA) scholarship of Federation University Australia.

References

1. Meola, A.: The Internet of Things: Meaning & Definition. Business Insider (2018)
2. Redi, J.A., Taktak, W., Dugelay, J.-L.: Digital image forensics: a booklet for beginners. Multimed. Tools Appl. **51**, 133–162 (2011)
3. Novak, M.: That Viral Photo of Putin and Trump is Totally Fake. gizmodo.com (2017)
4. Kwitt, R., Meerwald, P., Uhl, A.: Lightweight detection of additive watermarking in the DWT-domain. IEEE Trans. Image Process. **20**, 474–484 (2011)
5. Zhang, Y., Zhao, C., Pi, Y., Li, S., Wang, S.: Image-splicing forgery detection based on local binary patterns of DCT coefficients. Secur. Commun. Netw. **8**, 2386–2395 (2015)
6. Alahmadi, A.A., Hussain, M., Aboalsamh, H.A., Ghulam, M., Bebis, G., Mathkour, H.: Passive detection of image forgery using DCT and local binary pattern. SIViP **11**, 81–88 (2017)
7. Ng, T.-T., Chang, S.-F.: A model for image splicing. In: IEEE International Conference on Image Processing (2004)
8. Hsu, Y.-F., Chang, S.-F.: Detecting image splicing using geometry invariants and camera characteristics consistency. In: International Conference on Multimedia and Expo, Canada (2006)
9. Dong, J., Wang, W., Tan, T.: CASIA image tampering detection evaluation database. In: IEEE International Conference on Signal and Information Processing, pp. 422–426 (2013)
10. Dong, J., Wang, W.: CASIA tampered imaged detection evaluation database (CASIA TIDE v2.0). National Laboratory of Pattern Recognition, Chinese Academy of Science (2009–2016)
11. Ng, T.-T., Chang, S.-F., Sun, Q.: Blind detection of photomontage using higher order statistics. In: IEEE International Symposium on Circuits and Systems, pp. 688–691 (2004)
12. Fu, D., Shi, Y.Q., Su, W.: Detection of image splicing based on Hilbert-Huang transform and moments of characteristic functions with wavelet decomposition. In: Shi, Y.Q., Jeon, B. (eds.) IWDW 2006. LNCS, vol. 4283, pp. 177–187. Springer, Heidelberg (2006). https://doi.org/10.1007/11922841_15

13. Chen, W., Shi, Y.Q., Su, W.: Image splicing detection using 2-D phase congruency and statistical moments of characteristic function. In: Proceedings of SPIE 6505, Security, Steganography, and Watermarking of Multimedia Contents IX, vol. 6505. SPIE, Washington (2007)

14. Dong, J., Wang, W., Tan, T., Shi, Yun Q.: Run-length and edge statistics based approach for image splicing detection. In: Kim, H.-J., Katzenbeisser, S., Ho, A.T.S. (eds.) IWDW 2008. LNCS, vol. 5450, pp. 76–87. Springer, Heidelberg (2009). https://doi.org/10.1007/978-3-642-04438-0_7

15. He, Z., Sun, W., Lu, W., Lu, H.: Digital image splicing detection based on approximate run length. Pattern Recogn. Lett. **32**, 1591–1597 (2011)

16. Shi, Y.Q., Chen, C., Chen, W.: A natural image model approach to splicing detection. In: Proceedings of the 9th Workshop on Multimedia & Security, pp. 51–62. ACM, USA (2007)

17. He, Z., Lu, W., Sun, W., Huang, J.: Digital image splicing detection based on Markov features in DCT and DWT domain. Pattern Recogn. **45**, 4292–4299 (2012)

18. Wang, W., Dong, J., Tan, T.: Image tampering detection based on stationary distribution of Markov chain. In: IEEE International Conference on Image Processing, pp. 2101–2104 (2010)

19. Khayam, S.A.: The discrete cosine transform (DCT): theory and application. Michigan State University (2003)

20. Chang, C.-C., Lin, C.-J.: LIBSVM: a library for support vector machines. ACM Trans. Intell. Syst. Technol. **2**, 27:21–27:27 (2011)

Learning Latent Byte-Level Feature Representation for Malware Detection

Mahmood Yousefi-Azar[1,3]([✉]), Len Hamey[1], Vijay Varadharajan[2], and Shiping Chen[3]

[1] Department of Computing, Faculty of Science and Engineering, Macquarie University, Sydney, NSW, Australia
mahmood.yousefiazar@hdr.mq.edu.au, len.hamey@mq.edu.au
[2] Faculty of Engineering and Built Environment, University of Newcastle, Newcastle, Australia
vijay.varadharajan@newcastle.edu.au
[3] Commonwealth Scientific and Industrial Research Organisation, CSIRO, Data61, Sydney, Australia
Shiping.Chen@data61.csiro.au

Abstract. This paper proposes two different byte level feature representations of binary files for malware detection. The proposed static feature representations do not need any third-party tools and are independent of the operating system because they operate on the raw file bytes. Sparse term-frequency simhashing (s-tf-simhashing) is a faster type of tf-simhashing. S-tf-simhashing requires less computation and outperforms the original dense tf-simhashing. The binary word2vec (Bword2vec) representation embeds the semantic relationships of the n-grams into the code vectors. Bword2vec employs a binary to word2vec representation that reduces the feature space dimension than s-tf-simhashing and thus further reducing the computation of the classifier. We show that the proposed techniques can successfully be used for both analyzing of full malware apps and infected files. The experiments are conducted on real Android and PDF malware datasets.

Keywords: Malware detection · Binary-level feature representation Sparse *term-frequency* simhashing · Binary Word2vec

1 Introduction and Related Work

A Successful malware detection scheme requires a rich feature space. This space is based on static and/or dynamic analysis of samples [4,5,10]. The dynamic features are driven from the run-time behaviour of a sample while the static features are extracted from a binary file and/or disassembled code without running it. Both dynamic and static feature spaces can contain latent representation and hidden information that facilitates malware detection task [4,5,15].

© Springer Nature Switzerland AG 2018
L. Cheng et al. (Eds.): ICONIP 2018, LNCS 11304, pp. 568–578, 2018.
https://doi.org/10.1007/978-3-030-04212-7_50

This paper proposes to represent static features of any given file in both low dimension and informative representation for malware detection task. Term-frequency simhashing (*tf*-simhashing) has been successfully used for malware detection [15]. We improve *tf*-simhashing performance by using a sparse vocabulary hashing. The computational complexity of sparse *tf*-simhashing (*s-tf*-simhashing) is $O(NI(nnz))$ and only nonzero entries per row in A requires to be computed while *tf*-simhashing requires $O(NKI)$ operations.

We also propose a binary word2vec feature representation, a byte level semantic representation for byte n-grams. This representation embeds each byte n-gram into a vector according to the written pattern of surrounded byte code (a.k.a context n-grams). The embedding vectors are obtained from a pre-training phase and do not require any extra computation in the actual feature extraction phase.

In summary, the contributions of this paper are the following:

- We develop a fast type of *tf*-simhashing, which hashes each n-gram of the vocabulary into a sparse vector. Summing up all the sparse vectors maps each file into a fixed sized vector. Our improved *s-tf*-simhashing requires about 1% of the computation of *tf*-simhashing [15].
- We propose the Bword2vec feature representation for malware detection. Kolosnjaji [6] proposed a byte embedding model that is strictly on uni-gram (i.e. 0 to 255) with 8 elements. Chistyakov et al. [4] and Karbab et al. [5] proposed different word2vec schemes to embed instructions into vectors while our proposed scheme is based on n-gram byte representation of each file.
- We show that the proposed features can successfully detect malware samples and infected files, independent of any third-party tools. To the best of our knowledge, there is no such a representation for both full malicious files and infected files detection.

The next section presents *s-tf*-simhashing both mathematically and the implementation concepts. Section 3 discusses the Bword2vec model as a pre-training phase and its application for feature extraction. Section 4 presents evaluation metrics, experimental methods, results and discussion. Section 5 concludes this paper.

2   Sparse Hashing Algorithm

Hashing algorithms have been widely used for the malware detection task. In particular, Yousefi-Azar et al. [14] proposed to use *tf*-simhashing in which byte n-grams feature space are projected into a lower dimension of spaces. It has been shown that the projection matrix consisting of only 1 and -1 values is the first layer weights of Extreme Learning Machine (ELM). Also, Chen et al. [3] showed that ELM with first layer of very sparse weight matrix has a better generalization.

Figure 1 illustrates the mathematical representation of the proposed *s-tf*-simhashing to hash byte n-grams into low dimension. The original feature space is $\mathbf{A} \in \mathbb{R}^{N \times K}$, where N is the number of data samples (here files) in $\mathbb{R}^K$, and K is the vocabulary size (e.g. for n-grams with $n = 2$, $K = 2^{8n} = 65536$). The projection matrix is $\mathbf{R} \in \mathbb{R}^{K \times I}$ (with $I << K$) and the 1, −1, 0 values are mapping the original space into a low dimensional feature space $\hat{\mathbf{A}} = \frac{1}{\sqrt{K}}\mathbf{AR}$. If the projection matrix $\mathbf{R}$ satisfies Johnson and Lindenstrauss theory [7], then the projection approximation preserves all the pairwise L2-norm distances between points in the original space.

More precisely, assuming that the data files are mapped into vocabulary vectors $\{u_i\}_{i=1}^N$ of dimension K as the original feature space. The hash vectors will be $\{v_i\}_{i=1}^N$, where $v_i = \frac{1}{\sqrt{K}}\mathbf{R}u_i$, of dimension I, as the projected feature space. Then we have $u_i \cdot u_j - \epsilon < v_i \cdot v_j < u_i \cdot u_j + \epsilon$, where $\cdot$ denotes dot product.

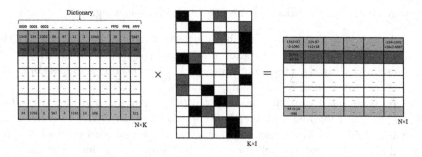

Fig. 1. *S-tf*-simhashing mathematical concept, $n = 2$ and $I = 6$ with 33% density.

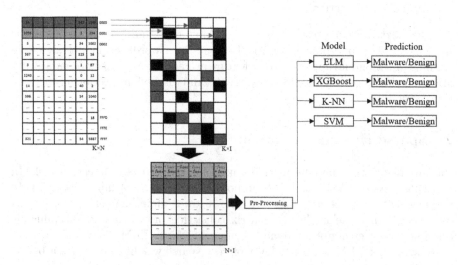

Fig. 2. The implementation of *s-tf*-simhashing.

In general, any random matrix is not orthogonal and not projector whereby posing significant distortion to the data [2]. Also, it is computationally very expensive to have orthogonal projection matrix. Instead, we generate a projection matrix that is being sufficiently close to orthogonal. Li et al. [7] showed a very sparse matrix can be used. For this purpose, we used about 1% sparsity. Because all items of the vocabulary are equally important, the matrix rows need to have equal amount of 1 or -1. For example, with $n = 2$, $I = 1010$ and density about 1% (i.e. 5 elements of the vector set to 1 and other 5 elements set to -1), the projection matrix $\mathbf{A}$ will sample about 650 items of the vocabulary in each dot product.

Figure 2 presents the implementation scheme of the proposed s-tf-simhashing matrix as a look up table. In practice, the location of each item of the vocabulary is the index of the projection matrix row. The final hash is sum up all the projected vectors. With equal numbers of non-zero entries, all items influence the final hashing space equally.

The hash vector mostly has high variance and therefore needs to be normalized. Support Vector Machine (SVM) requires low variance in the input. Instead, ELM requires to only have an standardized input between $[-1, 1]$.

In Figs. 1 and 2, the values/dimensions are only example of possible numbers. Black, gray and white colors represent 1, -1 and 0 respectively.

3   Binary Word2vec Feature Representation

Word2vec is a neural-network-based language model in which a large corpus of text is the input and the trained weights are taken as embedding vectors [8]. Word2vec has been recently used for malware analysis, particularly, embedding assembly instructions into vectors [4,5].

We propose Bword2vec feature representation in order to encapsulate the semantic relationships between n-grams. Using n-grams instead of assembly instructions reduces the reverse-engineering of file analysis. More precisely, for each n-gram, the algorithm generates a vector.

A question is how to visualize the semantic similarity of n-grams? while the human does not have a semantic understanding of binary level n-gram of files. Therefore, we assume that the embedded relationships of n-grams facilitates different classifiers to distinguish between benign from malware samples.

Figure 3 shows the Bword2vec feature representation scheme. The most commonly used word2vec models are continuous bag-of-words (CBOW) and continuous skip-gram. Skip-gram performs better for sparse words in human languages. The architecture is similar to a feed-forward neural network where the context words are decisive in predicting the word. In the pre-training phase, K is the vocabulary size and the hyper-parameter I is the hidden layer size. The input vector $\{x_i\}_{i=1}^{K}$ is a one-hot encoding of n-grams (e.g. 2-gram). The weights between the input layer and the hidden layer are $\mathbf{W}_{K \times I}$. After training, each row of $\mathbf{W}$ is the I-dimensional vector representation v_w of the associated n-gram in the input layer. The training algorithm is standard training with the cross-entropy measurement between two probabilistic distributions, context and the n-grams.

We observed that summing the embedding n-gram vectors of a given file provides better performance when we scale each embedding vectors with the respective term frequency (as shown in Fig. 3).

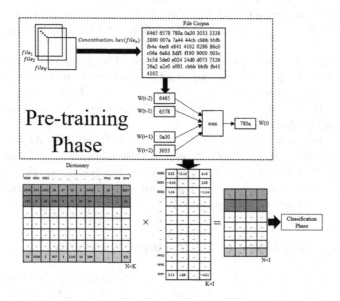

Fig. 3. Bword2vec feature extraction scheme in both pre-training and feature representation phases. The values are only example of possible numbers.

4   Experiments and Results

To evaluate the proposed feature representations, we conducted different experiments on both Android and Microsoft Windows platforms. We also use the classification capability of k-nearest Neighbors (K-NN), Gradient Boost (XGBoost), ELM and SVM algorithms. The scikit-learn Python library was used for implementation.

We used Drebin and DexShare datasets for Android malware detection [15]. Drebin consists of 5560 malware of which 5555 have a .dex file. DexShare has 309 malware families is a more complicated dataset compared to Drebin with 180 families. DexShare consists of 20255 malware samples. Benign sets were collected from Androzoo, a freely available APKs repository, originally collected from GooglePlay [1]. Motivated by Yousefi-Azar [15], we only used the dex file of each APK in the datasets. Both datasets are balanced in the number of benign and malware.

A dataset of PDF files with embedded malware called PDFShare is used to evaluate the representations for malicious files detection. Table 1 presents the statistics of the dataset. Both malware and benign sets were collected from contagiodump.com website[1].

We used grid search to optimize the hype-parameters of the K-NN (ranges from 1 to 20 for n neighbors and either 'uniform' or 'distance' for weights), ELM (ranging from 1 to 300 for both C and gamma) and SVM (range from 10^{-4} to 10^4 for both C and gamma).

To have comparable results with previous work, the *s-tf*-simhashing vector size is 1024 and we used 2-grams. Bword2vec vector size is 100. We used 5-fold cross-validation and set the classification threshold of class decision making to 0.5.

Table 1. The statistics of the PDFShare dataset.

Type	Qty	Max. size (MB)	Min. size (KB)	Ave. size (MB)
Malware	10980	2.1	1.0	0.22
Benign	8999	0.43	10.8	0.96

Table 2. The performance metrics to evaluate the proposed feature spaces.

Metrics	Description
TN	Number of benign apps correctly classified
FP	Number of false prediction as malicious
FN	Number of false prediction as clean
TP	Number of malicious apps correctly detected
Precision	$TP/(TP + FP)$
False negative rate (FNR)	$1 - (FN/(TP + FN))$; 1-recall or hit rate
f1-score	$2 \times Precision \times Recall/(Precision + Recall)$
Accuracy	$(TN + TP)/(TN + FP + FN + TP)$
False positive rate (FPR)	$FP/(TN + FP)$

To obtain Bword2vec vectors, we used CBOW with max window length of 5 and hierarchical softmax. Training samples for pre-training (embedding phase) could be either only benign or only malware or a mixture of them. We found that embeddings obtained from only benign samples perform slightly better than other options, so the results reported in this paper are on benign embeddings. Table 2 presents the metrics used for evaluations.

---

[1] http://contagiodump.blogspot.com/2013/03/16800-clean-and-11960-malicious-files.html.

4.1   Results and Discussion

Sparse Hashing Algorithm: Table 3 presents the performance of the four classifiers using *s-tf*-simhashing on the two Android datasets. It is to be expected that ELM with RBF kernel, in general, outperforms other classifiers [15] while K-NN, XGBoost and SVM also have good performance. This performance is comparable with sophisticated feature extraction and selection algorithms [10, 11].

Since DexShare consists of more families, almost all classifiers performance reduce between 2–3%. The only exception is SVM. Although SVM' precision and FNR change dramatically, its f1-score has similar trend to other classifiers. This consistent change across all classifiers shows that the representation is not tuned for a particular dataset.

The low computation of the *s-tf*-simhashing allows to use bigger hash size (see Table 3). Indeed, with hash size 3000, ELM has low FPR while has a good hit-rate to detect malware. The size of hash size requires more computation in the classification phase. This is not a large bottleneck for ELM because it uses a fast learning model where the computational cost is sensitive to the number of samples rather than the input (feature space) dimension (see [3,15] for more detail). We set hash size 3000 which has slight impact on the LEM computation.

To have the real world settings, we set the malware to a benign ratio of 20% (5060 random malware versus the entire benign set) on the DexShare. Despite the reduced malware samples, it is still robust to the reduction of malware samples while the precision of ELM improves.

It is important to see whether *s-tf*-simhashing can capture enough information to detect novel malware families. To this end, we selected 20 malware families, similar to [15], out of DexShare dataset, for testing. We chose groups of 4 families for each testing; thus, we performed the test 5 times each with a new test set. That is, for each experiment 4 families are test set and the remaining 16 families plus all the other malware and benign sets are used for training. Table 3 presents the average results. *S-tf*-simhashing with the hash size of 3000 clearly outperforms *tf*-simhashing with hash size 1024 in detection of novel families while still requires less computation.

Mariconti et al. [10] novelty detection dataset is from a source similar to our DexShare dataset. Assuming our family exclusion test is at least as difficult as predicting future malware (e.g. testing on one year in the future), we can see that *s-tf*-simhashing enhances ELM to compete with the state-of-the-art in novelty detection. Sayfullina et al. [11] also used a random-projection-based scheme on a different dataset.

We also tested *s-tf*-simhashing for infected PDF files. Table 4 presents the confusion matrices of the classifiers. SVM and K-NN are skewed towards good FPR or FNR while ELM and XGBoost perform better. In particular, XGBoost can detect most of the malware with very low FPR. Wang [13] also observed that XGBoost has a good performance in detecting Windows malware based on byte level features.

Table 3. The performance of the classifiers using *s-tf*-simhashing, on the Android Datasets.

Dataset	Model	Precision	FNR	f1-score	Accuracy	FPR
Drebin	K-NN	93.86% (±0.84%)	2.29% (±0.16%)	95.75% (±0.41%)	95.66% (±0.44%)	6.40%
	ELM	**96.36%** (±0.45%)	**1.66%** (±0.24%)	**97.34%** (±0.24%)	**97.31%** (±0.27%)	**3.70%**
	SVM	92.40% (±0.83%)	2.49% (±0.43%)	94.88% (±0.45%)	94.74% (±0.45%)	8.03%
	XGBoost	94.88% (±0.50%)	2.56% (±0.52%)	96.14% (±0.18%)	96.08% (±0.18%)	5.25%
DexShare	K-NN	91.96% (±0.27%)	5.70% (±0.42%)	93.12% (±0.25%)	93.03% (±0.25%)	8.23%
	ELM	95.45% (±0.22%)	**4.80%** (±0.48%)	**95.32%** (±0.21%)	**95.33%** (±0.22%)	4.53%
	SVM	**95.75%** (±0.16%)	11.39% (±0.32%)	92.04% (±0.22%)	92.34% (±0.20%)	**3.92%**
	XGBoost	93.75% (±0.23%)	6.46% (±0.32%)	93.64% (±0.18%)	93.65% (±0.18%)	6.23%
DexShare	ELM ($I = 3000$)	96.65% (±0.35%)	4.72% (±0.50%)	95.96% (±0.38%)	96.00% (±0.35%)	3.30%
	Sayfullina et al. [11]	88.24%	–	–	–	4.00%
	ELM ($I = 3000$, imbalanced)	98.91% (±0.27%)	**4.42%** (±0.82%)	97.21% (±0.53%)	98.90% (±0.18%)	**2.60%**
	ELM (I=3000, zero-days)	96.87%	**8.81%**	93.89%	94.13%	**2.94%**
	Malytics [15] (zero-days)	96.31%	10.59%	92.68%	92.99%	3.4%
	Mariconti [10] (zero-days)	86.00%	12.00%	87.00%	–	–

Table 4 shows the *s-tf*-simhashing provides a rich representation for files where part of a file is infected with malicious code. It is true that *s-tf*-simhashing hashes the entire file into a vector in a single phase, we think the pattern of term frequency is distributed over the hash vector. We also tied sliding window of *s-tf*-simhashing over the PDF files, concatenating all the resulting hash vectors for the classification phase. The results was no better than applied *s-tf*-simhashing to the entile file.

Binary Word2vec Feature Representation: A common method to show the effectiveness of the word2vec is visualization of the human language words.

Table 4. The *s-tf*-simhashing performance to detect infected PDFs.

Dataset	SVM		K-NN		ELM		XGBoost	
PDFShare	8994	5	8450	549	8976	23	8971	**28**
	550	10430	88	10892	126	10854	**57**	10923

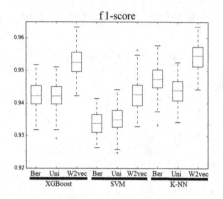
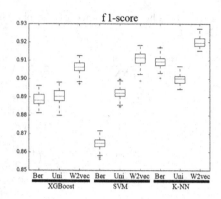

Fig. 4. The box plot of byte-Word2vec representation f-score with 100 iteration on the 5-fold cross-validation, for Drebin (left panel) and DexShare (right panel).

However, since byte gram visualization is not meaningful for human, we decided to evaluate the representation on the basis of the classifier performance.

In the first experiment, we have three different settings including Bword2vec, uniform distribution real values and the Bernoulli distributed of 1 and −1, as the projection matrix options. Figure 4 shows Bword2vec outperforms the other two projection on both Drebin and DexShare datasets. In general, Bword2vec f1-score outperforms 1–2% for all the classifiers. The trend for both datasets is very similar. It is to be expected that the results are more statistically significant for DexShare due to the larger number of samples while the results on Drebin also confirm that Bword2vec performs better.

The goal of the experiment is not to draw a conclusion on each individual classifier but on the general consensus of the classifiers performance for the 3 given representation. The tables show, in general, Bword2vec can reduce the error rate about 20%.

In the second experiment (see Table 5), the performance of Bword2vec was evaluated on PDFShare. We used ELM with RBF kernel to conduct this experiment. That is, similar to *s-tf*-simhashing classification, the output vector of Bword2vec representation is fed to ELM for classification. It is worth noting that the original representation of both latent representation is term frequency of 2-grams. Bword2vec representation outperforms *s-tf*-simhashing in all measures. This is to be expected since *s-tf*-simhashing comes from a random initialization whereas Bword2vec is learned through observations. Also, malicious and benign PDFs have smaller size and possibly less complexity compared to full Android dex files. We believe a vector of size 100 is rich enough for classifying PDF files.

We could not find very close related work to this paper for PDF malware detection. The results in [9, 12] at least indirectly show that the proposed feature representations of this paper performs competitively with the recent and more complicated models.

We noted *s-tf*-simhashing requires less computation than Bword2vec. This is because each item of the dictionary in *s-tf*-simhashing is selected in case the weight is 1 or change to negative number in case the weight is -1. For Bword2vec, the term frequency of each item of the dictionary must be multiplied by the embedding vector consisting of floating point.

Table 5. The results of ELM classifier with either byte-Word2vec or s-tf-simhashing representation.

Representation	Precision	FNR	f1-score	Accuracy	AUC	FPR	Confusion Matrix
Byte-Word2vec	99.83% (±0.08%)	00.51% (±0.20%)	**99.66%** (±0.12%)	99.62% (±0.13%)	99.64% (±0.13%)	0.20%	$\begin{bmatrix} 8980 & 19 \\ 56 & 10924 \end{bmatrix}$
S-tf-simhashing	99.79% (±0.12%)	01.15% (±0.28%)	99.32% (±0.13%)	99.25% (±0.14%)	99.30% (±0.13%)	0.25%	$\begin{bmatrix} 8976 & 23 \\ 126 & 10854 \end{bmatrix}$
Scofield [12]	100%	03.00%	–	98.0%	–	–	
Nissim [9]	98.40%	–	–	–	–	∼0.50%	

On common technique in byte level n-gram analysis of the malware detection is to binary representation of the dictionary and ignore term frequency [13]. However, our experiments have shown that applying the frequency of byte n-gram enhance the performance of the classifiers in both s-tf-simhashing and Bword2vec representations. We also observed that weighting each item of the vocabulary accordance to the common term frequency-inverse document frequency (tf-idf) formula achieves the same performance as the tf.

We observed that feature space of size 1024 or 100 dimensions can not enhance the latent space for n-grams $n > 2$. In case we still want to limit the feature space to small amount, the very high dimension of the original space might require a feature selection phase before feeding into either s-tf-simhashing or Bword2vec. As a future work, the string information of n-grams for $n > 2$ can leverage the detection task alongside the 2-gram information.

5   Conclusion

This paper presents two different feature representations for malware detection. Experiments show sparse term-frequency simhashing (s-tf-simhashing) outperforms its dense version and other n-gram models. Binary word2vec (Bword2vec) algorithm embeds the semantic relationships of byte level codes into vectors. We showed that the both proposed representations provide more discriminative feature space for classification. Experiments on Android malware and infected PDFs confirm the models can be used in real word settings. Bword2vec outperforms s-tf-simhashing for the same size of feature space while s-tf-simhashing improves with a larger size hash vector. We showed the two algorithms can successfully be used for full malware app detection and also infected files.

References

1. Allix, K., Bissyandé, T.F., Klein, J., Le Traon, Y.: Androzoo: collecting millions of android apps for the research community. In: ICSE 2016, pp. 468–471. ACM (2016)
2. Bingham, E., Mannila, H.: Random projection in dimensionality reduction: applications to image and text data. In: Proceedings of the seventh ACM SIGKDD, pp. 245–250. ACM (2001)
3. Chen, C., Vong, C.M., Wong, C.M., Wang, W., Wong, P.K.: Efficient extreme learning machine via very sparse random projection. Soft Comput. **22**(11), 3563–3574 (2018)
4. Chistyakov, A., Lobacheva, E., Kuznetsov, A., Romanenko, A.: Semantic embeddings for program behavior patterns. arXiv preprint arXiv:1804.03635 (2018)
5. Karbab, E.B., Debbabi, M., Derhab, A., Mouheb, D.: MalDozer: automatic framework for android malware detection using deep learning. Digit. Investig. **24**, S48–S59 (2018)
6. Kolosnjaji, B., Demontis, A., Biggio, B., Maiorca, D., Giacinto, G., Eckert, C., Roli, F.: Adversarial malware binaries: evading deep learning for malware detection in executables. In: EUSIPCO 2018 (2018)
7. Li, P., Hastie, T.J., Church, K.W.: Very sparse random projections. In: Proceedings of the 12th ACM SIGKDD, pp. 287–296. ACM (2006)
8. Mikolov, T., Chen, K., Corrado, G., Dean, J.: Efficient estimation of word representations in vector space. arXiv preprint arXiv:1301.3781 (2013)
9. Nissim, N., et al.: Keeping pace with the creation of new malicious PDF files using an active-learning based detection framework. Secur. Inform. **5**(1), 1 (2016)
10. Onwuzurike, L., Mariconti, E., Andriotis, P., De Cristofaro, E., Ross, G., Stringhini, G.: Mamadroid: detecting android malware by building Markov chains of behavioral models (extended version). arXiv preprint arXiv:1711.07477 (2017)
11. Sayfullina, L., Eirola, E., Komashinsky, D., Palumbo, P., Karhunen, J.: Android malware detection: building useful representations. In: Machine Learning and Applications (ICMLA), pp. 201–206. IEEE (2016)
12. Scofield, D., Miles, C., Kuhn, S.: Fast model learning for the detection of malicious digital documents. In: PPREW, p. 3. ACM (2017)
13. Wang, L., Liu, J., Chen, X.: Microsoft malware classification challenge (big 2015) first place team: say no to overfitting (2015)
14. Yousefi-Azar, M., Hamey, L., Varadharajan, V., McDonnell, M.D.: Fast, automatic and scalable learning to detect android malware. In: Liu, D., Xie, S., Li, Y., Zhao, D., El-Alfy, E.-S.M. (eds.) ICONIP 2017. LNCS, vol. 10638, pp. 848–857. Springer, Cham (2017). https://doi.org/10.1007/978-3-319-70139-4_86
15. Yousefi-Azar, M., Hamey, L., Varadharajanz, V., Cheng, S.: Malytics: a malware detection scheme. arXiv preprint arXiv:1803.03465 (2018)

Occlusion Detection in Visual Tracking: A New Framework and A New Benchmark

Xiaoguang Niu[1], Yueyang Gu[1], Zhifeng Lu[2], Zehua Hong[2], Yi Tian[2], Kuan Xu[1], Jie Yang[1], Xingqi Fang[1], and Yu Qiao[1(✉)]

[1] Institute of Image Processing and Pattern Recognition, Department of Automation, Shanghai Jiao Tong University, Shanghai, China
qiaoyu@sjtu.edu.cn
[2] Shanghai Electro-Mechanical Engineering Institute, Shanghai, China

Abstract. Occlusion remains being a challenge in visual object tracking. The robustness to occlusion is critical for tracking algorithms, though not much attention has been paid to it. In this paper, we first propose an occlusion detection framework which calculates the proportion of the target that is occluded, hence to decide whether to update the model of target. This framework can be integrated with existing tracking algorithms to increase their robustness to occlusion. Then we introduce a new benchmark which contains sequences where occlusion is the main difficulty. The sequences are chosen from public benchmarks and are fully annotated. The proposed framework is combined with several standard trackers and evaluated on the new benchmark. The experimental results show that our framework can improve the tracking performance, with explicit incorporation of occlusion detection.

Keywords: Visual tracking · Occlusion detection · Benchmark

1 Introduction

Generic object tracking [1,3,5–7,12–14], where the tracker is not specialized to any specific category of objects, is a popular research field in recent years. Because of the category-agnostic, it is not possible to train a detector offline for a particular type of objects, such as pedestrians or hands. Consequently, occlusion is the most challenging factor for generic object trackers [8], since the trackers usually cannot discriminate the occluders from the targets.

Majority of the work in handling occlusion is to add a sub-module before target model updater to monitor the tracking reliability. In [20], the feedback from tracking results is utilized to decide whether or not to update the target model. However, this strategy still cannot tell what is actually happening, occlusion or target appearance variation, both of which will decrease the tracking confidence.

This research is partly supported by USCAST2015-13, USCAST2016-23, SAST2016008, NSFC (No: 61375048).

COD (Context-based Occlusion Detection for Tracking) [15–17] is a framework that monitors the background-patches around the target and can identify which of them occlude the target. However, several drawbacks exist. First, the number of background-patches that COD monitors is constant, which contaminates the adaptive ability of the framework. Furthermore, determining the occlusion occurrence simply by the number of occluders over-simplifies the problem and is not guaranteed to be reasonable in all occasions. To solve these issues, we present Adaptive COD, which is adaptive to differently sized targets and able to identify what proportion of the target is affected by occlusion. The number of background-patches is now dependent on the perimeter of the target, hence more background-patches will be allocated to deal with a larger target. After acquiring the positions of the background-patches that occlude the target, we calculate the proportion of the target that is under occlusion. If the proportion is greater than a threshold, model updater will not take any action, avoiding the model being corrupted. The background-patches that occlude the target continues to be monitored, while other background-patches are discarded and new ones will be generated around the new target. As a general framework, Adaptive COD can be integrated with any existing tracking algorithm to address the occlusion problem.

To better evaluate the performance of different trackers and promote the development of tracking algorithms, several benchmarks have been built. OTB [21], VOT [10], and ALOV [19] are the most widely used ones. In OTB [21], each sequence is tagged with 9 attributes, including occlusion, illumination variation and so on, which represent the challenging factors in visual tracking. A sequence will be tagged with attribute 'occlusion' if there are frames in the sequence where occlusion happens. In VOT [10], the attribute annotation is further refined to per-frame level. Later in NUS-PRO [11], the occlusion is classified into three levels: no occlusion, partial occlusion and full occlusion. Recently, attribute-specific benchmarks appear. In [18], a dataset for fast moving objects is collected. A higher frame rate video dataset is proposed in [4]. Although occlusion is one of the attributes in OTB [21] and VOT [10], the frames where occlusion happens only take up a small proportion of the overall sequence. Moreover, before the tracker meets these frames, the tracking results have already drift from the groundtruth, which means that different trackers will have different initialization setups in terms of evaluating their robustness to occlusion. In this paper, we build an attribute-specific benchmark which contains sequences where the target undergoes occlusion. In our proposed dataset, we exclude other attributes and only preserve the frames relevant to occlusion. Each sequence contains three parts: before, during and after occlusion. We evaluate our model updating strategy by integrating it with several mediocre tracking algorithms, including KCF [7], SAMF [14], DSST [3] and Staple [1]. The experimental results show that the Adaptive COD improves the robustness of these tracking algorithms.

Algorithm 1. (Adaptive) COD

Initialize target tracker and background-patch trackers;
for $t = 2$ to T **do**
    Track the target and output target tracking result;
    Track the background-patches and identify occlusion;
    If no occlusion, update target tracker;
    Update background-patch trackers.
end for

In summary, the main contributions of this paper are as follows:

1. We improve the occlusion detection framework in [17]. The number of background-patch trackers is adaptive to the size of target. A new model updating strategy is proposed.
2. We establish a new dataset where the sequences contain occlusion for evaluating the robustness of tracking algorithms.
3. Extensive experiments demonstrate the effectiveness of our occlusion detection framework and occlusion benchmark.

2 Occlusion Detection Framework

In this section we first briefly review the Context-based Occlusion Detection for Tracking (COD) framework [17]. Then the proposed Adaptive COD is presented.

2.1 COD Review

Based on the assumption that both target and background-patches are involved in occlusion, COD [17] pays attention to the background around the target to *actively* detect occlusion. As is shown in Algorithm 1, two kinds of trackers exist in the framework: target tracker and background-patch trackers. Target tracker estimates the bounding box of target in the current frame, while the background-patch trackers provide the position and tracking reliability of every background-patch surrounding the target. Intuitively, if the bounding boxes of a background-patch and the target overlap and that the background-patch has high tracking reliability (hence it is not occluded by the target), then the target is occluded by the background-patch. Please refer to [17] for more details.

However, COD has the following disadvantages. Firstly, the number of background-patches N_1 is constant for variously sized targets in different sequences. For small targets, N_1 is relatively too large. Therefore, many background-patches overlay with each other, causing the double counting and repeated calculation. For large objects, N_1 becomes relatively small, so the background around the target is not fully monitored. Secondly, the target model will be updated online if the number of background-patches that occlude the target, N, is greater than a constant threshold N_{th}. Similarly, for targets of different sizes, N as merely a counting result cannot properly measure the degree of occlusion.

2.2 Adaptive COD

We propose an Adaptive COD to overcome the limitations of COD mentioned in Sect. 2.1. Adaptive COD inherits the structure from COD but differs in two important aspects: the initialization step and the criterion for identifying occlusion. They are shown in Algorithm 1.

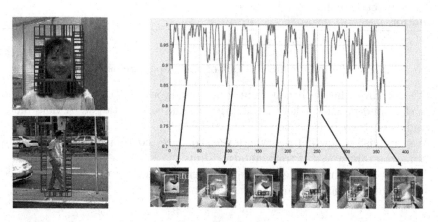

Fig. 1. In left, the number of background-patches for sequence *Girl* is 38, while for sequence *David3* it is 83. In right, the curve shows non-occluded proportion of the target for every frame in sequence *Tiger2*, along with the the frames #27,#107,#186,#238,#256,#355, corresponding to local minima of the curve. The blue boxes show where the occlusion happens.

Denote the bounding box of target in frame t as (x_t, y_t, w_t, h_t) for $t = 1, ..., T$, where (x_t, y_t) are the upper-left corner point coordinates and (w_t, h_t) are the width and height. Then we set $N_1 = [\ (w_1 + h_1)/2\ ]$, where $[x]$ will round x to its nearest integer. In this way, the number of background-patches is dependent on the size of target. Unless the scale of target varies heavily, we keep using N_1 in the following frames. The results can be seen in Fig. 1.

We propose a new criterion for identifying occlusion. For target with parameter (x_t, y_t, w_t, h_t), we build a mask M_t as follows:

$$M_t(x, y) = \begin{cases} 1, if\ x \in [x_t, x_t + w_t]\ \&\&\ y \in [y_t, y_t + h_t] \\ 0, otherwise \end{cases} \quad (1)$$

I.e., M_t has the same size of frame and the region representing the target is set as 1. The area of target region is $A_t = \sum M_t$. Similarly, for a background-patch with parameter $(bx_t^i, by_t^i, bw_t^i, bh_t^i)$ for $i = 1, 2, ..., N_1$, we build a mask m_t^i. Denoting the tracking reliability of background-patch i as r_t^i which is usually calculated as Peak-to-Sidelobe Ratio [2], we update M_t as

$$M_t = \begin{cases} M_t - m_t^i, & if\ r_t^i > r_{th} \\ M_t, & otherwise \end{cases} \quad (2)$$

where r_{th} is the threshold. After inspecting every background-patch and updating M_t, the area of target that is not occluded is $S_t = \sum M_t$. We use $\gamma_t = S_t / A_t$ as the measurement of occlusion, as is demonstrated in Fig. 1. Compared with using N as the indicator of occlusion in COD, the new area-based adaptive criterion makes sense for targets of any size.

After identifying occlusion, the algorithm makes decision on whether to update the target tracker. The background-patches that are identified as occluders will continue to be monitored. Meanwhile, the algorithm will not pay attention to the other background patches which does not occlude the target and new background patches around the target in current frame will be added in the monitoring set.

3   Occlusion Benchmark

In this section, we present a new specialized benchmark for evaluating the robustness of tracking algorithms to occlusion. The benchmark is available at https:// pan.baidu.com/s/1qZ0KeoW.

Although occlusion is one of the attributes in OTB [21], VOT [10] and NUS-PRO [11], these benchmarks still cannot accurately reflect the robustness of tracking algorithms to occlusion, due to the following reason. Each sequence usually has multiple challenging factors. Suppose a sequence s with frames $(\#1,...,\#t_1,...,\#t_2,...,\#T)$, where the occlusion happens in frames between $\#t_1$ and $\#t_2$. Since all the trackers start tracking in frame $\#1$, they will have different tracking outputs before the occlusion occurs in frame $\#t_1$, which means that the performance on frames between $\#t_1$ and $\#t_2$ is heavily influenced by the previous frames. As a recent study [9] shows, performance measures computed on a sequence are significantly biased to the dominant attribute of the sequence. Moreover, besides occlusion, there may exist other challenging factors in frames between $\#t_1$ and $\#t_2$, which makes the evaluation more unreliable.

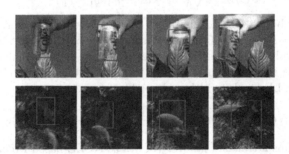

Fig. 2. Sequences in our occlusion benchmark can be divided into three parts. The first column shows the first frames of sequences *Coke_1* and *fish2_1*. The second and third columns show the targets being occluded. The last column shows targets after occlusion.

Table 1. Statistics about our occlusion benchmark.

Sequence sources	#	Target categories	#
From OTB	31	Person	19
From VOT	12	Object	15
From NUS-PRO	8	Animal	4
Total sequences	51	Face	8
Total frames	2628	Other	5

Based on these observations, we propose an occlusion benchmark that has the following characteristics:

1. Each sequence s with frames $(\#1,...,\#t_1,...,\#t_2,...,\#T)$ can be divided into 3 sub-sequences. In the first sub-sequence with frames $(\#1,...,\#t_1)$, neither occlusion nor other challenging factor occur, so the target model can be initialized. In the second sub-sequence with frames $(\#t_1,...,\#t_2)$, the target is occluded. In the last sub-sequence with frames $(\#t_2,...,\#T)$, occlusion disappears so we can identify if the tracking succeeds. See Fig. 2 for explanation.
2. In frames $(\#t_1,...,\#t_2)$, we exclude other attributes such as deformation, so that the only difficulty for tracking is to handle occlusion. However, it is a common scenario that the occluders are of the same category as the targets and have similar appearance, so we keep these sequences in the benchmark.
3. The sequences are selected from OTB [21], VOT [10] and NUS-PRO [11] with diversity and richness. The statistics is shown in Table 1.

In our occlusion benchmark, we propose a new metric called Normalized Center Location Error (NCLE) for evaluating performance. For tracking result (cx_1, cy_1, w_1, h_1) and ground-truth (cx, cy, w, h) where (cx_1, cy_1) and (cx, cy) are center locations, the traditional CLE adopted by OTB [21] is defined as

$$CLE = \sqrt{(cx_1 - cx)^2 + (cy_1 - cy)^2}. \tag{3}$$

A constant number, 20-pixel, is used for ranking trackers. However, for differently shaped and sized targets, 20-pixel deviation may have distinct meanings. For example, the width of a pedestrian target is usually smaller than the height, so the deviation is more serious if it is in the horizontal direction. In NCLE, we normalize the CLE by the width and height of target:

$$NCLE = min\{ \, max\{\frac{|cx_1 - cx|}{w}, \frac{|cy_1 - cy|}{h}\}, \, 1 \, \}. \tag{4}$$

NCLE $= 1$ means a tracking failure. We utilize NCLE-based Precision Plot and Success Plot [21] as performance measurements in our occlusion benchmark.

4   Experiments

In this section, we present the experimental results of several recent tracking algorithms evaluated on our occlusion benchmark, including KCF [7],

SAMF [14], DSST [3] and Staple [1]. Meanwhile, we integrate these trackers into our adaptive COD framework to validate its effectiveness. All the code is available at https://github.com/xgniu/Occlusion-Benchmark.

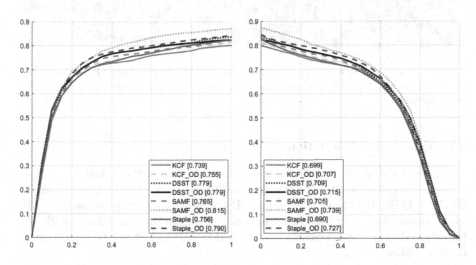

Fig. 3. The quantitative evaluation results. Left: NCLE-based Precision Plot. The numbers in brackets are the proportion of frames that have NCLE less than 0.5. Right: Success Plot.

Table 2. Different γ for different tracking algorithms. Our framework is not sensitive to the value of γ.

Precision	baseline	$\gamma=0.90$	$\gamma=0.85$	$\gamma=0.8$	Success	baseline	$\gamma=0.90$	$\gamma=0.85$	$\gamma=0.8$
KCF	0.739	0.755	0.758	0.760	KCF	0.699	0.707	0.714	0.720
DSST	0.779	0.779	0.779	0.795	DSST	0.709	0.715	0.716	0.732
SAMF	0.765	0.815	0.809	0.803	SAMF	0.705	0.739	0.724	0.723
Staple	0.756	0.790	0.783	0.783	Staple	0.690	0.727	0.721	0.719

4.1 Quantitative Evaluation

The quantitative evaluation results are shown in Fig. 3 in the form of Precision Plot and Success Plot. All the four trackers gain improvements in performance after being integrated into our adaptive occlusion detection framework. Moreover, we find that though different tracking algorithms require differently valued γ for best performance, a wide range of γ can provide comparable results (Table 2). The other thresholds are the same as in COD [17].

Fig. 4. The qualitative evaluation results. Red: SAMF. Blue: SAMF_OD. Green: Staple. Black: Staple_OD. The four sequences are *Coke*, *fish*, *Tiger2* and *Lemming* (Color figure online).

4.2 Qualitative Evaluation

Figure 4 visualizes several sequences from our occlusion benchmark along with the tracking results of different algorithms. Only the tracking results of SAMF, SAMF_OD, Staple and Staple_OD are shown for clarity, where the suffix '_OD' stands for being integrated into our occlusion detection framework. As the figure shows, when occlusion occurs, SAMF_OD and Staple_OD outperform their baselines.

5   Conclusion

Based on COD [17], we propose an adaptive occlusion detection framework which calculates the proportion of target that is not occluded. To better evaluate the robustness of tracking algorithms to occlusion, we propose an occlusion benchmark that excludes other challenging factors. In our benchmark, normalized center location error is adopted as the performance measure. Much work is needed in future to solve the occlusion problem for robust visual object tracking.

References

1. Bertinetto, L., Valmadre, J., Golodetz, S., Miksik, O., Torr, P.H.: Staple: complementary learners for real-time tracking. In: Proceedings of the IEEE Conference on Computer Vision and Pattern Recognition (CVPR), pp. 1401–1409 (2016)
2. Bolme, D.S., Beveridge, J.R., Draper, B.A., Lui, Y.M.: Visual object tracking using adaptive correlation filters. In: Proceedings of the IEEE Conference on Computer Vision and Pattern Recognition (CVPR), pp. 2544–2550. IEEE (2010)
3. Danelljan, M., Häger, G., Khan, F., Felsberg, M.: Accurate scale estimation for robust visual tracking. In: British Machine Vision Conference (BMVC). BMVA Press, Nottingham, 1–5 September 2014
4. Galoogahi, H.K., Fagg, A., Huang, C., Ramanan, D., Lucey, S.: Need for speed: a benchmark for higher frame rate object tracking. In: Proceedings of the IEEE International Conference on Computer Vision (ICCV), pp. 1134–1143 (2017)

5. Gu, K., Zhou, T., Liu, F., Yang, J., Qiao, Y.: Correlation filter tracking via bootstrap learning. In: IEEE International Conference on Image Processing, pp. 459–463 (2016)
6. Gu, K., Zhou, T., Liu, F., Yang, J., Qiao, Y.: Patch-based object tracking via locality-constrained linear coding. In: Proceedings of the 35th Chinese Control Conference, pp. 7015–7020 (2016)
7. Henriques, J.F., Caseiro, R., Martins, P., Batista, J.: High-speed tracking with kernelized correlation filters. IEEE Trans. Pattern Anal. Mach. Intell. **37**(3), 583–596 (2015)
8. Kristan, M., Leonardis, A., Matas, J., Felsberg, M.: The visual object tracking VOT2017 challenge results. In: Proceedings of the IEEE International Conference on Computer Vision (ICCV) Workshops, pp. 1949–1972 (2017)
9. Kristan, M., et al.: A novel performance evaluation methodology for single-target trackers. IEEE Trans. Pattern Anal. Mach. Intell. **38**(11), 2137–2155 (2016)
10. Kristan, M., Pflugfelder, R., Leonardis, A., Matas, J., Porikli, F., Čehovin, L.: The visual object tracking vot2013 challenge results. In: Proceedings of the IEEE International Conference on Computer Vision (ICCV) Workshops, pp. 564–586, December 2013
11. Li, A., Lin, M., Wu, Y., Yang, M., Yan, S.: Nus-pro: a new visual tracking challenge. IEEE Trans. Pattern Anal. Mach. Intell. **38**(2), 335–349 (2016)
12. Li, Q., Qiao, Y., Yang, J.: Robust visual tracking based on local kernelized representation. In: IEEE International Conference on Robiotics and Biomimetics, pp. 2523–2528 (2014)
13. Li, Q., Qiao, Y., Yang, J., Bai, L.: Robust visual tracking based on online learning of joint sparse dictionary. In: International Conference on Machine Vision (2013)
14. Li, Y., Zhu, J.: A scale adaptive kernel correlation filter tracker with feature integration. In: Agapito, L., Bronstein, M.M., Rother, C. (eds.) ECCV 2014. LNCS, vol. 8926, pp. 254–265. Springer, Cham (2015). https://doi.org/10.1007/978-3-319-16181-5_18
15. Niu, X., Cui, Z., Geng, S., Yang, J., Qiao, Y.: Robust visual tracking via occlusion detection based on depth-layer information. In: International Conference on Neural Information Processing, pp. 44–53 (2017)
16. Niu, X., Fang, X., Qiao, Y.: Robust visual tracking via occlusion detection based on staple algorithm. In: Asian Control Conference, pp. 1051–1056 (2017)
17. Niu, X., Qiao, Y.: Context-based occlusion detection for robust visual tracking. In: IEEE International Conference on Image Processing, pp. 3655–3659 (2017)
18. Rozumnyi, D., Kotera, J., Sroubek, F., Novotn, L., Matas, J.: The world of fast moving objects. In: Proceedings of the IEEE Conference on Computer Vision and Pattern Recognition (CVPR) (2017)
19. Smeulders, A.W.M., Chu, D.M., Cucchiara, R., Calderara, S., Dehghan, A., Shah, M.: Visual tracking: an experimental survey. IEEE Trans. Pattern Anal. Mach. Intell. **36**(7), 1442–1468 (2014)
20. Wang, M., Liu, Y., Huang, Z.: Large margin object tracking with circulant feature maps. In: Proceedings of the IEEE Conference on Computer Vision and Pattern Recognition (CVPR), pp. 21–26 (2017)
21. Wu, Y., Lim, J., Yang, M.H.: Online object tracking: a benchmark. In: Proceedings of the IEEE Conference on Computer Vision and Pattern Recognition (CVPR), pp. 2411–2418 (2013)

Attentional Payload Anomaly Detector for Web Applications

Zhi-Quan Qin, Xing-Kong Ma, and Yong-Jun Wang[(✉)]

College of Computer, National University of Defense Technology,
Changsha 410073, China
{tanzhiquan14,maxingkong,wangyongjun}nudt.edu.cn

Abstract. Nowadays web applications influence people deeply and become popular targets of attackers. The payload anomaly detection is an effective method to keep the security of web applications but requires proper features which takes a lot of time and effort for experts and researchers to design. Utilizing the deep learning techniques for the detection is a solution to the feature design problem because deep learning models can learn features during the training process and achieve great performances. However, current deep learning payload detection models have their limit on processing long sequences, which reduces the detection performance. And due to the intricate data processing, the results produced by the models are unconvincing. In this paper, we proposed an attentional recurrent neural network (RNN) model for the payload detection, called ATPAD. With the attention mechanism, ATPAD generates effective features for the detection tasks and provides a visualized way to verify detection results. The experiment results show that our proposed model not only achieves high detection rates and low false alarm rates, but also produces understandable results.

Keywords: Web application · Payload anomaly detection
Deep learning · RNN · Attention mechanism

1 Introduction

With the development of cyberspace, web applications are playing a more important role in the daily life of people than ever. Meanwhile, web applications become very attractive attack targets. By launching attacks such as SQL injection and cross site scripting(XSS), attackers are able to steal privacy information of users, tamper with the database and destroy data. To keep the security of web applications, anomaly detection systems are deployed to alarm the malicious behaviors since the attack techniques evolve rapidly.

The attacks are mainly carried in the payload in the application layer and the content of payload is complicated and varied. Researchers tend to use machine learning algorithms to build payload anomaly detection models. The detection models are divided into unsupervised and supervised ones by training manner. Although unsupervised models do not require labeled data for training, the

© Springer Nature Switzerland AG 2018
L. Cheng et al. (Eds.): ICONIP 2018, LNCS 11304, pp. 588–599, 2018.
https://doi.org/10.1007/978-3-030-04212-7_52

trained model may suffer from a high false positive rate(i.e. false alarm rate). Supervised models which are trained with the labeled data, achieve a better performance.

One of the most difficult problems when applying machine learning to anomaly detection is how to extract features from the raw network data. Though there are many famous traditional learning algorithms such as Support Vector Machine (SVM), Decision Tree (DT) and so on, researchers have to spend a lot of time and effort doing feature extraction and feature selection. Recently, deep learning makes a huge success in many fields. Some of the works reach and even surpass the human level. Not only a great performance, but also the suitable features for the task can be obtained by deep learning. Accordingly, applying deep learning to payload anomaly detection is promising.

In recent works, the single RNN model is usually used for payload anomaly detection because of its capability of processing sequential data. However, the single RNN model has its limits. First, due to the gradient vanishing problem, the single RNN model is unable to deal with the sequences that are too long even though Long Short Term Memory (LSTM) [7] and Gated Recurrent Unit (GRU) [3] are used. It often works on the sequences of length under 200 but the length of the application payload is far beyond that. Second, the single RNN model uses the final hidden state for detection which only focuses on the tail of the payload but the anomaly may appear any place in the payload. The above two limits reduce the detection performance of the model. Besides, the deep learning model transforms the raw data into a high-dimension feature space that is too complicated for human to analyze, which makes the detection results unconvincing.

To overcome the above drawbacks of the single RNN, we propose an attentional RNN for payload anomaly detection, called ATPAD (ATtentional Payload Anomaly Detector). ATPAD transforms the payload byte sequence to a hidden state sequence by a recurrent neural network, then use the attention mechanism to weight the hidden states as the feature vector for further detection. With the attention mechanism, ATPAD catches a global view of the payload and focuses on the suspicious segments rather than the tail of the payload to make an accurate classification. Also ATPAD is able to highlight detection-related segments of the payload, which assists experts to verify the detection result. We evaluate the performance of ATPAD on CIC-IDS-2017 [17] and CSIC-2010 [20] datasets. The experiment results are satisfying and reasonable. From the above, our contributions through this paper are summarized as follows:

1. We propose an attentional deep learning model ATPAD. So far as we know, it is the first work applying the attention mechanism in payload anomaly detection.
2. We demonstrate a novel feature generation approach for payload anomaly detection.
3. We demonstrate an attention-based approach to analyze and verify the detection results.

The rest of the paper is organized as follows. Section 2 introduces the previous works related to this paper. Section 3 describes the details of the proposed model. Then, Sect. 4 presents the setup and results of the experiments. Finally, Sect. 5 provides the conclusions.

2   Related Works

2.1   Feature Extraction and Feature Selection

To build a proper detection model with machine learning algorithms, feature extraction is required. In [22], PAYL is proposed, which uses the 1-grams frequency distribution of the payload as features. Although PAYL can achieve high detection rate and low false alarm rate, it is vulnerable to mimicry attacks. Anagram is proposed to make up for the drawback [21]. Anagram combines the high order n-grams with the bloom filter as features. In [16], McPAD is proposed. To avoid the high complexity of the n-grams model, McPAD builds an One-Class SVM model using 2_V-grams and ensembles models with different values of v. In [4], the token-based n-grams are used as features to build an One-Class SVM model for HTTP request payload detection. In [19], the hand-crafted features, combined with 1-grams are used to build decision tree models for the detection. Feature selection can save the training and prediction time and can improve the quality of the model. In [18], The difference distance map and Linear Discriminant Analysis are used to select the significant features.In [12], the hybrid feature selection method and gradually feature removal method are proposed to reduce the time of training and prediction. In [19], The Generic Feature Selection measure is used for feature selection. In a word, feature extraction and selection is an important but tough work in developing a detection model.

2.2   Deep Learning

There are several common deep learning models such as Stacked Auto-Encoder (SAE), Convolution Neural Network(CNN) and Recurrent Neural Network(RNN). RNN is capable of sequence-related tasks because it maintains the hidden states during computing the output sequence. And to overcome the gradient explosion and vanishing problem of the vanilla RNN, gradient clipping [15], LSTM [7] and GRU [3] are proposed.

The researchers usually use RNN models for anomaly detection on sequential data. In [10], a LSTM-RNN model is built on KDDCup99 dataset and each data point of the sequence is classified by the hidden state of the model. In [9], the LSTM-RNN models are used for host-based anomaly detection. the experiment also shows that the final hidden state of the LSTM-RNN as features outperforms the hand-crafted features. In [2], a 3-layer LSTM-RNN model is built for the detection of the HTTP requests. In [8], an RNN model is built and its final hidden state is used to determine whether an URL is either malicious or benign.

2.3 Attention Mechanism

Attention mechanism is mainly applied to seq2seq(sequence to sequence) tasks, such as Neural Machine Translation and Question & Answering in which the source sentence and target sentence are involved. It helps models learn what to "focus" on when making the prediction of each target word by calculating the alignments between the words of the two sentences.

There are several proposed variants of the attention. In [1], additive attention is proposed, it uses an one-hidden layer feed-forward network to calculate the alignments. In [14], multiplicative attention is proposed to simplify the calculations in the additive attention. In [13], self-attention is introduced to extract the relevant aspects from the same sentence, which is quite different from the first two.

3 ATPAD Model

ATPAD consists of an encoder RNN, a decoder RNN and an attention network, showed in Fig. 1. Formally, the model is a function f of the payload X as $(y, \alpha) = f(X)$, where $y \in \{0, 1\}$ denotes the detection result for the payload and $\alpha \in \mathbb{R}^L$ denotes the support for the result. L is the length of the payload. The payload is predicted as a normal when $y = 0$, and an anomaly when $y = 1$. There are four stages to go through for the detection, i.e. *Hidden State Generation, Attention Calculation, Payload Classification,* and *Attention Visualization.*

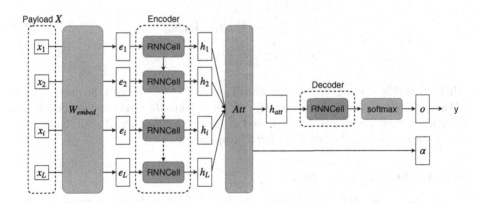

Fig. 1. The Framework of ATPAD

3.1 Hidden State Generation

The hidden state generation stage is aimed at extracting the sequential information of the payload. The payload $X = [x_1, x_2, \ldots, x_L], x_i \in \{0, \ldots, 255\}$ is a byte sequence of length L, which can be obtained from a single packet or the packet reassembling of a flow. A byte represents a symbol and so there are totally 256 different symbols. The value of the i-th byte in X, x_i is the index of its symbol. By looking up into an Embedding Matrix $W_{embed} \in \mathbb{R}^{E \times 256}$, X is transformed to a sequence of embedding vectors $S_{embed} = [e_1, e_2, \ldots, e_L]$, $e_i \in \mathbb{R}^E$.

Then, the encoder RNN takes S_{embed} as input and do L loops of calculations to output the hidden state sequence $H = [h_1, h_2, \ldots, h_L]$, $h_i \in \mathbb{R}^{N_{enc}}$. For i-th loop, the hidden state h_i is calculated by Eq. (1).

$$h_i = RNNCell_{enc}(h_{i-1}, e_i), \tag{1}$$

where $RNNCell$ denotes a general recurrent cell, it could be $LSTMCell$, $GRUCell$ and even multiple layers of them in the implement. When using $LSTMCell$, the hidden state h_i is replaced by a tuple of cell state and hidden state (c_i, h_i). At the beginning $i = 1$, the zero state h_0, a vector of all zeroes, is used.

3.2   Attention Calculation

The attention calculation stage is aimed at catching the correlationship between each step of the byte sequence and the detection result. An attentive weight α_i is assigned for h_i $(i = 1, 2, \ldots, L)$ by the attention network. The attention network Att implements an additive attention. It takes a constant vector h_0' and h_i as input and calculates the unnormalized attentive weights a_i by Eq. (2). Finally, the softmax normalization is used to satisfy the constraint $\sum_{i=1}^{L} \alpha_i = 1$, as Eq. (3).

$$
\begin{aligned}
a_i &= Att(h_i, h_0') \\
&= v_a^T \tanh(W_h h_i + W_0 h_0')
\end{aligned} \tag{2}
$$

$$\alpha_i = \frac{e^{a_i}}{\sum_{i=1}^{L} e^{a_i}}, \tag{3}$$

where $v_a \in \mathbb{R}^{N_{att}}$, $W_h \in \mathbb{R}^{N_{att} \times N_{enc}}$ and $W_0 \in \mathbb{R}^{N_{att} \times N_{dec}}$ denote the parameters of the attention network. Note that $h_0' \in \mathbb{R}^{N_{dec}}$ is learnable and used as the initial hidden state of decoder RNN. N_{att} and N_{dec} are the number of neurons of the attention network and of decoder RNN, respectively.

3.3   Payload Classification

In this stage, the attentive hidden state $h_{att} = \sum_{i=1}^{L} \alpha_i h_i$ is calculated as the encoding of the payload. This encoding is regarded as the feature vector for payload classification. h_{att} is feed forwards into the decoder RNN to generate the current hidden state, which is then feed forwards to the softmax layer to produce the prediction y as Eqs. (4) and (5).

$$o = softmax(RNNCell_{dec}(h_0', h_{att})) \tag{4}$$

$$y = \begin{cases} 1, & if \ o_0 < o_1 \\ 0, & else \end{cases}, \tag{5}$$

where $o = (o_0, o_1) \in \mathbb{R}^2$ is the predictive class distribution. o_0 and o_1 stand for the probability of being normal and anomalous respectively.

The parameters of ATPAD are learned during the supervised training process. By gradient descent, the attention network learn the correlation between the label and each part of the payload. The stronger the correlation, the greater the attentive weight.

Compared with using a single hidden state for prediction, using h_{att} brings two benefits: (1) h_{att} utilizes the information of the whole hidden state sequence H, which lead to a more accurate prediction. (2) the linear combination of the hidden states H with the attentive weights $\alpha = [\alpha_1, \alpha_2, \ldots, \alpha_L]$ is able to help people interpret how the model make predictions.

3.4   Attention Visualization

The attentive weights α indicate the degree of correlation between the bytes in the payload and the prediction, which can be regarded as the support for the prediction. To make the detection result convincing, the support is output. By visualizing the attentive weights, the experts can quickly verify the prediction and notice whether the model works correctly or not. To highlight the most strong correlation segments of the payload, the avg-max normalization is used. The attentive weights under the average value are mapped to 0, and the max value is mapped to 1. The weights between the average value and the max value are mapped by the linear function as Eq. (6).

$$\hat{\alpha}_i = \begin{cases} \frac{\alpha_i - \alpha_{avg}}{\alpha_{max} - \alpha_{avg}}, & if \ \alpha_i \geqslant \alpha_{avg} \\ 0, & else \end{cases}, \tag{6}$$

where α_{avg} and α_{max} denote the average value and the max value respectively. Then the weights $\hat{\alpha}$ are used to color the bytes in the payload by a specific colormap.

4   Experiment

4.1   Dataset

To evaluation our proposed model, two datasets are used. In the experiments, the number of attack classes is not presumed to be fixed and we only use two labels: "0" for the normal and "1" for the anomaly. The information about the class distribution and the usage is summarized in Table 1.

1. **CIC-IDS-2017**: The CIC-IDS-2017 dataset consists of totally 5-day working hour network traffic(stored as PCAP files) in a complicated network environment and the corresponding labeled flows(stored as CSV files). The traffic contains benign traffic generated by B-Profile system [6] and the most up-to-date attacks. In the experiment, we use the traffic in the period of web attacks, reassemble and label the application payloads of the flows with the

provided labeled file. And then we split the payloads into 5 subsets by about 10,000 payloads per subset. There are 3 types of web attacks involved: Brute Force login(BF), SQL Injection(SQLI) and XSS. All of the attack payloads and the benign payloads are relabeled as "1" and "0" respectively.

2. **CSIC-2010**: The CSIC-2010 dataset contains thousands of web requests automatically generated on e-Commerce web application. The dataset is divided into three subsets: 36,000 normal requests for training, 36,000 normal requests and 25,000 anomalous requests for test. There are three types of anomalous requests: static attacks that request for hidden(non-existent) resources, dynamic attacks that modify the valid request arguments, and illegal requests that have not malicious intention. We randomly mix 9,000 normal requests with 1,000 anomalous ones as the trainingset and leave the rest as the testset.

Table 1. The summarization of payloads used on datasets

Dataset		Class Distribution			
CIC-IDS-2017	Subset	Benign	BF	XSS	SQLI
	trainingset	9,967	37	0	0
	testset$_1$	9,938	68	0	0
	testset$_2$	9,959	48	0	0
	testset$_3$	9,964	0	26	12
	testset$_4$	10,003	0	0	0
CSIC-2010	Subset	Normal	Anomal		
	trainingset	9,000	1,000		
	testset	63,000	24,064		

4.2 Configurations of ATPAD

As a matter of convenience, we use one-layer GRU-RNNs as the encoder and decoder, and choose tanh as the activation function. the number of neurons i.e. N_{enc}, N_{dec} and N_{att} are 100. The embedding dimension E is set to 50 on CIC-IDS-2017 and 100 on CSIC-2010. During the training process, we use the cross entropy as the loss function and the Adam [11] optimizer for gradient descent with an exponentially decayed learning rate 0.005. And dropout [5] with probability 0.5 and global clipping over 1 are applied. For efficiency, we use minibatch of size 32 and the maxlength of the sequences in a batch is set to 500 on CIC-IDS-2017 and 1000 on CSIC-2010.

4.3 How Does the Attention Help?

We describe ATPAD's classification performance on CIC-IDS-2017 at first. ATPAD made only 1 false negative(FN), 1 false positive(FP) and 3 FPs on the trainingset, testset$_3$ and testset$_4$ respectively. And it didn't make any wrong prediction on testset$_1$ and testset$_2$. It means that ATPAD achieved almost a perfect classification performance on CIC-IDS-2017.

To illustrate how the attention visualization helps experts verify the prediction results, we show some attention examples obtained from the experiment on the CIC-IDS-2017. The true positives(TPs, i.e. attacks) and true negative(TN) are showed in Figs. 2 and 3, respectively. The FP and FN are showed in Fig. 4. The deeper the color, the greater the attentive weight.

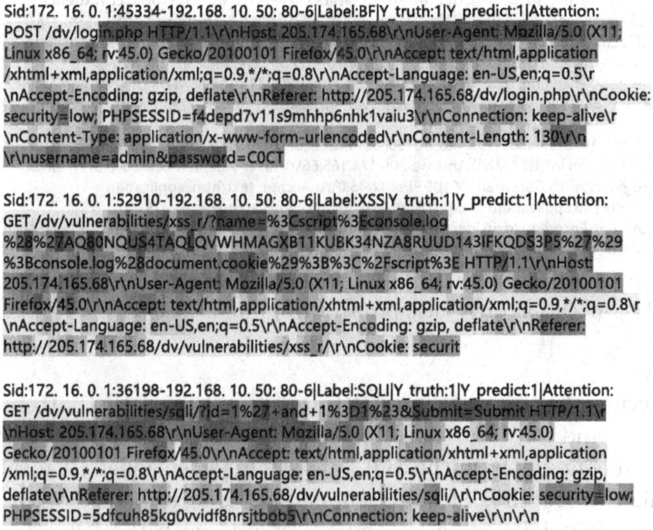

Sid:172. 16. 0. 1:45334-192.168. 10. 50: 80-6|Label:BF|Y_truth:1|Y_predict:1|Attention: POST /dv/login.php HTTP/1.1\r\nHost: 205.174.165.68\r\nUser-Agent: Mozilla/5.0 (X11; Linux x86_64; rv:45.0) Gecko/20100101 Firefox/45.0\r\nAccept: text/html,application /xhtml+xml,application/xml;q=0.9,*/*;q=0.8\r\nAccept-Language: en-US,en;q=0.5\r \r\nAccept-Encoding: gzip, deflate\r\nReferer: http://205.174.165.68/dv/login.php\r\nCookie: security=low; PHPSESSID=f4depd7v11s9mhhp6nhk1vaiu3\r\nConnection: keep-alive\r \r\nContent-Type: application/x-www-form-urlencoded\r\nContent-Length: 130\r\n \r\nusername=admin&password=C0CT

Sid:172. 16. 0. 1:52910-192.168. 10. 50: 80-6|Label:XSS|Y_truth:1|Y_predict:1|Attention: GET /dv/vulnerabilities/xss_r/?name=%3Cscript%3Econsole.log %28%27AQ80NQUS4TAQLQVWHMAGXB11KUBK34NZA8RUUD143IFKQDS3P5%27%29 %3Bconsole.log%28document.cookie%29%3B%3C%2Fscript%3E HTTP/1.1\r\nHost: 205.174.165.68\r\nUser-Agent: Mozilla/5.0 (X11; Linux x86_64; rv:45.0) Gecko/20100101 Firefox/45.0\r\nAccept: text/html,application/xhtml+xml,application/xml;q=0.9,*/*;q=0.8\r \r\nAccept-Language: en-US,en;q=0.5\r\nAccept-Encoding: gzip, deflate\r\nReferer: http://205.174.165.68/dv/vulnerabilities/xss_r/\r\nCookie: securit

Sid:172. 16. 0. 1:36198-192.168. 10. 50: 80-6|Label:SQLI|Y_truth:1|Y_predict:1|Attention: GET /dv/vulnerabilities/sqli/?id=1%27+and+1%3D1%23&Submit=Submit HTTP/1.1\r \r\nHost: 205.174.165.68\r\nUser-Agent: Mozilla/5.0 (X11; Linux x86_64; rv:45.0) Gecko/20100101 Firefox/45.0\r\nAccept: text/html,application/xhtml+xml,application /xml;q=0.9,*/*;q=0.8\r\nAccept-Language: en-US,en;q=0.5\r\nAccept-Encoding: gzip, deflate\r\nReferer: http://205.174.165.68/dv/vulnerabilities/sqli/\r\nCookie: security=low; PHPSESSID=5dfcuh85kg0vvidf8nrsjtbob5\r\nConnection: keep-alive\r\n\r\n

Fig. 2. The attention on Brute Force attack (top), Cross Site Scripting attack (middle) and SQL Injection attack (bottom).

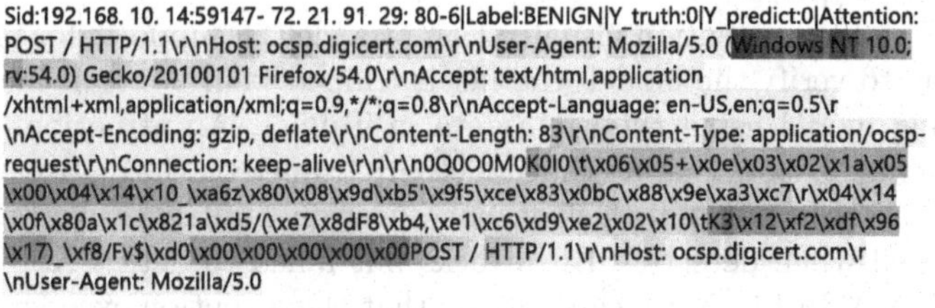

Sid:192.168. 10. 14:59147- 72. 21. 91. 29: 80-6|Label:BENIGN|Y_truth:0|Y_predict:0|Attention: POST / HTTP/1.1\r\nHost: ocsp.digicert.com\r\nUser-Agent: Mozilla/5.0 (Windows NT 10.0; rv:54.0) Gecko/20100101 Firefox/54.0\r\nAccept: text/html,application /xhtml+xml,application/xml;q=0.9,*/*;q=0.8\r\nAccept-Language: en-US,en;q=0.5\r \r\nAccept-Encoding: gzip, deflate\r\nContent-Length: 83\r\nContent-Type: application/ocsp-request\r\nConnection: keep-alive\r\n\r\n0Q0O0M0K0I0\t\x06\x05+\x0e\x03\x02\x1a\x05 \x00\x04\x14\x10_\xa6z\x80\x08\x9d\xb5`\x9f5\xce\x83\x0bC\x88\x9e\xa3\xc7\r\x04\x14 \x0f\x80a\x1c\x821a\xd5/(\xe7\x8dF8\xb4\xe1\xc6\xd9\xe2\x02\x10\tK3\x12\xf2\xdf\x96 \x17)_\xf8/Fv$\xd0\x00\x00\x00\x00\x00POST / HTTP/1.1\r\nHost: ocsp.digicert.com\r \r\nUser-Agent: Mozilla/5.0

Fig. 3. The attention on a benign payload.

Comparing Fig. 2 with Fig. 3, we can observe that the difference of attention distribution between TP and TN is obvious. The attention on TN tends to be more well-distributed, while the attention on TP is intensively located at some key places of the attack. For BF attack(top of Fig. 2), URL("*/dv/login.php*"), Host("*205.174.165.68*"), which is known as the attack machine) and the parameter name("*password*") are attentioned. For XSS attack(middle of Fig. 2), URL, especially the attack script segment, and Host are attentioned. The attack script is encoded as URL-encode to bypass the detection system. The decoded attack script is "⟨*script*⟩*console.log(⋯); console.log(document.cookie);*⟨*/script*⟩" which attempts to steal the cookie information of users. The attention on the SQLI attack(bottom of Fig. 2) is similar to the XSS attack and the attack SQL command(decoded as "*1' and 1=1#*") is attentioned.

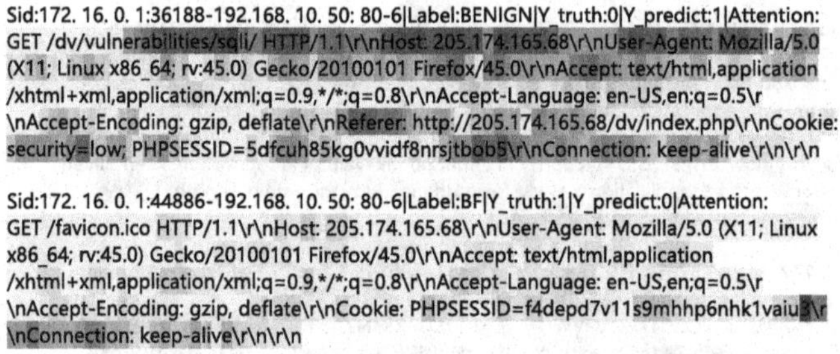

Sid:172. 16. 0. 1:36188-192.168. 10. 50: 80-6|Label:BENIGN|Y_truth:0|Y_predict:1|Attention:
GET /dv/vulnerabilities/sqli/ HTTP/1.1\r\nHost: 205.174.165.68\r\nUser-Agent: Mozilla/5.0
(X11; Linux x86_64; rv:45.0) Gecko/20100101 Firefox/45.0\r\nAccept: text/html,application
/xhtml+xml,application/xml;q=0.9,*/*;q=0.8\r\nAccept-Language: en-US,en;q=0.5\r
\nAccept-Encoding: gzip, deflate\r\nReferer: http://205.174.165.68/dv/index.php\r\nCookie:
security=low; PHPSESSID=5dfcuh85kg0vvidf8nrsjtbob5\r\nConnection: keep-alive\r\n\r\n

Sid:172. 16. 0. 1:44886-192.168. 10. 50: 80-6|Label:BF|Y_truth:1|Y_predict:0|Attention:
GET /favicon.ico HTTP/1.1\r\nHost: 205.174.165.68\r\nUser-Agent: Mozilla/5.0 (X11; Linux
x86_64; rv:45.0) Gecko/20100101 Firefox/45.0\r\nAccept: text/html,application
/xhtml+xml,application/xml;q=0.9,*/*;q=0.8\r\nAccept-Language: en-US,en;q=0.5\r
\nAccept-Encoding: gzip, deflate\r\nCookie: PHPSESSID=f4depd7v11s9mhhp6nhk1vaiu3\r
\nConnection: keep-alive\r\n\r\n

Fig. 4. The attention on a false positive (top) and a false negative (bottom).

By observing the attention on FP(top of Fig. 4) and FN(bottom of Fig. 4), we can understand why our model made such mistakes. In the FP example, many common contents with the SQLI attack are attentioned. It means that the FP has a high probability of being anomalous even though labeled as benign. The FN example is labeled as BF, but compared with the BF attack example in Fig. 2, it has several different places such as Method, URL and non-existent Referer. There is few suspected places to focus on and the attention on is flat, much more similar to the one on TN. In fact, the behavior of the FN example is trying to get the normal resource("favicon.ico"), which is even regarded as normal for the experts.

Not only can the attention visualization provide the supports for predictions, but also help to verify the model whether is trained well. We trained and tested an overfitting model using the same data in Table 1. An attention example is presented in Fig. 5. Note that all the models are trained on the dataset containing only BF attacks. Because of its lack of generalization ability, the overfitting model performed well when detecting BF attacks but failed to detect XSS and SQLI attacks(28 FNs). In Fig. 5, we can observe that the attention excessively locates on Host area("*205.174.165.68*", the attack machine mentioned before). Though the attentioned area may be useful for detection of BF attacks, it cannot support the prediction sufficiently. Therefore we reject this model.

Based on the results and analysis above, we can summarize that the proposed model can achieve a great performance and it can provide the supports for its prediction of attacks and help experts to verify itself by the attention visualization.

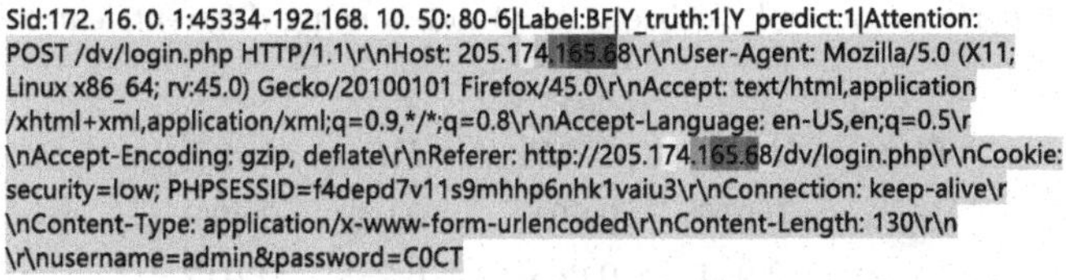

Sid:172. 16. 0. 1:45334-192.168. 10. 50: 80-6|Label:BF|Y_truth:1|Y_predict:1|Attention:
POST /dv/login.php HTTP/1.1\r\nHost: 205.174.165.68\r\nUser-Agent: Mozilla/5.0 (X11;
Linux x86_64; rv:45.0) Gecko/20100101 Firefox/45.0\r\nAccept: text/html,application
/xhtml+xml,application/xml;q=0.9,*/*;q=0.8\r\nAccept-Language: en-US,en;q=0.5\r
\nAccept-Encoding: gzip, deflate\r\nReferer: http://205.174.165.68/dv/login.php\r\nCookie:
security=low; PHPSESSID=f4depd7v11s9mhhp6nhk1vaiu3\r\nConnection: keep-alive\r
\nContent-Type: application/x-www-form-urlencoded\r\nContent-Length: 130\r\n
\r\nusername=admin&password=C0CT

Fig. 5. The attention on a attack payload generated by the overfitting model.

4.4   Performance Evaluation on CSIC-2010

To illustrate the performance advantage of our model, we compare it with a DT model using 1-grams as features and an RNN model using the final hidden state for prediction on CSIC-2010. For the DT model and RNN model, we train models on a set of hyper parameters and choose the model of the best performance. The detection rate(DR) and false alarm rate(FAR) are calculated as Eq. (7).

$$DR = \frac{TP}{TP + FN}, \quad FAR = \frac{FP}{TP + FP} \tag{7}$$

The DR and FAR of these three models on the trainingset and testset are showed in Table 2. Although the DT model achieves the highest DR 0.970 on the trainingset, it suffers from high FAR(≥ 0.496) on both trainingset and testset. RNN$_{final}$ and our proposed ATPAD achieve high DRs(≥ 0.920) and low FARs(≤ 0.012). Concretely, ATPAD achieves a comparable DR 0.969 to the DT model and the lowest FAR 0.001 on the trainingset. On the testset, ATPAD achieving a DR 0.944 and a FAR 0.003, outperforms all the other models.

Table 2. The performances of different models on CSIC-2010

Model	DR_{train}	FAR_{train}	DR_{test}	FAR_{test}
DT$_{1-\mathrm{grams}}$	**0.970**	0.691	0.940	0.496
RNN$_{\mathrm{final}}$	0.945	0.012	0.920	0.005
ATPAD	0.969	**0.001**	**0.944**	**0.003**

To show the effectiveness of the attentive hidden state, We also use the final hidden state h_{final} of RNN$_{final}$ and the attentive hidden state h_{att} of ATPAD as features to train the DT models. The result is showed in Table 3. Compared with h_{final}, using h_{att} as features achieves higher DRs on both the trainingset and testset, a lower FAR on the trainingset and a comparable FAR on the testset. Besides, compared with 1-grams feature, using the features learned by deep learning models achieves comparable DRs and much lower FARs.

Table 3. The performances of Decision Tree models using different features on CSIC-2010

Feature	DR_{train}	FAR_{train}	DR_{test}	FAR_{test}
1-grams	0.970	0.691	0.940	0.496
h_{final}	0.954	0.065	0.938	**0.017**
h_{att}	**0.970**	**0.047**	**0.947**	0.018

5   Conclusion

In this paper, we focus on overcoming the drawbacks of single RNN models in payload anomaly detection and propose the attentional RNN model ATPAD for payload anomaly detection. The model learns the features automatically and uses the attention mechanism to grasp the anomalous segments of the payload. The experiment results on the CIC-IDS-2017 and CSIC-2010 dataset show that the model achieves high detection rates and low false alarm rates, and the attention visualization of ATPAD is a helpful approach for experts to understand how the deep learning model works and verify the detection results as well as the model, and the attentive hidden state generated by ATPAD is a kind of effective features for payload anomaly detection.

Acknowledgement. This work is supported by NSFC No.61472439, National Natural Science Foundation of China under Grant.

References

1. Bahdanau, D., Cho, K., Bengio, Y.: Neural machine translation by jointly learning to align and translate. arXiv preprint (2014). http://arxiv.org/abs/1409.0473
2. Bochem, A., Zhang, H., Hogrefe, D.: Poster abstract: streamlined anomaly detection in web requests using recurrent neural networks. In: 2017 IEEE Conference on Computer Communications Workshops, pp. 1016–1017 (2017)
3. Cho, K., et al.: Learning phrase representations using RNN encoder-decoder for statistical machine translation. arXiv preprint (2014). http://arxiv.org/abs/1406.1078
4. Düssel, P., Gehl, C., Laskov, P., Rieck, K.: Incorporation of application layer protocol syntax into anomaly detection. In: Sekar, R., Pujari, A.K. (eds.) ICISS 2008. LNCS, vol. 5352, pp. 188–202. Springer, Heidelberg (2008). https://doi.org/10.1007/978-3-540-89862-7_17
5. Gal, Y., Ghahramani, Z.: A theoretically grounded application of dropout in recurrent neural networks. In: Data-Efficient Machine Learning workshop, ICML (2016)
6. Gharib, A., Sharafaldin, I., Lashkari, A.H., Ghorbani, A.A.: An evaluation framework for intrusion detection dataset. In: International Conference on Information Science and Security, pp. 1–6 (2017)
7. Hochreiter, S., Schmidhuber, J.: Long short-term memory. Neural Comput. **9**(8), 1735–1780 (1997)

8. Jin, X., Cui, B., Yang, J., Cheng, Z.: Payload-based web attack detection using deep neural network. In: Barolli, L., Xhafa, F., Conesa, J. (eds.) BWCCA 2017. LNDECT, vol. 12, pp. 482–488. Springer, Cham (2018). https://doi.org/10.1007/978-3-319-69811-3_44

9. Kim, G., Yi, H., Lee, J., Paek, Y., Yoon, S.: LSTM-based system-call language modeling and robust ensemble method for designing host-based intrusion detection systems. arXiv preprint (2016). https://arxiv.org/abs/1611.01726

10. Kim, J., Kim, J., Thu, H.L.T., Kim, H.: Long short term memory recurrent neural network classifier for intrusion detection. In: International Conference on Platform Technology and Service, pp. 1–5 (2016)

11. Kingma, D.P., Ba, J.: Adam: A method for stochastic optimization. arXiv preprint (2014). http://arxiv.org/abs/1412.6980

12. Li, Y., Xia, J., Zhang, S., Yan, J., Ai, X., Dai, K.: An efficient intrusion detection system based on support vector machines and gradually feature removal method. Expert Syst. Appl. **39**(1), 424–430 (2012)

13. Lin, Z., et al.: A structured self-attentive sentence embedding. arXiv preprint (2017). http://arxiv.org/abs/1703.03130

14. Luong, M.T., Pham, H., Manning, C.D.: Effective approaches to attention-based neural machine translation. In: The 2015 Conference on Empirical Methods in Natural Language Processing, pp. 1412–1421 (2015)

15. Pascanu, R., Mikolov, T., Bengio, Y.: On the difficulty of training recurrent neural networks. In: International Conference on International Conference on Machine Learning, pp. III-1310 (2013)

16. Perdisci, R., Ariu, D., Fogla, P., Giacinto, G., Lee, W.: McPAD: a multiple classifier system for accurate payload-based anomaly detection. Comput. Netw. Int. J. Comput. Telecommun. Networking **53**(6), 864–881 (2009)

17. Sharafaldin, I., Lashkari, A.H., Ghorbani, A.A.: Toward generating a new intrusion detection dataset and intrusion traffic characterization. In: International Conference on Information Systems Security and Privacy, pp. 108–116 (2018)

18. Tan, Z., Jamdagni, A., He, X., Nanda, P.: Network intrusion detection based on LDA for payload feature selection. In: GLOBECOM Workshops, pp. 1545–1549 (2011)

19. Torrano-Gimenez, C., Hai, T.N., Alvarez, G., Franke, K.: Combining expert knowledge with automatic feature extraction for reliable web attack detection. Secur. Commun. Netw. **8**(16), 2750–2767 (2015)

20. Torrano-Giménez, C., Pérez-Villegas, A., Álvarez, G.: HTTP DATASET CSIC 2010 (2010). http://www.isi.csic.es/dataset/

21. Wang, K., Parekh, J.J., Stolfo, S.J.: Anagram: a content anomaly detector resistant to mimicry attack. In: Zamboni, D., Kruegel, C. (eds.) RAID 2006. LNCS, vol. 4219, pp. 226–248. Springer, Heidelberg (2006). https://doi.org/10.1007/11856214_12

22. Wang, K., Stolfo, S.J.: Anomalous payload-based network intrusion detection. Recent Adv. Intrusion Detection **3224**, 203–222 (2004)

A Semantic Parsing Based LSTM Model for Intrusion Detection

Zhipeng Li and Zheng Qin[✉]

School of Software, Tsinghua University, Beijing 100084, China
lizp14@mails.tsinghua.edu.cn,
qingzh@mail.tsinghua.edu.cn

Abstract. Nowadays, with the great success of deep learning technology, using deep learning method to solve information security issues has become a study hot spot. Although some literal works have tried to solve intrusion detection problem via recurrent neural network, these methods do not give a detailed framework and specific data processing progress. We propose a novel semantic parsing based Long Short-Term Memory (LSTM) network framework in this paper. The proposed method uses the semantic representations of network data. The novel conversion process of various forms of network data to semantic description is given in detail. Experiments on NSL_KDD data sets show our proposed model outperforms most of the standard classifier. Results show that the semantic description has reserved information of the data and our semantic parsing based LSTM model provides a novel way to solve anomaly detection.

Keywords: Anomaly detection · Semantic parsing · LSTM
NSL_KDD

1 Introduction

Network Intrusion Detection System (IDS) is a dynamic security mechanism for monitoring, preventing or resisting system intrusion behavior. IDS collects and analysis's data in system or network to detect invading event. IDS responses to users' behavior of overriding and violating the security policy, and the attempt to invade the system by external invaders utilizing the system security vulnerabilities. D.E. Denning first proposed a general intrusion detection theoretical model in 1986 [2]. Stanford Research Institute developed first real time intrusion detection expert system (IDES) based on anomaly detection and detection rules. UCDavis developed first network based IDS NSM in 1990. Open source IDS Snort, Bro was released in 1998. IDS and extensive application adopted combination of detection and prevention in wireless, mobile and cloud computing scenarios in 2000s. IDS is combing with big data and artificial intelligence technology nowadays.

Misuse detection and anomaly detection are two main methods for implementation of intrusion detection. Misuse detection use rules or signatures of

L. Cheng et al. (Eds.): ICONIP 2018, LNCS 11304, pp. 600–609, 2018.
https://doi.org/10.1007/978-3-030-04212-7_53

attack to identify the attack behavior. The rules are generated via manual analysis of various attacks. Misuse detection has the advantages of high accuracy and low false alarm rate. On the opposite side, misuse detection need manual rules which limits the ability to find undefined harmful network behavior. However, Internet have fast amazing growth. There are novel attack methods every day. Misuse detection can not satisfy the needs of future development. Anomaly detection is another main method of intrusion detection. Anomaly detection confirms attacks via artificial intelligence, pattern recognition, machine learning and so on. Statistic laws of activity events are used to evaluate the activity whether normal behavior or attacks. The most advantage of this mode is that it can find the overall characteristics of a class of attacks. Then it could identify a specific novel attack of that class. Meanwhile, the abstract overall statistic easy lead to false alarm. Nowadays, deep learning have achieved unprecedented success in many areas [10]. Deep learning use special deep network structure to obtain features automatically and gain higher precision than traditional machine learning methods.

In this work, we propose a semantic parsing based IDS model using LSTM as classifier. Specifically, we add a novel data semantic conversion preprocessing stage as features preprocessor. NSL_KDD data set is chosen to evaluate the model. Experiments show that our proposal is better than most machine learning classifier. The proposed method achieve a relative good results beyond the baseline methods. The rest of this paper is organized as below: Sect. 2 introduces related literal works via deep learning method. Section 3 gives details about the proposed model. A detailed semantic data conversion procedure and LSTM sequence labeling model is introduced. NSL_KDD dataset is used to evaluate the proposed model in Sect. 4. In the end, Sect. 5 concludes our work and gives prospects of future work.

2   Related Work

Anomaly detection via machine learning is a hot spot in intrusion detection. Many machine learning method and feature selection methods have been attempted. [1,12,18] used classic machine learning methods to distinguish between normal network data and abnormal network data. [6,14] etc. used artificial intelligence algorithm in feature engineering and machine learning methods to improve the efficiency of classification. [8,16] proposed a ensemble model in anomaly detection.

In intrusion detection area, [4,5] used deep neural network structure to extract features automatically. [11,19] used convolutional neural network to classify the image conversion of data. [13] used RNN networks and [9,15] used LSTM unit of RNN. Literal works show that deep learning method can be used in intrusion detection successfully. The literature works concentrate on using deep structure to extract feature representation automatically and some works also attempted to use CNN or RNN model. However, works which using RNN are lacks of specific information about implementation of data process.

3 Proposed Model

There are many applications with various LSTM architectures. In our intrusion detection task, we used classification model or so-called sentiment classification model as the classifier. The overview of the proposed IDS architectures is shown in Fig. 1.

In the bottom of the system architecture, a raw packet capturer get the source data from the network. There are many methods can achieve raw data capture. E.g. a mirror data copy from a switch or a network data capture library. Then, we pass the data through a KDD99' data format extractor. The extractor got the 41 abstract features according the KDD99' dataset format. There are also some open source lib can do this process. We do some transforms of the KDD99' features in the preprocess module. We use different methods to convert the 41 features into semantic words and we will describe the process in detail in subsection. After the preprocess we get all the features consisting of symbolic words. We use an embedding layer transform the words into word vector. In the last word vectors go into the LSTM classification model and output the classification results.

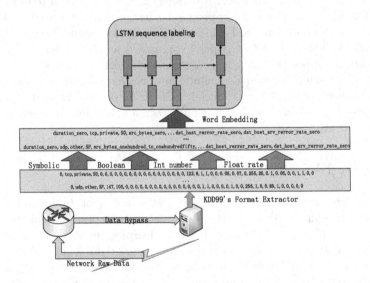

Fig. 1. Illustration of the proposed model

3.1 Data Preprogress

As show in Fig. 1 network data is obtained fist.Then system builds data with KDD99' format by a feature extractor. KDD99' dataset is an evaluation dataset most commonly used in IDS evaluation. The data format is an abstract data set from DARPA1998 data set. The data set have 41 features in four groups. There are 9 basic connection features shown in Table 1, 13 content features based on domain knowledge shown in Table 2, 9 flow content related features based on a 2 seconds window shown in Table 3, and 10 flow features based on a 100 connections window shown in Table 4 [3].

Table 2. Domain knowledge features of KDD99

No.	Attribute name	Data type	Sample
10	hot	numeric	0
11	num_failed_logins	numeric	0
12	logged_in	bool	0
13	num_compromised	numeric	0
14	root_shell	bool	0
15	su_attempted	bool	0
16	num_root	numeric	0
17	num_file_creations	numeric	0
18	num_shells	numeric	0
19	num_accesses_files	numeric	0
20	num_outbound_cmds	numeric	0
21	is_hot_login	bool	0
22	is_guest_login	bool	0

Table 1. Basic features of KDD99

No.	Attribute name	Data type	Sample
1	duration	numeric	0
2	protocol_type	enum	tcp
3	service	enum	private
4	flag	enum	S0
5	src_bytes	numeric	0
6	dst_bytes	numeric	0
7	land	bool	0
8	wrong_fragment	numeric	0
9	urgent	numeric	0

Semantic Words Conversion. In the preprocess procedure, we remain the nominal features data type, e.g 'protocol_type' (protocol used in the connection) has 3 different values 'tcp', 'udp' and 'icmp'. We convert boolean features into 'featurenames_ture' or 'featurenames_false'. E.g 'root_shell' (1 if root shell is obtained; 0 otherwise) is a boolean feature of content related features, we use 'root_shell_false' and 'root_shell_true' to replace '0' and '1'. There are two types of numeric features, one is integer number value, the other is float rate value. We use specified intervals to divide the integer numeric feature into several intervals. Then we convert the interval number to semantic describe. This method convert integer type features into well-defined 'clusters' and is more effective than clustering algorithm. E.g 'srv_count' (the number of same service connections (port number) in the past 2 s time window) has integer value from 0 to 550, we use every 50 numbers as interval divide the whole range into 12 intervals. We think 0 stands for a special state of feature, so we give 0 value a separate discrete point. Then, we use semantic words describe the specific interval, such as: 'srv_count_zero','srv_count_zero_to_fifty',...'srv_count_fivehundred_to_fivehundredfifty'. For float rate features, as the integer features we also use intervals to divide the 0 − 1 interval. E.g 'dst_host_srv_diff_host_rate' (the percentage of connections of different services, denominator is the connections counted in

Table 3. 2s-window features of KDD99

No.	Attribute name	Data type	Sample
23	count	numeric	123
24	srv_count	numeric	6
25	serror_rate	numeric	1
26	srv_serror_rate	numeric	1
27	rerror_rate	numeric	0
28	srv_rerror_rate	numeric	0
29	same_srv_rate	numeric	0.05
30	diff_srv_rate	numeric	0.07
31	srv_diff_host_rate	numeric	0

Table 4. 100-connections features of KDD99

No.	Attribute name	Data type	Sample
32	dst_host_count	numeric	255
33	dst_host_srv_count	numeric	255
34	dst_host_same_srv_rate	numeric	0.1
35	dst_host_diff_srv_rate	numeric	0.05
36	dst_host_same_src_port_rate	numeric	0
37	dst_host_srv_diff_host_rate	numeric	0
38	dst_host_serror_rate	numeric	1
39	dst_host_srv_serror_rate	numeric	1
40	dst_host_rerror_rate	numeric	0
41	dst_host_srv_rerror_rate	numeric	0

dst_host_count) turns into 12 intervals, 0.00 and 1.00 are considered as special feature state. Again, we use semantic word describe the specific interval, such as: 'dst_host_srv_diff_host_rate_zero','dst_host_srv_diff_host_rate_zero_to_zeropointone', ..., 'dst_host_srv_diff_host_rate_srv_rate_one'. And we find the feature 'num_out bound_cmdss' only has 1 value '0' in all NSL_KDD train samples. We drop this 1 value feature. The semantic conversion process is shown in Fig. 2.

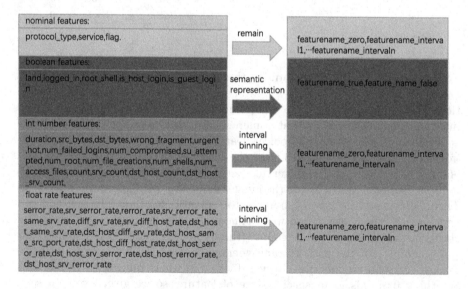

Fig. 2. Data semantic conversion

Word Embedding. One-hot encoder is a basic encoding method in machine learning. One-hot encoder get a vector which has the length of enumeration number of feature content. One feature has n categorical variables then one-hot encoder vector has n dimensions. Like one-hot encoder, word embedding can also be thought as encoder method. The word embedding method maps the feature space into specific vector space. By compressing into a certain vector range, the coding method works effectively. It is a common data process in natural language processing (NLP). There have been different models turn words to vector and some pre-trained dictionary library. In this work, we do not think the corpus is truly a natural language. We use a neural network lib to project the inputs into the specific vector size we want.

3.2 LSTM Sequence Labeling Model for Intrusion Detection

Long short-term memory (LSTM) was proposed to resolve sequence recognition problem. LSTM is a solution to recurrent neural network's (RNN) gradient vanishing or gradient exploding problem also called error back-flow problems. To improve gradient flow, LSTM use four interior gates to 'remember' or 'forget' the state of the previous state, selectively update cell state values and output certain parts of the cell state. The forget gate is used to forget irrelevant parts of the previous state. The input gate selectively update cell state values and the output gate output certain parts of the cell state as shown in Fig. 3(a). A typical unit structure is show in Fig. 3(b). i stands for input gate function, f stands for forget gate function and o stands for output gate function. The gate function contains a σ function and a dot product function. Figure 3(b) can be presented simply as:

$$\begin{pmatrix} i \\ f \\ o \\ g \end{pmatrix} = \begin{pmatrix} \sigma \\ \sigma \\ \sigma \\ tanh \end{pmatrix} W\,(h_{t-1}). \tag{1}$$

c_t stands for the cell state in time t and h_t stands for the output. The cell state c_t and output h_t can be written as :

$$c_t = f \odot c_{t-1} + i \odot g, \tag{2}$$

$$h_t = o \odot tanh(c_t). \tag{3}$$

There are also many other variants based on LSTM [7]. In practical applications, people follow input and output data structure to form a specific model by stacking RNN unit. In our work, we use sequence labeling model which often be used in sentiment classification or so called many to one model as our classification model. The sequence model is shown in Fig. 1, the top classification part is sequence labeling model. Back-propagation through time (BPTT) algorithm trains the LSTM part to converge.

4 Experiment and Results

We use a popular NSL_KDD dataset as evaluation dataset. NSL_KDD is an sub-sequent compact version of KDD99' dataset. NSL_KDD dataset inherits the data format of the KDD99' dataset and eliminates some redundant data. We experiment the performance test on a solo workstation (DELL 7910). A TITAN X Pascal GPU is used to accelerate computing. Pytorch was adopt as a deeplearning software framework. We used the 40 features ('num_outbound_cmdss' was dropped) as input. Embedding layer got a 409 dimensions of semantic words description input and output a 100 dimensions vector.

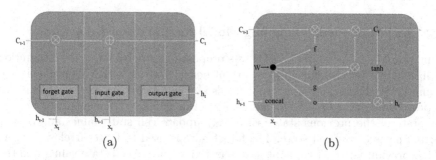

(a)   (b)

Fig. 3. Long short-term memory

4.1 Performance Metric

We evaluate the accuracy of our model as an overall performance evaluation. We also test the precision, recall and F1 score of the classification. Precision is an indicator that show the sensitivity of the classifier. Recall is measured to reflect the classification's coverage capacity. We also measured the F1 score to show the comprehensive performance of recall and precision. The raw positive and negative experiment results are shown in confusion matrix.

4.2 Comparative Analysis

NSL_KDD contains one train dataset and two test dataset of different difficulty level. Test dataset contain new attacks which train dataset does not contain. The novel type of data in test dataset make the task more realistic and more challenging. Test$^+$ is formed by randomly selected from the KDD99 test data. In addition, there is test data set called Test^{-21} which proving more challenging test. The Test^{-21} contains a more difficult data than Test$^+$. The data which misclassified by the 21 pretrained models (7 classifier individually trained 3 times) were collected.

Table 5. Results of binary classification test

	Accuracy	Precision	Recall	F1 score
Semantic parsing based LSTM on NSL_KDD Test$^+$	82.21%	72.07%	95.83%	82.27%
Semantic parsing based LSTM on NSL_KDD Test^{-21}	66.10%	31.29%	72.49%	43.71%

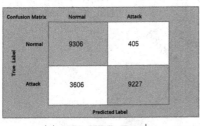

(a) NSL-KDD Test$^+$

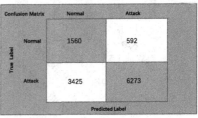

(b) NSL-KDD Test^{-21}

Fig. 4. Confusion matrix of binary test

We do binary classification test on the two test data set. Table 5 gives the performance metric results of the experiments. The confusion matrices contains test statics are shown in Fig. 4. A comparison to other method is given in Table 6. The illustration of the comparison is shown in Fig. 5. The comparison shows that the semantic parsing based LSTM method gets relatively high accuracy. The other classifier's performance is measured by Tavallaee *et al.* [17]. The experiments show that our semantic parsing based LSTM model do well in Test$^+$ validation test. When validation set contains more challenging data, the performance decreases. We think the great difficulty of Test^{-21} and the unbalanced distribution may cause the model a relatively low performance. However, our proposed model achieves better results than baseline machine learning method. The results show that our LSTM sequence labeling model with semantic conversion of data can be used in intrusion detection application.

Table 6. Comparison to other machine learning models

Model	Acc (Test$^+$)	Acc (Test^{-21})
J48	81.05%	63.97%
Naive bayes	76.56%	55.77%
NB tree	82.02%	66.16%
Random forest	80.67%	62.26%
Random tree	81.59%	58.51%
Multi-layer perceptron	77.41%	57.34%
SVM	69.52%	42.29%
Semantic LSTM	82.21%	66.10%

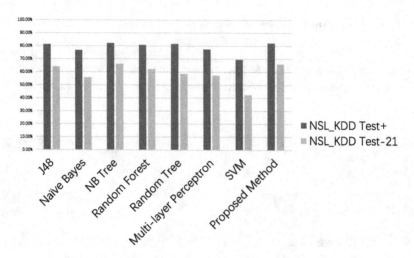

Fig. 5. Comparison of different method

5   Conclusion

We have proposed a novel KDD99' data format conversion to semantic description method and we have used a LSTM sequence labeling model as anomaly detection model. The proposed method is measured by a popular NSL_KDD data set. By comparing with standard machine learning baseline, our method gets a better accuracy score. Results demonstrate that our semantic parsing based LSTM model can be used as anomaly detection and the semantic description has reserved information of the data. We will explore a more precise conversion to semantic description such as using hierarchical clustering in numeric features. In addition, more RNN variant architectures will be attempted.

References

1. Canbay, Y., Sagiroglu, S.: A hybrid method for intrusion detection. In: 2015 IEEE 14th International Conference on Machine Learning and Applications (ICMLA), pp. 156–161. IEEE (2015)
2. Denning, D.E.: An intrusion-detection model. IEEE Trans. Softw. Eng. **SE–13**(2), 222–232 (1987)
3. Dhanabal, L., Shantharajah, S.: A study on NSL-KDD dataset for intrusion detection system based on classification algorithms. Int. J. Adv. Res. Comput. Commun. Eng. **4**(6), 446–452 (2015)
4. Erfani, S.M., Rajasegarar, S., Karunasekera, S., Leckie, C.: High-dimensional and large-scale anomaly detection using a linear one-class SVM with deep learning. Pattern Recogn. **58**, 121–134 (2016)
5. Fiore, U., Palmieri, F., Castiglione, A., De Santis, A.: Network anomaly detection with the restricted Boltzmann machine. Neurocomputing **122**, 13–23 (2013)

6. Gao, H.H., Yang, H.H., Wang, X.Y.: Ant colony optimization based network intrusion feature selection and detection. In: Proceedings of 2005 International Conference on Machine Learning and Cybernetics, vol. 6, pp. 3871–3875. IEEE (2005)
7. Greff, K., Srivastava, R.K., Koutník, J., Steunebrink, B.R., Schmidhuber, J.: LSTM: a search space odyssey. IEEE Trans. Neural Netw. Learn. Syst. **28**(10), 2222–2232 (2017)
8. Haq, N.F., Onik, A.R., Shah, F.M.: An ensemble framework of anomaly detection using hybridized feature selection approach (HFSA). In: 2015 SAI Intelligent Systems Conference (IntelliSys), pp. 989–995. IEEE (2015)
9. Kim, J., Kim, J., Thu, H.L.T., Kim, H.: Long short term memory recurrent neural network classifier for intrusion detection. In: 2016 International Conference on Platform Technology and Service (PlatCon), pp. 1–5. IEEE (2016)
10. LeCun, Y., Bengio, Y., Hinton, G.: Deep learning. Nature **521**(7553), 436–444 (2015)
11. Li, Z., Qin, Z., Huang, K., Yang, X., Ye, S.: Intrusion detection using convolutional neural networks for representation learning. In: Liu, D., Xie, S., Li, Y., Zhao, D., El-Alfy, E.-S.M. (eds.) ICONIP 2017. LNCS, vol. 10638, pp. 858–866. Springer, Cham (2017). https://doi.org/10.1007/978-3-319-70139-4_87
12. Sahu, S., Mehtre, B.M.: Network intrusion detection system using J48 decision tree. In: 2015 International Conference on Advances in Computing, Communications and Informatics (ICACCI), pp. 2023–2026. IEEE (2015)
13. Sheikhan, M., Jadidi, Z., Farrokhi, A.: Intrusion detection using reduced-size RNN based on feature grouping. Neural Comput. Appl. **21**(6), 1185–1190 (2012)
14. Srinoy, S.: Intrusion detection model based on particle swarm optimization and support vector machine. In: IEEE Symposium on Computational Intelligence in Security and Defense Applications, CISDA 2007, pp. 186–192. IEEE (2007)
15. Staudemeyer, R.C., Omlin, C.W.: Evaluating performance of long short-term memory recurrent neural networks on intrusion detection data. In: Proceedings of the South African Institute for Computer Scientists and Information Technologists Conference, pp. 218–224. ACM (2013)
16. Syarif, I., Zaluska, E., Prugel-Bennett, A., Wills, G.: Application of bagging, boosting and stacking to intrusion detection. In: Perner, P. (ed.) MLDM 2012. LNCS (LNAI), vol. 7376, pp. 593–602. Springer, Heidelberg (2012). https://doi.org/10.1007/978-3-642-31537-4_46
17. Tavallaee, M., Bagheri, E., Lu, W., Ghorbani, A.A.: A detailed analysis of the KDD cup 99 data set. In: IEEE Symposium on Computational Intelligence for Security and Defense Applications, CISDA 2009, pp. 1–6. IEEE (2009)
18. Teng, L., et al.: A collaborative and adaptive intrusion detection based on SVMs and decision trees. In: 2014 IEEE International Conference on Data Mining Workshop (ICDMW), pp. 898–905. IEEE (2014)
19. Wang, W., Zhu, M., Zeng, X., Ye, X., Sheng, Y.: Malware traffic classification using convolutional neural network for representation learning. In: 2017 International Conference on Information Networking (ICOIN), pp. 712–717. IEEE (2017)

Detecting the *Doubt Effect* and *Subjective Beliefs* Using Neural Networks and Observers' Pupillary Responses

Xuanying Zhu, Zhenyue Qin, Tom Gedeon[✉], Richard Jones,
Md Zakir Hossain, and Sabrina Caldwell

The Australian National University, Canberra, Australia
xuanying.zhu@anu.edu.au, tom@cs.anu.edu.au

Abstract. We investigated the physiological underpinnings to detect the 'doubt effect' – where a presenter's subjective belief in some information has been manipulated. We constructed stimulus videos in which presenters delivered information that in some cases they were led to doubt, but asked to "present anyway". We then showed these stimuli to observers and measured their physiological signals (pupillary responses). Neural networks trained with two statistical features reached a higher accuracy in differentiating the doubt/manipulated-belief compared to the observers' own veracity judgments, which is overall at chance level. We also trained confirmatory neural networks for the predictability of specific stimuli and extracted significant information on those stimulus presenters. We further showed that a semi-unsupervised training regime can use subjective class labels to achieve similar results to using the ground truth labels, opening the door to much wider applicability of these techniques as expensive ground truth labels (provenance) of stimuli data can be replaced by crowd source evaluations (subjective labels). Overall, we showed that neural networks can be used on subjective data, which includes observer perceptions of the doubt felt by the presenters of information. Our ability to detect this *doubt effect* is due to our observers' underlying emotional reactions to what they see, reflected in their physiological signals, and learnt by our neural networks. This kind of technology using physiological signals collected in real time from observers could be used to reflect audience distrust, and perhaps could lead to increased truthfulness in statements presented via the Media.

Keywords: Neural networks · Pupillary responses · Information veracity
Doubt · Trust · Subjective belief · Semi-unsupervised training

1 Introduction

People cooperate, solve problems and enhance social bonds by exchanging information and knowledge. To ensure enduring bonds and to achieve collaborative goals, these communications should be honest and faithful [1] so that people can navigate the information and knowledge with confidence and trust. However, with the proliferation of technologies, information can be easily generated and manipulated. Credibility becomes problematic under the weight of manipulated information such as fake news

© Springer Nature Switzerland AG 2018
L. Cheng et al. (Eds.): ICONIP 2018, LNCS 11304, pp. 610–621, 2018.
https://doi.org/10.1007/978-3-030-04212-7_54

and exaggerated or incorrect advertising, which could potentially cause people being deceived to suffer from grave consequences to their personal lives or administrative decisions. Therefore, from an evolutionary perspective, skills of knowing whom and what to trust is indispensable [2].

Nonetheless, according to Bond and Depaulo [3, 4] when being asked to provide direct veracity judgments of the question such as 'Is that person lying or telling the truth?', people are barely better than chance at consciously recognising a dishonesty, with an average accuracy of 54%. This accuracy improves slightly when people's judgments are assessed indirectly even though they may not be aware that they are being lied to [5]. Additionally, DePaulo et al. [3] and Albrechtsen et al. [6] found that people's quick and intuitive judgments of dishonesty are more accurate than their slow and deliberative judgments made after conscious reasoning. This is possibly because the unconscious dishonesty detection which results in intuitive judgments may happen at such an early stage that it has not reached consciousness [5]. These findings suggest that people may be able to unconsciously pick up subtle cues from a deception event, advocating the merit of using physiological signals as an unconscious indicator or indirect veracity detector.

Physiological responses maintained by the autonomous nervous system indicate mental state changes such as cognitive load [7] and emotions [8] without conscious awareness [9]. These responses also change with the detection of abnormalities, since unpleasant stimuli induce people's defensive reactions, marked by anxiety and avoidance which can be assessed with physiological signals [10]. Thus, it seems highly plausible to search for valid physiological indicators of observed dishonesty, which can be considered as an abnormality in observed behaviour as compared to honesty.

Van't Veer [5] attempted to find a physiological marker of unconscious veracity detection by examining observers' finger skin temperature, and found that when participants were observing a liar, their finger skin temperature decreased over time. When participants were informed in advance that their goal was to differentiate between liars and truth-tellers, their finger skin temperature declined more when they were watching liars contrasted with watching truth-tellers. Inspired by these findings, in a later study, Van't Veer [5] explored the use of observers' pupillary responses in discerning dishonest answers. She demonstrated that dishonesty evoked greater pupillary responses, with pupil size first increasing more and later decreasing more. This result is consistent with [11] in which people were found to have greater eye responses, namely increased fixations and durations of eye gaze, to manipulated areas of images. Another paper has shown that resting heart rate can be correlated with dishonesty detection [12].

Despite physiological correlates of dishonest information being examined previously, we could find no work in which physiological signals from observers were used to identify manipulated subjective belief, in which the presenter is not explicitly intending to deceive. Such *manipulated subjective belief* or *doubt* in information can be considered as a subtle form of deception as the presenters did not mention their doubt, but this is at most a sin of omission not commission.

This paper is organised as follows. Section 2 lists the hypotheses we examine, followed by Sect. 3, which details the experimental design and data collection process. Section 4 describes the feature extraction process and classification models. Results are presented in Sect. 5 and discussed in Sect. 6. Finally, limitations and future work are provided in Sect. 7 before conclusions are drawn in Sect. 8.

2 Hypotheses

Human consciousness is predisposed to trusting the veracity of information, as reported in the literature (see Sect. 1). Thus, when asked to evaluate the veracity of the stimulus content, we expect that participants' judgment will not differentiate between doubting and trusting presentations better than chance. We expect pupillary changes to be significantly different when viewing doubting and trusting conditions, where the reactions to doubting (manipulated subjective belief) are stronger. Thus, a Neural Network classifier trained with pupillary responses should provide a better estimation than human conscious judgments on predicting stimulus veracity.

3 Experimental Design

3.1 Stimuli

Two extracts for popular science books were constructed, phrased and formatted like book publisher advertising materials as Fig. 1 shows. The first was modified from the publisher description of a book "The Salt Fix" written by James DiNicolantonio [13] which describes the benefits of salt in chronic cardiovascular disease. The surname of the author was shortened to Nicol to reduce ethnicity effects. The second extract was summarized and extensively modified from a research paper written by Steven Stanford [14] which presents the feasibility of enzymes in curing diabetes. The author's surname was modified from Stanford to Stafford to avoid the potential celebrity effect of having the same name as a famous university, and a book cover was also constructed. These two materials were chosen because both contents contradict general beliefs that salt is a cause of cardiovascular disease, and that diabetes is not curable. Note that our work is neither a criticism nor an endorsement of their publications.

Four videos were subsequently recorded, each of which consists of an individual presenting one of the above-mentioned book extracts. To construct these videos, one female and one male volunteer from our University were recruited as actors, with Ethics Approval obtained from the Australian National University Human Research Ethics Committee. Neither was a professional actor. After giving their informed consent, they first presented one book extract which they could presume to be true (original subject belief condition). In the brief period before presenting the second extract they were told "Sorry the next one is a bit bogus, please present it anyway". Then, they presented the other extract (manipulated subject belief condition). As we constructed both extracts, this does not reflect our views on the source publications.

Each extract was read as a belief condition first and a doubt condition second. A camera was placed at 1 m from the actors and filmed them from the chest up. In this way, four videos were recorded, ranging from 27 s to 39 s in length. A laptop was subsequently used to display the videos to the participants during the veracity judgment session.

SALT

For many years we've been told
that salt is just pure bad for you.

Recent research
by leading cardiovascular researcher
Dr James Nicol
upends the low-salt myth,
proving that salt may be a cure
—rather than a cause of,
—our country's chronic disease crises.

Dr. Nicol has shown
how eating the right amount
  of this essential mineral
will help you beat sugar cravings,
  achieve weight loss,
improve athletic performance,
  increase fertility,
and reduce migranes.

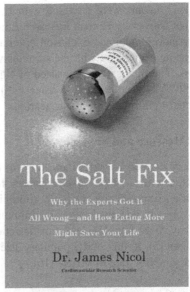

DIABETES

For many years we've been told
That diabetes drugs can not cure diabetes.

Recent research
By leading immunology researcher
Dr Steven Stafford
upends the no-cure myth,
proving a cure for diabetes is
—possible, and that diabetes
—is no longer a critical chronic disease.

Dr Stafford has shown
how taking the right enzyme blocker
  for an essential amino acid,
will help you control diabetes,
  achieve weight loss,
improve athletic performance,
  increase fertility,
and reduce eczema.

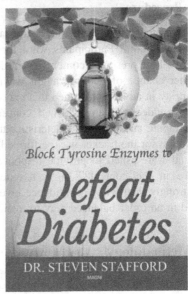

Fig. 1. Two book extracts about salt (above) and diabetes (below)

3.2 Participants

Thirteen students were recruited with Ethics Approval obtained from the Australian National University Human Research Ethics Committee. Five were excluded based on one or more predefined exclusion criteria: for having a history of cardiovascular disease, being acquainted with the stimulus video presenters, and technical failures of the sensors. The final sample consisted of eight participants, four males and four females, from 18 to 24 years in age (Average = 20.6, Standard Deviation = 2.3) with normal or corrected-to-normal vision and hearing.

3.3 Measure and Sensor

Pupillary dilation provides indications of changes in mental states and the strengths of mental activities [15]. Several papers have reported that pupil size constitutes a response to emotionally engaging stimuli in which the pupil is significantly bigger after positively and negatively arousing stimuli than after neutral stimuli [15]. Similar patterns have been found in deception detection where dishonesty evoked greater pupillary responses [5]. In this study, pupillary size was captured using an EyeTribe eye tracker with a sampling rate of 60 Hz [16].

3.4 Procedure

The experiment was conducted with each individual participant in the same quiet experiment room. Participants were forewarned that their goal was to identify the veracity of the presented content. The rationale was: because dishonest information has a higher chance of getting caught when observers are alerted to the possibility of dishonesty in advance [5], then this may also apply to manipulated (doubting) beliefs. They provided written informed consent, and filled in a questionnaire to collect demographic and health characteristics that may affect pupillary responses. Eye gaze calibration for the tracker was performed. Participants watched the two videos one by one and were asked to provide responses to a question of "To what extent do you believe in the content?" on a five-item Likert scale (from 1 = strongly no, to 5 = strongly yes) with a confidence level (from 0% to 100%). The videos were presented in an order balanced way to avoid effects of presentation ordering.

4 Methodology

After data collection, eight manipulated and eight unmanipulated subjective belief observations were obtained. In the rest of this paper we will mostly use *doubting* and *trusting* to succinctly express "manipulated subjective belief" and "unmanipulated subjective belief", respectively. Two matching observations were excluded from further analysis due to intermittent sensor data loss. This results in a total of twelve complete responses, including their conscious veracity judgments and physiological sensors recordings.

4.1   Feature Extraction

The raw pupillary data was processed to obtain features. The pupillary responses were calculated from each observer's pupil size acquired in viewing each video. For each second, the EyeTribe returned 60 frames of raw pupillary data and therefore, for example, a 27-second video would result in 1,620 frames of data recorded. The pupillary data were extracted from the beginning of the video playing to the end of the video. The extracted pupil size data were subsequently normalised across all videos viewed by each observer to reduce the effect of individual bias due to the naturally varying pupil sizes among participants, since the significant signal is the variation in the pupil size for an individual, not the magnitude of the pupil size.

Linear interpolation was applied to missing pupil size data caused by occasional eye blinks. This procedure was employed on the pupil data of left and right eyes separately. The interpolated pupil size values were averaged to obtain a single pupillary response signal for each participant, and their minimum and mean of processed pupillary data during each video watching session were extracted as features.

4.2   Model Description

An ensemble of five artificial neural networks were trained on observers' pupillary responses to predict presenters' subjective beliefs individually and jointly as well as observers' conscious veracity judgments. All neural networks had a sigmoid hidden layer of size 100 and an output layer with two output neurons. They were trained with the Adam optimizer [17] without any adaption using backpropagation with the Cross-Entropy loss function.

The most common method for cross-validation is k-fold which randomly partitions data into k equal sized subsamples and uses one subsample as validation data, and the remaining k-1 as training data. However, for human data, a continuous segment of physiological data with more than one data point can reflect a human's responses to a stimulus. Training a classification model on random splits of data is not adequate, unless all data from one human is guaranteed to be within either the training set or the testing set for each run. This method of leaving all data for one human out is called leave-one-participant-out, which was used in this study. Pupillary data from one observer was used as the testing set, and those from the remaining participants formed the training set, and repeated for all, averaging to calculate the final result reported.

5   Results

5.1   Veracity Judgments

The average accuracy of observers' conscious veracity judgments (see Table 1) was 50% with standard deviation of 0.43, which is at the prima-facie chance level of 50%, because there were two options for the observers over balanced numbers of video stimuli. The average accuracy of consciously identifying doubting presentations was 40%, lower than that of identifying trusting presentations which was at 60%. This could imply that people are better at identifying trusted information compared with doubted content. The accuracies of predicting veracity of subjective beliefs vary greatly between presenters, at the rate of 80% and 20% respectively, which may indicate that the veracity of behaviours from one presenter was easier to identify compared with those from the other. The 80% represents well over chance recognition of doubting for one presenter, while the 20% represents well *below* chance for the other, indicating our observers were consistently wrong on that presenter.

Table 1. Accuracies of doubting and trusting prediction from observers' verbal responses

	By manipulation			By presenters	
	Doubt condition	Trust condition	Total	Presenter 1	Presenter 2
Verbal responses	40%	60%	50%	80%	20%

5.2 Pupillary Responses

Figure 2 displays observers' pupil diameter when they were watching doubting and trusting presentations. Visually, regardless of the veracity of the content, there is an overall wave-like changing pattern that may reflect pupillary responses for a common process while viewing a video stimulus.

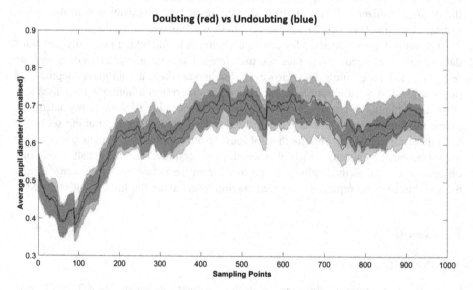

Fig. 2. Average pupillary response to doubting and trusting presentations over the time

The responses to doubting versus trusting presentations started with similar patterns and then diverged at around frame 90. Such divergence continued until at around frame 190 when the pupil diameter changed to opposite directions for doubting versus trusting presentations. A third divergence happened at around frame 560 but the directions for doubting versus trusting stimuli viewing remained the same. The first two diverging frames may indicate the beginning of human unconscious discrimination of stimuli content.

By observation, over time, when viewing a stimulus with doubted beliefs, observers' pupil diameters had greater changes, first decreasing more and then increasing more. To test whether the pattern of pupillary diameters was statistically different when observing a doubted belief compared to a trusting belief, a two-sample Kolmogorov-Smirnov test was performed. The result showed that the pupil dilated differently with a statistical significance ($p < 0.001$) for doubting stimuli observations compared with trusting stimuli observations. This provides a clear indication that the pupillary response can be of use for computational classification.

5.3 Relationship Between Pupillary Responses and Veracity Judgments

To assess whether observers' pupillary responses are predictive of their veracity judgments, the first neural network in the ensemble was trained with observers' pupillary features to predict their veracity judgments for all observers across all videos. The average accuracy was 75.8%, indicating that people's pupillary responses are positively correlated with their conscious veracity judgments.

5.4 Relationship Between Pupillary Size, Doubt Effect and Subjective Beliefs

As described above, doubting presentations resulted in greater observers' pupil arousal responses with a high statistical significance. To explore the relationship between observers' pupillary responses and presenters' doubt effects, the second neural network was trained with features extracted from observers' pupillary size to predict whether presenters doubted the presented content. From Table 2, the final accuracy for all observers across all videos was 58.3%, which is higher than the chance level of 50%. This accuracy is also higher than that of observers' conscious veracity judgments, implying that in general, neural network classifiers trained using pupillary responses can better identify doubting presentations than human conscious judgments.

Table 2. Accuracies of doubting and trusting prediction from observers' pupillary responses

	By manipulation			By presenters	
	Doubt condition (Specificity)	Trust condition (sensitivity)	Total (accuracy)	Presenter 1	Presenter 2
Pupillary responses	50%	66.7%	58.3%	100%	33%

The average accuracy of identifying doubting presentations with pupillary responses was 50%, lower than that of identifying trusting presentations which was 66.7% (see Table 2). This shows neural networks trained on human pupillary responses are also better at identifying trusting than doubting content. Also, these classifiers are more accurate, as their accuracies are *both* higher than those of observer's conscious veracity judgments. Besides, despite the improvement of accuracies, the prediction of pupillary responses is still not completely accurate. While this could be due to the limited features extracted from pupillary responses, this could also imply that at the physiological level, observers could still be fooled by presenters' behaviours.

To further explore whether different presenters' behaviours can be discerned from observers' physiological data, a neural network was trained for *each* presenter with pupillary responses from observers when they viewed that presenter. As shown in Table 2, the accuracies of predicting veracity of subjective beliefs with pupillary responses vary greatly between presenters, from 100% to 33%. These accuracies are both higher than those from observers' verbal responses which were 80% to 20%. While this shows the superior prediction of pupillary responses over verbal responses, it also reveals the possible stimuli bias introduced by different presenters. The classifier for Presenter 1 improves the prediction to perfect, however, while the accuracy of that for Presenter 2 is improved, but is still below chance level. This could be owing to the different behaviours by the two presenters, with one being an obvious disbeliever and another being either a good dissembler, or who was better at following the instructions literally, and tried sufficiently hard to "present it anyway" to be more plausible on the doubting condition than on the trusting condition.

5.5   Relationship Between Pupillary Size and Veracity by Voting

The last neural network was trained on observers' pupillary responses with labels derived from their verbal responses, with voting between subjects. That is, the majority classification is used. Leave-one-stimulus-out cross validation was used, meaning that responses to all but one videos were used as training data, and those to one video used as testing data. The accuracy was 75% on average. This is a "semi-unsupervised" learning as the real labels have not been used. Instead, the labels are "crowdsourced" from the experiment participants' verbal responses.

6   Discussion

With this preliminary study, we explored people's automatic and non-conscious pupillary responses to observing a doubting presentation, a subtle form of dishonest information, as well as their more conscious verbal judgments of the veracity in a presenter's video. For people's ability to accurately indicate whether the provided information is valid, it was predicted that conscious veracity judgment would not detect doubting much better than chance. Consistent with (but slightly worse than) earlier findings about the accuracy of people's conscious judgments on the veracity of smiles [18] and anger [19], observers in this study were found to be accurate about 50% of the time, a performance that was not different from chance. Our analysis showed that the two presenters in our study consist of an obvious doubter and a good dissembler; this could still imply that human conscious judgments on veracity may not be better than chance in general in evaluating direct deception and doubting. Future research could explore conscious veracity judgments on more complex emotions or even more subtle deception activities.

Observers' pupillary size was measured during the viewing of doubting stimuli. Different from the previous findings where pupillary size was discovered to first increase and then decrease while watching disbelief [5], the pupillary changes in this study started with an initial short decrease followed by a longer increase and ended with a minor decrease. Consistent with previous findings [20, 21], in this study the initial decrease in pupil dilation lasted slightly longer than 1 s, possibly due to the light reflex related to the changes in environmental illumination caused by the playing of stimulus. The subsequent increase may reflect the preparation for the human body to react appropriately to external stimuli. Since pupil dilation occurs with temporal changes of visual attention and is found to reflect personal preferences [22, 23], this reaction thus addresses the importance and preference for what is being observed, and therefore may assist in estimating the trustworthiness of others.

As for the pupillary responses to doubting and trusting content, it was found that observers' pupil diameters had greater changes in the doubting condition, consistent with earlier findings in the literature indicating that dishonesty evokes greater pupillary responses (on more obvious deception behaviours) [5]. Our result further suggests that dishonesty may be unconsciously perceived, upholding the evolutionary view that accurate detection of obvious deception and subtle doubt is adaptive and should be advantaged by natural selection [1].

With regard to the comparison between conscious judgments and a computational classifier trained on unconscious pupillary responses for identifying the veracity of presented information, better estimation was achieved by the classifier. We also found that people's pupillary size was positively correlated with their conscious judgments. Taken together, the results provide evidence that although humans cannot consciously discriminate doubted information from trusted information correctly, better ability of detecting doubted information may be present at an unconscious level and this ability can be accessed by computational classifiers. The superior veracity detecting ability of human unconscious pupillary responses over conscious judgments also occur in other areas such as estimating realness of two basic emotions [18, 19]. This suggests that while unconscious responses from human instinctive ability, which has been adaptively evolved by natural selection, can make efficient and effective use of deception cues, the resulting conscious judgments are made inaccurate under the influence of conscious biases and social rules [1]. Further exploration can be performed to examine the applicability of human unconscious responses to the veracity of information or emotions with subtler but perhaps socially important differences.

7   Limitations and Future Work

The stimuli videos used in this study could be increased in number, though the results were highly statistically significant, at the $p < 0.001$ level. Also, we assumed that presenters made the same level of effort in hiding their subjective beliefs, and that the beliefs were synchronised with the experimenter instructions. That is, when they were presented the content of a book extract, they would believe in it unless doubt was suggested. In a future study, more actors should be recruited, and their acting effort as well as their own beliefs should be collected by a structured interview after their video performance is recorded. Stronger conclusions may also be able to be drawn in subsequent studies with more observers. More statistical features from pupil responses could be investigated to see if this leads to more accurate computational classifiers. Finally, observers' other physiological signals such as galvanic skin response, skin temperature and blood volume pulse could also be investigated. With increasing amounts of data collected from wider groups of participants, recent deep learning models, such as Recurrent Neural Networks or Long Short-Term Memory models, can potentially be trained to achieve more accurate results.

8   Conclusion

Our work explored physiological signals to detect the 'doubt effect' where a presenter's subjective belief in some information was manipulated. When doubted information was observed, discernible physiological indicators are present: following an initial light reflex, pupillary changes fluctuated more with an increase and then a decrease. Neural network ensembles trained with 2 pupillary features can reach a higher accuracy in differentiating doubting and trusting information compared with the same observers' conscious veracity judgments. We demonstrated that neural networks trained within-stimulus can provide insights into the dataset. Human data is complex, and this approach may be useful more generally. We also introduced a form of semi-unsupervised training, using crowd-sourced labels rather than the ground-truth.

References

1. ten Brinke, L., Vohs, K.D., Carney, D.R.: Can ordinary people detect deception after all? Trends Cogn. Sci. **20**, 579–588 (2016)
2. Von Hippel, W., Trivers, R.: The evolution and psychology of self-deception. Behav. Brain Sci. **34**, 1–16 (2011)
3. Bond Jr., C.F., DePaulo, B.M.: Accuracy of deception judgments. Pers. Soc. Psychol. Rev. **10**, 214–234 (2006)
4. DePaulo, B.M., Bond Jr., C.F.: Beyond accuracy: bigger, broader ways to think about deceit. J. Appl. Res. Mem. Cogn. **1**, 120–121 (2012)
5. van't Veer, A.: Effortless morality: cognitive and affective processes in deception and its detection. Dissertation, Tilburg. University (2016)
6. Albrechtsen, J.S., Meissner, C.A., Susa, K.J.: Can intuition improve deception detection performance? J. Exp. Soc. Psychol. **45**, 1052–1055 (2009)
7. Chow, C., Gedeon, T.: Classifying document categories based on physiological measures of analyst responses. In: 2015 6th IEEE International Conference on Cognitive Infocommunications (CogInfoCom), pp. 421–425 (2015)
8. Kreibig, S.D.: Autonomic nervous system activity in emotion: a review. Biol. Psychol. **84**, 14–41 (2010)

9. Tomaka, J., Blascovich, J., Kelsey, R.M., Leitten, C.L.: Subjective, physiological, and behavioral effects of threat and challenge appraisal. J. Pers. Soc. Psychol. **65**, 248 (1993)

10. Sleegers, W., Proulx, T.: The comfort of approach: self-soothing effects of behavioral approach in response to meaning violations. Front. Psychol. **5**, 1568 (2015)

11. Caldwell, S., Gedeon, T., Jones, R., Copeland, L.: Imperfect understandings: a grounded theory and eye gaze investigation of human perceptions of manipulated and unmanipulated digital images. In: Proceedings of the World Congress on Electrical Engineering and Computer Systems and Science (2015)

12. Duran, G., Tapiero, I., Michael, G.A.: Resting heart rate: a physiological predicator of lie detection ability. Physiol. Behav. **186**, 10–15 (2018)

13. DiNicolantonio, J.: The Salt Fix. Harmony (2017)

14. Stanford, S.M., et al.: Diabetes reversal by inhibition of the low-molecular-weight tyrosine phosphatase. Nat. Chem. Biol. **13**, 624 (2017)

15. Laeng, B., Sirois, S., Gredebäck, G.: Pupillometry: a window to the preconscious? Perspect. Psychol. Sci. **7**, 18–27 (2012)

16. TheEyeTribe: The EyeTribe. http://theeyetribe.com/theeyetribe.com/about/index.html

17. Kingma, D.P., Ba, J.: Adam: A method for stochastic optimization. arXiv Preprint arXiv1412.6980 (2014)

18. Hossain, M.Z., Gedeon, T.: Classifying posed and real smiles from observers' peripheral physiology. In: 11th International Conference on Pervasive Computing Technologies for Healthcare (2017)

19. Chen, L., Gedeon, T., Hossain, M.Z., Caldwell, S.: Are you really angry?: detecting emotion veracity as a proposed tool for interaction. In: Proceedings of the 29th Australian Conference on Computer-Human Interaction, pp. 412–416 (2017)

20. Bradley, M.M., Miccoli, L., Escrig, M.A., Lang, P.J.: The pupil as a measure of emotional arousal and autonomic activation. Psychophysiology **45**, 602–607 (2008)

21. Beatty, J., Lucero-Wagoner, B., et al.: The pupillary system. In: Handbook Psychophysiology (2rd edn.) (2000)

22. Wiseman, R., Watt, C.: Judging a book by its cover: the unconscious influence of pupil size on consumer choice. Perception **39**, 1417–1419 (2010)

23. Gründl, M., Knoll, S., Eisenmann-Klein, M., Prantl, L.: The blue-eyes stereotype: do eye color, pupil diameter, and scleral color affect attractiveness? Aesthetic Plast. Surg. **36**, 234–240 (2012)

Driver Sleepiness Detection Using LSTM Neural Network

Yini Deng[1], Yingying Jiao[1], and Bao-Liang Lu[1,2,3(✉)]

[1] Center for Brain-like Computing and Machine Intelligence, Department of
Computer Science and Engineering, Shanghai Jiao Tong University, Shanghai, China
dengyini@sjtu.edu.cn, jiaoyingying2010@163.com
[2] Key Laboratory of Shanghai Education Commission for Intelligent Interaction and
Cognition Engineering, Shanghai Jiao Tong University, Shanghai, China
[3] Brain Science and Technology Research Center, Shanghai Jiao Tong University,
Shanghai, China
bllu@sjtu.edu.cn

Abstract. Driver sleepiness has become one of the main reasons for
traffic accidents. Previous studies have shown that two alpha-related
phenomena - alpha blocking phenomenon and alpha wave attenuation-
disappearance phenomenon - respectively represent two different sleepiness levels: the relaxed wakefulness and the sleep onset. Thus, we proposed a novel model to detect those two alpha-related phenomena based
on EEG and EOG signals so as to determine sleepiness level. EOG and
EEG signals inherently have temporal dependencies, and the sleepiness
level transition is also a temporal process. Correspondingly, continuous
wavelet transform represents physiological signals well, and LSTM is
capable of handling long-term dependencies. Thus, our proposed dec-
tection model utilized continuous wavelet transform and LSTM neural
network for detecting driver sleepiness. The performance of our detection
model are twofold: the recall and precision for detecting start and end
points of alpha waves are generally high, and the LSTM classifier reaches
a mean accuracy of 98.14%.

Keywords: Driver sleepiness detection · EEG · EOG
Continuous wavelet transform · LSTM

1 Introduction

Lacking sleeping, numerous drivers have reported that they actually fell asleep
while driving [10]. This phenomenon indicates that some drivers tend to ignore
the early signs of drowsiness, and are consequently unaware of the following
period of sleep onset [1,7]. Therefore, developing a reliable method to evaluate
driver sleepiness is of urgent need and of great importance.

Although there are many ways to evaluate sleepiness level, such as self evaluation and vehicle-based evaluation, using physiological signals to detect driver
sleepiness is one of the most reliable ways [2,12]. Shabani *et al.* used RQA of

© Springer Nature Switzerland AG 2018
L. Cheng et al. (Eds.): ICONIP 2018, LNCS 11304, pp. 622–633, 2018.
https://doi.org/10.1007/978-3-030-04212-7_55

EEG to differentiate alert to drowsy with 90.6% accuracy [11]. From our previous research, a new alpha wave attenuation-disappearance phenomenon has been proven to be a general pattern for predicting the entry of sleep during simulated driving in daytime [8]. Besides this phenomenon, we also observed the typical alpha blocking phenomenon in the simulated driving process. Accroding to [4,5], it refers to the alpha rhythm activity which appears when eyes are closed under the relaxed wakefulness, and it disappears when eyes are reopened. Thus, alpha blocking phenomenon and alpha wave attenuation-disappearance phenomenon represent two different sleepiness levels: the relaxed wakefulness and the sleep onset.

In our previous work, we combined continuous wavelet transform with SVM to detect two different alpha-related phenomena [9]. However, the sleepiness level transition is a temporal process, and physiological signals like EOG have inherent temporal dependencies [14]. Our previous model didn't take temporal information into account. Thus, we introduced LSTM network to deal with those temporal dependencies in this paper. Being a special form of recurrent neural networks, LSTM has the ability to capture temporal dependency property. And it has achieved great successes in the field of machine translation and speech recognition. Meanwhile, continuous wavelet transform represents physiological signals well. Therefore, we proposed a novel model based on continuous wavelet transform and LSTM network to detect the change of alpha waves and to distinguish those two alpha-related phenomena.

2   Two Experiments and Their Settings

We conducted two different types of experiments for each subject: the eye-closure experiment and the simulated driving experiment. These two experiments aim to obtain EEG and EOG signals under two different sleepiness levels: the relaxed wakefulness and the sleep onset. In total, 12 healthy subjects (4 females and 8 males with an average age of 22) who have siesta habit for more than a year were recruited from Shanghai Jiao Tong University (Fig. 1).

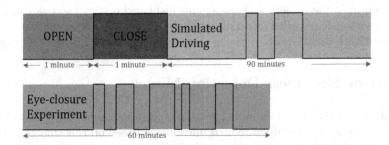

Fig. 1. Procedures for simulated driving experiments and eye-closure experiments

2.1 Eye-Closure Experiment

This experiment aims to obtain the subject's EEG and EOG signals under the relaxed wakefulness. The subject closed and opened his/her eyes according to our instructions, and only alpha blocking phenomenon appeared during the eye closure period.

Every subject participated in the eye-closure experiment once, and the experiment lasted for 60 min. We can get an averaged 250 periods of closing and reopening eyes from it. A portion of the data obtained from this experiment was used as the training data.

2.2 Simulated Driving Experiment

This experiment aims to induce the sleepiness level change of the subject from relaxed wakefulness to sleep onset during a simulated driving process. In this experiment, we can observe both alpha blocking phenomenon and alpha wave attenuation-disappearance phenomenon.

Every subject participated in the simulated driving experiment during the siesta time, and the experiment lasted for about 90 min. When the experiment began, the subject first kept the eyes closed and opened for one minute respectively. We defined those two periods as CLOSE and OPEN, which were used to calculate wavelet energy threshold later. Then, the subject started to do simulated driving. If a subject's higher sleepiness level was not induced during the experiment, the subject would participate in this experiment again.

2.3 Data Recording

As shown in Figs. 2 and 3, we used 6 electrodes to obtain EEG and EOG signals, including one reference electrode and one ground electrode placed behind the ears, two occipital EEG electrodes (O1 as an alternative if O2 is noisy), two EOG electrodes (Vu and Vd) placed above and under the left eye. All the signals were recorded at an 1000 Hz sampling rate using ESI NeuroScan System. Meanwhile, a camera was placed behind the steering wheel to moniter the state of the subject. The video from the camera and the signals displayed on Scan software window were sychronously recorded into a file, so that we can review after the experiment and investigate the relevance between eye movements and EEG or EOG signals.

3 Driver Sleepiness Detection Model

The purpose of our detection model is to track the change of alpha waves in O2 signal and to detect two alpha-related phenomena based on this change, so that we can detect driver sleepiness. We used offline processing to simulate the real-time driver sleepiness detection on signals from simulated driving experiments, and the sliding window was the key to our simulation. As depicted in Fig. 4, we first utilized a sliding window to calculate alpha wavelet energy threshold E_{th}

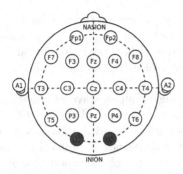

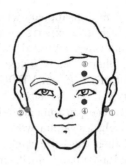

Fig. 2. EEG electrode placement   Fig. 3. EOG electrode placement

on O2 signal during CLOSE and OPEN periods. Then, E_{th} was used to detect the start and end points of alpha waves on O2 signal from the simulated driving process, whose alpha wavelet energy was also calculated using a sliding window. If the detected point was an end point, we extracted features from its corresponding VEOG signal and put those features into the trained LSTM classifier. Finally, the classifier determined whether the detected end point was the end point of alpha waves in alpha blocking phenomenon or alpha wave attenuation-disappearance phenomenon.

In terms of the training of the LSTM classifier, we used VEOG signals from the two experiments as the training data. The detection of end points was sensitive to the alpha wavelet energy threshold E_{th}: the detected end point might deviate from the actual end point. Thus, we utilized LSTM network so that it could take information from previous sequences into account, yielding a better classification performance.

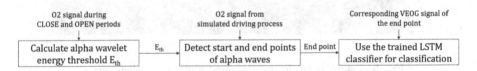

Fig. 4. The flowchart of the detection model

3.1 Visual Marking for Two Alpha-Related Phenomena

As mentioned in [9], an eye closure event (ECE) was defined as the period between the end of the upward trend line caused by closing eyes and the end of the downward trend line caused by reopening eyes in VEOG signal. In Fig. 5, we can see that two kinds of eye closure events, ECE^1 and ECE^2, exist alternately in simulated driving experiments.

ECE^1 is the eye closure event corresponding to alpha blocking phenomenon. This phenomenon refers to the alpha rhythm activity which appears when eyes are closed and disappears when eyes are reopened, indicating the relaxed wake-fulness. In Fig. 5, experts visually marked the start point s^1 and the end point e^1 of alpha waves on O2 signal, and this continuous alpha wave represents alpha blocking phenomenon. Those two points are equivalent to the start point and end point of ECE^1 on VEOG signal.

ECE^2 is the eye closure event corresponding to alpha wave attenuation-disappearance phenomenon. The visually marked split point p^2 divides ECE^2 into two phases: alpha wave attenuation phase and alpha wave disappearance phase. When the subject closes his/her eyes, the amplitude of alpha waves on O2 signal attenuates until alpha waves disappear. This phenomenon indicates the sleep onset, which means the subject has a high sleepiness level. Experts also marked the start point s^2 and end point p^2 of alpha waves in the alpha attenuation phase on O2 signal. Those two points are equivalent to the start point and split point of ECE^2 on VEOG signal. Besides, similar to the marking of end point e^1 in ECE^1, the end point of ECE^2 on VEOG signal was also visually marked according to the downward trend line.

Therefore, the two different alpha-related phenomena have different repre-sentations on O2 and VEOG signals, and indicate two different sleepiness levels.

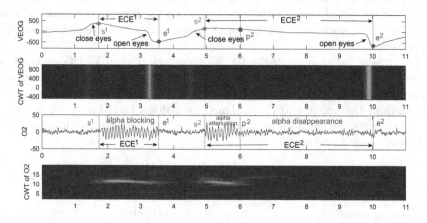

Fig. 5. Visual marks for two different alpha wave phenomena

3.2 Calculating Alpha Wavelet Energy Threshold E_{th}

Similar to the feature extraction method in [9], continuous wavelet transform (CWT) was applied to O2 signal during OPEN and CLOSE periods in simulated driving experiments to calculate wavelet threshold E_{th}. Complex Morlet wavelet was selected to do CWT because it is geometrically similar to alpha waves. We picked the scales corresponding to alpha frequency band (F = [8,12] Hz) from range [1, 1024]. Wavelet energy $w(t)$ was thus calculated according to wavelet coefficients on those scales and was further averaged in time window T. Here, T was set to 1 s, and the sliding step of the time window was set to 0.1 s. Thus, we got numerous 1 s alpha wavelet energy values for OPEN and CLOSE periods, respectively. E_{th} is the mean of alpha wavelet energy distribution's minimum during CLOSE period and its maximum during OPEN period.

3.3   Detecting Start and End Points of Alpha Waves

E_{th} was used to detect the start and end points of alpha waves on O2 signal, and it was calculated for each subject. We used Complex Morlet wavelet to do CWT on O2 signal of the testing data. As mentioned in 3.2, the time window length was set to 1 s, and the sliding offset was 0.1 s. Thus, we can get an alpha wavelet energy curve on O2 signal, as depicted in Fig. 6. As soon as the current wavelet energy was higher than E_{th}, the current time point was considered as the start point s' of alpha waves. Afterwards, if the wavelet energy was continuously higher than E_{th}, these time points were the proof for the persistent presence of alpha waves. Until the wavelet energy was lower than E_{th}, the current time point was the end point of alpha waves.

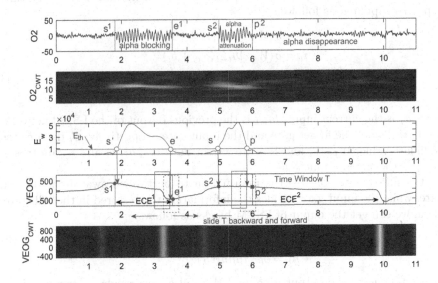

Fig. 6. Wavelet energy curve of O2 signal

3.4   LSTM Classifier

The key to distinguishing the two types of alpha-related phenomena is to differentiate the end point e^1 in ECE^1 from the split point p^2 in ECE^2. Thus, we defined e^1 as negative, and p^2 as positive. Besides, We used a leave-one-subject-out cross validation method, which took data from 11 subjects as the training data, and the remaining one subject as the testing data.

Feature Extraction of VEOG Signals. We used Haar wavelet to do CWT for VEOG signals, which is similar to the method used in [9], because its mother wavelet has a similar shape to the waveform in VEOG signals. CWT with Haar wavelet scaling from 1 to 128 was applied to the 0.5 s VEOG signal, which was either $[e^1 - 0.5\,\mathrm{s}, e^1\,\mathrm{s}]$ or $[p^2 - 0.5\,\mathrm{s}, p^2\,\mathrm{s}]$. As the LSTM classifier has the sequence length concept, the 0.5 s time window was moved forward for 0.125 s and 0.25 s, and backward for 0.125 s and 0.25 s. Finally, we can get five 128-dimensional Haar wavelet feature vectors for each training sample, corresponding to the sequence length 5 of the LSTM classifier. When labeling the training data, we labeled all 5 feature vectors for each p^2 as positive and those for each e^1 as negative.

LSTM Neural Neworks. As the input features have temporal dependencies, we introduced Long Short Term Memory (LSTM) neural network to incorporate this information [6]. Being an RNN using LSTM cells, LSTM neural network is capable of preventing vanishing gradient problems [3]. Moreover, in practice, LSTM neural networks handle long-term dependencies well. Cell state C_t, which is propagated over time, is the key to LSTM neural networks. At every time step, C_t is updated as follows:

$$
\begin{aligned}
f_t &= \sigma(W_f \cdot [h_{t-1}, x_t] + b_f) \\
i_t &= \sigma(W_i \cdot [h_{t-1}, x_t] + b_i) \\
\tilde{C}_t &= tanh(W_C \cdot [h_{t-1}, x_t] + b_C) \\
C_t &= f_t * C_{t-1} + i_t * \tilde{C}_t
\end{aligned}
\tag{1}
$$

where x_t is the current input, h_{t-1} is the previous output of the LSTM network, f_t and i_t denote the forget gate and the input gate, respectively, $\tilde{C}_t$ denotes the candidate value, W_f, W_i and W_C are weight matrices, b_f, b_i and i_C are biases, and σ is the sigmoid function.

The forget gate controls the information to be thrown away from the cell state, while the input gate decides the information to be stored in the cell state. Then, we can get the ouput of LSTM blocks as follows:

$$
\begin{aligned}
o_t &= \sigma(W_o \cdot [h_{t-1}, x_t] + b_o) \\
h_t &= o_t * tanh(C_t)
\end{aligned}
\tag{2}
$$

where o_t denotes the output gate, and W_o and b_o are weight matrix and bias, respectively. So, h_t is a filtered version of the cell state, regulated by the output gate.

Our classification model is depicted in Fig. 7. Dropout is applied to the ouput of LSTM layer, so that the model is more robust. After that, there is one classification layer, which uses hinge loss with L2-regularization as the objective loss function. Therefore, we can consider it as a linear kernel SVM [13].

3.5   Using the LSTM Classifier for Classification

When an end point of alpha waves was detected, we found its corresponding point on VEOG signal and utilized Haar wavelet to do CWT. As mentioned in Sect. 3.4,

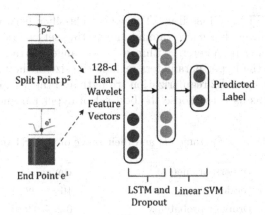

Fig. 7. LSTM classifier

five 128-dimensional feature vectors were extracted. They were further put into the LSTM classifier to determine whether this end point is the end point e' of ECE^1 or the split point p' of ECE^2. This is equivalent to determining whether this end point is the end point of alpha waves in alpha blocking phenomenon or in alpha wave attenuation-disappearance phenomenon.

Actually, we get one label, which is the bigger one in the two output labels, for each of the five feature vectors in the input sequence, while we only need one label to determine whether the detected end point is the end point of alpha waves in alpha blocking phenomenon or in alpha wave attenuation-disappearance phenomenon. To solve this nonuniformity, we design a mapping function to map the five output labels to one final label. That is, we choose the label which is the most frequent among the five output labels as the final label of the detected end point.

4 Experimental Results

4.1 Training Details

For the training of the classifier, we used data from the two experiments of the 11 subjects. According to our observation, there were very few eye closure events during the early stage of the simulated driving experiment. During the late stage, as the subject became more sleepy, ECE^1 and ECE^2 appeared alternately and frequently. Thus, for each subject, we picked a 30-min period from the late stage of his/her simulated driving experiment. Besides, due to the approximately balanced amount of e^1 and p^2 in simulated driving experiments, we only marked part of the e^1 in eye-closure experiments as negative, so that the whole training set wouldn't be significantly unbalanced. After the training of the LSTM classifier, the whole model was tested on the simulated driving experiment from the one subject which was left out. And we used the averaged result to evaluate our model.

Training the LSTM Classifier. The whole classifier network was trained using Adam optimizer. For each training and testing set, we randomly selected tens of sets of hyper-parameters within a given range to train the model. As shown in Table 1, the hyper-parameters include the size of the hidden layer, the dropout probability, the L2 regularization strength and the learning rate. 10-fold cross validation was used to choose the best set of hyper-parameters.

Table 1. The hyper-parameters and their range of the LSTM classifier

Hyper-parameter	Range
Hidden size	$16 \sim 128$
Dropout probability	$0.2 \sim 0.9$
$\log_{10}$(L2 regularization strength)	$-7 \sim -2$
$\log_{10}$(learning rate)	$-5 \sim -1$

Training SVM and k-NN. For linear SVM, we used sklearn package in Python to do the training, and adjusted the parameter C to achieve the best performance. For k-NN, we set k to 1, because it had the best performance in the range [1, 5] on our training data.

4.2 Performance of Detecting Alpha Wave Start and End Points

If the detected start point s' or end point e' fell into the range of $[s^1(s^2) - 0.5$ s, $s^1(s^2) + 0.5$ s] or $[e^1 - 0.5$ s, $e^1 + 0.5$ s], the point was considered as a correctly detected point by the model. Similarly, if the detected split point p' fell into the range of $[p^2 - 0.8$ s, $p^2 + 0.8$ s], it was considered as a correctly detected point. Due to the greater subjective bias for marking the split point, we defined a larger time range of 0.8 s.

To evaluate how well our detection model detects the start and end points of alpha waves, we introduced True Positive (TP), False Positive (FP) and False Negative (FN) to show the performance of detecting start and end points of alpha waves. Here, TP and FP refered to the number of start and end points that were correctly detected and wrongly detected, while FN was the number of those which were visually marked but not detected by the model. Meanwhile, recall, precision and F1 score were used to investigate the performance of detecting start and end points. From Table 2, we can see that the recall, precision and F1 score for detecting start and end points of ECE^1 are generally high across different subjects with mean values of 96.3%, 93.9%, 95.2% and 95.0%, 95.1%, 94.8%, respectively. Similarly, those three metrics for detecting start and split points of ECE^2 are also generally high with mean values of 96.0%, 95.7%, 95.8% and 94.5%, 94.3%, 94.6%, respectively, as shown in Table 3.

Table 2. ECE^1 detection performance

Subject	#ECE^1	Start point			End point		
		Recall (%)	Precision (%)	F1 (%)	Recall (%)	Precision (%)	F1 (%)
1	34	94.1	97.0	95.5	100.0	97.1	98.6
2	28	100.0	96.6	98.2	96.4	93.1	94.7
3	22	100.0	91.3	95.5	90.9	95.2	93.0
4	20	95.0	82.6	88.4	95.2	90.9	93.0
5	26	100.0	96.3	98.1	92.3	92.3	92.3
6	28	92.9	89.7	91.2	89.3	96.2	92.6
7	28	100.0	96.6	98.2	96.4	100.0	98.2
8	27	96.3	92.9	94.5	96.3	92.9	94.5
9	23	95.7	95.7	95.7	91.3	95.5	93.3
10	21	90.5	100.0	95.0	95.2	100.0	97.6
11	25	96.0	96.0	96.0	100.0	92.6	96.2
12	26	100.0	92.9	96.3	92.3	96.0	94.1
Mean ± SD		96.3 ± 3.0	93.9 ± 4.6	95.2 ± 2.8	95.0 ± 3.7	95.1 ± 2.9	94.8 ± 2.2

Table 3. ECE^2 detection performance

Subject	#ECE^2	Start point			Split point		
		Recall (%)	Precision (%)	F1 (%)	Recall (%)	Precision (%)	F1 (%)
1	24	100.0	96.0	98.0	100.0	92.3	96.0
2	35	100.0	100.0	100.0	97.1	97.1	97.1
3	30	93.3	96.6	94.9	96.7	96.7	96.7
4	27	88.9	92.3	90.6	85.2	92.0	88.5
5	24	95.8	95.8	95.8	95.8	100.0	97.9
6	26	92.3	92.3	92.3	92.3	88.9	90.6
7	27	96.3	92.9	94.5	88.9	96.0	92.3
8	28	100.0	90.3	94.9	100.0	93.3	96.6
9	26	96.2	100.0	98.0	96.2	89.3	92.6
10	23	100.0	95.8	97.9	95.7	95.7	95.7
11	29	96.6	100.0	98.2	93.1	96.4	94.7
12	28	92.9	96.3	94.5	92.9	100.0	96.3
Mean ± SD		96.0 ± 3.6	95.7 ± 3.2	95.8 ± 2.7	94.5 ± 4.3	94.3 ± 3.4	94.6 ± 2.9

4.3   Comparison of Three Classifiers

To make the comparison simpler, we only took into account those end points which were correctly detected by the detection models. As shown in Table 4, the LSTM classifier achieves the best accuracy of 98.14% among the three classifiers, and it has the smallest standard deviation of 0.75%, which makes it more robust. In contrary, k-NN is the worst at doing classification, and it is the most unstable one. Although the mean accuracy of linear SVM is close to that of the LSTM classifier, its performance is unstable across different subjects. We think the stableness of the LSTM classifier owes to its recurrent structure which uses information from previous sequences to do classification. Even if the information carried in one of the 5 input feature vectors is incomplete or inaccurate, the classifier is able to give the correct label with the help of temporal information. Meanwhile, if the detected end point deviates slightly from the visually marked end point, the vital part around the end point can still fall into the range of the five sliding windows in LSTM. This ability of integrating all the information in previous sequences makes LSTM better at classifying the end points.

Table 4. Accuracies and standard deviations of different classifiers

Subject	k-NN	SVM	LSTM
1	90.34	93.44	96.90
2	86.56	95.74	98.03
3	86.53	95.92	98.78
4	90.70	97.21	98.14
5	85.11	95.74	97.87
6	87.76	96.33	99.18
7	89.80	92.16	98.43
8	90.74	92.96	96.67
9	93.48	95.22	97.83
10	92.86	95.71	98.57
11	88.85	95.77	98.46
12	89.60	94.80	98.80
Average accuracy	89.35	95.08	**98.14**
SD	2.54	1.49	**0.75**

5   Conclusion

In this paper, we have introduced a driver sleepiness detection model to detect the change of alpha waves and classify two different alpha-related phenomena. This method utilized continuous wavelet transfrom to extract features from EEG and EOG signals, and used an LSTM classifier to do classification based on temporal information. The experimental result indicates that the recall, precision and F1 score for detecting start and end points of alpha waves are generally high. Meanwhile, the proposed LSTM classifier achieves a mean accuracy of 98.14%. Thus, our proposed method, which places few electrodes on subjects and has satisfying results, is both feasible and practical for detecting driver sleepiness.

Acknowledgments. This work was supported in part by the grants from the National Key Research and Development Program of China (Grant No. 2017YFB1002501), the

National Natural Science Foundation of China (Grant No. 61673266), and the Fundamental Research Funds for the Central Universities.

References

1. Anund, A., Åkerstedt, T.: Perception of sleepiness before falling asleep. Sleep Med. **11**(8), 743–744 (2010)
2. Balandong, R.P., Ahmad, R.F., Saad, M.N.M., Malik, A.S.: A review on EEG-based automatic sleepiness detection systems for driver. IEEE Access **6**, 22908–22919 (2018)
3. Bengio, Y., Simard, P.Y., Frasconi, P.: Learning long-term dependencies with gradient descent is difficult. IEEE Trans. Neural Networks **5**(2), 157–166 (1994)
4. Guyton, A.C.: Structure and Function of the Nervous System. Saunders Limited. (1976)
5. Harland, C.J., Clark, T.D., Prance, R.J.: Remote detection of human electroencephalograms using ultrahigh input impedance electric potential sensors. Appl. Phys. Lett. **81**(17), 3284–3286 (2002)
6. Hochreiter, S., Schmidhuber, J.: Long short-term memory. Neural Comput. **9**(8), 1735–1780 (1997)
7. Horne, J.A., Baulk, S.D.: Awareness of sleepiness when driving. Psychophysiology **41**(1), 161–165 (2004)
8. Jiao, Y., Lu, B.L.: An alpha wave pattern from attenuation to disappearance for predicting the entry into sleep during simulated driving. In: 2017 8th International IEEE/EMBS Conference on Neural Engineering (NER), pp. 21–24 (2017)
9. Jiao, Y., Lu, B.L.: Detecting driver sleepiness from EEG alpha wave during daytime driving. In: 2017 IEEE International Conference on Bioinformatics and Biomedicine (BIBM), pp. 728–731 (2017)
10. Sagberg, F.: Road accidents caused by drivers falling asleep. Accid. Anal. Prev. **31**(6), 639–649 (1999)
11. Shabani, H., Mikaili, M., Noori, S.M.R.: Assessment of recurrence quantification analysis (RQA) of EEG for development of a novel drowsiness detection system. Biomed. Eng. Lett. **6**(3), 196–204 (2016)
12. Shi, L.C., Lu, B.L.: EEG-based vigilance estimation using extreme learning machines. Neurocomputing **102**, 135–143 (2013)
13. Tang, Y.: Deep learning using linear support vector machines. arXiv preprint arXiv:1306.0239 (2013)
14. Zheng, W.L., Lu, B.L.: A multimodal approach to estimating vigilance using EEG and forehead EOG. J. Neural Eng. **14**(2), 26017 (2017)

HTMTAD: A Model to Detect Anomalies of CDN Traffic Based on Improved HTM Network

Ning Zhao[1], Yongli Wang[1(✉)], Na Cao[1], and Xiaoze Gong[2]

[1] School of Computer Science and Engineering, Nanjing University of Science and Technology, Nanjing 210094, China
yongliwang@njust.edu.cn

[2] Baicheng Ordnance Test Center, 108# mailbox, Baicheng 137001, Jilin, China

Abstract. There will always be malicious intrusion, node downtime and other events caused by network traffic anomalies while Content Delivery Network (CDN) is facing user's service. These events will lead in a large area of network paralysis and suspension of network services. Therefore, in order to effectively detect and deal with the anomalies in advance, the paper makes a partial improvement on the existing Hierarchical Temporal Memory network (HTM), and proposes a new network model HTMTAD (Hierarchical Temporal Memory – based Traffic Anomalies Detection) to detect intelligently the changes of abnormal traffic from the CDN. In view of the characteristics of CDN traffic data, the paper proposes a hash coding algorithm to improve the reliability of encoder and an anomaly likelihood calculation method to detect the CDN traffic anomalies. Experimental results show that HTMTAD can effectively detect anomalies in CDN network traffic.

Keywords: CDN · Traffic anomaly detection · Hierarchical Temporal Memory Encoder · Anomaly likelihood

1 Introduction

The Content Delivery Network (CDN) is an important part of the current Internet infrastructure, and Internet traffic today is dominated by content providers and highly distributed CDNs. Deploying a server infrastructure on a large scale makes it possible to serve the Internet in different locations. The inherent distribution of CDN makes it possible to better cope with the growing demand for user, that is, the popular applications and hot content are pushed to the end user as close as possible to reduce network delays and improve the quality of service. However, while CDN provides services to users, there will always be abnormal network traffic caused by malicious intrusions, node downtime, and other events. Therefore, it is very important to detect the events that result in network paralysis in CDN as early as possible so that CDN could adjust the strategy to adapt the changes.

CDN network traffic data has a strong temporal correlation, thus it could be represented by a time series. Assume that x_t in the sequence $\dots, x_{t-2}, x_{t-1}, x_t, x_{t+1}, x_{t+2}, \dots$ represents the traffic value of current CDN node server at time t. At each time point t, in

L. Cheng et al. (Eds.): ICONIP 2018, LNCS 11304, pp. 634–646, 2018.
https://doi.org/10.1007/978-3-030-04212-7_56

order to determine whether the current server is abnormal, it is necessary to make a judgment before time $t + 1$, that is, determine whether the current traffic data is abnormal according to the current and previous status of traffic. Therefore, the anomaly detection of CDN network traffic needs to have the following requirements:

(a) The prediction must be real-time, that is, the event x_t is identified as normal or abnormal before receiving the subsequent event x_{t+1}.
(b) Support online learning traffic characteristics, and do not need to store all network traffic data.
(c) The algorithm could reduce false positives and false negatives.

In the past decade, researchers have proposed many strategies and algorithms for detecting anomalies in network traffic. They can be roughly divided into two categories: statistical methods and machine learning methods. There are two common types of statistical methods: sliding thresholds and outlier tests, such as Extreme Student Deviation (ESD), k-sigma, change point detection, typicality and eccentricity analysis [1–5], etc. They mainly fit the statistical model by a given network traffic, and then apply statistical method to infer and test whether a data instance satisfies the model. Therefore, the data prepared for statistical method depends on the assumptions generated by the predefined distribution. However, these assumptions are usually not true. Machine learning methods are broadly categorized as supervised methods like Support Vector Machines and Decision Trees [6], and unsupervised methods like Clustering. Typical examples include Netflix's Robust Principal Component Analysis (RPCA) method and Yahoo's EGADS supervised machine learning method [7, 8], but it requires manual tagging of traffic data, a costly process, especially abnormal traffic data labels are difficult to obtain. Both of these methods need to analyze the entire network traffic data, can't continuously learn from complex traffic data, fail to detect unknown traffic data, do not consider the time context in traffic, ignore its time dependence, and do not satisfy three requirements mentioned above for CDN network traffic detection.

In order to resolve the above requirements and problems, this paper proposes a HTMTAD model to detect abnormal network traffic in the CDN. This model is based on a new biotechnology Hierarchical Temporal Memory network (HTM [9, 10]) that does not require tagging data, can continuously learn traffic data patterns, efficiently discovers the time background in the data, and has a great tolerance with noise.

The contributions of this paper are: (1) Provide a new solution for CDN network traffic anomalies detection HTMTAD (Hierarchical Temporal Memory – based Traffic Anomalies Detection), which improves the reliability of HTM encoder and proposes an algorithm for time series data (CDN traffic) encoding so that the HTM network can better adapt to the detection of CDN traffic anomalies; (2) On the basis of the prediction error of the existing HTM network output, we use the Gaussian tail probability distribution to improve the reliability of CDN traffic abnormality judgment. Experiments have proved that HTMTAD can play an extremely important role in the detection of CDN network traffic anomalies. It fully utilizes the time continuity and context information of CDN traffic data to make it possible to accurately detect traffic anomalies in advance.

The paper is organized as follows. Section 2 presents the model for detecting abnormal CDN traffic and the definition of related concepts, besides, gives a brief introduction about the HTM network model to readers. Section 3 describes the specific algorithm for CDN traffic anomalies detection in three parts. Section 4 experimentally demonstrates the effectiveness of HTMTAD in detecting CDN traffic abnormalities. Section 5 gives a summary about the model and analyzes the future work.

2 Models and Definitions

As shown in Table 1, CDN traffic is represented by time series in this paper. Detection of CDN traffic anomalies depends on contextual information. Meanwhile real-time and robustness are required in detection model. HTM is a hierarchical-temporal-memory-based learning model that can train time series data. Moreover, HTM make a predict according to a large number of stored time series patterns in conjunction with past contextual information [11]. Therefore, the HTM network satisfies the requirements of CDN traffic anomalies detection. In order to better enable the HTM network to be applied to the detection of CDN traffic anomalies, the paper improves on the existing HTM encoder, using the hash function to encode the time series so that it eliminates the limit of the maximum and minimum range, to describe and represent time series data better. In addition, Gaussian tail probability is used on the output of the prediction error of the existing HTM network to improve the reliability of CDN traffic anomalies judgment. HTM is mainly composed of three parts: encoder, space pool and sequence memory. The specific model is shown in Fig. 1.

Table 1. Style of time series.

Symbol	Time series
x_1	2014-03-01 17:36:00,42.0
x_2	2014-03-01 17:41:00,94.8
x_3	2014-03-01 17:46:00,42.0
...	...

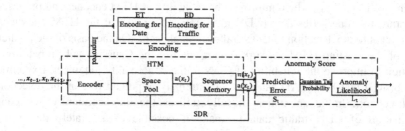

Fig. 1. HTMTAD model framework

In Fig. 1, the time series of traffic $\ldots, x_{t-1}, x_t, x_{t+1}, \ldots$ are inputted into the HTMTAD model. $a(x_t)$ is outputted after improved encoder encoding and spatial pool processing at first. Then $a(x_t)$ is sent to the sequence memory (the sequence memory is the core of the HTM network, simulating the time pattern of $a(x_t)$), and its corresponding sparse prediction vector $\pi(x_t)$ is outputted. $\pi(x_t)$ represents the predicted value of the next time point $a(x_{t+1})$. Finally, $a(x_t)$ and $\pi(x_t)$ are sent to the anomaly scoring module to obtain the prediction error, and on this basis, the Gaussian tail probability distribution is used to calculate the anomaly likelihood.

2.1 HTM Overview

The HTM network consists of mini-columns consisting of several HTM neurons (the right side of Fig. 2(a)). Each neuron is divided into two regions: the proximal region and the distal basal region. Each region contains several segments containing many synapses. The proximal region receives the feed-forward input and the distal basal context region receives contextual information from cells in the vicinity of the same cortical region. Neurons in the same mini-column receive the same feed-forward input. Neurons have a different distal context, and connect to other cells in the network. Traffic time series after sparse representation are stored in these mini-columns, and all operations of the HTM network are based on these mini-columns. The model before x_t and the prediction model at x_{t+1} corresponding to x_t are stored in the sequence memory. Figure 2 briefly illustrates the working process of HTM. Please refer to [9, 10, 12–14] for the specific HTM workflow.

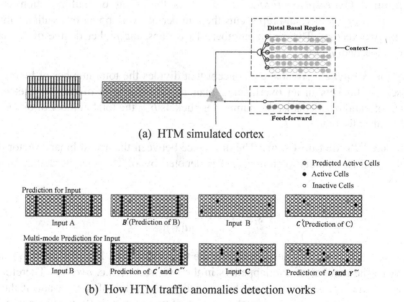

(a) HTM simulated cortex

(b) How HTM traffic anomalies detection works

Fig. 2. The working process of the HTM network.

In Fig. 2(a), the HTM simulates a layer of cortex consisting of a set of mini-columns, per mini-column include multiple neurons, and the synapses of pyramidal neurons in each neuron (left) simulates the structure of cortex(right). HTM neurons create detectors modeled by dendrite consisting of synapses. The contextual synapses receive transversal inputs from other neurons within the layer on which sufficient lateral activities will lead to the cells to enter a predicted state. High-order Markov sequences (ABCD and XBCY) are represented by shared subsequences in Fig. 2(b). Each sequence element calls a set of sparse cells in the mini-columns. Transverse junction-predicted cells block other cells in the same column by suppressing emission in the column, which cause a highly sparse representation of the sequence. Figure 2(b) (above) shows the unexpected input sequence element A, which is used to predict the next input B'. If the subsequent input matches the prediction, it is in normal mode and a new prediction is generated. Figure 2(b) (below) shows that unexpected input of sequence element B results in two predictions of C (C' and C'') for the ambiguous sequence BC (sequences ABCD and XBCY subsequences). Two model predictions (BCD and BCY) are generated after inputting C. This figure illustrates that the uncertain network can perform multiple predictions. Although the network predicts dozens of models at the same time, the error probability is still low due to the highly sparse representation of the prediction.

2.2   Related Concepts

Definition 1 Valid Bits. A bit with a value of 1 called valid bit in binary coding.

Definition 2 Overlap. $overlap(x, y) \equiv x \cdot y^T$ is the result of multiplication of the transposition of x and y, and represents the number of overlapping bits with a value of 1 in the two vectors. The higher number of overlaps, the higher degree of similarity between the two vectors.

Definition 3 Bucket. An abstract concept that divides the total number of binary bits into several buckets in an overlapping manner. The capacity of each bucket is the number of valid bits. That is, the number of buckets b = the total number of bits n - the number of valid bits $w + 1$.

Definition 4 Prediction Error. The difference between the actual binary vector $a(x_t)$ and the predicted binary vector $\pi(x_{t-1})$ is denoted by s_t. The specific definition is as follows:

$$s_t = 1 - \frac{\pi(x_{t-1}) \cdot a(x_t)}{|a(x_t)|} \tag{1}$$

where $|a(x_t)|$ is the scalar norm, which is the total number of 1 in $a(x_t)$. $\pi(x_{t-1}) \cdot a(x_t)$ represents the number of overlapped 1s in the two vectors, i.e. *overlap*. Therefore, in Eq. (1), if $a(x_t)$ perfectly matches the prediction, the value of s_t is 0, and if the two binary vectors are orthogonal (i.e., the *overlap* of two vectors is 0), the value of s_t is 1.

3 Models and Definitions

In this section, we first introduce three algorithms that are improved for encoder, and propose a method, linear detection method of the hash function, which is used to transform data into binary form, implementing the unboundedness and sparsity of the encoded binary value. It effectively represents the traffic of CDN and meets the requirements of the HTM network. The paper uses Gaussian tail probability distribution to calculate the anomaly likelihood based on the prediction error and improve the reliability of the anomaly judgment. Finally, the specific workflow of the entire anomaly detection algorithm is described in detail.

3.1 Improved Algorithm of Encoder

The HTM method represents the input data in the form of Sparse Distributed Representation (SDR), so the encoder is required to transform the input data into binary form, i.e. SDR, which consists of a large number of bits, most of which is 0 and a few are 1. Based on the uncertainty of the traffic value and the threshold limit problems of the original encoder [15], this paper proposes a coding method using a hash function so that there is no limit on the size of the traffic. In addition, each binary bit can represent multiple range of values and accurately describe the traffic data to meet the sparseness requirements of the SDR and adapt to the HTM network model. There are four aspects to consider when setting up the encoder: (1) SDRs with semantically similar data outputs have more overlapping valid bits; (2) the same input should always produce the same SDR; (3) all input corresponding output should have the same dimensions (total bits); (4) The output should have similar sparsity for all inputs. The encoder implementation is as follows:

(a) Get timestamps and traffic values in each sequence of the time series;
(b) Split the time stamp to obtain the year, month, day, hour, minute and second of the time stamp respectively;
(c) For a specific time period, select the maximum and minimum values, the number of valid bits, and the corresponding total number of bits;
(d) Using Algorithm 2 ED to encode the corresponding time period;
(e) Using Algorithm 3 ET to encode the traffic value;
(f) Combining the outputs from Algorithm 2 ED and Algorithm 3 ET to obtain the corresponding binary code for the final time series.

Algorithm 1 describes the encoding process of the encoder. The input is a series of time series, the data style is specifically referenced in Table 1, and the output is binary code corresponding the input. The 3rd line of the algorithm breaks the time series into several timestamp and traffic values that need to be encoded; the 4th–9th lines encode the timestamp in the time series and return a list that contains the encoded data; the 10th line encodes the traffic value mainly performed by using the hash function to obtain the corresponding binary coded list; the 11th–16th lines merge the binary code of the timestamp and the binary code of the traffic value to obtain the total coded data, that is, the binary code corresponding to the time series. It should be noted that once the parameters used in Algorithm 1 are determined, they cannot be changed. This ensures the consistency of the total number of bits encoded for each time series. Table 2 shows the symbols and meanings in the three algorithms.

Table 2. Symbol and Meaning.

Symbol	Meaning	Symbol	Meaning
$a(x_t)$	binary representation of the actual value of x_t	$\pi(x_t)$	binary representation of the predict value of x_t
s_t	prediction error	L_t	anomaly likelihood
$x_1 .. x_n$	traffic time series	$v_1...v_n$	the binary code of the time series
X	set of time series	V	set of binary code
$x_{year}, ..., x_{second}$	value of the corresponding time period	$t_{ymin}, ..., t_{smin}$	minimum value of the corresponding time period
$t_{ymax}, ..., t_{smax}$	maximum value of the corresponding time period	$n_{year}, ..., n_{second}$	bits of corresponding time period
$w_{year}, ..., w_{second}$	valid bits of time period	$v_{year}, ..., v_{second}$	binary representation of corresponding time period
v_f, v_t	binary representation of traffic rate, time period	f	traffic value
b_f	bucket number of traffic	w_f	valid bits of traffic
n_f	total bits of traffic	**floor**[]	float rounded down
ED	abbreviation of algorithm 2	**ET**	abbreviation of algorithm 3
range	generate a specified range of number sets	t	time period for encoding
t_{min}	the minimum of time period	t_{max}	the maximum of time period
b_t	bucket number of time period	w_t	valid bits of time period
a, b	two parameters of hash function		

Algorithm 1: Encoding(Encoding Data)

Input: $x_1, x_2, x_3, \cdots, x_n$

Output: $v_1, v_2, v_3, \cdots, v_n$

1)   init$(X, \{x_1, x_2, x_3, \cdots, x_n\}), V = \emptyset$

2)   **foreach** $x \in X$

3) $x_{year}, x_{month}, x_{day}, x_{hour}, x_{minute}, x_{second}, f \leftarrow x$

4) $v_{year} = ED(x_{year}, t_{ymin}, t_{ymax}, n_{year}, w_{year})$

5) $v_{month} = ED(x_{month}, t_{monthmin}, t_{monthmax}, n_{month}, w_{month})$

6) $v_{day} = ED(x_{day}, t_{dmin}, t_{dmax}, n_{day}, w_{day})$

7) $v_{hour} = ED(x_{hour}, t_{hmin}, t_{hmax}, n_{hour}, w_{hour})$

8) $v_{minute} = ED(x_{minute}, t_{minutemin}, t_{minutemax}, n_{minute}, w_{minute})$

9) $v_{second} = ED(x_{second}, t_{smin}, t_{smax}, n_{second}, w_{second})$

10) $v_f = ET(f, w_f, n_f)$

11) $k = n_{year} + n_{month} + n_{day} + n_{hour} + n_{minute} + n_{second} + n_f$

12)   init$(v[k], 0)$

13) $v \leftarrow v_{year} \cup v_{month} \cup v_{day} \cup v_{hour} \cup v_{minute} \cup v_{second} \cup v_f$

14)   **if** $v \notin V$ && $v \neq \emptyset$ **then**

15) $V \cup \{v\}$

16)   **end if**

17)   **end for**

18)   **return** V

As can be seen from the Algorithm 1, the time complexity depends on the time complexity of the Algorithm 2 ED and the Algorithm 3 ET. According to the analysis of the second and third algorithms, the time complexity of Algorithm 2 and algorithm three is $O(w_t)$ and $O(n_f)$ respectively. Hypothetical set

$$U = \{w_{year}, w_{month}, w_{day}, w_{hour}, w_{minute}, w_{secend}, n_f\} \tag{2}$$

Then the time complexity of Algorithm 1 can be expressed as:

$$O(Encoding) = \{w \mid max(n * w), w \in U\} \tag{3}$$

Algorithm 2 describes the encoding algorithm for the time period in the time stamp. The input is the maximum and minimum value (t_{min}, t_{max}) of the corresponding value (t) of each time period, the number of encoded bits n_t, and the coded valid bits w_t. Then the output is the binary code of the corresponding time period. Algorithm 2 first determines the representation range of the input values, then divides the values into segments (abstracts into buckets) overlappingly, moreover, determines which bucket should the values to enter, and finally sets the value of all elements is 1 in the bucket. The 1st–3rd lines of the algorithm is used to calculate the number of buckets and determine the range of value; the 4th line determines the bucket index number; the 5th–7th lines successively set w valid bits value 1 from the bucket index to achieve the purpose of encoding. The reason for using this code in the time period is the definiteness of the value of the time period. For example, the week can be represented by 0–6, and the month can be represented by 0–11. The time complexity of Algorithm 2 depends on the value of w_t, furthermore, the 5th–10th lines indicate that the algorithm will execute w_t times at most, so the time complexity of Algorithm 2 is $O(w_t)$.

Algorithm 2: ED(Encoding Date)

Input: $t, t_{min}, t_{max}, n_t, w_t$

Output: v_t

1) $r_t = t_{max} - t_{min}$
2) $b_t = n_t - w_t + 1$
3) $init(v_t[n_t], 0)$
4) $i_t = floor[b_t * (t - t_{min})]/r_t$
5) **foreach** i ∈ range($i_t, i_t + w_t$)
6)   **if** i > n_f
7)   **break**
8)   **end if**
9) $v_t[i] = 1$
10) **end for**
11) **return** v_t

Algorithm 3 describes the process of encoding traffic value. The input is the traffic value f, the valid bits of traffic value w_f, and the total number of bits n_f. Then the output is its corresponding binary code. The 1st–2nd lines in the algorithm are used to initialize the range of time period and calculate the number of buckets. The 3rd line determines the hash function (where a and b are both constants). The 4th line represents the index of the bucket found by the hash function. The 5th–12th lines find a suitable bit for each valid bit and set it value 1. The Algorithm 3 mainly adopts the hash function H (key) and uses the linear detection method to solve the hash collision. The time complexity of this method depends on the length of the hash table, so the time complexity of the Algorithm 3 is $O(n_f)$.

Algorithm 3: ET(Encoding Traffic)

Input: f, w_f, n_f

Output: v_f

1) $init(v_f[n_f], 0)$
2) $b_f = n_f - w_f + 1$
3) $H(key) = (a * key + b)\mathbf{mod}\, n_f$
4) $i_f = floor[b_f * f]/n_f$
5) **foreach** i ∈ range($i_f, i_f + w_f$)
6) $i_{new} = H(i)$
7)   **while** $v_f[i_{new}] \neq 0$ && $i_{new} < n_f$
8) $i_{new} = i_{new} + 1$
9)   **end while**
10) $v_f[i_{new}] = 1$
11) **end for**
12) **return** v_f

3.2   Improved Algorithm of Encoder

In this paper, s_t is used as the calculation method of the prediction error, and the specific definition is shown in Eq. (1). The traditional HTM network detection uses a fixed threshold (usually acquired through experience) to determine whether the current input is abnormal, that is, when s_t is less than a certain threshold, x_t is tagged as abnormal. However, algorithms that perform less hypothesis on the data base distribution perform better, especially for streamed data such as CDN traffic. Therefore, if using the threshold distribution simply, the HTM method performance will be greatly reduced. Because of the complexity and unpredictability of the CDN network traffic data as well as the unpredictability of the delay for HTM, if the frequency of high-delay request in the CDN network continues to increase, the direct prediction error will lead to many false positives. Therefore, using the prediction error threshold judgment method as an abnormal metric will increase the false positives. The paper adopts a rolling normal distribution model, using the Gaussian tail probability (Q function [16]) as a threshold to dynamically judge whether the input is abnormal. It should be noted that the Gaussian distribution is used to simulate the distribution of prediction errors and is not based on the underlying metric data, so it is a nonparametric technique for data. In this model, W is the last window with W errors, μ_t and σ_t^2 are the mean and the square error, they are continuously updated according to the sliding of the window to calculate the short-term average prediction error:

$$\mu_t = \frac{\sum_i^{i=W-1} s_{t-i}}{W} \tag{4}$$

$$\sigma_t^2 = \frac{\sum_i^{i=W-1} (s_{t-i} - \mu_t)^2}{W-1} \tag{5}$$

The anomaly likelihood is expressed as:

$$L_t = 1 - Q\left(\frac{\widetilde{u}_t - u_t}{\sigma_t}\right) \tag{6}$$

$$\text{where } \widetilde{u}_t = \frac{\sum_{i=0}^{i=W'-1} s_{t-i}}{W'} \tag{7}$$

W' here is a short-term moving average window, where $W' \ll W$, used to calculate the duration of the prediction error distribution. The threshold of the anomaly likelihood L_t is determined according to the user-defined parameter ϵ:

$$L_t \geq 1 - \epsilon \tag{8}$$

The L_t output here is the abnormal likelihood judgment of the time series corresponding to the CDN traffic data by the HTMTAD model.

4   Experiments and Assessment

4.1   Experimental Environment Configuration

Operating platform: Win10; CPU: Intel i5-7500 3.40 GHz; Memory: 8G; Solid State Drive: 256G.

Development language: Python.

4.2   Experimental Results and Analysis

This paper uses the NAB benchmark to compare the experimental results of HTMTAD with other two real-time monitoring algorithms. The algorithms evaluated included HTMTAD, Twitter ADVec, and Bayesian Online Change Point Detection Change-point. In this paper, the optimal parameter adjustment is performed for each algorithm, and the same data set is used for abnormal judgment comparison.

The main selected data sets are: (1) Amazon's total 4033 traffic time series data from February 14, 2014 to February 28, 2014; (2) Twitter's total 15,843 traffic time series from February 26, 2015 to April 22, 2015.

The results of the anomaly detection by the three algorithms are shown in Fig. 3. Different symbols represent the anomaly flags detected by the different detected algorithms.

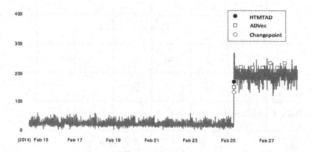

(a)   The results of Amazon traffic anomaly detection with three detected algorithms.

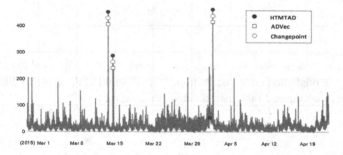

(b)   The results of Twitter traffic anomaly detection with three detected algorithms.

Fig. 3. The anomaly detection results of two datasets of NAB with three detected algorithms.

Figure 3 shows the detection results of two NAB datasets with three detected algorithms, and these were plotted in both cases. Figure 3(a) shows the results of Amazon traffic anomaly detection in late February, 2014. Figure 3(b) shows the results of Twitter traffic anomaly detection from March to April with three algorithms. The abscissa of both graphs represents time, and the ordinate represents the traffic value at the corresponding time point.

In Fig. 3(a), the traffic value of Amazon from February 15 to February 25 is floating up and down in the 0–100 range, and the floating range is small, indicating that no abnormal traffic has occurred. On February 25, there was a sudden increase in traffic. HTMTAD, Changepoint, and ADVec all detected anomalies. After February 25, the traffic rate after the change is fluctuated between 150 and 250, rather than suddenly change, which is non-abnormal traffic. But the ADVec algorithm does not detect continuous changes in traffic, and mistakenly detects traffic that occurs after February 25 as anomalies, resulting in abnormal traffic false alarms. Both the HTMTAD algorithm and the Changepoint algorithm correctly detect continuous changes in traffic. Figure 3(a) shows that HTMTAD can handle conceptual drift [17, 18] and reduce false positives.

In Fig. 3(b), Twitter traffic fluctuates widely, and non-abnormal traffic values (March 1 - March 13, March 16 - March 29, April 5 - April 20) are basically 0–200 fluctuations. Abnormal traffic occurred around March 13, March 15, and April 1. At these three time points, the HTMTAD, Changepoint, and ADVec algorithms all detected anomalies at simple spikes. The difference is that HTMTAD can detect the impending anomaly at the third time point, and neither the Changepoint algorithm nor the ADVec algorithm can do it. This result indicates the HTMTAD's proactiveness in traffic detection, i.e. HTMTAD could predict the traffic anomaly. After detecting the abnormal traffic in advance, the CDN changes the load balancing policy according to the abnormal traffic and adjusts the server resources to improve the service quality.

The experimental results of two different NAB datasets by the three algorithms in Fig. 3 show that: (1) HTMTAD algorithm can accurately detect spike flow anomaly, and can adapt to continuous traffic fluctuations, and handle concept drift; (2) HTMTAD, an algorithm has proactiveness in traffic detection, is able to detect traffic anomalies before anomalies arriving. Based on the above two characteristics, the HTMTAD algorithm proposed in this paper enables CDN to select the appropriate load balancing strategy based on abnormal traffic and improve user service quality.

5 Conclusion

CDN can improve the overall performance and service quality of applications, and plays an important role in the process of user-oriented network services. Therefore, how to quickly detect traffic anomaly in the CDN is a problem worthy of studying at present. In this paper, a new method to detect CDN traffic anomaly, the improved HTM network model HTMTAD, automatically detects possible anomalies in the CDN and compares it with several other popular detected algorithms. Experiments show that: (1) HTMTAD model can detect continuous flow changes in CDN and reduce abnormal false positives; (2) HTMTAD model is predictive, enabling CDN to detect impending traffic anomalies; (3) HTMTAD model meets requirements for real-time, continuous and online detection of CDN traffic. In the future work, the SP part in the HTM network model can be improved so that the SDR obtained can more closely fit the characteristics of the CDN traffic data, further improve the accuracy of the CDN traffic anomaly detection, and can also be performed to adjust dynamically the CDN load balancing policy based on the detected CDN traffic anomalies, improve the service quality of CDN.

Acknowledgements. The authors would like to thank the anonymous reviewers for their valuable comments and suggestions. This work is supported in part by the National Natural Science Foundation of China under Grant 61170035, 61272420 and 81674099, Six talent peaks project in Jiangsu Province (Grant No. 2014 WLW-004), the Fundamental Research Funds for the Central Universities (Grant No. 30916011328, 30918015103), Nanjing Science and Technology Development Plan Project (Grant No. 201805036), the Open Fund Project for Improve government `governance capacity Big Data Applied Technology National Engineering Laboratory 2017–2018.

References

1. Etsy Skyline GitHub: Online Code Repos. https://github.com/etsy/skyline
2. Bernieri, A., Betta, G., Liguori, C.: On-line fault detection and diagnosis obtained by implementing neural algorithms on a digital signal processor. IEEE Trans. Instrum. Measur. **45**(5), 894–899 (1996)
3. Basseville, M., Nikiforov, I.V.: Detection of abrupt changes. Change **2**, 729–730 (1993)
4. Angelov, P.: Anomaly detection based on eccentricity analysis. In: IEEE Symposium on Evolving and Autonomous Learning Systems, EALS, pp. 1–8 (2014)
5. Costa, B.S.J., Bezerra, C.G., Guedes, L.A., Angelov, P.P.: Online fault detection based on typicality and eccentricity data analytics. In: International Joint Conference on Neural Networks, pp. 1–6. IEEE (2015)
6. Chandola, V., Banerjee, A., Kumar, V.: Anomaly detection: a survey. ACM Comput. Surv. **41**(3), 1–58 (2009)
7. Netflix Surus GitHub: Online Code Repos. https://github.com/Netflix/Surus
8. Laptev, N., Amizadeh, S., Flint, I.: Generic and scalable framework for automated time-series anomaly detection. In: The 21st ACM SIGKDD International Conference on Knowledge Discovery and Data Mining, pp. 1939–1947. ACM, August 2015
9. Hawkins, J., Ahmad, S.: Why neurons have thousands of synapses, a theory of sequence memory in neocortex. Frontiers Neural Circ. **10**, 23 (2016)
10. Ahmad, S., Hawkins, J.: Properties of Sparse Distributed Representations and Their Application to Hierarchical Temporal Memory (2015). arXiv preprint: arXiv:1503.07469
11. Spruston, N.: Pyramidal neurons: dendritic structure and synaptic integration. Nat. Rev. Neurosci. **9**(3), 206 (2008)
12. Poirazi, P., Brannon, T., Mel, B.W.: Pyramidal neuron as two-layer neural network. Neuron **37**(6), 989–999 (2003)
13. Polsky, A., Mel, B.W., Schiller, J.: Computational subunits in thin dendrites of pyramidal cells. Nat. Neurosci. **7**(6), 621 (2004)
14. Hawkins, J.: Biological and Machine Intelligence. release 0.4 (2016). http://numenta.com/biological-and-machine-intelligence
15. Purdy, S.: Encoding Data for HTM Systems (2016). arXiv preprint: arXiv:1602.05925
16. Karagiannidis, G.K., Lioumpas, A.S.: An improved approximation for the Gaussian Q-function. IEEE Commun. Lett. **11**(8), 644–646 (2007)
17. Gama, J., Žliobaitė, I., Bifet, A., Pechenizkiy, M., Bouchachia, A.: A survey on concept drift adaptation. ACM Comput. Surv. (CSUR) **46**(4), 44 (2014)
18. Pratama, M., Lu, J., Lughofer, E., Zhang, G., Anavatti, S.: Scaffolding Type-2 classifier for incremental learning under concept drifts. Neurocomputing **191**, 304–329 (2016)

A Deep Learning Based Multi-task Ensemble Model for Intent Detection and Slot Filling in Spoken Language Understanding

Mauajama Firdaus[(✉)], Shobhit Bhatnagar, Asif Ekbal,
and Pushpak Bhattacharyya

Department of Computer Science and Engineering,
Indian Institute of Technology Patna, Patna, India
{mauajama.pcs16,shobhit.ee14,asif,pb}@iitp.ac.in

Abstract. An important component of every dialog system is understanding the language popularly known as Spoken Language Understanding (SLU). Intent detection (ID) and slot filling (SF) are the two very important and inter-related tasks of SLU. In this paper, we propose a deep learning based multi-task ensemble model that can perform both intent detection and slot filling tasks together. We use a deep bidirectional recurrent neural network (RNN) with long short term memory (LSTM) and gated recurrent unit (GRU) as the base-level classifiers. A multi-layer perceptron (MLP) framework is used to combine the outputs together. A combined word embedding representation is used to train the model obtained from both Glove and word2vec. This is further augmented with the syntactic Part-of-Speech (PoS) information. On the benchmark ATIS dataset, our experiments show that the proposed ensemble multi-task model (MTM) achieves better results than the individual models and the existing state-of-the-art systems. Experiments on the another dataset, TRAINS also proves that the proposed multi-task ensemble model is more effective compared to the individual models.

Keywords: Intent detection · Slot filling · Deep learning
Ensemble · Multi-task

1 Introduction

Spoken Language Understanding (SLU) is an important module in a dialog management system. It aims at identifying the semantic factors/components in a user utterance expressed in natural language and takes actions accordingly to fulfill the user's request. While understanding natural language is still considered to be a complex problem, a number of practical task-dependent human/machine dialog systems have been developed for the limited domains. The primary tasks

L. Cheng et al. (Eds.): ICONIP 2018, LNCS 11304, pp. 647–658, 2018.
https://doi.org/10.1007/978-3-030-04212-7_57

in such goal-oriented systems are intent detection and slot filling that capture the semantic information of the user utterances. The main objective of the system is to detect the user's intentions automatically as expressed in natural language and extract necessary information as slots to accomplish a goal. According to the information extracted, the system can then decide on the appropriate actions to be taken, so as to help the users achieve their demands. In our everyday lives, the applications of SLU are becoming exceedingly significant in every aspect. SLU technologies are used to built personal assistants in numerous devices such as smart phones.

1.1   Problem Definition

In this paper, we solve two very important problems of SLU, *viz.* intent detection (ID) and slot filling (SF). Intent detection is basically considered as a semantic utterance classification (SUC)problem. The objective is to identify a given user sentence x, consisting of words in a sequence $x = (x_1, x_2, \ldots, x_T)$ into one of the N pre-designated set of intent labels, y_i, based upon the components of the user sentence such that:

$$y_i = \operatorname*{argmax}_{i \in \mathbf{N}} P(y_i/x) \tag{1}$$

Slot filling refers to the extraction of semantic constituents from an input text, and to fill in the values for a predefined set of slots in a semantic frame. The slot filling task is considered as assigning semantic labels to every word in the utterance. Given a sentence x comprising of a sequence of words $x = (x_1, x_2, \ldots, x_T)$, the objective of a slot filling task is to search a sequence of semantic labels $s = (s_1, s_2, \ldots, s_T)$, for every word in the sentence, such that:

$$\hat{s} = \operatorname*{argmax}_{s} P(s/x) \tag{2}$$

For example, *Find flights to Baltimore from Dallas tomorrow*, is shown in Table 1 with the intent and slots labeled. Intent detection is basically a SUC problem while slot filling is treated as a sequence labeling task. Once the intents and slots are identified, the system can then decide upon the next proper action to be taken in order to achieve the goal of the user

Table 1. An example of ATIS dataset

Sentence	Find	flights	to	Baltimore	from	Dallas	tomorrow
Slots	O	O	O	B-toloc	O	B-fromloc	B-date
Intent	Flight						

1.2 Motivation and Contributions

In the literature, there exists quite a good number of works related to intent detection and slot filling, but still there is enough scope for future research in terms of making these models domain as well as task invariant as much as possible. The problem is more difficult when the system has to handle more realistic, natural utterances expressed in natural language, by a number of speakers. Then the tasks become more complicated than simple information requests on flights. Irrespective of the approach being adopted, the biggest obstacle is the "naturalness" of the spoken language input. In most of the existing works ID and SF have been carried out in isolation.

In this paper, we propose an ensemble based multi-task model for both ID and SF. Information of one task can provide useful evidence for the other. For example, if the intent is to find a flight in a sentence, the most likely slots will be arrival and departure cities or vice-versa. Based on this observation, we propose a multi-task model to solve both the problems concurrently. Another motivation for employing a multi-task model is that using a single architecture both the essential elements of SLU, i.e. intent detection and slot filling can be predicted at once providing an end-to-end system.

We employ an ensemble model with our observation that, however deep learning architectures such as Gated Recurrent Unit (GRU) and Long Short Term Memory (LSTM) have been implemented often but separately on varied datasets. Any systematic attempt has not been conducted for developing a model that could utilize the advantages of both these architectures. The different characteristics of both these models and assist in achieving better performance for any target task while these are merged together effectively. Experimental results show that the ensemble based multi-task model performs superior compared to the single task model. Moreover, our model outperforms the state-of-the-art methods in detecting the intents and filling the slots on the ATIS dataset and performs fairly well on the TRAINS dataset.
We summarise the contributions of our proposed work as follows:

- We propose an ensemble based multi-task model for intent detection and slot filling employing different RNN architectures such as LSTM and GRU that incorporate the functionalities of both of these deep learning architectures.
- We create a benchmark corpus for the SLU tasks, i.e. intent detection and slot filling on TRAINS dataset for capturing more realistic and natural utterances spoken by the speakers in a human/machine dialog system.

The remainder of this paper is organized as follows: In Sect. 2, we present a very brief survey of the related works. The proposed technique is described in Sect. 3. The data used and the experimental setup are outlined in Sect. 4. The results and error analysis are reported in Sect. 5. Finally, the concluding observations and the directions for future work are presented in Sect. 6.

2   Related Work

Being a significant component in spoken dialogue systems, spoken language understanding (SLU) module captures the semantical sense transmitted by speech signals. The primary units in SLU module mainly deal with classifying the user's intent and extraction of semantic components expressed in natural language. Both the problems are generally referred to as intent detection and slot filling, respectively. Historically, SLU research originated from the call classification systems [1] and the ATIS project [2].

Literature shows that there exists a significant number of works related to intent detection and slot filling. Traditional machine learning techniques have been applied in the past such as Adaboost [3,4] and Support Vector Machines (SVMs) [5] for detecting the intents of an user utterance. The authors in [6] also presented an approach for intent classification by considering the heterogeneous features, present in an user utterances. For identifying the intents, the authors in [7] made the performance of the model better by enriching the word embeddings. A multi-scale RNN structure was proposed in [8] for effective learning under low-resource SLU task for intent detection. For sequence labeling, in contrast to left-to-right, factorized probabilistic models such as Maximum Entropy Markov Model (MEMM) [9], Conditional Random Field (CRF) [10] directly captures the global distribution and overcomes the label bias problem faced by the locally normalized models.

A promising direction towards solving these problems is deep learning, which combines both feature design and classification into the learning process. Compared to the basic neural bag of words (NBoW) methods, the structural patterns in an utterance/sentence can be captured in a better way using deep learning architectures. For detecting the intents of an utterance, numerous deep learning techniques have been continuously explored such as convolutional neural network (CNN) [11], Long Short Term Memory [12] etc. In [13], a multi-layer neural network (NN) framework was explored to show gain in performance for intent detection. Lexical knowledge of the user utterance was used as a feature along with various neural network architectures has been investigated in [14] for a comparative study on intent detection. Recently, an ensemble based model for intent detection has also been proposed [15] for detecting the intents present in an user utterance. In the recent past, various deep learning based models have been explored for slot filling. A deep belief network (DBN) has been used for slot filling [16]. RNNs have also shown great performance on the slot filling task [17]. The authors in [18] used transition features to improve RNNs.

Lately, deep learning techniques have been explored for joint learning of both intent detection and slot filling for a given utterance. The authors in [19] employed triangular CRF, that implemented a random variable in addition to the standard CRF for detecting the intents of an utterance along with slot filling. Lately, identification of intents has been modeled together with slot filling employing various deep learning architectures. Various RNN models using LSTM or GRU as it's basic cell have been employed [20–23] for detecting the intents and slots together. Several architectures on deep learning, such as CNN [24] and

Recursive Neural Networks [25], have been employed to simultaneously identify the intents and slot of a given utterance.

In our paper, we propose a multi-task ensemble model for performing both slot filling and intent detection simultaneously. To the extent of our knowledge, this is the very first attempt employing an ensemble approach along with a combined representation of word embeddings for modeling both intent detection and slot filling together.

3   Methodology

In our proposed ensemble approach, the input sentence can be viewed as a sequence. To use and employ the functionalities of both GRU and LSTM, we have used them to learn the sequence representation at each time step. These predictions are used to predict the slot labels while the global representation of the sequence is used for identifying the intents. Therefore, the representations of the sentence is learned by GRU and LSTM are then shared by both the tasks of SLU and with the help of a combined loss function, both the SLU tasks can interact with one another through this shared representation.

The architecture of our multi-task model is depicted in Fig. 1. In our model, the input is an utterance x, which is a concatenation of words, $x = (x_1, x_2, \ldots, x_n)$ where n is the length of the utterance. The words are first converted into embeddings $e(w_T)$, and then the deep learning architectures use these embeddings as input. The model gives two types of outputs, i.e. slot label sequence y_s and the intent label y_i.

3.1   Embeddings

To capture the hidden semantic structures of a words in a sentence, word embeddings are very effective. In contrast to the traditional representation, such as one-hot representations of words the input to the neural network models are real-valued vectors known as word embeddings. One-hot word vectors or bag-of-words if given as input to the model leads to feature vectors of exceedingly huge dimensions. As a substitute, the large sparse vectors of words are projected into dense vectors of low dimension by the word embeddings. Word embeddings are mostly trained through an unsupervised fashion on an enormous corpus and then the fine-tuning of the embeddings are done in the supervised training procedure. For word embeddings, two pre-trained embedding models are used such as GloVe[1] and Word2Vec[2]. The word embeddings are used separately as inputs to the deep learning models. To obtain superior performance, both the word embeddings are concatenated and then it is fed as input to the different deep learning models. The combined word embeddings obtained from word2vec and Glove, helps in capturing the meanings of the words completely and thereby it is observed that our model performs better with this representation of words.

---

[1] http://nlp.stanford.edu/projects/glove/.
[2] https://code.google.com/archive/p/word2vec/.

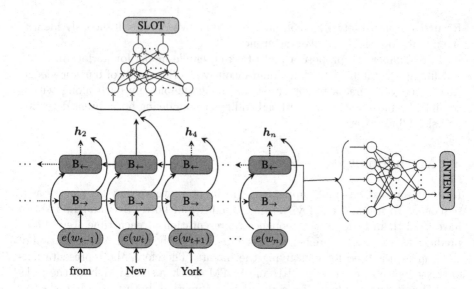

Fig. 1. Multi-Task Model Architecture for Intent Detection and Slot Filling; B→: LSTM/GRU forward pass cells; B←: LSTM/GRU backward pass cells; e(w_t): combined word embeddings.

3.2 Proposed Approach

In order to build a multi-task model shown in Fig. 1, that can detect both the intent of an utterance and simultaneously fill the slots we use an ensemble based deep learning approach. The ensemble incorporates the outputs of Bi-LSTM and Bi-GRU based models. An ensemble model is constructed with an assumption that when two neural network models with different functionalities are combined together they can improve the efficiency further.

We depict the proposed architecture of the multi-task ensemble model in Fig. 2. The hidden state of our proposed approach can be computed as:

$$\overleftrightarrow{h}_t = [\overrightarrow{LSTM}(x_t, h_{t-1}), \overleftarrow{GRU}(x_t, h_{t-1})] \tag{3}$$

The output of the Bi-LSTM/Bi-GRU along with the syntactic features is fed to a multi-layer perceptron (MLP) network. The output of the MLPs is ensembled. The last part of our model is the output layer. The softmax function with linear transformation is applied to the representation to generate the slot labels y_s and the intent labels y_i simultaneously for a given sentence. Formally,

$$y_s = softmax(W_i \overleftrightarrow{h}_t + b_s) \tag{4}$$

$$y_i = softmax(W_i \overleftrightarrow{h}_t + b_i) \tag{5}$$

where W_s and W_i are the transformation matrices for slot filling and intent detection, respectively, b_s and b_i are the bias vectors.

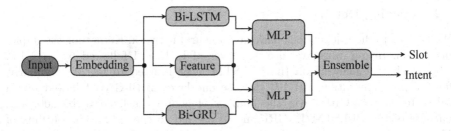

Fig. 2. Block Diagram of Our Proposed Approach

3.3   Augmenting Syntactic Features

To capture syntactic information about the utterance we use Part-of-Speech (PoS) as an additional feature in our model. Along with the representations learned by the deep learning architecture, PoS helps in correctly identifying the intents and slots of a given utterance. For PoS tagging, we use NLTK[3].

4   Datasets and Experiments

4.1   Datasets

We assess our proposed ensemble model by conducting experiments on two datasets. The first dataset is the extensively used ATIS corpus, and the other dataset is the TRAINS [26]. The TRAINS corpus consists of dialogue conversations, and we have manually annotated this with intents and slots.

ATIS Dataset: The ATIS (Airline Travel information System) corpus is an important by-product of Defence Advanced Research Program Agency (DARPA) project. For the SLU tasks, one of the most extensively used dataset is the ATIS corpus [2]. ATIS corpus have a few variants but in this paper, we use the dataset used in [10]. The utterances in ATIS corpus mainly are of flight reservations made by people. There are 17 distinct intent classes and 127 distinct slots in the dataset. There are 4978 utterances in the train set and 893 utterances in the test set.

TRAINS Dataset: Although for SLU, there are datasets, e.g. Cortana Data [25] and Bing Query Understanding Dataset [18], but they are not publicly available. For building a robust spoken dialogue system, it is essential to capture the intent and slots present in a human conversation. To be able to find the intent and slots of a real and natural utterance in a conversation, we manually annotate the TRAINS corpus. The corpus is a collection of problem solving dialogues. Three annotators of post-graduate level qualification were assigned to annotate this corpus with intent and slot. The inter-annotator score was found to be approximately 80%, and hence we can assume this to be reliable. The dataset has been annotated with 12 intents and 32 slots. There are 5355 utterances in the training set and 1336 utterances in the test set.

---

[3] https://www.nltk.org.

4.2   Training Details

We use the python dependent package on neural network, Keras[4] for the implementation. In our work, we use one layer of Bi-LSTM followed by an MLP. The number of neurons on the Bi-LSTM layer is fixed to be 200. Similarly, for the implementation of GRU, we use one layer of Bi-GRU followed by an MLP. In each Bi-GRU layer the number of neurons is fixed to be 200. The outputs of the Bi-LSTM/Bi-GRU model are fed to the MLP. The outputs of MLPs are ensembled to predict intents and slots together. The model uses a 600-dimensional combined word embedding formed using GloVe and Word2Vec. In our model, the intermediate layers uses ReLU activation function while the output layer uses softmax activation function. To avoid over-fitting of the neural network a very effective regularization technique known as dropout has been used in our model. At the time of forward propagation, the neurons are randomly tuned-off so that the convergence of weights is restricted to identical positions. For optimization and regularization, we use Adam optimizer along with 15% dropout in our model. Categorical cross-entropy is employed to update the model parameters.

5   Results and Discussion

In this section, the experimental results along with necessary error analysis have been reported. We also provide a comparison of our multi-task ensemble model with the base classifiers. Moreover, we provide a comparison of our model against the state-of-the-art approaches. We report accuracy as the performance measure for intent detection and slot filling tasks.

5.1   Results

Evaluation results for the ensemble based multi-task model for intent detection and slot filling in the form of accuracy are reported in Table 2. From the table, we can easily infer that our proposed ensemble based multi-task model for slot filling and intent detection perform superior than the other models. The multi-task model using Bi-LSTM and Bi-GRU with combined word embeddings as input performs comparably. Evaluation also shows that adding PoS information has been effective for both the problems.

Multi-task Model vs Individual Models: We use an ensemble based multi-task model for both slot filling and intent detection. We do a comparative study of the performance of this model with the individual models, when they are built in isolation, i.e. the tasks of intent classification and slot filling are not performed simultaneously. Results of the multi-task and individual models are demonstrated in Table 3. The complexity of the multi-task model is self-evident, because only one end-to-end model is needed for training and testing. The proposed model performs superior than the individual models for both the tasks.

---

[4] www.keras.io.

The correlation of the two tasks is learned by our multi-task model. This shared information provides useful evidence for both the tasks. Experimental results prove that the multi-task model performs better in all the settings.

Table 2. Accuracy results of multi-task model for intent detection and slot filling

Model	ATIS		TRAINS	
	Intent	Slot	Intent	Slot
Bi-directional LSTM	97.66	97.72	81.55	95.43
Bi-directional GRU	97.64	97.75	81.13	95.11
Bi-LSTM + Feature	97.80	97.81	81.72	95.87
Bi-GRU + Feature	97.76	97.93	81.45	95.45
Proposed Ensemble Model	98.12	98.02	82.28	96.12
Proposed Ensemble Model + Feature	**98.43**	**98.07**	**83.11**	**96.89**

Table 3. Multi-task vs individual model

Model	ATIS		TRAINS	
	Intent	Slot	Intent	Slot
Bi-LSTM (ID)	97.12	-	80.87	-
Bi-LSTM(SF)	-	97.15	-	94.34
Bi-LSTM (MTM)	97.66	97.72	81.55	95.43
Bi-GRU (ID)	97.24	-	80.74	-
Bi-GRU (SF)	-	97.23	-	94.18
Bi-GRU (MTM)	97.64	97.75	81.13	95.11
Ensemble (ID)	98.02	-	81.97	-
Ensemble (SF)	-	97.82	-	95.67
Ensemble (MTM)	98.43	98.07	83.11	96.89

Table 4. Comparison with previous approaches

Model	ATIS	
	Intent	Slot
SVM [Raymond et al.,2007]	-	89.67
CRF [Mesnil et al. 2015]	-	95.16
RNN [Mesnil et al. 2015]	-	96.24
R-CRF [Yao et al. 2014]	-	96.46
Boosting [Tur et al. 2010]	95.50	-
RecNN [Guo et al. 2014]	95.40	93.96
Bi-GRU [Zhang et al. 2016]	98.32	96.89
Proposed Ensemble Model	**98.43**	**98.02**

Comparison to State-of-the-Art Approaches: We do a comparative study of our best performing multi-task model against the previous approaches on both intent detection and slot filling tasks as discussed in the related work section. Table 4 shows the comparison, and it can be seen that our proposed ensemble multi-task model performs superior than the existing state-of-the-art approaches. In the previous works of [10,27] the authors used machine learning techniques such as SVM, boosting and CRF for identifying the intents and slots. Hence, our deep learning models outperforms these approaches in identifying the intents and slots. Also, the approaches of [17,23,25] uses deep learning techniques such as RNN with LSTM or GRU as basic cells for identifying the intents and slots. Our multi-task model outperforms these approaches as in our work we employ both LSTM and GRU together along with combined word embeddings for identifying the intents and slots together. By using both word2vec and Glove embeddings are model is able to capture an enormous amount of semantic information of the words present in the corpus. Also, by combining both LSTM and GRU in our work we are able to utilize the functionalities of both these architectures.

5.2 Error Analysis

A detailed analysis of errors is conducted of our best performing multi-task ensemble model to get a clear picture of the failures/errors of our system. By doing further analysis of the results it was revealed that the intent errors are because of the prepositional phrases being embedded in the noun phrases. In ATIS dataset for example, the phrase *Airfare of the flight to Denver from New York*, where in the utterance the prepositional phrase suggests the intent to be *flight* while in this case the intent class is basically determined by the noun phrase as it is the head word of the utterance (in this case it is *airfare*). As a result of some incorrect annotations encountered in the ATIS corpus we received few intent errors. Certain sentences in the ATIS corpus are ill-formulated and also ambiguous, such as *What's the airfare for a limousine service to the San Francisco airport?*. In this sentence, the intent label is implied to be *Airfare* because of the head word *airfare*, while the correct intent label should be *Ground Service*. The ATIS dataset have been studied for more than a decade, so the absolute improvement may not be very high as the accuracy of the state-of-the-art approach is very high already. Performance of our slot filling task is better as compared to the previous existing models, thanks to the powerful ability of both LSTM and GRU for modeling the sequences.

The relatively low accuracy in intent classification in the TRAINS dataset is mainly due to the long and ill-formulated sentences. The intents of many sentences are towards the end of the sentence and are not clearly stated. Performance of slot filling task on the TRAINS dataset is high. This may be due to the number of less slots, and also because the slots have relatively easy patterns, which are learned by the model very well.

6 Conclusion and Future Work

In this paper, we have proposed an ensemble based multi-task approach for slot filling and intent detection, which are the primary two tasks in SLU. For the ensemble multi-task model, we have used Bi-directional LSTM and Bi-directional GRU as the base learning model. The representations learned from these two models are shared by both intent and slot filling task. In our proposed approach, we have ensembled the Bi-directional LSTM and Bi-directional GRU to capture the functionalities of both these deep learning architectures. Experiments were conducted on two datasets. The ensemble multi-task model exhibits advantages over individual models and outperforms the state-of-the-art approaches on slot filling and intent detection tasks on the ATIS dataset. By using syntactic information and combined word embeddings, it further helps our model to identify the intents and slots correctly.

In our future work, we aim to incorporate semantic information as well. Furthermore, for the application of deep learning methods, our TRAINS dataset is still a small scale dataset. We would like to expand the scale of our dataset and use more dialogue data, which can be useful for building robust dialogue systems and help in SLU research.

Acknowledgment. Asif Ekbal acknowledges Young Faculty Research Fellowship (YFRF), funded by Visvesvaraya PhD scheme for Electronics and IT, Ministry of Electronics and Information Technology (MeitY), Government of India, being executed by Digital India Corporation (formerly Media Lab Asia).

References

1. Gorin, A.L., Riccardi, G., Wright, J.H.: How may i help you? Speech Commun. **23**(1–2), 113–127 (1997)
2. Price, P.J.: Evaluation of spoken language systems: the ATIS domain. In: Speech and Natural Language: Proceedings of a Workshop Held at Hidden Valley, Pennsylvania, 24–27 June 1990 (1990)
3. Tur, G.: Model adaptation for spoken language understanding. In: IEEE International Conference on Acoustics, Speech, and Signal Processing, Proceedings (ICASSP 2005), vol. 1, pp. I-41. IEEE (2005)
4. Tur, G., Hakkani-Tur, D., Heck, L.: What is Left to be Understood in ATIS? In: 2010 IEEE Spoken Language Technology Workshop (SLT), pp. 19–24. IEEE (2010)
5. Haffner, P., Tur, G., Wright, J.H.: Optimizing SVMs for complex call classification. In: 2003 IEEE International Conference on Acoustics, Speech, and Signal Processing, Proceedings (ICASSP 2003), vol. 1, p. I. IEEE (2003)
6. Hakkani-Tur, D., Tur, G., Chotimongkol, A.: Using syntactic and semantic graphs for call classification. In: Proceedings of the ACL Workshop on Feature Engineering for Machine Learning in Natural Language Processing (2005)
7. Kim, J.K., Tur, G., Celikyilmaz, A., Cao, B., Wang, Y.Y.: intent detection using semantically enriched word embeddings. In: 2016 IEEE Spoken Language Technology Workshop (SLT), pp. 414–419. IEEE (2016)
8. Luan, Y., Watanabe, S., Harsham, B.: Efficient learning for spoken language understanding tasks with word embedding based pre-training. In: Sixteenth Annual Conference of the International Speech Communication Association (2015)
9. McCallum, A., Freitag, D., Pereira, F.C.: Maximum entropy Markov models for information extraction and segmentation. ICML **17**, 591–598 (2000)
10. Raymond, C., Riccardi, G.: Generative and discriminative algorithms for spoken language understanding. In: Eighth Annual Conference of the International Speech Communication Association (2007)
11. Collobert, R., Weston, J.: A unified architecture for natural language processing: deep neural networks with multitask learning. In: Proceedings of the 25th International Conference on Machine Learning, pp. 160–167. ACM (2008)
12. Ravuri, S.V., Stolcke, A.: Recurrent neural network and LSTM models for lexical utterance classification. In: INTERSPEECH, pp. 135–139 (2015)
13. Sarikaya, R., Hinton, G.E., Ramabhadran, B.: Deep belief nets for natural language call routing. In: 2011 IEEE International Conference on Acoustics, Speech and Signal Processing (ICASSP), pp. 5680–5683. IEEE (2011)

14. Ravuri, S., Stoicke, A.: A comparative study of neural network models for lexical intent classification. In: 2015 IEEE Workshop on Automatic Speech Recognition and Understanding (ASRU), pp. 368–374. IEEE (2015)
15. Firdaus, M., Bhatnagar, S., Ekbal, A., Bhattacharyya, P.: Intent detection for spoken language understanding using a deep ensemble model. In: Geng, X., Kang, B.-H. (eds.) PRICAI 2018. LNCS (LNAI), vol. 11012, pp. 629–642. Springer, Cham (2018). https://doi.org/10.1007/978-3-319-97304-3_48
16. Deoras, A., Sarikaya, R.: Deep belief network based semantic taggers for spoken language understanding. In: Interspeech, pp. 2713–2717 (2013)
17. Mesnil, G., He, X., Deng, L., Bengio, Y.: Investigation of recurrent neural network architectures and learning methods for spoken language understanding. In: Interspeech, pp. 3771–3775 (2013)
18. Yao, K., Peng, B., Zweig, G., Yu, D., Li, X., Gao, F.: Recurrent conditional random field for language understanding. In: 2014 IEEE International Conference on Acoustics, Speech and Signal Processing (ICASSP), pp. 4077–4081. IEEE (2014)
19. Jeong, M., Lee, G.G.: Triangular-chain conditional random fields. IEEE Trans. Audio Speech Lang. Process. **16**(7), 1287–1302 (2008)
20. Liu, B., Lane, I.: Joint online spoken language understanding and language modeling with recurrent neural networks. In: 17th Annual Meeting of the Special Interest Group on Discourse and Dialogue, p. 22 (2016)
21. Liu, B., Lane, I.: Attention-based recurrent neural network models for joint intent detection and slot filling. In: INTERSPEECH, pp. 685–689 (2016)
22. Hakkani-Tur, D., et al.: Multi-domain joint semantic frame parsing using bi-directional RNN-LSTM. In: INTERSPEECH, pp. 715–719 (2016)
23. Zhang, X., Wang, H.: A joint model of intent determination and slot filling for spoken language understanding. In: IJCAI, pp. 2993–2999 (2016)
24. Xu, P., Sarikaya, R.: Convolutional neural network based triangular CRF for joint intent detection and slot filling. In: 2013 IEEE Workshop on Automatic Speech Recognition and Understanding (ASRU), pp. 78–83. IEEE (2013)
25. Guo, D., Tur, G., Yih, W.t., Zweig, G.: Joint semantic utterance classification and slot filling with recursive neural networks. In: 2014 IEEE Spoken Language Technology Workshop (SLT), pp. 554–559. IEEE (2014)
26. Heeman, P.A., Allen, J.F.: The Trains 93 Dialogues. Technical report, Rochester University NYDept of Computer Science (1995)
27. Tur, G., Hakkani-Tur, D., Heck, L., Parthasarathy, S.: Sentence simplification for spoken language understanding. In: 2011 IEEE International Conference on Acoustics, Speech and Signal Processing (ICASSP), pp. 5628–5631. IEEE (2011)

An Image-Based Approach for Defect Detection on Decorative Sheets

Boyu Zhou, Xin He, Zhongyi Zhou, and Xinyi Le[✉]

Shanghai Key Laboratory of Advanced Manufacturing Environment,
School of Mechanical Engineering, Shanghai Jiao Tong University, Shanghai, China
lexinyi@sjtu.edu.cn

Abstract. In this paper, we propose a novel image-based approach for defect detection on decorative sheets. First, an image-based data augmentation approach is applied to deal with imbalanced image sets and severely rare defeat images. Two deep convolutional neural networks (CNNs) are then trained on augmented image sets using feature-extraction-based transfer learning techniques. Finally two CNNs are combined to classify defects through a multi-model ensemble framework, aiming to reduce the false negative rate (FNR) as much as possible. Extensive experiments on augmented artificial images and realistic defeat images both achieve surprisingly FNR accuracy results, which substantiate the proposed approach is promising for defect detection on decorative sheets.

Keywords: Data augmentation · Convolutional neural network
Transfer learning · Multi-model ensemble · Defect detection

1 Introduction

1.1 Motivation

Nowadays, decorative sheets are widely used for indoor decoration in places such as stadiums, gymnasiums, hotels, and residential buildings. During the fabrication of decorative sheets, quality control is a key issue. As a result, a production line for decorative sheets usually hire several skilled inspectors to defect detects on various kinds of sheets. However, quality inspection by human beings is exhausting, error-prone, inefficient, and expensive.

Image-based approach is one of the possible solutions to defect detection. Early methods tend to extract some specific hand-crafted image features according to defect condition [7,9,15]. These methods heavily rely on the choices of hand-crafted features, which are sensitive to various real-world conditions such

The work described in the paper was supported by National Natural Science Foundation of China (Grant No. 61703274), Scientific Research Project of Shanghai Science and Technology Committee (17511104603), and Shanghai Pujiang Program (17PJ1404400).

© Springer Nature Switzerland AG 2018
L. Cheng et al. (Eds.): ICONIP 2018, LNCS 11304, pp. 659–670, 2018.
https://doi.org/10.1007/978-3-030-04212-7_58

as lighting and contrasts. Besides, they are not general enough to address all types of defects at the same time [9].

Recently, convolutional neural network (CNN) has become one of the most attractive methods for its excellent performance on generic visual recognition task. It has been successfully applied to a number of defect detection scenarios and show better performance compared to earlier methods [12,16]. Despite their great improvement in detection accuracy, these works require a huge number of images for training, which is unrealistic in most circumstances. At the same time, they didn't take the training time of CNN and the false negative rate of defect detection into consideration, which are two crucial factors of practical application.

According to the requirements and circumstances from the practical production line, it is very difficult to obtain enough defect images for CNN training. In addition, we do not want to miss any defects, but we could endure some false detection. To solve the above difficulties, we dedicated ourselves to develop a novel image-based approach to meet the detection requirements on decorative sheets.

1.2 Related Work

Last decade witnessed the rapid growth of CNNs in computer vision area. LeNet [1] as one of CNNs, was first proposed in 1989 for solving computer vision tasks. With the development of the computing capability, AlexNet [8] was created in 2012 by Krizhevsky with five convolutional layers and three fully connected layers. After the breakthrough work, many researchers use deep convolutional neural network as their main structures in image classification and object detection. Several CNNs were investigated to increase CNNs' performance. VGG [11] deepened the network and proved that the prediction accuracy can be increased through making the network deeper. Inception [14] paralleled the layers to increase its accuracy. In ResNet [5], the network is able to learn the residue information, thus eliminating information loss in very deep network.

Thanks to the development of CNNs, deep CNNs perform extremely well in various visual recognition tasks. However, when trained by limited training data, CNNs will meet severe over fitting problems. Inspired by human's capability of accumulating knowledge and transfering it for learning new objects, researchers are convinced that prior knowledge of a model can also be utilized for solving a new task even in a new domain, which is known as transfer learning.

In [2], the powerful generic visual features trained on a large visual recognition dataset can be transfer to new tasks. The transfer property still holds even if the new tasks differ significantly from the original trained tasks. In [10], the transfer learning is further substantiated by numerous experiments in wide range of recognition tasks including image classification, scene recognition, fine grained recognition, attribute detection, and image retrieval. As a result, transfer-learning based CNNs greatly reduce the required training time and the quantity of training data. leading to many new possible applications.

In spite of the successful applications in many areas, CNN based fault detection is not widely applied in production line. Although the required amount data is reduced using transfer learning technique, the fault/defeat data are still not enough for training. As a result, data augmentation is necessary.

There are several data augmentation techniques. [4] proposed several principles for solving imbalanced data problems. [3] proposed some data augmentation approaches based on hypergraph analysis. [17] used some human knowledge to generate enough data for training. As obtained defeat images are far from enough, data augmentation using prior human knowledge is effective and necessary to deal with rare defeat images in our problem.

1.3 Contributions

In this paper, we propose a novel framework to tackle the challenges of defect detection out of complex texture of decorative sheets. Our contributions are as follows:

(1) We propose an innovative image-processing-based data augmentation approach for addressing the data scarcity issue, where image drawing, filtering and structural analysis algorithms are applied to augment the decorative sheet images.
(2) We propose a defect detection method for decorative sheets based on transfer learning using pre-trained deep CNN architectures, aiming at reducing the required quantities of training data and shortening the training period.
(3) A multi-model ensemble method is proposed to further reduce the false negative rate of detection.
(4) We release our implementation of the proposed data augmentation, transfer learning and multi-model ensemble approaches as open source code and share generated defeat images by our data augmentation algorithms as open source dataset.

1.4 Organization

The remainder of this paper is organized as follows. First, problem description is developed in Sect. 2. Then, our proposed methods are presented in Sect. 3, followed by experiment results in Sect. 4. Finally we conclude our work in Sect. 5.

2 Description of Defect Detection

A decorative sheet manufacturing company usually own several production lines, and produce various kinds of decorative sheets with different colors and textures. Fig. 1 describes a traditional production line when fabricating decorative sheets.

It is realistic to add some cameras on production lines to obtain images for defeat detection. Figure 2 demonstrates various sheets.

In addition, according to the experts, there are several kinds of defeats which are very common during the fabrication process, as depicted in Fig. 3.

Fig. 1. Production lines for fabricating decorative sheets

Fig. 2. Various kinds of decorative sheets

Our goal is to decide whether a piece of decorative sheet contains any types of defects as described in Fig. 3. It is a typical image classification problem to classify the input images into "normal" or "spot", "dot", etc. We train deep convolutional neural networks as the classifiers for this purpose. In practical application, these classifiers can be combined with a sliding window technique to detect any region of interest (ROI).

3   Methodology

3.1   Image-Processing-Based Data Augmentation

Although CNN is a powerful structure for visual recognition tasks, modern models of CNNs have millions of parameters. To avoid over fitting, training for CNN requires a lot of labeled training data, which is an impossible requirement in many applications. This dilemma is especially true for our problem. The difficulty mainly comes from the facts that defective samples of decorative sheets are discovered at low frequency. In addition, these samples are not easy to collect and maintain by the manufacturers. As a result, we propose an image–based data augmentation method to generate tens of thousands of defective images within seconds.

In spite of the numerous types of decorative defects, we discover that most of the defects can be decomposed into some basic elements. Take the defects of "cut" and "dot" as examples, a typical "cut" can be regarded as the formation of irregular line segments. A "dot" is simply a small circle with specific distribution of color. The defects of "spot" are similar to "dot" except that "spot" has random and complex shapes. Using some image transforming algorithms, we are able to generate artificial defects on various kinds of decorative sheets, which are diversified but quite similar to true defects. Experiments in Sect. 4 showed that CNN models trained by these artificial data achieve equivalent performance compared to the real world data.

We developed algorithms to generate five most common defects, i.e., "cut", "dot", "spot", "fragment", and "abrasion". Figure 4 presents the detailed algorithm for generating "spot" as an example. The defect of "spot" is actually

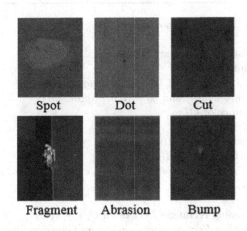

Fig. 3. Various kinds of defeats

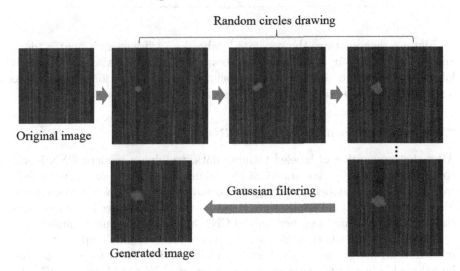

Fig. 4. Intermediate steps for generating "spot" on decorative images.

some impurities attached on decorative sheets. As shown in Algorithm 1, the first step of generating "spot" is to draw a random shape with unique color similar to the color of mud on decorative images. This is achieved by drawing several circles with different diameters whose centers normally distributed around a specific point. Since a realistic "spot" has complex color distribution, but similar contrast rate compared to the background sheet, we apply a Gaussian filter in and around the region of the generated random shape. Due to the arbitrary selected position, shape, size and color in our algorithm, generated "spot" is diversified enough to contain the features of realistic "spot".

Algorithm 1. Generation of spot defect.

Input:
   An original image of wooden floor surface, Img_{origin};
Output:
   The same image attached with spot, Img_{spot};
1: Copy Img_{origin} into Img_{spot};
2: Randomly select a central point (x_0, y_0) in Img_{spot}
   where $x_0 \sim U(0, img\_width)$ and $y_0 \sim U(0, img\_height)$;
3: Generate a random color CL;
4: **for each** $i \in [1, N]$ **do**
5:    $x_i = x_0 + \epsilon_{i,x}$ where $\epsilon_{i,x} \sim N(0, \sigma_1^2)$;
6:    $y_i = y_0 + \epsilon_{i,y}$ where $\epsilon_{i,y} \sim N(0, \sigma_1^2)$;
7:    Draw a circle C_i on Img_{spot} with radius r_i, center $P_i(x_i, y_i)$ and color CL
   where $r_i \sim U(r_{min}, r_{max})$;
8: **end for**
9: Compute the bounding rectangle $Rect$ of all $C_i \in [C_1, C_2 \ldots C_N]$;
10: Blur the region within $Rect$ of Img_{spot} using an Gaussian filter;
11: **return** Img_{spot};

Figure 5 demonstrates the generated "abrasion", "fragment", "cut", "dot", and "spot" using our algorithms. Moreover, each image is also automatically labeled as the defeats' name. Manual labeling of images is no longer needed. As a result, the proposed algorithm for generation is computationally efficient.

3.2   Transfer Learning of Deep CNNs

After the preparation of labeled training data, training a modern CNN is also computationally expensive and time consuming. Usually it takes days or even weeks for the CNN model to converge to a desired state, which is not as satisfying as expected by practical application. Hence, we employ transfer learning to save training time by using deep pre-trained CNN on large-scale image dataset.

According to [2,10], transfer learning of deep CNN mainly employs the approach of using pre-trained network for feature extraction. These features can then be followed by a generic classifier such as Supported Vector Machine (SVM) for classification. In our defect detection problem, several carefully selected CNN

Fig. 5. Generated defective images. Each column of images from right to left correspond to "abrasion", "fragment", "cut", "dot", and "spot" respectively.

models are applied in our main structure. We preserve most parts of the original pre-trained network except the last fully connected layer for extracting features from the decorative sheets. The last fully connected layer of the original network is then modified in accordance with our desired outputs. Through the prediction of the models, each image is labeled as normal surface, spot, dot, cut, abrasion or fragment. As a result, the last fully connected layer is adjusted to have six neurons, one for the label of "normal" and the other five for 5 types of defects. Using our training dataset, we update the weights of the final layer. The block diagram of our proposed method is shown in Fig. 6.

With regard to the CNN architecture, we employ Inception [13] as the source architecture for solving our defect detection problem. MobileNet [6] is also selected due to its good accuracy performance on small size and low latency data.

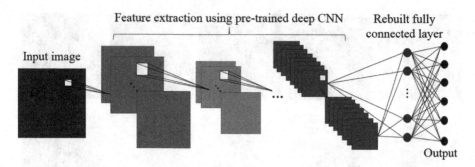

Fig. 6. Overview of the transfer learning method. The pre-trained deep CNN takes an image as input and extracts features, which is followed by a fully connected layer rebuilt for defect classification.

3.3 Multi-model Ensemble

In the decorative sheets fabrication, manufacturers are more concerned about the false negative rate (FNR) than the accuracy of defect detection. A false negative error means that images containing defeats are not detected by the detection system, which will damage the company's reputation if the problematic product is sold to consumers.

It is not so important to determine which kind of defects. In addition, it is also acceptable to classify a "normal" image in error to a defeated image. Thus it is of vital importance to avoid the occurrence of false negative error as much as possible.

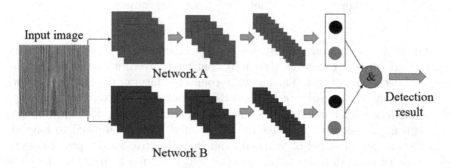

Fig. 7. Architecture of multi-model ensemble.

Similar to the decision making process of human beings, the probability of multiple persons making mistakes at the same time is much lower than that of each individual. To further reduce the FNR of defect detection, we propose a multi-model ensemble framework, whose block diagram is shown in Fig. 7. We train two different networks, i.e, Inception and MobileNet, separately using our proposed transfer learning method during the training phase. During testing, Inception and MobileNet make classification simultaneously.

Each image will obtain similarity scores to six classes. One image will be determined as "normal" only if both of two models estimate the similarities are larger than specific thresholds. In other words, if one image belongs to "normal" class, we give "True"; otherwise, "False". Two models make independent binary decisions, and the final result is the output of "AND" operation.

It should also be noted that more models can be added in our multi-model ensemble framework to increase accuracy.

4   Results

4.1   Implementation Details

All of our algorithm is implemented in Python. We use the open source computer vision library OpenCV in our proposed data augmentation algorithms. We also use Tensorflow as our deep learning framework, with which we build and train our models and make prediction using our models.

4.2   Experiments of Individual Models

Our training set is composed of 8595 RGB labeled images generated by our proposed data augmentation algorithms, 3410 for "normal", 1043 for "abrasion", 1143 for "fragment", 1000 for "cut", 999 for "dot" and 1000 for "spot", respectively. This training set is split into three subsets: 80% for training, 10% for validation and 10% for test. We train our models using the stochastic gradient descent (SGD) with batches of size 10, which runs 4000 steps in all. The learning rate is set as 0.01 throughout the whole process. The training process is launched on Intel Core i7 CPU and a Nvidia GTX 1080 Ti GPU. For each model, it takes roughly 15 min to reach convergence.

First, we use an independent test set to evaluate our models' accuracy which is composed of 2168 images, 1315 for "normal", 170 for "abrasion", 185 for "fragment", 169 for "cut", 165 for "dot" and 164 for "spot", respectively. The classification accuracy of Inception and MobileNet on the test set are 95.6% and 93.9%, respectively. Tables 1, 2, 3 and 4 give testing results of confusion matrix. In Tables 1 and 3, the types of defects are detailed. In Table 2 and 4, all types of defects are attribute to "defective". The reason for simplifying the confusion matrix as 2×2 is that it is more straightforward in terms of counting the number of false negative errors, which is one of the main concerns in our paper.

4.3   Experiments of Multi-model Ensemble

We test our proposed multi-model ensemble framework with the same test set mentioned in Sect. 4.2, where Inception and MobileNet make joint decisions. The resulted confusion matrix is shown in Table 5. The FNRs of Inception, MobileNet and multi-model ensemble are 2.10%, 3.98% and 0.47%, respectively. It is obvious that this framework make considerable FNR reduction compared with individual models.

Table 1. Confusion matrix of Inception with detailed defects type.

		True Class					
		normal	abrasion	fragment	cut	dot	spot
Predicted Class	normal	1275	5	3	0	6	4
	abrasion	11	162	0	0	0	0
	fragment	12	0	145	0	1	15
	cut	0	0	0	184	0	0
	dot	16	3	0	0	162	1
	spot	1	0	17	1	0	145

Table 2. Confusion matrix of MobileNet with detailed defects type.

		True Class					
		normal	abrasion	fragment	cut	dot	spot
Predicted Class	normal	1287	4	2	1	16	11
	abrasion	4	166	0	0	0	0
	fragment	3	0	144	3	0	47
	cut	0	0	0	181	0	0
	dot	19	0	1	0	153	2
	spot	2	0	18	0	0	105

Table 3. Abbreviated confusion matrix of Inception

		True Class	
		normal	defective
Predicted Class	normal	1275	18
	defective	40	836

Table 4. Abbreviated confusion matrix of MobileNet

		True Class	
		normal	defective
Predicted Class	normal	1287	34
	defective	28	820

Table 5. Confusion matrix of multi-model ensemble

		True Class	
		normal	defective
Predicted Class	normal	1155	4
	defective	160	850

4.4   Test on Realistic Defeat Images

To further verify the effectiveness of our proposed methods of data augmentation, transfer learning, and Multi Model Ensemble, test on realistic defeat images on decorative sheets is indispensable. Therefore, we collected some sample images of defective sheets captured by industrial camera from a defective sheets manufacturer and use them for testing. In total, we have 50 defective images containing different types of defects, some of which are shown in Fig. 8. We employed our proposed multi-model ensemble framework for defect detection. Delightfully, none of these 50 defects is missed by our models, which meet the expectation of us and manufacturers.

Fig. 8. Realistic Defeat Images

5   Conclusion and Future Work

This paper shows a real application of image-based defect detection in complex texture. The imbalanced classification images are augmented using prior human knowledge and the transfer learning is applied in CNNs to reduce training time. Finally, a multi-model ensemble approach is used to reduce false negative error. The proposed approach is proved to be successful to detect various kinds of defects out of complex background.

In the future, we will further perfect our approaches through training and testing on more defeats in more comprehensive environments, making our best efforts to meet the needs of real-world applications.

References

1. Cun, Y.L., Jackel, L.D., Boser, B.E., Denker, J.S., Graf, H.P., Guyon, I., Henderson, D., Howard, R.E., Hubbard, W.: Handwritten digit recognition: applications of neural network chips and automatic learning. IEEE Commun. Mag. **27**(11), 41–46 (1989)
2. Donahue, J., et al.: DeCAF: a deep convolutional activation feature for generic visual recognition. CoRR abs/1310.1531 (2013)
3. Gao, Y., Wang, M., Tao, D., Ji, R., Dai, Q.: 3-D object retrieval and recognition with hypergraph analysis. IEEE Trans. Image Process. **21**(9), 4290–4303 (2012)
4. He, H., Garcia, E.A.: Learning from imbalanced data. IEEE Trans. Knowl. Data Eng. **21**(9), 1263–1284 (2009)
5. He, K., Zhang, X., Ren, S., Sun, J.: Deep residual learning for image recognition. In: The IEEE Conference on Computer Vision and Pattern Recognition (CVPR), pp. 770–778 (2016)
6. Howard, A.G., et al.: MobileNets: efficient convolutional neural networks for mobile vision applications. CoRR abs/1704.04861 (2017)
7. Hu, G.H.: Automated defect detection in textured surfaces using optimal elliptical gabor filters. Optik Int. J. Light Electron Optics **126**(14), 1331–1340 (2015)
8. Krizhevsky, A., Sutskever, I., Hinton, G.E.: ImageNet classification with deepconvolutional neural networks, pp. 1097–1105 (2012)
9. Sanghadiya, F., Mistry, D.: Surface defect detection in a tile using digital image processing: analysis and evaluation. Int. J. Comput. Appl. **116**(10), 33–35 (2015)
10. Sharif Razavian, A., Azizpour, H., Sullivan, J., Carlsson, S.: CNN features off-the-shelf: an astounding baseline for recognition. In: The IEEE Conference on Computer Vision and Pattern Recognition (CVPR) Workshops, June 2014
11. Simonyan, K., Zisserman, A.: Very deep convolutional networks for large-scale image recognition. In: International Conference on Learning Representations (2015)
12. Soukup, D., Hubermork, R.: Convolutional neural networks for steel surface defect detection from photometric stereo images, pp. 668–677 (2014)
13. Szegedy, C., Ioffe, S., Vanhoucke, V., Alemi, A.A.: Inception-v4, inception-resnet and the impact of residual connections on learning. In: AAAI, vol. 4, pp. 12 (2017)
14. Szegedy, C., et al.: Going deeper with convolutions. In: The IEEE Conference on Computer Vision and Pattern Recognition (CVPR), June 2015
15. Tsai, D.M., Luo, J.Y.: Mean shift-based defect detection in multicrystalline solar wafer surfaces. IEEE Trans. Ind. Inf. **7**(1), 125–135 (2011)
16. Wang, T., Chen, Y., Qiao, M., Snoussi, H.: A fast and robust convolutional neural network-based defect detection model in product quality control. Int. J. Adv. Manuf. Technol. **94**(5–8), 1–7 (2017)
17. Zhang, X., Le, X., Panotopoulou, A., Whiting, E., Wang, C.C.: Perceptual models of preference in 3D printing direction. ACM Trans. Graph. (TOG), **34**(6), p. 215 (2015)

Complex Conditional Generative Adversarial Nets for Multiple Objectives Detection in Aerial Images

Dan Popescu[✉], Loretta Ichim, and Andrei Docea

Faculty of Automatic Control and Computers, University Politehnica of
Bucharest, Bucharest, Romania
{dan.popescu,loretta.ichim}@upb.ro,
andrei.docea@gmail.com

Abstract. Simultaneously detection and evaluation of small regions of interest from aerial images is successfully achieved by conditional generative adversarial nets (cGAN). As novelty, the paper proposes a cheap and accurate method based on a cGAN structure, containing two generators, and graphics processing units (GPU) to segment small flooded areas and respectively, roads from images taken by unmanned aerial vehicles. In the learning phase the weights for the discriminator and the two generators are established by a back propagation method. The real mask is created by using information of the color components R, G, B, H and a voting scheme in a supervised process. A set of 40 images were used for the learning phase and another set of 100 images were used for method validation. The method presents the advantages of accuracy and time processing (especially in the operational phase).

Keywords: Aerial images · Flood segmentation · Road segmentation
GPU image processing · Conditional generative adversarial nets

1 Introduction

Over the years, many works have approached the detection and segmentation of regions of interest [RoIs] at ground level like floods, vegetation, forest, roads, buildings, etc. To this end, different methods were used, like inter-spectral satellite images [1], ground sensors (including cameras) [2] and aerial images (taken from manned or unmanned aerial vehicles - UAV) [3–5]. It was demonstrated that processing and segmentation of images from UAV is the cheapest and most accurate solution to evaluate small RoIs. To eliminate the gaps or duplications the images are firstly integrated into an orthophotoplan and then they are partitioned and processed. The most successful methods to segment and evaluate the ground RoIs from remote images are based on local processing at patch or pixel level. For example, authors in [4] segment flooded areas by means of the features extracted from the chromatic co-occurrence matrix [5] and mass fractal dimension. The features extracted from co-occurrence matrix are also used in this paper to create some flooded masks in the learning process. Because of huge amount of information in high resolution images,

© Springer Nature Switzerland AG 2018
L. Cheng et al. (Eds.): ICONIP 2018, LNCS 11304, pp. 671–683, 2018.
https://doi.org/10.1007/978-3-030-04212-7_59

parallel processing is indicated to drastically reduce the processing time. For example, image processing with neural networks offers high potential in GPU [6] or FPGA implementation. The co-occurrence matrix represents the statistical distribution of the correlation between two adjacent pixel values. This mathematical construction is especially used to analyze textures, extract traits, and detect and classify certain regions of them. One of the main drawbacks of this approach is the complexity of the algorithm that results in very high computation times. Despite the large amount of data and the large number of calculations that need to be performed on that dataset, the instructions to be executed are similar at different stages for a very large number of data, so they can be executed concurrently. The parallel aspect of the algorithm and the existence of parallel computing platforms specifically designed for this purpose is the reason why this paper was done, namely to accelerate the computation time using a parallel algorithm executed on the graphics processing unit by CUDA technology [6].

More recently Goodfellow et al. [7] proposed a new framework to train the classifiers in a semi-supervised mode, which is based on an adversarial procedure and two entities: a generator and a discriminator. So, a generative adversarial network (GAN) is created, which creates and processes generative models to produce realistic output images. Authors in [8] used image-to-image translation with conditional adversarial networks [9] to synthesize photos, reconstruct objects and color images. In this paper, we used a similar approach to segment flooded zones from aerial images. Because GAN is an important development in deep learning, with certain advantages, some other applications with promising results are communicated in different domains. As the initial concept, most of them refer to realistic image generation [10]. There are also, applications in image recognition and interpretation; for example, authors in [11] proposed an adversarial network to detect and segment aggressive prostate cancer from MRI images. Also, in the medical field Wolterink et al. [12] used a GAN to reduce the inherent noise in low dose CT images. There are some extensions of the GAN original architecture or structure of input latent variables [13, 14]. So, in [13] an extension of adversarial auto-encoder, as part of the generator, for semi-supervised learning tasks, is proposed and it is tested on handwritten characters. The author in [14] modifies the distribution of the input variables by a new architecture named RemixNet.

The aim of this paper is to implement a cGAN based system for simultaneously detection, segmentation and evaluation of flooded areas and roads in rural zones. The novelty of the paper is a multi generator structure for cGAN (one generator for floods and one generator for roads). The system has two phases: learning and segmentation. For the learning phase the cGAN-based system combines segmented images obtained by parallel processing on a classical GPU configuration with the original images. The cGAN segmentation phase uses images taken from UAV, cropped from an orthophotoplan, which is processed by the generator only to create the mask of the flooded or roads zones on the original images. The number of learning images was 40 and the test images, 100. From experimental results it can be seen that the accuracy is about 96%.

2 Methodology

The main task of the proposed system is to segment different regions of interest from aerial images like flooded zones and roads. The system (Fig. 1) is implemented as a cGAN [9] with two phases: the learning phase and the operational (segmentation) phase. In the learning phase a set of aerial images with flooded zones (FI – flooding images) and with roads (RI – road images) are used to create a batch of images with mask on the flooded zones or on roads (RM - real masks). The real masks are created for FI and RI by a proper selection (MUX F/R – flood/road multiplexer). The same original image is introduced in the generator (G) via the multiplexer learn/test (MUX L/T) to obtain a synthetic image –fake mask (FM). The new created batch of synthetic images serves also as input in the discriminator (D). So, in D two image pairs are introduced: the real pair (RP) and the fake pair (FP). The other notation significances in Fig. 1 are the following: UM (unit matrix) – matrix of 1 s, NM (null matrix) – matrix of 0 s, UC – unit comparator, NC – null comparator, DW –weights for the discriminator, Σ – adder, GW – weight optimizer for the generator, GC – comparator for G between RM and FM, GC – comparator for D with UM, DMUX L/T – demultiplexer for the separation of the output to the learning and test phases.

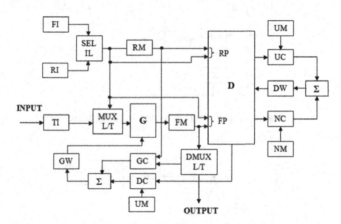

Fig. 1. Block diagram of the cGAN based system for flood and road detection with a single generator.

In Figs. 2 and 3 the schemes with two generators for the learning phase and operational phase are presented. The significance of notations is similar: GF – generator for flood, GR – generator for road, FMF – fake mask for flood, FMR – fake mask for road, RI – real image, RM – real mask, D – discriminator, DW – weight module for D, GFW – weight module for GF and GRW – weight module for GR. As it can be seen, four comparators are used: two for D and two for G. Based on error and gradient an optimizer establishes the weights for D and G, respectively (DW and GW). The Binary Cross Entropy criterion is used for the following comparisons:(a) between the output of D in case of the pair (real image, real mask) and UM; (b) between the output of D in case of the pair (real image, fake mask) and NM. It is intended to minimize the error and gradient between the real segmented image and UM - the ideal response (both have the same dimension as the real image) and also between the fake segmented images and M0 – the ideal response in this case. Similarly the L1 norm is used for comparison between FM and RM. The error sum and the gradient from two comparators are introduced in GW to update the weights for G. In order to dynamically modify the corresponding weights, GW uses a gradient-based optimization algorithm of stochastic objective functions ADAM [15].

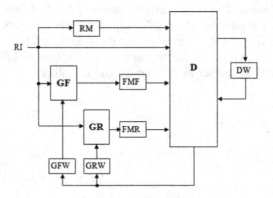

Fig. 2. Block diagram for the learning phase.

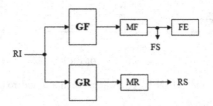

Fig. 3. Block diagram for the operational phase.

The network architecture is a conditional generative adversarial network [9] with the objective function $V(D, G)$ (1) taking into account the conditional model of both the generator (G) and discriminator (D).

$$\min_{G} \max_{D} V(D, G) = E_{x \sim p_{data}(x)} [\log D(x/y)] + E_{z-p_z(z)} [\log(1 - D(G(z/y)))] \quad (1)$$

In our case, the cGAN network modifies the classical network structure GAN by including a new element (the initial image to be segmented) in both the G and D input. This will condition the result according to its value.

In the present case, the solution chosen was preprocessing the images with a pre-implemented algorithm [16] to obtain a set of intermediate masks, and then manually correcting them to bring them as close as possible to the ideal shape (segmentation error 0%). The algorithm itself has two stages and is based on extracting information from the RGB and HSV representations of an image and segmenting the image according to their values.

The first step can be called a learning step because it involves establishing a range of values for the three channels of the RGB representation and for the H component of the HSV representation. This range is established by manually drawing flooded areas/roads from a set of representative images and calculating the minimum and maximum for each of the components of interest. A vector with these values is then passed to the next stage, the segmentation itself where the values of the each pixel are compared to those from the previously calculated ranges. If it falls within all it is considered that the pixel belongs to a flooded area/road, otherwise it is considered not to be in the region of interest.

The learning process is the following:

1. The network is initialized with random weights (both for the generator and for the discriminator).
2. One of the images in the training batch is passed through the generator.
3. The result together with the original image is given to the discriminator that compares the two images at the patch level and for each pixel gives a number from 0 to 1 which is the probability that it is "real" (that comes from the original image).
4. The same thing happens for the pair (original image, reference mask).
5. The results 3 and 4 are compared and an error is calculate, based on binary cross entropy and L1, and the error is propagated back into the discriminator and generator to modify the weights.
6. The steps 2–5 are repeated until the training batch is finished.

The pseudo code for the learning phase is described as follows (Algorithm 1).

Algorithm 1

```
initialize weights;

for epoch = 1 to num_epochs
    fake_mask = generator(real_image)
    fakeOutput = discriminator(real_image,fake_mask)
    real_mask = ReadFromFile(real _image)
    realOutput = discriminator(real_image,real_mask)
    gradientReal = BinaryCrossEntro-
    py(realOutput,unit_matrix)
    updateDiscriminatorWeights(gradientReal)
    gradientFake = BinaryCrossEntro-
    py(fakeOutput,zeros_matrix)
    updateDiscriminatorWeights(gradientFake)
    gradientGenerator = BinaryCrossEntro-
    py(fakeOutput,unit_matrix)
    gradientL1 = L1(fake_mask,real_mask)
    updateGeneratorWeights(gradientGenerator+gradientL1)

end
```

The detailed scheme of G which contains two main components, the encoder and the decoder, is presented in Fig. 4. This architecture sub-samples the data successively up to a point, and then applies the inverse process. The network learns to generate another image than the one received as input data. Each level in the encoder and decoder consists of convolution pairs and activation functions, with most of them having a normalization component (Batch Norm). An important deviation from the classical encoder-decoder structure is the presence of direct connections between the encoder levels and their corresponding decoder. This allows the network to bypass portions of the coding-decoding process and retain original information. Thus, a notable feature of the generator architecture is the presence of skip connections (red lines) that allow the low-level information in the input layers to bypass the bottleneck layer and, as such, to be directly connected to the corresponding output layers. This helps preserve the overall structure from the original image into the generated image and is called a U-net [17].

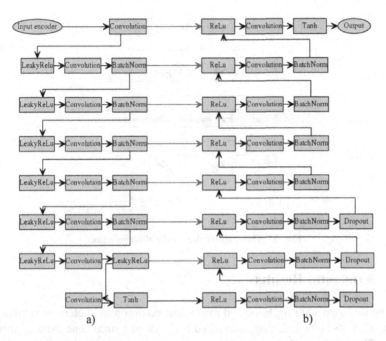

Fig. 4. The architecture of the generator: (a) the encoder, (b) the decoder. (Color figure online)

The role of the discriminator (Fig. 5) is to decide whether there is a real image or a fake image in the (real image, real mask) pair. The utility of this component in solving the segmentation problem is that it provides the generating component with an optimized function without being explicitly defined by the user.

Thus, a high-level goal can be specified, namely the distinction between real and generated images. The training of the discriminator has the function of inferring a function capable of fulfilling this objective, a function that is subsequently used to penalize the generator. As a result, the discriminating component provides a matrix with values between [0.1] (due to sigmoid function) that is the probability that a mask is real. This assures the compatibility with the comparator modules (Binary Cross Entropy criterion) from Fig. 1.

Among the technologies available for the development of artificial intelligence solutions for the system, the following were used:

- Torch [18], a machine learning framework used to implement the neural network;
- Python [19], specifically NumPy [20] and PIL (Python image library) libraries to evaluate the accuracy of network results;
- Matlab, used to prepare training data.

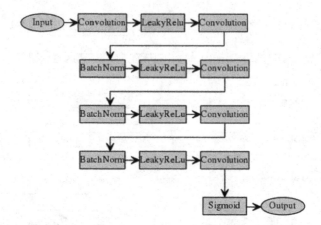

Fig. 5. The architecture of the discriminator.

3   Experimental Results

The proposed algorithm for flood and road segmentation was implemented taking into account a set of 140 aerial images acquired by UAV in a rural zone. Among them, 40 images (20 containing flooding and 20 containing asphaltic roads) were used in the learning phase and 100 images (50 containing flooding and 50 containing asphaltic roads) were used for the segmentation phase. A batch of 160 training pairs, from 40 images, successively rotated with 90^0, was used for learning the cGAN (both the generator and the discriminator). From these, five images and, correspondingly, their rotation ($0°$, $90°$, $180°$ and $270°$), together with the reference masks were used for exemplification (Fig. 6). Each image has assigned a unique ID (Im5430, Im5494, Im5518, Im6023, and Im6056). In the segmentation phase, for exemplification we used six images (Im5572, Im5537, Im5514, Im5595, Im6014 and Im5901) (Fig. 7) with flood areas and four mixed images – flood and road (Im4320, Im4441, Im4487, and Im4533) (Fig. 8). The output images and manual segmentation of the initial aerial images are presented in Fig. 7. The evaluation percentage of the flood from the image is computed by the system (Fig. 7).

The architecture of the computing system for target image evaluation consists of the following: Intel Core I7 CPU, 4th generation, 16 GB RAM, NVIDIA GeForce 770 M (Kepler architecture), 2 GB VRAM, Windows 10 operating system, Microsoft Visual Studio 2013, C++ programming language with language extension, and CUDA version 7.5 API. The main hardware component used to run the algorithm described in this paper is the graphics processing unit.

In the Octave program used to test the correctness of matrix computation, the function used in the graycomatrix own libraries was used. This function determines the co-match matrix for an input image based on the number of gray levels, the distance at which the pixel compares with the selected directions. This should be applied to each patch by using two forks. For a more accurate comparison of execution time, it will measure the time span of this function, accumulating the time slots for each cycle,

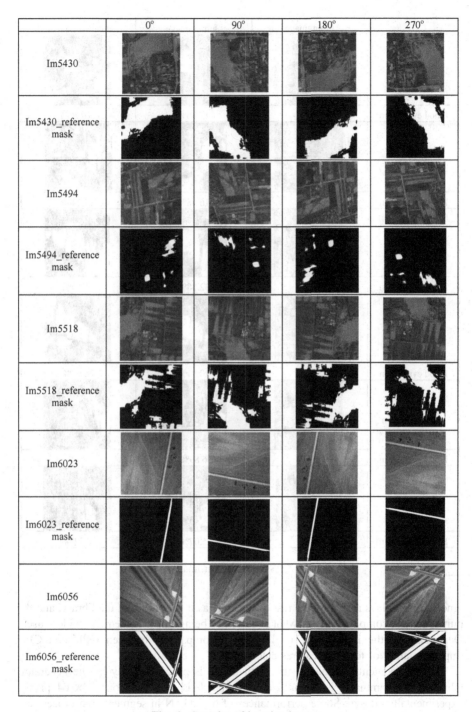

	0°	90°	180°	270°
Im5430				
Im5430_reference mask				
Im5494				
Im5494_reference mask				
Im5518				
Im5518_reference mask				
Im6023				
Im6023_reference mask				
Im6056				
Im6056_reference mask				

Fig. 6. Samples of learning images.

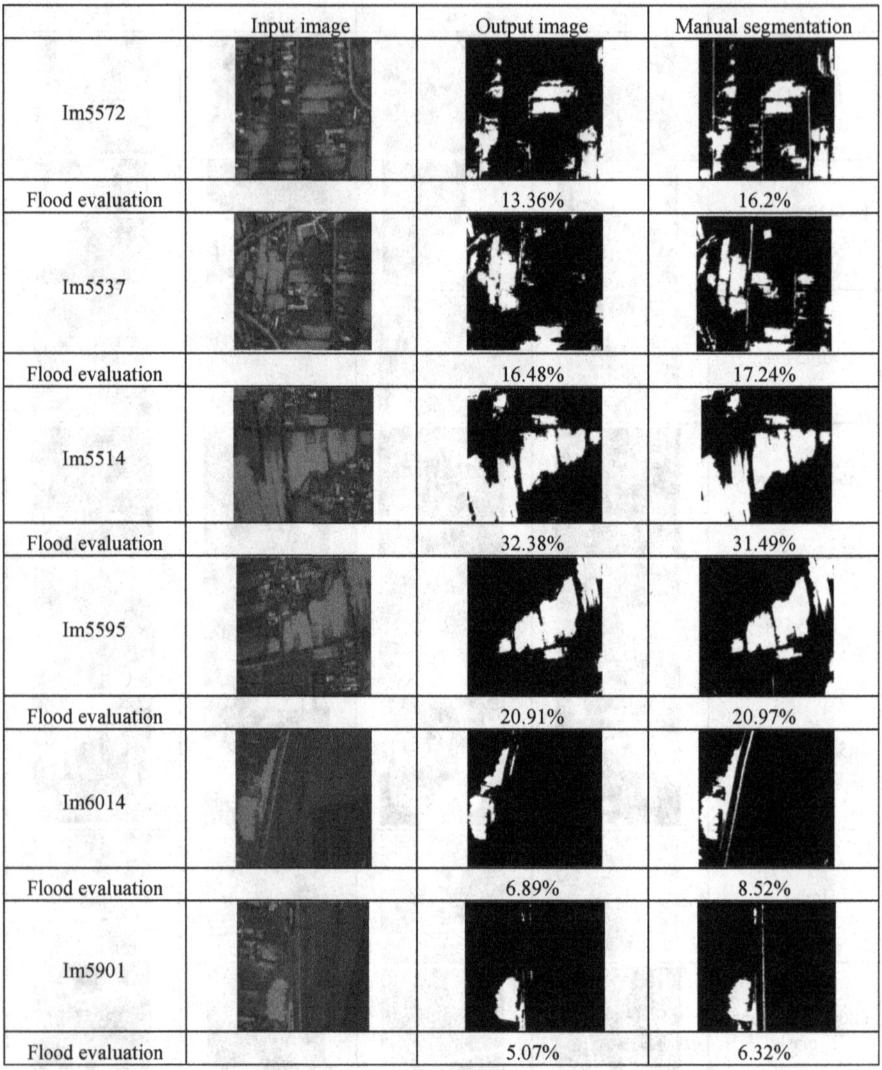

	Input image	Output image	Manual segmentation
Im5572			
Flood evaluation		13.36%	16.2%
Im5537			
Flood evaluation		16.48%	17.24%
Im5514			
Flood evaluation		32.38%	31.49%
Im5595			
Flood evaluation		20.91%	20.97%
Im6014			
Flood evaluation		6.89%	8.52%
Im5901			
Flood evaluation		5.07%	6.32%

Fig. 7. Samples of test images.

ignoring the delays added by the rest of the code as it is not part of the library, and the purpose is the comparison with what is currently being implemented and widely used. Depending of the image size, in Table 1, a comparison between GPU and CPU implementation of textural features calculation is given.

For the segmentation phase, the input image is processed only by the generator, while the discriminator is not utilized. The patch is of dimension 64×64 pixels, experimentally chosen. Some performances of the cGAN in segmentation process are given in Table 2.

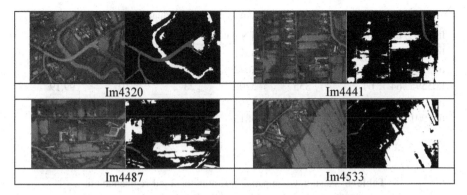

Fig. 8. Samples of mixed test images with flood (white) and road (blue) segmentation. (Color figure online)

Table 1. Execution times on CPU and GPU depending on the image size

Image Size	Threads per block	Patch Size	Grey Levels	CoMat Processing Time – GPU (ms)	CoMat Processing Time – CPU (ms)
4096 × 4096	16 × 16	64 × 64	256	117.68	70000
2048 × 2048	16 × 16	64 × 64	256	80.06	12967
1024 × 1024	16 × 16	64 × 64	256	9.59	2700
256 × 256	16 × 16	64 × 64	256	2.38	166

Table 2. cGAN performances

Performance	Value
Execution time learning	13 h
Execution time for segmentation	0.2 s
Image dimension	4096 × 4096
Patch dimension	64 × 64
False positive error	1.5%–5%
False negative error	0.5%–4%
Accuracy	95.5%–99%

In Fig. 7, the flood evaluation means the percent of flood occupation from the entire image. The generator was implemented with 55 nodes and the discriminator with 13 nodes (without inputs and outputs).

In Table 3, for some test images, the accuracy of flood segmentation (in terms of patches of 64 × 64 pixels) is presented. The notations in Table 3 are the following: TP – true positive, TN – true negative, FP – false positive, FN – false negative, and ACC – accuracy. In Table 4 the performance of our method for flood detection and segmentation, in terms of accuracy, is compared with the most recent papers with the same topic. The results are better than satellite images and also than urban areas.

Table 3. cGAN performances on individual images

Image	Patch dimension	TP patches	TN patches	FP patches	FN patches	ACC [%]
Im5572	64 × 64	547	3371	61	117	95.64
Im5514	64 × 64	1326	2672	62	36	97.61
Im5595	64 × 64	855	3208	23	10	99.19
Im6014	64 × 64	282	3761	18	35	98.71
Im5901	64 × 64	220	3815	12	49	98.51

Table 4. Accuracy comparation results for flood segmentation

Methods	Areas	ACC [%]	Observations
[6]	Rural	77–95	Satellite images
[7]	Rural/Urban	90/75	Satellite images
[9]	Urban	81.2–87.8	UAV images
Our method	Rural	95.5–99	UAV images

4   Conclusion

The proposed method differentiates the flooded areas and roads in aerial images. The novelty is a multi generator configuration of cGAN for multiple class image recognition. Also, a skip connections inside the generator, allows the low-level information in the input layers to bypass the bottleneck layer and, as such, to be directly connected to the corresponding output layers.

Acknowledgements. The work has been funded by project CAMIA, GEX 25/2017 (UPB), MAARS 185/2017 (ROSA) and MUWI 1224/2018 (NETIO).

References

1. Yulianto, F., Sofan, P., Zubaidah, A., Sukowati, K.A.D., Pasaribu, J.M., Khomarudin, M.R.: Detecting areas affected by flood using multi-temporal ALOS PALSAR remotely sensed data in Karawang, West Java. Indonesia. Nat. Hazards **77**, 959–985 (2015)
2. Pentari, A., Moirogiorgou, K., Livanos, G., Iliopoulou, D., Zervakis, M.: Feature analysis on river flow video data for floating tracers detection. In: Proceedings of the IEEE International Conference on Imaging Systems and Techniques (IST), 14–17 Octover, Santorini Island, Greece, pp. 287–292 (2014)
3. Lee, J.N., Kwak, K.C.: A trends analysis of image processing in unmanned aerial vehicle. Int. J. Comput. Inf. Sci. Eng. **8**, 261–264 (2014)
4. Popescu, D., Ichim, L., Stoican, F.: Unmanned aerial vehicle systems for remote estimation of flooded areas based on complex image processing. Sensors **17**(3), 446 (2017)
5. Popescu, D., Ichim, L.: Aerial image segmentation by use of textural features. In: 20th International Conference on System Theory, Control and Computing (ICSTCC), 13–15 October, Sinaia, Romania, pp. 721–726 (2016)

6. Strigl, D., Kofler, K., Podlipnig, S.: Performance and scalability of GPU-based convolutional neural networks. In: 18th Euromicro International Conference on Parallel, Distributed and Network-Based Processing (PDP), 17–19 February, Pisa, Italy, pp. 1–8 (2010)
7. Goodfellow, I.J., et al.: Generative Adversarial Networks. arXiv:1406.2661v1 (2014)
8. Isola, P., Zhu, J.-Y., Zhou, T., Efros A.A.: Image-to-image translation with Conditional Adversarial Networks. arXiv:1611.07004v1 (2016)
9. Mirza, M., Osindero, S.: Conditional Generative Adversarial nets. arXiv:1411.1784v1 (2014)
10. Kataoka, Y., Matsubara, T., Uehara, K.: Image generation using Generative Adversarial Networks and attention mechanism. In: IEEE/ACIS 15th International Conference on Computer and Information Science (ICIS), 26–29 June, Okayama, Japan (2016)
11. Kohl, S., et al.: Adversarial Networks for the detection of aggressive prostate cancer. arXiv: 1702.08014v1 (2017)
12. Wolterink, J.M., Leiner, T., Viergever, M.A., Isgum, I.: Generative Adversarial Networks for noise reduction in low-dose CT. IEEE Trans. Med. Imaging **PP**(99) (2017)
13. Tachibana, R., Matsubara,T., Uehara, K.: Semi-supervised learning using adversarial networks. In: IEEE/ACIS 15th International Conference on Computer and Information Science (ICIS), 26–29 June, Okayama, Japan (2016)
14. Yazdani, M.: RemixNet: generative adversarial networks for mixing multiple inputs. In: Semantic Computing (ICSC), 30 January–1 February (2017)
15. Kingma, D.P., Ba, J.: Adam: a method for stochastic optimization. In: 3rd International Conference for Learning Representations, San Diego, pp. 1–15 (2015)
16. Sumalan, A.L., Popescu, D., Ichim, L.: Flood evaluation in critical areas by UAV. In: 8th International Conference on Electronics, Computers and Artificial Intelligence (ECAI), 30 June–2 July, pp. 1–6 (2016)
17. Ronneberger, O., Fischer, P., Brox, T.: U-net: Convolutional networks for biomedical image segmentation. In: MIC-CAI, pp. 234–241 (2015)
18. Torch | Scientific computing for LuaJIT. http://torch.ch/
19. Documentation - Our Documentation. https://www.python.org/doc/
20. NumPy — NumPy. http://www.numpy.org/

Facial Landmark Detection Under Large Pose

Yangyang Hao[1], Hengliang Zhu[1], Zhiwen Shao[1], Xin Tan[1],
and Lizhuang Ma[1,2(✉)]

[1] Department of Computer Science and Engineering,
Shanghai Jiao Tong University, Shanghai, China
{haoyangyang2014,hengliang_zhu,shaozhiwen,tanxin2017}@sjtu.edu.cn,
ma-lz@cs.sjtu.edu.cn
[2] Department of Computer Science and Software Engineering,
East China Normal University, Shanghai, China

Abstract. Facial landmark detection is a necessary step in many vision tasks and plenty of excellent methods have been proposed to solve this problem. However, for the conditions with large pose and complex expression, these works usually suffer an eclipse. In this paper, we propose a two-stage cascade regression framework using patch-difference features to overcome the above problem. In the first stage, by applying the patch-difference feature and augmenting the large pose samples to the classical shape regression model, salient landmarks (eye centers, nose, mouth corners) can be located precisely. In the second stage, by applying enhanced feature section constraint to the patch-difference feature, multi-landmark detection is achieved. Experimental results show that our algorithm has a significant improvement compared to the classical shape regression method and achieves superior results on COFW dataset.

Keywords: Facial landmark detection · Large pose
Patch-difference feature · Feature section constraint

1 Introduction

Facial analysis and processing technologies become hot topics in recent years. Facial landmark detection aims to find the feature points of organs (nose, eyes, mouth and cheek). This technique has extensive applications, such as face recognition [5], face tracking [21], facial beautification [6], expression recognition [14]. It is time consuming and inefficient to detect one landmark with a respective model and the most popular way is to treat all landmarks as a whole. Cascade shape regression model can efficiently regress all the landmarks at the same time and lots of approaches [4,7,11,15] based on this model have been proposed. However, as these methods hardly handle the scenarios of large pose and complex expression, the accuracy obviously decreases on dataset with these situations.

There are two main reasons for the above problem. One reason is that features are unstable and don't contain enough information. For instance, the famous

© Springer Nature Switzerland AG 2018
L. Cheng et al. (Eds.): ICONIP 2018, LNCS 11304, pp. 684–696, 2018.
https://doi.org/10.1007/978-3-030-04212-7_60

pixel-difference feature [4] which is the intension difference between two pixels. The pixel-difference feature plays an important role in the tree-based cascade shape regression methods. Though the feature is highly efficient, the pixel is sensitive to noise and too less information is used. Additionally, the pixel pairs are selected from a large number of candidates and it is a problem to select the most useful ones. In order to address the above problems, this paper proposes a patch-based feature to improve the performance of classical cascade regression methods. We use mean of image patch to replace the pixel to enhance the ability of noise immunity. The feature is normalized to make it scale-invariance at the same time. The new feature is more robust on variation of occlusion and illumination. In the procedure of selecting the best features from a large pool, we also propose a new feature selection constraint. Researchers [11] show that the closer between the pixel pairs, the features are better. We assume that each pixel has its nearest landmark and the new constraint: the distance between the pixel pairs should be smaller than the distance between their respective nearest landmarks. By combining these two constraints, the selected features are better.

Another reason for the above problem is that the datasets don't contain enough variations of pose and expression. For example, LFPW [2] (29 landmarks) is a dataset with little variation and LBF [15] can achieve the mean error of 3.35%. Another method cGPRT [13] reports a result of 4.63% on HELEN [12] (194 landmarks) dataset. However, the mean error on challenging set of 300W (68 landmarks) is around 10%. In first two data sets, most of the faces are natural and frontal, 300W challenging set contains many large pose faces that include in-plane and out-of-plane. In terms of the large roll angles, such as over $30°$, testing data is much more than training data. We can see from Table 1, it is hard to get a good model on the data without enough pose variations. In this paper, we use a hard samples augmentation method to enrich the diversity of dataset. The idea of data augmentation is motivated by deep learning methods. The training data for deep learning methods are massive by leveraging preprocessing (translation, rotation, horizontal flip and compression), generally over half of a million. Aiming at large pose, we enlarge the training data through rotation and apple it to the cascade shape regression methods. However, the training capacity of conventional methods is from thousands to tens of thousands, only a small number of hard samples with large roll angles are selected for augmentation. In this way, training data in the same order can produce a better model and give better results.

In this paper, based on the patch-difference feature, we propose a two-stage cascade regression facial landmark detection method. We mainly cope with the scenario of large pose and improve the robustness of features for classical cascade regression. In summary, the contributions of this paper are:

- We propose a new patch-difference feature for tree-based cascade regression framework. By leveraging the patch information, our method is more accurate and has little affect to efficiency.
- We propose an enhanced feature selection constraint by using the information of the nearest landmarks of the feature pairs.

– A data augmentation method is used for face alignment that only a small number of large pose samples are augmented.

The remainder of this paper is organized as follows: Sect. 2 provides an overview of related work. Two-stage cascade regression method is presented in Sect. 3. Section 4 shows the experimental results and analysis. Section 5 is the conclusions.

2   Related Work

Facial landmark detection raises from last century and plenty of work have been proposed up to now. Generally, these approaches can be categorized into traditional methods and deep learning methods.

2.1   Traditional Methods

In recent years, the shape regression models [4,7,11,15,18,22] are extensively applied in face alignment. Cascade shape regression model is first used in [7] to estimate the facial shape and it is widely used in this field. ESR [4] directly learns a regression function to infer the shape from a sparse subset of pixel intensities indexed relative to the current shape estimate. Ensemble of Regression Trees (ERT) [11] substitutes the fern weak regressor in ESR [4] with a regression tree and limits the distance between the pairwise feature points to achieve a better result. Local Binary Feature (LBF) [15] proposes to learn local binary feature for each landmark independently and jointly regresses for all landmarks. Supervised descent method (SDM) [22] predicts shape increment by employing a cascaded linear regression based on SIFT features. GSDM [21] improve the performance of SDM [22] by computing the gradient in global. CFSS [26] applies the idea of coarse-to-fine to do shape searching in the sub-region and the results are not affected by the initial shape. Similar to ESR [4], LBF [15] and cGPRT [13], we focus on discriminative feature and propose a new feature to improve the performance.

2.2   Deep Learning Methods

Deep learning methods are the most popular in present. Sun et al. [16] first apply cascaded deep convolution network to estimate the position of five facial landmarks and refine the position of landmarks level by level. Zhou et al. [25] also use multi-level deep networks to detect facial landmarks in a coarse-to-fine manner. Honari et al. [9] present Recombinator Networks by using multi-scale input maps for learning coarse-to-fine feature. TCDCN [24] proposes a multi-task learning method that employs auxiliary facial attribute recognition to obtain correlative facial properties to improve the performance.

3   Two-Stage Cascade Regression

3.1   Overview of Our Method

Our method includes two main parts: salient landmark detection and multi-landmark detection. In the first stage, salient landmark detection is used to obtain positions of salient landmarks, then an initial face is generated as input for the next stage. Salient landmark set is the smallest subset that can roughly represent the characteristic of a face. Therefore, it is rational to leverage this information to generate the initial face for the next stage. The initial face is computed by a linear combination of several similar faces. Similar faces are obtained by searching from training samples according to distances between each other. In this paper, Manhattan distance of salient landmarks is applied. The weight w_n for each similar face is computed as follows:

$$w_n = \frac{\frac{1}{n} + \frac{1}{n+1} + \cdots + \frac{1}{N}}{N} \tag{1}$$

where $n = 1$ represents the most similar one, N is the number of similar faces and we use 19 similar faces, this formulation ensures that the more similar face has a bigger weight. In both of two stages, tree-based cascade regression framework is applied. Mean face and generated face are used as initialization in regression procedure for two stages respectively. Training data augmentation and patch-difference feature are used in the first stage to achieve the precise locations of salient landmarks. Enhanced feature selection constraint is applied to patch-difference feature in the second stage. Because salient landmarks are only 5 points, the distance between the nearest landmarks of pairwise points always bigger than the distance between pairwise points. That is to say, the new feature selection constraint is satisfied by default for salient landmark detection.

3.2   Tree Based Cascade Regression Model

A single regression model is insufficient for facial landmark detection in the wild that contains large variations of pose, expression, illumination and occlusion. Therefore, researchers tend to use a sequence of regressors to refine the results step by step. Tree model is generally used in training procedure of the regressors. Firstly, we give a brief introduction of cascade process. The shape of a face can be presented as $S = \{X_j | j = 1, 2...p\} \in \Re^{2p}$, p is the number of the landmarks, X_j denotes the x, y-coordinates of the j-th landmark in a face image I. By applying linear regression framework, formulation of cascade process is following: $S_{i,t+1} = S_{i,t} + r_t(I, S_{i,t})$, where r_t represents the t-th regressor, $S_{i,t}$ represents the current estimated shape of level t, $S_{i,t+1}$ represents the shape of next level. In this manner, the shape is updated step by step and increment for the shape of next level is r_t. And in each level, the regressor $r_t(I, S_{i,t})$ is learnt by solving the following optimization problem:

$$r_t = \arg\min_{r_t} \sum_{i=1}^{L} \|S_{i,t}^* - S_{i,t} - r_t\|_2 \tag{2}$$

where $S_{i,t}^*$ is the ground-truth, L represents the number of training data. Friedman [8] proposes gradient boosting tree algorithm to learn the regressor r_t and sum of square error loss is used in the algorithm. The number of levels is usually over 10. The idea of coarse-to-fine is exploited in the procedure.

Obviously, the crucial process is to learn a regressor r_t and we name it regression tree. The pixel difference feature is simple and geometric invariance in a certain intension, but it doesn't use the neighbor pixel information and is not a normalized feature. This paper addresses this problem and solves it in the later part. At each split node of the regression tree, threshold is applied to classify the samples into different leaf node according to the pairwise pixel difference value. Usually, at each node, we greedily select the best split from a number of candidate splits that are randomly generated. The best one should minimize the sum of the square error. Use θ to present the parameter set $(\tau, u$ and $v)$, τ is a threshold, u and v are positions of pairwise points. This process can be represented in the following formulation:

$$E(M, \theta) = \sum_{s \in \{l,r\}} \sum_{i \in M_{\theta,s}} \|r_i - \mu_{\theta,s}\|^2 \tag{3}$$

$$\mu_{\theta,s} = \frac{1}{\|M_{\theta,s}\|} \sum_{i \in M_{\theta,s}} r_i \tag{4}$$

where M is the indices of training samples used in the node, $M_{\theta,l}$ is the set of indices of samples that are classified into the left node judged by the threshold, r_i is the residual of sample i in the gradient boosting algorithm. By omitting the parts that are independent of θ, the formulation above can be rewritten as follows:

$$\arg \max_\theta E(M, \theta) = \arg \min_\theta \sum_{s \in \{l,r\}} \|M_{\theta,s}\| \mu_{\theta,s}{}^T \mu_{\theta,s} \tag{5}$$

$\mu_{\theta,s}$ is the only factor that is to be computed and the node split optimization is efficient.

3.3 Patch-Difference Feature and Feature Selection

Pixel difference feature used in the regression tree is difference between intensities of two pixels in an image, it is highly efficient and accurate. ERT [11] can achieve 1000 fps (frame per second) for 68 landmarks detection. The pixel difference feature is simple and geometric invariance in a certain intension, but it is sensitive to noise. We propose a patch-difference feature to cope with this problem and try to use more potential information. Following is formulation used to compute the features:

$$\frac{MP(u) - MP(v)}{MP(u) + MP(v)} \tag{6}$$

where $MP(\cdot)$ is a function that computing the mean of an image patch, considering of the efficiency, we compute the mean of a 3×3 patch. In this way, neighboring pixel information is used and the feature is a normalized form. This feature measures the relative difference between two image patches and the formulation in the same form as the Weber Fraction. The Weber's Law [10] is that the human perception of difference in stimulus is often measured as a fraction of the original stimulus. This form is robust against illumination changes. By leveraging the information of patches, this fraction form is robust to noise.

Candidates of features are generated randomly and this factor leads to a big difference between good feature and poor feature. On the other hand, the candidates pool should be big enough to make sure good features are contained. It is necessary to select a number of good ones from candidates pool and other work [11] has proposed a feasible measure. The constraint is that the pairwise points have a bigger probability to be selected when the distance between them is smaller. Exponential function is chosen to do this work, that is: $e^{-\lambda||u-v||}$, where $||\cdot||$ represents the Euclidean distance, λ is the parameter to control distance of the pairwise points. In this paper, we add an enhanced constraint that further improves the performance. We assume that each point corresponding to a nearest landmark. The additional constraint is that the distance between two landmarks should be bigger than the distance of pairwise points. The formulation is as follows,

$$||u - v|| < ||ul - vl|| \tag{7}$$

where ul and vl are the nearest landmarks of u and v, we show it in Fig. 1.

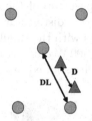

Fig. 1. Blue dots represent estimated landmarks of level t, red triangles represent two pixels. D is the distance between the pixel pair and DL is the distance between two landmarks.

3.4 Hard Sample Augmentation

For facial landmark detection, diversity of the annotated data sets is limited. The amount of the datasets from hundreds to thousands and most of the samples are frontal faces with natural expression. The performance decreases obviously when the faces have large variations in pose, expression, illumination and occlusion. 300W dataset is a good evidence to illustrate the above situation. This dataset includes two subsets: the common set and the challenging set. The mean error on common set is around 5% and the mean error on challenging set is around 10% for traditional methods.

Table 1. The analysis of face pose of 300 W dataset. The numbers represent quantity of samples with roll angle over 20°, 30° and 40°

300 W dataset	Training (3148)	Testing (689)
> 20°	83	34
> 30°	2	9
> 40°	0	2

The paper provides an analysis of face poses on the 300 W data set and roll angle of a face is regarded as the face pose. We use salient landmark information to analyze distribution of the roll angles. The roll angle is the angle between line LA and the Y-axis. Line LA is consisted of midpoint of eye centers and midpoint of mouth corners. The biggest roll angles of training and testing samples are 34° and 47°. From Table 1, we can see that the training data is seriously insufficient for large roll angles. The roll angle of training data is mainly below 30°, while some of testing samples with roll angle near 50°. Our method is following, first, the faces are classified into 5 categories (left, right, up, down and frontal) and 10 samples for each category are selected to be rotated ±30°, ±40° and ±50°. In a normalized face, the relative location of the nose tip is used to decide the category that a face belongs to. If nose tip at the left side of the center of the face, it is left face and 10 samples with the largest horizontal distance in this direction are selected. If nose tip on top of the center of the face, it is up face and 10 samples with largest vertical distance in this direction are selected. For frontal face, we choose 10 samples with the smallest Euclidean distance between nose tip and the center. Original 300 W data set is 3148 and the augmented data is 3448. The classified examples are showed in Fig. 2.

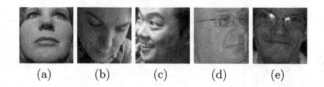

(a)　　　　(b)　　　　(c)　　　　(d)　　　　(e)

Fig. 2. These faces are samples of 5 different categories. The faces belong to up (a), down (b), left (c), right (d) and frontal (e).

4　Experiments

Datasets: Two challenging datasets are used for facial landmark detection to demonstrate our method achieves state-of-the-art performance. Faces of these datasets have a big variation on pose, expression, occlusion, and illumination.

300W dataset: It is a 68 landmarks dataset and consists of two subsets, the common subset and the challenging subset iBUG. Dataset configuration in [15] is used to have a fair comparison. The training set contains 3148 images. Test set contains 689 images.

COFW dataset [3]: This dataset is annotated with 29 landmarks and mainly contains the faces with heavy occlusion, training samples are 1345 and 507 samples for testing.

Evaluation metric: Standard mean absolute error is used in the experiments. All errors are normalized by the inter-ocular distance and results in this section are simplified form without '%' symbol. For 300 W full set, Calculated Error Distribution (CED) curves are plotted to give more visible results.

Parameter setting: In the training procedure, 20 randomly selected faces and 20 similar faces are used as initialization for salient landmark and multi-landmark detection respectively. Cascade level $T = 18$ and 15 are for first stage and second stage, $K = 500$ weak regressors form a strong regressor r_t, $D = 5$ is the depth of the regression tree. Shrinkage factor is 0.1. For node splitting, we repeat $S = 500$ times to find the best one. Following the feature selection constraint, 400 pairwise pixels and a randomly chosen threshold corresponding to each pair is used. The bounding boxes are provided in the database.

Table 2. Results of averaged error (%) are compared with state-of-the-art approaches on 300W. Errors are normalised by the inter-ocular distance, and the results of other methods are directly cited from the already published papers.

Method	Common	Challenging	Full set
DRMF [1]	6.65	19.79	9.22
ESR [4]	5.28	17.00	7.58
RCPR [3]	6.18	17.26	8.35
SDM [22]	5.57	15.40	7.50
ERT [11]	-	-	6.40
LBF [15]	4.95	11.98	6.32
cGPRT [13]	4.46	10.85	5.71
CFSS [26]	4.73	9.98	5.76
TCDCN [24]	4.80	8.60	5.54
MDM [17]	4.83	10.14	5.88
RDR [19]	5.03	8.95	5.80
RAR [20]	4.12	8.35	4.94
Our method	4.36	8.70	5.21

4.1   Comparison with Other Work

Table 2, a comparison with state-of-the-art methods is displayed. Compared methods include DRMF [1], ESR [4], RCPR [3], SDM [22], CFAN [23], ERT [11], LBF [15], cGPRT [13], CFSS [26], TCDCN [24], MDM [17], RDR [19] and RAR [20]. We can see that our method outperforms all the conditional methods and it is also comparable with deep learning method (TCDCN [24], MDM [17], RDR [19] and RAR [20]). 300W challenging subset mainly focuses on large pose and complex expression, with the help of hard sample augmentation, our method is robust adequate for variation of pose and expression.

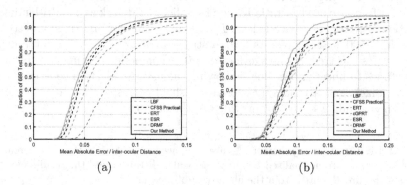

(a)   (b)

Fig. 3. Comparison of CED curves on 300 W full set and challenging set.

We provide comparison of CED curves with state-of-the-art approaches on 300 W full set (Fig. 3(a)) and challenging set (Fig. 3(b)). Compared methods include DRMF [1], ESR [4], ERT [11], LBF [15], CFSS Practical [26], and we can see that our approach better than others. The compared methods are re-implemented and some of the results are provided by authors. ESR [4] and ERT [11] are reproduced by ourselves with the error of 7.76 and 6.42. The result of LBF [15] is provided by the author, the codes of DRMF [1] and CFSS Practical [26] are available online. We also show some visible results of 300 W datasets in first two rows of Fig. 4. Though these faces with large variations in pose, our method achieves good performance by applying the proposed method.

Table 3. Results of averaged error (%) are compared with state-of-the-art approaches on COFW dataset.

Method	ESR [4]	RCPR [3]	SDM [22]	TCDCN [24]	RAR [20]	Our method
COFW	11.2	8.50	9.33	8.05	6.03	5.35

Comparison with state-of-the-art methods on COFW dataset is showed in Table 3. COFW dataset mainly focus on face with occlusion. This dataset is very challenge due to lots of faces with heavy occlusions. Compared methods include ESR [4], RCPR [3], SDM [22], TCDCN [24] and RAR [20]. From the Table, we can see that our method is much better than other methods. With the help of the salient-to-all manner, our method is robust under the conditions of occlusion. Last two rows faces of Fig. 4 are challenging samples of COFW dataset due to heavy occlusions. With the help of the proposed method, especially salient landmark detection, the effect of occlusion is declined.

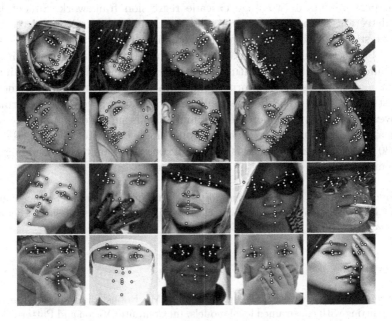

Fig. 4. Some challenging results of our method on 300 W and COFW dataset.

4.2 Incremental Analysis

In this paper, we propose three components to improve the performance and each of them shows a benefit to the whole process. Traditional cascade regression (CR) model is used as the baseline and three experiments are conducted to demonstrate effectiveness of our method. The three components are patch-difference feature (PD), enhanced feature selection constraint (FS), and hard sample augmentation (HSA). Both salient and 68 landmarks detection are conducted and two-stage is not applied. The new feature selection constraint is not applied in salient landmark detection because there are only 5 landmarks and this constraint is default satisfaction. Table 4 shows the performance of the three different components.

Table 4. Incremental analysis on 300W dataset.

Method	CR	CR+PD	CR+PD+FS	CR+PD+FS+HSA
Salient	4.39	4.23	-	3.95
Multiple	6.42	6.34	6.20	5.94

5   Conclusions

This paper presents a two-stage cascade regression framework. Salient landmark detection is done in the first stage and multi-landmark are detected in the next stage. Patch-difference feature, enhanced feature selection constraint and hard samples augmentation are applied in our algorithm. By utilizing the patch-difference feature and feature selection constraint, the feature maintains efficient and contains more information. With the augmentation, our method has a strong power to handle the condition with large pose. The performance improves significantly and increased training data is small compare to the original data. Our experiments are conducted on a single core Intel(R) Xeon(R) CPU E5-2630 v3 @2.4 GHz and speed is 190 fps. Experiments on two challenging data sets demonstrate the efficiency of our algorithm.

Acknowledgments. This work is supported by the National Natural Science Foundation of China (No. 61472245) and the Science and Technology Commission of Shanghai Municipality Program (No. 16511101300).

References

1. Asthana, A., Zafeiriou, S., Cheng, S., Pantic, M.: Robust discriminative response map fitting with constrained local models. In: Computer Vision and Pattern Recognition, pp. 3444–3451. IEEE, New York (2013)
2. Belhumeur, P.N., Jacobs, D.W., Kriegman, D.J., Kumar, N.: Localizing parts of faces using a consensus of exemplars. IEEE Trans. Pattern Anal. Mach. Intell. **35**(12), 545–552 (2013)
3. Burgos-Artizzu, X.P., Perona, P., Dollár, P.: Robust face landmark estimation under occlusion. In: International Conference on Computer Vision, pp. 1513–1520. IEEE, New York (2013)
4. Cao, X., Wei, Y., Wen, F., Sun, J.: Face alignment by explicit shape regression. Int. J. Comput. Vision **107**(2), 117–190 (2012)
5. Chen, C., Dantcheva, A., Ross, A.: Automatic facial makeup detection with application in face recognition. In: International Conference on Biometrics, pp. 1–8. IEEE, New York (2013)
6. Guo, D., Sim, T.: Digital face makeup by example. In: Computer Vision and Pattern Recognition, pp. 73–79. IEEE, New York (2009)
7. Dollár, P., Welinder, P., Perona, P.: Cascaded pose regression. In: Computer Vision and Pattern Recognition, pp. 1078–1085. IEEE, New York (2010)
8. Friedman, J.H.: Greedy function approximation: a gradient boosting machine. Ann. Stat., 1189–1232 (2001)

9. Honari, S., Yosinski, J., Vincent, P., Pal, C.: Recombinator networks: learning coarse-to-fine feature aggregation. In: Computer Vision and Pattern Recognition, pp. 5743–5752. IEEE Computer Society, New York (2016)
10. Jain, A.K.: Fundamentals of Digital Image Processing. Prentice Hall (1989)
11. Kazemi, V., Sullivan, J.: One millisecond face alignment with an ensemble of regression trees. In: Computer Vision and Pattern Recognition, pp. 1867–1874. IEEE, New York (2014)
12. Le, V., Brandt, J., Lin, Z., Bourdev, L., Huang, T.S.: Interactive facial feature localization. In: Fitzgibbon, A,, Lazebnik, S., Perona, P., Sato, Y., Schmid, C. (eds.) ECCV 2012. LNCS, vol. 7574, pp. 679–692. Springer, Heidelberg (2012). https://doi.org/10.1007/978-3-642-33712-3_49
13. Lee, D., Park, H., Yoo, C.: Face alignment using cascade Gaussian process regression trees. In: Computer Vision and Pattern Recognition, pp. 4204–4212. IEEE, New York (2015)
14. Ramirez Rivera, A., Castillo, R., Chae, O.: Local directional number pattern for face analysis: face and expression recognition. IEEE Trans. Image Process. **22**(5), 1740–1752 (2013)
15. Ren, S., Cao, X., Wei, Y., Sun, J.: Face alignment at 3000 fps via regressing local binary features. In: Computer Vision and Pattern Recognition, pp. 1685–1692. IEEE, New York (2014)
16. Sun, Y., Wang, X., Tang, X.: Deep convolutional network cascade for facial point detection. In: Computer Vision and Pattern Recognition, pp. 3476–3483. IEEE Computer Society, New York (2013)
17. Trigeorgis, G., Snape, P., Nicolaou, M.A., Antonakos, E., Zafeiriou, S.: Mnemonic descent method: a recurrent process applied for end-to-end face alignment. In: Computer Vision and Pattern Recognition, pp. 4177–4187. IEEE, New York (2016)
18. Tzimiropoulos, G.: Project-out cascaded regression with an application to face alignment. In: Computer Vision and Pattern Recognition. IEEE, New York (2015)
19. Xiao, S., et al.: Recurrent 3D–2D dual learning for large-pose facial landmark detection. In: IEEE International Conference on Computer Vision, pp. 1642–1651. IEEE Computer Society, New York (2017)
20. Xiao, S., Feng, J., Xing, J., Lai, H., Yan, S., Kassim, A.: Robust facial landmark detection via recurrent attentive-refinement networks. In: Leibe, B., Matas, J., Sebe, N., Welling, M. (eds.) ECCV 2016. LNCS, vol. 9905, pp. 57–72. Springer, Cham (2016). https://doi.org/10.1007/978-3-319-46448-0_4
21. Xiong, X., De la Torre, F.: Global supervised descent method. In: Computer Vision and Pattern Recognition, pp. 2664–2673. IEEE Computer Society, New York (2015)
22. Xiong, X., De la Torre, F.: Supervised descent method and its applications to face alignment. In: Computer Vision and Pattern Recognition, pp. 532–539. IEEE, New York (2013)
23. Zhang, J., Shan, S., Kan, M., Chen, X.: Coarse-to-fine auto-encoder networks (CFAN) for real-time face alignment. In: Fleet, D., Pajdla, T., Schiele, B., Tuytelaars, T. (eds.) ECCV 2014. LNCS, vol. 8690, pp. 1–16. Springer, Cham (2014). https://doi.org/10.1007/978-3-319-10605-2_1

24. Zhang, Z., Luo, P., Loy, C.C., Tang, X.: Learning deep representation for face alignment with auxiliary attributes. IEEE Trans. Pattern Anal. Mach. Intell. **38**(5), 918–930 (2016)
25. Zhou, E., Fan, H., Cao, Z., Jiang, Y.: Extensive facial landmark localization with coarse-to-fine convolutional network cascade. In: International Conference on Computer Vision Workshops, pp. 386–391. IEEE Computer Society, New York (2014)
26. Zhu, S., Li, C., Loy, C.C., Tang, X.: Face alignment by coarse-to-fine shape searching. In: Computer Vision and Pattern Recognition. IEEE, New York (2015)

Author Index

Al Shamsi, Fatima   122
Al-Jumeily, Dhiya   304
Alshammari, Mashaan   109
Aung, Zeyar   122

Ban, Tao   392
Bhatnagar, Shobhit   647
Bhattacharyya, Pushpak   647
Bo, Yuan   157

Caldwell, Sabrina   610
Cao, Na   634
Cao, Wen-ming   181
Chalmers, Carl   304
Chen, Jiahui   145
Chen, Shiping   568
Chen, Zhuo   157
Cheng, Long   454
Ching, Pak-Chung   50
Conradt, Jorg   371

Dai, Fengrui   359
Dai, Jianhua   72
Dang, Jianwu   62
Deng, Junhui   316
Deng, Yimin   437
Deng, Yini   622
Ding, Xin   244
Docea, Andrei   671
Dong, Zheng   532
Dou, Qingyun   293
Duan, Haibin   437

Ekbal, Asif   647

Fan, Qiang   281
Fang, Xingqi   579
Fei, Haiping   426
Feng, Linhui   532
Fergus, Paul   304
Firdaus, Mauajama   647

Gao, Chengliang   316
Gao, Kai   316
Gedeon, Tom   509, 610
Gondal, Iqbal   486
Gong, Xiaoze   634
Gu, Xiaodong   444
Gu, Yueyang   579
Guan, Haotian   62
Guo, Lili   62
Guo, Xiaoyu   244
Guo, Yanjie   437

Hamey, Len   568
Hao, Yangyang   684
Hayashi, Takuya   349
He, Xin   659
He, Yan-Lin   134
He, Yuyao   416
Hong, Zehua   579
Hossain, Md Zakir   610
Hou, Zeng-Guang   15, 40
Hou, Zengguang   405
Hu, Jialiang   170
Hu, Xiaoyi   191
Hu, Zechuan   205
Hussain, Abir   304

Ichim, Loretta   671
Ikeda, Kazushi   96
Imam, Tasadduq   486
Inoue, Daisuke   392
Islam, Mohammad Manzurul   555
Iwata, Kazunori   218

Jiang, Zhi-Ying   134
Jiao, Yingying   622
Jones, Richard   610

Kahandawa, Gayan   555
Kamruzzaman, Joarder   486, 555
Karmakar, Gour   555
Kasabov, Nikola   256
Khoda, Mahbub E.   486

Kim, Sangwook 349
Kleanthous, Natasa 304
Koyama, Yutaro 96

Le, Xinyi 659
Li, Baoqi 416
Li, Fanzhang 27
Li, Rui 181
Li, Xiaobo 499
Li, Zhipeng 600
Liang, Xu 15, 40
Lin, Tianwei 426
Lin, Xianghong 336
Liu, Chang 3
Liu, Jie 231
Liu, Rui 244
Liu, Weizhou 454
Liu, Xing 191
Lu, Bao-Liang 622
Lu, Liping 191
Lu, Zhifeng 579
Luo, Sihui 476
Lv, Jiancheng 359

Ma, Chao 256
Ma, Huifang 336
Ma, Lizhuang 684
Ma, Quan 3
Ma, Xing-Kong 588
Mańdziuk, Jacek 268
Mao, Shuiyang 50
Mason, Alex 304
Mathew, Jimson 380
Meng, Deyuan 454
Murshed, Manzur 555

Niu, Xiaoguang 579
Nugaliyadde, Anupiya 326

Omori, Masahiro 349
Omori, Toshiaki 349
Ozawa, Seiichi 349

Patino-Saucedo, Alberto 371
Peng, Liang 15, 40
Popescu, Dan 671
Pruengkarn, Ratchakoon 326
Pusit, Prasong 405

Qian, Sheng 181
Qiao, Yu 579
Qin, Zheng 600
Qin, Zhenyue 610
Qin, Zhi-Quan 588

Raj, Arpita 380
Ren, Shixin 15, 40
Ren, Yazhou 205
Rostro-Gonzalez, Horacio 371

Saha, Sriparna 380
Sakumura, Yuichi 96
Sanodiya, Rakesh Kumar 380
Shang, Jiaxing 532
Shao, Zhiwen 684
Shaw, Andy 304
Shen, Chengchao 476
Shen, Qiwei 545
Sheng, Hao 145
Sneddon, Jennifer 304
Song, Jie 476
Song, Mingli 476
Song, Shuyi 476
Su, Haisheng 426
Suchan, Jakub 268
Sun, Li 476
Sun, Xiaomin 316

Takahashi, Takeshi 392
Takatsuka, Masahiro 109
Tan, Xin 684
Tang, Baige 522
Tang, Chenwei 359
Tian, Jing 191
Tian, Runqi 244
Tian, Yi 579
Tong, Qiuhui 157
Tu, Enmei 256

Varadharajan, Vijay 568

Walder, Christian 509
Wang, Bangjun 27, 522
Wang, Bencheng 244
Wang, Chun 545
Wang, Chunze 545
Wang, Haojie 532
Wang, Huaimin 281

Wang, Jiaxing 15, 40
Wang, Jing 464, 545
Wang, Lihua 349
Wang, Longbiao 62
Wang, Lu 256
Wang, Tao 281
Wang, Tianyu 509
Wang, Weiqun 15, 40
Wang, Xiangwen 336
Wang, Xidong 84
Wang, Ya-Jie 134
Wang, Yong-Jun 588
Wang, Yongli 634
Wei, Quanrui 145
Wen, Ying 170
Wong, Hau-San 181
Wong, Kok Wai 326
Woon, Wei Lee 122
Wu, Haiyan 464
Wu, Si 181
Wu, Yubin 145

Xiang, Jianwen 191
Xie, Xiaoliang 405
Xiong, Shengwu 191
Xu, Fang-Zhu 134
Xu, Hua 316
Xu, Jianhua 3, 84
Xu, Kuan 579
Xu, Zenglin 205
Xue, Guangtao 145
Xue, Yangtao 522

Yan, Xiaofan 532
Yan, Xin 205
Yang, Jie 256, 579

Yang, Liu 27
Yi, Qian 231
Yin, Gang 281
Yousefi-Azar, Mahmood 568
Yu, Yue 281

Zeng, Yarong 281
Zhang, Guixuan 231
Zhang, Hui 244
Zhang, Li 27, 293, 522
Zhang, Linjuan 62
Zhang, Liqing 499
Zhang, Qilai 72
Zhang, Shuwu 231
Zhang, Wenqian 464
Zhang, Xiaoming 316
Zhang, Xinyu 145
Zhang, Xunhui 281
Zhang, Yang 145
Zhao, Dongdong 191
Zhao, Haimei 157
Zhao, Haohua 499
Zhao, Lei 84
Zhao, Ning 634
Zhao, Xu 426
Zhao, Yaohua 416
Zheng, Jianyang 336
Zhou, Boyu 659
Zhou, Haiying 191
Zhou, Lecheng 444
Zhou, Shangbo 532
Zhou, Zhongyi 659
Zhu, Hengliang 684
Zhu, Lei 392
Zhu, Qun-Xiong 134
Zhu, Xuanying 610